EMBRYOGENESIS

EMBRYOGENESIS

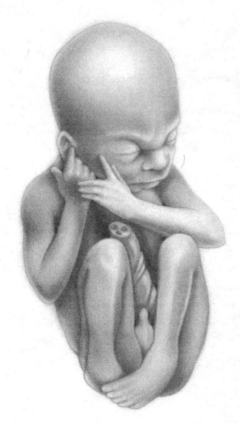

*Species, Gender,
and Identity*

**RICHARD
GROSSINGER**

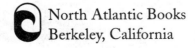
North Atlantic Books
Berkeley, California

For Robin

Published by
North Atlantic Books
P.O. Box 12327
Berkeley, California 94712

Cover art by Phoebe Gloeckner
Cover and book design by Paula Morrison

Printed in the United States of America

Embryogenesis: Species, Gender, and Identity is sponsored by the Society for the Study of Native Arts and Sciences, a nonprofit educational corporation whose goals are to develop an educational and crosscultural perspective linking various scientific, social, and artistic fields; to nurture a holistic view of arts, sciences, humanities, and healing; and to publish and distribute literature on the relationship of mind, body, and nature.

Library of Congress Cataloging-in-Publication Data
Grossinger, Richard, 1944–
    Embryogenesis : species, gender, and identity / Richard Grossinger.
      p. cm.
    Includes bibliographical references (p. 829).
    ISBN 1-55643-359-X
    1. Embryology. 2. Life—Origin. 3. Evolution (Biology)
  I. Title.
QL955.G76 2000
571.8'6—dc21   97-19749
                                   CIP

1 2 3 4 5 6 7 8 9 / 04 03 02 01 00

Table of Contents

Psyche and Soma

Applications

Acknowledgments

THE ACCOUNT IN THIS TEXT has been compiled from many sources (as spelled out in the *Notes* at the end). Without the complex, lucid accounts of hundreds of different biologists, a book of this sort could not have been written. Although I have organized, synthesized, and recontextualized their material, they are the ones who carried out the research, compiled it as information, and put it into language. I acknowledge them (collectively here, and again individually in the bibliographies at the beginning of the notes for each chapter).

The technical aspects of different drafts have been corrected in places by Dr. Stephen Black of the Department of Zoology, University of California at Berkeley (the version written in 1982–1984); Dr. R. Louis Schultz (retired) of the University of Colorado Medical School in Denver (the 1986 published version); Dr. Mary Tyler of the Department of Biology, University of Maine at Orono (an early 1998 draft of this edition); and Dr. Stuart A. Newman, Department of Cell Biology and Anatomy, New York Medical College at Valhalla. None of them read the 2000 published version, so they should not be held responsible for any mistakes that remain. Instead I encourage readers to send me corrections (care of the publisher) so that they may be incorporated in future editions.

I would also like to thank readers for help in particular areas of the text: Dr. Barry Coller of Mount Sinai Hospital in New York (the embryology of blood); Lynn Margulis, Department of Botany, University of Massachusetts at Amherst (symbiogenesis); Dr. Harvey Bialy, editor, *Nature Biotechnology* (genetics); Dr. Richard Strohman (retired), Department of Biology, University of California at Berkeley (the relationship between genetic determinism and epigenesis); and Charles Stein, editor, Station Hill Press (Jacques Derrida, mathematical theory, and Greek philosophy). Likewise, none of these people read the final draft; thus, none of them should be held responsible for misinterpretations.

My friend, the Rolfer and somatic theorist Michael Salveson, constantly pushed me into new territory around issues of morphogenesis and the relationship between morphogenesis and healing. Emilie Conrad showed me that embryogenesis can be lived, even after parturition, even by adults. Her colleague, anatomy- and movement-teacher Robert Litman, provided a wealth of metaphors, models, and possible universes.

During my three-year somatics training program at Deer Run House (1990–1993), instructor Randy Cherner and my fellow students taught me the limitations of intellect by demonstrating the compensatory intelligence of flesh and power of palpative touch.

From the researchers and physicians at the Upledger Institute I learned about the experiential relationship between matter and energy. Their therapeutic work with Vietnam veterans (post-traumatic stress syndrome), autistic children, victims of torture in Bosnia, sea mammals, etc., makes Palm Beach Gardens, Florida, a radiant point on an esoteric map of Earth 2000. My seminars and sessions (1992–1996) with Judith Bradley, Suzanne Scurlock-Durana, Jay Kain, Frank Lowen, Joann Easter, Francine Hammond, and, of course, the master, John Upledger himself, were breakthroughs into an alternative vision of the nature of being. Dr. John arrived at spirit by way of anatomy, without missing a single tissue complex, nerve, or bone; his opus alone balances the whole of materialistic biology and sets the terms for a survivable future.

I recognize here the lineage of those who taught me two interior modes of inquiry, t'ai chi chuan and meditation (1974–1997): Andy Shapiro, Carolyn Smithson, Paul Pitchford, Carol Lee, Benjamin Pang Jeng Lo, Martin Inn, Peter Ralston, Chris Flynn, Ron Sieh, Denise Forest, and Jeff Kitzes.

I was introduced to the weight and movement of my own body (1993–1999) by Breema and yoga teachers Kathy Vahsen, Manocher Movlai, Jon Schreiber, and Cybèle Tomlinson. Bob Frissell educated me in guided breathing, affirmation, and rebirthing. Barbara Thomas and Amini Peller showed me limitless and nameless domains of cosmic energy and improbable hope for the universe. Gene Alexander taught me by example and friendship. In our conversations over twenty-five years Ellias Lonsdale changed for me the context of the post-modern world (and everything in it) by putting us back in galactic time.

I AM GRATEFUL to my son Robin Grossinger, environmental scientist and Director of the Bay Area Historical Ecology Project at the San Francisco Estuary Institute, for our dialogues, beginning at a startlingly early age (for him), about the wonders of living creatures and oddities of scientific law. My daughter, performance artist Miranda July, betrayed the edges and radical segues of the ideas in this book by putting a small section of them on stage in her *Love Diamond*. She turned bacteria and cells into a galloping horse and, by the magic of transformation rather than syllogism, brought life forms onto Planet Sweet Chariot. My wife, Lindy

Hough, helped me get through big ideas and cut unnecessary affectations, though many (to her occasional chagrin) still remain.

I WOULD LIKE TO REMEMBER my graduate-school teacher at the University of Michigan, Dr. Frank B. Livingstone, who planted the initial seed for this book when he told our class in 1968 that we would cover two of the three keynotes of physical anthropology: population genetics and primate archaeology—but we would skip the third, embryology, because it was always left out. In fact, embryology (in particular, the relationship between ontogeny and phylogeny) is the vehicle, the energy, that joins the changing gene pools of hominid groups to the fossils of *Homo sapiens;* it alone provides an actual mechanism for biological continuity, mutation, and the emergence of culture. Without embryology, physical anthropology is stuck holding the air between two isolated paradigms (one genotypic, one palaeontological; both statistical and hypothetical). This book offers physical-anthropology students a glimpse at the hidden ontogenetic relationship between evolving societies and their biophysical roots.

Stanley Keleman revived my interest in the topic ten years later when he showed me movies of developing frog embryos while exclaiming in exasperated wonder at the metamorphoses unfolding before us, "Now what in hell is *that?*"

I THANK KATHY GLASS not only for her attentive editing but for grasping the essential concepts of this book, thus reflecting back to me the radical change in worldview that comes from seeing daily life through a mandala of embryogenesis.

Thanks to Paula Morrison for doing an elegant job of design.

Thanks to Victoria Baker for her thorough index.

I acknowledge the various artists for their work, notably medical illustrator Jillian Platt O'Malley who constructed a universe of embryos from my wish list (Jill7@bigfoot.com—or care of North Atlantic Books—for those who wish to commission work from her). Phoebe Gloeckner, Harry S. Robins, Dr. Jeremy Pickett-Heaps and Julie Pickett-Heaps of Adelaide University, Rudy Rucker of San Jose State University, and Bradley R. Smith of Duke University also either created or modified images for me.

Although many publishers and authors gave permission for illustrations, I would like to thank in particular W. B. Saunders Company, John O'Connor of Eastland Press, and Dr. Lynn Margulis for allowing me to use a large number of images each, and Dr. Donald Ingber (with the help of Jeanne Nisbet) and Dr. Stuart A. Newman for providing specific images to illustrate their concepts. Other images were culled from older books either in the public domain or for which a present copyright was unlocatable. Anyone owning such an image may contact me or the publisher for rectification.

Note on spelling and usage

SOME WORDS IN THIS TEXT appear in general scientific literature in more than one form—sometimes even with slightly different meanings for the same form. Many of them also have more than one spelling and are either capitalized or not, italicized or not, depending on the author and/or era of the source. I have tried to be more or less consistent but, given the length and range of subject matter in this book and a number of other, nontechnical priorities, I have made many decisions contextually rather than globally. Likewise, the copy-editor has put her energy into keeping the book faithful to its intended level of meaning and analysis rather than aiming for a flawless terminological accuracy.

For instance, I have italicized many Latin anatomical terms, while either leaving others in Roman in recognition of standardized usage in virtually all sources (despite my italicization of equivalent terms) or italicizing them inconsistently because of deviating usages in the sources I chose. Also, developing the text in layers, I have chosen usages and meanings from different eras and contexts. A frequent use of the glossary should guide the reader through this territory.

THE ILLUSTRATIONS HEREIN do not always match the accompanying text, and they are also not comprehensive. Some concepts have been illustrated either barely or not at all; others have been pictured in far greater detail than they have been discussed. There are also illustrations of concepts not described in the text.

This is basically a work of literature, not science. It is also not a textbook. Visual material occurs fluidly, its use depending on either aesthetics or what I was able to find and/or purchase permission for. I commissioned a certain amount of art specifically for this book; the potential topics for illustration, however, far outstripped my budget.

Preface

1.

Tʜɪs ʙᴏᴏᴋ ᴅᴇsᴄʀɪʙᴇs ᴛʜᴇ ɢᴇɴᴇsɪs of life beings on Earth; it brings together ontogeny and phylogeny, lineages on radically different intervals that converge in all plant and animal embodiment: the one chronicling the assemblage of a creature from a single cell, and the other tracing the evolution of myriad species from a primal cell.

Embryogenesis is also a text about being and mortality, matter and spirit, body and energy. The relationships among these plurally interchangeable terms (being and mortality, body and spirit, energy and matter are just as serviceable) map the contours of a great riddle of everything. Are these contrary manifestations of an elemental pith, a single agency (as of course they must be)? Do their apparent antitheses cast a mere semantic mirage? Are they not (from our vantage) independent entities seeking a mechanism for fusion, ceaselessly commuting antipodes of a process vastly more complex than we are, yet including us? Then how does energy conduct itself into matter; how do spirit and body rendezvous and alloy?

Embryogenesis (far more graphically than atomic synthesis) is the raw terrain, the visceral flank of the universe, for it is not only where spirit accosts substance but where the fact of being, the words of this text, meet the creation of stuff. Nowhere else do we see inert molecules becoming animate; nowhere else does undifferentiated (and, for all intent and purposes, infinitely dense) mass become differentiated and timebound. Embryogenesis demonstrates how spirit infuses matter (scrupulously and in full accord with the laws of both) or, if spirit and matter are one, how they come to be temporally segregated and their union mortal.

By definition this is a book about science and religion, not their vaunted paradigmicities and fundamentalisms in the late twentieth century but their deep seminal cores. I have laid open the modern (and post-modern) consensus of what we are and how we are made, while at the same time seeking the meaning of such an existence. I have searched not for the apparent revelation of orthodox science but its shadow. The truth supposedly revealed by the material world masks another truth concealed by that same world. The so-acclaimed spiritual realm is also a mask obscuring another spiritual domain.

If, on the one hand, we accept uncritically the empirical and statistical facts culled from nature, we forfeit the thread of existence and become epitomes of hollow machines. If, on the other hand, we adopt theosophical landscapes without experiencing their inception in our cultural biases and personalities, we are well on the road to becoming nihilists again in the end. In *Embryogenesis* I have taken a different path in place of either a choice between these two or a modernistic synthesis of them. I have tried to validate an experience that occurs outside the orderings of science and religion, yet with reference to each of their roles in forming our ideology.

"Stone by stone," sang an anonymous Sufi minstrel, "a structure takes shape beneath a sky full of mystery."[1]

2.

READERS MAY EXERCISE PRUDENCE in navigating through the embryology in this book. Skim shamelessly where you get bogged down. I wrote not to enforce terminology but to give a sense of our predicament and to depict our reality in the many morsels and layers of its implacable determinism. When I considered degrees of detail, I chose to err on the side of meticulousness. I refused to settle for homilies like "genes programming traits," "organs forming from fields," "the tao of biology," or "the self-organizing universe." After all, these are already abstractions.

I preferred instead to "make" actual guts, lungs, legs, and genitals, even if words are but another crook of abstraction. With layers of depiction piled atop one another from organelles to organisms, my saga offers a glimpse of the multidimensional grid underlying existence—a more textured one than could come from a mere assertion of the same event.

All languages, graphemic and mathematical as well, are metaphors. I have chosen metaphors of concrete things over metaphors of other metaphors. Yet the reader must break any passing spell that I am talking about real events. Biology falls smack in the heart of the modern fallacy that Alfred North Whitehead calls "misplaced concreteness" and René Guénon "the reign of quantity." Biological terminology is rife with anthropomorphic metonymies, misrepresentations of processes as things, and falsely linear narratives and historicisms. It is a drama of ideological patinas: the countervailing moralisms of creationism and survival of the fittest, the capitalist commoditization of the cell as a factory and nature as a productive business venture, the Marxist view of multicellularity and organism as universal stages of evolving communalism, the academic trademarking of cells and genes for initiates and customers, etc. Every metaphorical transposition of scale drags along with it a politics regarding such issues as genetic determinism, formal patterns underlying biological systems, the rule of order, and even the ontological status of mathematics and calculus presumed to underlie morphology and metabolism. Microbiology and embryology textbooks read like comic books of earnest little kingdoms, their solemn, odd-shaped and discontinuously scaled citizens always hard at work, loyal to their tasks and cell-masters, blind to the paradox of their eternal conscription.

The deeper I go into cell dynamics and organ making, the more deceptively my text conveys mirages of substantiality, sham domains in which cartoon organicisms masquerade as if the principalities of life. In the end we must set all these drawing boards aside. The real is far more vivid, mysterious, and ineluctable in its undeclared immediacy: *I y'am what I y'am what I y'am!*

3.

JEANNINE PARVATI, a spiritual midwife and herbalist, though a supporter of this text, believes I have been undiscriminating in my adherence to biological observations at the expense of intuition and worship. From her standpoint, to marginalize "woman's mysteries" in this way is to pretend to be objective overseers of the universe rather than life forms embodying transformations. She considers my narrative a dangerous fiction (for reasons other than those enumerated above), drawn from documents of researchers who tortured embryos and creatures for tainted facts.

Experiments famously squash complexity in attempts to unveil its proximal components. As one scientist remarked of the cyclotron, we smash Swiss watches to find out how they work. We crush atoms to classify matter; we dissect animals to see how they exist; we ablate tissue to pinpoint functions of the brain.

The "torturing" of worms and insects (and even cells) is an ongoing crime against the sacred fury of the universe. Though a scientist towers over a bee in generations of intelligence, he is puny against the fixed time of its species. (As a Generation X hipster in the North Atlantic Books warehouse remarked one day about the immediate fauna, "Those bees are into some *deep* shit.")

I would gladly support a more compassionate path to knowledge. However, to ignore the science of cells because it is based on the mutilation of creatures would be, for me, an ideologue's position that would prevent the writing of this book.

I agree that information gained from the severing of brain lobes of octopi, squirrel monkeys, etc., and the induction of tumors in helpless rabbits, chickens, and the like, must, in some way, be sullied by the experiments themselves. But to boycott such knowledge is to leave the twentieth century. The damage has already been done; we might as well examine the coinage for which we rent the goose.

4.

*E*mbryogenesis COMPLETES A REVISED TRILOGY. I began the project in 1977 with the first version of *Planet Medicine: From Stone-Age Shamanism to Post-Industrial Healing* (Doubleday/Anchor Books, 1979; revised edition, Shambhala Publications/Random House, 1983). Later I rewrote and expanded that book into three separate volumes. The first, *Homeopathy: An Introduction for Beginners and Skeptics* (North Atlantic Books, 1993), was itself revised again as *Homeopathy: The Great Riddle* (1998). *Planet Medicine: Origins* and *Planet Medicine: Modalities* (North Atlantic Books) were published together in 1995.

The second volume in the trilogy was *The Night Sky: The Science and Anthropology of the Stars and Planets* (Sierra Club Books/Random House, 1981; rewritten for J. P. Tarcher, 1988).

Embryogenesis, drafted between 1981 and 1984, was prepared for publication by Avon Books in 1985 but never released; it was published by North Atlantic Books in 1986. This version, its successor, was written from 1996 to 1999.

The trilogy is the dispatch of a twenty-two-year inquiry into origins and boundaries. I began with alternative medicine as the self-diagnosis of a civilizational disease. From there I addressed images of space-time and creation. The ways that we categorize extraterrestrial sparkles in the bottomless void reveal the character of our cultures. The present randomly explosive cosmos evinces an equally violent postmodern landscape.

5.

THE EMBRYOGENIC EPISODE IS OSTENSIBLY THE RESIDUE of an evolutionary process that began with raw inanimate elements billions of years ago and, without prompting—merely by interpolating and building upon itself—constructed creatures, thoughts, and symbols. A pearl was fashioned from mud—not just a pearl but an entire regime of swimming, walking, flying, breathing gems.

The notion that life made itself out of nothing and is but a random undulation within a sterile, godless universe is now the Rosetta Stone of contemporary logic. Yet it undermines virtually every moral order, every system of justice and equality, every plan to seed values (other than materialism, with prosperity for the fortunate).

If "being" is the prize in a lottery, there is no built-in requirement to behave well, to play any game other than survival of the fittest: exploit the weak; imprison the underclasses; defeat or exterminate enemies, rivals, and citizens of other tribes.

The traditional antidote to agnosticism has been to assert, in place of blind kinetic forces, a wise and rational deity behind creation. Unfortunately modern creationism (from Christian evangelicism to Shiite Islam) is mostly disingenuous, an anti-intellectual ploy to blame scientists for all the world's problems and, in renunciation of technological progress and its accompanying nihilism, an ostrichlike demand to return to an old world run by a strict and benign patriarchy.

If everything were reduced to only those two camps, this book would necessarily fall into the canon of evolution. However, polarity is a surface illusion. Evolution may be a brilliant explanation of appearances and their seeming mechanism, but it is a gross oversimplification of the real action of cell life. In recent years purely material accounts of the origin and continued gestation of species have continued to ensnarl at deeper and deeper levels until even statistical biology no longer measures an orderly march of creatures through a mesh of random environmental opportunities. Inscrutable factors weigh in at every stage; evolution is now pervaded with sleights that suggest a master alchemy (if not an alchemist), assaying incredibly devious and hidden qualities in base matter by a finesse of stirring it to life.

That a spheroid of molten stellar material could develop philosophies, laws, and religions out of raw atoms simply by trial, error, and accident defies common sense, and everyone knows it, even evolutionists. "The idea that perfect order 'evolved' from chaos, inanimate mud, or goo, without a Creator or blueprint," declares Pat Boone ("... *each time we saw the tide/take our love letters from the sand....*"), is so stupid that a six-year-old child would reject it. How can something come from nothing? How can incredible diversity and complexity 'evolve' mindlessly and randomly from one-celled slugs? And where did *they* come from?"[2]

Good questions. Questions I mean to address throughout this book without abandoning the big piece of the puzzle that evolution solves, at the same time without betraying the mystery of creation.

Being made of cells is not some secondary fact to be disputed or negotiated. It is *our* fact. Or, more precisely, its rendering in degrees of mathematical and semantic structures is the portiere through which we make ourselves real in a materialistic age. Cell existence has to be accounted for, historically and epistemologically—likewise, cell aggregation and coalescence. What does it mean to discover ourselves as this, to be this? Not: how do we evade this?

Embryogenesis lies at the heart of our riddle. It is the unadulterated text of both creation and evolution. Every time a creature forms anew out of raw atoms it makes a replica of the inception of life itself. The way in which it organizes its body and complexifies is the way in which meaning was invented.

All the pages that follow merely attempt to decipher how the cosmos literally writes itself on its own body.

6.

DESPITE APPEARANCES, this is not a biological text *(... to be repeated throughout reading....)*. It is an inquiry into being alive. The terms of existence are quite strange and haunting—in fact, totally implausible and unaccounted for. Our basis is not obvious. Cells and tissues merely saturate a deeper enigma.

Hypothetically, existence could take all manner of forms. Yet we are this. Whether "this" is the only thing we could be or whether we could as easily be disparate entities in myriad, quite alien domains (embodied in this kind of stuff, some other kind of stuff, unembodied and hyperdimensional, etc.—yet still alive and sentient) is the topic of a different speculation.

To study embryology is to meditate on the objectified language and microphotographic evidence for our formation in three dimensions. It is a factual commitment, though its facts teeter in absence of a context. From within "being," we presume to chart the tangibilities and material vectors of our becoming. We track the intricate maneuvers of proteins, protoplasm, and cells in coalescing and carving live mannequins. The sincerity of our observer status (plus an inner conviction of our own immediate presence) creates the illusion of a context.

Biology accumulates generations of collective empirical inquiry and critical analysis, perpetrating a mirage of systemically organized occurrences. Life science is a titanic cultural koan that makes the sound of one hand clapping by feeding off itself and a bottomless propagation of kicking-and-breathing chimeras. While developing more and more subtle and *bona fide* prerequisites with the advancement of the technologies that certify them, these (in every sense) model organisms have established their seemingly irrevocable concreteness over centuries. This book is then my answer.

BEING (in the manner in which biology depicts it) stands against two famous foils: "not being" and "being something (or somewhere) else."

Biology recognizes only one contrariety to itself: "not being." Embryology (insofar as it comprises both phylogeny and ontogeny) characterizes "being" as a purely physical event, perhaps one with epiphenomenal consequences. The only alternative to this "being" is nothingness—a nothingness of stone, fire, and water. Out of such inert parings, nature (acting as the chance material forces of the universe) fashioned life harum-scarum. Without this unpredictable accident the universe would have remained insensate and uninhabited: a maelstrom, void and eternally asleep. "Being" would never exist.

However, "being" also stands in antithesis to "being something else." If life forms have a primal essentiality, if consciousness precedes chemistry, then embryogenesis is not the inventor of existence, only a loom for one version of it. The embryo is the silt for getting consciousness into atoms. Without its issuance as an accident of evolution, "being" would still exist elsewhere, perhaps infinitely and multi-dimensionally.

Regardless of whether one believes that "being" stands in opposition to "not being" or in opposition to "being somewhere else" (or both), an embryogenic process is necessary. In the former instance an embryo must assemble matrices so complex and subtle that they metabolize, individuate, and then become sentient out of their own intrinsic circuitry. In the latter instance an embryo must weave a fabric rich enough and sympathetic enough to lure plumb existence into a molecular habitat.

Though I will go back and forth between these antipodes of embryogenesis, the process I am describing must always remain the same.

7.

MOST SCIENTISTS IGNORE THE GAP between a sense of being alive in a mysterious world and their own doctrinal explanations for it. Perhaps they fear the shadows collecting just outside the veil of law and its cavalcade of sanctioned reality. After all, despite a longstanding empire of rigorous fact, everything could change in an instant (taking with it the vaunted physical rules and their skeins of cause and effect)—and we could hardly protest or be surprised, since we don't know where any of this is happening or what imposes its rules and keeps them in place.

The illusion is that biology is dealing with something established and proven, but the molecular, cellular phenomena behind embryogenesis are so latent and old that there is no thread at all by which to get at them and their true agency. Only the outer edge of their mechanism is exposed, like frazzles of yarn; the rest has been subsumed in the transformational process itself. Biology's pictures of the making of life are ludicrous oversimplifications of an occasion denser than a neutron star. Over unimaginable epochs the infinite and cosmic has buried itself in the infinitesimal. Evolution has taken something as big and complex as the universe—in fact has taken the universe itself, its collective hieroglyph—and, over billions of years, stuffed it (along with billions of years of agglutinated, ensnarled events) into something as tiny as the nucleus of a cell. And there alone, to our knowledge any-way, under heaven and hell does it dwell. Collective events of unknown histories have encapsulated themselves in a timeless morphological contrivance, from whence they continue to unwind both back and forth.

Cells are the outcome not of genetic maps but planetary forces. Tidal ripples (and other ancient incidents) are incised in the backbones of genetic molecules as well as in the whorls of leopards' fur or tortoises' shells. Life is not a machine or blueprint/design; it is a radical event through which the invisible and profound energies of the universe find form and get themselves here.

Unknown meanings roil in dense layers across all of space-time to come to their single resolution. Each embryo unwraps the crypt of cosmic mystery while wrap-ping it in a new set of figurations—cells and tissues. The various embryonic shapes are the only message the void has for us.

We don't have to journey to the ends of the universe to find alien life forms, for the ends of the universe have travelled here to embody themselves. We don't have to go to another galaxy to see the haunted city of the strangers. The city, with its labyrinthine suburbs, is at hand. We don't have to imagine exotic beings spawned on remote moons. Anemones, hydras, bees, snakes, and beavers are real enough. Confucius came to this domain—so did Charlemagne. Joan of Arc, Bessie Smith, Crazy Horse, Jesse James are cell-beings. William Blake formed the same way; so did Sitting Bull, Jimi Hendrix, Esquimaux and their dogs. So did you.

Do you want to know what creatures sound like on planets in remote galaxies? Gregorian chant, Blind Lemon's blues, Pachelbel's Canon, the calls of whales, the squawks of birds, the squeals of raccoons are cosmic notes incubated in elemental strata. They are refined states of the stuff inside matter, signals uttering brogue. First, lattices self-assembled out of molecules; then they procured ghostlike shapes, culling patterns their properties allowed. Then they sang.

Each of the specters on any world is a form indigenous to the nebulae that forged its atoms. The artifacts made by those creatures, their nests, flags, and syllabaries, are faint replicas of structures and landscapes elsewhere. As the embryo unfolds (i.e., folds), unravels (i.e., ravels), the history of the universe is invented on Earth.

8.

THE EMBRYO MASQUERADES AS A BIOLOGICAL OBJECT. We swaddle it in aliases of chemistry and natural history. The same transmuting blob is a philosoph-ical object—a conduit between timelessness and time.

If creation (beyond "Big Bang" landscapes) goes on forever, what will occupy its duration? Mere linear things cannot populate it; even souls would get lost in eternity.

Embryos represent a radical process, a series of topological events shattering eternity and creating venues. As bodies evolve, more and more of consciousness is captured in a temporal mirage, a membrane. In fact, only through progressions of dense, divisible (indivisible) metamorphoses can meaning dodge timelessness and manifest to itself as alive, can it scrabble together creatures and chronologies.

THE NOTION THAT EXPERIMENTALLY DERIVED FACTS at all clarify our situation is science's foremost and fatal blunder. Most researchers not only apply a limited and predetermined hierarchy of knowledge (and naive epistemology) to everything they delve into and "discover"; they operate under the illusion that wherever they discern bigness and causation (galactic mass, elemental molecularity, relativity, subatomic charge, chromosomal code, and the like—seeming absolutes and boundary markers) ultimate reality must also inhere. Plato diagnosed this fallacy long ago: we mistake the dance of shadows for the "thing." Now technology has given us more acute tools for revealing apparitions at ever deeper and more superficially sophisticated levels.

Self-annointed mages and bigshots of secular ministries pronounce "complete descriptions" and working models of creation. Through their assorted field theories and origin cosmologies, they muddy the very waters they are attempting to sublime, projecting their own disillusion, noveau-riche cynicism, and spiritual shallowness onto nature's rare and diaphanous flow. The primal spring, the lucent object science seeks, is beyond reification or calibration and, at its true and actual heart, more likely a cosmic celebration than the death march we now observe.

Unconsciously we fabricate a merciless, tragic, and circumstantially materialistic universe to reflect and match the moral failure of Western civilization. Embryos are that universe's ugly ducklings.

Matter is not junk, molecular parings, or stellar debris; its relation to time and energy is beyond equation. Creatures are not things or events. Organisms are not the sums of samples of their modules. We are not even properly named in the dialects of our own chronicles; we are scions of a secret language spoken by nature itself—a language with remote and fragmentary residues in the calls of birds and trills of mammals (but also in the lapping of waves and rustle of wind). The domain professional biology prescribes is not, by ordination, the domain of life, nor does the word "life" even begin to account for what *this* is.

THE PRESENT-DAY STATE OF THE BIOSCIENCES is well portrayed by a Sufi parable. In one version of this tale a man, having lost his key in a dark alley, is searching for it in an adjacent lamp-filled courtyard.

When asked by a well-meaning passer-by what he is doing, he replies, "Looking for my key."

"Where did you lose it?"

He points back to the alley.

"Then why are you searching here?"

"Because that is where the light is."

9.

SOON AFTER FINISHING THE FIRST VERSION of this book in 1983, I went for a walk with my fourteen-year-old son. Reaching the top of a hill, we heard pigeons cooing on telephone wires. Sun etched their every stripe, ruffle, and hue. "That's what my embryology book is about," I suddenly realized. "How did those birds get there?"

We stared at the alien *pijons*. Then he remembered how his ten-year-old sister had asked him once if the universe went on forever. "I didn't know what to tell her."

The difference between *The Night Sky* and *Embryogenesis* lies between those questions. *The Night Sky* asks: "Why anything?" *Embryogenesis* asks: "How did those birds get there, in fluke of feather and flash of desire?"

These questions are a tar-baby. Go at them hard with fists and feet (and words) and you will find yourself stuck in them forever. Hit them again and again, book after book, and you will not get out alive. The riddles are unanswerable and will always be unanswerable. It is false profundity to devote one's life to them. They devour everyone and everything. When addressed by a Western mode of analysis, they merely double back with new mirages and perplexities.

For me this account more truly begins with Gene McDaniels singing "A Hundred Pounds of Clay." Each time I wondered why I was typing away, I put on my old 45: *"He took a hundred pounds of clay/and he said, "Hey listen...."*

Embryogenesis is not a warrant of facts and philosophies; it is a melody taken off the surface of America.

The obscurity of guileless feelings finally outweighs the most abstract taxonomy or labyrinthine metaphysics.

—Richard Grossinger, Berkeley, California, 1984, 2000

Part One

MECHANISM

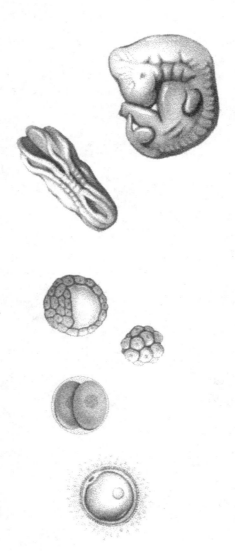

Illustrations by Phoebe Gloeckner

Embryogenesis

"For all we know, this may only be a dream...."

WHAT IS THE LIFE THAT BRINGS US HERE? How is stray energy of the cosmos snared in tissues and personalities? By what agency do entities, awake and aware, evade the vast anonymity?

Mostly, we distance ourselves from the intuition of our life and its mortal consequences. We project the mystery within to an artificially perceived outside. Then we placate it with metaphors and relativities, as if it did not swallow our destiny into its otherness. Not only our minds but our nerves, guts, lungs, and hearts prefer "business as usual," so it *is* business as usual, right up to the end.

We pass through world as shadows through fog. Life is around us, in us, inside our inside; yet we do not stave it and cannot grasp it.

As from nowhere we become alive, we encounter a remote apprehension of absolute existence; we sustain its fragile range all our days. A chorus sings, *"For all we know, this may only be a dream./We come and go just like ripples in a stream."*

There is also a spirit within us that approaches life as limitless possibility; that expects to be surprised, forever; and that labors to make us real to ourselves. We behave as if we had been here since the beginning of time and seen it all come to this.

Immortality is out of the question at this stage of things—the linear immortality of Western teleology—but everything we are, including the part of us that was "immortal" during the Middle Ages, before the ascent of science, arises in an embryonic process whose origin and principle lie outside the present economy of nature. It is to that process we must look both for meaning and the peril of no meaning at all.

FOR MOST OF SENTIENT HISTORY, human beings have indeed been considered finished and perfected creatures—final causes of deific agency. But nothing in the
quite different world depicted by science is complete or final. The physical basis of
life is a template of pulsating, transiting atoms. It takes but five years to replace
every one of them in us with another. Thus, bodies are made of stuff on shorter loan
than suits or cars. They are definitely not "ours." When friends meet after an interim,
they are new assemblages. One so closely resembles the other (and bears its memory) because prior atoms induce new ones in positions equivalent to their own.

Even the corpse that gets buried (or cremated) is just an atomic cell marker for
something invisible. Though it is truly the last remains, it contains nothing personal of the deceased, unique to him or her. It is not even as much a human artifact as other items in the last will. Its substance, decayed and saturating, will drain
out of specialized organs back into nature as dust and molecules, becoming soil,
air, sludge, bacteria, midges, and the like. Atoms themselves are common, undiscriminating pellets. They are so abundant and we use so many of them that each
of us contains dust that was part of Homer and Buddha, as well as billions of worms,
jellyfish, ancestral birds, crustaceans, corals, and Stone Age hunters.

At another level, life is a sequence of cellular fields, each nested upon a previous one, so that creatures emerge from drafts of antecedent species, from a prior
beginning in inanimate crystals which themselves originated in molecular clusters.
Life is also an abnormally organized zone of molecular debris or, as biologist Frederick Hopkins deduced, "a dynamic equilibrium in a polyphasic system."[1] Life is a
partial realization of the informational potential in atoms and molecules. But life
is something else altogether.

"... the more those origins take form retroactively, even as they recede from us...."

BY OUR MODERN WORLD VIEW, being is neither inherent nor inevitable and, if
circumstances had gone differently, there would be no one on Earth (and perhaps no one in the universe), not only now but forever. The molecular building
blocks of plants and animals are reputed to have even greater potential for lifelessness than for life, and there is nothing we know that predisposes them to make
hounds and hares.

What we have said about life at large is even more true (if that is possible) for
human life. Most scientists find it so unlikely that, to them, intelligence is a great
farce upon a lesser one. The universe should be a vacant cauldron, pure sound and
fury, no jolly coachmen anywhere.

Our own assessment of the odds against our coming into being, however, cannot undo the present fact.

WE ARE STUCK AT A CURIOUS PLACE: our search for origins (intended to bear solace and company) has left us more and more alone in an alien vortex. Meaning crumbles at our lightest touch, upon smudges of distant galaxies as upon ephemeral footprints of particles—upon interest rates and commodities likewise. Hamlet's "to be or not to be" has spread from the players to the audience to those not even holding tickets to the Super Bowl/World Cup of universal relativity and deconstruction. "We [now] invent our lost objects posthumously," chants postmodern scribe Steven Shaviro. "The more we brood over supposedly estranged origins, the more those origins take form retroactively, even as they recede from us. Melancholia . . . continually generates the very alienation of which it then complains."[2]

We expect the daily sun to operate normally, but we know it is only "the sun," a fallible stellar machine that may perform superbly through our lifetimes but will surely give some generation of our children (if our species endures that long) a barren red or indigo morning. We expect daylight to be safe, atmosphere to be breathable, fields and woods to flower and fructify, stormy weather to end; yet we pour the worst imaginable toxins into ocean and air, daily assaulting these functions as if they were guaranteed and indestructible. We expect to achieve something with our lives, to experience great truths; yet we smother the knowable with flagrantly symbolic realms, carrying out extravagant, maudlin battles of kings and clerks and estranging ourselves (down through the centuries) from the joys and sorrows of our own experiences. All that our many crusades have uncovered, both outside and inside us, is a bottomless cavity more void than thing.

Alienation is our species' proudest theology. We are enveloped by inflation, corruption, crooks, and vamps. Schizophrenias and nothingnesses riddle our hardwon mindedness. The fact that we are dissipating irreplaceable resources makes us an event not only without a meaning but without a future. We are free of the promises and threats the gods made through-

FIGURE 1A. Painted and carved woman-giving-birth door from Dutch New Guinea.

From Joseph Needham, *A History of Embryology* (New York: Abelard-Schuman, 1959).

out history, but it is only the same freedom that raw stellar elements had in the beginning, to make us or not.

We are algorithms. So, who would bother to speak for us or against us?

There is no conventional way out of this dilemma, so we barricade ourselves within tinsel hierarchies and merchandise gluts—the victims of fashionable histories and recreational regimes posing as statistical laws and controlled states. We were once the victims of the divine right of kings and proletarian revolutions.

Either way, the universe is not "a gigantic clockworks, brilliantly lit," as Puritan abolitionist John Brown proclaimed in Russell Banks' words in the novel *Cloudsplitter*. "It's an endless sea of darkness moving beneath a dark sky, between which, isolate bits of light, we constantly rise and fall."

We once imagined we could do God's work, or at least oversee the fastidious order of nature or, failing that, have a good time at the party. Now "we pass between sea and sky with unaccountable, humiliating ease, as if there were no firmament between the firmaments, no above or below, here or there, now or then, with only the feeble conventions of language, our contrived principles, and our love of one another's light to keep our own light from going out. . . ."[3]

WE HAVE REASONABLY ASSESSED the thermodynamic aspects of suns and atoms and of protein crystal. We have objectified our standing in the universe, removing ourselves hypothetically from our actual place and reconsidering our existence in the context of higher mathematics and its subset, the physiology of nervous systems.

Yet our fantasies and hungers do not square with the algebra of our being, except as we seem to find its forerunners as instincts in animals, its volatile aspects in the chemistry of carbon. We have not only missed the core of creation, we have missed ourselves missing it. *We* are the basis of existence—not as congeries in fragile, intelligent harmony but as single shock experiences. Scientific discourse cannot recover this raw clarity without first breaking from its own objectivity—which would be a fruitless detour inasmuch as it would lead only to the reinvention of science again (and again) until the cessation of our species.

The compulsion to salvage ourselves through objectification seems to be inherited with mind itself.

Once, there was no such plan of things, and all of this had to be invented.

EMBRYOGENESIS IS A NINETEENTH-CENTURY FUSION of two Hellenic stems: *enbruein* ("to grow in") and *genes* ("born"). Embryology is a branch of biology and social science that tracks the ongoing transformation of living and symbolic systems from their conception through their birth and dying. (Psychologists propose a thought-entity epiphenomenal to the flesh, but psyche must be anchored somewhere in the strata of cells.)

We do not *have* children; rather, they pass through our tissues uncognized and inalterable. We are the receptacle for their germ, a capacity imbedded in us by a prior receptacle in which we were seeds. Beyond this wheel of fortune, there is no fatherhood or motherhood, no bloodline or pedigree. The question: "Which came first: the chicken or the egg?" is more a dilemma of nomenclature than of real primacy, for the chicken never stops being a differentiating egg, and the egg is never more than a chicken gestating.

Early in the embryonic life of most complex creatures, germ cells providing a blueprint for the next generation migrate to a region of tissue which will become gonads. Codes residing in the nuclei of those cells are loaded into a zygote. New, almost identical gaggles of cells are spawned, each collocated out of prior ones with the consignment of their strings of genetic material.

Only one continuum is provided by the nucleus of any single cell—a microlith combining elements of both its parents. An embryo can become only the thing its parents are, or perish. It is a miracle that this law, which cannot be broken from one generation to the next, is broken cumulatively over many generations; otherwise, how would new species commence?

EACH CELL ARISES FROM A PREVIOUS CELL, by mitosis if it is a general bodily cell, or, if it is to become germinal, by meiosis (a fractionalizing process of fissioning). As cells are thus manufactured, they interact with the environment and with one another to mold an animate creature. A full-grown human consists of uncountable trillions of them, all interdependent for context and survival. A rotifer consists of several hundred cells, and an amoeba consists of one. All of these cells, no matter the tortuosity of their organism, are also individual life forms bearing germ nuclei for the continuation of their own lineages.

Not only are cells spawned everywhere on Earth and differentiated locally but they are then organized and reorganized in layers and layers of layers comprising creatures. No sooner are these layers formed than other layers are occurring within them. Structure is hidden inside structure, disassembling existing configurations as it barges its way through tissue that surrounds it. In a drama that has been recognized from ancient times, the winged rainbow in cocoon melts down the prior grub until its integrity evaporates into her own and she glides away with it.

"Is the butterfly 'at one' with the caterpillar?" asks Shaviro. "Is this housefly buzzing around my head 'the same' as the maggot it used to be? One genome, one continuously replenished body, one discretely bounded organism; and yet a radical discontinuity both of lived experience and physical form...."[4]

AFTER GESTATION, plants and animals continue to grow and change (usually more slowly), as their nuclei spew cells and tissues from a primordial jug. A multicellular organism exists as a yarn of tissue layers, each layer generating the next until the template withers or the creature is killed. Parturition is but a transition from one phase of embryogenesis to another. Grasshoppers among thistles and daisies, schools of eyespots surfing thermals, children in kindergartens, even aged eagles on crag summits are embryonic. That is why their tissues are able to heal.

Even catastrophic discontinuities do not snap the thread, not if a spore or two can escape. In principle, no plant or animal lives forever as itself, but all of them manufacture seeds with the potential to transmit their lineages through time.

ONCE, THERE WAS NO SUCH PLAN OF THINGS, and all of this had to be invented, to invent itself. The earliest one-celled animals, which no longer exist, were imbedded in life forms which also no longer exist. Along this tunnel of descent, creatures have continued to be imbedded billions of times over in sequences ancestral to every organism calling the Earth home. The vast majority have disappeared, their traces almost unrecognizably condensed and merged in later forms (which have since been condensed and combined). The plans of ancient animals incorporate the plans of primeval animals, and the plans of modern animals consolidate them all.

From the standpoint of denominations of animals, evolution is a death knell, for it is the extinction of species that provides germ plasm and niches for new species. Of course, species do not become extinct because their material is required for new ones. They become extinct for material and environmental reasons, and what genetic messages their last transmuting representatives can salvage become the building blocks for novel plants and animals, but only insofar as these species are able to thrive for equivalent material and environmental reasons. The changing macro- and microclimates of the Earth are merely the most obvious arbiter of evolution, for the tiniest variations in every current of nature, informed by fortuity, synchronicity, and quirky, ineffable fate, ultimately determine the origin and destiny of all species, ancestral and future. Traits and possible traits flow through nature from computers operated by not even the metaphorical equivalent of monkeys keyboarding nonsensically away. Yet mortality is utter, tragic to archivists of elephants and lizards, and irredeemable as well.

Any span of tissue touching one life always touches at its other end the genesis

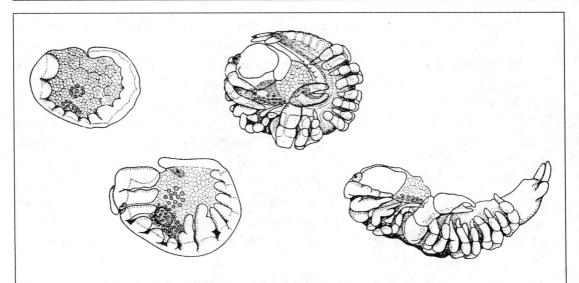

FIGURE 1B. Embryogenesis of crustacean showing development of its segmented pattern. From Rolf Siewing, *Lehrbuch Der Vergleichenden Entwicklungsgeschichte Der Tiere* (Hamburg: Verlag Paul Parey, 1969).

of protoplasm. It takes all of creation working from then to sculpt any one unique being. This delicate history is what we squeeze out of a fly when we crush its crystals. So William Beebe wrote: "... when the last individual of a race of living things breathes no more, another heaven and another earth must pass before such a one can be again."[5]

WHAT WE INHERIT TODAY is a series of designs that go on happening by precedent through a continuity of the basic thermodynamic frame in which they arose and became linked. They are energy swarms batting at their limits, herded always back within the membranous corral. They go on happening not because anything requires them (not on this plane of existence anyway); they go on because, at thresholds joined sequentially, they are bound biochemically to replicate and warp; they could not do otherwise.

A little ball splits and unzips; a tail rolls out of one end, curling up; a head swells at the other; a face gradually impresses itself; a torso etches out of a central cavity.

This sequence of images is deceiving because it presents embryogenesis as a problem solved in advance, a *fait accompli*. Although we know better, we tend to imagine that the reason a fetus finishes its own design is that a full-formed infant lies at its terminus. A woman may imagine her own character imbuing and shaping her baby; it will become human because she is human.

In truth, cells make no promises. They must labor furiously stage by stage to produce each child as though it were the first child ever made in the world. And all cells obedient to the nursery are potentially wild, antipathetic assassins, obedient to nothing greater than their own heat conversion and amoeboid sprawl. Every cell in a body is imprisoned there against its original "will."

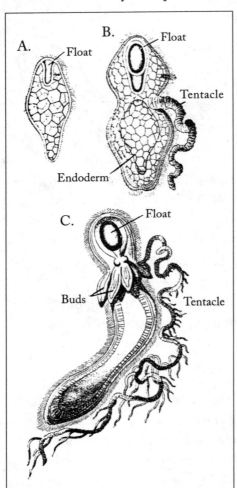

FIGURE IC. Stages in Development of Siphonophore. A. Planula with open invagination of ectoderm; B. Older larva; C. Developed larva with endoderm.

From Walter Howe, *Textbook of Embryology, Volume 1, Invertebrata* (London: MacMillan & Company, 1914).

If life continues because of the desire of living creatures, then that desire has found a way of being that does not require it for its own entelechy. For instance, we can claim that love draws conjugal pairs together, but this seeming romance is merely a symptom of life; it is really the fissioning and joining of cells that compel the act of being.

"To what green altar, O mysterious priest...?"

DESPITE THEIR RADICAL TRANSFORMATIONS, living forms and their artifacts are always continua; there is no alternative. Sequencing must recur from level to level within hierarchies, and from hierarchy to hierarchy.

Just as stages of embryos are absorbed in each other, species of plants and animals vanish into descendants. Staring back through our own chain of grandfathers and grandmothers, we find but a few thousand humanoid seers. Our ultimate great-grandparents are not even human. They look like monkeys, then like moles. Our ancestors lose speech, lose consciousness, ultimately receding from even countenance and contour. Speed up the retrograde heretofore universe, and their carapace becomes worms, then bacteria, eventually comet dust. Going forward, our existence is made of clustering, enveloping posses—first atomistically in the stellar alembic; then as cell tapestries heaped from an impregnated ovum.

The way in which embryogenesis has assembled us is also the way in which we are shaped within lives. When memories dissolve, what takes their place is aggregate meaning. Likewise, when tissue stages are absorbed, they are summarized by a generic organism. We continuously imbibe images, experiences, knowledge; yet we reduce and fuse them and make them part of collective identities in which they are singly obliterated.

Is there even one moment you remember from the third week of August 1958, the first of October 1989? Surely you lived it once, and believed it, and intended to stay loyal to its wonder and plan. Each day unfurled with sparkling clarity at the time. A photograph may suddenly bring back the ardor of an event now riddled by oblivion:

"She felt a surge of longing disbelief—where had that moment gone? Gazing into the camera, smiling at whoever was behind it—that memory was as lost to her as if she had never been there."[6]

The domain of a two-year-old is eclipsed in a six-year-old; her world turns opaque to the teenager. Shirley Temple came to experience her childhood roles as the performances of another little girl. In 1997 Bob Dylan wondered whose voice he heard on a tape from the early '60s: "What's this? I thought it was some obscure person. But it wasn't. It was me."[7]

William Faulkner had forgotten the characters of *Absalom, Absalom* by the time he reincarnated them in *The Hamlet*. Ultimately their English will become extinct too, but not before imbedding itself in a successor. Languages no longer spoken are cached within present tongues, but by now their every phoneme and syllable have changed. Shakespeare's comedies have already begun to erode; Chaucer's tales are part dinosaur; *Beowulf* is another species, a variant preceded by others ultimately as remote as Navaho and Maring.

We must speak unknowingly for the dead and about things we are not aware of, even as we must use the tissues of the dead and don the bodies of strangers.

The disappearance of discrete memories, like the tissues of embryonic layers, does not prevent their contribution to the whole. Even when events are forgotten, they continue to be wound in the fabric of life. The original Freudian insight was an embryological one: We are spackled and sublimated in layers of codes; most of them have become unconscious, inaccessible, but nothing has been lost. That is where the blue and red boat of the two-year-old went, and endures.

Palaeolithic bands awoke as Mesolithic clans, tribes as villages, kingdoms into empires, and from mediaeval cities arose the Flemish trade fairs. Just as a daughter jellyfish originates in its mother, so do pots, harpoons, and myths spawn offspring. Musical themes snarl, simplify, and entangle again like songbird plumage. The universe is a spectrum of fading continuities.

It is tragic that what was precious once must be eclipsed, both in psyche and in tissue. If it meant something, it should not be stolen. If it lived, it should not merely have its existence shorn and incorporated in subsequent creatures. Love should stay true and fuel the unfolding of the universe.

Yet if we remembered everything and lived millennia, time itself would become a jail. The universe needs some other method of preserving essence. Our memories are subsumed in each other not to sever but to protect the filament of existence.

"Today while the blossoms still cling to the vine . . ." We see the petals, the fragility of their connection. A tiara of white climbs bricks behind the Japanese restaurant. Our bodies arise bud by bud, flower, and fade *("I'll taste your strawberries, I'll drink your sweet wine.")*. The song will become a conventional ballad, the summer passes, but the connection to something eternal remains. There is no escape, either from being born or dying. *("A million tomorrows may all pass away,/ere I forget the joy that is mine today.")* The protozoans in the Precambrian mud may have felt that too, but they couldn't have expressed it. Not then, not yet.

"Who are these coming to the sacrifice?" asked John Keats. "To what green altar, O mysterious priest,/Lead'st thou that heifer lowing at the skies?"[8]

IN MOMENTS OF EXTREME HAPPINESS, in the calm of a summer day—golden blossoms, wild birds, the sound of a brook—it doesn't matter that it's not perfect because there is nothing to replace it, no other way for things to be.

In knowing what it is like to exist, to be curdled and then stamped in flesh, to live out desires of a lineage, all creatures, however abbreviated their lives, share an event. If the panther or hawk could articulate the thing they swear by, they would each of them tell you that they are the universe.

Leo Tolstoy's final words were: "I don't understand what I'm supposed to do."

But as Gertrude Stein told us in another way, that is not the question, that is the answer.

The Original Earth

Genesis

ALONG AND PURE SILENCE PRECEDED US. It would seem to be behind us forever. But we have distorted both time and space by our presence.

At one point in another time, there was no Sun; there was no Earth in this region of the Milky Way—only a gaseous, dust-laden cloud with motion and weight. Ancient stars elsewhere consumed initial fuels. Around 4,650 million years ago, shock waves from the spiralling arms of a migrating galaxy rippled through our anonymous heap, creating a vortex. Out of the cosmic spoor spun an unstable star giant. It was still disintegrating one hundred million years later when another spiral arm swept past, stirring its mass and debris into a Sun trailing a ripple zone of orbs. The Earth was one of the tinier inner lumps in the swirling porridge. By comparison the outer Jovian gas giants roared with the grandeur of junior suns. Lifeless though they may be, they monopolized the primal organic chemistry of this system.

An energetic young Sun pounded its new worlds with fluxes of ultraviolet light— a strong deterrent to the formation of subtle, indigenous webs. Today Sol is more spent and restrained, the Earth's atmosphere significantly thicker.

Over the next fifty or so million years, nickel and iron, still molten, sank to our planet's core, sorting lighter elements into a crust. Nitrogen, water vapor, and carbon dioxide hovered overhead. Soon water and hydrocarbons triturated. There was no one to listen to the rain. It rained for a thousand years, a hundred thousand, a million years, but it rained for only a second.

In Genesis, the "Book of Moses" in the Bible of the West, we are told, "In the beginning . . . the Earth was waste and void *[tohu wabohu]*; and darkness was upon

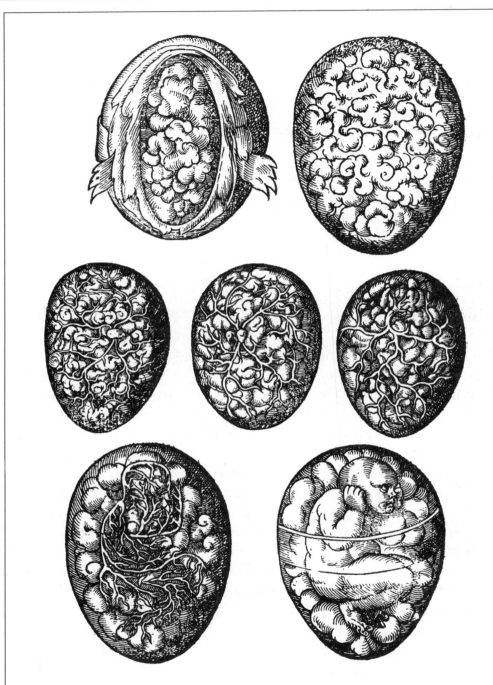

FIGURE 2A. Aristotelian coagulum of blood and seed in the uterus, 1554 drawing by Jacob Rueff entitled "De Conceptu et Generatione Hominis."

From Joseph Needham, *A History of Embryology* (New York: Abelard-Schuman, 1959).

the face of the deep."[1] This is before the Breath of God said, "'Let there be light,' and there was...."[2]

"Let there be light" preceded Sun, Moon, and stars, mere temporal candles. "Let the earth grow grass, plants yielding seed of each kind and trees bearing fruit of each kind."[3]

Beastlings followed. "Let the waters swarm." They did, down to the most infinitesimal droplet.

"... and let fowl fly over the earth across the vault of the heavens."[4] We see the dense flocks of their descendants above land and sea.

"Then the heavens and the earth were completed, and all their array."[5]

THERE WERE NO RELIGIONS ON THE PRIMAL EARTH—no Christianity, no Buddhism, no Taoism. Pre-Apache wind tore at a field of pre-Aranda dream-ghosts. Clouds of voodoo splattered volcanic mercurial Stone. A Dogon water *Nummo* bathed in liquors of Zoroastrian *xvarenah*. Vowels and consonants were all mumbo-jumbo. All gods and spirits commingled in a pan-Gaian pagan rite.

Elements of Creation

THE EARTH THAT CONDENSED FROM HYDROGEN AND HELIUM was uninhabitable. It was as much a sun as a land with a geography.

Stars may be colossal dense objects, but they are simple elementally. Their climates do not allow arrays of molecular edifices. Matter generally remains in its light single-proton state—hydrogen.

On cooler, temperate worlds, proton nets increase, with an abundance of buoyant elements and a fair portion of heavier gold, lead, and uranium.

Smaller planets like ours were unable to hold the bulk of their aery donations—neon, argon, xenon—from the original sidereal cloud; these simple elements, sifting upward in the Earth's atmosphere, have mostly fled our world along with a great percentage of medium-weight elements like oxygen, silicon, carbon, and phosphorus. Many other potentially fugitive gases have been retained by gradual molecular bondage into minerals.

The Earth's core remains nickel-iron, for the most part. Its crust now has the general composition of the mineral olivine—iron, magnesium, silicon, and oxygen.

UNIQUE LANDSCAPES CONGEAL from associations of different grids of atomic matter outside stellar furnaces. Although atoms' properties are determined by the protons and electrons in their shells, their activities are not predictable on a simple mathematical basis, either individually or in interactions with one another to form molecules. Unforeseeable compounds and phenomena originate place by place, indigenous to locales. Each irregular millimeter becomes its own habitat with a specific temperature, gravity, composition, and weather, all flowing together into greater habitats and climate zones. A planet is made up of trillions of such thermoclines and landscapes interacting with one another across obliquities of pressure and heat. The result is a seamless kaleidoscope of purlieus.

Similar phenomena may have developed (or be developing) on other worlds that share the Earth's size, composition, and distance from sun-stars of the same size (or proportionate distances from stars of greater or lesser size), but since random distribution of materials and chance events tilt microenvironmental gradients in unlikely directions, even these worlds will never be the same as ours (and may be exotically different in every aspect).

Apparently it is a dispersion of carbon reactions harnessed in the body-mass of primeval creatures that has given the Earth its present oxygen-rich atmosphere and climate. An identical planet without such a biochemistry would materialize in entirely strange ways. Venus, which is not wildly dissimilar from our world (in its size and distance from the Sun), is a metallic desert under thick acid cloud-cover, apparently lifeless for all of its history.

Water

THE GASEOUS EARTH EMBER was plunged at once into cosmic night, its wintry bath. Vicinities gnarled from temperate scarps. Even as crust formed, lava continued to pour from under the surface. Virginal oxygen bubbling out, hydrogen discovered its natural affinity with the gas' outer shell and shared electrons with it. Instead of escaping the planet, these molecules were held in their combine, water.

Water is a unique and talented substance. The structures of its individual molecules fluctuate chaotically, so that hydrogens are constantly displacing one another and bonding with new oxygens. This disorder gives water viscosity and electrical conductivity. Ionized atoms vibrate and trap energy, absorbing heat. Thus water resists easy melting or boiling; everywhere, it has a moderating influence and provides gentler environments.

The first terrestrial water condensed instantly into steam to fall again as rain. As soon as it hit the hot stones the hydrogen bonds between its molecules stretched and layers of it peeled back into atmosphere. This process continued for millennia, epic storms pounding and eroding the ground and filling its basins. The planet crackled with lightning and shook with thunder—the same weather report everywhere.

Some say that between four and five billion years ago a cataclysmic event occurred—the single most important incident in the Earth's history (other than its formation). An asteroid the size of Mars thudded into our planet, convulsing it to the core and blasting so much of its mantle into outer space that, as the debris gelled under the compaction of gravity, it became a full satellite, the Moon. This world circled the Earth in a much tighter orbit then, raising colossal tides and setting a lunar timing mechanism in terrestrial fluids.

Gradually the aftershocks ceased, landslides ended, and even the dust settled. Over epochs the rocky topography cooled and molded itself; low places filled with pools; water seeped beneath the ground.

About half a billion years ago various giant islands floating in the world-sea began to collide into one another and accrete as mega-continents. Along the seams of their thunderous impacts giant mounds were pushed up, including the most ancient mountain ranges on Earth, the Himalayas and their kin. Titan peaks interrupted the jetstream and sucked torrents out of thick carbon-dioxide cloud layers. All manner of waterfalls and cataracts permeated stone ledges and precipices; rivers dwarfing Mississippis and Amazons roared across mottled lands.

The Earth was no longer a hot ingot revolving and rotating alone. It had a landscape and a companion body made of its own stuff. Its clouds shone orange-yellow in the day and glimmered purple, violet, mauve at sundown.

The interstellar void had announced its regency. Now seasons spun their subtle theaters, each with native colors, storms, and tiers of humidity and ice.

"Only earth and sky matter," writes Annie Proulx. "Only the endlessly repeated flood of morning light. You begin to see that God does not owe us much beyond that."[6]

ANY WATER ON THE SURFACES of the Sun's tinier inner planets baked dry. No saps or herbal springs, their liquids are melted minerals. On Mars, the asteroids, Pluto, Charon, and the moons of the Jovian worlds all potential water was frozen into its metallic state (with the exception of seasonal melting on Mars and oceans heated volcanically by Jupiter's gravitational pull under the ice of its moons Europa and Callisto). Jupiter, Saturn, Uranus, and Neptune remained stellar and provided no crusts at all.

On Earth, copious waters oozed into stones, and there they displayed remarkable abilities. Denser as a solid than as a liquid, water froze on the surfaces of lakes and ponds, insulating itself. Later it fell as snow and sleet, piled up and accreted in glaciers, rolled as mists through valleys and as fog off bays; it barged north and south in icebergs and roared down rivulets into rivers and seas.

All of these talents were to take on a special meaning when water ultimately revealed its *tour de force:* it came alive.

IN THE BELLY OF THE EARTH, simmering plates shuddered and shifted; new basins and islands formed (and continue to form, for the core of the planet remains liquid). Mountain ranges rose and were swallowed, seas filled with Noachian volumes of rain which flooded inland after earthquakes, sundering new seas and leaving ancient seabeds dry plains which would no doubt be seas again. Whole geographies vanished and were replaced. This was an epoch of volcanoes, geysers, and gullies. In some regions lava poured uncontested from the crust. Elsewhere chill water was dammed, absorbing the salted sediments of the land and rushing jaggedly over abstract statuary into mineralized bays and oceans. As rivers ground their way through stone, they left identical patterns across wide-ranging areas, giving the planet its characteristic veined and ridged topography (visible likewise on photographs of Mars as a remnant of a wetter time).

Spreading waters further moderated the climate of the old Earth, evening temperatures, taming molten outposts, and spilling tepid broth into regions that were beginning to feel night's breath.

It is estimated that the original flooding of the Earth provided about a fifth of the water of the present oceans. The rest has been squeezed to the surface gradually over millennia, newly condensed, or spat directly up by submarine volcanoes.

THOUGH THERE ARE BILLIONS OF STARS in the universe, blue watery planets orbiting them are likely as rare as diamonds. Most worlds are probably Martian or Mercurial, bare rockscapes having forfeited their oceans or never having possessed standing water. Each sea-world among the galaxies, whether inhabited or not, will display a unique map—islands and isthmuses, alps and steppelands, archipelagos and signature coastlines.

An atlas of the ancient Earth would be equally unfamiliar to us—mere random shoals and subcontinents in ocean main. It could have been any planet anywhere. There were no Asias, Australias, or Americas. There were no mosses or shrubbery to cover bumps and craters, no diatoms or crabs. It was a majestic desolation, a lookalike of a world circling Antares.

Scientific logic assumes that at this point things could still have gone either way. The Earth could look much like that today but for the accident of life. No railroads or operas—just volcanoes and fog. Even scientists who argue that, given the chemistry of the primal Earth, life was inevitable would still not propose a prescience anywhere. No wild ghost on the wind, life would bring its own spirits onto the land.

Gradually something resembling our home geography began to emerge. The present continents were initially bunched together, with South America and Africa unriven to make up the ancient Gondwanaland (as geologists have named it). In this yolk, North American rock was in the process of being torn from Eurasia and Greenland. The Tethys Sea, the aboriginal Atlantic, lay to the south of Eurasia, breaking through a fissure of New World and Old. Oceanus lay bottled to the west—the natal Pacific.

As debris washed down from the atmosphere, the waters were sown. At the same time, the outer planetary membrane was charged by ultraviolet rays, electricity, and ion-molecule reactions.

Life probably began in the waters of the Earth between three-and-a-half and four billion years ago, only a billion years after the formation of the planet—long before oceans had settled to their present levels and lands receded into the continents of modern times ("And the waters surged mightily over the earth, and all the high mountains were covered"[7]). Early life was subjected to the full turmoil of subsequent rearrangement. Its descendants, even today, must withstand shock waves from the same forces: floods, volcanoes, earthquakes, tornados, waterspouts, perhaps even pole shifts, asteroid collisions, and radiation. Earth has never been a safe place to raise a family.

THE BIBLICAL AND SCIENTIFIC VERSIONS of the Creation differ in every conceivable way, but they depict the same mystery. The Biblical version, while attempting to herald vast externalizing acts, actually recounts an interior awakening, the birth of a sacred planet. The scientific version, although seeming to, cannot actually go back beyond the Word, for it uses language as an objectifying tool. It grasps a Gnostic element the literal Bible of the West omits—that the moment of creation continues to unfold, that the way in which the Void was originally breached recurs each instant as new creatures burst from darkness into a forest of sound and light.

(The conflict between creationism and evolutionism is a mirage. It is but a clash of extremist preachers of opposing modernisms. God and Nature could not be at war.)

The Earth's silence came to an end when the rain was heard, but it was not first heard by us; it was recorded by primordial cells. Without knowledge, without recall, without context, these sensings foreshadowed the zen gong whose resonance dissolves back into molecularity.

Life

ACCORDING TO CLASSICAL GREEK SCIENTIFIC THEORY, the four basic elements (iconicized in earth, air, fire, and water), by the dynamics of their combined natures, spawned living and nonliving substances alike. Aristotle's explanation of this etiology survived over two millennia:

"The hermit crab grows spontaneously out of soil and slime, and finds its way into untenanted shells," he declared in *History of Animals*. "Some insects are not derived from living parentage, but are generated spontaneously; some out of dew falling on leaves, ordinarily in springtime, but often in winter when there has been a stretch of fair weather and southerly winds; others grow in decaying mud or dung; others in timber, green or dry; some in the hair of animals, some in the flesh of animals, some from excrement after it has been voided; and some from excrement yet within the living animal. . . .

"Eels . . . grow spontaneously in mud and in humid ground; in fact, eels have at times been seen to emerge out of . . . earth guts, and on other occasions have been rendered visible when the earth's guts were laid open by either scraping or cutting."[8]

The vitalistic paradigm was presumed. In the seventeenth century the chemist Jan Van Helmont offered a favorite recipe for the assemblage of mice in a mere twenty-one days—from soiled clothes left in a dark, quiet place and sprinkled with wheat kernels. The subsequent discovery of animalcules through a lens only reinforced the conviction that animals were made of germs and could disintegrate into germinal components.

In 1859, with his publication of *On the Origin of Species,* Charles Darwin presented a thoroughly mechanical explanation for life, which stuck. Life is energy passing through dynamic physicochemical systems. Darwin couldn't use such terminology then, but that is what "common ancestry," "overproduction of offspring," "natural selection," "the struggle of males for females," "progression and continued divergence," and "survival of the fittest" add up to. Passenger pigeons with more powerful wings and antelopes with stronger legs replicate the divergences of microscopic creatures and their rudimentary shapes, all the way back to animalcules so simple they were indistinguishable from carbon dioxide bubbling through hot springs. Even among such primal fizzes, nature always favored the more vigorous, the more fertile, those entities with "the greatest facilities for seizing their prey."[9] There is no thermodynamic or elemental basis for any other mode of spontaneous generation of life forms.

Five years later, on April 1, 1864, the French scientist Louis Pasteur unsealed a number of test tubes in which he had incubated many of the popular "recipes" for infusoria, maggots, and rodents, including ample hay and dung. Because he had tightly stoppered the vessels, there was nothing alive in them—no fungus, no bacterium, no infusorian.

Pasteur did not deny the existence of "germs"; he showed that they were abundant in our midst beyond the wildest fantasy of fecundity and infestation. A cubic meter of air in Paris during the summer, Pasteur announced, held ten thousand viable germs. They float freely and invisibly about us and spawn under favorable conditions.[10] But life can arise only from other life; it is not elemental, or primal, or indestructible. "The impossibility of spontaneous generation at any time whatever," declared a contemporary physicist, "must be considered as firmly established as the law of universal gravitation."[11] In fact, even protozoa are so complicated and organized they could not spring from any random association of inanimate substances. But if life could not ever occur by spontaneous generation, how did it arise once upon a time, before there was previous biology?

Panspermia

A NOVEL NINETEENTH-CENTURY SOLUTION was to presume that, long ago, spores ascended through atmospheres of distant worlds; travelled, dormant and dehydrated, across the unbelievably vast acreage between solar systems; then seeded themselves in the oceans of new planets. These seeds, according to panspermia theory, would have to have been hardy to survive the cold of interstellar space, as well as airlessness and radiation, and they would then have had to pass through the thick atmospheres of their new homes without incinerating from friction.

The borderline theory of cosmozoan microbes was revived in the early eighties by Francis Crick and Leslie Orgel.[12] Their rationale was that life emerged on the Earth too suddenly to be indigenous and, additionally, that protoplasm itself varies in key ways from any terrestrial medium in which it might have evolved (for instance, cells use the rare element molybdenum in critical enzyme functions, an unlikely choice for terrestrial indigenes). Likewise, life is unitary; all life uses the same molecular codes and dialects to communicate with itself across species and kingdom barriers. This essentially random choice could not have been made the same way more than once, for there are trillions upon trillions of possible variants and equally viable alternatives.

Crick and Orgel's modernized extraterrestrial scenario involved anaerobic (nonoxygenating) cells bioengineered and packaged by intelligent aerobic scientists on some distant dying world concerned to preserve life itself (if not their own mode of it) elsewhere in the galaxy. Anaerobic organisms would stand the best chance not only of surviving the journey but adapting to any of a number of alien environments. Upon arrival millions of years later in a solar system unimagined by their senders they would spin a novel life chain, different from the one on their native world. Whole new bionts would spring from these nucleic seeds. The genetic code does in fact have suspiciously biotechnological aspects, but these have simpler explanations.

Panspermia theories require fantasies of super-scientists on remote worlds. For instance, Crick and Orgel imagine an original sun-star from the most ancient epoch of the universe, with billions more years of existence than our third-generation Sun.

This would give time for the origin and development of life and the maturation of a civilization, a process culminating in the approaching extinction of the star and a launch of seeds. Yet we do not even know the history of the European Middle Ages that well, so we can hardly speak for the ancient peoples of the Milky Way.

IN THE LATE 1990s, with the discovery of possible fossilized microbes in a meteorite that likely originated in the northern hemisphere of Mars, some scientists considered whether the missing link between molecular matter and life might lie on another, *nearby* world. Life would then have come a much shorter distance (and by pure accident) from an uncultivated biosphere.

For instance, the primal Earth might have been heavily enough bombarded by asteroids and comets that incipient organisms would have been exterminated many times, our ecosphere sterilized. If a primordial ocean-covered Mars had escaped fatal onslaught, microbial life evolving there could have travelled to Earth on a chunk blasted into space by an asteroid. In such a circumstance, if human beings were (billions of years later) to colonize Mars and terraform it, their offspring would not be foreigners propagating their templates by Martian molecules as much as terrestrials formed in Martian microbes originally, returned to Mars to give birth to primate offspring out of Martian stuff again—Martians who never could have evolved on Mars itself!

This plot is interesting more for its possibility than its likelihood.

THE INGREDIENTS OF BIOSPHERES originate not only on planets but throughout interstellar space. Snowballs form directly from galactic matter. Simple compounds of hydrogen, oxygen, and carbon, including sugars, glycerin, fatty acids, and amino acids, occur wherever the constituent materials are present: on comets, meteors, asteroids, and in cosmic dust. Among meteoric salt veins and microgeodes of iron-sulfide crystals nest purines, ethanol, and nitrogen-rich porphyrins, forerunners and building blocks of chlorophyll, hemoglobin, and other enzymes. Meteorites, cracked open, have revealed configurations resembling vacuoles, sea urchin spicules, double membranes, wriggly wormlings, and, in one case, the fossil of a cell in mitosis.

The vitalists were not wrong on one point: apparently, form precedes life. Biogenesis is morphologically and mysteriously grounded in cosmogenesis.

Gaia

THE MORE SERIOUS PROBLEM is that a panspermia doctrine does not solve the riddle of the origin of life; it simply transposes it to other worlds, one of which still must be the womb. The dilemma was never our raw material (which is abundant) but our DNA template, which is native to somewhere.

Another kind of solution was first advanced in 1924 in a paper by the Russian scientist A. I. Oparin entitled "The Origin of Life."[13] Oparin reasoned that the best remaining alternative was to violate the law of the impossibility of spontaneous generation with a single instance at the dawn of the Earth's history. One "spontaneously generated" organism could then have given rise to all subsequent plants and animals through natural selection.

Oparin described a previously unknown planet—the Earth before biological infestation. Carbides and heavy metals shot through the rocky crust and, from the superheated steam of the atmosphere, a steady downpour of hydrocarbons cascaded into the rising sea. J. B. S. Haldane later advertised the ocean of this world as "a hot dilute soup."[14] The brew was unique to its epoch and could occur only in the absence of life—a perception of Charles Darwin back in 1871:

"It has often been said that all the conditions for the first production of a living organism are now present which could ever have been present. But if (and oh! what a big if!) we could conceive in some warm little pond, with all sorts of ammonia and phosphoric salts, light, heat, electricity, etc., present, that a protein compound was chemically formed ready to undergo still more complex changes, at the present day such matter would be instantly devoured or absorbed, which would not have been the case before living creatures were formed."[15]

With this mitigating clue Haldane, Oparin, and others began to reconstruct the meteorology and incipient biochemistry of the abiotic Earth. In the absence of both photosynthesis and prior life forms, the atmosphere would have contained little or no oxygen but much stellar hydrogen that had not yet escaped into space or been bound into other compounds. Too much oxygen too soon would have consumed any primitive organic molecules and, as ozone, blocked ultraviolet radiation needed for protocellular energy. The most ancient life was thus likely anaerobic and produced its energy in a reducing (hydrogenizing) environment by fermentation rather than oxidation.

This hypothesis was experimentally tested in 1952 when Stanley L. Miller and Harold C. Urey recreated a version of primitive "soup" in their laboratory with a Jovian atmosphere and terrestrial gravity. Methane, water, ammonia, and hydrogen molecules were subjected to electrical discharges. The solution spewed many organic molecules, including some of the twenty amino acids used in our protein code. However, other amino acids and organic molecules *not* used by terrestrial life equally appeared. This was promising but hardly definitive.

More notably, the addition of oxygen to the original "atmosphere" of the flasks eliminated all molecules found today in living systems.

Obviously, days in a test tube cannot replicate hundreds of thousands of years in ancient seas and tidepools, so no experiment has spawned anything resembling even a poor subvirus. But the dawn planet probably shared much with these Miller-Urey jars that it does not with the contemporary planet that evolved from it.

The primitive Earth was the cradle of life; the modern Earth is a by-product of life. The first chlorophyll molecules transformed photons (quanta of electro-magnetic energy) into starches and sugars (quanta of carbohydrate energy), enabling cells to feed themselves and, ultimately, their predators, from our single proximate storehouse, the Sun. At the same time, they purloined carbon dioxide from the air, returning oxygen.

Under bombardment of ultraviolet radiation, oxygen fusing into ozone exuded a membrane at the outer border of the planet, absorbing the impact of subsequent cosmic rays, effectively shielding the planet from life-threatening wavelengths. By then there was enough energy in the pot not to require high cosmic voltage.

Paradoxically, the cells generating this skin lay under miles of ocean. As their own breath gradually shielded them, they rose slowly, generation by generation, to the surface. Water and ice scything across land ultimately eroded its new mantle, depositing fresh minerals (and molecular variations) for use in the assemblage of primitive DNA and its nascent life forms. Meteorites and asteroids—some gigan-tic, others little more than dust—delivered a ceaseless rain of exotic carbon and other mineral pellets from cosmic slag. At the same time, the transformation and breakdown of stone released more oxygen into the atmosphere. Between twenty-two hundred million and fifteen hundred million years ago, a surge of fertile photosyn-thetic organisms breathed out a new world habitable for oxygenating life forms. Together these organisms stabilized the atmosphere at its present equilibrium.

Prior to this epochal event, oxygen's participation in the Earth's atmosphere amounted to, at most, a part per hundred million. If plant life were suddenly to dis-appear, it is estimated that the weathering and oxidating of ferrous iron would remove all but that one part per million from the atmosphere in a mere million years.

The first plants and plant-animals were the unconscious purveyors of a design so elegant we wonder today if the Earth is not a single organism. It has already transformed its own atmosphere, developed a protective skin, adjusted its temper-ature, and scorched a nervous system across its crust. A pebble of the Sun that once laid cells of homespun algae among its waves now integrates telecommunications filaments from the computers and satellites of its primates.

3

The Materials of Life

The Molecular Lattice

A S MATTER SIFTED THROUGH HETEROGENOUS MESHES, the biosphere gradually differentiated itself from other zones, incorporating aspects of earth, air, and water, returning them only to borrow again. Jellyfish and whales are literally water; birds and insects are too. They are likewise air—large and small blubber balloons.

The biologist J. D. Bernal calls life "an epiphenomenon of the hydrosphere."[1] This is true even for creatures who have never seen or touched an ocean. Where life exists beyond water it does so only synthetically, by including sea within its membranes. Generations of flora and fauna have captured the aboriginal waters in their tissues through a series of embryogenic inversions, so their insides are a displaced replica of the pool in which life began. The alembics of our bodies continue to keep cell cultures alive within semi-permeable membranes.

The main ingredients of life are the lighter and more reactive elements, the first notes built by stars on a hydrogen scale. Four of the most abundant and simplest elements constitute 99.4% of the human body (and 99.9% of the biosphere). Hydrogen is the digit (1) on the periodic table; carbon, nitrogen, and oxygen are 6, 7, and 8, respectively. Life begins in such bare mathematics. The quantum numbers of the electrons of the elements and the kinds of multiple bonds they form (especially in the context of molecular water) lead to complex and varied macromolecules. These become the components of amino acids, nucleic acids, proteins, fats, and starches, fundamental to the embodiment of protoplasm.

If we broke the human body down into atoms, we would find that 63% of them are hydrogen, the most basic and abundant element in the universe (and on the primeval Earth). Sixty-six percent of seawater (H_2O) is also atomically hydrogen.

The next most abundant element in both the human body and seawater is oxygen—25.5% in us and 33% in the oceans. From hydrogen and oxygen atoms set loose in the original atmosphere, water vaporized and deliquesced into seas. The present-day atmosphere is 21% oxygen, but, as noted in the previous chapter, this is a by-product of photosynthesis and not the primal condition. The Earth's crust itself holds most first-generation oxygen atoms—scientists estimate 47% of the original complement—still trapped (oxidized) in rocks.

The next two most abundant elements in the human body are carbon (9.5%) and nitrogen (1.4%). The nitrogen of life originated atmospherically, a realm which is still over 78% nitrogen. Carbon atoms make up 3.5% of the Earth's crust, but seawater contains virtually no nitrogen and less than 0.01% carbon. Since the biosphere, as a whole, is almost 25% carbon, we assume that carbon was appropriated wholesale from the waters and atmosphere by nascent organisms. Living creatures are now the crust of the ocean—in the elemental sense, its "earth."

Much of life's carbon must have been subsequently cosmic, infused in stellar dust with silicates and metallic iron and nickel, and blown in the solar wind off the Sun's corona into the Earth's vicinity.

The Nature of Substance

WHAT ARE THE PRIMARY ELEMENTS? Do they have any essential qualities other than the geometry and mathematics of their lattices and valences that endow them with properties in their bonds with themselves and one another? Does anything dispose them to life?

Water, stone, plants, and animals are all composed of hydrogen, oxygen, carbon, and nitrogen; how and wherein their characters are engendered are, for the most part, a mystery. Yet if we search for elemental rudiments in substance, we discern their faint inklings—"an expression of the primary reality of cosmic shaping forces working in material condensations."[2]

In the 1940s German occult chemist Rudolf Hauschka, summarizing generations of vitalist biology, proposed intrinsic primal qualities that work their way through the webs and labyrinths of substance without losing their quintessential nature. Whether true atomic attributes can translate into molecular and cellular dispositions is a riddle that lies at the heart of another riddle, and we will explore both of them throughout this book. For now it is useful simply to consider the possible raw properties of elements as a means of understanding the basic characteristics of all physical systems in the universe. After all, nature builds in layers. Without a substratum of integral properties, more complex entities comprising subtle combinations of qualities cannot be assembled. And if attraction is the basis of life, it arises intrinsically at an atomic level.

Hydrogen is the lightest substance on Earth, and its compounds tend toward becoming gases. No matter how heavy the substance making a bond with hydrogen, the resulting entity usually takes to the air: methane (CH_4), phosphene (PH_3), hydrogen sulphide (SH_2). Water vapor (H_2O) easily turns atmospheric. Even heavy lead is aerobicized by hydrogen.

Hydrogen is heat-giving too. "It has the hottest of all flames. Iron and steel are welded by an oxyhydrogen torch that uses a mixture of hydrogen and oxygen, and hydrogen is the source of heat in all other autogenic welding processes.... Now the question arises whether this tendency is to be regarded as a purely physical phenomenon of anti-gravity, or as the last visible remnant of a cosmic fire-force that pervades the universe as a dissolving, de-materializing element?"[3] That depends on whether fire or gravity is more etiological but, either way, hydrogen appears to provide the fieriness and hot-bloodedness of life, conferring both its buoyancy and warmth. The heat and airiness of this element imbue the solubility of water and the blossoming of flowers; they provide animals with their fierceness and all creatures with courage and enthusiasm. Hydrogen is probably the underlying force expressed by the human heart.

In fact, apart from the whirlpool-like and lotus shapes of galaxies, hydrogen ("water originating element") is a poor name for this kernel; it was appended by French scientist Antoine Lavosier, the founder of modern chemistry, to an unseen gas that arose from his laboratory water. "If hydrogen were to be baptized with a name indicative of its inner nature," declares Hauschka, "we would have to call it 'pyrogen' ('fire-substance')."[4]

WE THINK OF OXYGEN AS AIRBORNE, but only about a fifth of the Earth's atmosphere is oxygen. Much more terrestrial oxygen is bonded to hydrogen in immense reservoirs of water. By chemical weight, pure water is almost 90% hydrogen.

Hauschka defines oxygen as the originator of "being," the carrier of life, or "the bearer of forces whereby 'being' becomes 'appearance....'" It combines with almost all substances and makes them capable of chemical reaction. Silicon, calcium, and other elements become chemically active only when they have [bonded] with oxygen, which enables them to become silicates, lime, and so on."[5] He thereby renames this element "biogen."

Water itself is pre-protoplasmic, pre-cellular, forming rhythmical funnels and complex vortices around obstacles, shifting molecularly within whirlpools of varying temperature, slowing itself with turbulence, percolating through rock and soil.

In organisms these will become bone and tissue. Of all inanimate compounds water is the most supple, self-regulating, and mindlike.

CARBON IS THE SMALLEST ATOM in the group of the periodic table that rests midway between the elements that give up electrons in their bonds and those that take them up. The carbon atom has equal proclivity to gain or lose the four electrons in its outer shell, so it forms a high variety of stable compounds, including ones using sodium, hydrogen, and chlorine—prime ingredients of ocean water. In addition, the smallness of the carbon atom allows intimacy with other elements, for carbon can bring its electron veil very close to the nucleus of an atom with a positive charge. Such bonds are fast ones.

Carbon spun the protoplasmic cloth. Its atoms have the same capacity to form bonds with one another, so, in their congenial company, they attach in extraordinarily long chains and, when the ends of the chains meet, carbon rings occur. Many of these diadems then collect secondary and tertiary chains, including ones attached to other rings. The bonding propensities of ancient carbon spun billions upon billions of exotic crystals. This choreography (performed as a ballet in 1939 at the national meeting of the American Chemical Society in Baltimore) filled the sea with lattices of carbon, hydrogen, nitrogen, oxygen, phosphorus, sulphur, and other elements.

If oxygen (biogen) provides being and appearance and hydrogen (pyrogen) contributes stellar expansion and heat, carbon is needed to balance and stabilize them into carbohydrates, grounding them in earthy matrices. Plants use carbon to convert ontological and fiery properties into fixed material substances in the form of starches and sugars (see Chapter 5).

"If no bounds were set to pyrogen the carbohydrates would become formless, as they do in sugar, color, scent, and pollen, and would be etherealized away into the cosmos."[6] However, without hydrogen, the life-force of oxygen would freeze in carbon crystals and convert into pure wood (cellulose).

CARBON WAS THE LOOM, but the tepid waters, which have ever after been a breeding ground, nurtured the fragile membranes. Water is by far the most abundant molecule yet found in the universe in a liquid state, so its proliferation on Earth is both fortunate and ordinary. Its own molecules continuously reorienting in relation to one another, water is a soft, chaotic balm, flexible enough not to freeze protein in crystals (as the hydrogen bonds of other hydrides like ammonia do). Water makes secondary hydrogen bonds with proteins and thus preserves their complex structure. H_2O molecules also help develop electrical charge in cytoplasm, saturating it with life energy.

Thus, life oscillates among form, fire, and being—between stability and volatility, petrification and dissipation, the ethereality of water and the configurability of carbon bonds.

NITROGEN, COMPRISING ALMOST 80% of the atmosphere, is true air-substance, bestowing dispersion, motion, and flux. Far from being the incendiary atom connoted by its name, nitrogen is a diluter and neutralizer of fire, a rhythmic pacer of substance and breath. Left to themselves, pyrogen and biogen would ignite the atmosphere in a sunlike conflagration. Nitrogen distributes and tempers their forces into breathable air. Likewise, hydrogen and oxygen alone in the nervous systems of organisms would explode in a paroxysm of directionless sensations. Nitrogen diffuses and sorts their neural sparks into feelings.

Hauschka credits nitrogen with introverting and permutating protoplasm into a system of functional organs in animals. From their fused expression of nitrogen's cosmic nature, pyrogen, carbon, and biogen are stretched, interiorized, and activated into heart, lungs, intestines, kidneys, and muscles. Organisms are deeded independent, autonomous mobility—freedom of action.

Secondary Elements and Trace Properties

THE OTHER ELEMENTS THAT ARE USED in plants and animals have been borrowed from the environment in more infinitesimal amounts for their specific silting and charging properties in enzymes and proteins. Both phosphorus (0.22% of the human body) and sulphur (0.05%) are essential in coenzyme molecules. In addition, phosphorus is a crucial component of ATP (adenosine triphosphate), which participates in the basic energy relations of cells and is a unit in the formation of nucleotides of DNA. Phosphorus is also used for support in creatures with bones. Sulphur is a component of many of the amino acids employed by the genetic code.

Innumerable other elements are required by creatures, though none of them singly compose even one-half of one percent of the atoms in the human body or the biosphere. Many of them are present in amounts less than one-hundredth of a percent. It may be that some elements became part of animate systems initially because they were travelling in just the right sector of the ancient waters and so reacted with existing carbon chains. Once their atoms were trapped in protein structure, their unique characteristics were incorporated and utilized.

Mutations and gene variations regularly patented new regimes. No pellets could hide, not even ones that just happened to be there. Their atomic properties and the properties of their compounds, when activated by emerging dynamic systems, were integrated into a variety of harvests: the structural resilience of developing skeletons; the chemical engines of respiration, digestion, reproduction; and the subtle biochemistry of enzymes, coenzymes, and hormones.

Cells are not electrically neutral; they maintain a charge on both sides of their membranes—negative within and positive without. Positive and negative ions—cations and anions—regulate this distribution and also the osmotic pressure within and without membranes. From the beginning, cells had to keep some ions out while including others, though all were part of their environment. It was the capacity of primitive entities to internalize selective elements and maintain their properties within membranes that marked the beginning of bounded life forms.

POTASSIUM MAY HAVE ADHERED to the more clayey parts of the organic broth when it was first forming, for its molecules have difficulty sticking to water. Once potassium is in a cell, it establishes its electrochemical properties as a cation (lacking an electron in its shell).

At their existential basis our cells have hoarded potassium and magnesium as cations and excluded sodium and calcium, other cations; this way bodily fluids maintained an electrochemical neutrality. Anions contributing to the balance include chlorine in the form of chlorides, sulphur in the form of sulphate ions, and phosphorus in the form of phosphate ions. The charges in these particles also help to regulate the chemistry of blood, lymph, hormones, and other internal liquids.

Elements were expressed again and again at different levels as creatures evolved. Extracellular calcium provided the basis for shells of the first invertebrates. Magnesium is a crucial component in the photosynthesis of plants and the derivation of necessary enzymes in animals.

The unique traits of metals, even in trace amounts, make them critical. Copper is used among invertebrates in hemocyanin for oxygen transport and in various enzymes for photosynthesis and skin pigmentation. Iron is integral for oxygen transport in hemoglobin and is incorporated in a wide variety of other enzymes. "[W]ere it not for iron's healing property we should be constantly poisoned by the cyanide compounds formed in the process of digestion. The iron in our blood, however, instantly transforms these compounds into harmless ones."[7]

Cobalt can replace iron in some functions of blood chemistry and is vital for DNA biosynthesis and amino acid metabolism. Nickel and manganese participate in the formation of red corpuscles. Zinc is indispensable for protein digestion in enzymes, and in the formation of carbon dioxide and the metabolism of alcohol. Molybdenum is a prerequisite for the metabolism of purines. Selenium enables liver function. Vanadium and niobium are used in the respiratory systems of sea squirts and other invertebrates.

"Why blood is red and why the grass is green are mysteries that none can reach unto,"[8] wrote Sir Walter Raleigh. But the physical basis, at least, of this reality has been disclosed. Bernal reminds us that the greenness of plants originates in the alternating single and double bonds of the magnesium-bearing chlorophyll molecule, and:

"The redness of blood ... is written into the molecule of haemin; this is to be found not only in the blood of vertebrates, but also in the larvae of some flies, the bloodworms in stagnant pools, and in the nitrogen-fixing nodules in the roots of peas. In all these cases the color is effectively due to the quantum states of the complex, partially filled electronic shells of ferric iron as modified by the porphyrin groups in which it is placed. Electron shells have existed as such ever since the first iron atoms were built inside a primitive supernova."[9]

Primordial matter enters into life in a multiplicity of ways, oblivious to partitions between vegetable and mineral, cosmic and terrestrial. We are made of star stuff and meteorite debris. Organisms are defined by their developing capacities to make functional structures out of metals and stones, wrapping their layering around them while retaining many of their key properties. Bacteria forge tiny internal loadstones out of magnetite; even today fungi assimilate the toxic by-products of industry and chemical warfare; oceanic invertebrates turn barium sulfate and calcium phosphate into sense organs (otoliths) and bones or shells, respectively. The boundary between geochemistry and biology is artificial and fluctuating.

JUST ABOUT ALL THE LIGHTER ELEMENTS are used in life, with the exceptions being those that are inert, like helium and neon, and those that are unambiguously poisonous like beryllium and arsenic. Fluorine has a structural role in bones and teeth. Silicon, similar to carbon, has the capacity to form chains (though not as extensive) and is used structurally as well, for instance, in diatoms. However, fluorine and silicon, as well as other elements, play unknown and conceivably subtle roles in living organisms. Though highly corrosive, iodine is a component of thyroid hormones; it is the heaviest element known to be essential to life, but its exact role is obscure (flocks of sheep become ragged and diseased without their minim of iodine).

If we take into account the possible roles of subdetectable quantities of substances, like homeopathic microdoses and the molecules identified by occult chemists, then no substance naturally occurring on the Earth can be excluded from suspicion of biopoesis. Gold and silver have been used medicinally, as has mercury. Perhaps the biosphere has developed "herbal" isotopes, microdoses of these toxic elements within coarser substance that make substances that stimulate form. About rare potions like ytterbium, cerium, and rhenium, we can only guess.

Life has formed from the beginning while being poisoned by intruders who, in most cases, were accidentally trapped within membranes. Where the intruders did not fatally disrupt the primitive organism, they got included in such a way that their toxicity became first neutralized and later incorporated in the framing of genetic or enzymatic messages. Their incorporation was also their neutralization. Since their presence automatically gives rise to new properties, selection ultimately was a matter of which organisms would creatively integrate properties and which be overwhelmed by them.

Oxygen was one of the early poisons, but without it, life could not have generated enough energy to maintain itself and diversify. No doubt iron and copper were toxic at first too. Our antecedents probably fought them off—and then magnesium and phosphorus—before accepting them, and becoming them. At every level of our emerging complexity we deny ourselves chemically, and then use that denial to continue our growth. No wonder on a psychic level we continue to deny and then reinvent ourselves from shadows. We were never made whole. We were birthed, as the pre-Socratic philosopher Heraclitus divined, from strife itself.

Metabolism and Reproduction

The elements of life did not, of course, assemble seamlessly in the ocean. Before there was an organization or plan, parts came together haphazardly and structures braided in random associations.

Any creature exists through the confluence of two events so remote from each other in scale that the life form is their only meeting point. One of these events is the developmental continuum of successive generations of replicating, mutating life forms over millions of years; the other is the succession of stages in a developing embryo, lasting anywhere from a few hours to a few thousand hours. Unicellular entities become multicellular, historically and again in each birth. Simple membranous animals develop deeper, denser structures and neuralized networks through accumulated adaptations within increasingly complex ecosystems made up of other plants and animals. Then they replicate that complexity embryogenically.

Phylogenesis is one long planetary embryogenesis which sprouted initially without seeds and without genetic material. Embryogenesis, conversely, carries out a brief, synthetic phylogenesis. If embryogenesis took even a billionth the duration of phylogenesis, creatures would not survive their gestation and, in fact, would not survive at all, for the succession of generations would occur too sluggishly for adaptation to changing environments.

EVERYTHING ABOUT LIFE ON EARTH suggests that it arose at a unique site and then spread. Life is a singular chemical event. There are not two kinds of life, or three kinds of life; there is one. All living cells resemble all other living cells: their asymmetrical molecules rotate the plane of polarized light in the same direction, they use the same chemical reactions to metabolize, and they reproduce from the same genetic molecule—they speak to one artisan.

The kinship of life is a more fundamental trademark than the divergence of species. That is why the cells of caterpillars can metabolize the cells of elm and apple leaves, why the anteater's tissues draw sustenance from morsels of ant. Even viruses and bacteria belong to our lineage; if they did not, they could not read our codes and appropriate them. All life on this world is the clone of a single cell, whether it was indigenous to the Earth or seeded itself from some other world where it arose presumably in a similar fashion.

TWO FUNDAMENTAL CHARACTERISTICS distinguish creatures from the environment around them—their capacity to assimilate other substances for their metabolic requirements without losing their identities in the mutual reactions, and their ability to reproduce themselves precisely. For a chemical composition to harbor these characteristics it must first be organismally self-maintaining (autopoietic); it must experience changes in the environment instantaneously and respond to them strategically (with organized motility).

Creatures with these capabilities are composed of proteins and genetic molecules, among the most fragile structures in a universe in which the hardest stones are worn to less than dust eventually. Their mere existence represents remarkable molecular pliancy combined with dynamic mutability. Although, without exception, life must be reincorporated into the general chemistry of the planet, its metabolizing fabric resists such degradation far longer than it would if it were an inanimate compound.

The survival of single minute organisms in the vast ocean was remarkable enough, but even the most complex macromolecule is "indifferent to existence: chemical systems have no priorities," hence no "genetic continuity."[10] They all would have been meaningless solo acts amidst pelagic anarchy were it not for their complementary capacity for replication. As Bernal points out, "without definite molecular reproduction it is very difficult to see what an organism means: if it is merely a piece cut out of an undetermined extension of metabolically active material, it has no *raison d'être* of its own."[11] If a fortuitous living creature came about and could not reproduce, then inertial dissipation would eliminate not only that zooid but its unique configuration. We would then have to await another chance creature for a mimicry of life.

But if any creature, by hook or by crook, were able to replicate itself, then its mortality would be incidental to its issuance of progeny. They would replace it, proliferate, and be replaced at their own deaths. Precise reproduction is, of course, distorted ceaselessly by mutations, but life also seizes this crisis, to change, evolve, and diversify.

LOOKING BACK ON THE UNLIKELY CIRCUMSTANCES that spawned us, the biologist Francis Crick wrote: "... it is impossible for us to decide whether the origin of life here was a very rare event or one almost certain to have occurred."[12] Conventional Western religion assumes the former but solves the problem with divine intervention. Western science invokes the vastness of the ancient seas and the large number of possible marine, tidal, and estuarine sites for molecular association and thus assumes that life was inevitable.

We can tilt the odds any way we want, but it seems strangely wonderful, indeed, that all this came from nothing and now sits contemplating its own event.

4

The First Beings

Polymerization

MOLECULES TAKE ON MYSTERIOUS PHENOMENOLOGIES from the configurations of their atoms and then, from their own bonds with one another, yield compounds with even more astonishing properties. Hue, resonance, abrasion, incandescence, phosphorescence, buoyancy, stickiness, fabric, heat, odor, symmetry, density, striation, and contour all arise from an invisible underworld and, tangling with one another, twist out and anneal into nodules we call "things." The geology room of any natural-history museum displays the myriad shapes and heterogeneous colorings of inanimate stone. The same ilk of compounds parades with deafening rumble and screech across the face of Neptune in swirling clouds.

Whenever entities are integrated into more comprehensive systems they lose a portion of their prior identity and gain radical new identities with novel qualities. Atoms make molecules; molecules assemble compounds. Cells aggregate into tissues, tissues into organs, organs into organisms. Since all matter, life, and consciousness derive ultimately from subatomic particles and their ostensible components (more akin to energy than matter), we can explain the transcendence of particle nature only by the introduction of unique characteristics—emergent properties—at each next level of synthesis. These characteristics are a lot thicker—more motley and agglomerated—than mere aggregate expressions of elemental quintessences.

The simplicity of carbon vanishes into the traits of primeval organic chains held together by carbon bonds; polymers then transubstantiate carbon-based monomers. With the passage of energy and heat from system to system, congeries invent their own avant-garde physics. Without any violation of the laws of thermodynamics sunlight floods and mutates through molecular lattices into cell life. There is no exogenous source for biological complexity, and in no way was it prefigured by atoms or subparticles before they arrived at it through radical interactions and bonds.

The tendency to shuffle, combine, and disperse energy is apparently an inherent propensity of molecules and their compounds, passed on to minute animals, mammals, and ultimately to thought itself which continuously sifts, associates, develops valences, and connects, and cannot stop this activity even in sleep. Associations established by the first cells continue as jellyfish colonies, termite nests, flocks of birds, tribes of primates, parasites and hosts, mating pairs, legislative bodies. Of course, these must elicit hydrogen, oxygen, carbon, nitrogen, and the like in some fashion, but they express them only as any edifice reflects the stone and mortar of its bricolage.

RANDOM CARBON-BASED MONOMERS apparently once teemed within the primeval "soup," but in order for these molecules to polymerize, two events had to coincide: concentration and energy, the former bringing together components of a potential membrane-enclosable system and the latter catalyzing its molecular reactions. We have no idea how rich the prebiological soup was or, for that matter, how rich it had to be. Harold Urey suggested that it consisted of as much as 25% organic material; however, other biologists surmise that as little as 0.1% would have been sufficient for biogenesis. After measuring the amount of organic stuff in chicken bouillon, Leslie Orgel accepted it as a rough equivalent of the primeval waters.

Only if all living things and their by-products were dissolved back into the sea could its original gelatinousness be restored.

Through aeons of transformation, the Earth's biosphere has been projected out of its hydrosphere. The bath, substantially thinned, is now bubbling with piscean gems of that alchemy.

THE OPEN WATERS WOULD HAVE PRESENTED twin obstacles to polymerization: a cold aqueous environment dissipates energy while, at the same time, waves tend to dash incipient chemical chains. However, Oparin felt that, with their sheer abundance of cosmic molecules, the ancient seas overwhelmed any such objections. Repeated bombardment by light mixed with ultraviolet rays would have charged and mutated reactions near oceanic surfaces. The earliest proto-life forms at the time were probably viscous droplets which were in the process of separating with colloidal particles from the general hydrosphere. The colloids would be differentially charged in layers but in states of equilibrium as they floated through similar, but more dilute molecules. These hypothetical coacervates (as Oparin named them) were made up of polymers including primitive proteins, albumin, and gum arabic.

Whole regions of sea came partially alive.

THE MAIN OBJECTION TO OPARIN'S CREATURES was that they seemed already too complex to precipitate from unorganized deliquescence. The least excursive way around this was for relatively small portions of broth to have become isolated in biochemically propitious microenvironments. For instance, thin layers of high organic concentration might be blown along as foam and then deposited intact onto shores. Organic tea might collect in shoreline cavities or be sequestered in inland ponds.

Lipidlike slicks deposited on beaches would likely retain many of their bubbles. They were cell motifs waiting to happen. The surface of each bubble is a prospective membrane, a pellicle for assimilating nutrients and turning an oil globule into a life form.

Crick presumed that the Moon was in fact once significantly closer to the Earth, its back-and-forth tug "produc[ing] continual wetting and drying . . . in pools near the margins of the oceans and seas."[1] Polymers were synthesized with the aid of

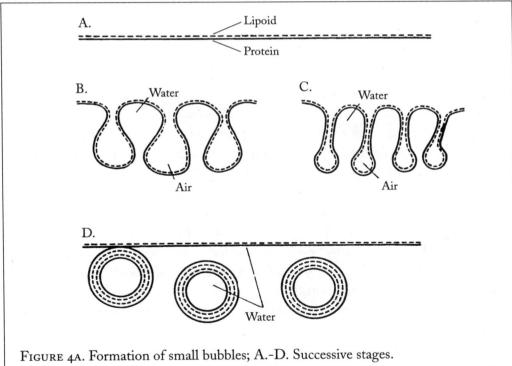

FIGURE 4A. Formation of small bubbles; A.-D. Successive stages.
From A. J. Oparin, *Genesis and the Evolutionary Development of Life* (New York: Academic Press, 1968).

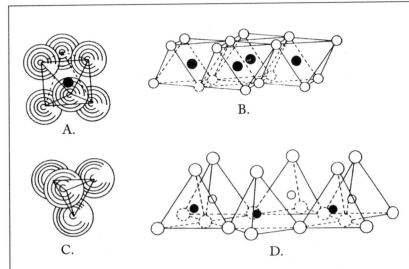

FIGURE 4B.
Precellular units and sheet structures.
A. Octahedral unit;
B. Octahedral sheet structure;
C. Tetrahedral unit;
D. Tetrahedral sheet structure.

From Mella Paecht-Horowitz, "The Possible Role of Clays in Prebiotic Peptide Synthesis" (*Origins of Life*, Volume 5, #1, 1974).

periodic changes in standing chemical composition and levels, recurrent dehydration of the substrate, and regeneration of reactive sites among monomers once they bonded to each other, leading to long, recurrent chains of them. Spring tides then distributed sun-dried hydrocarbons, ultimately globally.

Even at the present-day neap, preorganic material would have been deposited along estuaries in suspension with clay and water, forming a kind of ooze in which porous matrices of atoms assembled geodesically in octahedra and tetrahedra. These flexible grids incited catalysis of a wide range of chemical reactions. At ebb tide the colloidal wafers would have been imprinted on mud banks, the absorption of activated molecules simmering charges across sheets (for instance, between negative silicate layers and positive aluminum ones). The volts would bind molecules to sites and provide free energy for protein synthesis and polymerization.

Life arose from the mud, a golem quickened by a spell. The setting was Cytherean: a tidepool, a sunny beach, a cradle fed by surf foam. (If Europa's or Callisto's subglacial seas are heated by tidal forces, local equivalents of coacervates might have formed on the Jovian moons too.)

More recent experimenters have discovered that amino acids synthesize in polymers of two hundred or more at hot-spring temperatures between 160° F and 210° F. "Proteinic microspheres" are small spherical bodies manufactured in tepid environments, alternatives to coacervates. If some oceanic "soup" found its way into a volcanic cinder cone—perhaps in waves at high tides—subterranean heat might have polymerized the brew while it was evaporating; then microspheres would have washed back into the next high tide.

Crystals

CONDENSED ENVIRONMENTS ALLOW different molecules to "explore" possible relationships. Atoms of similar form (either identical ones such as gold or different ones with similar size, shape, and/or charge) draw together, forming crystal beds. Groups of molecules cluster into balls like those of copper in pipes from which substance has been worn.

In the absence of complex proteins serving as enzymes and coenzymes, such assemblages were probably catalyzed by minerals in equivalent roles, that is, transferring energy to maintain consonant sequences of bonds.

The analogy to crystals is striking. These molecular forms also repeat and restore themselves. Oparin was infatuated with "ice flowers" that formed on his windowpanes; they looked to him like tropical vegetation and suggested a transitional domain between minerals and cells.

Twinning, dimensionality, and regulation are properties of biological as well as of geological fields. Early polymers took crystalline form, and later DNA helices constructed themselves on series of gemlike spirally displaced axes.

PATROLLING THE QUASI-SYMMETRICAL transition between matter and life, an innate wave pattern gives structure to a variety of phenomena prone to "snap back" into stable patterns—from proteins to swarms of bees, from myth cycles to supergalaxies. Migrations of stones through the Earth's crust "mimic" tissue composing

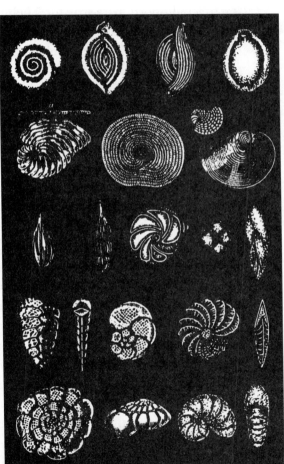

FIGURE 4C. Forms of Foraminifera.
From William Hanna Thomson, *What is Physical Life?* (New York: Dodd, Mead, and Company, 1909).

organs and dressing wounds; ice likewise knits winter's lakes. Collagen fibers of bone are similarly regenerative, sliding past one another to fill gaps (with a pressure great enough to generate an electrical current). The lens of the eye is a gigantic crystal consolidated from proteins—themselves elaborate micro-crystals.

Crystals have so long tantalized us with their lifelike characteristics that metaphors merge suspiciously with morphologies, subatomic and galactic events with transcultural ones. Aerial photography has revealed cities growing outward like amethysts, axis by axis. Houses are also crystals, as are cars, clocks, and cyclotrons. Computers have now exposed human symbol systems as vast quasi-symmetrical lattices mapping crystal-like syntaxes and morphologies. There is likewise an invisible crystal (deep proportionate structure) behind languages, laws, and concertos. The mathematics describing a crystal must also, in some sense, *be* a crystal.

The difference between geological and biological crystals seems to be the adamantine rigidity of stone. Life maintains a more supple, open-syntax congruity.

Anaerobic Bionts

THE HYPOTHETICAL DIVIDING LINE between a lifeless polymer and a life form is impossible to mark. Many chemical phenomena (fire, for instance) behave like animate beings, and some creatures seem as inert as stones (tiny tardigrades in diapause lie dormant for decades; viruses can remain latent seemingly forever). Less metaphysical but equally imponderable is the question of which came first: the cell metabolizing substances, or the gene copying itself? And how did they affiliate in a single mechanism? Did a sticky chemical reaction become surrounded by a membrane, or did a macromolecule begin somehow replicating itself? How did metabolism within a membrane survive the aeons it would have taken to develop genetics to preserve it, or how did the naked gene reproduce itself accurately from scratch without catalysts, or sustain itself without metabolism? There is only one possibility: they developed in concert, coevolutionarily.

After the miracle of their emergence 2.7 billion years ago, Earth's maiden creatures burbled and hissed, producing energy by fermentation, gobbling up hydrogen sulphide and other ubiquitous poisons. Anaerobic metabolism would be a sluggish and inefficient battery by present standards, but time was irrelevant in the Archean world. Rudimentary life forms would not rot, and there were no competing microorganisms to devour them.

Some contemporary bacteria use hydrogen to produce methane; others metabolize by sulphur. (Perhaps anaerobic beings permeating the soils of Mars greedily devoured food sent to them from America by spaceship in 1976 in order to test if they existed, or perhaps "they" were only a peroxide reaction. Not only do we always ask these same questions, but we get the same tantalizing answers.) Like creatures could have evolved in dark underground rivers, energized partly by radioactivity; they might occur (for that matter) even in the cores of meteors and asteroids. Without air, without light, without water, they would be torpid, mineral-like creatures. Although they would have very little evolutionary potential in our terms, intelligent microchip life forms may have developed somewhere in the universe.

Photosynthesis

AFTER MANY EPOCHS—a billion to two billion years (depending on your shale source)—a radical new chemistry tapped the Sun and, with its burst of energy, a matching kingdom of entities arose and spread across the planet, trumping the denizens of the anaerobic age. Photosynthesis is the ability of certain molecules (resonating lattices of carbon rings around central magnesium atoms) to split apart hydrogenous substances (usually water), freeing hydrogen to reduce carbon in various compounds, including carbon dioxide. A chlorophyll molecule fluctuates between configurations as its oscillating grids trap, store, and translate the energy of light quanta passing through them. Networks of three hundred or so chlorophyll entities operate as photosynthetic units, absorbing photons, jumping their electrons to higher orbits, and transferring them in these excited states (in a few trillionths of a second) from one molecule to another within the hive. The result is complex light-harvesting and amino-acid-transforming effects.

Although photoreceptive, chlorophyll is inert structurally so does not steal for itself; at the same time it is able to catalyze the bonding of hydrogen to carbon in the splitting of carbon dioxide. Groups of chlorophyll molecules later distribute the liberated quanta among themselves so that energy continues to be transferred in metabolism instead of being degraded into heat.

It is pointless to try to guess how carbon and magnesium discovered such quixotic properties in each other, but, once imbedded in biological systems, they dug in for the long haul, reversing the balance of energy in the microcosm. The carbon-magnesium grid produces a far greater surplus than the simple metabolism which it supports can use, so the rest is available for other reactions. These run the gamut, fueling bionts as small as rhodophytes and as large as whales.

Plants turn sunlight into organic compounds; animals without this capacity must avail themselves of nutrition in other life forms. Since nonphotosynthetic organisms cannot capture sun directly, they pilfer energy from plants or from animals that have consumed either plants or other animals that have consumed plants, etc. There is no separate pathway of energy, no subsidiary food chain. The voltage of this one grid, from its inception, has been tapped by all living systems since.

Life did not soar in a single beanstalk from Jack's seed; it had to be guided through eyes of needles, gardens and groves of mazes, at different scales. Proto-enzymes ensured that currents generated through one metabolism were transferred to another, or, to state it in neo-Darwinian terms, substances that simplified bonding became included within membranes in an organized way because membranes which captured such substances had a competitive advantage, thus thrived in the ancient brine at the expense of their rivals. Some early polymers may themselves have served a catalytic function for others, hence the birth of true protein enzymes, coenzymes, and the types of chemistry that assemble macromolecules like proteins and nucleic acids.

Through these primordial, epochal events the energy of the Sun has now been quarried and stored for billions of years, relocated to the bodies and activities of plants and animals, and to the tribes and civilizations of this planet.

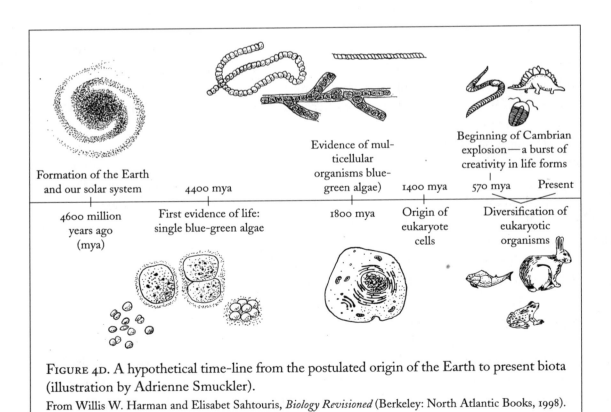

FIGURE 4D. A hypothetical time-line from the postulated origin of the Earth to present biota (illustration by Adrienne Smuckler).
From Willis W. Harman and Elisabet Sahtouris, *Biology Revisioned* (Berkeley: North Atlantic Books, 1998).

Somatacist Stanley Keleman concludes:

"Life is old, very old, and we are old in it. The continuum of existence we experience has no discrete beginning.

"A living process is an eternal process.

"A living event is committed by a continuum to billions of years of existence, an infinite chain of living events."[2]

The DNA and segmentation in the first cyanobacteria are the same hereditary molecule, the same episodic body motif found in a monkey. The shape of life begins in the ancient abyssal ocean with the manufacture of energy from glucose and a dance of light quanta in a weir of carbon rings about a chlorophyll snare.

Prokaryotes and Eukaryotes

THERE IS NOTHING IN PRINCIPLE requiring biological entities to be comprised of cells, but this is the manner in which plants and animals on Earth have

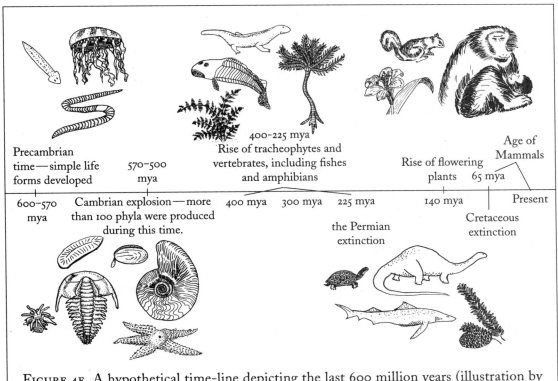

FIGURE 4E. A hypothetical time-line depicting the last 600 million years (illustration by Adrienne Smuckler).

From Willis W. Harman and Elisabet Sahtouris, *Biology Revisioned* (Berkeley: North Atlantic Books, 1998).

manufactured themselves. Elsewhere, among the vagrant galaxies, perhaps quite different types of structures have come to house oxidation, hydrolysis, glycolysis, phosphorylation, or other metabolic processes—even bionts not requiring embryogenesis for assembly. Down here, it is cells or oblivion.

The most primitive cell still in existence on our planet is the prokaryote, with no distinct membrane-bounded nucleus and a single genetic molecule. Nowadays, prokaryotes are represented solely by around 10,000 different kinds of bacterial microbes (including the blue-green bacteria once thought of as algae because they use sunlight to split carbon from carbon dioxide and nitrogen from nitrogen gas). Although bacteria are evolved contemporary organisms, not aboriginal remnants, they suggest what primitive simplified cells might have been. Their company includes fermenting organisms, spherical microbes, and thin, motile spirochetes in the shape of helices (as, for instance, are found in victims of syphilis). Lacking fractionalized subentities and deep overall structure, microbes do not have extensive internal membranes or enough molecules to assemble a mitotic spindle, so they do not divide by traditional mitosis. They regulate metabolism and transmit hereditary information in only one type of cell, which they already are, so they cannot join up and differentiate as tissue or develop specialized functions. Each prokaryote is an indelible monad.

Prokaryotes swarmed over the primal Earth. With no free oxygen in the atmosphere their metabolism was necessarily anaerobic, synthesizing chemical energy by degradation of sugar phosphates and other complex molecules.

The aggregated cells that constitute modern plants and animals, from paramecia to camels, are called eukaryotes. Either these life forms arise as single-celled algae and amoebas, and roam at liberty; or they consolidate in tightly integrated populations of one another that take on shapes of gophers, falcons, tulips, redwood trees, field mice, etc.

Each eukaryote is also composed of a diverse lot of heterologous monads ("homologous to a community of microorganisms"[3]); these differentiate and cohere in a variety of subcellular structures called organelles—in a sense, the organs of cells. Quite different from one another in history, design, and function, these components include entities that literally function as cells within cells (mitochondria, chloroplasts, and a nucleus); specialized sub-animalcules like cilia, microtubules, microfibrils, and microfilaments; primary structural elements (a membrane and intracellular membranes); and membrane-derived minions of uncertain origin (ribosomes, Golgi bodies, and lysosomes). Their various faculties come to contribute unique pathways of energy for aerobic life.

Endosymbiosis

THE MOST OBVIOUS ORIGIN for the differentiated cell lies in the population dynamics and commensalism of prokaryote beastlings whose life cycles became so entangled in one another that they merged and coevolved. This is the only way we can explain the remarkable intricacy of the simplest remaining true creatures, the protozoas, which could *not* have been assembled, *ex nihilo,* in the primal soup.

Far more rudimentary life forms than these (akin to simple bacteria or organelles) once wriggled from tidepools or volcanic cones and later fused together. The meshing of federated animals out of free critters is not so bizarre as it first seems. If parasitic bacteria can serve as actual flagella in protozoan parasites of termites, and nitrogen-fixing bacteria can regulate the digestion and protein synthesis of leguminous plants by attaching themselves to root hairs, then beings, ancestral to both organelles and bacteria, surely could have combined to form one-celled plants and animals.

In 1970, biologist Lynn Margulis hypothesized that a primordial eukaryote cell originated in an endosymbiotic fusion of prokaryotes (that is, a consortium of organ-

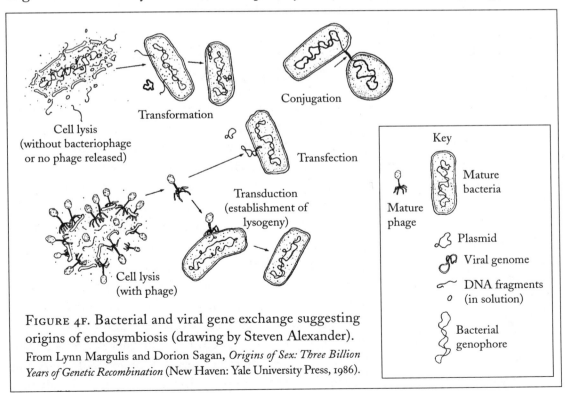

FIGURE 4F. Bacterial and viral gene exchange suggesting origins of endosymbiosis (drawing by Steven Alexander).
From Lynn Margulis and Dorion Sagan, *Origins of Sex: Three Billion Years of Genetic Recombination* (New Haven: Yale University Press, 1986).

isms of different species maintained throughout the life cyles and separate generations of all of them). By this etiology the mitochondria, the nucleus, and even the membranes of cells are each the descendants of disparate free-living organisms—assorted species of protobacteria, evolving mongrelly in bits of sea foam or long-standing puddles, and swimming independently for millions of generations before serially merging in cataclysmic episodes.

Throughout ocean worlds of the late anaerobic Earth, raffish zooids no doubt stalked promiscuously. These forerunners of cells hunted, collided, ravaged, and mixed. Alien plasmids, viruses, and protogenes were donated or appropriated. Such were the standard biology and sociology of the time. Then—to a large degree because of a mushrooming eukaryote population—the molecular balance of the atmosphere tilted in favor of oxygen metabolism.

Protoorganisms who historically fended for themselves gradually had their metabolisms spliced into a polymorphic, heterotropic colony. Thereafter, each of the components of the colony (the cell and its nucleus)—though still partially subject to differential selective pressures at deviant intervals and mutating and reproducing at asynchronous rates—had their destinies joined within the greater metabolism of an emerging system. We examine the eukaryote cell's evolution and assortment of monads in detail below and in the next chapter.

Organelles

Mitochondria

The conjectural first stage of cell amalgamation was the capturing, by other primitive organisms, of mitochondria on the high seas some 2200 million years ago. Without mitochondria, modern animal cells would have access to only old-fashioned methods of energy production: anaerobic splitting of glucose sugars and the like; they could not metabolize oxygen. Though no creature back then responded to oxygen as anything other than an unwelcome poison to be shunned, some life forms ancestral to mitochondria apparently developed an incipient ability to neutralize the exotic molecule and convert it into energy.

Oxygenation is a complicated series of chemical reactions following respiration: "Beginning usually with food molecules in the cytoplasm (carbohydrates like dicarboxylic acids, such as malic, succinic, and fumaric acids), carbon, oxygen, and hydrogen atoms are removed and the hydrogen reacts with oxygen so that the end products of eukaryote cell respiration are carbon dioxide and water. The energy released from food breakdown is stored in ATP nucleotide molecules, components of DNA and RNA."[4] Nowadays the mitochondrial factory is so streamlined and efficient that thirty-six molecules of ATP are generated from each molecule of glucose oxidized (compare this to the meager two-for-one performance of anaerobic glycolysis). Creatures able to complete the oxidative decathlon, even in a primitive or inefficient manner, stood ready to claim the biomass of this planet (see the next chapter, "The Cell," for more detail on ATP).

Today the descendants of these precocious intracellular breathers have their own membranes and proteins (distinct from the enveloping cytoplasm); their own DNA with its peculiar dialect of genetic language (among other things, lacking conventional "stop" codons); their own small ribosomes for transporting hereditary material; their own RNA with a different rhythm of replication from that in the cell nucleus; and a characteristic metabolism—all in keeping with a purported bacterial parentage. (For a more thorough account of mitochondria and their hereditary contributions, see pages 74–75.)

Changing shape plasmatically as they parade and undulate their bodies through cells, mitochondria resemble the stiff, oblong bacteria from which they apparently descended. Relatively large-sized for subcellular entities (0.5 to 1 micron), they sometimes fuse with one another only to part company again. As tiny rings of contemporary mitochondrial DNA synthesize their twenty or so proteins, the mitochondria themselves extend, fuse, and split in two, just like cells but at a higher speed and energized by their own intermembranous enzymes.

Mitochondria also adapt themselves to cellular environments. In sperms they wrap themselves around the tail (the flagellum) and participate in rotatory movement. In heart cells they align with the pulsating pump.

AMINO-ACID ANALYSES and X-ray crystallographic examinations of modern bacterial proteins offer circumstantial evidence that mitochondria arose from an ancient purple photosynthetic bacterium in which light-harvesting capacity had already deteriorated, leaving only a primitive respiratory pathway for metabolism.

A contemporary bacterium, named *Thermoplasma acidophila* for its tendency to thrive in torrid, acidic springs, boasts an eukaryote-like protein around its DNA, shielding it from environmental corrosion. If a hardy archaic form of such an organism, predisposed by mutations to oxygen tolerance or even preference, swallowed but did not assimilate an oxygen-spewing microbe, it might have developed its own oxygen-metabolizing properties. That is, the formative DNA of the zooid in the process of being cannibalized resisted digestion and continued manufacturing its proteins in foreign protoplasm, translating predation into symbiosis and trading nuclear material in genderless syzygy. Natural selection over many generations of increasing atmospheric oxygen could have regimentalized and preserved such contamination in orthodox scripts. Protomitochondria, when incorporated into the emerging metabolism of a cell, eventually evolved into true mitochondria.

By a different skein, a garrison of incipiently oxygen-metabolizing parasites could have functioned therapeutically within a bacterium if, by their accidental presence, they "cured" its oxygen toxicity. From their abode among its cytoplasm they could consume waste products and surplus nutrients and at the same time benefit their host by gobbling up oxygen molecules. If the hosts had already developed primitive membranes—in part, to protect their DNA from oxygen degradation—these membranes would gradually splice with those of the parasites, becoming continuous with them.

Hybrid creatures able to process the new abundance of toxic gas now had an immediate competitive advantage. They and their kin could seek out rich zones of molecular oxygen "pollution," flourishing at the expense of anaerobes. Roving and colonizing once uninhabitable environments, they became miniature hegemonies, enlisting more and more protomitochondria and membranes to protect and enhance their deep genetic components. Additional parasites of assorted ilk and skills were also invited harum-scarum into the party; a few stayed.

The descendants of oxygenating invaders breathe for modern cells. Whereas anaerobic metabolism occurs throughout eukaryote cells, oxygen exchange for these bionts is localized solely in the hundreds of respiratory enzymes of their mitochondria.

Protoctists

The first eukaryotes on Earth, loose confederacies of bacteria, gradually became distinct as miniature plants and animalcules. However, at the cell level, the difference between a plant and an animal is almost meaningless. Modern biologists tend to classify plants as photosynthesizing organisms containing plastids, and animals as hungrier, more mobile organisms without plastids. Yet *Euglena* is one of many active photosynthetic microorganisms whose existence suggests the need for a third kingdom of terrestrial life (or a fourth if the fungi are also enfranchised). Margulis has proposed the name Protoctista for this group—a kingdom embracing all known life forms with the exception of animals, plants, and fungi, thus comprising phyla of early eukaryotic creatures and some of their descendants (both unicellular and multicellular), including dinoflagellates; red, brown, and yellow-green seaweed; amoebae; ciliates; diatoms; and xenophyophores, mysterious inhabitants of the depths that are known only from their barium-sulfate skeletons.

The diets of protoctists are more similar to those of plants and animals than those of bacteria. No protoctists have embryos; some of them carry incomplete sets of organelles (lacking microtubules or mitochondria); others develop sophisticated structures made of microtubules. Some do not undergo mitosis. They range in size from barely more than a micrometer in diameter to hundred-meter-long seaweeds. Their earliest forms were likely the single eukaryotic cells Margulis calls "protists"— bacterial communities with genetic molecules, the hypothetical ancestors of every fungus, plant, and animal.

A protoctist kinship salvages water molds, slime molds, and chytrids from having to be classified as fungi, and recruits seaweeds out of the botanical kingdom, placing all of them among kindred entities that share a primitive morphology and unique life cycle.

Plastids

Like mitochondria, plastids of plant cells have their own genetic molecules, ribosomes, and membranes, and likely share heredity with the bacteria they resemble. Their nucleoplasm is cyanobacterial not algal. Green and blue-green bacteria, swallowed or invaded by primitive mitochondria-bearing life forms, might have lived on symbiotically in their harborers—hence incubating the forerunners of plants. The divergence of plant and animal lineages from each other was fated primevally by the former's acquisition of plastids. A horsetail is an entirely different entity from the snail crawling up it. The ledgers of botany and zoology began with a distinction of algae from amoebas.

Different-colored plastids include not only highly elaborate chloroplasts—the famous organelles of photosynthesis in plants—but also simple leucoplasts and amyloplasts (starch and lipid plastids *sans* both chlorophyll and full arrays of membranes), chromoplastids (organelles bearing yellow to red carotenoid pigments), and etioplasts (primitive plastids of plants grown without sunlight).

Simple light-activated pro-plastids lacking chlorophyll molecules most closely resemble the ancestral forms of photosynthetic bacteria that must have infiltrated cells. A variety of mutations compelled the inner envelopes of these bubblelike plastids to invaginate; flattened vesicles then formed. These later pinched off in folds, compressing and intensifying into a photo-chemical apparatus. For this metamorphosis to have consummated itself, the outer membranes of ancient chloroplast precursors must have gradually imported proteins (synthesized externally on cytoplasmic ribosomes) into their own molecular sequences. Thus they mixed very different lineages of DNA to concoct remarkable new structures. Ultimately they expressed superb electrical artistry in their manufacture of great botanical umbrellas and wreaths towering upwards toward the stars.

Whereas mitochondria infected sub-biological clusters that were neither plants nor animals nor even protists, chloroplasts were cell invaders of a much later aeon, giving rise (by their inclusion) to green algae, red seaweeds (inside which blue-green bacteria mutate into photosynthetic organelles known as rhodoplasts), and the entire botanical kingdom. The first chlorophyll organisms were ancestral to both plant and animal cells. The zoological kingdom then arose from a branch of early bionts able to develop separate lifestyles out of the oxygen and sulphur produced by their chemistry.

Peroxisomes

As the early cell developed, its survival was placed in immediate jeopardy by the plethora of toxic enzymes in the ocean (phenols, formaic acid, formaldehyde, and alcohol among them). For the conversion of pestilent chemistries, a visiting spherical zooid likely entered the cell and then evolved as a vesicle with a single membrane and a granular matrix slightly favoring electrons; the peroxisome is now an inhabitant of every eukaryote.

Modern peroxisomes remove hydrogen atoms from organic materials through an oxidative process yielding hydrogen peroxide, which they then convert by native enzymes into water (hence their name). Likely relics of the first primitive organelles that participated in the metabolism of cells after photosynthesis began to flood the Earth's atmosphere with oxygen, these microbodies are inefficient by comparison with high-powered mitochondria (because they produce no energy). Though peroxisomes became partially obsolete over time, they remain critical within liver and kidney cells whose role is to detoxify the bloodstream.

Peroxisomes resemble mitochondria and chloroplasts in their fissioning to produce daughter peroxisomes, but they carry no vestigial genetic molecules of their own, so their generations must be re-encoded solely by nuclear genes. They have archived their entire reproductive capacity outside their bodies; in this act of streamlining they are not alone.

Microtubules

The forerunners of these tiny hollow organelles might have been swift-swimming bacterial spirochetes that bore flagella inside the outer membranes of gram-negative cell walls. Such autopoietic creatures still procreate happily in muds, animal guts, and other anaerobic environments. Perhaps in a time of food scarcity, some microbes chose to imbed in mitochondria-bearing cells. Once inside, their proliferating colonies were absorbed into the cytoplasm; there they became individually streamlined, losing cell walls and plasma membranes, and ceding most of their biosynthetic functions to the nucleus.

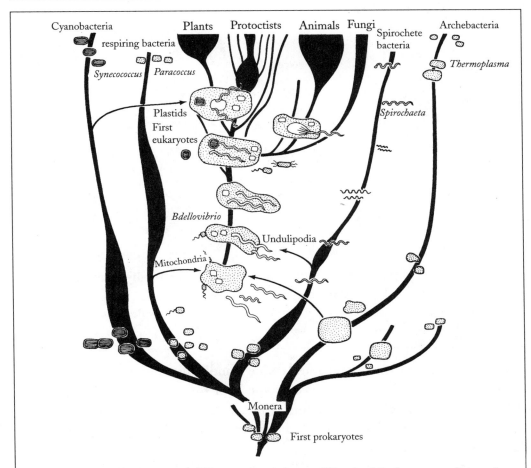

FIGURE 4G. Phylogeny of the five major kingdoms of Earth, showing organelles evolving from bacteria (drawing by Laszlo Meszoly).

From Lynn Margulis and Dorion Sagan, *Origins of Sex: Three Billion Years of Genetic Recombination* (New Haven: Yale University Press, 1986).

Fully symbiotic with their protoeukaryote hosts, spirochetes lost the ability to reproduce outside the cytoplasm. Their motility became transferred to (or differentiated as) locomotory organelles within their new cells. The vector-waves of their swimming became internal organ-like parts (organelles) and utility functions.

A universal dial-shaped core (the kinetosome) of nine microtubules suggests a common origin for a host of multitalented organelles; their initial metamorphoses may have been into undulipodia—rippling bundles assembled in either long axial threads (flagella) or short beating quills (cilia). Their common central shaft (the axoneme) biodynamically developed in a nine-plus-two microtubule arrangement.

Their intracellular lineage includes the propulsive tails of sperms as well as other

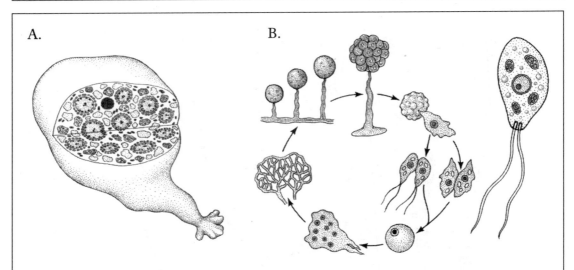

FIGURE 4H. Protoctista phyla. A. Karyoblastea; B. Amoebomastigotes life cycle and Myxomycotes (drawings by Laszlo Meszoly).

From Lynn Margulis and Dorion Sagan, *Origins of Sex: Three Billion Years of Genetic Recombination* (New Haven: Yale University Press, 1986).

exotic structures (see the next chapter). Nine-bundle microtubule clusters without axonemes formed centrioles (or basal bodies) for the mitotic spindles of animal cells. Thus, one spirochete-originating ultrastructure contributed both the prerequisites of motility and the fulcra of cell division to different lineages of its descendants inside tiny protoorganisms (see figure 5F, page 79).

As spirochetes lodged in host cells, they became subject to their metabolism and manner of reproduction. Many of them clustered in organizing centers to become involved later in mitosis and meiosis. Retaining only selected aspects of their structure and motility, these invaders gradually devolved from autonomous zooids into organelles dependent on host cytoplasm for reproduction and food. No longer free-living, the precursors of microtubules were compelled by hunger into general eukaryote metabolism, employing their formerly independent proliferative and agglutinative gifts to erect eclectic cylindrical structures that were then put to the service of cell differentiation.

The genetic message of the spirochete remnants became a source subcode for their polymorphous production of organelle derivatives; like that of peroxisomes it was transferred, Margulis presumes, over time to host RNA, leading to "many kinds of cell morphogeneses, most of which involved assembly of microtubule protein into microtubules...."[5]

Spirochete vestiges were ultimately replicated by surrogate templates rather than

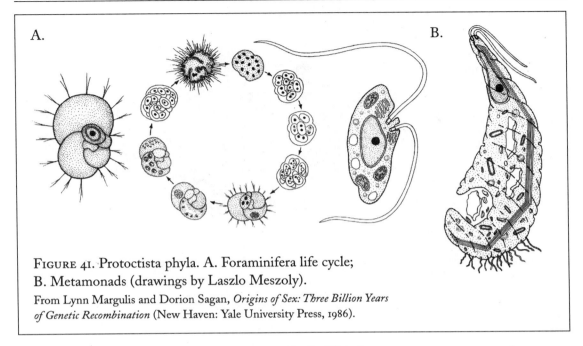

FIGURE 41. Protoctista phyla. A. Foraminifera life cycle;
B. Metamonads (drawings by Laszlo Meszoly).
From Lynn Margulis and Dorion Sagan, *Origins of Sex: Three Billion Years
of Genetic Recombination* (New Haven: Yale University Press, 1986).

their own DNA (which they no longer needed). This is a common energy-saving evolutionary concession (familiar in the acquiescing of the sperm and egg to the zygote), for " . . . as long as one complete copy of organellar DNA is present in the cell, wherever it may be, the DNA of other organelles may be safely lost."[6]

Where these spirochetes originally settled became, in the evolution of the eukaryotic cell, traditional organizing centers for microtubules—mechanisms of not only cell motility but mitosis and meiosis (see Chapter 7, "Sperm and Egg").

MEANWHILE, THE HOST, in order to preserve both the skills of its visitors and its own mitotic capability, came to fission not into two independent cells but into double cells, protists joined to each other; that way, neither information nor structure was lost, and the resulting creature was not only larger and more formidable but multiple. As this series of unifying replications continued, more and more cells were generated and linked, like soap bubbles popping out of each other and sticking together. Organicism progressed from monads to spiralling chains. These irregularly centripetal cavalcades provided fabric and versatility for later embryogenesis of multicellular creatures. Margulis summarizes:

"Cloned and differentiated, the 'microbes' become plants and animals. . . . Prokaryote sexuality was a preadaptation for tissue differentiation; cannibalism followed by indigestion (inability to digest conspecifics) was a preadaptation for meiotic sex."[7] Segregation became consolidation at a higher level.

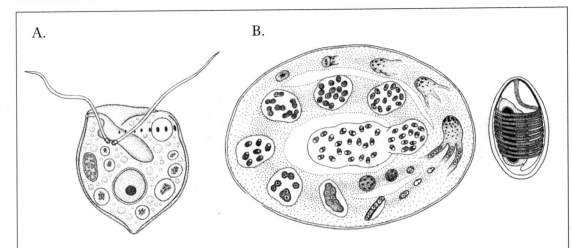

FIGURE 4J. Protoctista phyla. A. Cryptomonads; B. Microsporidians (drawings by Laszlo Meszoly).

From Lynn Margulis and Dorion Sagan, *Origins of Sex: Three Billion Years of Genetic Recombination* (New Haven: Yale University Press, 1986).

One type of energy, motion, and replicative principle—bestowed by spirochetes on their sanctum—was translated into another. External exploratory movement and shape-shifting—inverted and miniaturized biomechanically and morphologically—became intracellular engineering and, ultimately, raw material for multicellular diversification and development. A novel kind of collective motility emerged, as spirochete-infested, mitochondria-bearing creatures found themselves whipped along by swimming organelles. They were now super-predators, some with stabbing cells or poison darts: trichocysts, toxicysts, and nematocysts. Abridgment and confinement of function and energy became expansion and extension in a different dimension; monads, impressed into mere ancillary roles, contributed to the deployment of an exponentially greater whole. Eukaryotes developed a wider and more acrobatic repertoire than prokaryotes, including pinocytosis (cell drinking through narrow, deep channels inside the cytoplasm), phagocytosis (cell eating through engulfing particles), exocytosis (particle secretion and removal through a membrane), endocytosis (internal cell eating), and mitosis.

Margulis concludes:

"The undulating movements of the spirochetes conferred selective advantage on the early eukaryotic complex. The associated spirochetes developed permanent attachments to their hosts. With time they became entirely dependent on the metabolic products of their hosts. . . . The complex motility systems of eukaryotes hypothetically derived from this original merger. . . . The elaborate processing of nucleic

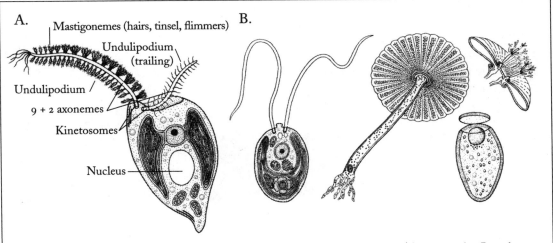

FIGURE 4K. Protoctista phyla. A. Chrysomonads; B. Chlorophytes (drawings by Laszlo Meszoly).

From Lynn Margulis and Dorion Sagan, *Origins of Sex: Three Billion Years of Genetic Recombination* (New Haven: Yale University Press, 1986).

acids, the large ribosomes, and the incessant internal activity of eukaryotes behave as products of composite ancestry."[8]

Microfilaments and Intermediate Filaments

Microfilaments are animalcules which function within cells like the contractile fibers of muscles. Confederations of actin-protein monomers, they may orient in bundles parallel to the long axis of the cell. Like microtubules, microfilaments have a status in cell differentiation and thus are integral to mitosis and tissue formation. They run in grids directly beneath membranes, organizing into belts before cell compression, and dispersing afterward. Microfilaments also form a stable meshwork in the cortex of cells.

INTERMEDIATE FILAMENTS are coiled protein assemblies differing from tissue to tissue. They include the keratins and neurofilaments of highly specialized cells. Their participation creates new classes of tissue with novel arenas of activity (such as neural transmission, horny coatings, or hormone synthesis). [See page 78.]

Cellular Membrane

From their mutual synthesis of proteins and lipid fats, all of the clustering organelles became engulfed by a great crystal—perhaps thickened sea spume, perhaps coagulating lava infusion. An outer envelope coalesced out of sheets of uniform thickness.

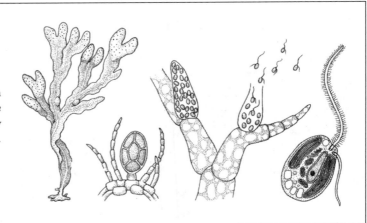

FIGURE 4L. Protoctista phyla. Phaeophytes (drawings by Laszlo Meszoly).

From Lynn Margulis and Dorion Sagan, *Origins of Sex: Three Billion Years of Genetic Recombination* (New Haven: Yale University Press, 1986).

This film trapped their endosymbiosis in a bounded zone—a super-enriched, metabolized pool. The membranous net within a cell resembles on a smaller scale the organism that forms the bright blue float of the Portuguese man-of-war. Its osmotic texture striated selectively to allow some molecules into the bubble while removing others. Membrane ceramics turned experience into microsculpture and microsculpture into experience. The inside of the inside of our bodies, hence, the inside of time, originates here.

The organelles were imprisoned but also protected and fed. They gave themselves—their bodies—to collective bodies, so were included profoundly as their autonomies were eradicated. Some were swallowed up in the dynamic relationships and fusions of others and vanished entirely (the enucleated red blood cells of mammals and dedifferentiated mitochondria in fermenting yeast are present-day relics that survived this process). Yet despite the loss of many phenotypes in cells, some genetic record—a copy of each heterologous genome—is preserved in the cell nucleus. Thus, the cell is only a partial expression of its full archived history and heredity, a mystery this text will ponder again and again.

THERE ARE NO special jurisdictions for a chaparral thicket, a lake, a spruce bog, or a communal microorganism except that the concentrated space of the latter and its intensified metabolic and genetic interactions exponentialize successions of generations and inculcate the effects of natural selection. Changes accumulate rapidly in dense hothouses. Though at first small and few, eukaryotes experimented with trillions of new concatenations, altering their internal ecologies (metabolisms) and teeming into millions of previously unimaginable niches within the microspheres of the Earth, altering planetary ecology in the process.

Differentiation is a radical conversion of bacterial and organelle chemistry with cosmogonic consequences extending throughout nature. Long ago, internalization and condensation of microbial plans deep in the cell bodies of our progenitors transmogrified into tissue differentiation. Symbiotically specialized, cells had no choice but to continue to express their inheritance—that is, impetuous "sperms, eggs, and spores are destined to repeat the incomplete cannibalistic encounters of their ancestors."[9] This results in lineages of flagella, cilia, and microtubules locked in an eternal dance of mitosis, meiosis, membrane manufacture, embryogenesis, commensalism, and predation—as long as there is a biosphere.

The violent origins of microbial hunting and mating may (in retrospect) supply some of the imperative of mammalian emotional life, the restlessness of our protoplasmic natures, the turmoil of our conquests. We inherit the fissionability, volatility, and adventurism of the early eukaryote universe.

The Nucleus

How replication of cells began is difficult to imagine. Crystals repeat structure, but they are inert and, in a sense, two-dimensional. They give rise to three-dimensional entities but only by operating repetitively on a sequence of surfaces. In biological replication, the genetic molecule must be able to "feel" what a preexisting crystal is like in *all its dimensions simultaneously* and then transmit the plan of its precise reassembly. Simple eukaryote cells have an innate predilection "to engulf and 'examine' many things: prey bacteria, particles of sediment, glass, and today even plastic beads. This tendency to engulf and examine is [an] aspect of the preadaptation of eukaryotes for what, in the end, became the meiotic sexual cycle."[10] The cycle later lodged in nucleic acids and found sanctuary in the cell nucleus.

Discovered in the 1830s, the nucleus first looked like a separate homogeneous mass. Its constancy and relatively large size (five to ten microns) were beguiling, like a *mare* on the Moon. By early in the next decade, biologists observed it coagulating into little hieroglyphic bundles. It was not long before better staining techniques revealed this granulation to be a by-product of cell division. In 1884 the bundles were named chromosomes (literally "colored bodies"). By the end of the century biologists knew that each organism had a set and even number of chromosomes in each of its cells and that, from cell to cell in an organism, chromosomes strongly resembled one another. Their critical role in heredity was not yet recognized.

The nucleus, like the eukaryote itself, is probably a gradually fused association of captive prokaryotes (within of course another, more disperse association of captive prokaryotes). The ribosomes—irregularly shaped hybrids of RNA and protein molecules—have a particularly independent role within the nucleus as they paddle back and forth like underwater snails between messenger RNA and amino acids. Once upon a time they invaded ancient organelles like viruses and, while attempting to steal their protoplasm, were instead impressed into using their latent genius to copy *it*.

Whereas all of the other organelles are concerned solely with immediate metabolism, the nucleus is the site of genetic transmission: the storage vat of heredity and the assembly plant for its replication. It conveys the chemical identity of the cell through time and space. Of course the nucleus has no such high ideal. Prior to its authorship of mitosis, it was detained in the cell by prehistoric symbiosis and must enact the process that its nucleotides now require. As it expends energy, new proteins are formed according to a blueprint notched along its body.

CELLULAR CLUSTERS COHERED and "accepted" their expanding hierarchies. The daily business of the internal milieu—its metabolism—took priority over any lingering independence in the organelles. The nucleus claimed reproductive control of all of them and rebuilt them by generations, like robots, always planting its trademark at their heart. Nevertheless, organelles continue to carry out aspects of the distinctive behavior that lured them into a community in the first place. That remains part of their value to the cell. They did not lose their identities or intrinsic functions; they merely componentialized them. Something in them remains liberated, forever outside the full reach of the cell "mind."

The nucleus is the one organelle that reverts to a feral state, shedding its membrane during cell division and enacting a round dance of ancient flight. Mitosis is the last throes of a trapped beast to fly the colony, to break the trance that somehow binds it. Instead of kicking free, the animalcule spins complexly in place, replicating its own dilemma to eternity and casting forth the creatures of subsequent time. Its gyrations win it a sort of freedom, if not *out of* the cell then *through* the cell into the multiplicity of plants and animals.

Imprinting scrolls through the corridor of its own componentiality, this chromosome-bearing organelle is a link between two utterly different types and scales of universes. In a suitable medium a wild nucleus can program the formation of trillions of cells, not only modelling them on prior cells but differentiating them from one another so that, even after millions of years, its offspring routinely compose gigantic creatures in three dimensions. Because of this activity cell endosymbiosis has been transformed into organic unity.

The true outcome of nuclear diligence is expressed finally in a superorganism transcending time and space, dwarfing both the original cell and its organelles, and its genes. DNA is, in truth, a single animal, unaware of the many costumes it wears. It is the Earth's only absolute, indivisible organism.

We are componential beings

So, WHERE IS OUR UNITY? From where does the singular existence we experience arise? For most of recorded history human beings have thought of themselves as completed entities with rational goals. They might be possessed by other entities or temporarily lose consciousness—regularly, in fact, in sleep—but such episodes are considered deviations. Even spirits and ghosts are perceived as whole and intentional.

The cell with its organelles is a serious breach of identity. It condemns us to being clusters and clumps of mites swarming into shapes of organs, swimming in cavities, nuzzling in the marble of bones. We exist as life forms only because cells collaborate blindly in our manufacture and mucilage. Each person is a wriggling, charged heap of defective amoeboid and paramecian zooids trapped in colonies.

Most of the events in this book rest on an incredible thing: we are made up of billions of separate cells themselves made up of subcellular autonomies. Though operating as a cohesive integrity in tissue, they are each existentially independent animals with lives of their own, using us merely as the medium of their survival and reproduction. Without them we don't exist. The baby cries, but the components of the baby are emotionless adults carrying out pond ecology.

Our existence as cells is both fantastic and ordinary. To the mind it makes no sense at all, but it is the sole reality. Perhaps it is a strange way to mold golems—plants and animals—but nature takes to it like a duck to water. Accretion of properties and enlistment of monads into other monads may in fact be the only way the universe can build complexity and consciousness. So this is who we are.

Beyond cells, we have zero epistemological or metaphysical claim. We are not offspring of gods at all; we are assemblages of microzoans—cell nations which have voices, however singly faint. Our philosophy and phenomenology are profoundly and quintessentially bacterial and cellular. The production of proteins and the birth and death of cells themselves—even though molecular rather than cerebral—contribute profoundly to our overall intelligence and sense of identity.

Where Platonists once hoped to find qualitative geometries regulating being, microbiologists discovered only indeterminate animals in search also of paternity.

HISTORICALLY, THE CELL is the forerunner of the unconscious mind—its prophecy and now its replica. Through his analysis of human behavior, Sigmund Freud demonstrated a powerful inaccessible vortex at the identity and core of every organism, far vaster in its domain and contents than all that is conscious. He did not mean it to be subcellular, but this is the direction in which we have looked for just about everything since his time, including the genesis of life. If all but the conscious aspects of our bodies were extinguished, everything we consider commonplace would instantly disappear; our world would either go totally blank or turn into an unimaginable nausea. Many psychologists still hope to prove that the unminded cellular substratum *is* the unconscious mind.

Post-Freudian philosophers openly acknowledge the fragmentation and multiplicity of their own personalities, the primacy of interior realms they will never know. We see their subcellular expressions in dadaist and cubist art; disjunctive, nonmelodic symphonies; and literature generated by algebras of collage.

OUR IMAGINED WHOLENESS is but affiliations of cells in cabals, our brain collections of nonthinking entities conducting thought. Each of these conspirators is a mosaic of vestigial bacteria stalking their own food, drawing their own breaths. Yet their coalescence and synergy have somehow usurped the reality of their separate existences. This is the power-rush felt by the cart-drawing ox, the descending hawk, the breaching whale. Associations of bacterial creatures make up oak leaf and lizard. They were established among protists, and there was nothing snails or octopi could do but live them and pass them on. By the time proto-human apes formed tribes and shone the first symbols on Earth, they were already doomed to discover in some twentieth after twentieth century minute predators occupying every shred of their being and pseudo-wholeness.

Alienation may be a modern symptom, but the identity crisis that it discloses has existed from the dawn of our lineage.

Self is our invention, not our heritage.

5

The Cell

Infusoria

FREE-LIVING CELLS WERE DISCOVERED IN 1674 through a microscope that was little more than a bead of glass mounted in metal and held up to the eye by a focusing pin. When the Dutch lens-grinder and amateur naturalist Anton van Leeuwenhoek examined standing pond-water under a magnification greater than a hundred diameters, he saw, to his astonishment, a horde of tiny luminous creatures swimming about. Some spun like wheels, running into each other and retreating. Others crawled like shapeless snails, engulfing their unfortunate neighbors. In ocean water he watched little "fleas" hopping great distances (for their size) within "the compass of a coarse sand grain."[1] In rainwater he found himself looking at a vorticella whose eye-stalks he described as "two little horns." He called it "the most wretched creature I've ever seen for when, with its tail, it touched any particle it stuck entangled in it, then pulled itself into an oval and did struggle by strongly stretching itself to free its tail, whereupon the whole body snapped together again leaving the tail coiled up serpent-wise...."[2]

Van Leeuwenhoek soon found that these creatures dwelled also in staggering numbers in hay, dust, dried mud, dirt from roof gutters, and moss. Since they arose spontaneously from infusions of those substances, they were named "infusoria."

It took more than a generation for people to get used to the news that the interstices were more densely inhabited than forest or sea. Van Leeuwenhoek "dug some stuff out of the roots of one of my teeth ... and in it I found an unbelievably great company of living animalcules, moving more nimbly than any I had seen up to now.... Indeed all the people living in our United Netherlands are not as many as the living animals I carry in my own mouth this very day."[3]

At first this microscopic world might have seemed a Nereid kingdom, but a century later, biologists classified van Leeuwenhoek's infusoria as elemental seeds out of which "real" animals are made. A renowned taxonomist, Count Georges Buffon, described them as organic molecules that did not have the vitality to reproduce as plants and animals do but which contributed components to their formation. Albeit "through a glass darkly" and without realizing the implications, he had deciphered an essential aspect of animalcules; they *were* the building blocks of plants and animals, not in their present but their ancestral form as the forerunners of dependent cells. What Buffon gazed upon were not seeds but the relatively unaltered descendants of ancient microbes.

The modern world comprises two distinct orders of cells. Van Leeuwenhoek's autonomous organisms prey, carry out chemical reactions, and reproduce by dividing. Dependent eukaryotes within living tissue also have membranes, derive their energy from phosphate reactions, and breed by mitosis or meiosis. Both originate from proto-cells in the primordial ocean, but the lifestyle of one lineage has barely changed since then, while the other's has undergone extraordinary transformations of scale and habitat. Since *we* precede van Leeuwenhoek's entities in our anthropocentric unravelling of natural history, we call them (anachronistically) "cells functioning as organisms," though we are more accurately "organisms confederating cells." In the context of our own embryogenesis, protozoa are "zygotes" in which the potential biological energy for multicellular cohesion and differentiation has been translated instead into motility and predation.

Dependent Cells

EARLY MAGNIFYING-GLASS BIOLOGISTS saw at least the fuzzy outlines of constituent cells when they looked at ant eggs, fly brains, larvae, and human blood. In 1665, when viewing faint divisions in cork, Robert Hooke named them by their resemblance to the "cells" in a monastery. Their actual character was discerned in 1838 by Matthias Jakob Schleiden, a German lawyer and naturalist. After close examination of botanical material under a microscope Schleiden declared that "plants consist of an aggregate of individual, self-contained, organic molecules—cells. The cell is the elementary organ, the one essential constituent element of all plants without which a plant does not exist."[4] At roughly the same time, physiologist Theodor Schwann crowned the cell as the sole organizing principle of tissues and organs— a status it continues to hold.

With greater magnification and improved context, observers began to recognize the similarity between cells and infusoria: they each had membranes and nuclei;

they divided into pairs from a crisp star-like pattern suggesting a spider's web or frost; they housed the same honeycombed structures with nuclei. An analysis of the cell's jelly showed it to be a kind of protein, one of the complex compounds of carbon.

By the mid-nineteenth century, the modern conceit was in full control of science: cells had become an unchallenged first principle of being; biology was an extension of cytology, life a cellular mechanism—the consummation of the qualities and relationships of cells. Medical materialists were explaining that "the brain secretes thought as the kidney secretes urine," and that "genius is a question of phosphorus."[5] In the industrial age the cell was reborn as nature's machine, and (in the twentieth century) the machine was updated again into a computer. We would have trouble today even explaining what "thought" is other than a "secretion" or current of the brain. That swiftly have we moved from an intimation of spirit to a certainty of our existence as molecular processors.

Their metaphorical reality grafted from cubicles with stony boundaries, the cells of the seventeenth and eighteenth century were discrete, physical formations, literally storerooms or chambers. Later they operated as adroit, multidimensional factories. Our recent unmasking of the deep structure and functions of cells owes less to visual surveying of actual cell space and more (like our knowledge of quasars and pulsing stars) to radiation and number patterns spat from machines. The twentieth-century cell is a confluence of data streams from electron microscopy, X-ray diffraction, application of biochemical and immunochemical methods such as fractionation and centrifugation of proteins, and recombinant DNA technology to map the organelles' individual components at a genetic level. The entity which modern biologists inherit is hardly an object at all in the seventeenth-century sense; it is a field through which a dynamic transformation of polymers dances, an eddy of "organic components that grow, shrink, and flow as needed."[6]

Hooke merely casts a cell-like shadow from a prior century.

DEPENDENT CELLS STILL LIVE "CELL LIVES." If they develop a coherence in tissue, this is an incidental matter to the habits of the solitary creatures. Despite the seeming homogeneity of plants and animalcules, cells do not melt and coalesce to create organisms; they continue to act very much like little beasties, each concerned with its personal survival.

When a tiny section of tissue from our body (like lung scrapings) is placed in a suitable medium, its component cells slowly stir to their discrete existences. They wriggle out of their lung adhesions (established and reinforced by millions of lineal ancestors) and snoop about curiously in their new environment, implicitly ask-

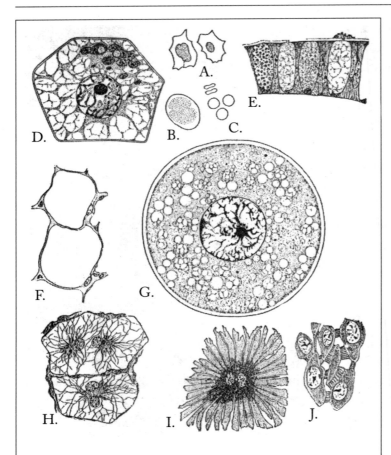

FIGURE 5A. Different types of cells. A., B. Human leukocyte; C. Human red blood corpuscle; D. Cell from root cap of calla lily; E. Four mucous cells in different stages of secretion, from epidermis of an earthworm; F. Fat cells in skin of chicken; G. Oocyte of cat; H. Connective tissue cells from lobster; I. Pigment cell from peritoneum of the fish *Ammodytes;* J. Stratified epithelium from human pharynx, with intercellular connections.

From William E. Kellicott, *A Textbook of General Embryology* (New York: Henry Holt & Company, 1913).

ing the questions any independent organism asks: who am I and what I am supposed to do? They are no longer lung at all; they are selves.

CELLS DIFFER RADICALLY in size, shape, and lifestyle. The largest non-protist cell is an ostrich egg; the smallest are yeasts ten microns each in diameter. Like one-celled animals, constituent cells may be spherical, ovoid, flattened, pointed, biconcave (blood), highly elliptical (neurons), ciliated, etc. Plant cells have especially dense walls (to support their girth) and large internal vacuoles surrounded by membranes. Their spores are occasionally protected inside thick-walled proteins—a configuration that allows them to survive harsh temperatures and periods of dessication or starvation.

Vertebrates contain more than two hundred distinct types of cells and hundreds more of subtly varying subgenera. While certain immune-system cells flourish as long as their bearer, erythrocytes in the bloodstream survive for only four months.

In a number of species, free-living cells gather in colonies which are ecologically, if not morphologically, the equivalents of multicellular plants and animals. They "pretend" to be organisms, sticking their bodies together and coordinating the stroking of their flagella in one swimming motion. Their ancestors did this long before anyone ran across a full-fledged multicellular beast. A volvox is a globular association of flagellates in which each cell feeds independently while wed to its siblings. Despite its resemblance to a blastula (noted frequently by early recapitulationist biologists), the volvox is a collaboration of separate cells, not a genetic unit. Acrasiomycota are aggregations of separate amoeboid slime molds drawn together by chemotaxis (see page 232).

Cell Biochemistry

THE MORPHOGENETIC ACTIVITIES and biosynthesis of cells are a collaborative result of their organelles, in terms of both the original nexuses brought into the emerging endosymbiotic community and the later coevolution of intracellular spirochetes and zooids into unique interfunctional units. Organelles are in effect the result of two suc-

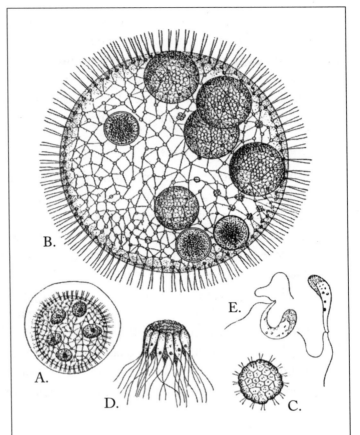

FIGURE 5B. Reproduction in volvox. A. Young colony showing distinction between somatic and reproductive cells; B. Older colony showing parthenogonidia, oogonidia, and spermagonidia in various stages of formation; C. Spermagonidium consisting of thirty-two spermagametes; D. Side view of C; and E. Spermagametes. From William E. Kellicott, *A Textbook of General Embryology* (New York: Henry Holt & Company, 1913).

cessive and distinct selective environments: their original diffuse oceanic abode with its predators, followed by a highly concentrated endogenous, symbiotic domain. Each organelle is a primitive ocean template modified by subsequent intracellular dynamics. The templates were likely trim, prolific molds with rudimentary specializations, sharing a singular talent for getting inside nascent cell-stuff. Most of their complex morphology and biochemical traits are probably an accretion upon this simple chassis through generations of mutational modification within the cell.

Their animate behavior notwithstanding, cells are physically no more than circumscribed concentrations of organic chemicals in aqueous solution. As noted, four elements alone—carbon, hydrogen, nitrogen, and oxygen—constitute more than ninety-eight percent of any cell's weight. These amalgamate in small organic molecules (intracellular polymers) comprising four major families: simple sugars, fatty acids, amino acids, and nucleotides.

Sugars and Fatty Acids

Sugars provide a goodly portion of cells' food molecules and some of their extracellular structural materials (like cellulose); they also combine with fatty acids and amino acids to form more complex molecular chains.

Fatty acids are excellent foodstuffs (providing twice as much usable energy as glucose sugar), and they participate in the construction of all cellular membranes (see below).

Polysaccharides and fats are structurally part of the cell and agents of energy transfer. Polysaccharides are polymers of sugar molecules. Fats are formed by single glycerin molecules bonded to three fatty-acid molecules.

By standards of inorganic chemistry these are complex substances, but they are eclipsed by proteins and nucleic acids, which are spun in macromolecules of unprecedented size and intricacy and with molecular weights often in the millions.

Nucleotides

Nucleotides, as building blocks for the assemblage of nucleic acids, are fundamental to the storage of biological information. As bases of chemical rings, they are transmitters of energy in hundreds of discrete cellular reactions involving ATP. The energy arises from a universal process involving the breakdown of adenosine triphosphate (ATP) to its diphosphate (ADP).

By a machine analogy, adenosine phosphates are the batteries of cells. Adenosines are members of a class of sugars bonded to nitrogen bases in units called nucleosides. Specifically, they are ribose ($C5$) sugars with nitrogen-carbon bases called pyrimidines. When inorganic phosphorus chains (phosphates) are in their environment, adenosines can form mono-, di-, or tri-phosphates depending on the lengths of their chains. When adenosine diphosphate already exists in a locale containing phosphates, adenosine triphosphate will be formed if there is energy to attach an additional phosphate to the chain. This is no small undertaking, and the amount of energy required is not generally available. However (and this is the key to the motility of life forms from parading ants to jetting squids as well as the basis of hunger and predation), horsepower can be provided by the decomposition of organic stuff, for instance in digestion of proteins or sugars. The energy of diges-tive reactions is stored in ATP bonds and released, along with ADP, when they are broken.

We eat for energy. Watch as a spider hastens along its web to get a fly into the top of its digestive funnel. Cells are chemical reactions that assimilate other chem-ical reactions or they die. Life survives only as cells, and dies likewise as cells. When the wolf consumes the rabbit, the intimacy of that relationship lies in the fact that their cells are virtually identical, and the carbons and hydrogens move from one chain to another. The owl adores the marmot for the same reason.

Amino Acids and Proteins

Proteins are polymers of the otherwise common and unpromising amino acids, often comprising a hundred thousand or more in a single protein. Just twenty amino acids, when linked in different combinations head to tail by peptide bonds, func-tion as subunits for the synthesis of proteins, which are themselves the agents for the growth and repair of plant and animal tissues. Elaborate crystals whose sepa-rate parts bend around and through one another to form multiple bonds, proteins themselves twist and attach and fold and grow until they fill and even seem to dis-tort the three dimensions of ordinary space. They dwarf the molecules that assem-ble them, and their fabrications are chemically and topologically innovative, providing raw material for all the artifacts and epiphenomena of life.

According to their underlying amino-acid sequences, proteins can be arranged in trillions of ways, so the potential diversity of attributes conforming to their shapes and surface topographies is astronomical, enough to express our singular multi-plicity. It has been noted by biologists that (as hard as it is to believe) there are more *possible* proteins than there are atoms in the known universe.

Enzymes

The reactions of proteins are catalyzed by substances that are also proteins—enzymes and coenzymes, which replaced minerals in early life chains.

Enzymes change molecular structure transiently to make certain things happen more easily. They form intermediate congeries that facilitate more enduring struc-tures. For instance, if the energy to enact a particular bond is not available, an enzyme provides sites for each of the molecules of the potential bond, and it brings them to a place where it will take less energy to complete their combination. For instance, the enzyme ATP synthetase is a cellular shaman, catalyzing the synthe-sis of ATP and converting one type of energy (chemical reaction) into another along an electrochemical gradient of protons in the mitochondria.

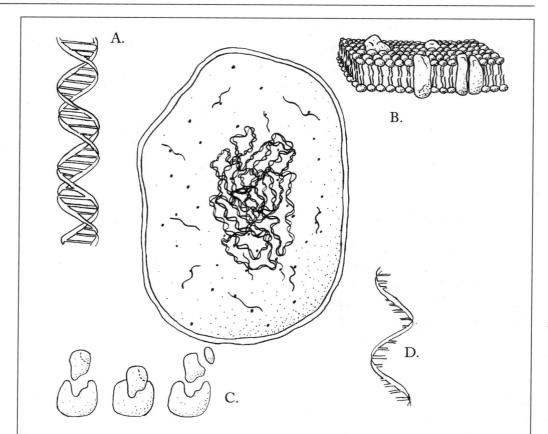

FIGURE 5C. The four essential components to any living cell. A. DNA; B. Phospholipid bilayer membrane; C. Enzyme; D. RNA (illustration by Adrienne Smuckler). From Willis W. Harman and Elisabet Sahtouris, *Biology Revisioned* (Berkeley: North Atlantic Books, 1998).

Enzymes act by changing the way substances lie in space. They use the natural shapes and spaces in molecules (which fit their own bodies) to twist the molecules into new positions, each enzyme catalyzing a singular effect.

They are architects not pharmacists, though at a subcellular level this is almost the same thing.

COENZYMES ARE ATTACHED to the surfaces of certain enzymes and contribute to their activities. Even as some chemical reactions cannot occur auspiciously without enzymes, some enzymes cannot execute their enzymatic functions without coenzymes. Many coenzymes are trace substances unsynthesizable by the animals that need them; these must be supplied by consumption of plants and microorganisms that contain them.

Essential life reactions happening almost instantaneously or in minutes would take years or even centuries without such help. Yet enzymes and coenzymes are not predesigned for their roles; their occasions are accidental until they happen. Within prebiotic material their unexplored properties changed relations among other molecules without the enzymes being altered themselves, so they were incorporated within compounds. That is, only those "cells" with magicians survived and reproduced.

Volcanic action might bond proteins at high speed, but it would also destroy their delicate configurations. Enzymes allow living stuff to congeal by providing the necessary energy without catastrophic heat or pressure.

It is both the limitation and versatility of enzymes that they do not alter or augment the charge differential between the raw substrates on which they act and the final products of their activity. They accelerate chemical reactions; they do not (and cannot) improve them or their energy balance. Since enzymes do not burn with chemical activity, they are tools, available again and again for the same reaction. There is nothing to stop them and nothing to wear them out. Once they are integrated into the embryogenic dance they must go on catalyzing, for every time they release their partner a new one will grab either hand.

Insofar as "enzymes catalyze all physiological reactions in organisms . . . they are absolutely essential to life."[7] Without these intruders biology would have waited forever to happen, and by then all its other preparations would have fallen into rubble, into the vast waters, to occur again, if at all, at random isolated sites. The whole possibility for integrated animate systems lies in seizing the evanescent moment when the manifold pieces are present in one place.

Cell Membranes

MODERN CELLS MAINTAIN THEIR INDIVIDUALITIES by outer membranes composed of lipid and protein molecules held together noncovalently (without sharing electrons); the constituent molecules remain mobile within the plane of the membrane. The coherent structure of the membrane is a continuous double layer of lipids, a pellicle as thin, in a relative sense, as the corona of the Sun or the film of life around the Earth's crust. A combination of a gate and a barrier, this plasma envelope is sensitive enough to transmit information about substances in its environs without letting those substances in, yet selectively admitting welcome molecules.

CELLS ARE ADDITIONALLY PERMEATED with membranous material separating their cytoplasm into regions and resulting in vesicles, vacuoles, and other structures. Submembranes partition microscopic space into compartments such that any organelle is cordoned off from its neighbors by at least a single barrier. The membranes are penetrated by transmembrane channels which filter protein molecules from blood and transmit other intercellular fluids and metabolic resources for cell function.

The overall cell membranes are a continuous ribbon of interconnected structures that arise from one another and lead one to another—a soft maze through which a particle could wander for miles without coming to an end or crossing its own path. The various membranes are one membrane, but that membrane *is* the cell structure. The cell as a whole is a functioning series of boundaries not inside or outside one another but arranged as layers of atmosphere are—by the chemistry that takes place among them. Since membranes that develop within cells combine with adjacent extracellular counterparts, there is also no concrete boundary between a cell's internality and its externality.

The variegated intracellular membrane (known as the endoplasmic reticulum) likely evolved long ago from a complex association of separately originating, vestigial zooids (see the previous chapter). Its intricacy and labyrinthine density provide a theater of biosynthetic function in the cell. It also translocates newly assembled proteins from the cytoplasm into pertinent organelles.

The dynamic environment within a cell is the result and also the source of its metabolism. The particular internal milieu that made cells possible continues to flourish and spread because cells must persist in synthesizing their environment, or, more precisely, come into being as it is synthesized.

RAMIFYING IN CISTERNAE (fluid-filled spaces) and tubules throughout the cytoplasm, the inner reticulum makes up as much as half of the cell's total membrane system and ten percent of the cell's volume. Its network is divided into rough and smooth mesh.

Identified by the many ribosomes attached to its surface after their expulsion from the nucleus, the rough endoplasmic reticulum is the bailiwick for the synthesis of secretory proteins. The smooth ER manufactures most of the lipids used in structuring cell membranes (including mitochondrial and peroxisomal membranes); it also synthesizes steroids in glandular cells, sequesters calcium ions for regulation of contractile tubules in skeletal-muscle cells, and detoxifies a variety of other tissues. For this latter reason it is particularly capacious in the hepatocytes of the liver and the cortex of the kidney.

THE CELL MEMBRANE INGESTS macromolecules by endocytosis. Small bilayered portions of the membrane invaginate to enclose a substance; they then pinch themselves off in a separate vesicle containing the hostage. Kept separate henceforth from the other contents of the cell, the engulfed molecule binds randomly to the surface of the cell. In crawling cells, however, the internalized membrane is returned solely to the leading edge, propelling it forward.

Conversely, in exocytosis the molecule is launched into extracellular space.

These vesicles are also the mechanisms whereby newly synthesized molecules are transported from the endoplasmic reticulum to the Golgi apparatus, then from Golgi chambers to other regions of the cell (see below). The same basic modes distribute polypeptide hormones such as insulin outside the cells of their origin to receptors elsewhere in the body.

The Golgi Apparatus

As proteins are manufactured within the rough endoplasmic reticulum, transport vesicles convey them into the Golgi apparatus, a cluster of flattened sacs of membranes (cisternae) surrounded by small membranous tubules and vesicles. Discovered by the Italian biologist Camillo Golgi in 1899, this complex congery of structures both modifies and sorts newly synthesized proteins and dispatches "predators" known as lysosomes; in the form of pinched-off membranes around vesicles, these latter circumscribe and degrade digestible materials. Both organelles (Golgi bodies and lysosomes) are present in all animal cells but are more prominent in cells whose tissue carries out their respective specializations, for instance the mucus-producing goblet cells of the intestinal epithelium (see figure 19v, page 496).

The Golgi complex's sophisticated, compounded morphology is highlighted by the curved stack of four to six parallel cisternae, each a micron in diameter. The overall structure presents two Gordian faces, a convex region known as *cis,* a concave one called *trans.* As the *cis,* or entry, face merges with transitional elements of the rough ER, newly synthesized glycoproteins and lipids enter the curved stack. The whole stack then functions as a series of processing compartments, from *cis* to *trans,* each with its own unique protein environment.

Carried by transport vesicles through one cisterna after another, substances are alchemized many times, with the particular outcome dependent on the present location of a travelling protein within the stack. The transport vesicles apparently dock at those membrane sites on the cisternae that display so-called SNARE proteins. Glycosylation (tacking complex series of sugar chains onto transmembrane proteins) is completed in the *trans* chamber. After synthesis the compounds are sorted according to final destination. Final products exit through the tubuluar *trans*-Golgi reticulum, ferried by locally budding transport bladders to their ports of operation: the plasma membrane, lysosomes, or secretory vesicles. Those departing the cell by secretion will

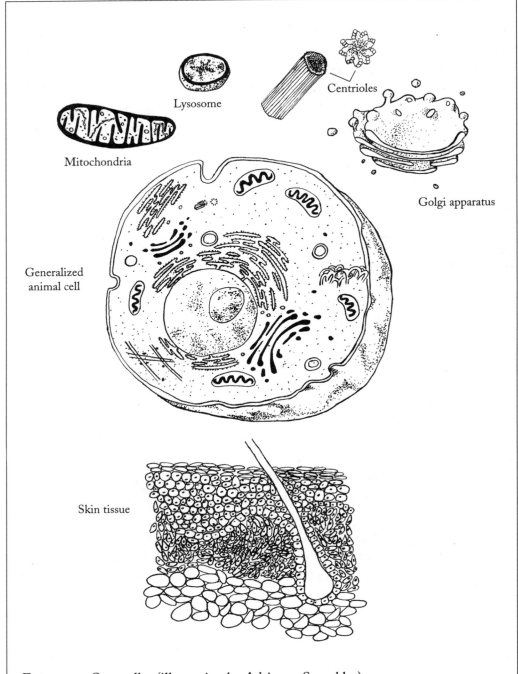

Mitochondria

Lysosome

Centrioles

Golgi apparatus

Generalized
animal cell

Skin tissue

FIGURE 5D. Organelles (illustration by Adrienne Smuckler).
From Willis W. Harman and Elisabet Sahtouris, *Biology Revisioned* (Berkeley: North Atlantic Books, 1998).

become digestive enzymes, blood-plasma proteins, and hormones; those remaining within the cell will participate in the membranous compartments.

Initially the Golgi compartments were thought of as "static warehouses, receiving and dispatching cargo in the shuttling vesicles."[8] The contemporarily reconceived Golgi system is more "fluid, self-correcting, and evolving. . . . The cisternae themselves may move forward while the vesicles actually move backward to recycle components of the ER and Golgi compartments to their sites of origin."[9] The cargo line moving one way intercepts the flow of enzymes going the other way.

The *trans*-Golgi network is richer in cholesterol than the *cis* and also has a thicker reticulum of tubules, secretory vesicles, and storage condensing vacuoles. It is where serious metabolic processing and designating take place, biosynthetic reactions too elaborate for description in this text (fatty acylation, glycosylation, collagen and glycolipid assemblage, oligosaccharide trimming, sulphation, etc.). Many of these have embryogenic and even epistomological implications.

Sometimes metaphorically referred to as the "post office of the cell," the Golgi apparatus probably does more than receive packets and redirect them. It may be the operational "intelligence center," providing an interface between mechanisms of protein synthesis inside a protein-lipid envelope and diverse activity in the outer world comprising, first of all, other cells (with their own Golgi bodies) and, secondly, the organism as a whole. Golgis in fact appear to move about cells, shifting their functions and the nature of their protein products, changing their behaviors in response to cues flooding across outer membranes. In their mobile, dynamic sacs they are more fluid protoplasmic artifacts than fixed machines.

Lysosomes

LYSOSOMES ARE LATENT ORGANELLES that were not discovered until the 1950s in liver cells; diverse and elusive in size and shape—in fact the most heterogeneous of organelles—they were identified first histochemically by centrifugation. Products of *trans*-Golgi synthesis, these organelles are less likely to be progeny of ancient zooid invaders; instead they may be virgin organelles hatched not in primordial but intracellular waters. Raw lysosome proteins synthesized in the endoplasmic reticulum arrive at the *cis* of the Golgi apparatus; after processing in both *cis* and *trans* compartments, they depart as small membranous bags—enzyme-rich vesicles—with a surrounding membrane bearing transport proteins.

Emerging from the Golgi sphere, lysosomes fuse repeatedly with vacuoles containing material needing degradation. The macromolecular material they metabolize is contributed by their membrane's transport proteins to the general biosynthesis of the cell. A cadre of acidic (hydrolytic) enzymes allows lysosomes (as a cell samples and cleans its matrix) to carry out intracellular digestion and "lyse" cellular debris into a structureless fluid. They conduct endocytosis in most cells and, in specialized cells, phagocytosis of very large particles.

Lysosomes that accumulate more material then they can dissolve become defunct.

The diversity of individual lysosome structure probably reflects the long-term evolution of distinct digestive functions to process a variety of intracellular and extracellular debris. There are at least three hundred total lysosomes in almost all eukaryotes, the exception being their absence in red blood cells. Greater numbers of them occur of course in cells involved in phagocytosis.

Lysosomes are crucial embryogenic artisans. In many specialized cells, their carnivorous actions (autophagy) have direct developmental consequences, as selected structures (including mitochondria and secretory vesicles) are destroyed to accommodate cell remodelling. Lysosomes thereby contribute to the differentiation of tissue, the regression of defunct organs (as the tadpole tail during frog metamorphosis), the degradation of old bone by osteoclasts, and the regulation of hormones and kidney proteins.

The Roles of Mitochondria and Chloroplasts (and Peroxisomes)

THE TWO MOST HIGH-POWERED and profoundly adapted organelles, chloroplasts and mitochondria, appear either to descend from the same lineage of parasitic zooids or to have converged developmentally after endosymbiosis. They have similar smooth, permeable outer membranes; inner membranes; equally specialized complements of DNA and RNA; and comparable energy-yielding chemistries. While chloroplasts synthesize their own lipids, mitochondria are dependent upon the endoplasmic reticulum.

Prokaryote parasites by heritage, mitochondria fused with the precellular unit in such a way that the nature of both was transformed and they became not only interdependent but one (see the previous chapter). At the same time, mitochondria withheld an aspect of their autonomy from their new surroundings. While exchanging proteins and genes with the cell and its nucleus as a whole, they maintained a license and exemptive barrier within the eukaryote, remaining metabolically and genetically separate at some energetic cost to the system as a whole. While the predominance of mitochondrial (and chloroplast) DNA originates in the nucleus and is imported into the cytoplasm by the cell's ribosomes, some of their protein continues to be encoded by organellar DNA and manufactured locally on mitochondrial ribosomes.

Mutating at ten times the rate of the nuclear genome, genes encoded by mito-chondrial DNA became crucial to the formation of complex amino-acid compo-nents in the developing cell. They were already present and in use and could not be arrived at independently by the nucleus. As unique sites of ATP energy pro-duction, mitochondria are now involved in the synthesis of nucleic acids and pro-teins, cell division, food intake, enzyme production, and overall motility. No longer expendable, these onetime outlaws have evolved into "cellular power stations."[10]

THE CHLOROPLAST IS the primary plant-energy organelle. These descendants of plastids emanate an elaborate membranous matrix for the annexation of light energy and metamorphosis of carbon dioxide into carbohydrates, discharging oxygen in the process. Flattened sacs of internal membranes called thylakoids radiate through-out the inner domain of the chloroplast, stacking in closely pressed bundles (grana) linked to one another by loose membrane sections (lamellae) to extrude a multi-layered architecture around a topography of gaps and lumens. This convoluted maze is the outcome of unique interactions among proteins and sugar- and phosphorus-bearing fats (glycolipids and phospholipids). The extensive space between the inner external membrane of the chloroplast (which is far smoother than its mitochon-drial counterpart) and the thylakoids is called the stroma. Water, oxygen, carbon dioxide, and other gases flow freely through it and the membrane itself.

In the transmembranous complexes of the thylakoids, polypeptides (amino-acid chains) and pigments harvest sunlight. Chloroplast energy production ensues with the absorption of light energy, the assemblage of an electron transport chain, and

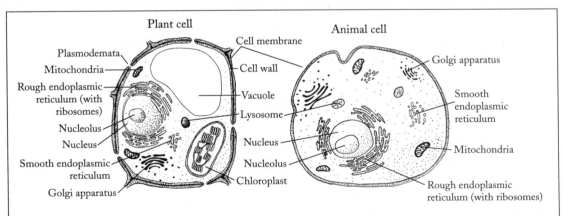

FIGURE 5E. Comparison of plant and animal cells. The animal cell lacks both a rigid cell wall and chloroplasts (illustration by Adrienne Smuckler).

From Willis W. Harman and Elisabet Sahtouris, *Biology Revisioned* (Berkeley: North Atlantic Books, 1998).

the concomitant scaling of a proton gradient in the membrane lumen for the ATP synthesis of photophosphates. Light-capturing antenna pigments funnel the energy of excited electrons through the thylakoids' asymmetrical protein geometry, its hexameric lattices, to the lair of the specialized chlorophyll molecule (as noted earlier). Complementing this process, an electron-deficient gradient pulls electrons out of water, releasing molecular oxygen. With the flight of electrons, protons are driven into the intermembrane space, a gradient siphoning off some of the free energy for ATP synthesis. Other protons are propelled from the stroma across the thylakoid membrane, regenerating the photosynthetic cycle.

Clearly this elaborate pharmacy and tiny electrical storm were not carried full-blown into the cell by bacterial invaders; they accrued step by step over millennia of generations, each successive innovation grounded in a deeper and more stably intricate network. Such is the history of all complex appliances.

PLANTS ARE OPAQUE MIRRORS, combining sunlight, air (carbon dioxide), and water into a rainbow of starches and sugars that are then metabolized by other creatures into a riotous spectrum of animal shapes and personalities. Carbohydrates are literally carbon and water; these comprise the basic material residue of botanical life. Yet no machine can turn them back into leaves and flowers; we cannot back-engineer true thylakoids. Chloroplasts are unique weavers of vital matter from the rawest ingredients in the Solar System.

"[S]tarch is a product of the plant's process of assimilation. This process takes place in the plant's middle zone, the leaf, when sunlight acts upon it in the presence of water and carbon dioxides. Plant physiologists express this . . . interplay of light [sun] with darkness [underground roots] in the following formula: $6CO_2$ [carbon dioxide] + $5H_2O$ [water] = $C_6H_{10}O_5$ (starch) + $6O_2$ [oxygen]. . . .

"Starch is subject to many metamorphoses in the plant organism. The most important one is the etherealizing of it into sugar as the sun's warmth draws it upward. Sugar is found in the nectars, but it is also present higher up, in the still more refined form of glucosides, in the blossom colors. When our 'enchanted rainbow' gleams in a field alight with flowers, it is as though heaven itself were greeting us."[11]

Colloidal, fluid starches manufactured and assimilated in leaves are transferred to the roots, flowers, and fruits of various trees, grains, herbs, bushes, groundcovers, etc., where they are cached. Molecularly heavy starches in foliage factories are warmed by the sun and gradually sublimated into lighter dextrins and sugars which are then "stored in blossom nectaries"[12] and fruits as well as leaves and roots. In "a ceaseless harmonization of the living polarities of earth and heaven, giving rise to an endless range of metamorphoses, . . . the static world of atoms and calculable happenings"[13] is translated into flowering and fruiting canopies. As botanical templates breathe sun and air, their starchy matter is continuously subtilized into colors, scents, aroma essences, etheric oils, foods, and medicines. The mineralized by-products of this process, cellulose and coal tar, provide the tensile strength and form for different botanical species from larches and ginkgos to clover and rye.

Beet sugar, cane sugar, grape sugar, fructose, and honey comprise similar carbohydrate chains. They vary subtly in that their solutions "behave differently in the presence of polarized light. Grape sugar turns the plane of polarization to the right, fructose to the left, thereby earning the names dextrose and levulose [literally 'right-turning' and 'left-turning']. Fruit sugar is like honey in being a mixture of the two."[14]

Honey was the singular sweetener of the Mediterranean until Alexander "led his armies through India and there discovered a 'reed that produces honey without the aid of bees'"[15]—the sugar cane.

Armies and bees are no longer necessary. Human laboratories and factories "conjure forth a synthetic mirror-image of the natural world: synthetic colors, scents, saccharine and other sweeteners, mineral oils, and therapeutic substances."[16]

OUTSIDE THE REALMS OF CONVENTIONAL BIOPHYSICS, experimenters quantifying chemical composition of soils and the plants growing in them have deduced that organelles carry out not only ordinary molecular reactions but effect actual transmutations of elements at an atomic level. Measuring increases in some mineral contents and decreases in others, they concluded that, somewhere among the hexameric lattices and electronic gradients of chloroplasts, subatomic particles are shot between atoms on a regular basis, leading to the creation of *new basic elements of matter.*

At an exoteric level, carbon dioxide is transformed into carbon and oxygen; at an esoteric level, molecules of carbon and oxygen flow through a living forge turning them into magnesium, then calcium, then phosphorus, then sulphur. Another series of experiments showed plants alchemizing virgin potassium out of nitrogen. Thus, the cell would seem to serve a cosmological as well as a pure biological function. "The soil," proclaimed the German vitalist botanist Baron von Herzeele of Hanover, "does not produce plants; plants produce soil."[17]

THE UNIQUE CHEMICAL PATHWAY of another organelle, the peroxisome, enables it to degrade very-long-chain fatty acids and phytanic fatty acids produced by the oxidation of chlorophyll. This microbody collaborates in the synthesis of ether lipids (some of which protect the cell from oxidation) and (in the livers of animals) bile acids. Specialized peroxisomes in plants (glyoxysomes) convert fatty acids in photo-respiration.

The Roles of Intermediate Filaments, Microtubules, and Microfilaments

MICROFILAMENTS, MICROTUBULES, AND INTERMEDIATE FILAMENTS provide mechanical strength and structural and tension-bearing forces within the cell.

Because their chief ingredient, actin, is as abundant as any protein in the cell, dense meshes of microfilaments arise just below the plasma membrane, linked in a stiff three-dimensional grid. Complex chemical reactions within this network introduce cytoplasmic mobility. The filamentous networks also differentially attach to the cell's plasma membrane and, through it, develop tension with the extracellular matrix. By pulling selectively, they can instantaneously change cell shape, squeeze out protrusions, or cause cell migration.

While microtubules and microfilaments are comprised of globular proteins, intermediate filaments are forged by fibrous ones. The latter are much more variable in size and shape, with both ropey sections that participate in tension resistance and structure-bearing and other chemically diverse segments that have exotic functions. The mechanical properties conferred by intermediate filaments underlie differentiation into highly specialized cell types; these include neurofilaments of nerve-cell axons, keratin appendages of epithelial (notably epidermal) cells, desmin protein strands of muscle cells, and the acidic glial lace of assorted neural components.

MICROTUBULES WERE FIRST RECOGNIZED in the nineteenth century when biologists using simple microscopes became aware of zones of increased activity within cells. These had a fibrous or granulated quality and appeared to viewers much like the fuzz of distant galaxies. Twentieth-century preparation techniques under electron microscopes revealed caches of tall, thin, hollow cylinders. A constant 240 ångstroms in diameter though of assorted lengths, the units organize themselves in idiosyncratic patterns (like a child's kit of thousands or even tens of thousands of adhesive straws).

Although once self-manufactured spirochetes, modern microtubules are assembled in cells by two distinct subunits of tubulin protein; these form in aggregations of duplicate globular subunits stacked in parallel circumferences around a hollow center. The minute tubes are not part of the cell reticulum but appear to be assembled within the cytoplasm, organized by the centrioles and basal bodies (centrioles of flagella or cilia). Microtubules each have fast-growing and slow-growing poles. The minus end is anchored at the centrosome, an area of differentiated cytoplasm bearing the centriole. We will discuss its activity in Chapter 7, "Sperm and Egg."

The walls of microtubules are constructed usually of thirteen subunits of an acidic dimer protein rich in glutamic acid. In cilia these shafts arrange themselves in ninefold double pairs with two singlets in the center. The uniform diameters and characteristic patterns of microtubules suggest not only common ancestry but long-standing symbiotic association within cells and stable organizing locales (see the previous chapter). As other proteins interact with tubulin, specialized varieties of microtubules also emerge.

ZONES OF MICROTUBULE ORIGINATION in the modern cell include nucleic-acid attachment sites—either centrioles or kinetosomes (depending on the subsequent configurations of their organelles). As these centers of production and organization spread, highly characteristic, irregular structures develop—each determined by the particular lineage, site, and stage of development of the cell.

As they can be induced to provide tension along different parameters, tubulin organelles once upon a time distorted cells in ways that gave them totally new mean-

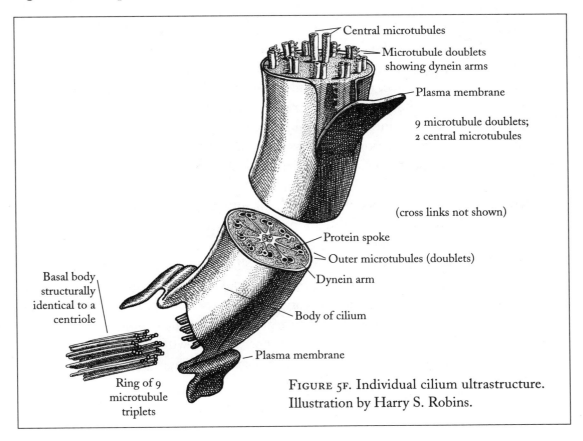

Central microtubules

Microtubule doublets showing dynein arms

Plasma membrane

9 microtubule doublets; 2 central microtubules

(cross links not shown)

Protein spoke

Outer microtubules (doublets)

Dynein arm

Body of cilium

Plasma membrane

Basal body structurally identical to a centriole

Ring of 9 microtubule triplets

FIGURE 5F. Individual cilium ultrastructure. Illustration by Harry S. Robins.

ings and functions in the formation of tissue. Microtubules continue to play a crucial structural role in the modern cell, maintaining and changing its overall shape. In such processes they may adjoin membranes or line up parallel to cell walls—activities critical in embryogenesis and cell specialization. Microtubules not only change cell shape, contribute to cell movement, and regulate the plane of cell division (in concert with microfilaments); they also afford a "structural basis for dynamic instability" within the cell, providing "an organizing principle for cell morphogenesis."[18]

Although their milling centers are the sole factory for the structures of mitosis, microtubules paradoxically customize some classes of cells into distinctive morphologies incompatible with future replication. That is, they provide the axles of fission in one domain and also the extinction of the capacity to divide in another. As so often happens in phylogenesis, opposite uses of the same information are not only complementary but interdependent in developing complexity. Margulis explains the "motive" behind the sacrifice of fission:

"The essence of animal-style differentiation is the formation of cells capable of nonmitotic internal motility (such as the growth of dendrites and axons in nerve cells, melanocytes in pigmented tissue, and cilia in epithelium). All these cells simultaneously lose the ability to divide by mitosis."[19]

Additional microtubule-composed organelles (introduced in the previous chapter) include flagella (locomotory whips), cilia (shorter locomotory strands), sensory cilia for taste and smell, mitotic spindles (arranged in bundles), sperm tails, asters (rosette-shaped mitotic axes), suctorian tentacles and pharyngeal baskets (among protoctists), tactile spines, and the movement-generating and feeding axopods of heliozoans. In combination with lysosomes, microtubules form phagocytes which engulf and transport food and capture small organisms.

MICROFILAMENTS ACT IN CONCERT with microtubules, though differently. Because actin is more prevalent in cells than tubulin and because microtubules are considerably thicker than their filamentous counterparts, microfilaments are generally thirty times longer than tubulin organelles. Unlike the cross-linking microfilaments, microtubules maintain individuality, radiating out from the approximate site of the cell nucleus and furnishing fibers for organelles to travel along throughout the cytoplasm, tracks that also orient and guide the location of the endoplasmic reticulum and polarize the Golgi apparatus in an opposite vector within the ER (see the discussion of tensegrity in Chapter 11, "Morphogenesis," pages 246–250).

The Nucleus, Nucleoplasm, and Nucleolus

THE CELL NUCLEUS HAS ITS OWN TWO-LAYERED MEMBRANE, each layer supported by intermediate filaments; inside this delicate shell nucleic acids and associated proteins are maintained in nucleoplasm and protected from oxygen contamination. The nuclear sap carries out anaerobic metabolism and maintains a different ionic state from the cytoplasm.

The outer nuclear element, bearing some ribosomes, is continuous with the endoplasmic reticulum, though of somewhat different protein chemistry. The inner membrane is overlaid by a lamina, a mesh of fibrous proteins. The entire double envelope, pocked with protein rings, is a fearsomely complex structure and, like other membranes, a selective barrier preserving a chemical and electrical differential, in this case between the nucleus and the cytoplasm.

Despite a strong continuity, the inner and outer nuclear membranes are chemically segregated. Their pore complex is fashioned from chunky protein granules arranged in octagonal grids. Aqueous conduits in the granules allow passage of water-soluble molecules between the nucleoplasm and cytoplasm. Ribosomes expelled in baby states grow too large in the cytoplasm to get back in and thereby distract the nucleus with their rigmarole of protein synthesis. Conversely, proteins, enzymes, coenzymes, and ATP find their way through the membrane into the nucleus, entering upon announcement of their "personal" identity sequences (nuclear import signals in the forms of peptides of four to eight amino-acid units). Larger polymerases require the assistance of receptor proteins that actively ferry them through the membrane while expanding its pores. The molecular basis of this process remains an enigma.

The nucleoplasm harbors a variety of regulatory proteins, enzymes, as well as DNA bundled with histones (amino acid-rich proteins) into chromosomes, so it provides a matrix for genetic chains.

The nucleolus is the nucleus' organelle, impressed at some point during endosymbiosis, perhaps in the incorporation of a simpler proto-cell by a more complex one for which the simpler cell became the hereditary molecule over time. A membraneless, fibrillar structure, the nucleolus is the nuclear site for the biogenesis of ribosomes, tantamount to a ribosome machine; its granularity represents ribosomal chromatin, ribosomal precursors, and ribosomes at various stages of maturity. The catalytic machinery for protein synthesis, most mature ribosomes (as noted above) eventually migrate out of the nucleus and are quartered in the rough endoplasmic reticulum. Their role and fate will be discussed in the next chapter.

Almost everything that goes on inside the cell nucleus involves production of nucleic acids—DNA replication during cell division, RNA synthesis the rest of the time. The nucleus is the only organelle that manufactures chromosomes and fissions mitotically; all the other organelles in a cell that bear DNA or RNA expand and twain by a bacterial nonmitotic mode.

The Origin of Independent Cells

ANCIENT, FREE-LIVING PROTOZOA do not have separate organs or specialized parts under the control of their own nuclei. They lack the cytoplasmic substance to exist as anything but a kind of crystalline by-product of water. Independent cells are protean integers, descendants of precellular globs in primordial scud. The early twentieth-century mathematician D'Arcy Thompson pointed out that almost all protozoan body shapes, from oblong tubes to floating bells and pears to swimming ciliate balls and assemblages of balls, can be derived from surface tensions between fluids. Such coagulating ripples are virtually independent of gravity. Their high surface-to-mass ratio makes specialized respiration unnecessary; each one breathes throughout its fabric. Each is, equally, lung and gut. As with dependent cells their energy originates from the conversion of ATP to ADP, though it is unclear how that energy is translated all the way to the filamentous protein molecules of their locomotory fibrils.

Protozoans are entirely aquatic, but they have adapted to a diversity of semi-moist environments over millions of years. Some live in damp soils—the little bits of water clinging to dirt particles serve as their lakes. Others imbed as parasites in the organs of animals and plants.

PARAMECIA ARE HIGHLY COORDINATED one-celled creatures, able to retreat quickly, for example, after striking objects, as their cilia reverse. Threads, shot from their microtubule-constructed basal bodies, can capture prey or hold to a spot like an anchor. Paramecia show clear "hunting" strategies and apparently "learn" from their errors, which is remarkable in a creature of this modest size.

We are seeing the dawn of a hidden power like gravity, an impulse of individuation which comes into being even before there is apparently enough neural surface to sustain it. Paramecia are "egos," not oil slicks.

The locomotory organelles of protoctists are critical to their independent identity and survival as predators. Movement is existential and ontological. Getting from "here" to "there" is one of the most profound vectors of existence. Simple flagellates like *Euglena* lash their single protein whip back and forth so that their body is thrown to one side and another in an overall spiral progression. The plane of motion lies at an angle to the animal so *Euglena* rotates as it swims. The flagellum also strokes food into its gullet. Ciliates, like paramecia, have a more linearly controlled gait generated by waves of tiny cilia. Amoebas are mobile guts. The outer part of their endoplasm is a gel and the inner part a sol which becomes a gel along the advancing lobes of the pseudopod. Meanwhile endoplasmic vacuoles secrete enzymes and absorb the food in their protoplasm.

Similar methods of locomotion have evolved in specialized descendants within organisms. Ciliate and flagellate cells move particles in digestive and excretory tracts, and amoeboid lymph cells and antibodies behave like pond-scavengers as they circumscribe trespassers.

WITHIN THE LIMITS OF THEIR DIMENSIONALITY, unicellular organisms complexify solely through their organelles. As animals they may be simple, but as cells they are not. They form membrane-bounded vesicles, some of which are excretory. Other membranes contract to squeeze out water. Ciliates develop cavities, funnels, and cell mouths—elaborations of compound cilia fibrils. Their behavior "can be photosynthetic or carnivorous, motile or sedentary. Their anatomy ... includes such structures as sensory bristles, photoreceptors, flagella, leglike appendages, mouth parts, fringes of cilia, snouts, stinging darts, and musclelike contractile bundles."[20] Although these elaborate organelles are functionally the forerunners of the organs of animals, it is impossible to see how they could share any homologous continuity with forms made *of* cells. Yet somehow, when the protists transcended their atomicity, the organelles of their bodies (mitochondria, microtubules, Golgi bodies, etc.) were revived in the organs (mouths, tails, guts, limbs) they composed. Heedless of scale or componentiality, a blueprint of organicism was passed on.

Single-celled animals are cloned in their offspring, dying only to recur. They go on fissioning and copying themselves, hypothetically forever. This is not the fate of the metazoan zygote, which will magically fission into millions of *diverse* connected cells assembling a gigantic creature that in no way resembles them; only its germ cells will live past it, and then only insofar as they generate their own entire, separate organisms that will also mature and die, each of them spawning a small number of like germs to carry on their lineage.

The Origin of Simple Worms

IF A SINGLE-CELLED CREATURE grew much larger within its membranes, it could not provide enough protoplasmic surface for absorption of either oxygen or food. Its evolving descendants would also be unable to snowball new protoplasm into a fattening pellet without choking and starving the substance at the core. Only if their unidimensionality were projected in a line (so that no new cell were more than a few microns from the surface) could they occupy three dimensions of increasing space. This linearization is a flatworm.

A better way to fill space with cells is to increase the complexity of intercellular surfaces (much as *intracellular* surfaces were complexified) through layering, branching, and folding. All large animals do this in tissues and organs. They breathe and eat through a sinuous topography of lungs and guts that packs matter and spatial pockets in spiralling interiors. Their bronchial pleats and twisting esophagi record the long and irregularly internalized events through which they evolved and shaped themselves surface-to-surface with our planet's liquid currents and volatile atmosphere. Creatures are chemicals tracked in mazes of thickening fabrics.

Multicellularity represents the discovery and occupation of space, of planet-scale landscapes—a movement across one decimal of the great ladder between microcosm and macrocosm. Increase in size and density led protoplasmic clusters to new habitats and ranges and also to a deeper dependence of parts. Worms compress their innards in fulcra to scrabble across the pittedness of the Earth. But it is not simply a matter of densification and growth; connection, communication, and coordination sprout, cell by cell. Synchronization of cells molds intricate internal milieus—central cavities through which organs are distended and their functions linked.

Ultimately this swelling cytoplasmic blubber will propagate a skeleton, a girth by which to thrust itself out of droplet life into a cosmic zone.

Multicellular Dependence

ONCE CELLS HAVE BEEN CONSOLIDATED in organisms and their activities subsumed, they are effectively symbionts and cannot survive outside the environment they have collectively created. Symbiosis is a variant of cell life, and if (as has happened to all dependent cells, except germinal cells in brief episodes) their capacity for independent survival has been lost, it is only because this capacity became a luxury. As we saw in the previous chapter, parts of their organelle physiology degraded as they were no longer required. Other aspects were transformed into mechanisms of shared metabolism and collaboration.

Dependent cells have substantial functional connections that must have evolved at the onset of their lineage. These include protein molecules in their coats that recognize and interpret one another, filaments and seals between membranes, and gap junctions through whose minute channels ions and other materials pass. The junctions lead to electrical coupling and symbiotic metabolism. These are not external connections like those of the volvox; they are deep, internal networks by which the cells make contact, eat and breathe in a unity, and send one another chemical messages (see Chapter 11).

Junctions, though they originate within the morphology of cells, are not fixed or permanent. They are specializations that arose relatively late in the history of cellular development when multicellularity became a popular lifestyle, and they disassemble instantly with treatment by certain chemicals and are quickly restored when cells are put in contact with one another, even if the cells are from different regions of tissue or from different animals of widely separated species. In cell cultures made of human cancer tissue and chicken embryo, the cells soon recognize one another as kin and behave accordingly even though they can't construct viable organs. They are more intrinsically tissue-knitting than autonomous. This reservoir of potential connectivity is crucial in embryogenesis because groups of cells often cluster together suddenly and move as a single field, whereas other cells that do not participate must be able to detach and define themselves in a different context.

It is because the cell is paradoxically capable of both independent and integrative activity that it can be the building block of life.

THAT WE ARE FASHIONED by cells in linked sheets is an experience we can probe anew by running fingers along face—fingers separated by gaps between cell clusters, lips and eye sockets formed where migrating cells turned on the edge of their own unperceived termini and shaped the relativized skin of their universe from the inside. The body is apparently the only part of the universe that feels itself from within. As fingers run through cat's fur, two separated galaxies made of exactly the same stuff meet.

Although it is an anthropomorphism, we might say that cells remain cells to themselves even as they form organisms; they are not "aware" of giving up liberty or of participating in a group effort. It simply happened that their matrix was transferred to a new environment; they went on metabolizing, growing, and dividing like their free siblings, providing our metabolism through their own. It would be a surprise to them indeed to find out that in conjunction with one another they had such elaborate and exotic by-products. We ourselves would hardly be interesting to the cells that embody us, for they are still discrete animalcules in salty pools. They are not subtle enough or cognizant enough to perceive that the surrounding waters have moved and complexified. And then again, perhaps it *is* still the same sea, and its curious plasma packaging is little more than another current or tidepool. (How would *we* know if our galaxy were a patch in the extracellular matrix of a mega-gigantic being?)

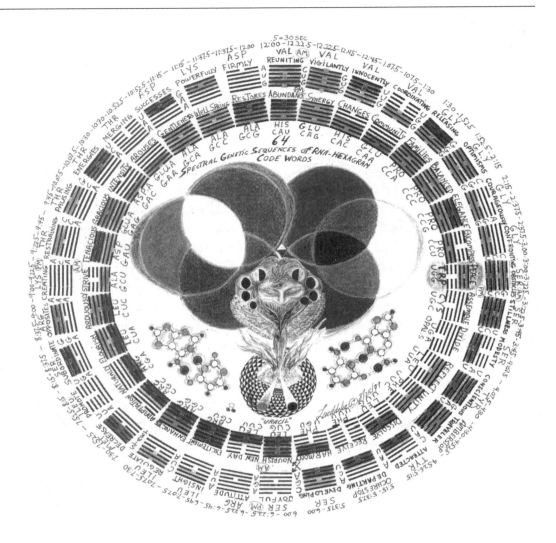

Spectral Genetic Ecology: The Circadian Light Tuner by Drs. Edward Lueddeke and Steven Leonard, and Patricia Lueddeke.

This is an illustration of one of the many late twentieth-century models for the integration of the 64 memes of the 6,000-year-old I Ching with the 64 triplet nucleotide bases of RNA (the transcription bar code for the 20 amino acids, building blocks of planetary biology). Practice with the "Circadian Light Tuner" is intended to generate an electromagnetic field of invisible complementary light of less than 100 nanometers. According to the authors, "the mathematical coefficient of the corresponding frequencies elicits a state of dynamic equilibrium between the nucleotide base pairs, synchronizing the individual with the [unknown] universal intelligence [behind them]."

For more information, write inwardbound@acadia.net, or Acadia School of Advanced Therapy, 104 George's Pond Road, Franklin, Maine 04634.

6

The Genetic Code

DNA

PROTEINS WERE SUCH CREATIVE AND INGENIOUS MOLECULES that it was assumed, until 1944, they were also the basis for heredity. In an experiment that year, Oswald Avery and his colleagues, working with two strains of the same species of lung bacteria (one with smooth, shiny colonies; the other with rough ones), managed to get some of the rough strains transformed into smooth ones by adding smooth extract to rough colonies. The change was later inherited.

The scientists then destroyed selective macromolecules in treated bacteria cells and discovered that not a protein at all but DNA, a nucleic acid (*deoxyribo*nucleic acid), was the *only* molecule required for a new generation of the microorganisms.

Nucleic acids are chains of polymerized nucleotides, regularly hundreds and often thousands of units joined in diverse and intricate ways. Their stability and the consistency of their arrangements endow them with reproductive potential. They originate in distinct structures—DNA in the chromosomes and RNA (ribonucleic acid) in the nucleoli in association with protein.

Stored within the rod-like chromosomes, DNA is pictured as an extremely long, dense polymer spun in recurrent fibers chemically resembling rayon or hair (chromosomes contain by weight approximately forty percent DNA; the rest is protein). Proliferating strands of nucleic acid are arranged in such a way that sugar and phosphate groups alternate on their backbones, with nitrogen bases attached to each sugar. The sugars and nitrogen bases form nucleosides which combine with the phosphorus units in compounds called nucleotides (see previous chapter). The chemical basis of these protein-DNA complexes is so ancient that hybrids of yeast and squirrel DNA, or rose and vulture, bind as tightly to each other as DNA from the same species.

The Double Helix

REPRODUCTION HAS ITS MOLECULAR EXPLANATION in the double-helix model of DNA proposed in 1953 by Francis Crick and James Watson. Their predecessors had already discovered that each DNA nucleotide is composed of the same sugar and phosphate and a changing nitrogenous base. Four different bases are attached to the nucleic-acid sugars. Two bases (cytosine and thymine) are of the pyrimidine type, bonding carbon and nitrogen in a single hexagon; and two are double-hexagon purines (guanine and adenine). Chemical analysis had shown that adenine content equalled thymine content, and that guanines and cytosines were in similar balance. Using X-ray diffraction studies, Crick and Watson arrived at their icon: two chains twisted about a shared axis, or a spiral rope ladder with rungs of bases. The sugar and phosphate backbones were on the outside, and the bases were turned in toward their common axis and periodically joined—adenine to thymine, guanine to cytosine—at regular intervals on each strand, jutting out transversely to meet their counterparts on the other like the cross-ties of railroad tracks. With each chain tracing a right-handed helix but running in an opposite direction from the other, DNA resembled a torqued ladder: two parallel ribbons joined by purine and pyrimidine steps.

Purines and pyrimidines are poles of a complementarity. Their hydrogen bonds express the creationary force at the heart of the Sun that also fuses water molecules. Molecular biologist Harvey Bialy compares them to the yin and yang of Chinese physics, opposing primal forces meeting in a vortex through which the universe manifests.[1]

By making replicas of themselves DNA strands pass on information about their own chemical structure, divulging (in essence) how they are made. This transcription became more than just a stencil or emblem; read holographically, it delivered the source code of life. So basic was the transmission that it lodged hieratically at the heart of the cell and became the oldest tongue spoken on the Earth.

INSOFAR AS THE GENETIC MOLECULE must stack its database in a reproducible, multidimensional

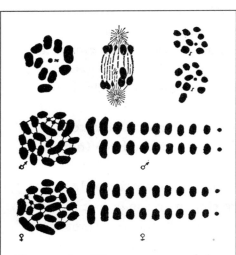

FIGURE 6A. Chromosomes of the squash bug.

From H. L. Wieman, *An Introduction to Vertebrate Embryology* (New York: McGraw Hill, Inc., 1949).

state, a double helix is an ideal geometry. If DNA were solid it could not be read; if it were a plane, it would hammer out sterilely redundant images. That it is essentially a line, a linear strand, was suspected before Crick and Watson. The surprise was that it was a bent line, a helix, with another helix, a twin, twisted around it. This is now as much a seminal image for the latter half of the twentieth century as the mushroom of the split atom and the blue-white mandala of Earth against night. The twin spirals suggest the mystery of life structure, an interminable winding staircase of growth and form, its molecules spinning section out of section, half of them always upside-down.

Because DNA is a flowing displacement we can portray it statically only through reflections of its appearances from different vantage points—twisted ribbons, parallel lines curving in three-space, a flowing ladder, or a wave phenomenon created by the polymerization of nucleotides. If we look at the ladder from the steps to the backbone, then the nitrogen bases combine with pentose sugars to form nucleosides, which are held in the backbone as nucleotides by the phosphates. Each nucleotide actually starts with a sugar and a base joined to three phosphates in a row, but two of them are always consumed in the polymerization of the chain.

The ribbons themselves are long and predictable. They go: phosphate-sugar, phosphate-sugar, phosphate-sugar millions of times, with the nitrogen bases strung on them as close as molecular forces will allow and each ribbon making a complete circuit of the axis every ten bases.

As we extricate its image from deep matter, the double helix strikingly resembles the entwined serpents of Mesopotamian legend as well as Hermes/Thoth's caduceus of healing. To an informal sect of New Age cosmologists this betrays that our template was fashioned by extraterrestrial visitors and incorporated by them in not only our cells but our myths. Letters of Egyptian and Hebrew alphabets, time units of the Mayan calendar, Druid tree codes, Biblical text, and twelve-sign zodiacs are each seen to correspond to human chromosomes along different parameters.[2] The double helix is viewed as a paraphysical grid, like runes of a tree alphabet or a microcosmic transdimensional coil.

DNA is so basic and elusive that it defies the complicacies of algebra and higher mathematics. Superficially it reads like a techno-masterpiece from the Pleiades or a supergalactic corporation's design for populating the cosmos. Something mysterious and sublime is hidden inside it, something we are still approaching, something far less elaborate than the entirety of modern science (or science fiction), yet at the same time exponentially more complex in a whole other way. No wonder it can seem proto-biblical and meta-hieroglyphic.

DNA Packaging

Human chromosomes contain approximately 10^8 base pairs of DNA; a cell holds about three billion bases. Scaled up to railroad-track size, the DNA in that cell would run about twenty million miles.[3] If all the DNA in any one of us were disentangled from the helices inside the nucleus of one of our cells and then stretched out, a single strand would span the nucleus many thousands of times. This comprises a much deeper and broader information base than is needed for the assemblage of any particular biont.

Strands of such length also do not compress into packets easily. The double helix is inherently rigid and negatively charged, thus resists pleating and stuffing into tiny capsules. Yet, with the aid of the abundant structural proteins called histones, this "thin but stiff cable is somehow wrapped, looped and folded to fit within a container whose linear dimensions are several times smaller. The packaging of DNA is, without exaggeration, an engineering feat of staggering proportions."[4]

It was once thought that genetic polymers maintained a precise thirty-six-degree helical twist between all adjacent base pairs, completing a constant ten pairs per turn. Yet the helices admit quite different lengths, widths, tilts, and degrees of flexibility and rotation, with their unique geometries leading to gradations of compression and a subsequent heterogeneity of gene activity and protein dynamics. Some helices have areas of deeper folding than others, for instance, "two turns of a left-handed superhelix, wound around an octamer of histone proteins ... [corresponding] to a roughly seven-fold condensation."[5] Tightly binding histones crimp and plait DNA, swathing it in tight coils, rendering it not only spatially concise but tidy. The beaded and

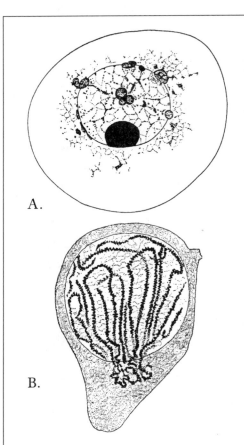

FIGURE 6B. A. Chromatin extrusion from the nucleus into oocyte of *Pelagia noctiluca;* B. Extrusion of chromatin into the cytoplasm during the maturation of the oocyte of *Proteus anguineus.*
From William E. Kellicott, *A Textbook of General Embryology* (New York: Henry Holt & Company, 1913).

coiled chromatin structure adds another forty-fold condensation of DNA. Looping of chromatin (an irregularly fibrous complex of histones, nonhistone chromosomal proteins, and nuclear DNA) deepens compaction, giving a furlike, squiggly appearance to the lampbrush chromosomes of oocytes that have an unusually high rate of transcription.

Without the chromosomes for its scaffolding, DNA in the nucleus would be "strewn about in tangles" like film in a cutting room; yet the chromosomes organize DNA like a "reel and shelving system. The DNA strands are complexed with proteins which serve as spools. The spooling is dynamic, so that the DNA is practically all put away when the cell is dividing . . . while during the normal activity cycle of the cell, filaments of the chromosomes are partially unraveled and float out in the nucleoplasm, making the genes relevant for activity in a given cell exposed and available, and the ones which are not are suppressed."[6]

During the critical metaphase of mitosis the chromosomes are even more densely stuffed. Throughout biosynthesis and morphogenesis, such concentrated "folding of DNA generates successive high-order structures."[7]

The degree of miniaturization and intercalation necessary for the tabulating and embryogenic potentiating of nucleic acids, cramming and organizing DNA into chromosomes eight thousand times more concentrated, is incomprehensible—the Library of Congress stuffed into a thimble without tearing a page. Yet the mechanism is so deeply entrenched and meticulous that it brings a register of stability and structure to an entropic cosmos and then replicates, maintains, and even improves its edict over time. This is irreducible to physics as we now practice it.

THE SAME LOOM THAT ORIGINALLY POLYMERIZED all the other macromolecules spun the antecedents of nucleic acids. Phosphates were already abundant in the primal "soup"; ribose sugars were likewise plentiful among cosmic debris. But a Spider God (or E.T.) would have bumbled and fumbled many times in sticking together bases and then knitting sugars both to them and phosphates. Perhaps sections of the code assembled separately, a babble of chance words that came together when their strings proved meaningful afterward. No doubt innumerable "wrong" chains appeared in the same environment, were able to contribute nothing to the emerging message, and so were randomly obliterated as they were concocted (or stowed in a null state for later reconsideration). Once fully replicating sequences developed, their assemblage was likely regularized by a mineral or a primitive enzyme. This reinforced an iterative pattern and kept meaningless units out of potentially functional codes. Although there was no terrestrial mathematician present, DNA reflects a deep awareness of the cosmological (and even numerological) capacities of integers and primes.

"Nature chose simple natural numbers (quantum numbers) for elementary non-living particles. It seemed to favor mostly prime numbers (amino acid numbers) for elementary units of living molecules. Numbers appear to 'breathe life' into the genetic code. Biology seems to remind scientists how to count with natural numbers. . . ."[8]

REPRODUCTION WAS THE OUTCOME of hundreds of millions of years of experimentation. "The microtubule organizing centers and the chromatin system stabilized and coevolved in a coordinated fashion . . . before the common ancestor of yeast, cows, and peas appeared, probably over 700 but certainly prior to 500 million years ago."[9] Ribonucleic acid was likely the first genetic molecule on Earth, an accidental polymer assembled from stray sections of organic material with the capacity and compulsion to transmit repetitious chemical instructions. Alone it was not able to do much, but if a chain of RNA were to associate with other organelles, perhaps ribosomes and primitive mitochondria, and display its proclivity for mimicry, then rudimentary protein synthesis could begin.

DNA was a later refinement, a full *bibliothèque* of genetic lore set in the nuclei of cells. DNA never actually leaves the chromosomes and does not participate in protein synthesis. Only its scions of RNA continue to transfer warrants from the nucleus to the cytoplasm where proteins are manufactured. There are slight chemical differences between the two, like dialects of the same mother tongue (RNA uses simple ribose instead of deoxyribose sugar, and uracil instead of thymine), but the complementarity and underlying informational integrity of the message do not change when it is transferred from one template to another.

IN 1998 WALL STREET-ORIENTED SCIENTISTS projected the ultimate marriage of biology and cybernetics to be a molecular computer—that is, an information processor constructed of genetic strands in place of silicon. Simple DNA chips baptized by microbiologists in 1996 can already rudimentarily "read the reams of genetic information in the genomes of living organisms. . . . [See Chapter 15, "Biotechnology," for elaboration.] Unlike most conventional computers, which are sequential and can only handle one thing at a time, DNA is a massive parallel computing machine and can theoretically compute a hundred million billion things at once. One scientist recently quipped that a small jug of DNA can compute more arithmetic than all of the computers currently in use."[10]

This anthropocentric view of nature, though intentionally ironic, is still self-congratulatory and upside-down. The dilemma is less quantitative than epistemological. We ourselves, the inventors of both arithmetic and cybernetic computation, are, of course, the one true molecular computer—the most complex DNA chip on

Earth. The holographic capacity of cell nuclei, while presently conceived of in cybernetic terms, is truly virtual and old. Our computers are Goliaths by comparison.

We are not inventors of DNA. Its bare shadow is what we "cyborg" with gross filaments and circuits.

DNA Synthesis

During episodes of cell division the double helix sears down its midline, the strands detaching as if a zipper had held them. DNA polymerase, a large enzyme formed from multiple polypeptide chains, follows the cleavage along the sides of the split track "like a repair locomotive, and duplicates each missing half."[11] The splitting of the double helix by unzipping its base pairs leads to functional replication of the unique pattern of its rungs, as each attracts "free bases from adjoining material in corresponding numbers, type, and sequences, thus providing the mechanism for the transmission of genetic material."[12] The synthesis is not foolproof, but it is a close enough "one-to-one matching of base for base [that] the DNA in the proliferating cells . . . guides the individual throughout its lifetime, and its progeny down through the generations."[13]

Unwound at the appropriate cadence, each DNA strand transmits its code to an assembling sugar-phosphate chain. Maiden double helices occur spontaneously—one old strand, one new strand matched in antiparallel fashion and separable too at the proper impulse from the environment. Crick describes the transient bonding of the strands as "like two lovers, held tightly in an intimate embrace, but separable because however closely they fit together each has a unity which is stronger than the bonds which unite them."[14]

DNA replication requires dozens of discrete proteins to catalyze the unwinding of the helices as well as the rearrangement of the chains in space, their rotation about one another, and the legible marking of the backbones. All of this must occur while the helices are separating at one end and being copied at the other; transcription is going backward on one as it is going forward on the other.

Genetic information is transferred in a series of events that suspiciously resembles an assembly line (even as the information resembles computer code). Our machines augur, after the fact, the synthesis of proteins, but this does not mean the cell is a factory (any more than DNA is a computer). Rather the cell has an aspect which, in a sequence of anachronistically industrialized paradigms, suggests a factory.

RNA

DNA COMPOSES RNA from its building blocks in the nuclear sap. RNA enzymes locate precise points of contact on DNA molecules for transcription; nucleotide by nucleotide they proceed through the message as they hold the chain in place. Because of the prochronic mechanics of replication, a terminating nucleotide sequence (called a telomere) repeats itself in order to avoid the loss of a few nucleotides of genetic information at the end of each strand, a defect which would otherwise successively shorten generations of chromosomes.

The RNA molecules are of course complementary, not identical to their DNA molds; they are anticodons—but then the cell knows of the existence of DNA only through its mirrored structure in RNA. In fact, RNA is the dynamic reality of the genetic molecule, its strands coming into being as the nucleoside triphosphates copy the DNA chain, uracil aligning with adenine, guanine with cytosine.

In combination with a squadron of protein molecules RNA is used structurally in the formation of ribosomes; but, most fundamentally, RNA is the working copy of the DNA in the nucleus. In this form, as messenger RNA, it carries genetic information to the ribosomes which are manufactured continuously in the nucleoli.

RNA is usually found in single strands, though it may initiate double helices with itself or a strand of DNA. Whereas once it served as the actual genetic material (and still does in some small viruses), its function is highly refined in the cells of most contemporary organisms.

Free polymerase molecules ordinarily bump willy-nilly along the chromosomes, adhering weakly at best. However, upon contact with the specific DNA sequence that marks the beginning of RNA synthesis (a site known as "the promoter"), the molecules lock and tightly bind. After the polymerase hooks to the promoter, the strands of the helix are uncoupled in such a way that the template is bared for nucleic scrutiny.

There are three distinct RNA polymerases in eukaryote cells, one of which makes all the precursors of the RNA genetic code (known as messenger RNA); the other two synthesize variants such as ribosomal and transfer RNAs, which have structural and catalytic roles (see below).

Transcription

CHAPERONED WITH THE AID of their polyadenylate tails, ribosomal subunits escape the nucleoplasm in traffic with transfer and messenger RNAs. All mature ribosomes involved in protein synthesis ultimately come to dwell in the cytoplasm outside the nucleus.

While itself still in the nucleus, messenger RNA may be acted upon by various enzymes and spliced in such a way that certain nucleotide sequences are expunged. Splicing empowers single genes to encode many different proteins and provides context for the random production of novel proteins and new relationships among existing ones. (We will discuss selective transcription in Chapters 11 and 12.)

This linguistic coup occurs without rebellion by the ribosomes, which are not permitted back through the nuclear membranes after maturing outside (see discussion of the nucleus in the previous chapter). Because the membranes are a true barrier, transcription and translation are separated from each other in space and time and meaning; thus, mRNA can undergo significant innovation and customizing before it acts in any final embryogenic fashion. Intervention adds subtlety to the biosynthetic process.

The nuclear membrane also restricts what proteins are actually allowed to come into contact with nuclear DNA. Control of transcription is the key to a healthy cell and the likely reason for the primordial quarantine of DNA as well as the deep structure of the nucleus' plasma envelopes.

FULLY EQUIPPED, MESSENGER RNA must go find the ribosomes in order to deliver its message. Then the protein-synthesizing machinery can swing into action.

MRNA molecules are first translated into amino acids with the aid of enzymes which recognize the shape of only one amino acid and so convey its message. They serve as the active bond between the static nucleotide sequence and the emerging polymers.

A POSSIBLE ANCIENT PREDECESSOR OF DNA, transfer RNA (tRNA) plays a critical role in transmission; both structural and informational, it (not its amino acid) determines where each amino acid will attach during protein synthesis. Relatively small (seventy to ninety nucleotides per molecule), tRNA is meticulously pleated into a distinctive three-dimensional configuration resembling a double helix. Activated by enzymes, it functions by attaching one end of itself to a codon of mRNA, the other to the amino acid stipulated by that codon, welding these sequences together. Afterwards, the amino acids array in a perfect match of the mRNA nucleotides. One end of each kind of tRNA supplies the codon for which it carries the complementary message. Meanwhile the molecule is folding such that its other end can be fastened to the ribosome at the point of protein synthesis.

Seen from a different angle, tRNA leads the reinless amino acids to the correct codons at the ribosomes where they are assembled into proteins.

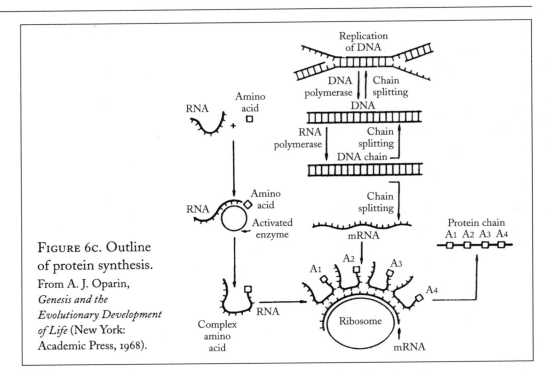

FIGURE 6c. Outline of protein synthesis. From A. J. Oparin, *Genesis and the Evolutionary Development of Life* (New York: Academic Press, 1968).

In its intermediary role tRNA is another anticodon, bearing the sequence complementary to the one on the formative strand of mRNA; they fit together like lock and key or adjacent puzzle pieces. Through tRNA the genetic code turns nucleotide sequences into protein sequences. Molecular surfaces forge one another with mirror specificity, three nucleotides of mRNA per amino acid. The process also creates high-energy linkages, simultaneously potentiating each amino acid to extend a peptide bond to the next one brought by the next tRNA. Without this continual activation, amino acids would have no capacity to form polypeptides. As it is, they roll along, electrons flowing between their atoms, each break in covalent energy restored by the arrival of the ensuing unit, each bond impelling another, the chain growing vast and curling into space.

Protein Synthesis

A RIBOSOME IS LIKE THE LOCK in the zipper. Its RNA component allows it to attach the sequences of codons from mRNA to the amino acids of tRNA. After responding to the start codon on an mRNA molecule, this zipper travels the length of the codons picking up successive transfer molecules with their amino acids. A ribosome releases a protein chain upon reading one of the three mRNA "stop" codons (see later). Numerous ribosomes toil at the same time along an mRNA strand, catalyzed at sites of protein synthesis. Ribosome by ribosome, growing peptide chains bind as a result of the mutual reactivities of the amino acids and the emerging configuration of the protein. When the last amino acid has been attached at the ribosome, another RNA molecule arrives at the site, and the process recurs. When the full polypeptide chain has been fashioned, the protein molecule is released and a new one begun. Through the agencies of mRNA, ribosomes, amino acids, and their cohorts, the fine, threadlike structures of the gene, with its primevally incised microalphabets, becomes the dense, globular fabric of multidimensional protein conformations. Ancient megalinear space is translated into modern tuberosities and chemically charged distensions.

Protein synthesis is the staple of cell life. From the instant following fertilization until the death of the organism, mRNA is selectively led into the polyribosomes for translation into amino-acid language. It is important to remember that this embryogenic process is a lifelong mechanism of multiform and heterogenous assemblage and reassemblage, never ceasing, resembling but removed (temporally and algebraically) from the singular drama of the germ cells and their onetime meiotic alteration and transmission of the hereditary plan into a new organism. Naturally the two processes are linked. No embryogenesis can occur without the creation of a seed, and no seed can be cultivated without prior embryogenic activity. The making of the seed—by selective copying of ancestral plans—resets protein synthesis at its beginning and initiates an enormous variety of different possible (nonsexual) protein-replicating events in the template creature's offspring. Thus, time moves forward while scale goes back and forth between the hieroglyphic coding of the "whole universe" in one indivisible DNA curd to the actual recreation of that universe from complex and varied information in the knot. The latter (embryogenesis) requires many separate and quite different RNA episodes occurring both simultaneously and sequentially along zones joining each microcosm to the same macrocosm.

PROTEINS ARE FABRICATED IN STRAIGHT LINES—the trajectories of their amino-acid sequences—but they travel into much more intricate space as amino acids interact and bend their chain around to join itself in loops and twists. Some of these are actual chemical bonds; some are electrostatic interactions; others are hydrophobic groupings. The latter crystallize when those amino acids that do not associate with water gather in the oily center of the protein. Then, those amino acids attracted to water form a ring around the core and fold together with electrostatic and hydrogen bonds. Covalent bonds anchor the lattice. This same process occurs of course at myriad sites throughout an organism.

Some proteins are quartered with the bound ribosomes inside the rough endoplasmic reticulum where they are routed for insertion into the cell's membranes, or for secretion into the local microenvironment or bloodstream. Others are produced on strings of unbound or "free" ribosomes and are hastened on their way to one of the cell's aqueous compartments upon their completion.

Certain proteins are DNA-binding; they act to control either their own synthesis or that of some other protein by directly contacting the DNA site from which RNA messages emanate.

Any one transcriptional event must bear differential but consistent relationships to others in time and space—to those within the same cell but also to those in other cells in the same tissue and ultimately to those in other parts of the same organism. Each local environment is transformed by and transforms the transcriptions it elicits. Despite this complexity and boundless opportunity for error, anatomies quicken and pulsate in unison; new environments emerge seamlessly from prior ones. The coordination is incredible, almost preposterous, as an organism quickens and self-chisels texture and functional depth across its geography. How could the cells working on a leg know their organ's relation to a kidney or ear, and get it right again and again? How could nature have orchestrated such coordination without a compass? How does the reality of the organism annex and supersede trillions of cell realities? How does multiplicity become unity?

PROTEINS ARE NOT "LIFE," but they are able to mold animate figures from their own intrinsic shapes and the dynamic chemistry of their components, improvising a startling array of differentiated structures from the same lifeless raw material: tissue, skin, blood, silk, feathers, nerves, stinging cysts, phosphorescent bodies, bone, scales, horns, and so on.

The approximate rate of translation is one amino acid per half-second. Hemoglobin molecules, with 150 amino acids, take one-and-a-half minutes to synthesize.

The transition from cell to organism may be colossal, but it happens only by single synchronized steps. The genetic code is not a set of instructions played back; it is an improvisation in which cells in radically changing environments continually revise their interpretations and uses of the same information to match their new surroundings. The self-assembly of functional forms from unstructured stock is known as epigenesis; it will be discussed again in Chapter 7, "Sperm and Egg," and Chapter 15, "Biotechnology."

GENETIC MESSAGES, GENES, are incorporated in sequences of nucleotides formed by the seriality of nitrogen bases: they have no other physical reality—an enigma we will revisit throughout this book. The nucleotide sequences program the assemblage of twenty amino acids; from this seemingly small number, arranged in different combinations, 10^{23} different proteins are possible, each an average of a hundred amino acids in length. Apparently, this is enough to turn clams into octopi, and worms into blue jays. However, without precise timing, synchronization, and the capacity for pause and termination, the code would be merely a prescription for chaos. Given the exquisite subtlety of carbon's bonding ability, such muck may abound in swamps bigger than the Earth on Jupiter, Saturn, or Uranus—raw organic sludge lacking a genetic molecule or membranes.

Gaia is special because of the rigorous artisanry of DNA. The precise strands ensure that proteins are reproduced and differentiated according to their ancestral plan. A coxswain would be superfluous. There is no other way for them to form, and there is also no way for them not to form, so they continue to reinvent themselves out of their intrinsic chemistry and the sequential fields they bring into being.

EXCEPT IN RARE CASES a cell contains the DNA for the construction of its whole organism, but it will translate only those portions of the hereditary code that pertain to its position in the developing embryo; that is, the fate of a cell is determined by selective translation of its genes (this index of embryogenic gradients will be examined in detail in Chapters 9 through 12). A lot of protein-assembling information is transcribed but never translated into stuff; instead its RNA is destroyed in the nucleus after being synthesized. The relative contributions of different genes thus rest upon their selective catalysis by enzymes and the length of time their messenger RNA survives in the cytoplasm before it is degraded. The more stable it is, the more protein chains a gene contributes.

Degree of folding and unfolding of compressed DNA may also affect the gross expression of certain genes, for unless RNA polymerase can get at base sequences, it cannot relay their amino-acid instructions. Conversely, naked or otherwise amplified loops of DNA may get duplicated many times.

The genetic code is degenerate and redundant.

EACH THREE CONSECUTIVE SITES on a strand of RNA constitute a codon for a particular amino acid. Since uracil, cytosine, adenine, and guanine can form sixty-four (4 x 4 x 4) triplets, there are sixty-four possible amino acids in the code. However, three (and possibly four) of the triplets simply punctuate the end of a polypeptide chain; the confirmed "stop" sequences (like gaps between words or periods after sentences) are uracil-adenine-adenine, uracil-adenine-guanine, and uracil-guanine-adenine. Adenine-uracil-guanine begins chains but is always excised before protein synthesis. Many of the other triplets signal identical amino acids (for instance, phenylalanine is specified by two sequences).

Some tRNA molecules are constructed tautologically so that they require only the first two positions of the codon to spell out the intended amino acid, leading to many functional mismatches in the third position, hence extensive duplication of matches.

Only the two least abundant amino acids (tryptophan and methionine) are generated by one codon each.

There are finally only twenty different amino acids transcribed by the multiplicity of nucleotide triplets.

And there is no fundamental relationship between these triplets and their amino acids. Utterly different nucleotides can carry the same final message.

MOST NUCLEOTIDE SEQUENCES are broken by interceding non-coding sequences called introns that will later be cut out of precursor mRNA. Introns can be altered internally without any noticeable effect on the genetic message, giving the impression that they are nucleic junk. The only functional introns are (ironically) those that code for intron excision.

ABOUT ONE PERCENT of most cells' DNA is located outside the nucleus in mitochondria, and those programming sequences are in a different language from that of the nucleus. Yeast and animal mitochondria use codons that mean only "Stop!" in the nucleus to fabricate perfectly good amino acids; conversely, they employ (as their own "stops") codons which yield amino acids in the nucleus. They generally violate the "universal" code of the nucleus to produce "wrong" amino acids from codons. They even jumble their own mitochondrial sequences from yeasts to mammals.

It would seem that the inventor of the genetic code cared less about memorable or indelible arias and more about highly pliant street melodies that interchange with and give rise to one another.

PHYSICAL CHEMISTRY OF GENES tells nothing of the proteins that will be synthesized, nor whether one, two, three, four, or five base pairs are involved.

THE PROCESS OF TRANSCRIPTION can also run backwards: RNA-directed DNA polymerase can spin DNA off an RNA template.

IT IS NOT A CODE made by a computer genius; it is a code made by an ocean. Yet it brings ontologies into being.

The code requires blind intermediaries
between itself and living structure.

GENES SPEAK THROUGH AMINO ACIDS; that is the extent of their "intelligence." The creativity of the genes lies only in the potential of complex structures assembled from amino acids. There is no evidence that genes have the capacity for innovating structure or responding instantaneously to function. They do not have a clue as to what they are doing. In fact, they are not producing proteins; they are engaged in a sewing bee.

Their mostly faithful translations are later jumbled by splicing and other enzymatic activities without their knowing it; their transcriptions are then commoditized by foreigners who do not speak their language.

One level does not predict the next. Genes do not directly program eye color, sex, blood, or instincts. They do not even choose between mammal and amphibian, starfish and sunflower. Amino acids are their sole voice.

By the time elephants stir and starlings quicken, the genetic code is a distant memory of a forgotten dream.

Genetic events are also one-dimensional and linear.

IT IS POSSIBLE that scientists assign so much of heredity to genes not because genes are the singular mechanical cause of organisms but because their participation in the creation of life is the *only* aspect simple enough for them to track and understand. The factory metaphor provides our sole level of comprehension. What happens after RNA improvises and proteins start dancing with one another remains a mystery.

In truth, all that DNA and RNA can do is translate one linear sequence of codons into another or into an equally linear chain of amino acids. Proteins then actuate the play of complex three-dimensional surfaces, crystalline axes, and dynamic stress planes that makes up embryogenesis.

The journey into three-dimensionality is not genetic at all. But then what is it?

DNA is designed for easy repair rather than economy or precision.

IF LIFE ON EARTH WAS FABRICATED by extraterrestrial masters of biotechnology, they clearly intended to be away for a long time. DNA has been left in charge of itself.

In all cells the fragile genetic message regularly undergoes potentially lethal changes from thermal fluctuations alone. Of the thousands of accidental errors in base pairs, only 0.1 percent result in actual mutations. This is because enzymes (DNA repair nucleases) recognize altered portions of DNA and remove them by hydrolysis. The gaps are then filled by DNA polymerase copying the correct information off the undamaged strand of the double helix.

Many different types of enzymes can excise and replace incorrect nucleotide sequences. Some of the remediators are specifically summoned by cells responding to their own lesions.

Genes have no set meaning or numerical value.

A GENE IS A NUCLEOTIDE SEQUENCE that serves as a functional template for the generation of an RNA molecule—no more. Chromosomes are extraordinarily long DNA molecules bearing a sequence of genes. The notion that each gene routinely encodes a single polypeptide chain is naive. Only some blocks of RNA (exons) will actually be translated into protein; introns will be spliced out. The splicing is conducted in a variety of mechanical and chemical exercises within the confines of the nucleus, so that, as noted, a single gene may end up programming many different functional proteins, hundreds in fact. These are called splice variants.

Sequences of DNA that are spatially dispersed on chromosomes may also program lone units of messenger RNA while collaborating on other units elsewhere.

THERE IS NO CORRELATION between the code and traits in bionts. In fact, in different contexts genes "program" utterly contrary characteristics and outcomes, some of which may be identical to outcomes programmed by discrepant genes. Pleiotropy is the formal name for the regulation of more than one attribute or function by a single gene.

Genes that get transported between chromosomes can have radically divergent effects on development from their new positions. These "jumping genes" bear transposable genetic elements that retain no consistent content or systematic morphological expression.

WHILE HUMAN CELLS CONTAIN a mere seven hundred times more DNA than most bacteria, there are plant cells with thirty times more DNA than human ones. Likewise, one amphibian can possess a hundred times more DNA than a close relative. This makes no quantitative or statistical sense. From a purely genetic standpoint, humans are only ten times more complex than fruit flies!

IN 1998, AFTER A DECADE OF EXPERIMENTS, a British-American team of researchers at Sanger Centre in Cambridge, England, and Washington University in St. Louis, respectively, were able to identify and place in functional order the 97 million genetic units comprising a blueprint for *Caenorhabditis elegans,* a silvery, translucent, soil-dwelling worm of 959 cells, barely a millimeter in length. Remarkably, a full seventy percent of the thousands of known genes in the human genome occurred either identically or in similar form in this simple creature.[15]

Clearly nature can use the same genes to write radically different scripts, to create entities of strikingly divergent scale and complexity. Mammals do not require the invention of a whole new repertoire of genes beyond those in worms and trilobites; their architecture simply redeploys ancient, proven genetic letters under radically new coefficients. Where these coefficients arise is the skeleton in the closet of both epigenesis and phylogenesis.

The code is mutable.

THE TRANSCRIPTIVE PROCESS is not only fallible but subject to mutagenic revision, usually from interaction between cosmic radiation and DNA. Physiological alterations occur at a deep nucleic level and, displacing and reattaching elements, cause lethal knots in nucleotide sequences. Most of these are corrected internally and routinely; many are not.

Mutations are spontaneous, ordinary, ubiquitous, and heritable. The body of life is slowly changed by motiveless billiard balls; whole creatures are lost, stage by stage, and replaced by others. "No objects, spaces or bodies are sacred in themselves; any component can be interfaced with any other...."[16] This is not surprising, given the egregious rewriting of worms as horses.

If perfect transcriptions of genetic material were biochemically guaranteed, life would remain static. There would be no parade of living creatures, only repetitive crystalline forms. The fact that the code is written in amino acids, rather than in fixed blueprints for organs, protects life. Because the code does not contain final-stage information, it is a variable fountainhead, giving rise to the pliant rubrics of form rather than concrete molds. It can be altered randomly without being totally degraded; thus, it is fluent and even disposable. Genes are relational networks that "work only when they break down, and by continually breaking down."[17]

A mutagen rearranges amino-acid sequences and relationships, but the creature bearing this "error" will spawn no lineage unless the newly selected proteins can be organized functionally during embryogenesis and the altered offspring are able to survive in a competitive ecosphere. Severe "noise" leads to total reproductive failure, hence does not sully the gene pool. Other errors are incorporated because they are not fatal and can be overridden by later cellular contexts. These deepen the informational potential of the chromosomes.

The choreography is subject to the random play of the universe. This is not a problem. This is our sole hope.

The code is arbitrary.

THE GENETIC CODE has no regular feature that could not be attributed to accident. The most important proteins are not written in the simplest sequences. There is *no* generic relationship between proteins and nucleotide triplets.

If we fear that such a code is too simple and random for such elephants and geese as populate the planet, we must remember another equally arbitrary code—human language. The fact that the phonemes and morphemes of this code have no intrinsic meaning does not stop their chains from bearing philosophical systems, laws, poems, and sacred and judicial concepts.

Both human languages and amino-acid chains begin with units that lack signification in and of themselves. Organisms transcend their genetic codes in much the way words transcend the nonsense syllables in which they are written. The initial randomness becomes irrelevant once the systems are operational.

In fact, our languages may come into being as distant echoes of subcellular codes generating us. The various alphabets of the Earth would then be hieroglyphic ciphers, through a glass darkly, of the incipient and creationary scripts of macromolecules. Their poems and songs would bear some of the primeval mantra of nature itself as it shuttles vowels among tiny instruments of the deep.

As noted above, not only is randomness not a hindrance, it is a source of creativity. An arbitrary basis allows novel properties and radical structures to originate from haphazard occurrences. Known words turn into unknown words, ideas into their opposites and then into whole different ideas; exotic species of plants and animals arise. Puns, homonyms, and onomatopoeia joke, tease, and allude with bottomless innuendo. How else could we explain Apache and Basque as dialects of the same proto-language? Walruses and orchids are likewise idioms of the same "speech." The play between one level of code and another allows infinite variety, "innovations far stranger and more radical than anything we can conceive on our own."[18] No logical nonarbitrary system could elicit such divergences.

Life on other worlds, if formed in our manner, will no doubt use carbon and other elements in entirely different morphophonemic codes.

Life forms can be colonized, cannibalized, kidnapped, and/or ravaged at a nucleic level.

THE MUTATIONS AND TRANSPOSITIONS underwriting any genome are themselves unstable. They disintegrate, contaminate, bastardize, and even maim one another, both within an organism and across organismic boundaries (as varied as fungi and plants, nematodes and fish, protoctists and insects); yet throughout their battleground they maintain life properties—the biological equivalents of zombis and loas. Genes sabotage and suppress each other's expressions in inconsistent fashion (epistasis) and, in general, behave as though they represent a snake and a bouga toad "buried together in [a] jar until they died from rage. Then ground millipedes and tarantulas [are] mixed [in]."[19] In genetic fields such malefic pharmacy is not mortuarial but necromantic. The voodoo of nature's biotechnology is to fuse extermination with satellization, hypothetical events with actual ones, and simulation of landscapes with xylographies. Thus does a crisis of possession and plague of tissue spread across the planet's domain. Anything could be here in our place, but what is—a grotesque carnival of extinct and mythological beasts prancing through moments in the sun—has totally colonized its own reality.

THE ONLY CONTEMPORARY UNITS smaller than cells that maintain life are parasitic upon cells. Their name taken from the Latin for "poison," viruses are extraordinarily simple nonautopoietic organisms consisting only of a protective sheath of protein around a strand of DNA. They were discovered near the end of the nineteenth century when unfortunately elected organisms developed the same diseases as other organisms from whom they received an injection of a plasma extract passed through a bacteria-trapping filter. Invisible under even extremely high magnification (until the electron microscope revealed the tobacco mosaic virus in 1939), completely inert except when they come into contact with a living cell, viruses outside a host are akin to crystals or mere chemicals. In a nutrient medium they still remain dormant. Yet once they get into a cell's storehouse, they literally steal its energy and metabolism. Resembling native chromosomes at this stage, they have been described as "wandering genomes, or parasites at a genetic level."[20] But they are apparently neither ancient nor primitive. They know too much about advanced organisms to have evolved before cells. Their predecessors were likely plasmids, small fragments of nucleic acids that developed the capacity to replicate themselves, to propagate indefinitely with autonomy from the chromosomes of their host cells. Thus, genetic information can travel independently of organisms (even as organic chemistry travelled once on meteors independently of planets).

A separate kingdom of nature that has spawned and mutated rapidly in response to life, viruses are an indication of the inherent autonomy and vitality of DNA—able to package itself as an sub-animal with little else, vital enough to awaken violently from centuries (and probably millennia) of mineral-like dormancy.

Virus' reproductive acts occurring only inside the cells of other creatures, their entire life mechanism involves placing their own genetic material in one of the cells of a plant or animal (or even a bacterium) and inducing it to make viruses instead of themselves. The viral molecules trick the transmembrane channels into opening and admitting them into a network of tubules that conveys their DNA or RNA right to the host cell nucleus—a feat which prequalifies them for delivery of recombinant DNA (see Chapter 15, "Biotechnology," pages 356–357). The virus then combines with the host's DNA in order to make modifications that further its own procreation. By-products of this takeover pour into the bloodstream at a rate approaching 500,000 new viruses per minute.

As a viral invasion spreads, more and more cells are destroyed and more and more viruses are synthesized, the collective effect often resulting in diseases such as polio or rabies. In between such flurries viruses are imperceptible, inert pebbles without metabolism.

Historically, evolutionarily, viruses transport information between systems; however alien and unpopular their gift may be upon receipt, over time it may lead to successful adaptations and wonderfully exotic variations of the host, giving rise to entire new lineages. DNA promiscuity lies at the heart of biospheric creativity and species mutability.

We continue to embody, in plagues and epidemics, shadows that underlie our creation. We learn all too ardently through our unquiet cells that we are not finished works immune to revision.

VIRAL NUMBERS INCLUDE HIV, herpes, ebola, bird and swine flus, and innumerable creatures whose hosts are plants. They provide ever new generations of offspring with ingenious variations in order to bluff their way into cells (the difference of just one nucleotide in 1700 can turn a minor flu into a global killer).

Viruses share a method of recombination with prokaryotes whereby genomes, instead of being integrated into the linear sequence of a cell's DNA, are borne outside the genophore. Thus, even prior to viral mayhem, DNA fragments replicate at different rates; some are exchanged between parents and spliced "virally" into existing sequences with care not to repeat or destroy necessary information (yet to excise duplications or contradictions). Bacterial and viral gene exchange and recombination are complex, subtle, primordial operations indicative of long histories of interspecies hybridization.

Viral existence "is a message—encoded in nucleic acid—whose only content is an order to repeat itself.... Here the medium really is the message: for the virus doesn't enunciate any command, so much as the virus is itself the command. It is a machine for reproduction, but without any reference or referential content to be reproduced. A virus is a simulacrum: a copy for which there is no original, emptily duplicating itself to infinity.... Marx's famous description of capital applies perfectly to viruses: 'dead labor which, vampire-like, lives only by sucking living labor, and lives the more, the more labor it sucks.'"[21]

Given this "free market" nucleic legacy, it is no wonder that plant and animal kingdoms are as mercantile and hegemonous as they are.

The full message is unexpressed.

THE BIOLOGICAL UNIVERSE exists hologrammatically in each of us. Eukaryote cells are stuffed with a great deal of DNA not used in protein synthesis— about ten to a thousand times what is needed to form the proteins of individual creatures. In the mammal genome, perhaps sixty thousand essential proteins are synthesized from over three million potential DNA nucleotide sequences. Most DNA either is not transcribed into RNA or does not survive RNA processing. Some of the surplus no doubt represents the early promiscuity of DNA passing from one microbe or organelle to another. These excess genes then provided thousands if not millions of different protein combinations for the diverse environments in which early life originated. The untranslated surplus continues to bear an evolutionary, organizational meaning in contemporary bionts.

Biologists traditionally look at "junk DNA," see that it can't be mapped or quantified, so dismiss it, calling it an "artifact." Yet this library of information is what makes the astonishing variability of life forms possible. Organisms are fluid bodies, splicing themselves from both immediately conferred codons (which themselves have multiple possibilities of configuration) and archived nucleic stuff (some of it unused for perhaps millions of years, yet faithfully transmitted in "junk" form). Harvesting their "memory," cells move into their own latent probability, reconfiguring and resonating with synergistically mutable transcriptions, expanding fields of expression. What's alive, what gets to live is what the environment naturally selects from this merry-go-round. Thus, species may be a collaboration between a collective DNA library and the ecosphere.

Biological manifestation (like consciousness) is now a matter of hierarchical selectivity: What the cells collectively do not *suppress* is what the organism finally becomes. Each somatic nucleus must withhold its potential replica of the whole organism. Insofar as every cell of our bodies—whether skin, hair, or liver—was at one time able to make another one of us, billions of our twins lie dormant in our flesh. On the simplest mechanical level, suppression can be as effective an evolutionary force as excision or replacement. If a mutation is able to muffle or, on the other hand, unveil the expressions of particular genes, either by changes in enzymes or more subtle transfigurations of the entire embryogenic field, previously unknown creatures march forth from new codons without wholescale changes in DNA consignment. Over generations, global morphogenesis succeeds—dormant loci sinking beneath like Sleeping Beauties, to be awakened in subsequent aeons.

We, as well as the other creatures on this world, contain vast documents of information about life itself, but most of it is inaccessible. In 1982 futurist biologist John Todd told me of a discussion he had with another futurist, Lyall Watson. They were talking about how species have been reinvading one another through viruses for millennia so that parts of plants, fungi, animals are continuously transferred back and forth between creatures and stored in viral nucleic acids. Everything is not only promiscuous but promiscuous at different tiers and levels of abridgment of the same code. Watson's theory, in Todd's words, is

"... that the silence in you represents the genetic imprint of all other beings. The silence in the oak is the genetic imprint of beings other than the oak. Even extinct creatures continue to exist. They're carried in some way in other creatures. Can you imagine! We could dance back that which is gone. What a project for civilization! We could bring back the pterodactyl or some ancient armored fish! Watson thought it might be easier to recreate a species that left recently. The animal he would elect to dance back is Stellar's sea cow, which was last seen about 1886 off Alaska."[22]

If so much information lies buried within, life is a singularity at incredible depth. This is no simple matter of light and darkness, consciousness and unconsciousness. We are the collective imprint of trillions of individual creatures, each of which has had its full manifestation suppressed to allow us. We are deceptively autonomous mud paintings, kaleidoscopic phantoms arising and dissolving underwater in our own elemental, algebraic seas. If we hear voices, some of them may be very ancient indeed. If it takes an effort to summon unity from our many origins, it is not just because we are made of parts. It is because we are made of other unities—sprites who would no doubt stir to life if awakened, and would manifest as we do. Some of them do, in fact, manifest, and their collective songs are us.

7

Sperm and Egg

Mitosis

FROM THE MOMENT IT IS MINTED from another cell, a cell—any cell—knows where it is and either divides again or specializes. It may specialize permanently like nerve cells or red blood cells and never divide again, or it may replicate its specialization.

General cell division, known as mitosis, is framed by biologists in four distinct but overlapping phases plus a fifth interphase during which the cell sits at relative rest. In most cells, prophase begins as the outer membranes of both the nucleus and the nucleolus as well as the nuclear lamina and pores disintegrate. The Golgi apparatus fragments; its components disperse in the endoplasmic reticulum. Loosely packed chromatin fibers coil and swell with condensed, discrete chromosomes.

Hundreds of microtubules and associated proteins construct a spindle at the site of two centriole pairs just outside opposite edges of the nuclear membrane (each centriole, as described in Chapters 4 and 5, is a cylinder of nine triplet microtubules in a ring). Around the centriole couplets other microtubules are busily assembling radial bundles (asters). The bundles suddenly lengthen, pushing the cytoplasmic centrioles apart and propelling them along the surface of the nucleus.

The nuclear membrane deteriorates; the mitotic spindle penetrates the nucleoplasm and, invading the altar of the chromosomes, snares them at the centromeres (constrictions of their DNA-binding nucleotide sequences). Meanwhile strands of other microtubules stretch from the cell's poles to its equator, hitching kinetochore fibers to the centromeres. The interaction of this web of connecting micro-fibers agitates the chromosomes and pulls them into alignment at the midpoint of a plane perpendicular to the spindle axis.

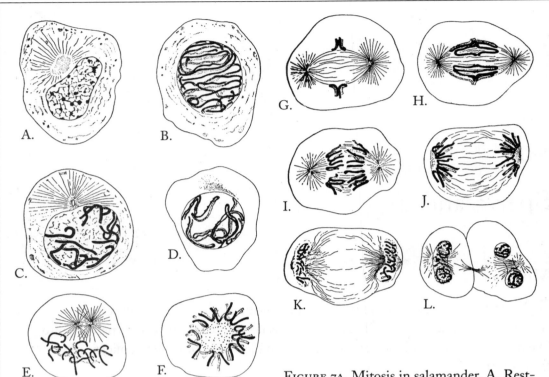

FIGURE 7A. Mitosis in salamander. A. Resting stage; B. Early prophase; C. Prophase, centrosomes diverging, spindle forming; D. Splitting of chromosomes; E. Disappearance of nuclear membrane, continued divergence of chromosomes and asters; F. Mesophase, formation of equatorial plate; G. Side view of F; H. Anaphase, diverging daughter chromosomes united at ends; I. Anaphase, chromosomes separated; J. Late anaphase, complete divergence of chromosomes; K. Telophase, beginning of reconstruction of daughter nuclei, chromosomes disintegrating; L. Late telophase, division completed, nuclei reconstructed, cell walls completed.

From William E. Kellicott, *A Textbook of General Embryology* (New York: Henry Holt & Company, 1913).

The nucleolus shrinks and then dissolves; the cell's components melt and merge. The chromosomes continue to condense; RNA production ceases.

During metaphase, ribosomal and other nucleolar proteins are released; they cling to their chromosomes in order to regain their places in the post-mitotic nucleus. Eventually they flow to the poles of the fissioning cell along with tiny vesicles from the disintegration of the nuclear envelope—the raw material of new nuclear envelopes.

The chromosomes, which have been copying themselves in the nucleoplasm, convene in pairs of new and old strands and coil along the equator of the nuclear core, their diffuse filaments now bunched tightly. The single complementary strands are joined like shocks of wheat at each centromere, their long axes at near right angles to the spindle axis.

Anaphase commences when the paired centromeres disengage, sundering the sister chromatids, each of which becomes a nubile chromosome. The chromatids pop apart, and the pairs journey poleward as if repelled by each other. The mitotic spindles are passive during the splitting; the actual cleavage is in the centromeres affixed to the microtubules of the spindles. The microtubules do not yank the chromatids apart—it is thought that molecular "motors" in the centromeres chug the chromosomes along the microtubules and that the microtubules themselves depolymerize in the wake of moving chromosomes. If the centromeres become detached from the mitotic spindle, the chromosomes lag or drift directionlessly.

The daughter centrioles then sever and migrate around the nucleus to opposite sides of the cell, fine microtubules lying down behind them. Transpiring at the speed of about a micron a second, the circuit of activity follows the mitotic spindle, centromere clips advancing first, telomere tails streaming toward the far pole. The kinetochore fibers shorten as the new chromosomes draw near the poles—sites which are already drifting apart—bearing their completed sets of chromosomes with them.

As the polar fibers continue to lengthen, new nuclei begin to materialize in the vicinity of the chromosomes. The original cell now resembles a three-dimensional figure-eight, a microscopic schmo or beenie baby with its head as big as its body. Each half will become a full-fledged daughter cell. Assembly of daughter cells (telophase) occurs as the lamina reunite, activating the fusion of the membrane vesicles and nuclear fragments into new membranes around the chromosomes. Protein-bearing vesicles bud off the reconstituted endoplasmic reticulum and fuse to compose the cisternae of new Golgi bodies.

A thin furrow forms in the surface of the cell and encircles it, cutting through the spindle, tightening like a knot, and squeezing out two separate cells with their nuclear membranes and nucleoli freshly reconstituted. Each of these is genetically identical to the parent cell.

These events are semi-visible to us as representations of motion in another dimension. Their coordination suggests hidden complexity. A current rips upward through a unit and twins it. The centrioles are cracks in the mirror of time as they travel across the cytoplasm, but they are also mirrors through the crack in time, for they reflect interminable generations as they pull apart cell after cell to reveal only new centrioles extruding.

A population of ordinary (nonsexual) cells divides like this, synthesizes fresh

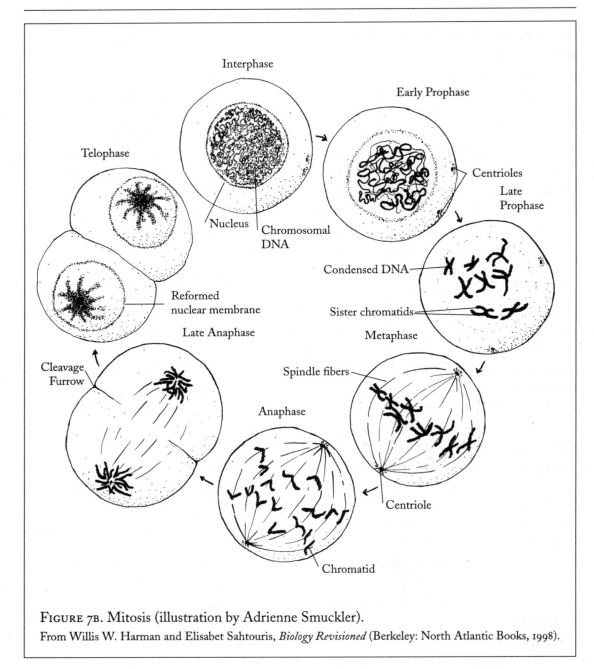

Figure 7b. Mitosis (illustration by Adrienne Smuckler).
From Willis W. Harman and Elisabet Sahtouris, *Biology Revisioned* (Berkeley: North Atlantic Books, 1998).

DNA, and divides again. Mitosis is the sole mechanical basis of tissue growth, autopoiesis, and maintenance. Without mitosis, there would be no fresh skin to cover wounds on a body—in fact, no body.

In early photomicrographs biologists watched this timelapse progression of ordinary cells in sheer wonderment. Even in cases where a nucleus was irradiated, the cells kept on fissioning—splitting and reorganizing their damaged chromosomal rods, salvaging the vestiges of organization from chaos, division after division, in harmonic unision, making sense out of their plight until they could manage the dissonance no longer and fell silent.

This is the relentless activity of life itself, contesting all obstacles and impediments, pulling organizational principles out of its unseen interior, improvising bits of sense out of nonsense. Mitosis is obviously many layers deep. Even though protoplasm was apparently constructed accidentally and out of nothing, its fissioning perseveres and keeps making bright new cells.

THE RATE OF MITOTIC FISSION is highest in the embryo, slackening with age. But age itself does not render cells completely decrepit: wounds still heal, if more slowly, in elderly creatures. Eventually, though, the life-span of each cell is exhausted (one by one), and mitosis ceases altogether; scar tissue knits as slowly as molasses. Cellular aging (cytogerentology) was first demonstrated by Leonard Hayflick during the 1970s[1]; the roughly fifty-replication lifespan of cells has since been known as the Hayflick limit.

The upshot is that cells wear down; they can divide only a certain number of times before they die.

Some cell cultures do, however, survive indefinitely. The mere fact that the cells of an African-American woman's tumor, preserved in the 1950s and adapted to plastic dishes, have continued to spread through the world in laboratories indicates the raw potential that is suppressed by tissue context in normal circumstances and let loose in the uncoordinated milieu of malignancies.

Meiosis

IN SOME PLANTS AND ANIMALS simple cell division hatches full-fledged gametes. In order for human (and, in fact, most animal) germ cells to function as gametes another division of chromosomes must occur—a resynchronization of mitosis such that the cell divides twice while its chromosomes are replicated only the first time.

Fissioning with bisection is called meiosis and first occurred in protists. Whereas mitosis is enacted throughout the tissues of an organism, meiosis is a special event affecting only two consecutive divisions of the preformative cells of sperms and eggs.

WHEN CHROMOSOMES are densely packed in cells that are to become sexualized, the strands come together along their length as if drawn to each other from head to toe. During meiotic prophase they actually break and rejoin so that homologous sections of genetic material are exchanged between original maternal and paternal helices; segments of adjacent nonsister chromatids splinter off and reaffix themselves to the other chromatid. This shuffle traditionally takes place without disrupting the arrangement of genes in each chromosome—an episode called "crossing over." At the same time, genomic contents are sorted and checked and, in some instances, errors are corrected—a kind of cybernetic utility process carried out by DNA strands.

During metaphase and anaphase of meiosis the sister chromatids do not separate. Adhering at their centromeres (which likewise do not divide), they travel as a single entity toward one pole while the homologous chromosome migrates toward the opposite pole (in mitosis the sister chromosomes proceed to opposite poles). Each pole now receives only half a chromosome set, and each new cell formed by the subsequent cleavage is likewise halved (or haploid). Afterwards prophase, metaphase, anaphase, and telophase recur, yielding four haploid daughter cells from two.

"Crossing over" allows recombination of gene-coded traits so that a virtually unlimited variety of gametes bearing unique heredities can arise from a single organism. In some instances enzymes can even alter the relative positions of nucleotide sequences, causing chromatid mismatches. Nonhomologous recombination, recognized first in bacterial viruses and fruit flies, provides mobile genetic elements for more complex animals as well. The incidence of genetic variants, with long-term evolutionary consequences, is abetted and accelerated by meiotic reassortment of genes and redistribution of DNA sequences. In a human being, with twenty-three separate chromosomes and forty-six pairs, there are already ten million different kinds of sperms and eggs possible, *before* crossing-over.

Sexualization

MITOSIS REPRESENTS TWO HISTORICALLY DISTINCT, chronologically divergent events in the cytoplasm which became linked over millennia—microtubule construction of a mitotic apparatus and nuclear synthesis of chromosomal DNA. Once free-swimming bacteria (see Chapter 4), microtubules were enlisted by their protist colonies to assemble mechanisms for the distribution of genes between creatures of different ancestries. They now operate a versatile, autonomic, close-to-foolproof micro-machinery in every cell.

The primacy of sexualization is demonstrated by its ordinary presence in modern protoctists. The colonial volvox divides meiotically to produce large egg-like cells and clusters of sperm-like gametes. In *Euglena* a new basal body arises from the centriole and generates a flagellum. As the nucleus splits, a global severing slides lengthwise down the body. Bisexual paramecia exchange nuclei from mouth to mouth in an act foreshadowing copulation. The gullet is the first organelle to divide, the parent then dissolving transversely and sharing its vacuoles with its fission-self. The cytoplasm always restores missing organelles after mitosis, so paramecia continue to splinter, coalescing their nuclei and reconstituting.

Meiosis has meaning only in lineages that produce zygotes from male and female partners. Using mitosis and meiosis, the lineal descendants of protists continue to mate, fuse, exchange products, splice, combine, replicate, and differentiate into organisms. This is how the meaning of "animal" (or "plant") is translated from microcosmic dimensions to such fullblown tuberous entities as now inhabit land and sea. Embryogenically—each time, each new creature—a series of metamorphoses has to occur both within and among its cells, but that can transpire only as their subcellular units "revisit" the microcosm (hundreds of millions of years in the past) and enact the ancient cyberneticized script of tissue assemblage.

Meiosis is a variant of mitosis that potentiates cells in some species for sex (other species clone germ cells for asexual reproduction). The halving of its chromosomes represents a moment of crisis for a contemporary cell. Suddenly, while carrying out standard mitosis, it is jolted; its parts become entangled; its breath speeds up; it divides quickly, and then finds itself depleted. Most germ cells perish without becoming whole again. The few that replace the missing chromosomes do so by fusing with their benefactors, i.e., with each other, turning their trysts into unions that swallow them individually and generate offspring. Thus, reproduction of multicellular creatures requires a reversion to an act like the one that fashioned the original cell.

As we have seen time and again, desperate and brilliant leaps reinterpret while miniaturizing primordial events. Gaps and their resolutions, dichotomies and their syntheses, represent crises that life and reproduction overcame in its infancy in order to survive. Now they are standard operating procedure.

Meiosis was a gulp, an unlikely mathematical solution to a cosmological problem, also a lick and a promise. Then it propelled evolution forward.

THE PROBLEM SOLVED by meiosis was nucleic fusion.

Adult creatures cannot combine bodies to exchange cells and rejuvenate their metabolisms. The blending of whole organisms (like great energy lattices) would make a certain thermodynamic sense but is not indigenous to this planet; terrestrial tissues are thick and impermeable. Life here neither advances generationally nor refreshes and alters itself by organisms intermingling their substances and incorporating each other. The merging of single incidental cells from each creature is a far shrewder method nature chose. But even that was not simple or obvious before it happened. Now it is a completed masterpiece of transposed scales, inverted surfaces, topological projections, and pigeons out of colored handkerchiefs. Back then it was a last-ditch algebra to salvage an incredible proposition.

Fusion and regeneration of bionts through their tiny extraneous sex cells proceeds only by the amalgamation of deep and hidden genotypes, never by actual phenotypes and never by expression of egos or the awarenesses of the creatures themselves (except insofar as sex acts beget their own symbolic imagination). Syzygy is not an amalgamation of creatures at all; it is a re-sorting of the elements of a cipher (that generated each of them) to make a different entity, or more precisely the mixing and sorting of the near-identical elements of both their ciphers to produce an independent synthesis of them as well as their underlying ciphers. Because the succession of generations occurs by regression to code rather than directly out of things, an indexing mechanism for the potentially unruly alphabet is provided. This is meiotic fission with chromosomal splitting and crossing over.

Meiosis prepares cells for nucleic sorting and fusion; intercourse and other forms of fertilization bring the outfitted cells into proximity for coalescence. Without chromosome halving, sexual exchange of nucleic material would produce mostly monsters and likely would never have originated on Earth. Cells could not otherwise fuse and transmit bilineal genetic material between generations, for they would already be too "whole."

Pre-meiotic spermatogonia and oogonia carry the normal chromosome number of their species. A cell spawned by their unreduced merger would have double chromosomes and its offspring would have four times the functional number. Meiotic variation "sexualizes" cells by halving their chromosomes and, in species in which propagation was established this way historically, it must happen again each generation.

The Advent of Meiosis

THERE ARE MANY WAYS meiosis could have originated, all involving intracellular mitotic asynchronies. If some ancient spirochetes began to replicate faster than their chromosomes (or vice versa), or if the replication of centromeres independently quickened or (conversely) was aborted, half or double chromosomes might have been the result, with genetic material lost in the former case and garbled in the latter. As halved creatures fed on one another, some might have resisted enzymatic digestion and been included in new bionts, a condition ultimately transposed into the genes.

Loss of somatic structure is one of nature's commonest ploys. We see how contemporary parasitic lineages become depleted; for instance, some insects lose hindguts during periods of their hosts' molting (when the food supply dwindles). Requiring only half the amount of DNA as diploids, haploid cells generally tend to be larger, hardier, and more adaptable to changing environments. On the other hand, diploid or tetraploid cells might have other selective advantages (extra genetic information) for tolerating drought or starvation, though their superfluous chromosomes eventually lead to birth defects. Oocytes in many contemporary species actually delay the completion of their first meiotic division in order to double the amount of DNA they have on hand for RNA synthesis; others make surplus copies of some nucleotides.

Diploid or tetraploid generations could also have alternated seasonally with haploid ones in some creatures (a process observed among fungal zygospores), or millennially (from deep, dormant genetic vats) as climates changed. Creatures that retained both potentials somewhere in their genome would have had the greatest selective advantage. This is demonstrated by their ubiquity and fecundity on the present Earth.

In any case, the reproductive plight of halved (haploid) germ cells could be resolved genetically only by fusions of their deficient "hosts" with each other. The plight of cells with doubled chromosomes could be resolved only by a primitive form of meiosis. Somewhere from among these various (and even antithetical) possibilities, sexualized gametes emerged.

THE ORIGIN OF MEIOSIS thus incorporates a variety of imponderable factors combining errors of spirochete replication, algebraic transpositions, Earth changes putting selective pressure on microbes, and primitive microbial colonies invading, contaminating, and agglutinating with one another. They engaged in virus-like transfer of genetic material. Multicellular evolution was their sexually transmitted disease.

"Everywhere poisonous mixtures seethed in the depths of bodies; abominable necromancies, incests, and feelings were elaborated."[2] These inhabitants of a great world ocean swam together, bacteria and protists, long before their progeny became plants and animals. Their "universal feeding frenzy was transformed into a delirious erotic intermingling: cannibalism became fertilization, and meiosis was forced to evolve. And that's why plants and animals have gonads today. It's also why our cells are stuffed with organelles: mitochondria that let us breathe oxygen, chloroplasts that plants use to photosynthesize. These are all contingent effects of unplanned, miscegenetic encounters: the evolutionary fallout of prokaryotic sex."[3]

It was spirochete motility, synchronization and asynchronization of life cycles, and meiosis that turned cannibalism into gonads and fertilization in surviving lineages of ciliates, plants, fungi, and animals. Ever since the primeval bacchanalia, sexuality has, to one degree or another, ruled reproduction and, though there are instances of cells escaping sexual modes of generation, they have done so only in terms of it and by carrying its inactivated template at their core.

Germ Cells

EARLY IN EMBRYOGENESIS certain cells become dichotomized and their role henceforth is the union of their polar material with each other. The "decision" to forge new gametes having already been made at the moment of a prior syzygy, these zooids arise seemingly independently of structural layers of tissue. The large human germ cells destined for meiosis first appear in the endodermal epithelium of the yolk sac, in the region where the placenta and umbilical cord will later form. From there they migrate to the gonads—a tribe crossing a subcontinent—not with the blood or lymph streams but by amoeboid crawling, arriving in a mass of undifferentiated cells which will form genital ridges. The British neurologist Sir Charles Sherrington refers to them as "old ancestral cells ... one narrow derivative line of descendants, nested in the rest of the specialized collateral progeny."[4]

The two animals that meet as sperm and egg in each syzygy of complex creatures are similar but not identical. If they were not similar their blueprints would not cohere in protein assembly (obviously, there is no mermaid compromise between a fin and an arm). They are not identical because they have accumulated different histories since their last point of congruity in a common ancestor of their lines. Actually individual genes share many points of commonality ranging from the basic amino acids that are present in all multicellular life to Ordovician marine skeletons to Jurassic placental organs to Pleistocene bones and blood types to Neolithic skin pigmentation. Millions of present-day humans can claim Alfred the Great as their ancestor, and there is nothing special about the ninth-century King of Wessex in this regard. "Six degrees of separation" is the current fabled limit to the number of relatives segregating two human beings from anywhere on this planet. Only six cousins of cousins stand between a European American and any Zulu, Cambodian, or Comanche. We are not only related to everyone; strangers may share our identical traits from correlative mutations or different remote ancestors even as our own siblings may suddenly diverge through mutations and meiotic crossing over.

Where common ancestors (or convergences) meet in the fertilized egg, there is simply identity of chromosomes. Wherever there is a difference, some characteristics from each lineage must be selected and their rivals eliminated in the surviving reproductive cell.

Preformation and Epigenesis

AS THE ROLES OF THE GERM CELLS became known during the seventeenth century, some scientists surmised that species of plants and animals were imprinted in the cell stuff and that successive generations already existed, nested one within another within each egg. As an egg developed into a mature organism the next mutating egg within it became tangible, but it still held in reserve a bottomless lineage of ever-smaller eggs, encased in one another until the end of time. Scientists and philosophers could not otherwise explain the ready development of embryos— they must have been present, preformed in eggs, concealed and condensed, waiting to unfurl.

During the sixth century B.C.E. the concept of preformation emerged in the West via Leucippus' and Democritus' cosmology of atoms. Democritus held that "nothing can come-to-be from not-being nor pass away into not-being ... [that] the atoms are infinite in number and in differences of size and, as they are borne along in the universe they form vortices, and thereby they generate all composite things."[5] Aristotle reports, somewhat enigmatically, that Democritus (and Anaxagoras) believed in "an infinitude of contacts ... among qualitative similars according to the one, contacts among the spermatic universal shapes according to the other."[6]

The first concrete reference to a preformationist mode of procreation is found in Seneca's *Naturales Quaestiones* from the first century B.C.E.:

"In the seed are enclosed all the parts of the body of the man that shall be formed. The infant that is borne in his mother's womb has the roots of the beard and the hair that he shall wear one day. In this little mass likewise are all the lineaments of the body and all which posterity shall discover in him."[7] Full development of this theory would await the discovery of actual spermatic shapes, seeds within seeds.

IN 1672, LOOKING THROUGH A NEW MICROSCOPE, Reineer de Graaf discovered the mammalian egg, though he mistook the ovarian follicle for the ovum itself.

Sperm was disclosed in 1677 when a student at the University of Leyden brought van Leeuwenhoek a bottle of semen from a man who complained of too-frequent nocturnal emissions. The student had his own preliminary diagnosis: After viewing the semen through a microscope and seeing thousands of small animals swimming

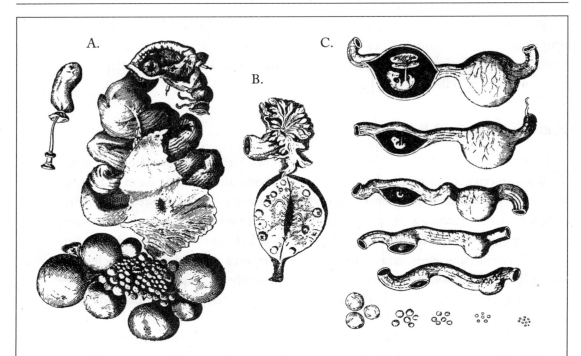

FIGURE 7C. Illustrations from *Regneri de Graaf Opera Omnio* (Lagdani Batavorum Ex officina Hackiana, 1677). A. *Exhibet ova prout in cuniculorum utero reperiuntur;* B. *Exhibet testiculum seu Ovarium Murlieris cum annexio Tuborum extremo;* C. *Gallinae Partes Genitales exhibet.* From Joseph Needham, *A History of Embryology* (New York: Abelard-Schuman, 1959).

with tails, he concluded that the man was infested with some infusorian. Van Leeuwenhoek confirmed the existence of these small animals under a more powerful microscope, but he found them later in the sperm of healthy men, men suffering from a variety of ailments, and animals ranging from rabbits and dogs to pike and cod. He wrote:

"I have seen so excessively great a quantity of living animalcules that I am astonished by it. I can say without exaggeration that in a bit of matter no longer than a grain of sand more than fifty thousand animalcules were present, whose shape I can compare with naught better than our river eel. These animalcules move about with uncommon vigor and in some places clustered so thickly together that they formed a single dark mass."[8] He added that there were ten times as many animalcules in the milt of a large male cod than there were human beings on the Earth.

A new model of preformation now arose: animalculism. Although van Leeuwenhoek did not advance it himself, others used his published findings to argue that future generations were preformed in sperms and that the womb served only as an incubator, a matrix for this seed.

Soon scientists were seeing imprints of tiny embryos in either sperms or unfertilized eggs. One Dutch biologist even sketched the angelic homunculi he "observed" in his spermatozoa, their umbilical cords wound into their tails.

Another early Dutch researcher, Jan Swammerdam, claimed that, just as the butterfly was hidden *(larvatus)* in the caterpillar and the caterpillar in the egg, so were "all men contained in the organs of Adam and Eve."[9] This explained original sin to him. "When their store of eggs is exhausted, the human race will cease to exist."[10] The Italian Aromatari later imagined he saw the complex rudiments of adult plant shapes engraved on bulbs like fossils.

So the argument was joined between the proto-feminists and early male supremacists over whether Adam or Eve bore the future human race.

Both sides, though, shared the belief that the germ-plasm was fixed from the beginning to the end of time and that either Adam's or Eve's body contained every man or woman that could ever be born. "This precaution of Nature," wrote Immanuel Kant, "to equip all her creatures for all kinds of future conditions by means of hidden inner predispositions, by the help of which they may ... be adapted to diversities of climate or soil, is truly marvelous. It gives rise, in the course of the migration and change of environment of animals and plants, to what seem to be new species; but these are nothing more than races of the same species, the germs and natural predispositions for which have developed themselves in different ways as occasion arose in the course of long ages."[11]

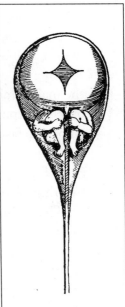

FIGURE 7D. Drawing of preformed male embryo inside spermatozoon. From Joseph Needham, *A History of Embryology* (New York: Abelard-Schuman, 1959).

THE PREFORMATIONISTS WERE OPPOSED by the epigenesists who contended that the sperm and egg each contained raw unorganized material, particles which came together interactively to form each individual. In 1776 J. T. Needham cited "the numerous absurdities which exist in the opinion of preexistent germs together with the impossibility of explaining on that ground the birth of monsters and hybrids."[12] His contemporary Karl Friedrich Wolff noted that if organs were preformed they would have to appear complete. The embryo would then develop by whole stages emerging out of one another like clowns changing costumes. Yet, instead, a growing plant under a microscrope revealed only homogeneous, undifferentiated tissue—simple prominences and swellings, no leaves, blossoms, or other organs. Yolk membranes likewise progressed by gradual vascularization.[13]

The epigenesists were backed in their point of view by no less an authority than Aristotle who, in *On the Generation of Animals,* had described "the catamenia, the menstrual blood, as the passive, plastic substance to which the dynamic, creative male semen gives shape and structure as the sculptor shapes the marble. The working of the male semen on the female blood in the uterus is likened here, too, to the action of rennet upon milk, solidifying the humid elements into drier, solid shapes."[14] The embryo is fashioned by the *eidos,* a primordial form-conferring essence that, although it leaves no hard tattoo in the emergent tissues, transmits their template and a program for their material development. (The Greek philosopher had foreseen DNA two thousand years before it could be located and topologically identified.)

The second-century anatomist Galen espoused a humoral version of the same thesis:

"The seed having been cast into the womb or into the earth—for there is no difference—then after a certain definite period a great number of parts become constituted in the substance which is being generated; these differ as regards moisture, dryness, coldness, and warmth, and in all other qualities which naturally derive therefrom."[15]

English biologist William Harvey takes up the same approximate theme, using a scientific metaphor common to 1653:

"The woman does seem, after spermatical contact, to be . . . rendered prolifical, as the iron touched by the loadstone and presently endowed with [its] virtue."[16]

In his book *Venus Physique,* the early eighteenth-century French natural philosopher Pierre-Louis de Maupertuis wrote: "The elements suitable for forming the foetus swim in the semens of the father and mother animals."[17] Chemical attractions between particles lead to the formation of heart, intestines, limbs, etc.

In *De Formatione Intestorium* (1768) Karl Friedrich Wolff "demonstrated that the chick intestine is formed by the folding back of a sheet of tissue which is detached from the ventral surface of the embryo, and that the folds produce a gutter which transforms itself into a closed tube."[18] Insofar as all organs seemed to emerge in this dynamic fashion, they could not be preexistent.

Wolff was challenged by his colleague Albrecht von Haller, a former epigenesist, who said of the vascularizing membranes: "I don't believe that any new vessels arise at all, but that the blood which enters them makes them more obvious because of the color it gives them."[19] He also wondered how epigenesis of one and the same raw material, the *vis essentialis,* could make a hen always give birth to a chicken, a peacock to a peacock.[20]

This riddle chased itself in bewildering circles because each school had intuited a polar aspect of biological reality: The trait-bearing gametes *are* preformed, but

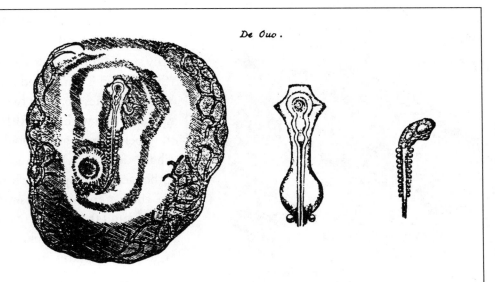

FIGURE 7E. Marcello Malpighi's drawings of the early stages of development in a chick embryo.

From Joseph Needham, *A History of Embryology* (New York: Abelard-Schuman, 1959).

the trait-organizing embryo is not. Organization begins from scratch each generation. The only design that passes between parents and offspring is a code for their lineage that must be retranslated and reenacted each time from elemental proteins. Sperm and egg are also not different kingdoms; they are different packaging for neuter gametes, their sexuality arising from minor variations in the amino-acid sequences of some genes.

And it remains a mystery how a peacock always breeds a peacock despite the lack of a fixed multidimensional plan in the nucleus of either the sperm or the egg. Epigenesis is a description not an explanation of biological form.

Spermatogenesis and Oogenesis

SPERMATOGENESIS OCCURS CYCLICALLY in the male gonads throughout a creature's life. Spermatogonia are generated by mitosis and remain dormant in humans from the fetal period until puberty at roughly thirteen to sixteen years of age, when they begin to multiply. They undergo an average of seven divisions before growing in volume and then, as primary spermatocytes, they enter the meiotic phase. The first division during meiosis yields two secondary spermatocytes, and the second fissions these into four spermatids containing the nuclear material of the four sex cells. The overall series of divisions strips most of the cytoplasm from the relatively chubby

Residual spermatogonium   Follicle cell of spermatocyst

Figure 7f. Spermatogenesis in a frog, some cells in mitotic division (enlarged 1600 times); at upper left, 26 metaphasic chromosomes of one spermatogonium. From Emil Witschi, *Development of Vertebrates* (Philadelphia: W. B. Saunders & Company, 1956).

spermatogonia, leaving almost solely nuclear substance in the tiny spermatozoa; it also separates their X and Y (male and female) chromosomes, leaving two with X's and two with Y's. Although the sperm nuclei are haploid, they share their DNA with one another across cytoplasmic bridges, so their genome is essentially diploid and they are able to repair mutated genes by checking them against templates from other nuclei.

The cores of these spermatids, having lost their RNA and almost all of their reserves of protein, are now dehydrated. While they have no Golgi apparatus, endoplasmic reticulum, or ribosomes, they pack numerous specialized mitochondria that power (by hydrolysis of ATP) a long, motile flagellum. This tail emanates from a centriole just outside the nucleus and is constructed (like all undulipodia) of two central single microtubules within a cylinder of nine doublets.

A granule sprouts in an idiosyncratically enlarged Golgi vacuole of the sperm head; it swells with a pointed tip called an acrosome. This specialization declares the sperm a radically different animal from all the other cells sharing its genome. The vacuole gradually dehydrates and spreads as a double-walled sheath over the sperm body. Inside the head the sperms' single strands of DNA are packed into a highly compressed haploid nucleus by idiosyncratic protamines (simple, water-soluble proteins); more complex histones are dispensed with. Only the primary plan is protected, imbedded in a "spaceship" for intergalactic flight. Microcosmic entities (like the imaginary panspermia launchers) are preparing a projectile for a journey they cannot undertake themselves.

Virtually all the living substance of the cell is exhausted in making this little animal. It has only enough matter to keep it alive while it swims toward the egg. If it does not reach the egg it has no mode of feeding, and dies. If it impacts the egg it fuses with the gigantic ovarian cell.

In human sperm, the acrosome is ovoid and flattened at the sides; a long flagellum-tail grows out of the filaments of a distal centriole. A proximal centriole forms at right angles to the distal one; it will enter the egg behind the nucleus and set up the spindle apparatus for the first cleavage of the egg.

Rodents and frogs have bent "sword" sperms with points, birds have corkscrew sperm-heads. Even plants have sperm-like male germ cells. In the genitalized cones of ginkgo trees male gametes develop fishlike tails. It is astonishing that a fern and a monkey could arise from such similar mites. The sperm is obviously an ancient, atavistic creature.

THE REDUCTION IN CELL SIZE, outfitting a sperm for its journey to the egg, likely replicates, in condensed and metamorphosed form, a series of episodes from the aeon of the earliest multicellular organisms. As one microbial mutant ovified, it gradually packed and stored nutrients while becoming immobilized; the other gave up anything that impeded its motility to reach the storehouse. At a subcellular level, which is where maleness and femaleness originate, a male is by definition an organism manufacturing small highly mobile gametes, usually in great numbers, and a female is an organism breeding a few plump, sedentary gametes. The sperm is lightened to accommodate speed and range; the female is loaded to serve as the site of multicellular development. Each sacrifices a significant microbial characteristic (either nutritional reserves or motility) to enhance the complementary aspect and thereby accomplish its procreative role with the least expenditure of biological energy.

The only requirement of both the sperm and the egg (as we have seen) is to bear a haploid copy of each of the genes of their lineages. Though eggs carry the same vintage of genomic information as sperms, they do not have to transport it, so they are

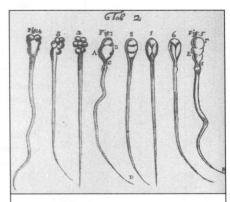

FIGURE 7G. Van Leeuwenhoek's illustrations of spermatozoa of a dog and a rabbit, March 18, 1678. From Joseph Needham, *A History of Embryology* (New York: Abelard-Schuman, 1959).

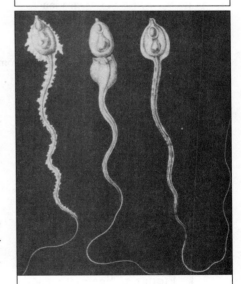

FIGURE 7H. Human spermatozoa in a nineteenth-century drawing. From Arthur William Meyer, *The Rise of Embryology* (Stanford University Press, 1939).

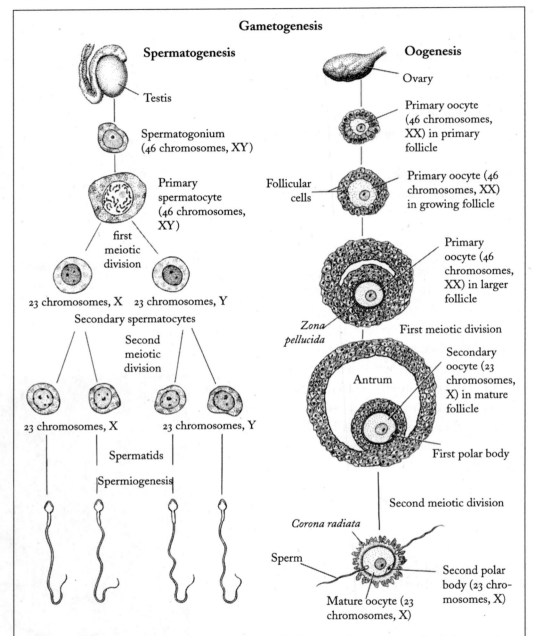

FIGURE 71. Spermatogenesis and Oogenesis. Following the two meiotic divisions, the diploid number of chromosomes (46) is reduced to the haploid number, 23. While one primary spermatocyte gives rise to four sperms, a single large mature egg cell forms from the maturation of a primary oocyte, as cytoplasm is conserved.

From Keith L. Moore, *The Developing Human: Clinically Oriented Embryology* (Philadelphia: Saunders College Publishing, 1977).

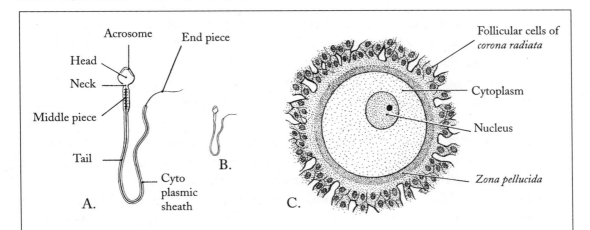

FIGURE 7J. A. Parts of the human sperm (enlarged by 1250 times). The head, composed mostly of the nucleus, is covered by the acrosome, an organelle containing enzymes critical to fertilization. B. The sperm drawn to the same scale as the oocyte alongside it. C. The secondary human oocyte (enlarged by 200 times), surrounded by the *zona pellucida* and the *corona radiata*.

From Keith L. Moore, *The Developing Human: Clinically Oriented Embryology* (Philadelphia: Saunders College Publishing, 1977).

not streamlined. Instead they are pantries of proteins and fats, for once the sperm arrives, the ovum becomes the abode of the newborn. It must provide all the raw material for the embryo's initial metabolism and development.

In species where the egg matriculates in a shell sealed off from maternal tissue, the full complement of nutrients for the embryogenic process must be stored from the beginning—the substances needed to assemble a whole creature after its kind. So the frog's egg grows by twenty-seven thousand times its original size over three years before maturation whereas the human egg, which can draw on the mother till birth, remains a tiny thing.

EGGS ARE BIPOLAR. A seed is wrapped in a nutritive liquid with particles of protein and fat suspended in it. The sphere of the egg that has the least of this yolk is called the animal pole; where yolk accumulates is the vegetal pole. The yolk is made up of elegantly textured platelets. Its raw material comes from a variety of maternal tissues; in birds and humans the liver supplies a large portion, but in insect eggs the blood provides the main nutrient.

Embryogenesis in placental mammals like us is carried out entirely by female tissue, which (of course) fashions males as naturally as it does females. Behind every creature stands a woman beyond simple gender, the Great Mother inside the ovum and placenta. In her breath and glands arise not only oxygen to drive the heart and feed its amassing cells but the cells themselves. Even the penis and scrotum are made from female substance.

The Great Mother is a chain of matrilineal tissue enveloping all embryos back to the dawn of our species and, on a cellular level, the beginning of life.

EVEN BEFORE IT IS FERTILIZED, the egg is organizing itself into a living shape and beginning to foreshadow organs. Brilliant reds and yellows in some species are a natural staining of the otherwise invisible texture of presumptive muscles and intestines. Although yolk does not contribute to the structures of the embryo other than providing nutrient, its relationship to cytoplasm determines the contexts of cell layers, tissues, and organs which succeed it.

Outside the egg, following fertilization, a membrane gradually coagulates to form a superficial but fixed cortex to prevent further sperm from entering. The shell of the robin and the jelly of the newt harden atop this membrane.

Most eggs are round even if their cortices are oval or tubular (but insect eggs are elongated).

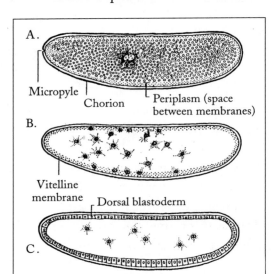

FIGURE 7K. A. Diagrammatic unfertilized insect ovum with nucleus centrally situated in cytoplasm; B. Cleavage nuclei centrally located, blastodermic nuclei in peripheral cytoplasm; C. Yolk cells centrally located, peripheral blastoderm complete, ventral plate forming.
From Harold R. Hagan, *Embryology of the Viviparous Insects* (New York: Ronald Press Company, 1951).

Labels in figure: A. / Micropyle / Chorion / Periplasm (space between membranes) / B. / Vitelline membrane / Dorsal blastoderm / C.

To BECOME EGGS the original germ cells acquire a simple asymmetry during meiosis. The spindle of the oogonium tilts sharply into the animal region with the result that only one fully formed oocyte survives division, receiving the potential larder of four. Ripe with nuclear sap, its chromosomes bristle out in lamp-brushes and become despiralized.

The egg matures in an envelope of protective follicle cells which also nourish it. Although the source of this residual tissue differs by species, some of it comes from aborted eggs. Insects, mollusks, and annelid worms all synthesize nurse cells from unused oogonia. As many as fifteen siblings may contribute to the growth of one functional egg.

Once again, parasitism and cannibalism underlie reproduction.

In most invertebrates the yolk reserves are small and evenly distributed through the cytoplasm. Yolk takes up a relatively larger portion of the amphibian egg and is regionalized toward the lower vegetal zone, its protein concentrated in large oval granules. Birds and reptiles, as noted, have enormous eggs because of their stores of vegetalized yolk. The cytoplasm is but a thin layer around the yolk, with a thicker cap at the animal pole where the nucleic material resides. As noted, the consumables in the yolk of shell-enclosed eggs limit the possible duration of embryogenesis.

A chicken egg cracked open reveals a single giant cell bloated with yolk, so much that the functional part of the cell is not immediately apparent. "While the ovum is the precursor of all subsequent development, it is nevertheless regarded as the least specialized cell type, holding the potential of differentiating form-creation in its huge cytoplasm like a void, while the directions for individuation are concentrated in the relatively small nucleus."[21] The animal pole and genetic substance are both squeezed into a cloudy nuclear film that looks at first like mucus in the yolk. Though it is never dealt with in cookbooks, it is the only true "egg."

The white (albumin composed of protein and water) is simply a protective and nutritive substance secreted around the egg as it passes down the oviduct. Of course, these yolk-and-albumin containers are among the juiciest free lunches in nature, not only for embryos of their species but poachers and gourmet chefs.

Arthropods, notably insects, have evolved centrolecithal eggs with yolk-filled cytoplasm in their interior surrounded by a thin layer of peripheral cytoplasm which is relatively yolk-free. The emerging ant or spider must develop in polar relationship to its cytoplasmic components.

The simpler eggs of water-dwelling invertebrates are neuter gametes, not unlike sperms. Seas, lakes, and rivers provide their wombs. The planula larva of the jellyfish is shot from the mouth of the adult soon after fertilization.

In general, where the mother has few layers of complicated cells, any connection between the tissues of the egg and the tissues of the "womb" is brief and minimal.

The development of maternal tissue in more advanced multicellular animals summarizes a gradual cumulative internalization of watery habitats, as land animals became mobile ponds in which their cells thrived.

But evolution does not proceed along simple linear parameters or unidirectionally. For instance, mammals were not the first creatures to develop a placental relationship. Some species of fish and sea squirts nourish their unborn babies in this way (although in mammalian embryos a new kind of profound bodily intimacy between two organisms was established in the womb). The yolk reduction of mammals (except monotremes) is a reversal laid down atop a prior system of abundant yolk manufacture (unlike the original invertebrate yolk sparsity which was an unadorned consequence of the first multicellular animals). Yolk was required in greater amounts as embryos became more complicated, then in exponentially greater amounts as a particular lineage transferred its embryogenesis to external receptacles in shells (biospheres) on land, then in drastically smaller amounts as that same lineage developed a uterine, placental environment for its young, "reverting" to the simpler invertebrate condition where the embryo can draw on a mother's organs or the external environment for food. Abandoning the shell made the mother's whole body the egg's receptacle.

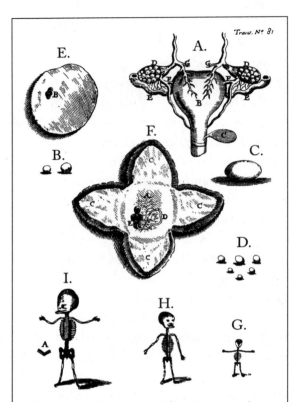

FIGURE 7L. Human fetal development from Theodore Kerckring (1670). A. "A matrix with its chief dependances"; B. "Eggs of different bigness … from the testicles (sic) of a woman"; C. "A bigger egg"; D. "Smaller eggs from the testicles of a cow"; E. "An egg opened three or four days after it was fallen into the matrix of a woman"; F. "A bigger egg opened a fortnight after conception"; G., H., and I. "The skeletons of infants three weeks, four weeks, and six weeks after conception." (original language corrected and simplified)

From Joseph Needham, *A History of Embryology* (New York: Abelard-Schuman, 1959).

DURING HUMAN EMBRYOGENESIS two million primary oocytes are bred and remain dormant in the ovary of a newborn female; no more will be manufactured during her life time. Most of these oocytes regress during childhood, leaving some thirty to forty thousand at puberty. Only about four hundred of them will ripen in the span between female puberty and menopause; ideally those will be dispatched one by one during monthly ovulation, a trigger ancient cells submitted to the lunation cycle, the gravitational pull and light of the Moon. Some signal was necessary, and the largest proximate body sent one rippling through membranous waters. Then cells wrapped themselves upon the many microcosmic "moons" and their phases.

Sperms are newly manufactured by the hour and discarded easily. Eggs are finite in number and may be fifty years old when they are fertilized.

The egg like the sperm is a whole animal. The illusion that it is only one-half of a yet-to-be-formed organism has nothing to do with its moment-to-moment crisis of survival. No sentimentalist, nature requires that eggs and sperms live by the same rules as other animals. They must be able to metabolize organic substances efficiently to generate energy. After all, eggs don't know they are mere eggs with a grand future. They behave like little myxozoa. They are just as hungry, just as anxious to breathe as any jellyfish or snake. There is no special dispensation for being embryonic; an ovum is one more protoctist in the sea. Even an animal parasitic upon its yolk or its mother's tissue is effectively competing for planetary real estate.

The zygote, counselled Ross Harrison, a pioneer American embryologist, is not just "a developing organism, in which the parts are important potentially, but also an organism in which each stage of development has functions to perform that are important for that particular stage.... Organic form is a product of protoplasmic activity and must, therefore, find its explanation in the dynamics of living matter...."[22]

For the egg itself, there is no embryo; there is only an egg. As it passes through stages of ontogeny it adds organs, but not as a means of assembling the functions of its adult destiny. At every phase it is another animal; its tissues are added gradually and only in relationship to one another. The embryo cannot give up its unity or history to accommodate even the most promising modification. If a structure has become critical for the adult form, then it will have been achieved through a chronological sequence of living phases, all of which must be repeated in traditional order. A modern blueprint already contains the most radical deviations, but then only through a long-established and seamless parade of mutating creatures, any one of which could well have been a terminus of development.

Just as an artisan does not produce incomplete works in his youth because he knows masterpieces are forthcoming, so the differentiating organism cannot postpone its survival until it has more complex organs. An artist composes each symphony (or canvas) as if it were his last, or he does not progress and encompass new stages. If he dies suddenly in youth (like John Keats or Percy Bysshe Shelley), his final works must be considered his most advanced. The same is true for the starfish embryo swallowed by a clam.

SPERM AND EGG are both animals *and* embryonic seeds. As animals they are protozoa; as seeds they are metazoa. In truth, all animals are protozoan and metazoan in different phases: organisms are seeds and the husks of seeds in anachronic stages of both evolution and development. What uniquely distinguishes sperms and eggs from adults is that they are haploid and potentiated. All of embryogenesis lies before them and hidden deep within their deceptive simplicity. They are an algebra—

pristine, nascent, and underived. In becoming what they already "are," sperms and eggs forfeit their latency, yet only while expressing it in full-blown creatures and also archiving its matrix in freshly honed algorithms (near clones) of themselves. Thus do organisms sort their components, differentiate, evolve, manifest ... vitiate, recur, redifferentiate, and mutate—through alternately synopsized and exponentialized versions of the same body.

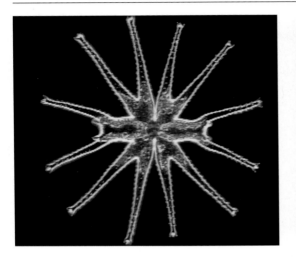

FIGURE 1.

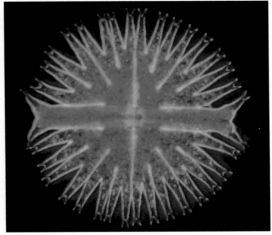

FIGURE 2.

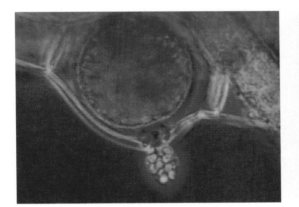

FIGURE 3.

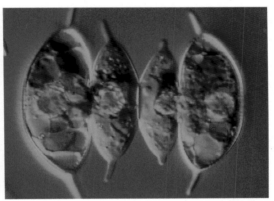

FIGURE 4.

FIGURE 5.

FIGURES 1 and 2. Two desmids (green algae). These beautiful cells are always symmetrical and several thousand species are known, all different in morphology and ornamentation. FIGURE 3. Fertilization in the green alga *Oedogonium*. The tiny sperm cell is just about to enter the much larger female cell. FIGURE 4. The desmid *Arthrodemus* about halfway through cell division. FIGURE 5. Colonies of the golden brown alga *Synura*, actively swimming. Images (and captions) supplied by Jeremy Pickett-Heaps from video microscopy of living material.

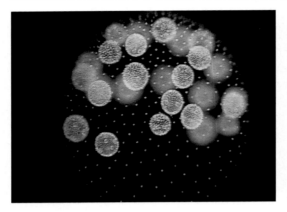

FIGURE 6.

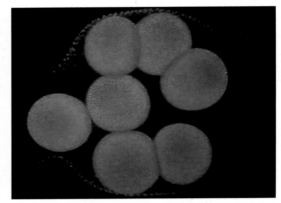

FIGURE 7.

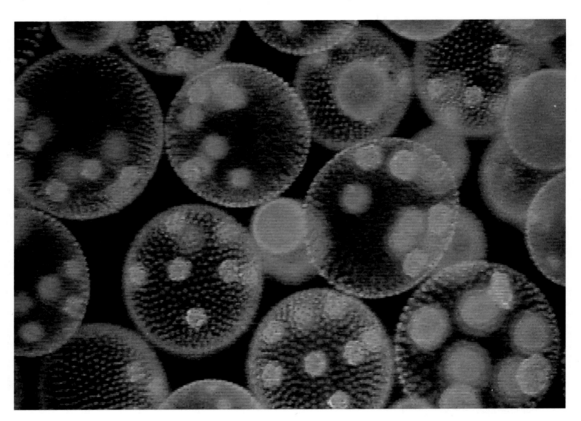

FIGURE 8.

Images of the colonial green alga *Volvox*. FIGURE 6. Numerous small colonies ready for release. FIGURE 7. Colonies being released from the parent colony. FIGURE 8. Colonies containing colonies at different stages of maturation, and embryonic cells starting cycles of cell divisions. Images (and captions) supplied by Jeremy Pickett-Heaps from video microscopy of living material.

FIGURE 9. This large single-celled protist is a heliozoan ("sun animal"), visualized with dark-field illumination; the very long, fine extensions act like a three-dimensional spider web, trapping prey organisms, which are pulled inwards and engulfed. FIGURE 10. Another large protist, *Stentor amethystinus;* the curved region is the mouth lined with cilia. The cell is filled with bright green algal symbionts. Images (and captions) supplied by Jeremy Pickett-Heaps from video microscopy of living material. (Many images similar to those on these pages may be viewed at www.cytographics.com.)

FIGURE 9.

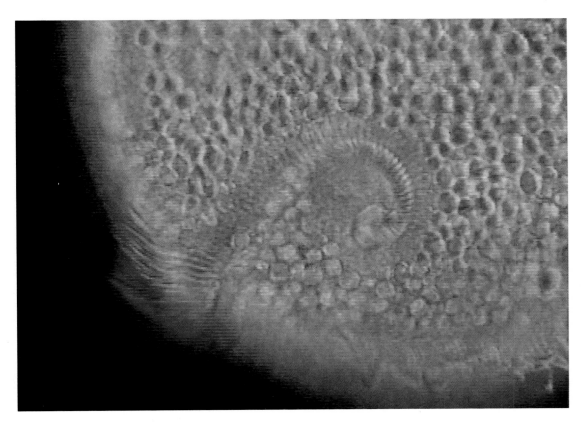

FIGURE 10.

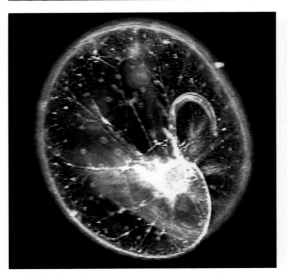
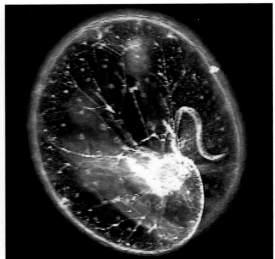

FIGURE II.

FIGURE II. The marine dinoflagellate *Noctiluca*. As its name implies, this cell often generates phosphorescence when growing in large numbers in the ocean. It is a fierce predator and uses the sticky, whip-like "peduncle" to catch its prey and carry the unfortunate organism into its mouth region. FIGURE 12. The diatom *Striatella*; all diatoms are enclosed within walls composed of silica (glass). The wall is composed of many segments which slide apart to allow the cell to grow. This cell has just divided. Images (and captions) supplied by Jeremy Pickett-Heaps from video microscopy of living material.

FIGURE 12.

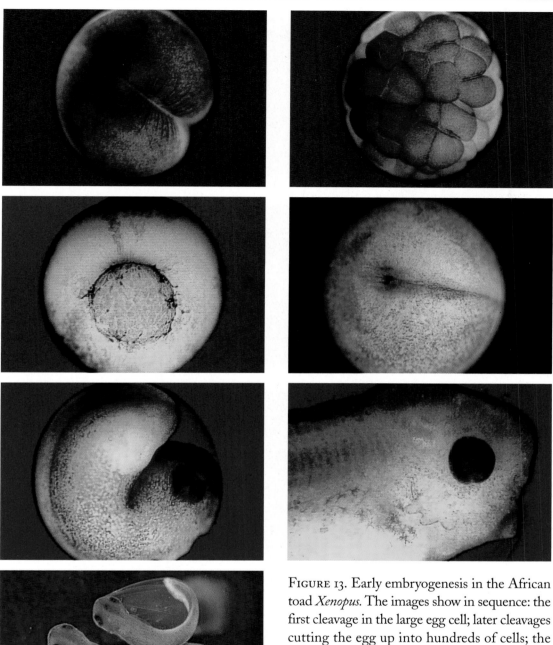

FIGURE 13. Early embryogenesis in the African toad *Xenopus*. The images show in sequence: the first cleavage in the large egg cell; later cleavages cutting the egg up into hundreds of cells; the blastula undergoing gastrulation; closing of the neural tube; elongation of the embryo, still inside the egg membrane; differentiation of the eye in the growing embryo; release of embryos. Images (and captions) supplied by Jeremy Pickett-Heaps from video microscopy of living material.

FIGURE 14. Mitosis. Illustration by Jillian O'Malley.

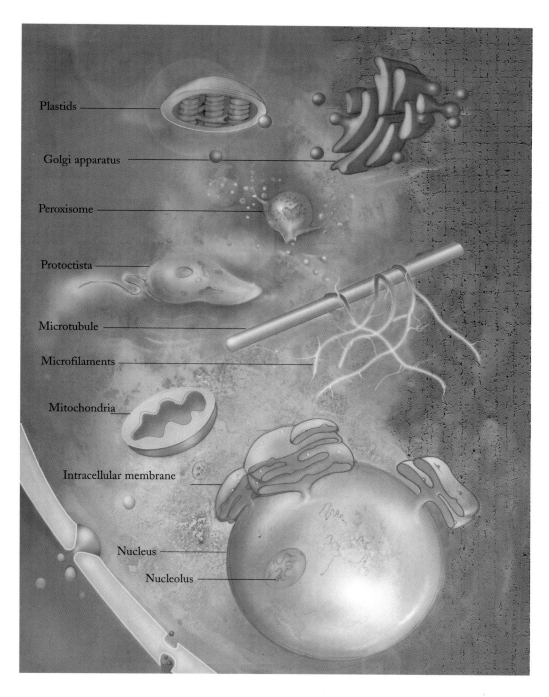

Plastids

Golgi apparatus

Peroxisome

Protoctista

Microtubule

Microfilaments

Mitochondria

Intracellular membrane

Nucleus

Nucleolus

FIGURE 15. Organelles. Illustration by Jillian O'Malley.

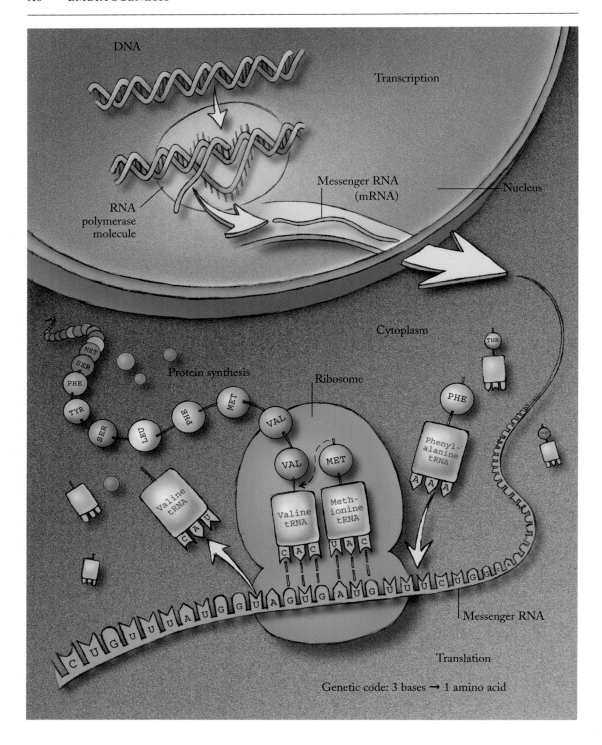

FIGURE 16. DNA transcription and amino acid formation. Illustration by Phoebe Gloeckner.

8

Fertilization

The Meeting of Sperm and Egg

FERTILIZATION OCCURS IN AN AQUATIC ENVIRONMENT, whether externally in a body of water or internally in the tissue ducts of animals. Most water-dwelling species, especially simpler invertebrates, spread sperms and eggs through the hydrosphere in the same way that plants cast seeds into the lower atmosphere. Random dispersal would seem doomed without a species attraction between a sperm and an egg; yet many scientists have believed that sperms are capable only of undirected swimming and that fertile collisions are, in essence, accidental. Biologists who challenge this assessment claim to have observed spermatozoa speeding up as they approach eggs of their own species or species with which they are interfertile. Recently a sperm-attracting chemical has been identified in the jelly of sea-urchin eggs.[1]

Selective fertilization occurs because eggs respond exclusively to their complementary sperms (from the thousands with which they might come into contact in an average aquatic zone); only such a sperm can normally penetrate the environment of the jelly coat around the egg. Foreign sperms slipping through may be sterilized by indigenous protein. Some eggs periodically repulse sperms from their own species, a vigilance perhaps necessary to prevent pathologies from multiple fertilizations.

Even a ripe egg is not easily penetrable. The wriggling sperms would be rebuffed at the outer extracellular coats if they did not bear acrosomal enzymes which break down the proteins in their path. Aquatic eggs bond by proteins to the head of an eligible sperm. In the disturbed vitelline jelly the acrosome dissolves and releases its granule, and the components of the sperm are delivered into the egg's surface.

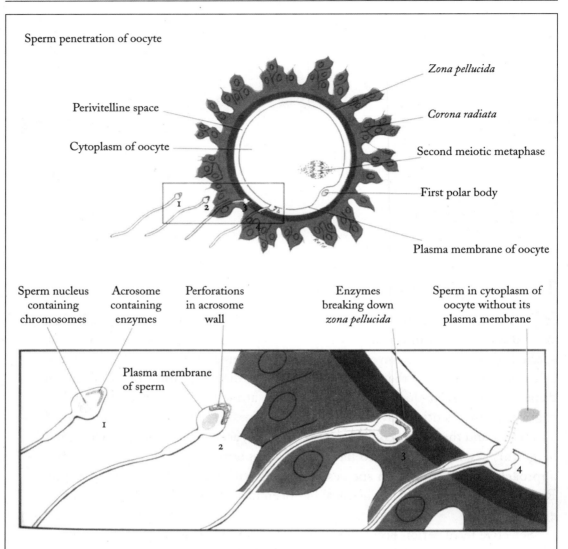

FIGURE 8A. 1. Sperm during capacitation; 2. Acrosome reaction; 3. Enzymes from acrosome digest a path through *zona pellucida*; 4. Sperm enters oocyte. Plasma membranes of sperm and oocyte fuse. As the head and tail of the sperm enter the oocyte, the sperm's plasma membrane remains attached to the plasma membrane of the oocyte.

From Keith L. Moore, *The Developing Human: Clinically Oriented Embryology* (Philadelphia: Saunders College Publishing, 1977).

The plasma membranes of the sperm and egg make contact with each other, fuse, and dissolve into minute vesicles. The continuous membranous structures internal to both become one membrane. The male nucleus proceeds into the cytoplasm of the cell; its journey will end at the center of the oocyte.

The German Oscar Hertwig was the first recorder of this event. In 1875, after fertilizing a sea-urchin egg, he turned the glass of Galileo the other way and spied a single large moon in the center of a meteor shower. One of the meteors suddenly burst through the atmosphere; a lunar shell formed, keeping out the rest. The sky changed color, became stormy, and pulsated irregularly. The point of contact bulged. The tail disintegrated but the head collided with the cell center and there formed "an extraordinarily characteristic starlike figure."[2]

As IT IS FINISHING the first meiotic division, the human egg is swept out of the ovary into the oviduct where repeated passes of the uterine tube's feathery fingers propel it along. The male genital sends five hundred million spermatozoa into the cervical canal. With wriggling tails they beat their way along, aided by the muscular contractions of the uterus.

"The ovum is about a hundred thousand times larger in volume than the sperm and for any one ovum, two or three hundred million sperm will vie....

"[It] is completely round and enclosed in a highly light-reflective thick membrane, the *zona pellucida*, and surrounded by a bright halo of follicular cells known as the radiating crown or the *corona radiata*. For a moment, the ovum hovers in suspense before being taken up by the wafting finger-like fimbria of one of the Fallopian tubes. Semen having been introduced into the vagina, millions of spermatozoa travel in a great stream upwards through the uterus and into the Fallopian tubes.

"The spermatozoon is the smallest cell in the body, only 0.01 to 0.02 mm. in diameter, and consists almost solely of a small torpedo-shaped nucleus and a fine flagellating tail. The head of the spermatozoon is [also] highly light-reflective so that the rapidly oscillating streams of sperm rushing toward the ovum appear like a flash of lightning when seen under the microscope. Fertilization occurs through the penetration of the ovum by one spermatozoon which sheds its tail, enlarges its head, and fuses with the nucleus of the ovum."[3]

Although all but the spermatozoon that penetrates the egg will expire quickly, the end of their individualities has come anyway, for the successful sperm is no longer a discrete being. It merges with the egg and its nuclei fuse with the egg's nuclei to make a new nucleus. The extinction of both individual cells supplies the critical disjunction for the renewal of life.

If we were to try to locate ourselves just before the moment of our conception we would find that, at least on the genetic plane, we were two creatures. This is almost as strange as being congeries of fifty trillion changing cells. Whereas our lattices are ultimately fused into a cellular hive with hundreds of millions of years of cohesion behind it, sperm and egg, bearing our potential embodiment of traits, arise anew from different organisms and are fully distinct, independent animals right up to the instant of their merger.

"As a sperm ... you and millions of your companions, all of you brightly shining there in that river of elixir, were drawn out into the ocean it seemed, or suddenly scattered like flying ants released from the mud hive to chase the queen. All [five hundred] million or more of your companions died that night. And you— how is it possible?—you out of all those millions were permitted to be metamorphosed so that you could contemplate the Mystery as we are presently doing it."[4]

So the Master Adi Da Samraj told his disciples. Death and birth, extinction and desire are indissolubly joined and, though some cells perish before life, all perish eventually:

"What happened to all those millions of others that night? How terrible that every last one of them died! How terrible that there are billions of us here now flying toward who knows what...!"[5]

IN HUMAN BEINGS fertilization occurs in the female, high in the oviduct, when ovulation and sexual intercourse more or less coincide—sperms can live for up to five days before fertilizing eggs. Women can have dizygotic fraternal twins (from two fathers) if two eggs are ovulated.

Even before the era of frozen gametes and artificial insemination, live sperms got transported unsuspectingly into uterine tubes. In one reported case a Caucasian woman gave birth to a part-Black child, and her husband sued for divorce on grounds of infidelity. Since the woman had not had intercourse with anyone but her husband, she hired a detective to solve the mystery. It turned out that the husband had visited a prostitute hours before having intercourse with his wife, and he came away with the residual semen of a previous client of hers adhering inside his penis. One hardy exogenous sperm was then used by him unintentionally to impregnate his wife.

The Zygote

WITH THE FUSION OF THEIR MEMBRANES and nuclei the gametes become one cell, a zygote. They now function chemically as a unity; their unique identities dissipate, and they embark upon a journey that will transform the contents of the egg into an organism. The female pronucleus must migrate to the center of the egg where it contacts the male pronucleus. When the pronuclei approach, they respond to the nearness of one another. In some species they send out protuberances which embrace and pull them together. The male pronucleus provides a

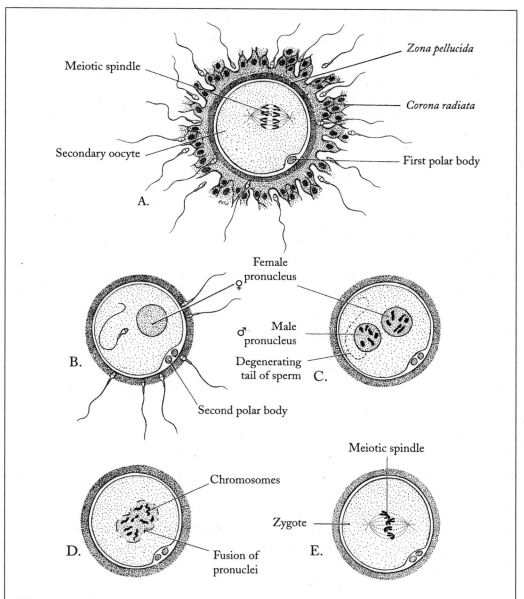

FIGURE 8B. A. Secondary oocyte (only four of 23 chromosome pairs depicted); B. The sperm contacts the oocyte's plasma membrane and enters the oocyte; the second meiotic division has occurred, and the *corona radiata* has disappeared; C. The sperm head has formed the male pronucleus; the cell, no longer an egg and not yet a zygote, is an ootid; D. The pronuclei fuse; E. The chromosomes of the zygote are aligning on a mitotic spindle in preparation for the first cleavage division.

From Keith L. Moore, *The Developing Human: Clinically Oriented Embryology* (Philadelphia: Saunders College Publishing, 1977).

centriole for the ensuing mitotic divisions. Then, as the nuclear membranes break down, male and female chromosomes mingle freely.

The penetration of the sperm does many things. It unmasks enzyme activity so that messenger RNA replication and protein synthesis are intensified, and it contributes DNA for the specific hereditary traits of the male lineage to the fused pronuclei, but it does not participate mechanically or structurally in the new organism.

The overall combination of spermatic changes also prevents other sperms from entering the egg.

A haploid egg must extract its lost genetic material from an impregnating sperm if it is to develop. As we saw in the previous chapter, this "reunion" of cells, restoring the necessary chromosome number for the species, imbeds sex ancestrally in fungal, plant, and animal lineages. The sperm body may dissolve back into the egg's cytoplasm, but its genetic component is incorporated in the nucleic "brain."

The oocyte accommodates the full heredity of the sperm—not just its male characteristics but all the allotted chromosomes of the lineages of males and females through which it has been incarnated from prehistoric times; likewise, the egg contributes male and female lineages both.

IN A NUCLEIC SENSE an egg already has what a sperm provides: a set of genes. Meiotic halving in sexualized species has made sperms essential for the zygote to have a full working complement of genes, but meiosis is only the germ cells' secondary adaptation to sexuality. If this heritage were artificially reversed by a suppression of the second polar body, the egg would then carry the full (diploid) number of chromosomes through oogenesis and require no father.

Sexuality may be a genetic requirement following meiosis, but asexual methods of fertilization and self-fertilization can also activate development. As noted, mature slime molds crawl together in damp soil and merge into a complete animal, a slug. Bacterial spores wall themselves off within their own self-created chambers and develop therein.

Eggs in nature have regularly been activated by sperms from other species. We know from experiments that in some cases of related species, such as the toad and the frog, the nuclei may even fuse and cleavage begin, but development is aborted soon after because of unmatchable genes. When the species are from different classes or even different phyla, an alien sperm can sometimes activate a egg and then die without damaging the female pronucleus, in which case a normal offspring of the mother may occur (without any paternal genetic material). Sea-urchin eggs have been fertilized in this way by sperms from sea lilies and mollusks. Usually alien-sperm development does not occur or progresses to the blastula stage only.

Although meiosis and fertilization are normally required for ontogenesis, full haploid or nondiploid organisms are possible in rare instances. DNA is, as we have seen, a volatile, unpredictable animal. It can also decide to replicate itself without nuclear division, supplying an unfertilized egg with necessary developmental genes from its own vast repository. Undoubtedly such indecorous episodes permeate the histories of species.

Unfertilized eggs can self-activate without any sperm. This process (known as parthenogenesis) has been regularly observed in rotifers, nematodes, and various species of insects. Pure sunlight has initiated development in turkey eggs.

When such virgin reproduction was first observed in plant lice and myriapods a few years after the discovery of the sperm's role, it was an unexpected triumph for the ovulists who had been all but defeated. If the egg could develop without the male, then the embryo must preexist in entirety in the egg. Throughout the eighteenth century a woman might conveniently claim that her pregnancy had occurred from a day in the garden.

Laboratory experimentations have produced parthenogenic offspring in mammals. Rabbit eggs activated by an embryo extract after suppression of the second meiotic division were then inserted in the womb of a virgin female rabbit where early stages of an organism formed, and, in one instance, a complete rabbit (which stands fat and healthy beside its rather diminutive mother on the pages of many embryology textbooks).[6] Parthenogenesis is also a regular stage of the life cycle of many species, alternating with generations of sperm activation to produce offspring of different sexes, respectively. The sperm that the queen bee collects on her mating flight can be used to fertilize her eggs, with only females being born from this process. If she lets them develop on their own, they will become males. Although there is no known case of human parthenogenesis, it is at least theoretically possible. Our eggs are not fundamentally different from those of rabbits, and they contain the necessary genetic information and organization to replicate an entire organism without any paternal contribution. The offspring would not, however, be a clone of the mother, since meiosis causes genetic shuffling (the two pronuclei from the second meiotic division are *not* her genetic equivalent).

Cloning

CELLS RETAIN A LATENT POTENTIAL for modes of reproduction not originally selected or phylogenetically inculcated. In 1996 Ian Wilmut and colleagues at the Roslin Institute in Scotland used ordinary mitotic cells, extracted from the udder of a Finn Dorset ewe and preserved in a freezer for six years after the ewe's

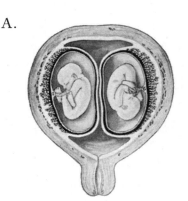
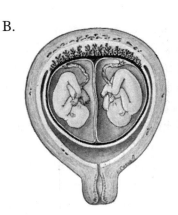

FIGURE 8c. Two types of human twins. A. Ordinary, double-egg twins with individual chorions and placentas; B. True twins with a single chorion and placenta.
From Leslie Brainerd Arey, *Developmental Anatomy: A Textbook and Laboratory Manual of Embryology* (Philadelphia: W. B. Saunders & Company, 1946).

demise, to breed a new sheep identical to the original—the famous Dolly. This was the first time a mammal had been cloned from a fully differentiated cell (it was approximately six years and forty mitoses down the road from its sexual origin). The cultured udder-cell placed on the outskirts of an enucleated ovum, the two cells were then fused, first by an AC-pulse electric shock to penetrate the membranes of donor and ovum, and then by successive DC pulses to combine them. Since 1996, variations of this experiment have been carried out with many different mammalian species, the most notable menagerie being Ryuzo Yanagimachi's fifty scurrying offspring of the mouse Cumulina in a lab at the University of Hawaii, a good number of them clones. In 1998 a Japanese team made eight replicas of a prize milk cow, and Korean scientists even claimed to have cloned a human. (See Chapter 15, "Biotechnology," for a continuation of this discussion.)

The unmasking of the organismic potential of nonsexual nuclei shows that the difference between the nucleus of a gamete and of any other cell is a slight and temporary one, spawned relatively late in the annals of micro-speciation and actuated by local somatic fields rather than by any intrinsic characteristic of cells themselves. Nature made all cells potentially sex cells and then, after the fact, groomed only certain of these for reproduction.

Identical twins formed by the splitting of a zygote into two embryonic primordia clearly have distinct minds and bodies. Environmental factors initiated as early as differential blood supply from the placenta (if not earlier) consolidate in each organism until even monozygotic multiple-birth siblings do not precisely resemble one another. They then go on to live completely separate lives.

Likewise, the lineages of any sperm and egg merge and invent a unique person.

Views of Sexuality: Reichian, Taoist, and Neo-Darwinian

THE FERTILIZATION OF GAMETES in the microcosm redeems the vagrant history of sexuality. Yet it would seem that erotic ritual far exceeds the requirements of reproduction. From a sheer biomechanical point of view, if the purpose of sexuality is to bring together germ cells expediently and at low cost, the efficiency of different gonadal tissues and their ongoing adaptability to changing environmental niches should be the sole baseline of sexual innovation. Yet eros has a million exotic seductions and masks.

Cloned germ cells are very efficient, requiring only themselves. However, neo-Darwinian biologists have long suspected that the sexual differentiation of plants and animals, despite the perilous adventures male and female cells must complete in order for embryogenesis to occur, is a secondarily favorable adaptation. Through meiosis and syzygy, sexuality breaks the repetitive fixity of the genotype, inventorying, correcting, and shuffling the chromosome deck. The resulting variations are the basis for new lineages of plants and animals, etc., the combination of traits available for innovation through natural selection continuously sorted through breeding populations, from males to females, from females to males. Likewise, damaged genes are replaced by their equivalents from other genomes. If originally maladaptive, sex became a vehicle for the myriad potential of life's expression in protein.

IN ANY CASE, we cannot know whether eros itself arose locally in the aftermath of accidental spirochete encounters and mutations or is an expression of a universal force. If the latter, then gamete fertilization (and ensuing embryogenesis) might be a local manifestation of a cosmic archetype. On remote worlds it would be embodied by other kinds of molecular crystals translating libido into the local equivalents of bodies and organs.

The psychologist Wilhelm Reich reified universal sexual energy in his thesis of orgone—a current exchanged between lovers but also among galaxies, planets and suns, thunderclouds, and organic and inorganic material—in other words, a stream of particles impelling all other physical motion, including gravity. In Reichian terms, sex cells form organisms because they translate orgone (primordial world-stuff) into zygotes. The charge itself is inherent and ancient and is realized in seeds of plants and animals only because it was already present in stardust.[7] This gives erotic desire a far loftier destiny than reproduction, and it also explains why it can fuel syzygy.

Creatures are highly concentrated biological stars. (See Chapter 24, "Healing," pages 629–632, for further discussion of this topic.)

Generally biologists have *not* considered that the method behind the union of germ cells might have a salubrious in addition to a pleasurable effect on the conjugating creatures; its utility has been deemed solely to transfer genetic potential into extrinsic zygotes. In Reich's model, mating creatures send waves of nourishment to each other. Or, if sentient and philosophically inclined, they can attempt to hoard the vital seed. Practitioners of tantric and Taoist yoga suppress orgastic release in order to draw and assimilate seminal and ovarian energies throughout their own bodies. They literally intend to pull prana or ch'i back out of their gametes into their organs in order to achieve improved health and longer life rather than offspring. Erotic charge is thus used medicinally and as a trigger of religious trance.

Stimulating and then denying orgasm replaces the goal of pleasure with that of yogic virtuosity, and syzygy with immortality. The attempt alone should alert us to the fact that sexuality masks another ritual.

MARGULIS MAKES HER OWN far more mundane case for meiosis and sexual interchange as a happenstance but successful survival mode during the epoch of heavy radiation when simple creatures were evolving. Photosynthetic organisms required light for their metabolism, but its accompanying ultraviolet radiation knotted and degraded their DNA. Since exposure was unavoidable, the ancient microbes that survived to launch the eukaryotic revolution must have developed methods of repairing their compromised DNA, first by checking and correcting it against their own templates (likely another strand), then (if there were unbridgeable gaps and snarls) by using sequences received from other organisms and splicing those into their own memories. The actual repairs were probably carried out by endogenous or exogenous enzymes that were themselves photoreactive. That was the invention of sex.

DNA continuity ultimately came to depend on a second, external source for fidelity—"the borrowing of an undamaged DNA."[8] This complement, to be functional ("recognizably similar to the first"[9]), had to originate in a related animal. Thus, sexual contact between nascently interfertile organisms not only protected heredity but allowed lineages to miscegenate, mutate, and enter new niches without losing reproductive capacity in the ultraviolet mayhem. "Sex has been preserved," Margulis concludes, "not because it is 'adaptive' but because the organisms in which it was coupled to reproduction reproduced. Human lovers, male and female, are evolutionary permutations, living reminders of the ancient microbial events comprising the origin of sex."[10]

Initially, creature fusions with genetic exchanges may have been infrequent—one in a zillion—but the offspring of such a process would have unparalleled flexibility in the face of environmental perils to their genetic material. The benefits included more lifetime and resources with which to procreate and spread their traits.

Once cut-and-patch mechanisms developed and spread, they were coopted by life itself. Their simple exercise is now

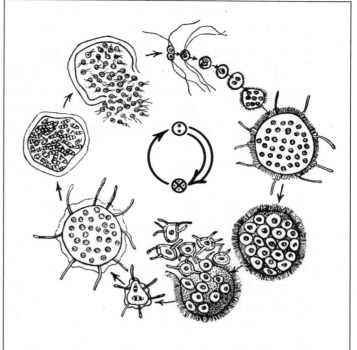

FIGURE 8D. Life cycle of protozoan *Trichosphaerium*. From Emil Witschi, *Development of Vertebrates* (Philadelphia: W. B. Saunders & Company, 1956).

reenacted trillions of times of day on the modern Earth, spewing forth plants and animals everywhere. Without it, the global economy would be ruled by bacteria alone.

It is possible also that sexuality had complementary and mutually integrative causes on more than one cosmological level. Perhaps biogenetic necessity in pelagic environments merged with deep archetypal imprinting on vitalized matter to produce shapes and structures reflecting the needs of both and conducive to their simultaneous resolution. In fact, major innovations may only occur when there are multiple agents. [See Chapter 22, "The Origin of Sexuality and Gender," for a continuation of this discussion.]

Sexuality, Eros, and Meaning

ALTHOUGH THE DISJUNCTION INTO SEXES occurred transhistorically (through successive creatures), the union of male and female is an explicit temporal event, rhythmic at a number of levels and expressing a synchrony of gametes. Release of sperm is a link between the life of the individual and the history of cells. That is why psyche reaches blindly for eros, as an affirmation of its own existence. One's sexuality, as the old poets divined, is an antidote to the portent of the corpse.

"But at my back I alwaies hear Times winged Charriot hurrying near,"[11] wrote seventeenth-century poet Andrew Marvell ("to his coy mistress").

Orgasm scrapes the membrane between existence and nonexistence, the gap of individuation and egolessness where everything (or nothing) is possible, now and forever. . . .

". . . and Yonder all before us lye/Desarts of vast Eternity."[12]

The lover may not become immortal, but his deed resonates with a timeless sensation and gives him a momentary vision of eternal gardens rather than "desarts."

A WOMAN'S EROS has a subtler and more profound relationship to primordial substance. She is preparing her eggs for syzygy and then sacrificing them (when they go unfertilized) in a tidal cycle, always on the verge of contacting the ovum's nature, of moving from her time to its time.

But part of her is phallus too, meeting the sperm's prana and completing the ring of desire. What for a man is a single rush toward tinder, for a woman is a deepening spiral of gyres, each one touching closer to the origin of life. While the male is tied to his act of galactic discharge, the female has a capacity for orgasm that can go on for hours. [For a further discussion of this topic, see Chapter 22, "The Origin of Sexuality and Gender."]

LOVE-MAKING AND FERTILIZATION are primordially and fundamentally cellular. The intimacy of sperm and egg is much greater and more profound than the passion of whole creatures. Like Plato's archetypal males and females (who have been rent from their "eternal" unions in passage from an ideal realm to embodiment in nature), the germ cells have been truly riven. So when they find their Other they require it forever. Gametes give themselves up without regard for race, beauty, or any actual love exchanged between partners in the multicellular world that juggles their fates. Our deviousness cannot damage their essential innocence (apparently they are just as avid for union in cases of rape and unwilling or artificial insemination).

The germ cells have a basic sexuality that transcends our temporal attractions and seductivities. They are pagan gnomes with sacraments and rites, a pale representation of which appears in pornography and erotic art. The sperm and egg are far too old to have human tastes.

When these animals are released in love-making, man and woman experience accumulated desires built up in layers of tissue around an original deliquescence.

Through their own eros they feel and complete a long cellular migration. The Japanese philosopher Michio Kushi imagines this cosmic odyssey:

"The reproductive cells which are active in the mother's womb are the result of a journey of life-transmutations of hundreds of billions of years, which began from the infinite ocean of the universe an almost unknown time ago. They have reached the stage of organic life as the primary constitution of animal life, highly charged electromagnetic vibration....

"When these antagonistic and complementary reproductive cells, the *yang* egg and the *yin* sperm, combine and fuse with each other, it is the beginning of the returning course of a hundred billion years toward the infinite ocean of life. Each of them carries its past memories and the vision of its future. When they fuse with each other—though their recent journeys separated them into different species of vegetables, different kinds of molecules and different blood cells of the father and mother—their memory that they have come from the same origin, one infinity, and their vision toward the future when they shall become one infinity, has never been forgotten."[13]

Lovers impart billions upon billions of years of history beside which their separateness is a flicker. Their kidneys and lungs are roots of one body, long ago twained by meiosis. They have embraced millions of times, as spores, protists, polyps, zooids, mudfish, frogs, snakes. They have shared blue sky and nubilous snows via cells in other bodies, as worms, moles, and lemurs—a brisk wind and icy grains on their epidermal pads. The synthesis of their germ cells is made possible by a primal and indelible singularity. Their kiss really does last forever. Their congress is all we have.

"The memories and visions, which were carried by the parents' reproductive cells and fused into one by fertilization, are distributed to each of the rapidly growing individual cells. Each cell carries the same memories and visions, and all cells comprehensively carry the same memory and vision as a whole."[14]

Sexuality is meiosis and syzygy internalized and writ large. In it we experience glimmerings of the fundamental act of the biosphere, rendering sentient (and even romantic) what cells merely do: making love. Of course if they could speak they might not judge the lugubrious gyrations of such massive creatures as anything like their own graceful fission-dance.

Love and sex are probably two separate drives that become associated (and confused) by our manner of embodiment.

FROM THE COALESCENCE OF SPERM AND EGG, human experience is constituted granule by granule—layers retained, layers erased, layers that turn into something else. Within this cocoon the phenomenology of life detonates in geysers of fine, iridescent chaff. Nuances of breeze trickle across cells; a straw day glistens with jasmine and dust. As fragments (neurons) detonate in disturbed rapture, pings of meaning pour from squalls across the savannah—an impregnated air, purple radiance, intimations of jazz, lightning, a soft voodoo drum. And then what?

Humans interpret the profundity and passion of this lifeform as love. Dragged through epigrams, social codes, kitsch epiphanies, "love" parades as a bewitching trope that holds all creation in its thrall. By invocations from Sappho to Shakespeare, Charlotte Brontë to D. H. Lawrence, troubadours grasp at what is both inexpressible and inexpressibly ancient: a prior, more intimate being of which men and women were etiologically shorn.

In her 1998 novel *Evening* Susan Minot traces Anne Grant to a moment forty years past—a lover who came from nowhere and vanished afterwards: "... she could not get enough of what he was doing to her. Where was he taking her? She didn't want to know. Who was she? Who cared.... The sky was an example of how far distance could go. I go on forever, it said, nothing can be contained. She was the same, she went on forever. She felt everything in her. Good and bad were not so different, she inhabited them equally. She was never more herself and yet never so altered *this is what you were made for....*"[15]

The episode is sex and (perhaps) love, but it is also a realization of being in the universe, of possessing a bottomless, inscrutable cell-hologram: a self. The Sufi poets of the thirteenth century recognized the same intoxication as a shadow of the prime number, the cosmic dance. Where inviolable chemical signature meets angelic charisma, a demi-god appears; circumference becomes center: "You experience a vehement love, a sympathy, an ardent desire, an emotional agitation so great as to provoke physical weakness, total insomnia, disgust at all food, and yet you do not know *for* whom or *by* whom.... Then this love attaches itself ...... you meet a certain person...; you recognize that this person was the object of your love, though you were unaware of it."[16]

Love is what impels and redeems the loneliness and isolation of body, what torques mass through becoming. One's own essence, a point of origin no longer extant or attainable, is suddenly realized in a lone figure, drifting in the same void:

"This is one of the most secret and subtle presentiments that souls have of things, divining them through veils of Mystery, while knowing nothing of their mode of being, without even knowing whom they are in love with, in whom their love will repose, or even what the love they feel is in reality."[17]

We hardly know where *we* are in reality.

Signified as guru, vagabond, swain, no matter how remote and apostate, the lover is always present: "For if you approach me,/It is because I have approached you. I am nearer to you than yourself,/Than your soul, than your breath."[18]

Getting made is a journey across infinity, dimensionalized in the microcosm by the activation of a deceptively modest spore or egg. Each creature then seeks to be remade, etched again in the deep cellular mirror, beguiled by a nameless replica *(thou shalt not see me)* that ignites abyss into form.

9

The Blastula

The Universal Cellular Mold

BEFORE OVULATION AN EGG IS DOR-
MANT. Oogenesis rouses it, and it
pumps out voluminous RNA. After the
entry of the sperm, the zygote is briefly
lethargic again, as if reconsidering. Then
the RNA sings the saga of multicellular
life recorded in its filaments, an odyssey
with many ports of call and long-extinct
isles and cyclopses. It divides by mitosis.
Its daughters each split, and their daughters split. The microplanet is engulfed by
motion and change. Cytoplasm begins to
flow, electrical potentials of membranes
fluctuate, stored mRNA starts assembling
the proteins necessary for development,
and (in many species) great gulps of oxygen are swallowed throughout.

While a sperm initiates blastulation, it
does not contribute to the choreography.
The egg was already building toward
cleavage at the time it was fertilized. Even
an enucleated cell, if stimulated, will
develop a pattern of serial divisions, for

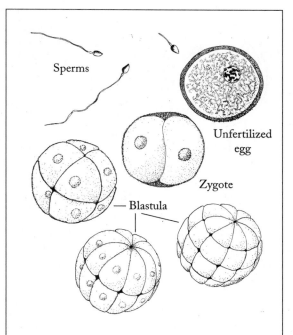

FIGURE 9A. Formation of a blastula from
sperm and unfertilized egg (illustration by
Adrienne Smuckler).
From Willis W. Harman and Elisabet Sahtouris,
Biology Revisioned (Berkeley: North Atlantic Books,
1998).

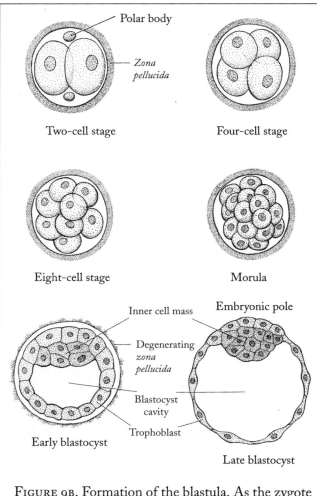

Two-cell stage

Four-cell stage

Eight-cell stage

Morula

Polar body

Zona pellucida

Inner cell mass

Embryonic pole

Degenerating *zona pellucida*

Blastocyst cavity

Trophoblast

Early blastocyst

Late blastocyst

FIGURE 9B. Formation of the blastula. As the zygote cleaves, the blastocyst forms.

From Keith L. Moore, *The Developing Human: Clinically Oriented Embryology* (Philadelphia: Saunders College Publishing, 1977).

the rhythmicity of the blastomeres is solely under the influence of its cytoplasm.

With each division, the stock of cytoplasm is distributed among more and more cells in a fissioning ball. Every succeeding generation receives less cytoplasm and more nuclear material, so its cubicles become smaller and smaller (each complement of cells is roughly half the size of the ones from which they came) until they shrink to beneath the size of the cells of normal tissue.

This is not just ordinary compartmentalization in an everyday spheroid; blastulation is a cosmic event like those attending the birth of hydrogen, the molding of galaxies and planets. It denotes creation of primal matter—an ushering of one into many, unity into multiplicity, timelessness into time, inertia into mutability, information into form. Events funnelled over millennia into a deep singularity bubble back into matter.

The swarming is not mere aggregation and expansion. It is transatomic metamorphosis, macromolecular alchemy, thaumaturgy arising deep within a sorcerer's crystal globe. Though it occupies a sphere, the sphere itself does not grow.

THE BLASTULA IS a universal subanimal formed by simple cell division, a volvox-like ball. Its organization is ancient, primary, and panspecific, affecting all eggs, from invertebrates to mammals, in much the same way. Regardless of ultimate destiny, the component cells first arrange themselves in a hive of sibling balls which is not quite a creature but a mold from which individual animals can be shaped.

The reason the general appearance of a blastula is spherical (or ovoid) is that the free surfaces of its constituent cells continue to be globular (like plastic bubbles) while the pressure of their surfaces against one another keeps compressing them inward. The fertilized zygote, a single round cell, fizzes outward into many smaller round cells, keeping the topology of a ball.

The pattern of emergence of this colony is similar throughout the animal kingdom, for the initiatory phase of multicellularity represents the epoch of history when replicating protists stuck together after fissioning, instead of swimming apart. The blastula, in essence, restages the drama: "No, we are not going to swim off in our own solitary directions. We need to feel each others' presences in order to be safe and get fed. We are going to flex our stickiness and protrude little hooks." This was a monumental feat once, taking generations upon generations of mutating protists to accomplish successfully. Now it is pure instinct—cow and calf, mother duck and duckling. The baby cells don't even have a chance of straying; it is no longer in their makeup.

If we view the embryo as a ball in space, its cleavage furrows are a ring developing simultaneously around the globe and pressing inward at all points. The sphere breathes, pauses, throbs, and then breathes again; at each breath the number of chambers has doubled, surficially and within. The pauses between divisions can range from a few minutes (as in a sea urchin) to several hours in a mammal.

The cells go on fissioning, yielding two for every one, until there are enough of them for the embryo to differentiate according to its historical plan.

The exotic variety of rotifers, sponges, algae, worms, and polyps in the world today unambiguously demonstrates the degree of experimentation that took place at a simple multispherical level through organelle dynamics. Primitive blastulation, even at a stage of just a few cells, provided enormous leverage and potential for a hodgepodge of plant and animal shapes once the microtubules and assorted filaments and cisternae did their dances in separate cell clusters.

IN HEALTHY BLASTULATION the periodicity and duration of cleavage of the cells are coordinated, forging a comprehensive unity. The cells also become more deeply joined and intimate, developing communication junctures across the minute gaps between them. In this form they are known as "blastomeres." Synchronous with cleavage and the formation of junctures is the transcription of DNA molecules of the specific animal design and suppression of all others: each of the daughter cells receives a complete working set of one kind of genes; otherwise, successful assembly could not occur. The nuclear material, spindle, and outer membrane of the blastula are solely

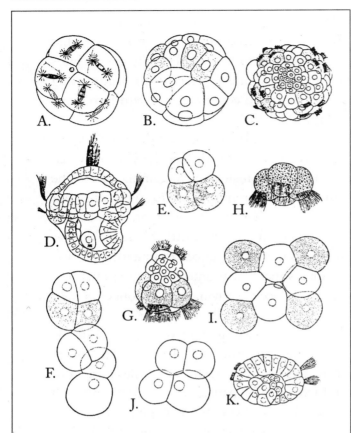

FIGURE 9C. Cleavage of isolated blastomeres of the mollusk *Patella* (the common European limpet). A. Normal eight-cell stage viewed from upper pole with fourth cleavage in process; B. Eight-cell stage from side; C. Ciliated trochoblasts of ctenophore-like stage; D. Trochophore; E. Second cleavage of an isolated micromere of the first quartet; F. Products of first and second quartet cells, forming holographically like the whole egg; G. Larva of twenty-four hours from one of eight micromeres; H. Product of isolated primary trochoblast at sixteen-cell stage; I. First division of isolated first quartet; J. Division of isolated basal cell of eight-cell stage; K. Larva of twenty-four hours, developed from a group similar to I, showing two secondary trochoblasts and two feebly ciliated pre-anal cells.

From William E. Kellicott, *A Textbook of General Embryology* (New York: Henry Holt & Company, 1913).

inherited from the egg. The sperm, having fused with the ovum, stitches its image on every blastomere; the blastula incorporates and checks the signal. It is as though a ship, just after leaving port, were given a secret mission, its new instructions having been tappiced in logs loaded on from another vessel.

The compound spherical creature swells as if a Zeiss projector were continuing to fill dome after dome of nested planetaria with the imprint of a single night sky, yet all the planetaria are hemispheres of phosphorescent jellyfish stuck to one another as they swim apart.

Seen differently, a hall of mirrors picks up the minute two-dimensional engraving of genes and radiates it outward—first, by expanding its actual space, and second, by creating an enormous variety of contexts and dimensions for its potential read-out. Yet a blastula is not a sterile algebraic function; it expands and contracts violently like a star pulsating, and each cryptic face of cells it reveals is the changing mask of a shaman in trance.

Types of Blastulas

IN MANY SPECIES the fissioning cells remain at the outer wall. Their inner edges (as we have seen) are pushed in by the surface tension of their membranes against one another, so a cavernous region forms at their collective interior, a hollow cavity called a blastocoel. Blastocoel landscape is underwritten by secretions of cells pushing one another away. Surface clusters of blastomeres, often of different sizes, maintain the spherical illusion, but the packed cells disguise a hollow center. In some species, contacts between the outer cell membranes seal into a permanent layer, an epithelium or skin.

In holoblastic radial cleavage, the pattern employed by sea urchins, the plane of the first cleavage is usually vertical, cutting through the animal-vegetal axis; the second division is also vertical but at a right angle to the first so that four hemispheres stretch from the top of the embryo to the base, each crossing the equator. The third division then slices the equator, separating animal and vegetal spheres. Four upper blastomeres rest perfectly atop four lower ones.

In holoblastic spiral cleavage, common to most mollusks, the blastomeres may be tilted slightly to the left or right. Because the cells originate in mitosis, their off-center rotations reflect oblique arrangements of the spindle fibers twisting the axes

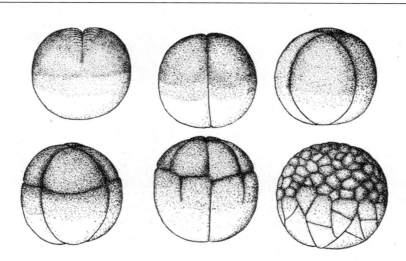

FIGURE 9D. Cleavage in formation of frog blastula.

From B. I. Balinsky, *An Introduction to Embryology*, 5th edition (Philadelphia: Saunders College Publishing, 1981).

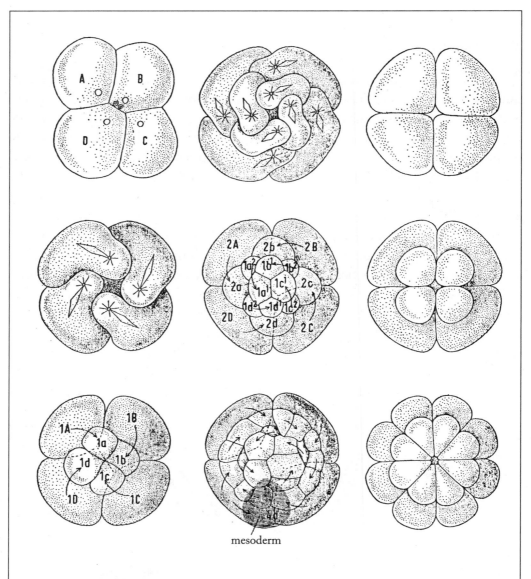

FIGURE 9E. Spiral cleavage.

From Rolf Siewing, *Lehrbuch Der Vergleichenden Entwicklungsgeschichte Der Tiere* (Hamburg: Verlag Paul Parey, 1969).

of their cleavage furrows; spirality is projected from the angle of fissioning and guides organization.

These are distinguished from meroblastic (incomplete) cleavage found in highly yolky eggs such as those of birds, fish, and insects.

In general, where the sizes of blastomeres differ, tiny micromeres form at the

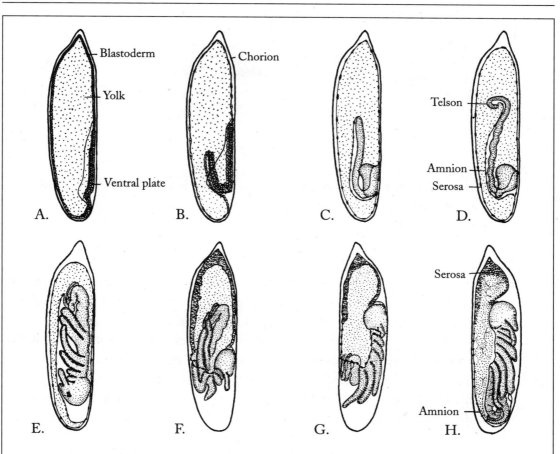

FIGURE 9F. Diagrammatic stages of blastokinesis in sagittal sections of damselfly *(Calopteryx)* egg. A. Posterior end of ventral plates sinks into the yolk; B. and C. Ventral plate migrates anteriorly and dorsally; blastoderm thins out to form the serosa; the junction between the embryo and serosa is the amniotic rudiment; C. Germ band invaginated except for anterior portion of the head; D. Metamerism visible; E. Embryonic appendages developed; amniotic cavity is ventral to embryo; F.-H. Well-developed embryo migrating to ventral surface of the egg; amnion ruptured and abandoned by embryo as serosa contracts dorsally.

This shows the characteristic embryogenic segmentation of a long-germ-band insect. In such insects the blastula (blastoderm) forms not by cell multiplication (as in other animals, including short-germ-band insects), but by nuclear multiplication within a common egg cytoplasm, the units joined in a syncytium. After a thousand or so nuclei coalesce, their cluster moves out to the egg periphery, assembling a monolayer. Then the layer of nuclei becomes cellularized by the formation of membranes around them. The result is the blastoderm. It should be noted that blastokinesis more properly belongs under gastrulation (next chapter), as it involves rearrangements of an already-formed blastoderm.

From Harold R. Hagan, *Embryology of the Viviparous Insects* (New York: Ronald Press Company, 1951).

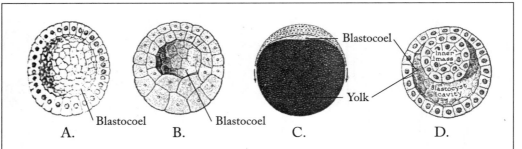

FIGURE 9G. Blastula types among chordates, shown as hemisections. A. Amphioxus; B. Amphibians; C. Reptiles and Birds; D. Mammals.

From William E. Kellicott, *A Textbook of General Embryology* (New York: Henry Holt & Company, 1913).

animal pole, macromeres at the vegetal end. However, among echinoderms (starfish and sea cucumbers as well as sea urchins), the micromeres cluster at the vegetal pole and become the primary mesenchyme (migrating) cells. At the beginning of gastrulation they are the first cells to do anything. They move inward individually, busting through the basement membrane and entering the blastocoel, eventually to form the skeletal spicules of the larva (see the next chapter).

Yolk and cellular material in insect eggs are partitioned in such a way that cytoplasm does not initially shear with its nuclei. After the nuclei fission they cluster in a central mass which migrates toward the embryo's surface, each nucleus trailing a gown of cytoplasm. As these line up in the peripheral cytoplasm, furrows roll inward and blastomeres form. The royal class of segmented, frenetic robots has an early oogenetic origin, a basis in anatomy distinct from the other zoological classes and orders.

In animals with a great deal of yolk (or segregated yolk), the rich proteins and fats hinder cell division. While the upper hemisphere of the blastula divides rapidly into micromeres, the yolky vegetal zone remains smooth and aloof, the embryo having the overall appearance of miniature crystals on a stone. The "crystals" are continuously displaced upward, away from the margin they share with the uncleaved stone.

Cleavage in birds and fish is discoidal. In chick embryos, for instance, a disc of cells situates atop the yolk. This layer is called the blastoderm, in nomenclatural preparation for three layers of body tissue, i.e., ectoderm, mesoderm, and endoderm. Bony fish have convex disc blastulas. Prior cell size generally determines regional thickness. The yolk near the vegetal pole of the frog embryo makes the blastoderm a puffy mass of jelly, whereas the animal layer is a mere lining of cells. Although the presumptive frog is everywhere (to appearances) a hollow sphere, the hidden blastocoel reaches almost to its surface at the animal pole. On the other hand, insect yolk fills its entire central area, obliterating any potential cavity.

Despite these species differences the underlying topology and meaning of the blastula are similar, and the phases of very diverse phyla overlap. In general among animals, blastomeres are presegregated into three tissue types which will become bodily layers.

The Morula

IN HUMANS IT TAKES ABOUT THREE DAYS to form a solid ball of sixteen blastomeres—called a morula from the Latin name for the mulberry it resembles. Although the morula contains an increasing company of cells, like all blastulas it remains no larger than the ovum in which it originates.

The cells of mammalian morulae are packed through to the interior. Blastulation occurs as the outer aspect of the cell mass, called the trophoblast, begins to secrete fluid, filling the intercellular spaces. Contacting one another, the spaces run together, eventually composing a fluid-filled blastocoel. The cells then continue to separate into the outer trophoblast (which will participate in the formation of the placenta) and a distinct inner cell mass, or embryoblast (which will provide the material for the organism and its sex cells).

This "cell division can be regarded not only as a quantitative process leading from one larger fertilized ovum to a cluster of smaller, similar cells, but also as a qualitative division of those cells which will continue to differentiate ever further into all the many tissues and organs from those other cells which will retain the general potential of procreation of the whole embryo. The former group consisting of the cells which will form bone, nerve, and the other tissues as well as liver, heart, lung, and other organs, are referred to as *soma,* meaning 'body' as distinct from the germ cells [sperm and egg]. Another aspect of the qualitative division is the separation into those cells that later form the embryonic sheaths or membranes."[1]

Placental attachment of the embryo requires this specialized anatomy. If mammals were directly derived from ancestors shared with fish and birds, it would be as if the potential density of their nearly nonexistent yolk were transmuted into trophoblast. Perhaps even without such an ancestry a disjunctive series of yolk-displacing (or other cytoplasm-displacing) mutations lies at the root of the entire mammalian lineage.

Though the morula resembles a sponge, it of course is not an adult multicellular organism; undifferentiated, it has no capacity for independent behavior. If we think of the emerging mammalian outside as communication, protection, and feeding, and the inside as digestion, transformation, and assimilation, then skin and intestines, nerves and gut must still be polarized, respectively, away from each other to form a real inside and outside. In all blastulas these lie in linear cotangency. There is no layer of muscles, no skeleton, no motility; there is not even a real central metabolic cavity (the blastocoel is more circumstantial than structural). With no true interior and exterior zones, cells merely occupy different positions in a uniform grid. In order to become an organism the blastula must develop a functional interior and express the layers of its three-dimensional destiny.

Determination and Capacity in the Cells of the Blastula

A SCIENTIZED VERSION OF PREFORMATIONISM persisted into the later nineteenth century in a belief that the body parts of any future organism were dictated by granules already in the egg—discrete, visible particles that were qualitatively different from one another and translated ultimately into biological traits. The first significant laboratory experiment in embryology was a test of preformationism.

In 1888, Wilhelm Roux used a hot needle to destroy one of the first two cleavage cells of a frog embryo. The surviving member went on to develop a half-embryo, apparently lacking just those parts that would have been provided by the eliminated blastomere. "Eureka!" thought Roux, "life is an assembly kit." Yet a missed observation at the time made this trial only a dramatic (and incorrect) confirmation of preformation. Expecting the same results, three years later Hans Driesch mutilated a sea-urchin egg in a corresponding way. To his astonishment he found in his dish the next morning not a half-organization but a whole gastrula. The embryo had mysteriously resupplied its own missing half.

At the turn of the century, the contradictory results of these experiments were still unresolved, and prominent embryologists and geneticists continued to argue about whether or not genetic determinants were passed on discretely from nuclei to individual blastomeres. In 1928, Hans Spemann undertook a definitive inquiry: he split the newly fertilized egg of a newt in such a way that the nucleus was isolated in one sphere. A thin bridge of cytoplasm was left between the spheres. Up to the stage of, roughly, sixteen blastomeres, only the nucleated half developed; at that point a single nucleus slipped through the bridge. The distorted cell was then severed. Soon the other half began cleaving. Both produced normal newts. Spemann had allowed the potentially functionless half of the divided egg to recover the component necessary for its reconstruction—the nucleus.

By leaving the dead blastomere forty years earlier, Roux had unintentionally placed the living one in an unfortunate situation it interpreted as: "My other half's still there, so I'll go about my usual business." It didn't repair its wound. Without understanding these factors, Driesch had then done a different thing; he made a clean division and so gave each blastomere autonomy to become whole anew.

On other occasions Spemann inserted his slip-noose in such a way that one section of the egg received all of the material of the gray crescent on the egg's surface; then the deprived section—though it continued to divide and subdivide—produced only an amorphous heap of tissue. Spemann called it a "belly piece" because it was at least part of a newt; it generated cells for liver, lung, intestine, and other abdominal material, but no axial skeleton, nervous system, or organizing principle. This indicated to him that the gray crescent had a major role in predetermining anatomical structure. Unfortunately, it was not a universal landmark; it dissipated entirely during cell multiplication, and it did not occur in all types of salamander.

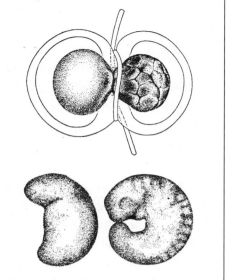

FIGURE 9H. Twinning of combined newt embryos, with nucleation and development delayed in one half. From Paul Weiss, *Principles of Development: A Text in Experimental Embryology* (New York: Henry Holt & Company, 1939).

Interpretation of these three experiments in the context of one another established both the primacy of the nucleus and gray crescent and the transcendence of the particularism of the genes by an epigenetic field.

MORE SOPHISTICATED VARIATIONS of Spemann's experiment were undertaken during ensuing decades. Eggs depleted by unequal divisions of genetic material adroitly restored themselves as well; their seemingly too advanced nuclei were able to backtrack and restart embryogenesis from the beginning. As techniques for transplantation improved, genetic material was extracted from later and later stages of organisms. In the 1950s, cell nuclei taken from blastulas with roughly sixteen thousand cells organized healthy animals. In 1968, nuclei from cells of the neural plate of a tadpole—cells well on their way to developing specialties within a nervous system—were transferred back to an enucleated ovum and assembled a normal frog larva. Gut, lung, and kidney cells have also functioned nucleically on rare occasions, though all of them more often produced either monstrosities or nonfunctional blastulas than healthy creatures. The fact that they succeeded at all demonstrates the nucleic potential that can be realized in ideal circumstances. Even the cell of a tumor growing in a frog, when isolated from the malignancy and placed in a healthy tadpole, participated in normal development, for its DNA contained the template of a regular, functional amphibian. It became a normal cell in a normal frog, demonstrating that its organismic context alone made it malignant.

In another set of experiments, scientists transplanted nuclei from various fully differentiated cell types with advanced nuclei into enucleated frog eggs. Those cells were too sophisticated to go back to infancy in one fell swoop; yet they could be helped along gradually to regain a full potential. After one formerly enucleated egg began development it was arrested at the blastula stage. A nucleus was taken from it and transplanted into a fresh enucleated egg. The new egg developed to roughly the same point; then the process of excision and transplantation into yet another enucleated egg was repeated.

By serially displacing genetic material many times in this manner, scientists were eventually able to deprogram formerly differentiated nuclei enough to support development of normal larvae.

These experiments do not, however, prove that kidney cells, for instance, are *the same* as ova, but in an emergency imposed by either intelligent-species interference or cosmic rays a kidney cell can stop becoming kidney and call up its latent encyclopedia of instructions for assembling the organism of which it is a differentiated part. Where such experiments fail, the problem, likely, is in tempo of cell metabolism more than in lack of information. Upon finding themselves in a new zygote, relocated nuclei must grow as much as thirty times their size. Endodermal and neural nuclei in particular have difficulty regaining the pace of early cytoplasmic division.

It is important also to realize that each of the original blastomeres does not necessarily retain the capacity of the ovum. During cleavage the cytoplasm, which is rich in information that will be directing the development of the germ layers, is divided unequally among them; the potential of a blastomere is therefore based on the type of cytoplasm it receives. Depending upon when the genome has been turned on, the nuclei of particular blastomeres may, however, remain equivalent in potential.

The degree of determination differs by species. In some embryos (mammals, for instance) the genome is turned on very early. In frogs it is turned on quite late — in the middle of the blastula stage. Still, mammalian embryos can be surprisingly supple; relatively advanced blastula cells, separated off, may still supply data for complete embryos. Twins are sometimes spawned in this way: an inner cell mass gets segregated into two masses and each propagates an embryo.

There are no simple rules in systems with this many factors. Embryos are complex entities that defy consistent experimental results. The same method of blastomere isolation favors one outcome with sea urchins, another with mollusks and sea squirts. (This discussion will continue in Chapter 12, "Biological Fields.")

Mosaic and Regulative Development

THE TERMS "MOSAIC" AND "REGULATIVE" have been applied by embryologists to the ostensible extremes of determination in development. Spirally cleaving eggs exhibit mosaic development; that is, each of the cells formed by cleavage furrows in the blastula has a determined fate in the adult organism. "The initial cleavage planes are oriented in a strict relationship to the chemical pattern so that each blastomere inherits a specific, predictable set of molecules."[2] It is possible that chemical determinants in the cytoskeletons of the cells regulate the planes of division and distribution of cytoplasm.

Nematodes are typically mosaic; their early blastomeres already correspond inalterably to endoderm, alimentary tract, reproductive organs, etc. Homologous traits and organs of many flatworms, segmented worms, mollusks, and insects develop along the same mosaic pattern, blastomere by blastomere, respectively, despite the fact that these groups have been separated evolutionarily for half a billion years.

In contrast, the cells formed by the divisions of regulative mammalian blastulas retain equal potential throughout early cleavage.

The difference between mosaic (fixed) and regulative (equipotential) development is a subtle one. Even in regulative development the genome is turned on and gradients form in the blastulated embryo (and sometimes as early as in the unfertilized egg).

It is unclear whether the quick determination of the blastomeres is a primitive trait—diagnostic of simpler and older species—or an idiosyncratic pattern maintained conservatively in diverse lineages. If all simple creatures were formed determinatively, then development by equipotential blastomeres would be a kind of higher, holographically redundant system (like the brain), representing more complex transpositions of information—but this is not the case. Ascidians (including sea squirts) are an example of a complex animal showing fairly determinative development.

"Primitive" and "complex" come to seem secondary variations employing the same underlying themes. Any creature initially mosaic can still become multidimensionally complex; likewise, high complexity can reach a mysterious limit at any stage, show its cards, and fold in. It is finally hard to categorize the difference between dragonflies, whose eggs can be divided to form two equivalent organisms, and houseflies, in which final development of the oocyte occurs regionally even before cleavage. Are these real and fundamental dichotomies, circumstantial embryogenic adaptations, or even experimentally induced quirks?

In addition, more sensitive laboratory procedures have revealed greater poten-

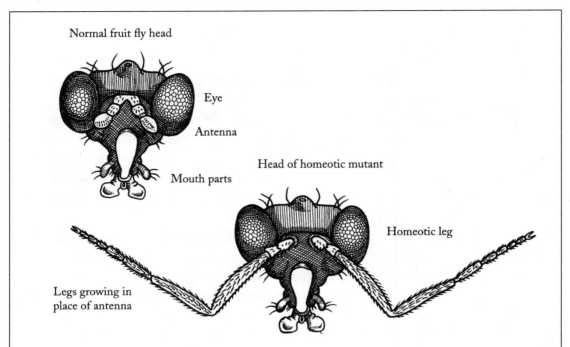

Normal fruit fly head

Eye

Antenna

Head of homeotic mutant

Homeotic leg

Mouth parts

Legs growing in
place of antenna

FIGURE 91. Artificial homeotic transposition. In nature displacements far more complex and useful than this occur over long periods of time. Illustration by Harry S. Robins

tial than thought possible in the blastomeres of mosaic species. Houseflies also show the importance of context and latency of heterogeneity even in a determinative framework. The discs of the adult fly are completely suppressed at the larval stage. The result is a squat worm. Only in dormancy within this flightless, immobile pupa does the soma read its own genes in such a drastically different manner that, from a chrysalis, a new animal unwraps sylphy wings and propels itself into a previously unexplored habitat. Metamorphosis becomes literally a whole second embryogenesis.

If potentiating discs are transplanted in a laboratory generation after generation, from fly to fly but in different segments from those in which they originated, cells arising from them can eventually change too and form the "wrong" organs—though only in consistent sequences. A genital disc can become an antenna or a leg, but not a wing. An antenna or a leg disc, including one that was formerly genital, can become a wing; a wing disc can give rise to thorax. This shows us again the latent potentiality in each organism and how, millennially, structures (whether simple or complex) may have been transformed to create new species. Organs are not only "invented" by simple bits of data but by series of changed contexts, leading eventually to whole new animals. Unused and presumably useless DNA may

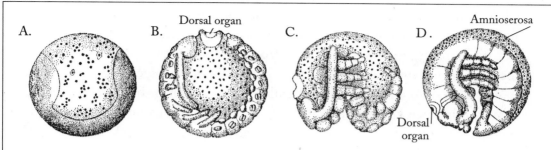

FIGURE 9J. Blastokinesis and early developmental pattern of *Campodea;* this embryo will become a slim six-millimeter-long, wormlike, eyeless insect bearing long, thin antennae, dwelling in damp places under stones or among rotten wood or leaves. A. Germ band; B. Embryo with dorsal organ; C. Blastokinesis of the embryo; D. Embryo after completion of blastokinesis. (See also FIGURE 9F.)

From Oskar A. Johannsen and Ferdinand H. Butt, *Embryology of Insects and Myriapods* (New York: McGraw-Hill Book Company, 1941).

be raided for ingenious solutions to anatomical and embryogenic traps. Homologous discs are transduced and transdetermined along continua over epochs and generations. An unadorned worm becomes a menagerie of idiosyncratically sculpted insects and crabs, their various mutagenic wings, claws, and feelers transposing segments into organs and organs into systems. But this is also how, in a single summer, a worm-like grub becomes a gliding ladybug.

At their beginning, human beings and sea urchins are surprisingly alike, but then ancestral patterns interweave in each in slightly different ways, and small divergences accumulate until whole discrete structures emerge and creatures utterly foreign to one another and at inconsonant scales and degrees of complexity appear. Where is the mosaic? Where is complexity introduced? How is regulation maintained across scales?

In a sense, modern biologists identify animals from their embryos not because they can discern the certain forerunners of specific organs but because they have learned after the fact which kinds of embryos give rise to which creatures. Once upon a time the ancestors of mammals and octopi diverged from the same primordial blastula without a single violation of growth and form and with a subtlety that does not reveal precise pathways or detours.

Cosmic and Therapeutic Gradients

THE BLASTULA IS A MEDIATION of two fundamental polarities: the deep vegetal pole provides gut material; and the more externalizing animal pole propagates skin, feathers, fur, hair, and claws (depending on the species). So the opposing zones of the blastula embody extremes of interiority and surface elaboration.

If too much cytoplasm is removed from the animal hemisphere the blastula will become vegetalized, i.e., interiorized, only a stomach; whereas a dominant animal pole will lead to a hollow ciliated ball. The two anterior blastomeres of the sea squirt, if isolated at the four-cell stage, will form only ectoderm, nervous system, notochord, and a trace of endoderm, but will not produce muscle tissue or cells for a circulatory system.

Textures of multiple organs originate in gradients of cytoplasm between poles. Prior to their actual formation these body parts preexist only as potentialities in the blastula's field. The unformed liver emerges closer to the vegetal pole. The unformed brain and eyes manifest in the animal plasm. Functional tissue thickens somehow from interference patterns in the currents between opposing morphogenetic fields— a biogenesis adumbrated in the most ancient Asian theories of embryology and medicine. Stanley Keleman fathoms this developmental principle behind acupuncture needles and osteopathic adjustment:

"Early cell proliferation, going from two to four, from four to eight to sixteen to twenty-four on up, keeps the surfaces in direct contact with each other. Cells touch one another. This direct contact clearly demonstrates the principle of tissue connectivity. In early embryological development all tissues and organs are intimately connected; the heart and the brain are just two surfaces apart. The heartbeat is tattooed right on the brain. No nerves are necessary. As development continues there remain vestiges of remembered contact.... We are a sheet of cells, twisted, bent, curved, rolled into organ systems and tubes and then into an organism."[3] An organism is also a microcosm, a dense subplanet.

IN SYSTEMS WHICH PRESUME a relationship between cosmic and life energies, the gap between universality and individuality, macrocosm and microcosm, is itself superficial and illusory. Changes in climate and the heavens are reflected in movements of cells and tissues. The Moon waxes; the Sun strums the Earth's Van Allen fields; mammoth Jupiter approaches; nearby Mars retreats—all these shifts (plus faraway stars and galaxies) send large and small ripples through embryonic fields.

The Japanese philosopher and physician Michio Kushi describes the blastula as "a replica of the earth, which is rotating, therefore producing electromagnetic belts around itself and periodically undergoing axis shifts."[4] The blastula is a nodule in a galactic field, an ingot of meteoric carbon robing itself in gravity's well—a very small, simmering star.

The body's pulse, like the microwave background of the cosmos, is a record of creation, and it continues to time the spatialization of body/mind in a universe from which it can never be disjoined—not in the least of its parts and not as a disarticulated whole. Cells and molecules are bodies with magnetic fields spinning in deeply impacted zones of the heavens; likewise, stars and galaxies are "eggs" and "blastulas" strung out along gravitational waves. Where could we begin to separate our skin from the universe's skin, one end of its pellicle from another?

From prehistoric times traditional Chinese herbs and fine metal thorns have been directed not mosaically at organs but at courses and inductions of energy through channels from which organs once emerged and in terms of which they continue to function. The dynamic vitality of these remedies is transduced to germinal paths (meridians) and originary dynamics of tissue formation. Western pharmacy by contrast targets treatments by their specific chemical effects on fully formed organs. (See Chapter 24, "Healing," for a further discussion of these topics.)

Metabolic and meridian cycles are thus truly multidimensional; they do not begin in the ears or lungs as such or drain in the lower guts, or vice versa; they emerge from all directions at once with differentiating matter and tissue, even as nebulae and intergalactic fields spring from nowhere and create space and time by their convergence. An obscure but indelible unity binds each cell both to the birth of the organism and the origin of the cosmos.

This is the hidden astrophysics of the embryo.

"Looking up at the blue sky and then at your mother's face you break the silence asking her if it is not in reality much more distant than it appears."[5]

We cannot hope to recall what it felt like to become. The sense of being born emerges anew from life itself, again and again. We awake each morning on a planet with bees swarming at the window, cloud caravans on the horizon. "The sky that is. The blue sky."[6] We see ourselves reflected a billion times through the mind of the universe, each one about to happen.

" . . . living is one constant and perpetual instant when the arras-veil before what-is-to-be hangs docile and even glad to the lightest naked thrust if we had dared"[7]: the words of William Faulkner in the voice of the mind of Rosa in *Absalom, Absalom!*

Our sense of our own oneness and completeness is tethered to a fissioning egg, a hollow ball, " . . . the globy and complete instant of its freedom . . . a fragile evanescent iridescent sphere."[8]

IN A DREAM:

I feel strong waves of gravity. I awake suddenly. I have been in hibernation with the others as our spaceship passed inertially from galaxy to galaxy. Now our body has been perturbed by a sun-star and we cannot escape this system. We are sucked down onto a huge planet. I see light flying out beneath me and widening as I sit up staring through the rear viewport. I am materializing through the sky onto this world. Giants, crooked and bent, stand on hills and look up at us as we fly past them. We are a UFO, but we are moving so fast they do not understand what we are. The strange topography looks familiar. Our ship is caught in the branches of a tree. I wait, hanging there, hoping that we will be able to take off again. The ship becomes smaller until it is only me perched there with a blanket wrapped around me. I throw off the blanket and climb down out of the tree. Now I am on this planet, and its inhabitants are dwindling to normal size and shape. I join a crowd in a village on the edge of a forest, and I walk with them until we blend in with the others on the city streets.

IN HER NOVEL *Dinner at the Homesick Restaurant* Anne Tyler describes how an elderly Pearl Tull, near death, instructs her son Ezra to read her own childhood diary entries to her. She has him go through the pages, day by day, week by week, "not because it had been so wonderful," Kim Stanley Robinson reminds us in his tale of two-hundred-dred-year-old Earthlings on Mars in 2128, "but simply because it had *been,* and now was gone"[9]: "*purchased ten yards of heliotrope brilliantine and made chocolate blanc-mange for the Girls' Culture Circle.*"[10] It was 1908; she was fourteen. He continues through "*a flaxseed poultice on my finger . . . some gartlets of pale pink ribbon.*"[11] Another day (another year) he reads: "*Washed my yellow gown, made salt-rising bread, played Basket Ball.*"[12] Finally he comes to the entry she was seeking (1910):

"*Early this morning I went out behind the house to weed. Was kneeling in the dirt by the stable with my pinafore a mess and the perspiration rolling down my back, wiped my face on my sleeve, reached for the trowel, and all at once thought, Why I believe that at just this moment I am absolutely happy . . .*

"*The Bedloe girl's piano scales were floating out her window, and a bottle fly was buzzing in the grass, and I saw that I was kneeling on such a beautiful green little planet. I don't care what else might come about, I have had this moment. It belongs to me.*"[13]

Her adult life had been filled with abandonments and broken dreams. She had stood most of her years behind a cash register in a grocery market. She was thoroughly disappointed by what had become of her; yet her existence, the soul of her being, was undiminished.

Psychic unity happens once, at the beginning, and from there radiates through the entirety of life.

Gastrulation

The Emergence of Complex Form

SEEN THROUGH A MICROSCOPE frog blastulae are wee puffballs, more like gem gravel than animals. With each spasm their cells are more and tinier. By the first or second morning a dramatic change has taken place: tadpole mites have unravelled from the beads and lie as little lines on the bottom of the jar.

Occasionally a nervous flutter activates one; it kicks up for an instant and settles back.

By the second or third morning the sun radiating through their water jar shows almost completely transparent bodies stretched up like gymnasts hanging in space, quivering slightly. The stream of light through them is interrupted only by the little black dots of their eyes and a translucent band of gut tissue from tail to cranium along their central axes. Rootless plants dancing in the sun, they are not awake but are negotiating gravity.

"Why do you want me so quickly?" they ask in neither Hopi nor English, like the primal creatures of a Pueblo creation myth.

THE LIFE OF THE BLASTULA is brief: its fission gradually slows and its chemistry changes; its cells start synthesizing proteins distinctive to

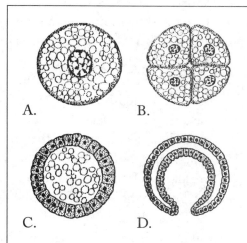

FIGURE 10A. Animal egg. A. Unfertilized ovum; B. Four-cell stage; C. Blastula; D. Gastrula.

From Oskar A. Johannsen and Ferdinand H. Butt, *Embryology of Insects and Myriapods* (New York: McGraw-Hill Book Company, 1941).

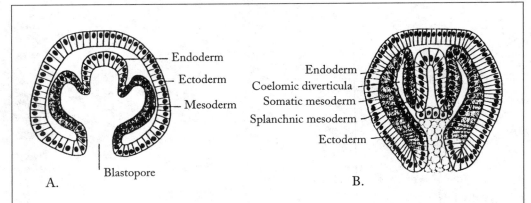

FIGURE 10B. Gastrulation of arrowworm. A. (Section) Mesoderm formation; B. Folds developing from the bottom of the gastrula giving rise to mid-gut and coelomic diverticula.

From Oskar A. Johannsen and Ferdinand H. Butt, *Embryology of Insects and Myriapods* (New York: McGraw-Hill Book Company, 1941).

the animal of their genotype. While the blastula was an undifferentiated animal, the gastrula is a living sculpture that will mold and carve itself from the inside out. Blastulation was primarily the breath cycle of the cytoplasm; gastrulation is the outcome of massive RNA transcription. As the nuclei stir, maternal and paternal chromosomes transmit their full holograms. A remarkable transition occurs: substance that was formerly exterior turns inward, and a mass of tissue stretches over its surface to cover it.

First, the inner realm of the blastula thickens as its cells accumulate. The surface of the hollow sphere caves in and back around itself, sucking the interior of the emerging adult body inside. "The blastula starts to 'turn outside in,' first becoming indented and then, as the process continues, ending up with an outside sheath which completely closes off an inner space, thus creating an inner and an outer world.... [U]p to this point animal development was plantlike. The gastrula stage, however, is the start of a wholly new turn of events."[1]

The only reason gastrulation does not look like a finger stuck in a spongy ball is that the blastula is under no pressure from the outside. It collapses into itself with extraordinary grace and precision, putting the organs in historical and functional relationship to one another and obliterating the blastocoel (which is physiologically meaningless) in order to create the archenteron, the forerunner of the alimentary system.

Gradually the mock volvox becomes a dense, organized ball with a fine structure of laminae, tubes, pouches, cavities, and curved membranes. These will develop into passageways for air and food and sources of neuromusculature. Multiple layers of insides and outsides crinkle into anal and oral openings, cerebral pleats, and other organ templates. At one pole an inside is tucking itself; at the other an outside is sliding over it like a smooth fingerless glove.

Protoplasm rushes toward itself. Margins plunge into contours; free edges meet and seal; linings close over them. Pouches swell out across pouches, creating multiple interiorities.

Later, as a flattened sheet of cells stretches into a tube for the spinal cord, the gastrula will become a cylindroid creature, a neurula. Its outer, ectodermal layer will develop skin and nerves. In its middle, mesodermal zone, muscles and blood vessels will striate. Deep inside, organs and viscera will further crimp and deepen the endoderm, opening furrows for nutrition. The distinct character and degree of complexity of the substrata will be determined solely by the lineage borne in the nucleus. But how is so much history regurgitated and integrated so rapidly and profoundly? How does it "conceive" architecture and function simultaneously?

GASTRULATION CONTINUES TO astonish even the most sophisticated witnesses. It doesn't seem possible; yet it occurs brazenly like a ringmaster showing off. Gastrulation is more magical than the most incredible Industrial Light and Magic effects because it happens in real time without an animator.

In conversation while writing this book I have used this event as an example of why embryology is such a powerful rubric for exploring the mysteries of form and being. Otherwise-hardened skeptics bow before the gastrula; their eyes grow misty.

At a 1998 publishers' Christmas party in Berkeley, a divorced doctor, escorting a sprightly female publicist, was letting his disdain for alternative medicine be known to the "naive" advocates in his vicinity. His voice was edgy, righteously piqued concerning Chinese herbs, homeopathic microdoses, chiropractic adjustments. As the author of *Planet Medicine*, I was clearly culpable. Then I cited my embryology book and mentioned gastrulation. His heart filled with good humor, his voice with awe. "You have these separate cells," he exclaimed suddenly, "all the same, going about their business. Then the goddamn thing just invaginates, folds in. Suddenly a liver cell knows it's a liver cell; a brain cell knows it's a brain cell; they start doing their things. Yet you take a liver cell and put it where a brain cell is, it stops being a liver cell and acts like all the other brain cells; up to a point—then it's stuck being a liver cell. How does that happen? It's uncanny."

Everyone was now on the same wavelength, thinking about the strangeness of their bodies, the wonder of being alive and gabbing away like this as if we were anything other than miracles.

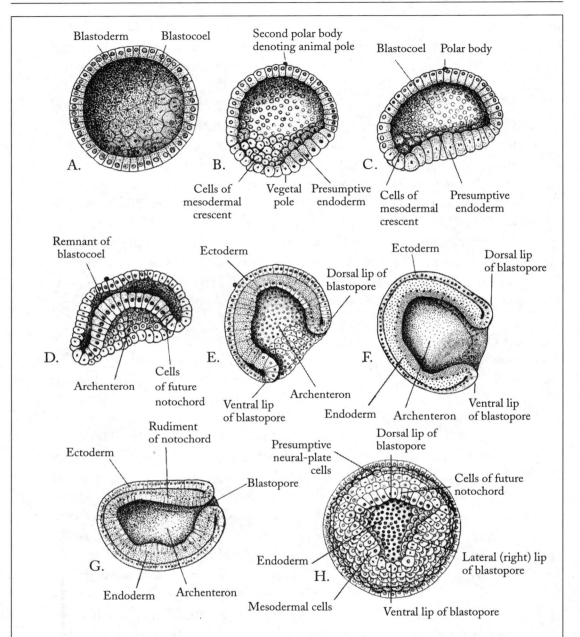

FIGURE 10C. Stages of gastrulation of *Amphioxus*. The embryos in A-G are represented as cut in the median plane. A. Blastula; B. and C. Beginning of invagination; D. Invagination advanced, the embryo attaining the structure of a double-walled cup with a broad opening to the exterior; E. and F. Constriction of the blastopore; G. Completed gastrula; H. Middle gastrula, whole, viewed from side of blastopore.

From E. G. Conklin, "The Embryology of *Amphioxus*" (*Journal of Morphology*, Volume 54, 1932).

Gastrulation is the original transformation of one kind of animate tissue and space into another. It differentiates and rearranges groups of cells and turns configurations inside-out within other configurations being turned inside-out, all the time not only keeping coordination with one another but deepening their structural and functional cohesivenesses so that utterly new meanings and performances arise. So many histories are reenacted and synchronized as they fold into one another it seems as though a gastrula could morph any shape at all if given enough time. Phylogenetically, this is what has happened; repeated gastrulations of prior gastrulas have derived worms from jellyfish, snails from worms, and monkeys from lemurs. Of course, these precise animals did not proceed from one another, but gastrulas at their approximate stages of configuration mutated site by site and gave rise to one another's topologies.

Although gastrulation differs in detail from order to order even within closely related groups, there are fundamental similarities which confirm that, like blastulation, it is an ancient event shared by every terrestrial animal. Relative amounts of yolk, larval and fetal adaptations, and specializations that work their way back to earlier phases of ontogeny camouflage to one degree or another the universal organization of creatures which must all eat, void, breathe, swim, reproduce, have insides and outsides (no matter what machinery and ornamentation surround these processes). They must all create an internal milieu (cavity) with conduits to the external environment. This means degrees of internalization, subtilization, and sanctuary (endoderm), of structure and locomotion (mesoderm), and an external sensitized shell, a globalized protozoan layer to communicate with the world (ectoderm). One needle sews these together in multiple planes.

IT IS DIFFICULT (ontogenetically as well as phylogenetically) to draw meaningful lines between stages that morph out of each other—oogenesis, blastulation, gastrulation, neurulation, and organogenesis. It is true that the egg must be ovulated if blastulation is to occur (the protozoa must become multicellular); but, as we have seen, a blastula is already beginning to form within the primeval egg. Blastulation likewise does not end during gastrulation—the cells continue to divide in phase, though their synchronized "breathing" is incorporated in a new series of disjunctions and differentiations.

Gastrulation takes separate loci of movement and independent animal origin and binds them in a functional whole. Protean intestines, glands, blood islands, and spleens, whatever their genetic histories and original existential contexts, cease being autonomous tissue nexuses and zooids and fold into an organism.

Three Germinal Layers

CELLS ARE SOLO-DIMENSIONAL FORMS; when they fission and stick together they are still discrete points. Protozoas are pure surface. A blastula is little more than a primordial association of protozoa.

When layers of cells move as units and undulate, they become tissues. We no longer notice the single "cell points"; structure predominates.

Original living tissue, before any layering, cavitation, or introversion, was all ectoderm—blastula. Protozoans, rotifers, and flatworms are essentially bidimensional animals—cellular and linear. (Of course, cells and tissues are topologically three-dimensional too, but the three dimensions of a cell are confined functionally to one of our dimensions, and the three dimensions of tissue operate as plane vectors in our bodies.)

Gastrulation universally transforms a blastula into three germinal layers and a central cavity, the coelom, the confederation of which thickens and interiorizes multicellular animal life. Primordial ectoderm is thereby deepened, pitted, and dimensionalized. A zoogenic field is created. The only exceptions are those creatures which make up the transition from one-celled existence to full three-layered girth: the twin-layered sponges, jellyfish, and related species.

In the following pages we will explore gastrulation both ontogenetically and evolutionarily, for the epochal series of events leading to classes of animals lodged itself in the nuclear templates of those bionts in such a way that they must complexify anew from a protozoan stage each time in order to become. Tissue itself is the textile of gastrulating looms.

Gastrulation as an Evolutionary Event
Recruiting New Species for Vacant Niches

Jellyfish: Two Layers

The Coelenterate phylum includes spheroids of dilated ectoderm around amorphous endodermal swellings. These are diploblastic jellyfish, sea anemones, corals, and hydroids. Their ectoderm is lined with nerve fibers and their inside is a gastrovascular cavity blind at one end. Between the exterior cytoplasm and the endodermal lining of the gut is a gel layer (called the mesogloea) made up mostly of water and inorganic salts. Undifferentiated cells lying close to the mesogloea give rise to amoeboid cnidoblasts whose Golgi bodies specialize as stinging capsules, harpoons laden with protein poisons. These nematocysts discharge barbed threads

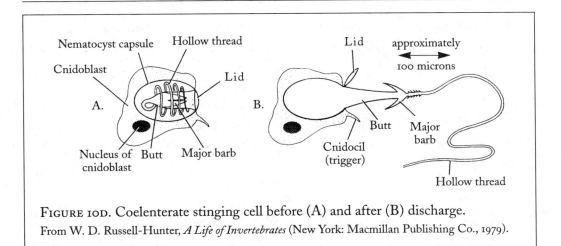

FIGURE 10D. Coelenterate stinging cell before (A) and after (B) discharge.
From W. D. Russell-Hunter, *A Life of Invertebrates* (New York: Macmillan Publishing Co., 1979).

with unfolding pleats which hone in on prey and harpoon it at high speed.

To a certain degree, jellyfish are enormous, simplistically endodermalized protozoa, thickened along one axis. Like one-celled animals they digest food in the vacuoles of their cells, though some enzymatic breakdown of large morsels occurs in their interior cavities. Epidermis and gastrodermis absorb oxygen directly and diffuse it to tissues, excreting carbon dioxide and nitrogenous wastes. Certain of the more complex jellyfish have branching guts with canals and pockets, but no anal opening; amoeboid cells carry larger particles of indigestible material out of their mouths.

Endoderm, by definition, must have developed in distinction to a surface; it is interior. From this primal polarity animals derive the most basic parameters of identity, shape, propriety, and position.

The diversity of creatures at this level represents millions of years of satisfactory adaptive radiation of a basic animal type without much adjunct experimentation. Jellyfish and their kin are polymorphic gastrulas; they include polyps that grow from the sea floor like plants and swimming medusae that bud from these strands like seeds from flowers. Some species sprout connective tissue that binds different zooids in a superorganism like the Portuguese man-of-war, a cluster which includes swimming bells that propel the colony through the water, gonad buds producing medusae, and individuals that spend their lives as oily bags and fill with gases secreted by ectoderm. These bright blue symbiotic floats, exposed to the winds, chug along like sailboats.

While ancestors of such creatures explored the endless varieties of diploblastic organization, expanding into niches in the seas, others were affected by different mutations and continued to "gastrulate," to add on tissues. These "pathologies" exponentialized the range, mobility, structural potential, and dimensional profundity of protoplasm. As tissue assembled in layers around cavities, an exquisitely dense, differentiated array of creatures emerged.

Because a thicker, meshier animal can exert greater force per unit volume of its body, it is not confined to a droplet or puddle; it can explore and manipulate external space. One small step for a jellyfish was one giant leap for protoplasmic life. Tractioning layer by layer, protuberance by protuberance, ultimately in cyborg form it climbed all the way to the Moon.

Worms and Larvae: The Origin of Mesoderm

The descendants of creatures who survived and integrated cataclysmic mutations became worms, fishes, and crustaceans, and later, insects, sand dollars, and vertebrates. But at each of these subsequent stages of development, there were always both countless conservative variations on a single theme and radical disjunctions into new arrays and configurations of tissue. Each layer of organization had its epoch. By our unrealistic industrial standards, evolution progressed excruciatingly slowly.

Worms are one fundamental expression of a gastrula stage. They are abundant and heterogeneous because they embody a distinct, simple, and fatally attractive level of morphology, like living "primitive streaks" that have adapted over aeons to exquisitely distinct niches at scales throughout the ecosphere. Large numbers of worms in all phyla, having become parasitic, have abandoned their ability to move about or feed. Their guts as well as their locomotory structures wither, and their metabolisms are imbedded in the life cycles of various mollusks, fishes, birds, and mammals. In these instances the original predisposition of cells to specialize and merge crosses the boundaries of creatures, blending coembryology with coevolution.

Natural selection fills niches, however obscure; it does not blindly march toward biomass and complexity. In parasitic forms, as noted, development eliminates stages, returning the body to an embryonic lifestyle.

THE PLANULA, a larval Coelenterate, is closely related to the free-swimming larvae of primitive worms, hence all multilayered animals. In ancient times some of these ciliated embryos (or allied forms) began seafloor crawling and matured sexually in a fetal state. This is how they altered the dynamics and layering of their tissue, hence potentiated themselves. Subsequent differentiation established separate tissue surfaces, including contractile layers for propulsion. No doubt locomotion stimulated development, as new classes of mobile mutated animals were favored by natural selection; these crawled, fluttered, and inched their way along.

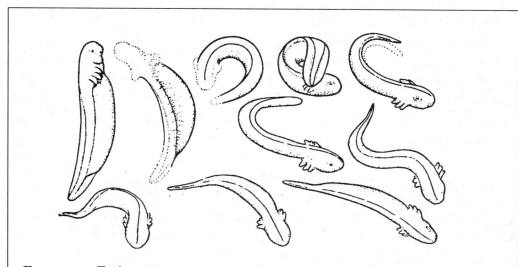

FIGURE IOE. Early swimming patterns of the salamander embryo, showing contractile layers used for propulsion.
From G. E. Coghill, *Anatomy and the Problem of Behaviour* (Cambridge University Press, 1929).

The larvae of modern parasitic Coelenterates all begin as free-swimming embryos—a vestige of their historical state and an indication of shared phylogeny; then their organs degenerate.

FLATWORMS ARE TECHNICALLY TRIPLOBLASTIC (three-layered) but lack a true coelom; mesodermal tissue arises between their epithelium and the endoderm of their gut lining. The opening in the gut is embryologically a mouth, but it must serve also as an anus. Simple folding and tubing of tissue forms muscle protrusions in the pharyngeal region. The flatworm moves by cilia and as peristaltic waves cross its body in loose coordination with contractions of mesodermal tissue. Internal flow of liquids is controlled by clumps of flagella (flame cells) which arise from fleshy bulbs in the mesoderm along with penis, ovary, and excretory organs. Some anterior concentrations of nerve cells suggest cephalization, but there is nothing like a real head, a seat of identity.

In most lines of animals, growing cell colonies could reach only a certain size and complexity without exponential increase in volume and pouching of internal surfaces and tissues. In a singular exception, ribbon worms may protract to two meters of length lacking gut walls, muscles, or segments, simply by adding spirally crisscrossing fibers around a tubular gut; but these tidal animals are an evolutionary dead end on this planet (though their analogs—up to thousands of meters long—may fill oceans on Europa or Jupiter itself).

In truly complex living systems, internal organs must be supported by folds or mesenteries in a mesoderm. Thick texturing occurs as a polarization away from surfaces across multiple struts. Segmented worms, mollusks, echinoderms, insects, though ancient creatures with genetically fixed behaviors, embody fully gastrulated three-layered stages with central cavities and ingeniously devised organs, including ectodermalized sensoria, nephridia (for removing nitrogenous waste), and mesodermal muscles and vessels (for movement and circulation).

Gastrulation as an Ontogenetic Event Varying by Lineage

SMALL FULLY-CLEAVING EGGS such as those of starfish and sea urchins develop cilia after their tenth division, and a day later the vegetal pole of the free-swimming creature flattens and cells migrate into the blastocoel and forge a prism-shaped larva. Lancelets gastrulate between the ninth and tenth division when they have about eight hundred cells. The vegetal pole suddenly sags horizontally and collapses inward. The sphere hollows into a cup, its external wall still facing outward as another wall lines the new cavity. The opening into the archenteron, called the blastopore, remains throughout adult life.

In species simpler than starfish and sea urchins—the *Spiralia* (or Protostomes)— the blastopore usually becomes a mouth; in the lineage descending from the echinoderms (the Deuterostomes), the opening becomes an anus and the mouth must develop from a secondary perforation. This distinction between the more primitive blueprint and the plan underlying complex animals is the most basic in the animal kingdom, though it is overlaid and often obscured by adaptive cleavage patterns of individual species that mix traditional and modern characteristics secondarily.

In amphibian embryos in which the vegetal wall is yolky enough to resist collapse, the cells of the marginal zone (between animal and vegetal hemispheres) initiate inversion by migrating inside and sinking. As the initial cells disappear, new cells follow them over the lip; this "inrolling" through the blastopore creates the archenteron wall. The migration of course has a mechanical basis: the cells in the region of the lip elongate in the shape of upside-down bottles. This causes them to fold inward collectively in a thin surface groove which severs the vegetal from the marginal zone and continues to spread until it meets itself on the ventral side of the embryo. Ultimately the dorsal, ventral, and lateral lips of the blastopore almost converge (as when a caterpillar brings mouth and hind tip together). Most of the infolded material travels from the dorsal region where gastrulation began; some subsequent tissue flows over the ventral and lateral lips. Even though the bottle necks stretch out, the cells remain connected at their surfaces and move in concert. As the inrolling ring spreads through the vegetal hemisphere, yolky endoderm plunging downward is sucked into the interior.

Once over the lips, the vegetalized cytoplasm moves curiously in a direction opposite to its migration on the surface, so, by the end of gastrulation, the vegetal region, which entered the archenteron at the lateral and ventral lips, has rotated ventrally through the embryo. While the prospective endoderm and mesoderm are involuting, the animal hemisphere must now expand to cover the deformed sphere. The part of the animal hemisphere that is to become the nervous system moves toward the blastopore, and the rest of the ectodermal sheet spreads (as its cells divide) to cover the entire surface of the embryo except for the yolk plug, which sticks out as the underbelly of an interior sphere. The covering of the surface with animal tissue is called epiboly, literally a "throwing forward."

The germinal cells of the notochord, gut, and brain form on the surface of the blastula with some of the eventual skin (other dermis awaits somites that will arise interiorly). If we project a full-grown newt back onto its blastula, the epidermis and nervous system flow into the upper animal hemisphere— an irregular island marking parts of the nose, ears, and the skin of the mouth; the eyes diverge to its opposite sides. The notochord bunches up at the equator, strands of connective tissue falling toward the vegetal pole. As the mouth comes apart, its endoderm sluices down into the vegetal zone with the gills and pharynx. Segments of internal organs pull like yarn and twist indistinguishably about one another into the yolk of the vegetal pole. The mesoderm of the

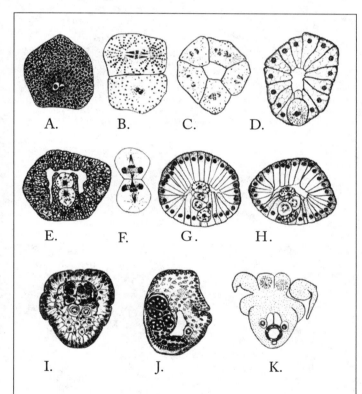

FIGURE 10F. Separation of germ plasm and soma in the early development of a crustacean. A. to C. Cleavage; D. Blastula with large stem cell; E. to H. Gastrulation; I. to K. Differentiation of the nauplius larva.

From Emil Witschi, *Development of Vertebrates* (Philadelphia: W. B. Saunders & Company, 1956).

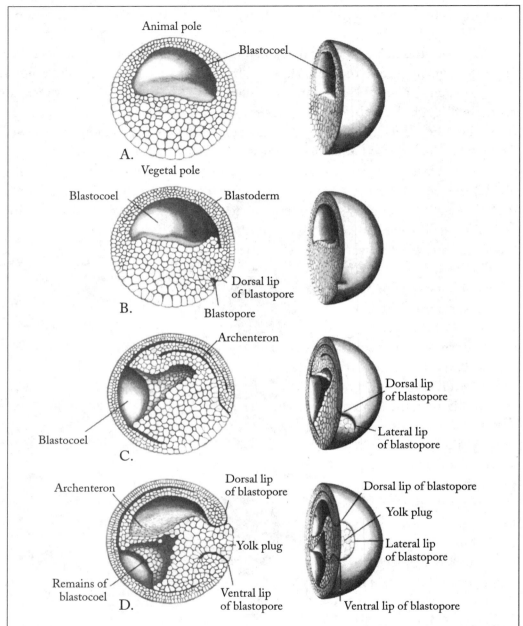

FIGURE 10G. Four stages of development of a frog embryo. A. Late blastula; B. Beginnings of gastrulation; C. Middle gastrula phase; D. Late gastrula phase. Embryos on the left are cut in the median plane; the same embryos on the right are viewed at an angle from the dorsal side, or, in D, from the posterior end.

From B. I. Balinsky, *An Introduction to Embryology,* 5th edition (Philadelphia: Saunders College Publishing, 1981).

body cavity and the kidneys unravels onto the other side of the embryo with the rear gut.

When we view this same process again with time flowing forward, a complex creature morphs itself in three dimensions out of separate lumps and stains.

Different organs will ultimately emerge along a gradient of invagination (or relative distance of migration). The notochord enters the interior from the marginal zone and becomes concentrated along the dorsal side of the archenteron roof. Other mesoderm originates as a sheet of cells travelling forward between the ectoderm and endoderm. While one edge of it rolls freely, the other remains connected to the notochord. As the archenteron lengthens, a minute anterior invagination of the ectoderm fuses with the cavity to form the stomodeum (the mouth rudiment). Contiguous endoderm becomes the pharynx. Since the mesoderm does not quite reach the anterior end of the gastrula, the mouth at the front of the gut puckers in a knitting of ectoderm and endoderm.

The deepest endodermal cells become duodenal, contacting the alimentary canal. Surface endoderm becomes pharyngeal and oral. Intermediate sections merge into the foregut.

IN INSECT EMBRYOS a median groove deepens from a thickening of blastoderm along the ventral side of the gastrula. As the groove advances inward toward the yolk, it spreads to become mesoderm and endoderm. Furrows develop across the germinal band, dividing the organism into metameres (segments) and enclosing it in an extraembryonic membrane.

The embryos of birds and reptiles (among others) feed and breathe through organs which will not be part of their adult bodies. Early cleavage in such eggs is relatively superficial and, as we have seen, limited to a germinal disc at the animal pole. Even before a chicken egg is laid, cells with greater amounts of yolk begin falling into a subgerminal cavity beneath the blastoderm. The actual body cells are located in the upper region of transparent tissue (the *area pellucida*); yolkier cells surround these in an *area opaca* (the cells may be less yolky than they look—there is no space separating their opacity from yolk).

Gastrulation begins as a thin layer detaches from the lower blastoderm; this hypoblast is the source of the yolk sac whereas the upper layer, the epiblast, will generate all three germinal tissue layers of the body and parts of the tissues of the extraembryonic organs. As the hypoblast sinks beneath the epiblast, the subgerminal cavity is displaced between the hypoblast and the yolk; and a new cavity, the blastocoel, opens between the blastoderm layers.

Once the egg has been laid, the blastoderm margin advances to enclose the yolk in

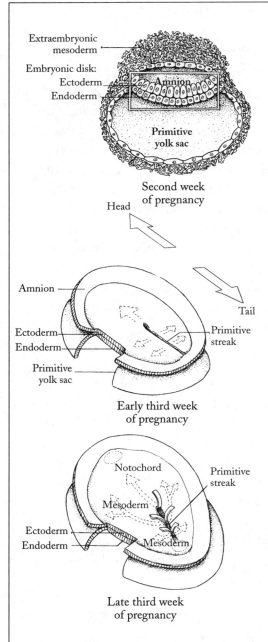

Extraembryonic mesoderm

Embryonic disk:
Ectoderm
Endoderm

Amnion

Primitive yolk sac

Second week of pregnancy

Head

Tail

Amnion

Ectoderm
Endoderm

Primitive streak

Primitive yolk sac

Early third week of pregnancy

Notochord

Primitive streak

Mesoderm

Ectoderm
Endoderm

Mesoderm

Late third week of pregnancy

FIGURE 10H. Initial embryonic differentiation of tissue layers and notochord.
From R. Louis Schultz and Rosemary Feitis, *The Endless Web: Fascial Anatomy and Physical Reality* (Berkeley, California: North Atlantic Books, 1996).

a sac; meanwhile a strip of blastoderm at the mid rear of the *area pellucida* thickens and stretches forward, pushing out a groove behind itself. The whole stirred-up area is progressively sucked into this strip, the posterior part to become mesoderm, most of it for extraembryonic organs, and the anterior zone to become both mesoderm and endoderm.

The "primitive streak" (as the strip is called) is not so much a structure as a disturbed region through which the surface of the blastoderm migrates into the interior without infolding. As the cells near the streak bunch together, other cells converge behind them, creating multidirectional motion. Tissue sinks down the center and then shifts sideways and both laterally and forward. The thickened knot at the head of the primitive streak (Hensen's node) contains a single plate of prospective notochord and mesoderm which extends between the migrating ectoderm and endoderm. Material for the heart and kidneys moves inward and then laterally. As the hypoblast develops into the yolk sac it is covered by a continued displacement of cells. When there are not enough cells left to feed the primitive streak, it shrinks and moves backward, disappearing into the cloaca and tailbud.

The recession of Hensen's node leaves in its wake the germinal neural plate, which is stretched backward (having occupied the region anterior and lateral to the streak) and drawn in toward the center. The front of the plate then folds around itself to form the neural tube which is subsequently enclosed by ectodermal folds knitting together in a smooth sheet. Between the upper ectoderm and the lower endoderm, the mesoderm subdivides. That portion lying next to the notochord breaks into prospective skeletal muscle and dermis units, called somites.

The alimentary canal of birds and reptiles does not delineate directly by involution as in amphibians; rather, folds of the epiblast roll downward and curve in along a narrow median strip of endoderm, the mid-section of this gut remaining open to tissue forming the yolk sac.

Birds and reptiles develop extraembryonic organs from upward folds of the extraembryonic region of ectoderm and mesoderm. One fold bends backward over the head and then along the body, its free edges fusing in an anterioposterior direction. The fold is originally ectodermal but incorporates mesoderm as it moves backward. The embryo is now enclosed in a cavity within a membrane. The inner surface of fused folds, the lining of the cavity, is called the amnion, and the connecting membrane of the outer surface is the chorion. As this amniotic cavity is packed with secretion, the unborn animal floats freely in fluid, buffered from shocks and without friction from its shell.

Between the embryo and the "external" organs only a compact endodermal stalk of tissue remains, the yolk stalk. This structure, rich in blood vessels, gradually encloses the neck of another outgrowth, an extraembryonic bladder (the allantois), which is hewn from endoderm of the hindgut and mesodermal lining. As the storage chamber for the embryo's urine, the allantois spreads out beneath the chorion,

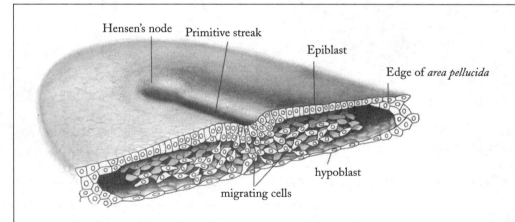

FIGURE 101. Anterior half of the *area pellucida* of a chick embryo cut transversely to show the migration of mesodermal and endodermal cells from the primitive streak. From B. I. Balinsky, *An Introduction to Embryology,* 5th edition (Philadelphia: Saunders College Publishing, 1981).

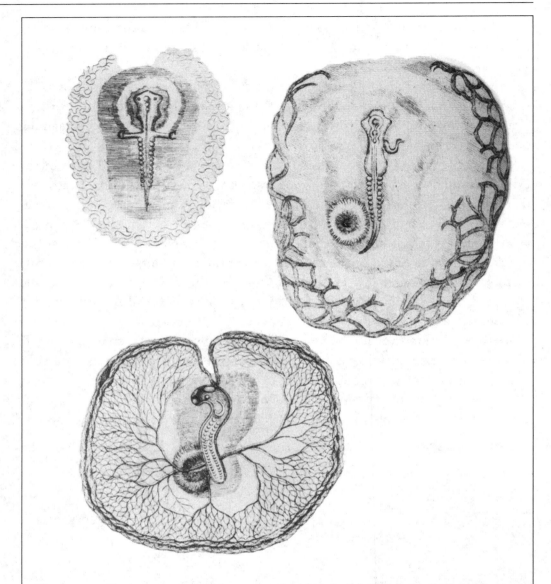

FIGURE 10J. Early stages in development of the chick, from Marcello Malpighi's 1673 drawings.

From Arthur William Meyer, *The Rise of Embryology* (Stanford University Press, 1939).

developing its own blood network to the heart (blood comes from the dorsal aorta into the allantois, then back through veins to vessels leading to the heart). It too surrounds the organism. The incipient eagle or crocodile breathes through external membranes made of their own tissue—an ecosphere in an eggshell.

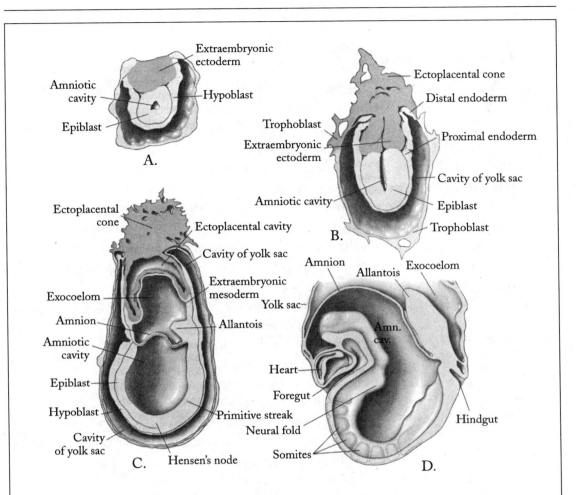

FIGURE 10K. Development of a mouse embryo. A. Early egg; the inner cell mass has become subdivided into the rudiment of the embryo proper (epiblast) and extraembryonic ectoderm; the endodermal (hypoblast) layer has been formed; B. The amniotic cavity has been formed by cavitation in both the embryo rudiment and the extraembryonic ectoderm. The latter has produced an ectoplacental cone, which is invading the maternal tissues; the endoderm has started spreading to cover the inner surface of the trophoblast (distal endoderm); C. This is the primitive-streak stage; the cavity inside the egg cylinder has become subdivided into the amniotic cavity proper and the cavity of the ectoplacental cone, the extraembryonic coelom (exocoelom) having developed between; the endoderm has lined the inner surface of the trophoblast, thus completely surrounding the yolk-sac cavity; D. Embryo with open neural plate and seven pairs of somites; the embryo is curved around the amniotic cavity and is concave dorsally. The allantois is growing from the posterior end of the embryo into the extraembryonic coelom.

From B. I. Balinsky, *An Introduction to Embryology,* 5th edition (Philadelphia: Saunders College Publishing, 1981).

Human Gastrulation

AFTER CONCEPTION THE DEVELOPING HOMINID MORULA travels down the Fallopian tube into the uterus. Around the fifth day, the outer cell-layer of the mulberry begins to secrete fluid; the seepage fills the intercellular spaces, washing out a cavity (the blastocoel) and saturating it with mucus. A blastocyst, a hollow sphere filled with fluid, now floats within uterine waters. Like the firmament of Genesis, it is a spherical layer separating waters above from waters below.

As the pellucid membrane between oceans dissipates, the outer cells are flattened by pressure from within. The inner cell-mass of the blastocyst "becomes attached eccentrically to the inner side of the outer layer of flattened cells which, because of the part [these cells] are going to play in embryonic nutrition, is called the trophoblast."[2]

Sometime between the seventh and tenth days, the trophoblast spreads outward in a manner more pronounced on the side of its inner cell-mass, thus losing some of its cell structure. Leading with its proliferating side, the blastocyst contacts the uterine wall and, as it begins to infest maternal tissue, the wall closes over it.

Because mammals have nearly yolkless cytoplasm, the young blastula stays alive by implanting its tissues in the uterine lining like a parasite. Placental tissue is then formed by a combination of maternal and fetal cells. This is no incidental event; it is a major excursion, a detour initially overwhelming any other tissue consideration, as though the main purpose of syzygy were to house and accommodate a pygmy inside a womb. The amniotic sac and placenta serve as a matrix and a cellular bridge between the embryo's and mother's bodies across which metabolites can be exchanged. These extraembryonic organs, like those of other mammals and birds, will be relinquished at birth and play no role in adult existence, yet their sanctuary allows for a nine-month development and complex layering of internal structure without exceeding uterine and maternal resources.

Meanwhile the inner cell-mass begins to segregate in layers of etiologically dissimilar cells. Within the ectodermal layer, a second, less distinct group—roundish and slightly flattened—foreshadows the entry of endoderm. Between these layers and the trophoblast, fluid accumulating in the amniotic cavity opens a hollowed zone out of dense cell-mass, the blastocoel. "As a second step in this stage, loose fibrous mesenchyme structures appear inside the blastocoel which, disappearing again, will give rise to the extraembryonic coelom (or primary yolk sac). Into this area, the yolk sac will develop from the inner layer of the cell-mass."[3]

By the thirteenth day the implanted trophoblast has become fully enclosed in the uterine wall; tiny villi sprout across its surface, larger in the cylindrical quarter where the amnion is attached. This is where the placenta will form. As the blastocyst sinks beneath the maternal tissue, it obliterates areas of the uterine wall with trophoblast cells and replaces them with a network of tissues and blood vessels. Chorionic mesoderm swells between the amniotic cavity and the trophoblast, ultimately contriving an umbilical stalk which, as the amnion and yolk sac rotate counterclockwise, comes to connect to the region of the embryonic disc where they also meet and join. A sickle-shaped allantoic bud extends from the yolk sac into the rudimentary umbilical cord. An umbilical cord will connect the embryo to the yolk sac and extraembryonic organs.

As the embryo matriculates through the fifteenth and sixteenth days of its life, little throbbing creatures infuse the umbilical stalk and surround the allantois and yolk sac. Hearkening to a more ancient epoch, these independent bionts in colonies—blood islands—come to life before an aortic pump even exists; they may in fact have evolved prior to the heart-driven circulatory system—amoeboid barges responsible for their own commerce in simple organisms. Eventually they will coalesce into primordial blood vessels and submit their regimes to the heart and placenta.

The Phenomenology of Extraembryonic Organs

THE MAMMALIAN EMBRYO DIFFUSES WASTE PRODUCTS from the fetal side of the placenta into maternal blood vessels. The endoderm of the mammalian allantois is stunted (in hominids it does not even reach the placenta); however, the mesodermal blood system of the allantois is part of the placenta and ferries sustenance from maternal tissues to the implanted fetus. Metabolic exchange is established between the two organisms even though the blood of each circulates independently. Water and oxygen pass into the fetal blood by diffusion; carbon dioxide, urea, and various salts and proteins are eliminated likewise. While the mother's antibodies immunize the offspring against certain diseases, others, like rubella and chicken pox, can infect embryonically through the placenta. The mother also passes somatic and likely even psychosomatic components to the embryo—diseases, toxins, neurotransmitter by-products, perhaps even heart pulse, rhythm of breath (or lack thereof), other vibrations, and rudiments of emotions.

Though we each build new lives we contain the residues of past generations: their immunities, poisons, and even perhaps their predilections. According to esoteric homeopathic philosophy, congenital syphilis and psoric miasms are transmitted from generation to generation through tissues weakened at a genetic or cellular level by drugs and suppressed illnesses. Whether we accept this paraphysical prognosis or not, we assume that the diet, constitution, and emotional ambiance of the mother radiate into the field of her offspring. Antibiotics, tranquilizers, hallucinogens, alcohol, and other drugs ingested by a pregnant woman unavoidably perturb the human embryo. Thalidomide caused thousands of industrial tragedies—infants born with weakened and irregular hearts and without full limbs because pharmaceutical firms chose to promote certain sedatives to gastrulating women.

The corruptions of cultures are incarnated in displacements of fledgling cells, though not necessarily in the way the people expect.

THE INTIMACY BETWEEN MOTHER AND EMBRYO, seminal to mammalian existence, develops at biological levels far beneath what language can later touch. The Mother is such a powerful figure humans barely know which parts of themselves come from her and which are metaphors for the deed of primal cell-sharing.

"When the fetus grows, the placenta is, in most cases, located close to the mother's navel. This close proximity ... allows the baby's umbilical cord to receive energy directly.... Within the chord, the umbilical arteries and umbilical vein provide the vital energetic circulation of blood.... The nutrients that enter the umbilical cord flow first through the eight holes in the upper four sacrum bones, and then up to the brain, providing the entire growth needed for the body. In this way the spine is like a river that allows the water [ch'i] to flow along its bed and to nourish the organs, muscles, tissues, bones, and other bodily parts. As the ears become the mouth of this river, a vibration is generated between the fetus and the mother. A 'ringing connection' is made in this process. This 'ringing connection' goes beyond mere skin contact with amniotic fluid; it alters the world."[4]

We experience incarnation first through our mother's impressions ("... natives of [her] body ... [our] bloodstream comingling with hers"[5]). Forming in the tissues of another being is an annointment, a ceremony. A body within a body, innards and fluids joined to innards and fluids, is erotic in a way that no adult carnality can match. But it is also a great danger.

The Mother is an archetype, a succession of originless images illuminated from within. Her manifestations bubble up spontaneously throughout life: mama, queen, femme, movie star, matriarch, hag. Beasts of prey lurk behind her beatific faces. The flesh of a young princess peels away to reveal a rotting skeleton; a belly dancer becomes a rabid wolf-woman. A Japanese female Ninja displays her breasts in order to startle her opponent. While he is distracted, she unleashes a concealed dagger. She seductively offers a flower which explodes in his face with blinding powder.[6] Statues of the Great Mother reveal vaginal teeth. Heinrich Zimmer portrays her as a carnivorous witch:

"In her 'hideous aspect' the goddess, as Kali, the 'dark one,' she raises the skull full of seething blood to her lips; her devotional image shows her dressed in blood red, standing in a boat floating on a sea of blood."[7] At the foot of her altar flows the blood of sacrificial animals, their heads piled in mud. She is stroking her cobras while gorging on entrails, and her own innards are connected by an umbilicus to the innards of the corpse of a beast.

The newborn is first attracted to the Mother as a projection of its own libido and a source of nourishment (the psychologist Erich Neumann names her three transformation-mysteries: the blood of menstruation, the life-blood of pregnancy, and the blood of the breast converted to shining milk). In this form the Mother is not a person but the personification of biological reality in warmth, incense, and folds of flesh. Her acts are numinous events providing seeds of consciousness for shaping body and psyche: "In the early phases of intrauterine existence and the first years of life, every maternal gesture and emotional reaction is instantaneously somaticized by the infant."[8] The Mother is the source of the self.

Eventually the ego realizes that it has barely escaped oblivion. The womb may have been life-giving to the body, but psychically it is a zone of death—the death of consciousness in preconsciousness, the death of cells in cells, the death of individuality in lust. The Mother and her minions and their disguises must be externalized and dispelled. Ultimate survival depends on establishing a self-correcting oscillation between the charismatic phantoms of the universe and the deep biological fields of the self. The organism must claim its own reality from the transformation-mystery of gastrulation, must create selfhood out of mere cells.

By making the Great Mother palpable in figurines and myths we experience the pain and paradox of her more directly and so are in less actual danger. Yet how many times must her emanation be played out in maudlin wars, grim executions? A terrible nostalgia for the Mother afflicts whole cultures, setting loose her craven and bloody myrmidons. She was a dark and bloody place to begin with, long before the blood of the womb.

The Great Mother is not a whim of surrealism or an idol of pagan culture; she is a personification of an unknown force that impels us toward change, birth, and reproduction. She is an invisible aspect of cell life raised to the luminosity and countenance of an ikon.

Neurulation

DURING THE TIME THE HUMAN EMBRYO is grafting itself into maternal tissue, the primitive streak originates as a furrow in the ectodermal layer at the hind part of the dorsal region of the blastodisc, in its approximate middle, and with the addition of cells at its rear, pushes forward as a ridge of epiblast. The streak thickens in front and a primitive knot develops. Other cells flow between the extraembryonic mesoderm in the umbilical stalk and the streak. In response to mesodermal cells migrating across the periphery, ectodermal cells sprout, meet them, and then flow downward. When the embryo is about sixteen days old, this group of cells migrates beyond the primitive knot and travels anteriorly and laterally between the ectoderm and endoderm until it contacts the prochordal plate of the endoderm, which is attached to the ectoderm. Some of these cells begin forming the notochord and others enter the trophoblast, which is still implanting, and participate in placentation.

Through the agitated site of the blastopore the notochord invaginates directly between ectoderm and endoderm and forges the longitudinal axis of the creature, the channel along which a spinal column will later unwind. Working its way down through the endoderm, the notochordal process fuses with it to form a flattened plate which rolls from the cranial to caudal end of the embryo to mold the actual notochord. The notochord then contacts tissue which is thickening into horizontal columns. Twenty days into the embryo's life the columns begin to split into somites which are pushed up like mountains within the mesodermal plane. Thirty-eight of these paired "backbone" bodies arise in a craniocaudal sequence along the germinal spine during the next ten days, and two to four pairs develop later in embryogenesis. These will generate the axial skeleton, its musculature, and layers of skin.

The mesoderm has separated into two integuments: an external body wall integrated with the ectoderm and a primitive gut wall connected to the endoderm; the space between them becomes the coelom, or the central body cavity of the creature. Initially, the mesoderm is solid, with only small isolated discontinuities, but these coalesce into a horseshoe-shaped cavity. Above and below this coelom, the mesodermal layers remain continuous with the extraembryonic mesoderm.

The spreading mesoderm, cutting a zone between ectoderm and endoderm and surrounding the notochord, establishes the bodily basis of up and down, cranial and caudal. It is also the compass of dorsal and ventral in its relation to the amnion and yolk sac and to the outermost and innermost aspects of the other two cell layers. Left and right are defined by additional mesoderm budding into primordial somites.

Human identity remains buried, a ghost in mud.

Twin ridges arising from the amniotic floor gradually close over the central groove and innervate its channel with a tube, the matrix of the nervous system. The neural tube seals shut in its mid section opposite the somites like a valley flattening into a plain between mountain ranges. Its head and tail sections maintain deep, thin crevasses. They will close (first head, then tail) during the fourth week. The lateral walls of the neural tube then thicken as the root of the brain and spinal cord; its residual cavity becomes the brain's ventricular system and the spinal cord's central canal. Nerve cell bodies then disperse axons through the brain, spinal cord, and dorsal root ganglia.

Through the fourth week of embryogenesis the primitive streak continues to invaginate mesoderm; then it gradually disappears into the base of the tailbud. In rare cases where it is not fully absorbed, tumors of partially unintegrated tissue occur. The difference between coordinated growth and shapeless malignancy here is a subtle one: the primitive streak is a global tumor under precise systemic control, whereas cancer is decentralized cell production, bad information, and unsupervised migration. (See the discussion of the relationship between trophoblast and malignant tumors in Chapter 20, pages 544–545.)

Subsequent cells migrating anteriorly flow around the notochordal process and prochordal plate and meet on the other side to begin the construction of the heart. At the junction of the amnion and the yolk sac the mesodermal cardiac plate and pre-cardiac cavity billow, flutter, and fold in prologue to a great organ. Pharyngeal and cloacal membranes swelling from the top layer of the yolk sac will plunge inward as the whole layer collapses and invaginates during the fourth week. The interpolation of membranes will forge a primitive foregut and hindgut, harbingers of intestines.

During this process of rapid growth and change the embryo, now head-down within its sheaths inside the womb, has reoriented itself by almost 180 degrees; its umbilical cord has been twisted from the caudal to ventral side.

"Complex growth occurs through multiplication, densification, layering and then cell specialization into such components as heart muscle or bone. With the development of tubes and their pouches, pulsation begins to take place vertically as well as horizontally and circumferentially. This new step permits anti-gravitational organization. To prevent collapse and ejection of our internal contents, expansion and contraction need support. Chambers and valves are needed to maintain the peristaltic rhythms against the forces of gravity.

"This is our metamorphosis from rhythmic pulsating cells into a multi-rhythmed pulsating organism."[9]

There may be other paths to the same point, but by now we are stuck with our one terrestrial method of getting our guts inside our skin and our organs in functional and historical relationship to one another.

Cellular Architecture: Layers and Tubes

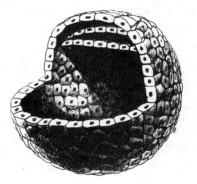

The Layering of Inner Space

The Birth of Pouches and Diaphragms

FIGURE 10K. From Stanley Keleman, *Emotional Anatomy: The Structure of Experience* (Berkeley: Center Press, 1985).

The Phenomenology of Space

"INSIDE" AND "OUTSIDE" ARE FUNDAMENTAL NOTIONS—not only inside and outside our bodies but inside our minds, in our memories. We find ourselves in bodies in families in cultures in history. Sensations and experiences continuing to flow inside us are internalized deeper and deeper. "The inside is in contact with the outside through the mediating mesodermal level. The outside is the boundary, the social self. The inside is the secret, deep, ancient past and present. The middle is the volitional self, modulation between inner and outer."[10] The social self subjects all its meanings and august occasions to the mysterious, complicated depth of its own abyss.

Symbolically and mechanically we live among enveloping structures. We inhabit caves, teepees, kivas, and hive-like apartment complexes, transporting ourselves through elevator shafts up buildings into rooms where information is stored in files and DNA-like programs. We locomote in cars and trains and, packed into elongated metal cans, are jetted by burning oil up from one zone down onto the membrane of another. We watch flattened filmstrips through which light is projected in animated replicas of three-dimensional shapes.

Our whole science has become a "degastrulation," not only of the bodies but of epistemologies and behavior patterns, of zodiacs on buffalo hide and optic glass, of fields of stars and matter wrapped in waves of gravitation and light around our world.

The surgery carried out by physicians (even from primitive times) is a scant and surficial defoliation of a labyrinth of shells housing organs formed by waves of introverting cells.

Palaeolithic peoples tried to get back inside sacred space. They embroidered star maps with spirits and pigmented cave walls with kangaroos and elk set dancing by firelight. The plank dwellings and canoes of Northwest American coastal peoples were bodies of totem whales and ravens, the masks of their sorcerers likewise an experience of self evaginated and introverted simultaneously.

We have finally managed to get outside the Earth, first in our imagination of the depth of the heavens and then as a consequence of designing mineral engines and carbon propellants, sealed vessels fueled by candles. Incarnation never was abstract, and the screech of metal against gravity is apparently as critical to the destiny of our species as the visceral cavitation of tissue. The astronaut Joseph Allen described floating outside and then returning to his membranous world:

"You know the Earth is round because you see the roundness and then you realize there's another dimension because you see layers of things as you look down. You see clouds towering up and you see their shadows on sunlit plains and you see a ship's wake in the Indian Ocean and brushfires in Africa and a lightning storm walking its way 1000 miles across Australia. It's like a stereoscopic view of all of nature...."[11]

And then reentry:

"All of a sudden you begin to hear the sound. Until now, *Columbia* has been a very silent ship, but there is a roar or a rush that builds and builds and builds. It's the rush of air.

"The next thing you're aware of is a color on the windows that starts out with just a faint tint of rose red that gets brighter and brighter, then changes to a whiter red, then an orange pink and ultimately a white that flickers around the windows and is the fiery heat of reentry. It's like being inside a neon light bulb."[12]

IN THE GASTRULA we are privileged to observe cosmic matter moving out of the universal into the temporal—a lotus mind embodying itself in soundless waves.

In viewing this from the outside we are watching time and scale created out of themselves. We are assembled in the deep underneath, in the great dark, layer by layer by layer; the mesh of it all is who we are. As thresholds of cells are absorbed and translated into new structures, regions of tissue fuse into other tissue; images arise, change, and disappear. A wall of flesh becomes wrapped around the universe; its interstices become lines of time and space, stretching across seas and mountain ranges and dreamscapes across galaxies into cosmogenesis.

"It is not birth, marriage, or death," wrote Lewis Wolpert, "but gastrulation, which is truly the most important time in your life."[13]

GASTRULATION SHOWS US how we are put into bodies. They are not rigid things of fixed parts; they are active reorganizing layers. Their folds and introversions of tissue are the grounds of our experience. Our excitement and our desire emerge in the same way as our gut cavity formed. Sadness and joy, as well as rage, have an original shape, divorced from any object or goal.

"There are some insects," writes the novelist Janet Frame, "that carry a bulge of seed outside their body as the intelligence of the universe carries its planets and stars. A spider has its milky house strung *fragilely* between two stalks of grass; and so God has pitched his worlds; and we who are replicas and live in the house of replicas cannot exist until we have shaped what we have discovered within the manifold; and know in the repeated shaping that we are not Gods, and not avoid knowing that we ourselves have been shaped and patterned not by a shadow of light or a twin intelligence but an original the sum of all equals and unequals and cubes and squares...."[14]

WE MAY REEXPERIENCE our formation as dreams of enormous masses of matter wedged up against us, passages which go out interminably and change the size and scale of our organs. We feel ourselves being twisted and rotated in sleep, but we cannot see the background of the motion. It is as though we are lying in a wall and the wall is moving. In other dreams we may burrow through long passageways and suddenly emerge under a field of stars, or find ourselves climbing through tiers of broken planks beneath a house, unable to reach the basement or make our way out.

In the key dream we stand beneath the entire starry heavens. The sky slips—not by much, just a fraction of an inch. The lynchpin surely is close at hand.

Part Two

THEORIES

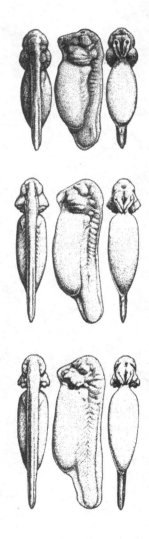

Embryology of the tadpole.
From Emil Witschi, *Development of Vertebrates*
(Philadelphia: W. B. Saunders & Company, 1956.

Morphogenesis

"How do the cells know which of them are going to form those organs?"

EMBRYOGENESIS HAPPENS TRILLIONS OF TIMES a day throughout the Earth's biosphere; yet development remains a mystery. Experiments have disclosed that cells are individual creatures compelled to specialization by nuclei; tissues are aggregates of these cells locked together to become organs. Yet nature has no persona seeking the most elegant arrangements of substance, and the innate intelligence of cells and tissues is merely the collective outcome of mechanical and chemical events regulating and regulated by each other through hierarchies established in the evolution of membrane-enclosed systems.

Cells themselves are motiveless; they react only to differential degrees of free energy and adhesiveness as their surfaces collide. Their nuclei transcribe proteins, and these events are coordinated by hormones released from other cells. Elaborate matrices (plants, animals, etc.) emerge not because protoplasm knows how to behave but because it cannot escape the homeostasis of its self-made fields of membranes. Despite this lack of enthusiasm and initiative, gestation proceeds with ease and at an amazing rate of success.

"What is development?" ask the authors in the introduction to a contemporary textbook.

"No one has completely defined it, any more than the organism itself has been fully defined. In all cases a cell or a group of cells becomes separate from the organism as a whole, either physically or physiologically, and progressively becomes a new complete organism or a new part thereof. A fern spore settles and develops into a fern gametophyte. An insect egg may become a caterpillar which transforms into a pupa and emerges as a butterfly. The stump of a salamander leg regenerates a new limb. A microscopic cell, the human egg, proliferates, and develops into a giant creature able to contemplate its own nature and origin."[1]

An embryologist, after waiting for his child to become old enough to learn how animals are formed, showed him (at age six) pictures of cells gathered in different planes as tissues, then cross-sections of those tissues assembled in organs. He had taught this rigorous sequence to graduate students for years, and no one had ever challenged its essential adequacy. But the perplexed child said: "Daddy, how do the cells know which of them are going to form those organs?"[2]

What could the professor say? Initial questions have not been answered without evasion while we have moved on to secondary matters. It takes a child to see that the "emperor" is naked.

Scale

WHEN WE JOURNEY THROUGH A RAINSTORM, droplets are driven against us by wind. If we take a plane above the storm we realize that it is actually one coherent layer of clouds. Astronauts see weather fronts as aspects of larger systems in which clouds arise in continuous cyclonic patterns from shifting centers. The pink and purple plumes of an orchid and the speckled skin of a butterfly are the same to us as the banded orb of Jupiter—organized patterns of substance. Scale is irrelevant; yet it is crucial in a way that defies simple explanation.

When we look at cells through a microscope we understand that they are earnest animalcules working like ants to manufacture structures. But we do not live among cells; we cannot communicate with them; we cannot relate our existence to theirs on a mundane basis; we cannot even find them without elaborate artifice. The gap between cells and people is unbridgeable; still we have established a scientific neutral zone in which we pretend that they and we are entities in the same realm— nouns in the same language.

They are tiny enough that we can dismiss their actual experiences, pretending that—though incremental to us—they cannot interfere with our agenda. Our meaning aloof from them is taken for granted. Yet, if cell reality is not the stuff of human existence—and civilization—what is?

We are lulled by wholeness because we stand either so close or far enough away that resonant waves of matter and material are invisible. Having no context, they merge with background, like the motion of the Earth itself.

The lines on the tiger's fur are etched there beyond correction, beyond error. Every mark is ingrained by millions of cells. Every thorn and hair on the beetle's leg as well as the flecks in its eyes, the signals from its ganglia, the junctions of its shell require the same unmonitored care. They are hewn, with a rugged specificity that defines intelligence even as it denies it.

To William Blake too, in the "fearful symmetry" of "The Tyger":

"What the Hammer? what the chain?/In what furnace was thy brain?"

And then the question that will haunt us till the end of time: "Did he who made the Lamb make thee?"[3]

Neo-Darwinian Theory

The Omneity of Genes

By a combination of genetic theory and Darwinian logic, anatomy and function are explained as chance by-products of ancient configurations of genes and their subsequent mutations. Genes program the assemblage of animate entities to feed and preserve one another and supply enzymes and occasions for their fusion into the carapaces of new beings. Organisms are "complex molecular machines controlled by the genes carried within them, bearers of the historical record of the species to which the organism belongs."[4] They are the manifestation of the genetic record. Even highly complex structures and functions that seem to transcend genetic wiring are presumed to have an origin in the hard copy of DNA at some level.

In our post-Darwinian universe, genes not organisms rule biological destiny; they sire offspring, exploit environments, and prosper at the expense of less successful rivals. In the wake of gene annointment the botany and zoology of adult life forms have become as antiquated and superfluous to modernistic cybernetic science as the species themselves are ephemeral. Biota may hog the landscape, but genes are the masters, patiently awaiting the demise of each generation to play their next round. The remarkable creatures stalking a terrestrial bestiary are summarily manufactured, then erased, every shred of their anatomies and somatic memories annihilated. Genes alone endure. True biological explanation lies in continuities of subcellular ciphers not Brobdingnagians.

There is but one set of rules: the cards (codes) that program "better" game-pieces have themselves better chances of surviving and being projected through time and space in the armature of subsequent creatures. The lone contribution of macroorganisms is to "test" the genes' experiments by bodies, either by surviving long enough to donate cells to a next generation or by dying without offspring and excising a link from the chain of life. The incentive to biology is thus a transfer of the raw fluvia of sunlight, gravity, and molecular activity through the fragile metabolisms of membrane-forming creatures into the nuclei of their cells where, given entirely different contexts and meanings from the physics in which they originate, they program the manufacture of others of their ilk. "[F]rom a relativistic point of view … individuals can be thought of as vehicles for their genetic material. It is passed from generation to generation as the darling of the evolutionary process. We have evolved [merely] to be 'good enough' vehicles for its transmission…."[5]

We are a grand if brief florescence in a regulated exercise, as the peaches of a single tree in a summer are to the Primordial Root of all trees.

FOR GENES TO HAVE such priority, two things must be true: first, they must control the entire process of embryogenesis, down to the minutest detail of content and sequence; and second, there can be no direct flow of information from the lives and bodies of organisms into their chromosomes (other than random, soma-based mutations and the sheer binary effects of differential fertility and survival, neither of which are strictly hereditary contributions). Valid proofs of both of these statuses have been axiomatic in biology for the last century.

They must control the entire process of embryogenesis.

Biologists claim that "the instructions for the assembly of a living system are embodied in the DNA molecules contained within the living cell…."[6] Most of them deem nucleotide sequences—genes—actual computer programs for organisms. If we could extract their source codes, they believe, we would find compilations of the ingredient molecules and exact stages for assembling (i.e., computing) complete functional organisms of their particular lineages. "[P]rotein is hardware," quips physicist Freeman Dyson, "and nucleic acid is software."[7] According to a typical contemporary geneticist, the "collection of chromosomes in the fertilized egg constitutes the complete set of instructions for development, determining the timing and details of the formation of the heart, the central nervous system, the immune system, and every other organ and tissue required for life."[8]

Plants and animals thus exist because of nuclear trait-maps that contain operational plans for organs and organisms. Their descendants continue to evolve through a process of natural selection. We honor this hierarchy despite the fact that we do not know how genes transpose matter from mere codons into integrated biological entities or why humans and worms, for example, are assembled from such similar programs.

The virtuosity of the organismal plan, displayed in each embryogenic episode, is little more than the bias that only certain carbon-based holograms survive—those matter-and-energy entities able to translate other matter and energy into serial, isochronal, relatively error-free playbacks of their own holography. The glue holding tissues together may be thermodynamic and photosynthetic, but its tenure is sheer accident and inertia, the outcome of trial and error in the Olympiad of free energy.

"Every force desires rest and movement," grieves psychologist Charles Poncé, as he and other neo-Platonist sentinels watch the old universe of gods and heavens fly by the arriving one of atoms and slot machines: "the flower, Fall, the bitch, Summer, each prompted by the turning of what bears us, an obeisance to the gyre of an eternal schedule, its wobble threatening the fear of our imaginations. We are not even ourselves. We are the residue of an accident, an exhalation of the earth, an itinerant pulse readied for flatline. Our hubris keeps us from the strangest thought: we are a celebration of the essence in every accident.

"There is no turning: no left or right, no above or below. There is just this being of awe we are, choking for breath even as breath leaves us, blinded by promise where there never was promise...where there is simply this: this day, this moment, this presence, this brightness. What more does one need to fashion a God, to scrape out of what we are an idea of our brief epiphany?"[9]

The answer is: nothing more; nothing more will be granted in our lifetimes. It is enough to acknowledge the corporate liquidation of resources and unimpeded flow of currency through global marketplaces. The descendants of medicine men stand in suits before the prosperity of the casino glow.

There can be no direct flow of information from the lives and bodies of organisms into their chromosomes.

As for the impenetrability of the genes by the organism, this condition was firmly established over a hundred years ago by the German zoologist August Weismann. In experiments on developing insect eggs, Weismann noticed that germ cells (sperm and egg) originate separate from the rest of the organism, remain separate in reproductive organs throughout its lifetime, and assemble new organisms without aid from or contamination by bodily cells. Genes live in their own distinct precinct and, by comparison with the mortal structures they make, are deathless, bar a planetary cataclysm.

What is now known as "Weismann's barrier" is less a rampart or membrane than an historically established segregation between differentiating, microtubule-organized organelles and reproductive ones, between soma and chromosomes—a gap that nature never overcame (and perhaps never even "tried" to overcome) despite the competing assertion of French biologist Jean Baptiste Pierre Lamarck (some hundred years before Weismann) that events in the lives of organisms fundamentally alter their reproductive contributions to the next generation, and alter them as explicitly and productively as an architect would amend his plans upon discovery that modifications were needed for perpetuation of his individual edifices.

Darwin also took for granted that adult tissue imprints modifications on germ cells. His theory of pangenesis "held that each cell in the body produced, and was represented by, invisible particles called 'gemmules' which circulated freely throughout the system and accumulated in the reproductive organs. The mixing of the gemmules would result in offspring that were in part similar to the parents and in part novel.... [Darwin] therefore proposed that the tissues of the body, upon being affected by 'changed conditions' such as use and disuse, will 'consequently throw off modified gemmules, which are transmitted with their newly acquired peculiarities to their offspring.'"[10]

Most previous nineteenth-century biologists concurred that nothing but inheritance of acquired characteristics could account for evolution by survival of the fittest (how else to test the fittest except by allowing nature to have spontaneous and detailed feedback as to the viability of each of its extant designs?). However, Weismann demonstrated, seemingly incontrovertibly, that the "germ plasm" (as he named it) is structurally and ontologically immured from the somatoplasm. Not only can no intelligent feedback ("we need a longer neck plus a bit more joint mobility") penetrate it, but no information at all can get through except "yes" or "no" at the level not of single alleles bearing patches of genetic designs but whole organisms surviving to reproductive age and producing sufficient germ plasm for transmission of their entire plan—or not.

THE SELECTIVE CONSTRUCTION of creatures is *even more random* than the above suggests, for the premature death of any one organism may have absolutely no relationship to most of the genes thereby penalized, an injustice clearly incidental to nature. "Natural selection is never aware of the long-term future. It is not aware of anything. Improvements come about not through foresight but by genes coming to outnumber their rivals in gene pools."[11]

Intelligence simply cannot insert itself into the casual sorting, even to salvage elegant sections of dysfunctional blueprints. Conversely, lethal genes that do not take effect until after reproductive age (for instance, those favoring catastrophic diseases of old age) continue to be unduly protected by the fertility of adolescents.

The Darwinian supposition is that—despite Murphy's Law ("everything that *can* go wrong *will* go wrong") and a heartbreaking waste of deserving organisms—over long periods of time, mainly favorable genes will proliferate and survive, even if far more wonderful ones are lost forever. While aimless mutations cause populations of organisms to vary in every imaginable characteristic, under conditions of limited resources, ultimately the luckiest of the best-adapted will survive to sexual maturity and produce enough offspring to maintain their hold on a portion of the planet's biosphere. The surviving genes will then manufacture new generations of similar entities to take advantage of prevailing thermodynamics and populate the changing landscape. Animals that flourish will transmit genes that flourish, not on a gene-by-gene basis, but in "complex arrays with other genes (just as many stars don't make a good team). The process selects team players rather than stars. Adaptations that are selected can work only with the finite assembly of traits existing at any one time. Adaptations are *intermediate structures,* selected to *indirectly* bring about reproductive success, and then only in a particular environment. So this process does not generate *optimum* solutions to reproductive problems, but, very slowly, finds the best ones that the material at hand can offer."[12]

The durability and overall functional cohesiveness of particular designs alone charge them with enough energy to go on replicating. Natural selection and history become, by default, their own explanation, and no one is the wiser about what really causes plants and animals to occur.

As DARWINISM HAS SETTLED IN as a civilizational religion, genes have become more and more our grail of life—and the justification for disingenuously confident answers to the riddle of form. Yet, despite this, the jury on genetic authority is still out.

Later in this chapter and throughout the next two chapters, I will consider scientific alternatives to a gene-ruled biosphere. In the last three chapters of this book I will consider still other alternatives, outside the realm of science. However, from the standpoint of charting morphogenesis, we must take for granted, for the moment, that the differentiations of cells—and thus tissues and organisms—originate totally and uniquely in the germ plasm of cells.

Laws of Growth and Form

UNDER GENETIC REGENCY THERE ARE TWO KNOWN WAYS in which nature imparts patterning—raw thermodynamic activity transmitted through atoms and molecules (choreographing nonliving as well as living entities) and the peculiar alchemy of cells on Earth giving rise to animate entities. As the laws of physics conduct themselves through gene-controlled spheres, the tight terrains and microtopographies of protoplasm—networks coordinated by genetic molecules—curdle in a landscape of intricately harmonized shapes, a realm of metabolism, heredity, knowledge, and sentience quite different from the rest of the universe.

The quandary then becomes: how much design information is transmitted directly by genes through molecules into metabolic structures, and (contrarily) to what degree are living shapes generated by other forces? While genetic determinists now consider the chromosomes almost total dictators of animate form, at the other extreme, a handful of renegade cell biologists read the patterning of life as essentially natural eddies later modified by genes.

In truth, the genetic imperative in our culture is so inflated that there is a tendency to assume that any natural process that reliably replicates itself must have an informational basis, must "be specified by a set of instructions."[13] Yet series of waves propagate through Earth's oceans, Mars and Neptune show up on schedule century after century, pulsing stars maintain cadences, and huge cyclones on Jupiter and Saturn recruit fresh molecules into fixed patterns over millennia. Atomic particles vibrate so precisely and reliably that we set the master Greenwich clock by them. Recurrent forces likely operate even more meticulously in tight cellular environments.

THE SAME COSMIC AND SUBATOMIC FORCES (nuclear, electrical, gravitational) that mold stars and planets and draw tides are the real basis of all cellular form. Organic fields begin in the electromagnetic and gravitational fields of atoms and stars. Although the astrophysical laws of matter and energy do not predict or explain animate creatures, those creatures also cannot exist without being rooted in a mechanical, geophysical substratum. Biology will never escape physics, for "both organic and inorganic matter are made of the same building blocks: atoms of carbon, hydrogen, oxygen, nitrogen, and phosphorus. The only difference is how the atoms are arranged in three-dimensional space."[14]

Identical mathematical properties are embodied in phenomena of very different origin and duration. The deformations and anisotropies of the protoctists *Vorticellae* and *Rhodophytes* could be a formation of spits and tombolos in the crosscurrents around an island, a gust of snowflakes, or hydrogen swirling into a star. A flower is a slow splash supported

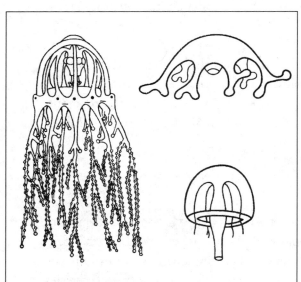

FIGURE 11A. Medusoids showing shapes of falling drops.
From D'Arcy Thompson, *On Growth and Form* (Cambridge: Cambridge University Press, 1966).

in membranes. Falling droplets from spouts recur as hydrozoan polyps in seas. Tatters of gas coalesce into planets even as bits of protoplasm are drawn into the nucleus of the cell. There are differences, but they don't nullify the similarities.

Gravity is bottomless and pervasive. Surface tension has its finger in every chink and cranny, so the earliest cells and groups of cells conserved energy by presenting the greatest possible volume with the least area of surface—a sphere. This is how an egg manifests embryogenically and, despite its cleavage divisions, the blastula remains a sphere, a sphere composed of spheres. Even cells artificially transplanted from distal regions of a gastrula will come together in a ball.

Mathematician D'Arcy Thompson contended a sphere "is strictly and absolutely the surface of minimal area, and it is, *ipso facto,* the form which will be assumed by a unicellular organism (just as by a raindrop), if it be practically homogenous and if, like *Orbulina* floating in the ocean, its surroundings be likewise homogenous and its field of force symmetrical."[15] Additionally, "the phenomenon of division of the growing cell, however it be brought about, will be precisely what is wanted to keep fairly constant the ratio between surface and mass, and to retain or restore the balance between surface-energy and the other forces of the system."[16] In this sense, a cell is no different from a molecule or a quark; it is subject to the basement laws of cosmogenesis. Lines of force rolling around sites of cellular coalescence mold simple bodies and confer discretion and symmetry on loose molecular coalitions.

Thompson derived most of the core shapes of living things simply by applying the rules of Cartesian transformation to protoplasm. In his book *On Growth and Form,* he showed that both animate and inanimate configurations arise identically from the relationships among surfaces, contours, and volumes. As geometries of different size interact and distort one another, forces propagating symmetry, polarity,

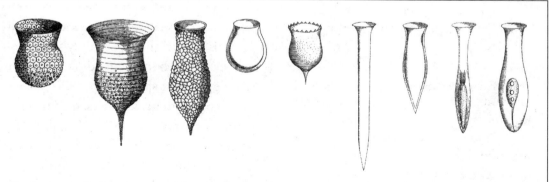

FIGURE 11B. Unduloid constricted cylinder shapes among one-celled life forms.
From D'Arcy Thompson, *On Growth and Form* (Cambridge: Cambridge University Press, 1966).

and gradient rearrange and newly order them.

If there were no other internal or external influences, life forms would remain as round as most stars and planets. But spiralling and longitudinalizing forces have also been present from the beginning. Even the tilt of the Earth on its axis and the positioning of the Moon provide series of oblique and irregular fulcra for living tissue to orient around. The environment converts these along winds, ripples, hard boundaries, weak electromagnetic fields, and the textures themselves of emerging organs into more complex grids. Waves of energy become displaced and fragmented in all directions as, right from the beginning, cells struggled to keep in contact with the environment from which they were metabolized and fed, and prevent themselves from getting trapped in solid cores. For every unit of density, there must be corresponding perforations and microtunnelling to allow breath and assimilation. Thus, shapes wind, twist, and cavitate as they are somaticized under the legislation of Archimedes' renowned edict: "... in similar figures the surface increases as the square, and the volume as the cube, of the linear dimensions."[17]

By the time an elaborate entity such as an animal egg has evolved multiple layers through uncountable generations its intrinsic dynamics are deeply aspherical; it "wants" to develop an axis. Stimulation by calcium or potassium ions, electricity, and/or light may hasten or localize axis-formation, but the internal dynamics of a cell, even in a homogeneous environment, will destabilize sphericity.

Thompson also proposed that transitions of surface tension, irregular cusps within heterogeneous fluids, bursting bubbles, and the like might serve as energetic reservoirs and morphological templates for the forerunners of tissues and organisms. With their freely gliding films the interi-

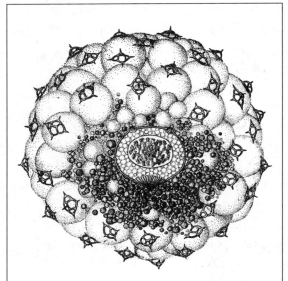

FIGURE 11C. An aggregating amoeboid form showing a bubblelike structure with basket-shaped spicules developing about each cluster of four cells and their intervening furrows. This is how four contiguous cells lying side by side in one plane normally arrange themselves. The structure of the organism displays a symmetry of forces arising from the play and interplay of surface tensions in the whole system. This creature replicates a roughly spherical mass of froth or an aggregation of vesicles.

From D'Arcy Thompson, *On Growth and Form* (Cambridge: Cambridge University Press, 1966).

ors of a heap of frothing soap-bubbles suggested to him nothing less than the dividing cells of a blastula.

THE FORMATIVE FABRIC of initial cytoplasm and tissues was modified by their simultaneous resistance to and movement through currents, so ancient polyps and crustacea, and even prototype fishes, were compromises between the emerging constitutionality of protoplasm and environmental tensions in which membranes congealed and performed. The gullet of the minute ciliate *Stentor* arises, according to Thompson, from a combination of membrane tension and its own cilia beating currents of water around it, so its "curved contour seems to enter, re-enter, and disappear within the substance of the body ... bounding a deep and twisted shape or passage which merges with the fluid contents and vanishes within the cell."[18] The ripples, density, and gravity of water and the simultaneous granular resistance to them mold the umbrellas of jellyfish, the arms of starfish, and the fins of fishes. "[D]ifferences in body shapes of related fish could be derived from one another by simple coordinate transformations that were plausibly tied to different strategies of accommodation to external forces, such as friction."[19]

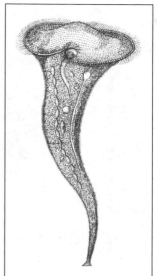

FIGURE 11D. A ciliated *Stentor*.
From D'Arcy Thompson, *On Growth and Form* (Cambridge: Cambridge University Press, 1966).

Insofar as our own bodies are derived through the phylogeny of aquatic creatures, the brunt of ancient waves and frictions is retained in our protracted, sinewy shapes (though we have tilted the lines as far upward as we could against gravity, leveraging it along a ladder of vertebrae). The vertebrate skeletal system uses the same method to distribute its load-bearing elements as a cantilever bridge, and this plan of assemblage has been reapplied at each level from "the gross anatomy of the spinal column to the microscopic configuration of trabeculae in spongy bone."[20]

The helical configurations of separate florets within a sunflower and the winding course of a snail's shell are likewise orchestrated by infinitesimal, composite forces working over time. Complex muscular oppositions sustain the kink in a chameleon's tail and generate the intrinsic spiral paths of insects proceeding toward light. That a spiral geometry is an historically attractive resolution for a variety of structures is shown by its ubiquity in the "transitory spirals in a lock of hair, in a staple of wool, in the coil of an elephant's trunk, in the 'circling spires' of a snake, in the coils of a cuttle-fish's arm, or of a monkey's or chameleon's tail."[21]

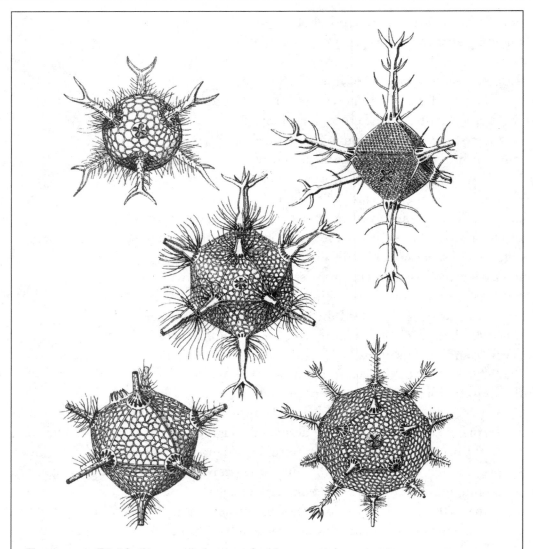

FIGURE IIE. Rigid siliceous skeletons of various marine protozoans.
From D'Arcy Thompson, *On Growth and Form* (Cambridge: Cambridge University Press, 1966).

AT THE CENTER of all large creatures lie small creatures; at the center of small creatures are primeval whorls, boundaries, and shifting densities, similitudes progressing out from nuclei, twirling around graded asymmetries, chemical variations of surface tensions, degrees of relative tensible restraint, and isobars of gravity, electromagnetism, and mass.

All other epigenetic elements of growth and form arise from cell and organelle mechanics in the context of these micro- and macroenvironments.

THIS ANALYSIS OF RELATIONSHIPS between physical and biological vectors can serve as a starting point for any discussion of cell and tissue patterning. However, because Thompson blindly inherited the nineteenth century's geometric idealism and linear kinetics, his models are only rough outlines of living systems. Minimizing developmental and embryogenic processes (and slighting Mendellian hereditary factors), he almost entirely missed the *actual* mechanics behind the phylogenesis of species, one out of another. Despite his agitated bubbles and splattering drops, he gradualized evolution, mapping each three-dimensional plant or animal, point by point, onto a subsequent anatomy as if the form of one organism could bend into the mature shape of a successor by mere gravitational sculpting of its soft medium. Artificially contrived by partial restraint rings and complicated cages of wire, his creatures were geometric toys. Their manufacture did not heed the nonlinear dynamics of genetically based systems with their raveled and mutually reinforcing feedback between mechanical and genetic factors—plus, as we shall see in Chapter 13, Thompson's "nature" was rounder, straighter, and spatiotemporally smoother than the so-called "realer" universe of contemporary physics.

A Repertoire of Morphogenesis

Base State of the Embryogenic System

A fertilized egg is a protozoan. Every multicellular animal assembles itself from clones of one prototype cell, each clone bearing the same genotype. Yet somehow these identical cells are differentiated phenotypically into interdependent entities with remarkably diverse shapes and functions, all of which have to be improvised on the fly. Furthermore, their floods of progeny are regimented solely to compose the biont blueprinted in the DNA of their shared genome, a dynamic pattern in which they will take on lifelong subsidiary roles.

All of this intricate transformation and stabilization is accomplished from a limited repertoire of molecular routines. Cells grow, split by mitosis, and die. As their nucleotide transcriptions are turned on and off, they also secrete singular proteins, develop topologically and chemically distinctive surfaces, change shape by juggling their microtubules and microfilaments, extrude unique artifacts (lamellipodia, filopodia, microspikes, etc.), and in general exert structural and dynamic forces on their neighbors, establishing and breaking adhesions with them, causing them also to change shape and locomote. The ongoing transformations of individual cells ordain subsequent gene transcription and protein synthesis, which impose additional complex vectors on the structure and dynamics of emerging tissue.

Throughout gastrulation and organogenesis, every plane of the congealing

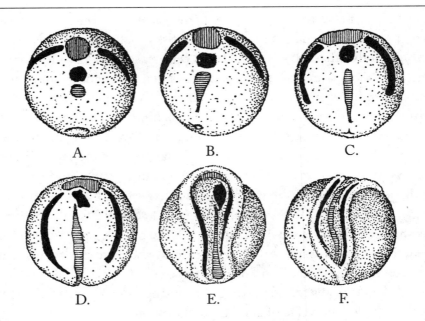

FIGURE 11F. Gastrula stage of *Triton* (newt), showing pliant morphing shape and sequential hieroglyphs. A. The anterior edge of the presumptive medullary material is marked by a shaded circle and two solid black streams on either side; B. to D. Shifting of medullary material during succeeding stages to closure of the blastopore; E. and F. Medullary fold stages, showing final resting place of black marks within the folds.

From H. L. Wieman, *An Introduction to Vertebrate Embryology* (New York: McGraw Hill, Inc., 1949).

embryo reveals a pliant morphing shape. Each three-dimensional hieroglyph squeezes itself out of the global dynamics of a prior one: a more detailed and exotic calligraphy and in deeper relief. All individual velocities and expressions of mass are subsumed in one oscillating harmony. Not only does the left side know what the right side is doing (and top distinguishes itself peremptorily from bottom), but every subpatch in every subregion "knows" where it is and what every other patch in every other region is doing.

How is this done? What role is played by a chorus of genetic dialects dispatching their primeval blueprint through assembly lines? What additional part is played by feedback between evolving structure and its underlying hereditary imperative? What architectural gamut is supplied by the indigenous forces of the universe, the children of heat, adhesion, and gravity? And, most importantly, how do nongenetic (structural and environmental) factors mesh on multiple levels of structure and function with the rigorously serial outcomes of chromosome transcription? That is, how do unbridled universal forces come to resonate stably within conscripted nucleic motifs over time? How do they derive from and augment each other in highly precise, generationally transmittable, anisotropic vortices such that plants and animals actually cohere and operate while evolving restlessly toward both greater complexity and simplicity?

In order to explain biological form we must travel back and forth between two regimes, both cosmic—one cosmic and biomolecular and the other cosmic and physicomechanical, the former a subtilized subset of the latter. Descriptions in the following sections will be difficult to follow serially and connect to one another. Aside from the simple fact that the author, who is not a biologist, is trying to coordinate cytological explanations and chronologies from a number of different scientific and quasi-scientific sources, there is the problem that even the most rigorously scientific authors do not always agree with one another as to what causes morphogenesis. Perhaps they do not diametrically disagree either, but researchers and theoreticians place decidedly different emphases and relative priorities on mechanics, shear forces, chemistry, genes, cell contexts, so-called fields, scalar effects, and a variety of other subtle intracellular, intercellular, and environmental factors.

The key events include: genetic control of amino-acid and protein manufacture and topology; cellular specification of which inherited genes are active across particular temporal and spatial milieus; intercellular communication across burgeoning geometries and geographies by a variety of chemicomechanical means; structural and textural behavior of materials making up cells and their matrices; structural and textural behavior of tissues comprised of cells and matrices; determination and expression of cell and organelle position within microanatomies (of a few microns), whole tissue regions, and an entire organism; macroenvironmental and electromagnetic fields; and the many different levels of interaction, coordination, and integration among these separate agencies.

In the narrative systematizing this information I have tried less to give a single comprehensive picture of how cells make organisms (which would be impossible anyway) and more to answer the question, "How do the cells know which of them are going to form those organs?" in a way that reflects the paradoxical and even contradictory inquiries that have gone on during this past century. There is no simple explanation. The variety of different commentaries suggests the imponderable complexity and delicacy of life. We know a great deal about its mechanics, but we do not know in what order these factors arose, how they trigger and compel one another, and how (from myriad, infinitesimal incidents) such singular cohesive entities as organisms are achieved.

That is, we don't really know how life works—not by our standards of proof, nor *can we know* because it did not organize itself using traceable objects within a time frame we can grasp. It is simply too big and too much in too compact a space. We cannot penetrate so much density in terms of its own temporality and organizational etiology. It will have to remain a wonder and a mystery.

We are not used to solving puzzles tangled and packed at billions of years of depth and diminishing exponents of scale. We are capable only of temporally and spatially finite experiments (even those of a bottomless uncertainty nature). We understand machines, so we explain cells and tissues as complex biomechanical cycles, arriving always at partial resolutions that sound good and almost always more or less "work." Yet none of the equations that we claim must hold cells together are capable of the deed. Other, competing machineries are equally viable (and equally deficient) because organic events and outcomes are jumbled and integrated both imperceptibly and millennially such that different hypothetical pathways to them are virtually interchangeable. Every explanation, when applied to actual biology, yields imperfect or incomplete results. They are all parables, metaphors, academic propaganda, and studies in hypothetical raw material.

It is likely that the manifold events (molecular, cytological, and intercellular) that comprise embryogenesis are multiply and holographically coded at a number of different levels simultaneously so that influences very different in terrestrial origin, meaning, materiality, and scale synergistically intersect and reinforce the same (or a consistent) outcome. These individual factors sometimes run in identical, reinforcing grids such that chemistry, genetics, dynamics, and scalar forces all collaborate to the same effect (as if programmed intelligently and extrinsically); at other times the same factors will impose quite separate itineraries and meanings.

The dance of embryogenesis and evolution has packed its choreography so deep while concentrating it so densely and delicately that everything is buried indecipherably in everything else. Yet with such a traffic jam in such tight quarters, there is no gridlock and nothing gets in anything else's way. In fact, the very act of cramming interlocks separate and uniquely arising meanings so as to make their collective magic possible.

Cell Biodynamics and Interactions

Cells are chemically and tactilely aware of one another's existences and define their own natures by the properties and locations of the cells around them. As they mingle in their closely packed world they read their neighbors' shapes, charges, and the notches and chemistry of their surfaces. Receptor molecules protruding from the membranes of adjacent cells provide sites for bonding. Each family of cells apparently brews a specific collection of adhesion molecules which determine what type of cells they can adhere to. The electrical aspects of cells likewise enable them to recognize one another and stick together after incidental contact (the divalent cations calcium and magnesium establish the underlying charge).

But in this regard cells are *not* planets; they do not attract from a distance; they respond only to direct touch. If their coats are removed chemically they will not aggregate.

THE PHYSICAL AND MORPHOGENETIC TRAITS of cells are grounded in innate, archaic attributes: their production of distinctive proteins such as tubulin, actin, kinesin, and myosin; their individual faculties to differentiate in response to signals in the form of molecules from other cells or elsewhere; their ability to secrete idiosyncratic macromolecules themselves; and their collective capacity to produce and maintain fields that catch and transmit metabolic energy. All of these properties are more or less rendered genetically.

With its minimal tolerances of variation, the intracellular environment is the main precinct of uncompromising genetic influence with, correspondingly, fewer and weaker exogenous and extraneous effects. Inside the cell is a hothouse for close interactions between macromolecules, resulting, over epochal time, in prototype "cellular machines" such as cytoplasmic motors and regulated membrane channels conferring shape and motility on their housing. Under hegemony of physicochemical signalling, exotic macromolecules coevolved into genetically sensitive structures that mingled their individual properties in collective tasks. These trenchant relics emerge anew each cycle of modern cells.[22]

The deep cellular environment is also physicomechanical. Nucleophilic displacements and hydrophobic interactions take place within membranes; as we move further from the arena of direct protein by-products, additional agents impose themselves. The extracellular matrix, the first major sphere outside a cell, is not as susceptible to molecularly mechanized colonization, thus is correspondingly more receptive to mechanical and dynamic action. Various universal influences (gravity, gravity-driven convection, adhesion, elasticity, surface tension, interfacial tension, phase separation, striation, buoyancy, reaction-diffusion, etc.) ply their forces principally beyond the range of macromolecular and cellular determination, increasing along a gradient out from the cell nucleus. This transition is considered either quaint or trivial by a science infatuated with genes. Yet, quite obviously, all the products of genes, from the core of the nucleus to the boundary of the organism, are subject to some physical and mechanical sanction in their morphogenesis—the further out, the more extrinsic forces enter the fray. At the same time, these invaders are never anarchic; they interact profoundly with genetic and enzymatic factors in translocating cells and assembling tissues.[23]

Just as a journey from the Earth's molten core out to its ionosphere (and beyond) undergoes a progression from geologic jurisdiction to cosmic influences, so movement from the heart of a cell through its extracellular matrix into tissue incrementally trades chromosomal authority for macroenvironmental fluxes. Though beyond the range of that cell's DNA intervention, the ecosphere nourishing the cell is a true part of its morphogenetic milieu, providing novel evolutionary pathways.

As tissues are formed through subtle interactions of intracellular structures, extracellular matrices, and external environments, there is no limit to the inward nesting of subcellular domains or outward flux of instigating waves. Far from being confined to genes, influences on morphogenesis come from the Sun, subatomic particles, and—either way—beyond, to quarks and galaxies in unknown ways. The gravitational and mechanical forces described by D'Arcy Thompson permute and snarl cellular algebra, as the immediate environment within and without cell envelopes changes in thickness, composition, and texture. Coagulation and diversifying chemistry and architecture, in turn, incite much more complex, interactional cell behavior. At a certain point, physics almost seems to be replaced by morphogenetic intention and organismic strategy. Whether or not this is the case (for reasons beyond the realm of traditional physics), local thermodynamic factors generated by cells themselves and their genes continue to orchestrate and guide all cellular motifs and the ranges of possible innovation, including those that seem superorganic in origin.

Molecular machines engineer tiny circumscribed regions while synthesizing precise tissue relics—tubules, cisternae, laminae, phalanges, and the like. On the opposite scale, hydrodynamic and environmental forces (rendered by Sun, Moon, electromagnetic fields, wind, water, etc.) allot overall shapes to self-sufficient bionts. Natural forces operate in concert with genetic factors to improvise such widely divergent metabolisms as sessile photosynthetic collector-pipelines and modular, mobile factories processing elemental fuels into zoogenetic gels.

Tissue Formation

Most tissues originate when so-called "founder cells" adhere to their progeny and restrain their natural proclivity to set off on their own. Instead, a connection is established either between their cell surfaces or macromolecules in their extracellular matrices. Tissues can also form when one cell population infringes upon another and mingles with it to form a mass of hybrid origin (see below).

Cells in tissues become bound together across the little bits of space between them by tight junctions "composed of an anastomosing network of strands that completely encircles the apical end of each cell in the epithelial sheet . . . a series of focal connections between the outer layers of two interacting plasma membranes.... The ability of junctions to restrict the passage of ions through the spaces between cells increases logarithmically with increasing numbers of strands in the network, as if each strand acts as an independent barrier. The strands are thought to be composed of long rows of specific transmembrane proteins in each of the two interacting plama membranes, which join directly to each other to occlude the intercellular space."[24]

As CELLS OF MULTICELLULAR ORGANISMS secrete their proteins, they also construct the extracellular matrix to which they become anchored and within which they associate and travel. Cells attach by affixing "to surface-bound molecules in the extracellular matrix; ... [they] are not evenly 'glued' to the matrix; rather they are 'spot welded' in localized sites...."[25] Cell walls impinging, their microtubules and microfilaments realign, "transmitting mechanical forces and chemical influences ... across the morphogenetic field."[26]

The extracellular matrix is predominantly a network of macromolecules (poly-saccharides and fibrous proteins) synthesized by the cells located within the matrix. The polysaccharides provide an aqueous component for diffusion of nutrients, hormones, and metabolites from the main bloodstream to units within the matrix. Some fibrous proteins impart durable girth and rubbery resilience to tissue; others imbue an adhesive component enabling cells to attach and connect in sheets of integrated layers of tissue (epithelia). In connective tissue fibroblasts themselves generate most of the macromolecules. (For further discussion of the extracellular matrix, see the description of collagen in Chapter 20, pages 518–519.)

Individual cells and sheets of cells are insulated from the tissue that materializes under and around them by thin, tough mats of a specialized component of the extracellular matrix. This layer is synthesized by the cells themselves out of adhesive proteins and a different class of collagen from that used in forming the musculoskeletal system. Muscle, fat, and nerve cells are all wrapped in these sheets.

ANCHORING TIES IN THE FORM of adherens junctions, desmosomes, and hemidesmosomes connect cytoskeletal components of each cell in tissue either to the extracellular matrix or another cell. These junctional complexes are forged from attachment proteins or linker glycoproteins which develop either chemically bonding or mechanically coupling sites within either the cytoskeleton and external matrix or in the transmembrane complex of a neighboring cell. Adherens junctions provide brooches for actin filaments, themselves linked in bundles by transmembrane glycoproteins. Together the links can assemble a continuous belt of calcium bonds circulating around each of the adjacent cells in an epithelial sheet. This enables such sheets to

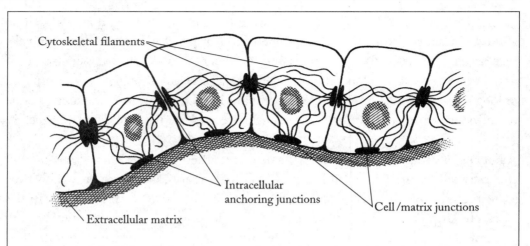

FIGURE 11G. Anchoring junctions join cytoskeletal filaments from cell to cell, from cell to extracellular matrix. Illustration by Harry S. Robins.

fold uniformly into tubes. During embryogenesis, as an adhesion belt tightens the apical ridge of epithelial cells in the neural plate, every cell narrows at its constricted apex and the plate rolls up into a pipe.

Desmosomes are pivotal anchors for intermediate filaments. Tiny rivet points composed of linker glycoproteins yoked to cytoplasmic plaque, they stick cells together or tie them to intermediate filaments. Hemidesmosomes, despite their name and resemblance to desmosomes, are chemically and functionally distinct; they affix the basal surface of epithelial cells to a specialized lamina of the extra-cellular matrix spreading between the epithelium and a region of connective tissue. Both types of attachment buttons help to strengthen groups of cells within an epithelium and enable them to operate as structural unities; serving as bolts or pins, they disseminate tensile and shearing forces through the epithelium and its under-lying connective tissue, reapportioning and deforming their topology. Anchoring junctions are more common in tissues under mechanical stress—cardiac muscle, skin epithelium, the neck of the uterus, etc.

In general, the extracellular matrix guides the movements of cells and helps to allot their specialized character—their shape and polarity, the organization of their cytoskeletons, what chemicals they synthesize, and how they differentiate. Collec-tively, the bonds between cells and between cells and their matrix receptors com-prise gradients of affinity. The overall propensity and repugnance between any two adjoining cells contribute to a sort of morphogenetic code that ultimately disposes either their formation into one genre of tissue or another, or their permanent dis-engagement from each other.

Gap Junctions

Cells are also segregated and linked by narrow permeable gaps of approximately three nanometers' width. These spaces are formed when transmembrane proteins in adjacent cells merge into an aqueous estuary that, in a sense, confederates their interiors. Yet the border proteins impinge in such a way that a complex interrupted chink occurs between their plasma membranes. Because the chink is continually interrupted by thin clusters of connections and because so much activity is generated across it, it is given an oxymoron for a name: gap junction.

Although there is no full anatomical connection between adjacent cells across the gap junction, they communicate by passing small water-soluble molecules,

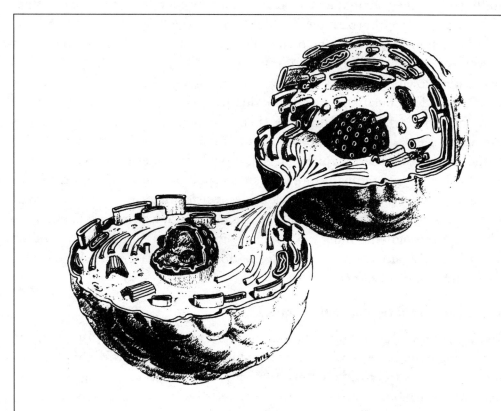

FIGURE 11H. This could represent two cells with a junction; two pouches connected by a passsageway, each pouch (like a cell) with a nucleus; two viscera with a sphincter between them; or the head and the body joined by the nervous system. The principle of formation and connection is similar: expansion, fission, separation, junction.
From Stanley Keleman, *Emotional Anatomy: The Structure of Experience* (Berkeley: Center Press, 1985).

including inorganic ions, directly through their cytoplasms. Thus, the cells, while not morphologically hinged, are electrically and metabolically allied. As their intracellular metabolites are shared, they cohere and collaborate.

Electrical couplings synchronize heart-muscle cells in contraction and smooth-muscle intestinal cells in peristalsis; they also cause cilia to beat in unison along epithelia and propel action potentials from axon to axon, enabling fish and spiders to evade and baffle predators, likewise capacitating predators to spring and pursue.

IN THE BLASTULA, cells are held together by the basic mechanical geography of their situation but also by the passage of ions and other small molecules through gap junctions. In gastrulation, as groups of cells disjoin from surrounding clusters and differentiate, they retain their cooperative unity by continuing to signal through gap junctions. This authorizes rows of cells to develop as a layer rather than individually. The homogeneity of the neural plate, after it uncouples from overlying ectoderm, is partially gap-junction-coordinated.

Molecules flowing through gap junctions establish gradients from high to low concentration. This fluctuating set of boundaries gives cells highly discrete positional information which not only helps them stay together in a moving layer but differentiates them within that layer by dint of their individual placements.

Adhesion and surface recognition in cells provide an underlying organizational plan that guides tissues through complex metamorphoses. As the crumpled epithelium of a larval fruit-fly's imaginal disk elongates, twists, stretches, and subtly refolds into a wing, position-specific molecules at the surfaces of cells maintain and project the shape through its point-by-point morphing process. When scientists introduce antibodies against gap-junction formation, all organized development in eight-cell fly blastulas ceases.

Adhesivity, Cohesivity, and Sedimentation

Morphodynamic and morphogenetic factors interact with molecular products from the beginning of embryogenesis even as they must have mingled with them in Precambrian tidepools to provide templates for life. Organisms are configurations of clusters of cells (bubbles), which gather in spots, disaffiliate selectively from one another, migrate to new sites either singly or in layers, and reattach to tissue elsewhere by adhesion molecules. In this manner shapes emerge across three-dimensional grids.

At early stages of their development cell clusters differed in the proclivities of the molecular components they secreted or appended. Some of their ingredients became denser than the cytoplasm in which they floated. As clusters evolved, gravity pressed asymmetries into them, and asymmetries stippled different chemistries and arrays of adhesion molecules.

Adhesivity is a primeval and fundamental building block of metazoan design, biophysically prior to junctioning. Sprawling over zones of emerging internal space, adhesion provides the essential cohesivity of tissue, hence the morphogenetic force of multicellularity. By modulations of adhesivity and de-adhesivity, cells form aggregates and linearized segments, carve out compartments and lumens. Over long epochs a few simple motifs emerged: "A tissue layer can engulf or be engulfed by its neighbor layer, undergo segmentation, form a hollow ball, or roll up into a tube."[27] These were preanimate features of multicellular systems; their recitation required little or no genetic regulation and occurred independently in separately evolving protoplasmic clumps, as inevitably as in any lineage regulated by genetic molecules. All by themselves, variations in molecular sedimentation led to pattern formations, for, as they divided, founder cells passed on idiosyncratic components (primitive yolk platelets, stored mRNAs, etc.) and thus sired asymmetrical masses. Similar chemicomechanical factors now organize each new egg according to the same principles.

Gastrulation is a highly complex phenomenon transpiring at multiple levels with many vectors. Biologists have generally tried to explain it by genetics and natural selection: protist cells, once upon a time trapped between competing imperatives to fission or ciliate (that is, to improve their status either by becoming colonies or by adding useful organelles), in a stroke of biokinetic genius used their predicament (and surplus energy) to turn themselves into gastrulae, thereby resolving the cosmic riddle by transcending it. A crossroads thus became a labyrinth.

Something like this may well have occurred phylogenetically, but the raw choreography of gastrulation—its trademark deed of engulfing compartmentalization—may also be triggered dynamically simply by adjacent companies of cells of differential density in a blastula resolving their adhesivity-cohesivity gradients such that one group sinks into the other and is engulfed by it. In laboratory experiments spectrums of inconsonant relative adhesivity and cohesivity reliably initiate gastrulation-like phenomena in a variety of cellular and quasi-cellular substances.

Hence, gastrulation, while genetically a brilliant resolution, was already dynamically inevitable given the quantitative adhesive differentials in solid and hollow balls of precellular bubbles. By fortuity the solution to a morphogenetic and ecological dilemma was accorded millennia before the dilemma even arose. Like sexuality, gastrulation may be the result of a convergence of events on two totally different levels—one nucleic and intracellular, the other extracellular and mechanical.

Several key blastomere digressions initiate gastrulation and the subsequent development of multicellular organisms—sometimes one mode per organism, sometimes

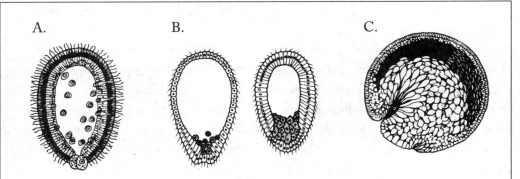

FIGURE III. Different modes of gastrulation: A. Multipolar introgression in a sponge; B. Two stages of unipolar introgression in a hydroid; C. Emboly (invagination) in an amphibian.

From Stuart A. Newman, "Generic physical mechanisms of morphogenesis and pattern formation as determinants in the evolution of multicellular organization" *(Journal of Biosciences,* Volume 17, Number 3, September 1992).

two or more. All of these "could, in principle, be achieved simply by modulation of adhesivity in different regions of the embryo. In *invagination,* a ball of cells moves into the cavity of a hollow blastula. *Epiboly* and *involution* both involve the envelopment of a distinct vegetal or marginal population of cells by the spreading ectodermal layer. *Delamination* occurs in solid blastulae when an outer layer of cells separates from the inner cell mass, and *ingression* involves the independent detachment of cells from the blastula wall and their accumulation within the cavity."[28]

All of these initiatives yield constellations of multiple germ layers subject to subsequent permutation, providing bases for tissue boundaries and profusions of new cell and body types. They invented themselves in free space during the Precambrian and Cambrian periods and, under genetic regulation, they recur in each modern embryo. These are mechanical, generic pathways to interiorization and tissue structure. However, as we shall see in a few pages, they are abetted by nucleically assembled organelles.

FOLLOWING GASTRULATION some embryonic regions are circumscribed into serial reiterations of linear patterns. Though (like involution and epiboly) subject to molecular refinement, this segmentation can also occur purely physicomechanically from chemical oscillation and differential adhesion. Compartments with alternating ranges of adhesivity provide the kinetic equivalents of a segmental body plan. This sort of formation occurs routinely in the subdivisioning of insect bodies, the gathering of mesoderm into somites among vertebrates (leading to the formation of a musculoskeletal axis), and the periodic swellings (rhombomeres) across the vertebrate hindbrain.[29] It is a rippling of terrestrial gruel.

The Physics of Soft Matter

Cells (like electrons) are launched in a sphere of quantal forces (strong and weak); then they enter the boundless domain they share with puddles, stones, breezes, and crystals. Though tissues are assembled mostly genetically, they emerge into a bailiwick of textural and tensional factors like adhesivity. Shear and compressive forces arising within their immediate fabric and abetted by surface tension conduct stress patterns tangential to protoplasmic planes, pushing adjoining layers past one another. Yet it is not a pure slide. While liquids flow passively, cells and tissues propagate both motile and contractile forces; within their epithelia they spread molecular bonds. Molecularity restores links among them.

Epithelia operate as tightly packed elastic sheets with strains from interfacial tensions accruing and diminishing between adjoining tissues. Morphogenesis in a sense represents the shifting balance sheet of deformations between them.

Depending on their relative strengths, heterotypic and homotypic cell types maintain degrees of fusion and separation; to differential degrees they adhere and avert between similar and dissimilar denominations. Cells and their extracellular matrices also constitute distinct physical phases with interfacial tension between them. Thus, most raw tissue fluctuates in a semi-chaotic fashion, molded by its own inherent rubberiness.

Liquid tissues (like blood) shear in relatively unimpeded currents; similarly, within semiliquid layers cells slip past one another, dissipating strains before they warp beyond recognition. Most tissues are neither solid nor liquid but have viscoelastic and semisolid components with proclivities of both. Fluid characteristics allow groups of cells to travel along adhesive substrata by interfacial tension. This process (haptotaxis) likely contributes, in differing contexts and degrees, to the unidirectional migration of neural-crest cells, the spreading of blastoderm on the inner surface of the chick vitelline membrane, the protraction of embryonic epicardium over pigeon myocardium, the passage of one cellular zone over and around another during gastrulation of bony fishes, the extension of the pronephric duct over the lateral mesoderm of the salamander, and the congregation of mesenchymal cells prior to the formation of vertebrate cartilage.[30]

The viscosity and elasticity of each soft tissue domain will determine the likely proportions and scope of its deformation.

By contrast, within solid tissue (bone, for instance) shear stresses yield minimal cellular movement.

UNDER THE NOMADIC INFLUENCE of their own viscoelastic topographies tissues regularly develop secondary nonuniformities of density and chemistry. Some regions dissipate nonequilibria by mixing across zones; others remain chemically immiscible. In amphibian oogenesis cytoplasm and yolk support distinct, separable fluid phases like oil and water, leading to heterogeneous developmental mechanics.

Over time tissues sustained heterogeneity by emergent metabolic processes consuming free energy. As barriers developed between immiscible regions, interfacial tension accumulated an imbalance between adhesive and deforming forces. Energy was built up in the local system; contours and boundaries sideslipped. Stable idiosyncratic shapes emerged along interfaces with both transient and far-reaching morphogenetic consequences. The algebra of each fluid-phase boundary contributed to the morphology of the tissue layers and organs formed within and contiguous to the boundary.[31]

IN MULTICOMPONENT SYSTEMS convection arises from tension and viscosity. Subregions stipple and interdigitate. Striated countercurrents form, gravity influencing density gradients along boundaries of oppositely moving flows. Droplets follow lines of strain, stirring motion across interfacing regimes (not unlike streams at the air-liquid interface in a glass of dinner wine). In the context of interfacial stress semifluid tissues launched different relative motions. Their convective flows converted metastable states into adhesive equilibria. Microfingers formed at scales of ten to hundreds of microns, coopting particles and cells until (over vast periods of time) they translated surface-tension inhomogeneities and metabolic processes and flows into lasting morphologies.[32]

Streamers extend from the downward flow of cytoplasm after fertilization of an ascidian egg as well as in the migration of the embryonic neural-crest cells in the axolotl; likewise the developing kidney-collecting tubules in mammals reflect prior microfinger morphology, albeit in a more semipermanent guise.[33]

Chemical Waves from Reaction-Diffusion Coupling

The role of feedback interactions between reacting and diffusing ingredients was chronicled by the mathematician Alan Mathison Turing in a 1952 paper, "The Chemical Basis of Morphogenesis." In purely chemical reactions distributing positive and negative feedback loops, components diffuse at different rates, leading to lifelike self-organization and spatiotemporal patterning. Such ancient reaction-diffusion mechanisms underlie a range of dynamic processes that lead to molecular inhomogeneity and could have provided the underlying basis for features like stripes, spots, dapples, calico, plaid, and bars in permeable tissues. "Organisms with a wide spectrum of body plans (*e.g.* segmented, 'checkerboard,' annular, and 'pinwheel')

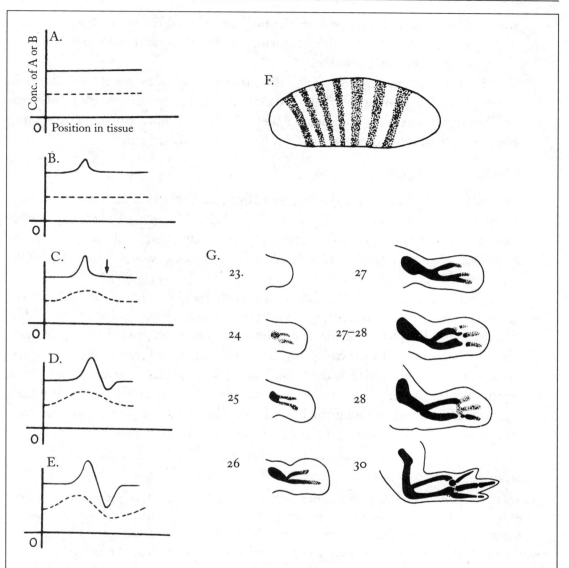

FIGURE 11J. Chemical wave generation by a reaction-diffusion mechanism, and examples of stripe patterns during development. A.-E. Graphical representation of chemical-wave formation. It is assumed that two substances, A. and B., which influence each other's synthesis, are produced throughout a row of cells, and that there is a balance in the rates of synthesis and utilization of A. and B. (i.e., they are at a steady state). The steady state shown in panel A is *spatially uniform:* the concentrations of both A. and B. are unvarying along the row of cells. Under certain conditions a *spatially nonuniform* stationary state (panel E.) can be achieved by the growth and stabilization of fluctuation (panels B.-D.). The following conditions are

(Continued on page 222)

would all be expected to arise from the interaction of differential adhesion effects with gravity-driven and reaction-diffusion mechanisms, and to have proliferated during early metazoan evolution."[34]

The distinctive patterning seen on the coats of today's zebras and leopards might reflect the diffusion of primordial substances within dawn-time puddles. Likewise, as Thompson proposed, the grids for skeletal motifs (limbs, hands, feet, etc.) may have originated as gradients within fluids and viscoelastic fields.[35]

Intracellular Mechanics

Nowadays cells change their shapes internally through the production or rearrangement of organelles (such as microtubules and microfilaments)—a mostly genetically orchestrated phenomenon. A mesh of these talented organelles throughout the cytoplasm gives a cell functional and structural coherence as well as inherent mechanical properties.

When the microtubules become ionized into parallel sheaths, the affected cell must elongate. This is a decisive factor in the modern separation of epithelial layers, for when clusters of cells form bottleheads they contract inward homogeneously, the outer edge of their layer collapsing. Since the extenuated ends of the cells adhere to one another, contraction is then transmitted across an entire layer. Such a global event happens dramatically during gastrulation when the epithelium of the marginal zone invaginates along the blastopore. The surface is pulled into the cytoplasm in a smooth concave pocket which becomes the archenteron.

FIGURE 11J. *(Continued from page 221)*

sufficient to bring about this phenomenon: substance A has a positive effect on the synthesis of both itself and substance B; substance B has an inhibitory effect on the synthesis of A; the diffusion rate of B is greater than that of A. Arrow in panel C indicates the point at which a reduction in the concentration of A to below its uniform steady-state level will be initiated on the basis of the assumptions above. The number of peaks and valleys of A and B that will be in place when the system finally reaches the new steady state will depend on reaction and diffusion rates, the size and shape of the spatial domain in which these events are occurring, and the modes of utilization of A and B at the boundaries of the domain. F. *Drosophila* (fruit-fly) blastoderm-stage embryo showing early *even-skipped* protein pattern (stippled stripes). G. Progress of chondrogenesis in the chick wing bud between four and seven days of development. Solid black regions represent definitive cartilage; stippled areas represent early cartilage.

From Stuart A. Newman, "Generic physical mechanisms of morphogenesis and pattern formation as determinants in the evolution of multicellular organization" *(Journal of Biosciences,* Volume 17, Number 3, September, 1992).

The unremitting power of microfilaments is demonstrated in the high-speed beats of the wings of insects and tails of rattlesnakes as these organelles expand and contract in alternation. In morphogenesis microfilaments commonly tighten like purse strings and contract or narrow the cells of which they are part. They are a major driving force in neurulation, squeezing out the necks of prospective neurons while intervening cells flatten into epidermis. In the formation of glands microfilaments pull the inner open surfaces of cells together so that a cone imbeds in a cavity, for instance, the anterior pituitary in the stomodeum.

Cells also morph through the accentuation of some aspect of cellular anatomy. For instance, the expanding vacuoles cluster together to make the notochord.

The Formation and Dynamics of Epithelia

Epithelia arise, as seen, from a variety of structural and biochemical episodes. Once formed, they function (like cell membranes) as selective permeability barriers. Fluids on either side of them maintain different aqueous constituents. The linings of the small intestine and the urinary bladder *must* seal most of their organs' contents within, allowing only predesignated molecules to slip through the cell sheet.

Nutrients released in the intestines are pumped into the extracellular fluid of neighboring connective tissue; from there they spill into the bloodstream. Carrier proteins within the membrane transport chosen molecules across the epithelial barriers. Even these proteins have localized precincts. For instance, an apical set of carriers is confined to the intestinal surface facing the lumen of the gut and must under no circumstances carry their booty into the basolateral surface of the cell (which has its own patrols). This differentiation of carrier proteins into regional spheres ensures a directional flow of nutrients (one way) and wastes (the other). When proteins select appropriate molecules and keep them moving at the desired heading across their zone, the molecules are prevented from diffusing accross the membrane, back down the gradient, and the junctions between the epithelial cells keep them from leaking by default into other cytoplasm.

ALTHOUGH THE EMBRYOGENIC DYNAMICS of epithelia are inextricably complicated, there are a few basic mechanical possibilities that are propagated in endlessly novel contexts. Tissue essentially moves where there is least resistance or where it is drawn by other tissue. Plates of cells thicken, thin, separate, buckle up or down, or break apart. As these movements follow one another in intricate algebraic juxtaposition, organs are molded.

Sections of tissue become similarly folded and refolded in the construction of separate fields. In the cerebral cortex of the brain and the duodenal and intestinal

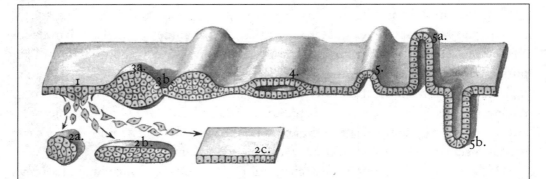

FIGURE 11K. Morphogenetic processes. 1. Cell migration; 2. Cell aggregation, forming: a. masses, b. cords, and c. sheets; 3. Localized growth, resulting in a. enlargements, b. constrictions; 4. Splitting, including the delamination of single sheets into separate layers, the cavitation of cell masses, and the forking of cords; 5. Folding, including circumscribed folds which produce a. evaginations, or out-pocketings, b. invaginations, or in-pocketings.

From Leslie Brainerd Arey, *Developmental Anatomy: A Textbook and Laboratory Manual of Embryology* (Philadelphia: W. B. Saunders & Company, 1946).

loops of the alimentary canal, pouches and cavities take on distinctly different meanings in terms of their relationships to adjacent tissues and the functional unity of the organism. Ears and pharynx are both twisted labyrinths. Lungs and kidneys emanate from repetitive branching. Heart, eyes, and tear ducts all originate as hollowed-out regions. There is no biomechanical rule for even the formation of homologous organs. Within the same lineage of creatures an ear may either be a pocket in an evagination or a solid mass of cells subsequently hollowed. The human archenteron is a cavitated morula; the salamander archenteron is an involuted cavity. The embryogenic origin and enactment of a basic motif can change radically and seamlessly without loss of function. Once again, there are multiple paths through the labyrinth.

Inhibition of Movement

When the leading edges of two inimical cells come into contact with each other, they are simultaneously paralyzed as each cortical cytoskeleton receives an alarming signal in the area of impingement. Then the cells halt their microspike and lamellipodia production at the affected sites and begin synthesizing the same organelles along other edges. The new microspikes and lamellipodia provide each of the colliding cells with a trajectory away from the other. This is probably the method by which neurons from the central nervous system shun their peripheral counterparts. The axons in each of these systems creep happily along axons from their own system but retract from axons of the alien system.

In wound healing, epithelial cells extend lamellipodia and move in sheets over a damaged area. But when the epithelial patches begin to overlap and cells from different margins impinge across the gash, the edges retreat, making the repair neat and precise, with relatively scant scarring. Such behavior suggests primeval packs of cells, stalking other packs and protecting their own territories.

It should also be noted that physicist Robert Becker has shown electrical and electromagnetic forces playing a role in limb regeneration and tissue healing among salamanders.[36] Waves are likely generated by microtubules, cell membranes, and collagen fibers connecting cells to the extracellular matrix. Perineural Schwann cells carry these signals, stimulating regional activity.

The Cohesive Movements of Cells

With their pliable walls and tight links, cells migrate either singly or in sheets.

When cells recognize their kinship they unite in epithelia. At the same time, they move away from other epithelia they do not recognize. As groups of cells lose contact with one another (no longer acknowledging each other as part of the same layer), crevices develop between their tissue; for instance, the coelomic cavity between the parietal and visceral layers of mesoderm. Many of these gaps are simultaneously filled with secretions that aid in the separation and equilibrium (as in the relationship between the mammalian blastocoel and surrounding tissue layers).

The same pattern dynamics repeat along a hierarchy of scales. Gastrulating neural plate and mesoderm sink inward together, away from the epidermal and endodermal cells with which they were previously fused. The blastopore forms by the invagination of cells. It spreads, as inwardly travelling cells radiate over the blastula's inner surface to produce deep tissue layers. During neurulation, an inner sheet of already gastrulated cells furrows again along the neural plate to cut a cylindrical tube. Later, the section of the tube that becomes the optic vesicle itself caves in to provide an optic cup; nearby flat epidermal cells respond to the existence of the vesicle by swelling and invaginating along an inward gradient to mold a transparent circular lens, a primitive image-transmitting device.

Somites in the mesoderm are also formed by loss of contact between groups of cells. As a series of clefts develops perpendicular to the surface of the epithelium, any individual cell must gravitate to one somite clump or another.

The Thickening of Epithelia

A regional surge in mitosis (or a migration of cells to a single site) provides material for the formation of new organs.

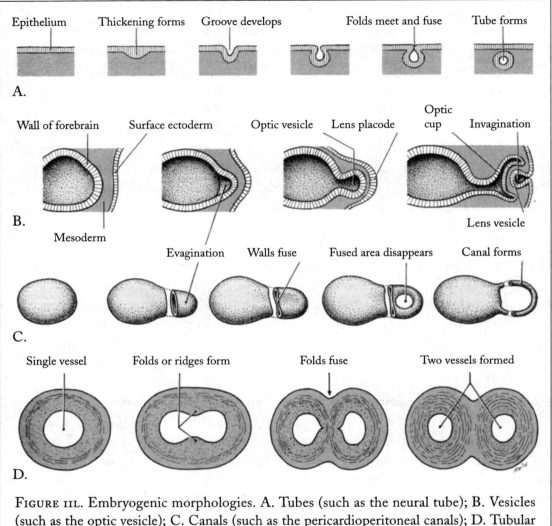

FIGURE IIL. Embryogenic morphologies. A. Tubes (such as the neural tube); B. Vesicles (such as the optic vesicle); C. Canals (such as the pericardioperitoneal canals); D. Tubular structures (such as large blood vessels).

From Keith L. Moore, *The Developing Human: Clinically Oriented Embryology* (Philadelphia: Saunders College Publishing, 1977).

The neural plate appears first in humans as a thickening of ectodermal tissue toward the mid-dorsal region of the gastrula. Neural-crest cells split off the plate when the neural tube is being formed and later colligate into a variety of structures, e.g., the spinal ganglia and ganglia of the autonomic nervous system, pigment cells, bones of the face, pharyngeal skeleton, adrenal medulla, odontoblasts of teeth, etc.

The brain begins as a series of thickenings at the front of the neural plate.

The gonads of the vertebrates arise as carbuncles on mesodermal tissue lining the back of the body cavity. As this germinal ridge extends into the coelom it is squeezed laterally so as to become suspended and brought into contact with emerging excretory organs.

The lungs and pancreas originate in regional evaginations, thickenings of endodermal epithelia which continue to branch out and fuse in individual themes, in some places forming secondary ridges.

The heart rudiment swells when free edges of mesodermal mantle come together in an area outside the mesoderm and consolidate, creating an endocardial tube with a cavity inside.

Hair, feathers, scales, and glands all start as little round pegs—raised placodes. If they continue to buckle out, scales and feathers emerge, but they can also invaginate to produce glands and teeth. Thickening of epidermal cells generates hair. In fishes longitudinal thickening of epidermis becomes fins.

The Thinning of Epithelia

Cells can also degenerate (sometimes through the participation of lysosomes and other organelles).

Regional cell death causes the separation of bones at the joints (such as between the radius and ulna) and occurs in the skin between digits of feet; among ducks and other water birds a webbing is left behind. Interestingly, once cells have received the signal to expire they will deteriorate even if the tissue they are part of is transplanted to a blossoming region.

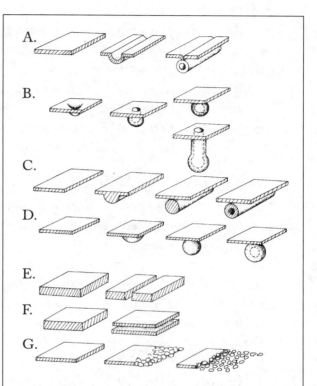

FIGURE 11M. Different types of transformations occurring in embryonic epithelium in the course of development. A. Groove giving rise to a tube; B. Pit giving rise either to a vesicle or tubule open to the exterior; C. Development of a tube from a solid elongated thickening of the epithelium; D. Development of a vesicle from a solid circumscribed thickening of the epithelium; E. Splitting of epithelium perpendicular to surface in sections; F. Splitting of epithelium tangentially into two layers with a space in between; G. Breaking up of an epithelium to form separated mesenchyme-like cells.

From B. I. Balinsky, *An Introduction to Embryology*, 5th edition (Philadelphia: Saunders College Publishing, 1981).

Conventionally, epithelia thin out when the number of layers of cells in them is reduced. For instance, during epiboly the epidermis expands and covers the yolky mass of the embryo while cell fission simply does not keep pace with the extension of the layer's territory.

As a vertebrate elongates from head to tail, its epidermis continues to stretch, though it will eventually fold and thicken in spots as it forms organ rudiments in combination with and by induction from underlying tissues.

In fishes and amphibians the endoderm at the front of the alimentary canal is dilated. This region later becomes the epithelial lining of the pharynx and propagates a series of pouches just behind the mandibular arches—the rudiments of gills (converted to glands in land vertebrates).

Mesenchyme and Cell Migration

Epithelia may also become uncoordinated and break apart; this process simultaneously opens cavities and provides raw material—free cells for a new generation of tissue. As these rampaging zooids mingle with tissue from disparate sources, they differentiate into a wide assortment of tissues, including those comprising the peripheral nervous system. Neural-crest cells, having individually detached themselves from the edges of their neural folds, migrate throughout the body, participating in the remarkable diversity of structures described on the previous page; at the same time they leave a gap between the epidermis and the partially fused neural tube. (See Glossary, pages 804–805.)

In many species invagination of the endoderm and mesoderm occurs through the systematic disassembly of layers of tissue. The cells of the primitive streak of the chick slip away from the surface one by one. Because they do not adhere to one another they cannot form a smooth-walled cavity. An archenteron must be created secondarily (by folding).

CELLS ARISING FROM THE BREAK-UP of epithelia make up a swarm called mesenchyme and, unlike epithelial cells, travel on their own outside any homogeneous layer. The tissue from which they arise is usually left intact though somewhat depleted of cells.

Mesenchyme cells often accumulate in a colloidal substratum, as they are unable to navigate through fluid. If the fibers of the substratum are disperse, the cells are likewise scattered but, if the fibers are organized, the loose cells are led to a specific site, oriented by epithelia, and held in place. (Not only does the primordial gastrula provide the shape and inductive grids for the organism, its vintage is the unhewn material from which continuously more subtle, interstitial tissues are assembled.) Mesenchyme is a biocosmological substance, continuously supplying raw tissue for vessels, bones, and nerves throughout the body.

Cartilage begins as mesenchyme on the inner wall of somites and later captures some of the free-moving neural-crest cells. The rudiments of blood appear first as angioblasts, mesenchymal aggregations in the mesoderm. The limb-buds of the vertebrates originate in dense packs of mesenchyme. Others of these peripatetic cells participate in sense organs, sex cords, neural arches, and the heart.

Although mesenchyme movements are amoeboid in appearance they do not use the amoeba's internal streaming of protoplasm to fill a pseudopod. Instead the ruffled cell edge of a pseudopodium makes contact with a surface and pulls the body along, then relaxes. For instance, in orienting toward the newly formed interior of the body, cells of the primitive streak send out pseudopods which are withdrawn if they contact blastocoel fluid. New edges are cast until they finally locate the cavity.

Travelling zooids are not only able to find their way to their ancestral assemblage sites but they immediately recognize the cells to which they will bond in tissue. Whole armies surge nostalgically into appropriate zones where they are welcomed and incorporated. Remember, these cells retain both an independent and subordinate nature. They are simultaneously autonomous and componential, alien and familiar. And they bring radically new characteristics to indigenous tissue masses.

Their ritual odyssey resembles, before the fact, turtles and birds in their instinctual nesting journeys. How it got internalized in membranes long ago and scripted so rigorously at a zooid level is one of the mysteries of development. But mesenchyme is primordial cellular intelligence—the migrating life-connector of the body and its web of consciousness.

Protein Chemistry

Individual cell chemistry is obviously the language in which many of the instructions of organismic development are coded. Clear signposts of this include the variety of tissues responding to female sex hormones during puberty and the effect of extraneous thyroid on a tadpole (which turns it progressively and irreversibly into a frog).

Chemical signals tell cells how and when to differentiate. Some of these signals are brief; others are slow and usually long-lasting in their impacts. The actions of some substances may be biphasic: Ordinary or strong doses inhibit activity stimulated by minute doses of the same chemical. Even larger doses may be toxic. If greater initial stimuli instigate negligible changes (or recognition) when lesser initial stimuli induce quite powerful effects, it is no wonder that the embryogenic outcomes of discrete chemical signals may be alternate, opposite, and/or paradoxical across ranges of doses and contexts, and that microdoses lead to macrostructures.

As we have seen, protein molecules are continually dispersed and interpreted at fresh hierarchical levels. Sites of synthesis and distribution must first be blueprinted in mRNA and amino acids, then assembled by ribosomes. After enzymes are disseminated, they activate other biochemical centers throughout the organism, factories that not only provide new materials and inducers but have secondary effects on already existing structures and activities. These myriad reciprocal networks are joined in a complex web of activity.

In organisms as widely varying as human beings and fungi, a small number of steroid hormones synthesized from cholesterol molecules regulate many critical developmental and physiological processes. Whereas water-soluble molecules cannot pass through the lipid bilayers of a cell's plasma membranes (hence, must link with receptor proteins on the cell's surface to deliver their instructions), the tiny, hydrophobic steroids are fully lipid soluble. Transported through the blood by carrier proteins to which they bind reversibly (rather than by hydrophilic dilution), at signalled sites they break free of their accompanists and sail unimpeded through the membrane of a target cell, binding tightly (though again reversibly) to one of ten thousand or so steroid receptors inside the cell, either in the cytoplasm or nucleus. Ensuing receptor activation kindles specific DNA sequences so that nearby genes are bound and begin to transcribe—an extremely subtle and fundamental choreography that was not identified until 1983.

The products of transcription regulate not only the spectrum of proteins synthesized but the transcription of other genes, for secondary DNA responses to steroids are triggered by the initial output. The signalling is stable and discrete, transcription stimulated even when its enhancers are separated by thousands of nucleotides from the promoter site of RNA synthesis. The system depends on the inherent character of a cell itself—its emerging chemistry—for one kind of receptor protein can supervise the transcription of different genes in quite disparate cells. Thus diverse proteins can be targeted by the same steroid or the same proteins by dissimilar steroids. Though signalling dialects may vary phylogenetically from cell to cell, coevolving steroids remain able to coordinate them into consistent organismic landscapes.

As sequences of proteins are synthesized, they establish internal chemistries. Proliferation of organelles like Golgi bodies and peroxisomes, and arrangements of microtubules, microfilaments, and intermediate filaments then particularize within concordances of cell networks. Keratin is manufactured in epidermal cells, crystallins in lens cells, digestive enzymes in cells that are to become part of the gut. This happens not, for instance, because an eye has lost the genes for digestion or because skin does not contain nucleotide sequences that could result in blood chemistry, but because proteins bind specific DNA sequences in each nucleus; these perform as enhancers and silencers such that only some genes are activated in each cell—a feature we will discuss at length in the next chapter.

Individual spirochete/organelle specializations become organismic ecologies. The pancreas emerges as a distinct organ participating in carbohydrate metabolism as its cells assemble insulin molecules. Other cells become cartilaginous as their nuclei impel them to manufacture collagen. Muscles form when the proteins myosin, actin, tropomyosin, and troponin are synthesized by a fused alliance of cells and woven into force-producing filaments that span this giant multinucleated fiber. When the genetic message calls for red blood, the appropriate cells must suppress their own nuclei, attract iron, and design oxygen-bonding hemoglobin molecules. When the field requires neural fibers, large numbers of cells exaggerate the electrochemical properties of their membranes and extend them grotesquely in fibers. In the region of the retina some cells fold their plasma membranes and become rods. Where pigment is specified, a complement of cells specializes in the production of appropriate enzymes like melanin. Others must secrete large amounts of fluid, so their Golgi complexes swell and multiply while their endoplasm is roughened.

CELL MIGRATION IS PROBABLY FACILITATED by an organellar release of chemicals, including various polysaccharides. In a variety of tissues, as cells migrate, a molecule of repeating sugar residues—hyaluronic acid—is produced in voluminous quantities, attracting water and thereby swelling the extracellular matrix and encouraging cells to break free of it; hyaluronic acid is naturally degraded by hyaluronidase enzyme, ending the period of migration.

Migrating cells are propelled directionally by organelles. Lamellipodia formed from actin filaments bulge outward, generating tension in the cell cortex. As noted, cells in motion also preferentially relocate bits of the ingested plasma envelope secreted during their endocytic cycles. Actin filaments and microtubules transport recycled membranes to their leading edge, thus polarizing cell vagrancy in a uniform direction.

Migration is also mediated by adhesive, fibril-forming glycoproteins in the extracellular matrix that bind cells to macromolecules in the matrix. These proteins must be sticky enough that the migrating cells cling to their pathway along the matrix and follow it like blind travellers honoring a railing, but not so sticky that the cells get bogged down. For instance, telltale concentrations of fibronectin mark the trail followed by formative mesodermal cells during gastrulation of amphibians. This is

FIGURE 11N. Fusion of slime mold cells to form slug (drawing by Christie Lyons).
From Lynn Margulis and Dorion Sagan, *Origins of Sex: Three Billion Years of Genetic Recombination* (New Haven: Yale University Press, 1986).

strikingly similar to the process of chemotaxis whereby as many as one hundred thousand free-living amoeboid slime molds indigenous to the forest floor suddenly come together when their food supply is exhausted and (in a span of thirty hours) aggregate within sticky extracellular matrices and crawl away as a coherent, motile slug. This snail-like entity elicits fruiting bodies able to emit fresh spores.

Position

Any organism is the partial outcome of gene expression into form, but this cannot happen unless genes give cells means of locating themselves.

Large-scale tissue configurations and body plans are organized as global pantheons of signals inculcate cells with multidimensional grid information. Some of the identity tag remains in the cell memory as a "positional value," and it continues to affect how a cell responds to new environments. Usually a series of positional values builds up in a single cell from bits of geographical chemistry conferred chronologically. As a cell's sense of its purpose and spatial requirements becomes more fixed, it incorporates not only a location but a history, a path. While organizing its own autonomous material, it responds to signals from other cells.

Each cell may contain the same genome but, as parallel cells move into different states and explore alternative paths, every cell is (at least in principle) specified in a new way. This constant interplay within an amalgamated array leads both to great diversity and highly organized complexification.

The collective molecular states within cells and adhesive properties of their surfaces result in not only specialization and differentiation of tissue but adherence to an overall organismic pattern and cognizance (at some level) of their unique relationship to it. Gradients of intercellular landmarks and concentrations of proteins diffusing through tissues serve as morphogens, guiding fields of tissues into sequences and motifs. Positional information and molecular addresses likely originate from feedback loops. First, weak asymmetries are reinforced by positive feedback; then they become polarized into tissues and layers of tissues.

Feedback loops impose their body-plans on a region-by-region basis, with each region comprising perhaps only a hundred cell diameters (a millimeter). The micro-plan gets translated outward as the organism grows; coarse positional values are incised into organ rudiments. Burgeoning mass and intricate patterning within each region generate positional information at ever deeper hierarchical levels. The organism moves forward in time while coiling across space.

In fruit-flies the basic segmentation of the larva—the body plan—is apparently conferred by three gap genes; disruption of these kingpins leads to malformation of the overall insect. Next in line, eight pair-rule genes count and calibrate the segments provided by the gap genes; mutations in these usually lead to only half the genome's body units being expressed and other geographical anomalies. Mutations in the ten segment-polarity genes cause portions of a segment to duplicate themselves in sterile mirror images replacing the remainder of the unit.

Each set of genes thus provides underlying positional values that are organized by the succeeding set. Global positioning makes tiered positioning possible; products of pair-rule genes then lay down the templates on which organ crafting can begin. A historical lineage that made itself thick and complex reenacts its mechanical assemblage from nucleic memory, creating grids that allow subsequent grids to form.

In the blastula initial asymmetries include head and tail, back and belly, but these are not as fervidly differentiated as they will be after their deviations are augmented by episodes of molecular activity incrementally encouraging and deepening headness, tailness, etc. Furthermore, at progressive sublevels within each of these dichotomies (head and tail, back and belly, inside and outside) equivalent dichotomies continue to bifurcate and position each other, texturing the entire entity.

The specifications in the blastula and early gastrula are rough and vague; functional details of viscera are etched during organogenesis. Only after the organismal plan has emerged can (for example) vertebrate fore limbs differentiate from hind limbs; prior to that they are homologous. Likewise, the gut is one homogeneous lumen until separate organs are induced within it and proximal to it.

Feedback between some zones so enhances polarities that initially minor distinctions between groups of cells accrete and intensify their signals, taking off into different tissue states. Eventually there is no smooth gradient left between them, only a sharp distinction such as the one between cartilage and muscle.

As groups of cells bud from existing ones, newcomers may inherit partially isomorphic, partially drifting destinies from their predecessors. The new cells can main-

FIGURE 110. Part of a caterpillar with the skin dissected away to show the buds from which a butterfly's wings will arise.
From C. H. Waddington, *How Animals Develop* (London: George Allen and Unwin, 1935).

tain the continuity of the emerging pattern only by adopting intermediate positional values, i.e., by filling in the sizes between the largest and smallest or shapes between the straightest and most bent. By this "rule of intercalation," cells proliferating in a given region continue to churn out progeny until the entire gap between initial and final polar states has been filled in by a continuous gradient of cell types. This is a common organizing principle in molting segments and limbs of highly determined insects.

Pulsations and Electromagnetic Fields

IN A MORE MYSTERIOUS MANNER than bare positioning in grids, pulsations of viscera enlist membranes into global flutters, their spontaneous rhythmic cohesion reenacted in intervals from single tissues to a whole organism. Heart cells, in proximity with one another, synchronize a single rhythm to beat in unison. Even without touching, they incite one another's oscillations. When isolated, they cease fibrillation and die.

The cell membrane and extracellular matrix are apparently tuners and amplifiers of weak magnetic fields and, in concert with other morphogenetic factors, such fields can stimulate or suppress collective and regional tissue activity. Thus, grids and channels joining the movements of cells (like the meridians in traditional Chinese medicine) may be, in part, electromagnetically (and astrophysically) coordinated.

Scalar fields and their waves have also been postulated as powerful morphogenetic organizers. Scalar electromagnetics "is based on the notion of a vast, unseen background of scalar energies (as opposed to *vector* energies) which underlie all physical reality."[37] Nineteenth-century Scottish physicist James Clerk Maxwell originally postulated four equations of electromagnetism that are still current today. However, the form in which they have been inherited has been significantly abridged, the scalar domain of complex numbers excised from the four-dimensional (quaternion) system. This pruning was carried out by Oliver Heaviside soon after Maxwell derived the equations. His well-intentioned goal was to divorce a simple vector universe from its far more complex scalar dimensions, hence making it accessible to scientists. The strategy worked: the electronics industry, including telephone and radio technologies, grew out of Heaviside's simplification.

However, "when Heaviside threw out the scalar part of the quaternionic electromagnetic equation, he unknowingly threw out the possibility of *unifying gravitation with electromagnetism*—which has been a holy grail for scientists since Einstein himself wrestled with the problem." Maxwell's scalar mathematics "modelled the 'stress on the aether'—which leads to curving/warping space-time...."[38]

Scalar fields hypothetically can be produced by vector fields interfering with one another in a nonlinear medium; rhythmic variances in such fields generate unique forces—ripples in space-time. These scalar waves change the structure of pure space and/or mass in a tiny area (such as a field of cells or a cell), sending influences and signals across a space-time curvature. A physics of this order would enable cells to collaborate instantaneously with one another in the assemblage of organisms. Tissue would have to have a scalar as well as a mechanical basis, making its protoplasmic field more bioelectrically "alive." (Someday in the future an actual technology unifying mechanical vectors with scalars, and gravity with electromagnetism, might yield conversions back and forth between them, enabling the gravity, time, inertia, and apparent mass of an object to be altered.)

On the other hand, to suggest that organisms actually bend space as per general relativity smacks more of New Age myth and science fiction than terrestrial biology. It would require a whole *X-Files* universe operating at the level of organelles and dimers.

THERE ARE CONCEIVABLY OTHER FACTORS systematizing cell activity and movement: unknown subtle energies, paraphysical and telekinetic forces, microdoses and biphasic stimuli of enzymes, and psychic and telepathic signals at a cellular level. Organisms somehow capture and entrain each other's rhythms and attentions. Healers, shamans, and lovers are all pulsating sympathetically in some enhanced manner. If human beings can share feelings and emotional connections globally and mysteriously, the cells that comprise them probably do too. At a microscopic level, cell "meanings" likely enlist one another by modes of oscillation and attraction. Charisma and eros are cytogenic.

These varying pulsations resolve finally as layers, tubes, and compartments. Structure and location become function. Compartments balloon into pouches which provide milieus for active body cavities (oral, thoracic, and abdominal). Tubes become pumps which regulate a generalized motility of fluids. Coordinated cell positional value and pulsation harmonize and unify levels of structure which continue to expand and differentiate into more elaborate pulsating, finely stippled structures.

Layering

Cells retain the imprint of their original layer of organization. When migrants from different layers of tissue are combined artificially in a laboratory, those from the same layer find one another and adhere. Ectodermal cells automatically gather on the surface of the culture while endodermal cells retreat to the interior, but only if ectoderm is present; otherwise, they spread externally to mesoderm. Mesodermal cells meanwhile form a layer between them and begin organizing themselves around coelomic cavities. Mixed liver and retinal cells revert to some basic cell-to-cell recognition system and (like slime molds spreading beneath a redwood canopy) hurry to sort themselves out and reassociate according to their tissue of origin. Whatever disorder they find themselves in (and experimenters make their plights as exotic as possible), cells seek their counterparts and attempt a gastrulation. If true introversion is impossible they will still form a series of layers resembling the rudiments of organs.

The affinity among the cells of layers transcends even species. If the cells of a mouse and a chicken at the same stages of development are centrifuged together, they will fuse according to their origins in ectoderm, endoderm, and mesoderm rather than according to their immediate parent. Chicken and mouse disappear altogether, and there arise strange chimeric organs that look like parts of bones or dissociated tubules and abortive kidneys and hearts. In a breakdown of the natural order of species, cells return to the cosmogonic layers from which their collective ancestors long ago arose. The universal template overrides all conflicting derivations, thereby showing organs to be relatively superficial variations on primal themes.

Except for gap junctions, cells in actively developing tissue are not tightly linked. When removed from embryos and set in culture disks, they immediately show amoeboid independence, casting out microspikes and lamellipodia as they crawl around. It is possible that these embryonic cells may recognize one another by different receptor systems and gather by adhesion into loose tissue layers which later become stabilized by actual junctional contacts. Only then do they lose most of their atavistic character. For these vestigial protozoa, bound to one another by nooses of their own making, freeborn life is permanently curtailed.

A Variety of Perspectives on Morphogenesis

Induction

The movements of cells in general are induced by solid surfaces such as the extracellular matrix. In the early part of the twentieth century, Ross Harrison learned

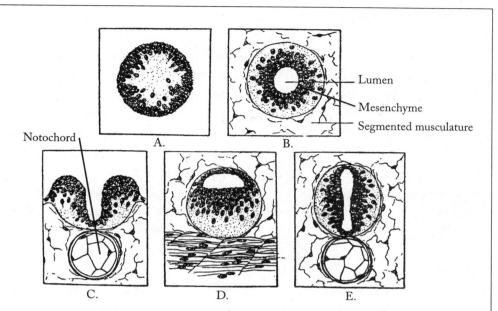

FIGURE 11P. Effect of extrinsic conditions on the morphology of the neural tube in amphibians. A. Solid neural mass, nuclei of gray matter clustered near the surface, white matter in the interior; B. Neural tube surrounded by mesenchyme, nuclei massed at inner (free) surface; C. Failure of tube to close, thinning of the floor of the tube in contact with the notochord, gray matter along the free surface; D. Neural tube underlain by musculature, lumen eccentric at far side, white matter at near side; E. Neural tube underlain by notochord, normal appearance, slit-shaped lumen oriented towards notochord.

From Paul Weiss, *Principles of Development: A Text in Experimental Embryology* (New York: Henry Holt & Company, 1939).

this while studying protoplasmic activity at the ends of nerve fibers grown on spiderweb frames. These fibers migrated along paths created by solids (even along the glass covers of laboratory dishes).

As noted, cells in morphogenesis use previously differentiated cells for their orientation; stimuli are thus transmitted from one embryonically emerging structure to another. In some organs, epithelial buds cannot form secreting structures without induction by surrounding mesoderm. The neural tube will not develop without the presence of a notochord underneath. Induction may be the *sole* function of a vestigial structure, so certain primordial tissues and structures must live and then die for appropriate development to occur around them. For instance, although a notochord is necessary at an early stage of human development, it must later degenerate for normal development. But if a piece of it is transplanted under the epithelium in another area, a neural tube will also develop there.

In an interpretation of embryology with phylogenetic significance, development is said to begin by inducing all other tissues out of primordial ectoderm and then specific organs from those tissues, often in the context of ectoderm. Uninduced ectoderm merely synthesizes mucus-secreting epithelial cells.

Life once began as immaculate ectoderm—even the vegetation of plants is a kind of skin—and then induced deeper structures from its epidermis. Protozoas are ectodermal creatures. The original orienting context in a paramecium is its cell cortex. When a whole cortex is scraped off, the cell dies, but if just a tiny bit is left, it is able to sprout again along the edge and restore itself. Thus, it is likely that organelle replication in mitosis was induced by cytoskeleton—the probable predecessor of all inductors.

Environment

There is a tendency to regard the world inside embryos (and inside organisms) as qualitatively different from the world outside them and thus to imagine that cells' inheritances are permanent and inalterable. Not only, however, is there a semi-permeable boundary everywhere between interior and exterior, but embryos must develop in accordance with their environment. Nutrients, weather, and gravity all play significant embryogenic as well as ecological roles. Gastrulating and organ-forming cells respond not just to other cells but to all contextualizing forces within their domain, from the mineral content of their habitat to the gravitational fields of external bodies.

At the end of the 1900s Gaston Bonnier conducted a series of experiments to determine the effects of vicinity and climate on botanical growth. He took single young specimens of dandelions and scabious, cut them each in half lengthwise, and re-rooted one cutting in an alpine garden (two thousand meters above sea level) and the other in a garden outside Paris at thirty-two meters.

As they developed, maturing halves of the plants hardly seemed to be of the same species, let alone from the same organism. They deviated from each other in characteristics right down to the individual cell structure. The lowland plants, with relatively sparse root mesh, unfurled luxuriant stems, leaves, and flowers; the alpine plants grew deep and tenacious roots crowned with tiny leaves and few flowers. The mountain scabious developed no side stems at all while its Parisian twin seemed almost boastful in its verdure and florescences. In each case, the same genes led to radically different somas.[39]

Botanists have found similar variations between plants grown in mixtures of dense, compacted sand and loam and those rooted in loose, composted garden soil.

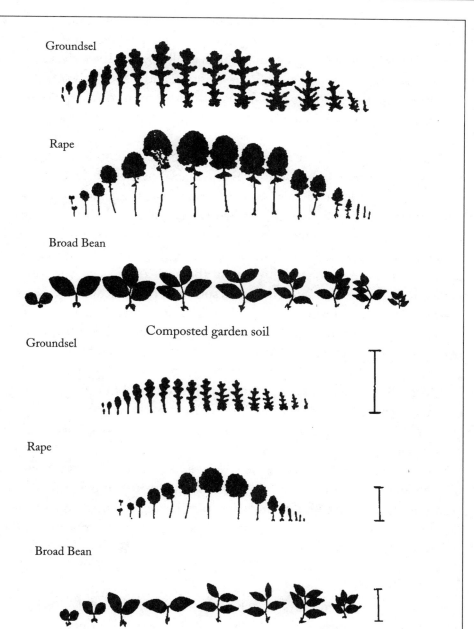

Groundsel

Rape

Broad Bean

Composted garden soil

Groundsel

Rape

Broad Bean

Sand-Loam soil

FIGURE 11Q. Each of six rows of leaves from the main stems of plants grown in two different kinds of soil.

From Craig Holdrege, *Genetics and the Manipulation of Life: The Forgotten Factor of Context* (Hudson, New York: Lindisfarne Press, 1996).

Right from germination garden specimens emit deep forceful roots and tower in bushy, robust vegetation; those in sand and loam send out single unforked roots capped by unbranched stems with isolated small flowers. Even the individual leaves and flowers in garden plants are more complex and elaborately lobed.

Other experiments have shown that the medium in which bacteria grow (type of sugar, for instance) affects key traits such as antibiotic resistance of individual colonies.

Environmental cellular influence is not limited to embryogenic organisms. In outer space, under conditions of weightlessness, astronauts lose skeleton—up to 6% of heel-bone density in just twelve days during the Apollo and Skylab missions. Equivalent decrease in density occurred in the radius and ulna. Conversely, soccer players develop larger femur bones, tennis players correspondingly expanded radii.

Just from standing up and walking during the first years of life, all hominids undergo narrowing and straightening of their calcaneous and talus bones. If these same animals were born into gravityless surroundings, it is unlikely that any amount of genetics could deform the original roundness of their bones.

The implication of these episodes is that the genes are only one aspect of the overall morphogenetic context of an organism. Not only do the cells and tissues emerging from the nucleotides create an immediate field for subsequent growth and form, but external and environmental factors which differ from organism to organism have effects which, though they may not be as critical as those of intrinsic biochemistry, help to determine size, texture, complexity, hue, and hardiness of organisms.

Yes, cellular mechanics impose absolute limitations on morphology; clearly a dandelion cannot turn into a rose bush or a hedgehog into a pig, no matter their surroundings, but the vitality and even the character and personality of a dandelion or hedgehog may be determined by its ecology—the morphogenetic "membrane" fields of the larger universe.

Symbiogenesis

Symbiogenesis, a term introduced by Konstantin Sergeivich Merezhkovsky at the turn of the nineteenth century, "refers to the appearance of new physiologies, tissues, organs, and even new species of organisms as a direct consequence of symbiosis (... different species living in close contact)."[40] Such morphologies arise when interactional cell induction occurs between species, for tissues of exogenous organisms can instigate cell movement and specialization even as endogenous tissues do.

In incidents recalling the endosymbiosis of the first cells, microbes dwelling in states of cyclical symbiosis with plants entice new botanical structures as well as augment existing ones; likewise, the hosts impel biochemical deviations in their microbial guests. Specialized morphologies arising cyanobacterially from less complex rudiments include leaf cavities in hornworts, coral-like roots in cycads, and stem glands in the various species comprising *Gunnera* (succulent creeping ground cover in New Zealand and giant herbs called "poor man's umbrella" in South America). In all cases the symbiont involved is one of the jellylike creatures christened nostoc by Paracelsus (from his belief that they were deposited by starlight).

Symbiont-induced histogenesis among cycads includes "small specialized root cortical cells that degenerate, leaving mucus-filled spaces that accommodate the cyanobacteria [plus] . . . other cells of the root cortex differentiat[ing] into elongated, fingerlike shapes thought to facilitate nutrient exchange."[41] In *Gunnera* new plasma membranes form around the visiting nostocs and develop into nonmotile photosynthetic cells and nutritive mucus-coated bracts in succulent and umbrella forms, respectively. Something more global than species integrity is at work.

Heterotrophic bacteria also engage in combinatorial induction with various leguminous angiosperms (beans, clovers, alfalfas), eliciting curled root hairs in the "shepherd's crook" structure underlying the plants. As the bacteria compel the cortical cells of the roots to divide, plant nodules and bacteria proliferate together, the latter producing idiosyncratic varieties of polysaccharides. Gradually, as the bacteria become nondividing, nitrogen-fixing cells, they are submersed in the plant's membrane where they contribute nutritionally and structurally; meanwhile the nodules of the legume synthesize a special leghemoglobin (an oxygen-binding protein) that buffers the so-called bacteroids from oxygen contamination and allows them to continue to metabolize. Plants and bacteria thus encode proteins and enzymes for each other's use. Information travels across creature demarcations.

Biochemical crosstalk with cyanobacteria is not limited to plants. On the threadlike filaments (hyphae) of green zygomycete fungi spreading along the banks of the River Salze in Germany, nostoc microbes induce blue-black ovoid bladders of lipidrich vacuoles through which nutrients are exchanged between symbionts.

Insects also partake in symbiogenesis. In tropical forests ant acacias and other trees and legumes develop symbio-organs called Müllerian bodies—red or yellow polyps in dense hairy mats at the base of their petioles. This tissue, opulent in oils, protein, and glycogen, is as beneficial to the trees as to the resident *Azteca* ants.

So much for the law of the jungle and a planet run solely by commandments of the fittest. Cooperation and communalization seem at least as fundamental laws of nature. DNA is fundamentally a trans-species system.

Although symbiogenesis does not require the transfer of nucleic-acid sequences over species boundaries, such evolutionary activity (known as DNA transduction) is not only possible but (over aeons) quite common. Many viruses "recombine frequently with their host cell genome and with one another and in this way can pick up small pieces of host chromosome at random and carry them to different cells or organisms. Moreover, integrated copies of viral DNA (proviruses) have become a normal part of the genome of most organisms."[42] These arbitrary chunks of genetic messages can take on surprising new meanings when integrated into fresh genomes; they encode distinctive proteins as if they had been present from the beginning, and they participate in morphologies quite alien to their source.

WHEN LYNN MARGULIS MAILED ME her 1998 co-authored article on symbiogenesis, I read it and then wrote her back, asking whether symbiogenesis involved exchange of actual genes between organisms or mutual induction of structures in each other, without chromosomal transfer, by bionts whose life cycles had become permanently interlodged.

She replied by hand: "As to your question, both are true—and more. Genes are transferred (and acquired) as genes (transduction, transferring principle DNA, bacterial conjugation, etc.); as genomes (including protein synthesis, i.e., bacterial symbionts that become organelles); and as cyclical bionts inducing morphogenetic change whether they are sperm/pollen tube/haustoria [specialized branches of hyphae by which fungi obtain nutrients from their hosts]/rhizobia [nitrogen-fixing bacteria] of legumes/dinomastigotes of coral/chorella in hydra/schistosomes [parasitic trematode worms] in snails/insects that induce galls, etc., etc."[43]

I could make out her notes except for the phrase "transferring principle DNA," which I had misinterpreted as "transferring principally DNA"; when that didn't make sense and I was unable to reach her, I decided to leave it off the list altogether. When she returned from a trip and saw both my faxed versions, she wrote at once to say that I had it wrong and should restore the phrase because "transferring principle DNA" was a seminal genetic mechanism of the cell. She went on to explain:

As demonstrated by bacteriologist Oswald Avery, heritable changes in bacteria can be brought about by exposing groups of them with different characteristics to each other's extracts. The key to the exchange "turned out to be a phosphoric acid macromolecular substance that [was] named 'transforming principle....' [Thus], the bacterial genetic system is global and has been for at least two billion years, and bacteria [are not] ... assignable to species that are comparable stable categories like those of eukaryotes."[44]

Margulis added that this notion was quite unpopular with "card-carrying microbiologists."[45]

If her view is correct, bacteria are more like syntactic units or interchangeable codes than regular organisms. Like ancestral spirochetes, they exchange DNA promiscuously and participate in chemotactic networks with each other, inducing new morphologies. Even rogue protein particles lacking nucleic acids (prions, implicated in the neural degeneration of scrapie in sheep and goats, "mad cow disease," and "laughing sickness" among headhunters of New Guinea) may transmit molecular and scalar information between organisms by inducing proteins in neurons to mimic their structure and multiply through the "alien" brain and nervous system, altering generations of behavior with their itches and euphorias. Perhaps parasites likewise steal code along with nutrition, *becoming* as well as collaborating with dormant aspects of their "victims."

WHEN WE LOOK BEYOND the ordinary organism-based meanings of morphogenesis, we see that the frameworks for the synthesis of protein structures, cell specialization, and tissue induction are far broader than milieus within individual embryos. Vectors of morphogenesis can incorporate tissues and chemistry of other organisms; they can respond to and coopt foreign DNA.

Symbiotic and chemotactic partnerships may well have been causative agents in some of the planet's most momentous evolutionary events.

Shear Force and other Geometries

Theories of dynamic tension conducting embryogeny are as old as the Chinese primordial yin and yang generating, by their polar interactions, all things among the stars and on planets—likewise the four elements of Western cosmology. By the same reasoning, some Greek and Mediaeval naturalist philosophers thought that the three germ layers of the body might represent primary principles of form and motion.

The fundamental tripartite division of the embryo was recognized as early as the second century when Graeco-Roman physician Galen described a triad of "organs" emerging out of an unformed tissue mass. Though he did not specify ectoderm, endoderm, and mesoderm, he intuited their rough innateness.

To early Platonic embryologists the triunity of tissues suggested that mesoderm comes into being through a dialectic of ectoderm and endoderm. Others considered mesoderm a ground state, preexisting and undifferentiated as extraembryonic mesenchyme, and generating ectoderm and endoderm by polarity with the embryonic disc itself.

Projective geometry and atomism provided a mathematical, elemental model for immanent speculation. Democritus proposed that the space-restricting womb imposed a shape on the embryo directing its unfolding. His etiology was revived in the early seventeenth century, as Pierre Gassendi and René Descartes each attempted "to explain embryology as mechanical processes of movements and pressures in a mathematical-geometric theory against the background of an atomistic, iatromathematical approach."[46] At the same time, Marci of Kronland explained "the manifold and complex development of form from seed to organism by means of the analogy of lenses which produce complicated rays and patterns of light from a simple light source. The formative power of light is imagined as radiating from the geometrical center of the foetal body, creating the differentiated complexity of form without losing any of its own power."[47]

A more sophisticated version of archetypal morphology was developed by Lawrence Edwards during the 1970s; he tried to demonstrate that "projective geometrical transformations between vortex forms and averaged forms of the human uterus create two-dimensional projections of successive forms of early human embryonic development. The same set of transformations provides a fully three-dimensional model of the general gesture which the embryo makes in forming the neural canal across the neural plate immediately after the disappearance of the primitive streak."[48]

DURING THE EARLY 1980s George Oster, an entomologist at the University of California in Berkeley, offered a purely mechanical explanation for morphogenesis. Combining elements of nineteenth-century mechanics with computer technology, he derived fertilization, gastrulation, and other embryogenic motions from abstract models of cell interaction within the high-viscosity cytoplasm. Tension, compression, and shear force, transmitted incrementally from cell to cell in the form of waves and globally coordinated over an embryo, generated structures in computer projections similar to those impelled by fertilization and gastrulation in actual embryos. This process begins when the sperm functions as a winch polarizing the egg's contents back through the cell, the acrosome acting as a moving hydrodynamic boundary encountering drag, initially at the extracellular coats of the egg.[49]

In the thick, turgid environment of cytoplasm, motion follows geometrically determined paths, stopping only for natural boundaries or from the redistribution of forces. The laws of low Reynolds numbers apply, leading to a brand of Archimedean physics in which floating bodies stop instantly when impetus is suspended. The archenteron, mouth, and teeth (in their downward undulations) and the rudiments of the liver, pharyngeal pouches, and pancreas (in their folding upward) are all generated by the "same" computer wave constricted at different points and in different directions.

Similar forces may operate on a subcellular level too, for instance, the invagination of the inner envelopes of plastids to form vesicles in plant cells and the infiltration of the nucleoplasm by the mitotic spindle.

By such a principle, waves of tension (not genes *per se*) set herds of antelopes galloping and fish squiggling. We can imagine all biological motion as the inertial bumps and taps of ancient creatures echoing and attenuated across time and space—a hologram in which touch alone imparts kinesis, and invagination is propagated by a compression wave across timeless skeins of high-viscosity fluid. In a kind of originless "Rube Goldberg" machine, the motions of extinct worms are captured in the pulsations of gulls over the city, their hearts beating, as blood islands disperse and hemoglobin contacts their cells.

This etiology requires tissue that is simultaneously dynamic and chemical, fluid and hard, agitatable yet cohesive. Influences must continue to radiate from level to level while becoming organized according to precedent and plan. Resonating forces cannot produce bedlam, and they cannot stop dead in their tracks. As we shall see in Chapter 24 ("Healing"), the osteopathic medicines of compression and shear force transmit palpations through the surface of the body deep into viscera and perhaps even into cells, effecting global psychosomatic changes by redistribution of forces.

Phyllotaxis

Phyllotaxis (the arrangement of leaves on a plant stem) is controlled in irregular periodicity by combinations of parallel and perpendicular forces along the outer margin of the stem. The direction and structural proportioning of these budding and flowering ratios is not fixed by the genes of any one species but follows from the environment of plant tissues it forms. The bias of a twist is determined by the same class of stimuli as cause water to ripple distinctively or flow in one direction. A plant is a semi-fluid river sluiced in a simple, often spiral geometry.

In the region of rapid mitosis at the tip of plant roots and branches (the meristem), epidermal cells manufacture cellulose microfibrils whose walls lie parallel to one another, molding a pressure-resisting elastic cap against the swelling tissue beneath. The cap stretches much as any fabric realigns in response to vectors of stress. As clusters of cells continue to segment and bifurcate, newly sprouting leaves initiate lines of tension on epidermal cells opposite them, and cellulose is deposited to support the meristem. Perpendicular to the reinforced cells, "the tissue will buckle laterally and fold outward under the pressure, producing a leaf primordium on the opposite side to the closest growing leaf."[50] Alternate schemes are thereby available, stored in almost quantal fashion within each plant's uncertainty state.

The behavior of cellulose microfibrils produces the whorled and distichous patterns seen on plants in nature. Furthermore, "plants with spiral phyllotaxis tend to locate successive leaves at an angle that divides the circle of the meristem in the proportions of the Golden Section"[51] (the same ratio on which ancient Greek architects designed temples). Sunflowers and pineapples are natural Fibonacci gargoyles and dowsing nodes.

Tensegrity

Genes probably do not invent form. Their reign of structure and superstructure does not begin until a repertoire of motifs has been furnished by purely dynamic, inanimate events in nature. Later even quite small physicomechanical shifts rearranging cells and extracellular matrices can result in transformations of both the size and shape of a tissue domain, with lasting morphogenetic consequences.

This is how complex organisms are built and regulated—not solely by cells but by cells under forces. Most of the major embryogenic processes have initial physicochemical components. Newtonian equations underlie life, providing the rudiments of gastrulation, epiboly, invagination, delamination, segmentation, involution, neural-crest migration, and assorted episodes in which populations of cells invade free territory.[52] Such activity is as dynamic as it is genetic.

In fact, if it were not for genetic control developed through aeons of evolutionary attunement, mechanical processes and gravity-mass relationships would run riots of useless shapes through protocellular material (as they do throughout nature). The gaudy diversity of morphologies arising from inanimate landscapes must be subtilized, sculpted, and metabolized by genes. As large shapes and currents press in (from natural forces), fine detail and integration push out (from chromosomal regulation and consequent organelle tectonics).

DONALD E. INGBER, in a paper on morphogenesis and biological structure in the January 1998 *Scientific American*, summarized his own recent work and that of Ning Wang of the Harvard School of Public Health, along with prior experiments conducted at Yale University, Michigan State University, Johns Hopkins University, and the Friedrich Miescher Institute in Basel, Switzerland. Ingber's conclusion was that cell shape and subcellular and extracellular mechanical forces have significant determinative influences on the activities and expressions of genes.

Ingber first reviews the obvious: Contrary to the assertions of genetic determinists, DNA does not contain the complete plan for organisms; it only produces messenger RNA, which, in turn, programs the synthesis of single proteins in sequence. The fabrics created by these proteins then tell the genes what proteins and enzymes to synthesize next. Morphology is subject to complex emerging influences which

Molecular structure of microfilament.
Each sphere represents an actin monomer.

Microfilaments (bundle)

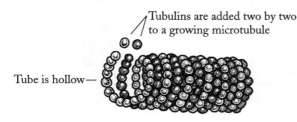

Tubulins are added two by two
to a growing microtubule

Tube is hollow—

Microtubule (assembly showing tubulin proteins)

End shows seven-fold weave

Intermediate filaments in section
These connect microtubules and contractible microfilaments
to surface membrane of cell and nucleus of cell.

Key to photographs:
1. Actin (a section of microfilament)
2. Actin (a section of microfilament)
3. Microtubules
4. Microtubules
5. Entire cytoskeleton

FIGURE 11R. Microfilaments, microtubules, intermediate filaments, and cytoskeleton. First four photographs by Donald E. Ingber; last photograph by Kate Nobes and Alan Hall. Illustrations by Harry S. Robins.

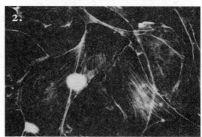

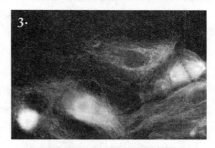

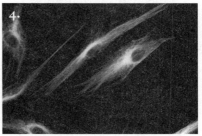

not only cannot be predicted from the genetic code but translate similar ground states into radically different architectures. This is not to deny that genetic information is at the core of organismic development, but it is to switch the emphasis from an exclusive priority of genes to a series of uncertainty states and feedback circuits that, though requiring genes in order to emerge, come to contain information that is nowhere in the chromosomes. This suggests a biological universe in which complex form is generated out of confederations of organelles and cytoplasm at one level and tissues and organs at another—not only generated spontaneously but maintained through progeny such that protein structures as well as genes bear and transmit consistent, inheritable morphologies.

Ingber underscores the dynamic arrangements of microfilaments and microtubules in differentiating cells. He proposes that these organelles are regulated and stabilized in geometries similar to the tensegrity-maintained structures of architects and artists (R. Buckminster Fuller's geodesic domes and Kenneth Snelson's sculptures among them).

Tensegrity is a series of configurations whereby cells not only counterbalance themselves mechanically but reorganize and transmit "instructions" from one to another, causing epithelia to form, spread, and likewise reorient. In Ingber's words, it is a method of structure and equilibration through which "tension is continuously transmitted across all structural members" such that "... an increase in tension in one of the members results in increased tension in members throughout the structure—even ones on the opposite side.... [T]he structure stabilizes itself through a mechanism that Fuller described as continuous tension and local compression.... Tensional forces naturally transmit themselves over the shortest distance between two points, so the members of a tensegrity structure are precisely positioned to best withstand stress. For this reason, tensegrity structures offer a maximum amount of strength for a given amount of building material."[53] This is true even if the building material is carbon-based as well as alive.

In mature vertebrates, bones (serving as compression struts) and tension-bearing muscles, tendons, and ligaments (counterbalancing the bones) make up a musculoskeletal tensegrity. Likewise, the extracellular matrix together with microfilaments, intermediate filaments, and microtubules compose tension-compression structures among cells. While microfilaments exert tension throughout each cell and pull its components and membranes toward the nucleus, microtubules and bundles of other microfilaments within the cytoskeleton function as compressive girders; the extracellular matrix is a strut.

Anchored through their cell's membrane to the matrix, contractile microfilaments shorten and increase as isometric tensions vary within their amalgams. Meanwhile, intermediate filaments link "microtubules and contractile microfilaments to one another as well as to the surface membrane and the cell's nucleus. In addition, they act as guy wires, stiffening the central nucleus and securing it in place."[54] Elsewhere, microfilaments unanchored to the matrix pull and are tugged back against one another.

With microfilaments spreading and establishing themselves (or crumbling) under the influences of stress patterns and changes in calcium, potassium, or other cytoplasmic molecules, cells shift shape and position, sometimes extending filopodia and lammelipodia which grip their counterparts on adjoining cells and adhere tightly (as in the movement of mesenchyme). When cells become flattened out and stretched, that context agitates increased cell division and tells new cells to thicken the layer, i.e., to heal a wound. If cells swell up and crowd together, selected ones among them die to forestall malignancy. When microtubules redistribute stress patterns from a cell's poles to its spindle at its centromeres, the chromosomes are pulled into alignment and then migrate apart, turning two cells into one—a procedure fundamental to all morphogenesis.

Ingber and Wang found that by increasing and reducing stress on the molecules that penetrate the cellular membrane and bind the extracellular matrix to the cytoskeleton inside the cell, they could cause the cells to stiffen or become more flexible as more or fewer microtubules and/or microfilaments came to orient along the axis of the tension—an activity crucial to the formation of tissues such as tendons and muscles. These tissues then induce visceral and neural structures around themselves.

As forces are conducted along discrete molecular paths, overall cellular sculptures flow through and among internal and external environments of each cell and across consortia of cells. Cell colonies follow and spread along their mutual surfaces as contractile microtubules align in bundles and tug between adhesion sites. Systems of heterogeneous and enveloping texture and turgor imbed in one another. Layers of tissue achieve complexity and functional integration through continuous interpenetration and coalescence within those miniaturized realms of surface tension, density, and gravity.

THE CHANGING GEOMETRIES and mechanical restructurings of cells and tissues inform genes as to where they are, which ones among them are next to be activated, and what materials they are to produce from their repertoire. The behavior of microscopic straws, strands, and bundles sets the uncertainty states of morphogenesis. Cytoskeletal geometry and mechanics influence and are influenced by the biochemistry and emerging infrastructure of the cell. Structure changes biochemistry which alters structure. Chemistry generates tensegrities which program chemicals which realign space. From uniform stasis in the egg, chemistry and structure unfold reciprocally, their interactions manifesting in multidimensional patterns. Identical dynamics (as noted) underlie flowers, worms, limbs, roots, jellyfish, eyes, and brains.

Who is to say such resonant vibrations do not originate on atomic and subatomic levels and work their way through molecules into carbon edifices? Proteins providing structure and chemistry for cells might themselves be arranged in tensegrities orchestrating strings of amino acids, folded-up zones of backbone, and equilibria between the pulls of hydrogen bonds and the resistance of coiled protein to shortening. Cascades of molecular restructurings would radiate outward into cell matrices and then tissue. As symmetries are broken, novel themes emerge and rearrange their liberated elements. Dimensionality is colonized; energy and mass are distributed and conserved. Morphogenesis becomes ontogenesis; ontogenesis becomes culture.

From another viewpoint electromagnetic and scalar-wave vibrations may flow from the external world to the intracellular world, pulling the universe taut in a single pulse, a grand exponential avenue of helices. Energies and shapes originating in helices of galaxies and planetary systems thereby stream ceaselessly across space, down through world atmospheres and tissues, through collagenous helices of extracellular matrices into cells themselves, all the way into the DNA helices of their nuclei.

As the poet Michael McClure reminds us:

"All life is a single unitary surge, a single giant organism—even a single spectacular protein molecule. In the four billion years life has grown on this planet it is not possible to imagine (in view of its whirlwind energy and delicate complexity) that there have not developed interacting fields, forces, auras, within the behemoth topology of it.... A ribosome in a liver cell in a salmon might relate to a field of energies or a 'position' within a quasar or in a distant sun...."[55]

Perhaps gravity fields anchored by black holes transmit morphogenetic-like forces through interstellar space along mammoth transgalactic fibers. Planetary systems strung on the fulcra of their orbital rings attach (locally) to sun nuclei and globally to the extraplanetary hydrogen matrix. By this grand astrology operating through invisible zodiacs nested in one another at harmonic scales, proteins and stars mutually induce each other (and everything else in between and above and below).

"The Universe IS indeed an aura of trillionically multiplex interrelations ... primarily comprised of natures of matter that we do not, certainly not consciously, contact."[56]

12

Biological Fields

Is there such a thing as a gene?

THE CHARACTERISTICS OF CELLS supposedly originate in the sequencings of
their nucleotides, transmitted to the cytoplasm by RNA molecules and converted there into amino acids. Yet chromosomes do not tell us how or why macromolecules express themselves in structure.

One biologist compares morphogenesis to the scene that hypothetical observers
from another dimension might see as they look down on a city and try to figure out
the pattern of people-flow. In the morning, there is heavy traffic. In the evening,
the traffic returns but its units travel in opposite directions. For five days, there is
one general stream; for two days a different one. Then the five-day motif returns.
There are also unexpected, irregularly spaced intervals in the five-day sequence:
Christmas, New Year's, July 4th. Some of these changes are longstanding. Others,
like Martin Luther King Day and the transposition of two-day cycles to three- and
four-day cycles, are relatively new. The observers remove people and cars from the
pattern at random and dissect them; this reveals nothing at all. Finally they notice

FIGURE 12A. Diploid sets of human chromosomes, arranged in pairs according to size
and shape. At the end of the male series is an uneven pair (XY).
From Emil Witschi, *Development of Vertebrates* (Philadelphia: W. B. Saunders & Company, 1956).

the traffic lights. When these break down, things go to hell. So, theories are built around the sequences of lights. In reality, however, until they discover human communications and the protocols of social life, they'll never really get it.[1]

The traffic lights of this extended metaphor are, of course, genes. We overendow them because they are the most consistent and dramatic markers. Yet our knowledge of their domains and blueprints is derived almost solely from breeding experiments tracking singularities—singularities like shape, color, size, and immunity which coalesce in the life of an organism and exist in isolation only semantically.

RESEARCHERS DURING THE EARLY DECADES of the twentieth century identified distinct hereditary traits; these were presumably implanted once upon a time by biophysics, natural selection, and mutations. Breeding mutant flies with normal ones, scientists were able to discern the existence of two thousand different such traits, hence ostensibly two thousand different genes. They also deduced that linked traits (traits inherited as a set) represented pairs of homologous chromosomes—for example, blackness, reduced body hair, and dwarf wings in one group of fruit flies; sepia eye color, curled wings, and stubbled body hair in another; they based their early gene/chromosome identifications on these concrete associations. However, the same traits—i.e., black body and reduced hair—also become separated from each other in anywhere from two to ten percent of all flies. Instead of abandoning their hypothesis, though, the scientists attributed the dislocation of traits to the mechanics of crossing-over during meiosis: i.e., so-called nearby traits *more often* remain linked in offspring. This premise was the basis of the first chromosome maps of the 1930s—and the communiqué of the existence of genes.

But are genes real things? The inventor of the term, Wilhelm Ludwig Johannsen, wrote in 1909:

"By no means have we the right to define the gene as a morphological structure in the sense of Darwin's gemmules or biophores or determinants or speculative morphological concepts of that kind. Nor have we any right to conceive that each special gene (or a special kind of genes) corresponds to a particular phenotypic unit-character or (as morphologists like to say) a 'trait' of the developed organism.... The word *gene* is completely free of any hypothesis; it expresses only the evident fact that, in any case, many characteristics of the organism are specified in the germ cells by means of special conditions, foundations, and determiners which are present in unique, separate, and thereby independent ways...."[2]

These characteristics are not of course specified in jukebox or rebus form—a trait per unit. "[A] gene constitutes nothing more than the cell's replicable record of the primary sequence of an RNA molecule, or, indirectly, a protein."[3] It is a strip of macromolecular terrain masquerading (in our culture) as a cybernetic quantifier. Its informational content is redundant, promiscuous, nonlinear, nonsemantic.

At the wellspring of biological form lie archaic genelike algebraic bits. Heredity is a series of alphabets locked inside one another. There is no consistent index or causative chain, merely a syntax of potentiation. And our dogmatically optimistic rendition of genetic language looks suspiciously like something a biotechnician might have written to assemble bionts from protoplasm if he were not already written *in it,* i.e., if someone did not clearly have a prior patent (this, remember, is the plot of the great cosmic conspiracy theories, *X-files,* and whodunits of the late twentieth century).

A code lies between us and life, another code between us and every thought (including every theory we expound to explain both life and the codes). The scenario is far too suspicious to be an extraterrestrial ruse. It is more likely that we unveil what we already know, and we know it because it expresses itself in us at every emergent layer along the way.

In 1999 "gene" remains a statistical bluff rather than a thing in reality. The resemblance of nucleotides, codons, and amino acids to alphabets needing a Rosetta Stone may have more to do with the fact that *our* languages emerge primevally from phonemes and runes than that genetic reality is itself runic and alphabetical. Yes, there *is* a code-like aspect to the molecular transitions between levels of intracellular phenomena, but the phantom source code may equally be an archetypal cipher of nature, beyond algebra as we know it, or a mirage of mind in the mirror of nature.

And this is but one of many dead ends to which the search for the abode of the genes leads us.

GENES "EXIST" FROM THE ASSUMPTION that something must lie behind, for instance, a violet-flowered generation of hybrid peas with a white-flowered parent. The unwitting discoverer of genetics, Austrian biologist Gregor Mendel did not surmise his trait-sources were concrete things; he called them *Anlagen* (factors or mathematical elements) of heredity. He extrapolated them solely from the statistical distribution of anatomical properties. Once concretized into genes by a later generation, they became the posited material causes of traits; yet neither experiments nor microscopic observations reveal their precise biochemical make-up. In fact, we speculate their existence predominantly by their expressions — or more precisely we presume that there are elementary particles of heredity because damaging or disrupting chromosomes alters development in a statistically consistent fashion, or completely prevents it, and because clipping fragments of chromosomes from one organism and splicing them into another transfers traits from the first biont to the second.

Even the physically demonstrable chromosomes are not hard traits—or things. A chromosome is not the spininess of a horse-chestnut shell or the red of a rose. It is a transient form in the nucleus of a cell that lives its life and perishes in its own time, quite separate from the duration of phenomena to which it supposedly gives rise. In assigning substantial forms to genes, we are making an intellectual reduction from a field of appearances. We are imagining that the way living forms emerge and develop bestows the same materiality on the codes and elements at their origination.

How do genes express themselves?

WE LEARN BY EXPERIMENTAL DEDUCTION that a specific genetic allele is responsible for blue eyes in a cat. But this does not mean that the gene *colors* the eyes blue. It has no farseeing sentinels to direct how creatures are formed. Its only link with the world outside the nucleus is its transcription of itself onto messenger RNA. Neither nucleic nor amino acids colors the eyes. By the time that a substance reflecting blue light has located in the iris the singular expression of the gene has been "lost" in the complexity of the system. The genes underlying neural and optic aspects, like the genes for components of the blood and the cellular integument of the heart, are scattered dormantly through many other organs too.

"Failure to realize ... that a single factor may have several effects, and that a single character may depend on many factors, has led to much confusion between factors and characters...."[4]

Even though biologists have identified instances in which a mutation in one gene can produce identical results in different bionts (for instance, double paws on different species of raccoons or phosphorescence in fireflies and plants), genetic space—as a whole—does not translate in any simple or consistent fashion into organismic space. Understanding genes no more discloses the complexities of organisms than fathoming molecules yields the meanings and properties of compounds, or reading traffic lights explains cars. Genes are neither unicausal nor representational, nor are they (as the current metaphor proclaims) selfish.

DNA HAS A SINGLE UNIQUE ABILITY—to replicate and partition itself accurately into daughter cells. This anchors embryogenesis; it does not invent it, for no amount of self-copying will turn a part into a more complex whole. Left to itself, DNA can at best simplify and degrade.

It is the cellular and global context of DNA that seizes the genetic message and transposes it into a system of integrated meanings and functions. Biological structure in its totality may routinely be credited to genes but, without cells, genes do nothing. They can't send a message; they can't program structure; they can't organize life forms; they can't even duplicate themselves. If we indulge the deceptively functional metaphor of the cell as a factory, the genes need that factory in order to be genes. Alone, a gene is a blank statuary, the architect of nothing.

In the words of a contemporary geneticist:

"The world's most boring book will be the complete sequence of the human genome: three-thousand-million letters long, with no discernible plot, thousands of repeats of the same sentence, page after page of meaningless rambling, and an occasional nugget of sense—usually signifying nothing in particular.... The gene sequencers are pursuing the ultimate reductionistic program: to understand the message, we just have to put all the letters in order. There is an opposing view which suggests that, having sequenced the genome we may be in the position of a nonmusician faced with the score of Wagner's Ring cycle: information, apparently making no sense at all, but in fact containing an amazing tale—if only we knew what it meant."[5]

Molecules, proteins, and cells by themselves are also vacuous pebbles. Outside of the exquisite orderings of tissue themes they mean nothing. Eye cells do not by themselves see. Brain cells do not think. They live typical protozoan lives—metabolize, manufacture proteins, divide or die. The organs they compose are neither self-sufficient modules nor operable segments of a machine.

TO THE DEGREE THAT CODONS provide the elements of form and function, they do so nonlinearly and nonreductively. Collectively, genes are like musicians in an orchestra. Each of them plays a single instrument. Some play once, some twice, some many times; some play on and off, some continuously. It is the harmonized array of all of them that defines the "meaning" of a particular clarinet or piano note at any given time, a meaning which changes as the symphony progresses and the fly and the albatross are composed.

One gene is like a flashlight beam, which is most discretely identifiable closest to its source and disperses from there over an increasingly wide area. As new genes are activated, areas of tissue executing the biochemistry of previously synthesized proteins are also changing their relative positions, and groups of cells once remote from one another are being brought into proximity, as others separate. The beams crisscross at many different levels of dispersal, their expressions altering one another and surpassing any original orthodoxy.

If the cross-eyedness, eye color, and shades of fur pigmentation of the cat are linked in a single gene (as researchers have indicated), this would be a salient example of the capacity of multidimensional expression that exists in all genes. One field may read a gene chromatically, another in terms of gross structure, another in terms of function or behavior. The cells generated and conducted by genes are also multipotent and can follow a variety of different trajectories depending upon the situation in which they find themselves. As noted in previous chapters, they also have a degree of flexibility and freedom, being able to differentiate one way, stop suddenly, and then differentiate in a different direction as the morphogenetic field changes.

We have a classical paradox: the cells are genetically under the control of their nuclei (and have no other source for independent information), but the nuclei are unmasked by substances in the cytoplasm (which could not become regionalized without the nuclei). Apparently, differentiation is determined by both the genes and the cytoplasm in such a manner that each requires the other to express itself in an overall design. Form literally brings its own shaping context into being. How genes can create fields which alter the later expressions of the same genes is a mystery. The outcome, before it even occurs, would seem to be influencing the source. Is such deep feedback possible? Is genetic density so redundant as to defy simple chronology? Or is it that complexity intrinsically seeks embodiment, so overrides all impediments to its becoming?

THE MUSICIANS IN the human half of the metaphor can at least hear and appreciate their own playing, but the "genetic" musicians are deaf, numb, and blind and do not even know what music is.

Despite the fact that morphogenesis is defined by regimented precedent, neither genes nor cells have scores. Their so-called programmed molecules meet one another anew each time. There is no sentience here, only ocean brine caught in eddies, compiling neuron cups that somehow see:

" . . . the whole structure, with its prescience and all its efficiency, is produced by and out of specks of granular slime arranging themselves as of their own accord in sheets and layers, and acting seemingly on an agreed plan . . . two eyeballs built and finished to one standard so that the mind can read their two pictures together as one. . . . That done, and their organ complete, they abide by what they have accomplished. They lapse into relative quietude and change no more."[6]

The Search for the Basis of Organization

Organizers

Eighteenth-century biologists once assumed that somatic qualities arose from the elemental constituents of tissues. But this is mere redescription, like saying a rose is a rose because it exudes redness and sweetness. A century and a half later, geneticists shifted the developmental bias from abstract organic determinism to individual cellular potentials. However, when severed blastomeres still gestated into whole organisms and dorsal skin cells became brain tissue after being transplanted from one embryonic frog to the neural plate of another, hopes for a linear geometry of development were disappointed. The determinant was located neither in the body parts (as eighteenth-century mechanists had thought) nor in the blueprint from which they were assembled (as late nineteenth-century geneticists had proposed to demonstrate).

Embryological experiments since have charted the role of prior tissue configurations in the formation of subsequent ones. As early as 1921, Hans Spemann grafted small sections of epidermis from a newly formed amphibian gastrula into the neural region of one more advanced and found that they differentiated according to their surroundings—as neural plate.

Trying various combinations of tissue in an ultimately unsuccessful attempt to pinpoint causal factors, Spemann discovered that a piece of potential belly skin taken from above the blastopore of one salamander became nerve tissue when transplanted in the area below the blastopore of another salamander that would have become nerve tissue. The same process worked in reverse: nerve tissue transplanted into the belly area became skin. However, notochordal mesoderm grafted to the ventral region of another embryo formed a whole second embryo.

The regulative ability of cells to change their commitment if their position in the field changes (known as prospective potency) gradually narrows as the organism develops. Neural ectoderm will not behave as epidermal ectoderm if transplanted at the end of gastrulation; it will sink from the surface and develop a vesicle with thickened walls—a functionless brain and spinal cord adjacent to the "true" integrated one. Competence is a pliancy of early embryogenesis. As the body matures, its underlying field rigidifies and becomes more circumscribed and fated.

In 1924, Hilde Mangold, a student of Spemann's, transplanted the dorsal lip of a blastopore of a young newt gastrula onto the ventral surface of an equivalent newt gastrula of a different species. The graft invaginated and ultimately developed a whole second set of organs (notochord, ear rudiments, kidney tubules, gut lumen, etc., missing only the anterior section of the head).

The dorsal lip when transplanted to a second embryo had induced this embryo to hatch yet another embryo around the graft. The result was a very strange being indeed: twins, joined at the belly facing each other, each with its own notochord around which the rest of it was organized. The only contributed material from the transplant was the notochord of one of the twins. Although development proceeded

FIGURE 12B.
Multiple embryo
formation resulting
in triplet trout.
From Emil Witschi,
*Development of
Vertebrates* (Philadelphia:
W. B. Saunders &
Company, 1956).

FIGURE 12C. Young
victimized toad
(Bombinator) on
which an additional
limb has been grafted
in the head region.
From J. Graham Kerr,
*Textbook of Embryology,
Volume II, Vertebrata*
(London: Macmillan
and Company, 1919).

no further, the central role of the dorsal lip of the blastopore was conclusively demonstrated. Because it could induce a virtually complete second organism when transplanted, Spemann called it "the primary organizer." Its forerunner was identified as the gray crescent in the oocyte of the newts of the earlier experiments (see Chapter 9); later in development this becomes the notochord. The Precambrian precursor of the crescent may have been some sort of contamination or infection within a primordial membrane to which other molecules responded and in which (aeons later) organelles congregated.

Biologists had great initial confidence in the chemistry of the organizer as the final solution to the developmental riddle, but by 1932, experiments by C. H. Waddington and others revealed that even a dead organizer, one that had been boiled or treated with alcohol, induced organ rudiments in chicks and newts. In addition, when the dorsal lip was replaced with certain other tissues, induction still occurred and the same structures formed. Subsequent experiments showed that reptilian, insect, and human cells (including human cancer cells) induced whole organs in newts. In many unlikely species guinea-pig bone marrow is an excellent inductor of mesodermal organs and spinal cord. The liver of a guinea pig, on the other hand, instigates brain vesicles and eyes in an equal variety of creatures. Kidney tissue from an adder will induce newt hindbrain followed by ear vesicles.

The variety of potential inducing substances and their inconsistent relationships frustrated early promises that simple proteins guide tissue along historical paths, so experimenters reasoned anew. If inducers do not create tissue patterns, then they might trigger preprogrammed sequences of cell activity. Something like this indeed appears to happen. Regions of tissue arise in juxtaposition to each other, providing framework and timing for development. Thus, stand-ins can take the role of native tissues.

Ross Harrison concluded: "The organizer, itself a complex system with different regional capabilities, merely activates or releases certain possible qualities which the material acted upon already possesses. The orderly arrangement which results depends not only upon the topography of the organizer but also upon that of the system with which it reacts."[7]

As noted in the previous chapter, the system seems to have evolved in such a way that contextualizing tissues create fields to which they as well as their neighbors respond: "... the emphasis upon 'determiner' and 'determined' leads to a very lopsided and often erroneous view of the process, for it is questionable whether one factor can influence another without itself being changed."[8]

Induction is not solely a mechanical or chemical event. It is a property of relationship among fields of proteins and the changing potentials of nuclei and cells as embryogenesis proceeds. An embryo "creates itself" by moving from one unified state to another in developmental order, taking even experimentally introduced tissue into itself and redetermining it (activating its nuclei). Organisms "grow internally and are made up of many different components with different rates of increase."[9] Embryogenesis is not static, linear, or conventionally chronological. Prior form and ongoing dynamics interact, producing mobile boundaries in space and time.

There can be no embryogenic activity without structure and no structure without embryogenic activity.

The Mystery of Context

In the cat, every cell has at its heart the same die, the same effigy; now some are fur, some twitch as whiskers, others track in eyes—most sustain an electro-elastic, metabolic wetsuit inside an epidermal wrap, a grumbling churn that gives a creature life and purpose. The same is true of the cells in Humphrey Bogart and Ingrid Bergman, their icons preserved on illuminated celluloid. Yes, they have faces and bodies, but these congealed only after a central cell matrix bevelled out multiple interlocking fields.

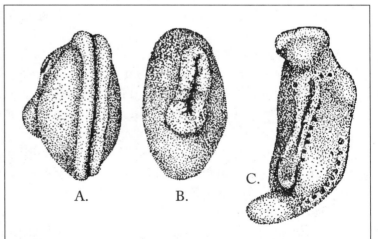

FIGURE 12D. Artificial induction. A. Neural fold of a *Triton* (newt) embryo in which an organizer from another newt species has been implanted in the gastrula phase; B. Secondary neural folds produced by organizer; C. Later stage, showing secondary embryonic axis with its tail, neural fold, somites, and otic sacs.

From H. L. Wieman, *An Introduction to Vertebrate Embryology* (New York: McGraw Hill, Inc., 1949).

We start out homogeneous and undifferentiated, a carbuncular hive, a singularity. All our integers are synonomous. Then the mirror is smashed. Cells which are the same make one another different. Context arises from sameness. The one becomes many; the singular, multiple; the multiple, whole. The surface becomes interior; the interior, surface. Symmetry crumbles into sheaths of asymmetries from which a new gyre of symmetries arises.

We are inverted, folded, and twisted into grids of nucleic swarms, splayed and winnowed into a living sculpture robed in regional protein microtextures, dispatched from an invaginating vortex. Once, these were indivisible mites with common histories. Now they are *Casablanca,* Ringling Brothers, Ford Motor Company.

How does this make any sense at all?

The Gray Crescent

Despite their enormous differences, a range of creatures, from houseflies to vertebrates, emerges from embryos that divide into body segments at right angles to their germinal tissue layers. This portends underlying sequence homologies in the animal plan. Initial asymmetries, as we have seen, are rearranged through their own by-products and patterns into consistent designs. Complexity originates from a limited number of simple elements (summarized as "genes") and expresses those elements again and again in structures built atop other structures. This occurred chronologically and sequentially, layer by layer and genome by genome through the evolution of phyla and orders, but it is organized in a single genome in each modern biont, thus must reemerge during ontogeny in tracks of synchronized interactive motifs until the historical level of complexity is achieved—no less, no more. The longer and deeper the embryogenesis, the more extensive and enfolded the evolutionary history that can be disgorged. All this differentiation requires a beginning point, a seam where ontogeny meets evolution and arises anew from it—ideally some early blemish marking the projection of a specific biological configuration into the blastula.

Experiments with newt eggs in Spemann's era and, later in the twentieth century, with fly eggs gradually allowed embryologists to identify primary influences on organization. During oogenesis regional separations of charged molecules at the animal and vegetal poles express differential properties in a gradient of cytoplasm and yolk that becomes discernible in the gray crescent. Thus, original egg polarity is provided by maternal-effect genes prior to any impartment of nucleic elements from the sperm.

In frog zygotes the gray crescent develops on the side of the egg opposite sperm penetration. Becoming the dorsal lip of the blastopore (made of notochordal mesoderm) and later the roof of the archenteron, it functions as an organizing center for other tissue layers. In sea squirts yellow cytoplasm flows down to the vegetal pole and a crescent forms just below the equator of the egg in a region that is to become muscles and mesodermal tissue. During gastrulation of more complex animals, the crescent/gradient also becomes the dorsal lip of the blastopore and sinks to form the roof of the archenteron which becomes the notochord; there it establishes the primary axis of the embryo, inducing a series of organs from head to the tailbud. In birds and mammals, Henson's node of the primitive streak is the primary organizer and becomes the notochord. It is induced by underlying hypoblast.

Cell-to-cell contact in a mouse blastula appears to contribute to gradients among cells. Where cell surfaces break free of each other for tiny stretches, microvilli form on the external surfaces of the blastomeres, marking asymmetrical chemical distribution beneath. This irregularity tilts the next cleavage plane: one daughter cell leans toward becoming part of a prospective inner cell mass and the other faces outward—predispositions that are inherited and enhanced by descendant blastomeres. Tiny irregularities become whole tissue plexuses.

The nature and origin of the gray crescent were elusive to Spemann and his associates, but later twentieth-century experiments have revealed that the egg itself manufactures morphogens which, after fertilization, trickle throughout the blastula, establishing faint chemical gradients. These tracks awaken primary developmental genes, the dawn-time alleles that inaugurate embryogenesis in all phyla. Molecules released at this phase then journey throughout the cells in the prospective organism, instructing them as to who they are, where they are to go, and what they must make.

As proteins and enzymes stream and are read in terms of existing (and constantly changing) meanings, patterns converge and diverge, resynchronize and resyncopate. Regional differences in the cytoplasm "tell" the nuclei which genes to activate and how to differentiate proteins once they are synthesized.

Competence

We now understand some of the concrete factors that separate pure genetic space from more complex organismic space.

During gastrulation the moving center of organization spawns regional centers, each one itself engendering hierarchies of subsystems. For instance, the first material to invaginate induces head organs which in turn induce brain ventricles, eye and nose rudiments, and ear vesicles. The last material to enter becomes the trunk and induces posterior organs.

As noted, the same genetic datum may be interpreted uniquely in different regions of the organism and at different stages of development. It may have one expression in neural tissue and another, entirely original one in cartilage. Invaginating mesoderm may be morphogenetically similar to invaginating neural plate, but they have distinct organismic meanings. Arms and legs are likewise regional interpretations of identical underlying instructions.

Specialization occurs because either only some genes in each differentiating cell are copied by the RNA, or, after transcription, a portion of the information is destroyed in the nucleoplasm and thus does not contribute to protein synthesis. That is, embryogenesis comprises either differential transcription of genes or differential passage of RNA into the cytoplasm. As relatively minor variations accrue along their gradients, groups of cells are induced chemically and mechanically to deviate from one another and begin to behave idiosyncratically, spurring further deviation. Regional heterogeneity then provides the context for new cycles of differential expression by some cells and lack of expression of others.

As a creature grows and complexifies, structure continues to be regulated by surrounding hierarchies in matrices of tissue and various extracellular secretions.

THE CAPACITY OF TISSUE to respond to induction in a morphogenetically specific way is determined by its location and present degree of specialization. Neural competence becomes spinocaudal competence under the influence of induction by the notochord. Ectoderm which can no longer join the neural plate has secondary competence as ear vesicles if induced by the hindbrain, and nasal pits if induced by the forebrain.

Functional organs emerge along gradients of induction. In insect embryos, for instance, a neuralizing gradient forms brain and sense organs, then mouth, appendages, and subesophageal ganglia; a mesodermalizing gradient originating caudally induces legs, wings, claspers, and copulation parts. Where neuralizing and mesodermalizing influences meet, trunk structures and spinal cord develop. The nerve cord itself is the thinning out of a neuralizing gradient through mesoderm.

More delicate branching organs radiate from the vortices of central organs. Limbs are induced by a thickening of ectoderm; fingers and toes by limbs. Harrison pointed out that a limb rudiment "may not be regarded as a definitely circumscribed area, like a stone in a mosaic, but as a center of differentiation in which the intensity of the process gradually diminishes."[10] Blood vessels are induced in mesoderm by endoderm; blood cells by other blood cells. The lens of the eye is induced by the optic vesicle, and secondary nerve branches are induced by primary nerve branches. "The egg and early embryo consist of fields — topologies or differentiation centers in which the specific properties drop off in intensity as the distance from the field center increases, but in which any part within limits may represent any other."[11]

Though it is obviously an oversimplification to see embryogenesis in this way, it is the best recipe we have. Cells start off as meiotic DNA, global records of organisms (hence, the sustained forensic viability of even hair follicles and dead skin in identifying the perpetrator of a crime). Later, individual cells differentiate in fields, then subfields, at the same time changing their nucleic character—i.e., genes generate fields; fields divide and disperse; their topologies reprogram the genes. Embryologists now speak of differential gene expression and inhibition as the mechanism behind *all* cell discrimination and development. Post-transcriptional control is at least as important as transcription in gene expression. A gene may be "on," but because of post-transcriptional controls, no protein is made. Many, many genes transcribe without expression; in fact, up to ninety percent of the genes in a differentiated cell are likely to be permanently turned off. Their DNA is reprogrammed to such a degree that each has lost the entire memory of its origin in meiosis as a germ cell; it "thinks" it was always kidney (or skin, or blood).

During cloning, differentiated cells taken from mature animals somehow become deprogrammed, lose their memory of having been udders and guts, and become totipotent all over again. Though no one has found them yet, it is possible that microdoses of ontogenetic chemicals in the cytoplasm of ova and early blastulas have seminal enough power to retrodifferentiate DNA.

Morphogenetic Fields and Subfields

By the 1930s physics had provided biology with a new paradigm—the particle (or position) within the field. Location of any one entity in a biological field affects the positions of all other entities (just like bodies in a gravitational field).

In 1931 Joseph Needham wrote: "Determination or chemo-differentiation takes place with reference to the whole organism; what any given part will develop into depends upon its position with reference to the whole."[12]

Fields split into smaller fields until the embryo becomes "a system of equilibrated spheres of coordinated action."[13] Each field is simultaneously whole and partible, and each subfield retains those same characteristics. Hence, brain tissue can develop within its general neural field and simultaneously give rise to progressive sets of individual sense organs. This series of synecdoches makes the whole and the part equivalent, distinct only in terms of spatial and chronological scale and context-derived function.

Limb regeneration and wound-healing can be explained as fields compensating for deficiencies. A split-off blastomere recreates a whole blastula for similar reasons.

IF CELLS ARE DIFFERENTIATED by tissues which arise from differentiating cells, we can no more assign priority here than to call light a wave instead of a particle, or hold matter separate of energy. Every organic shape comes to exist not as a carving of external space to fit a puzzle but as a coalescence of initially amorphous qualities, transformed through the intrinsic geometries and algebras of substance into larger, mathematically rigorous entities. DNA may lie at the basis of life, but the innate curiosity and mutuality of fields leads to their confederation and fusion, translating indeterminate polyps and filaments into elaborate organisms with deep internal integrity.

A SEED CONTAINS CERTAIN ASPECTS of a plant which the phenotype cannot escape but, as we saw in the previous chapter, not all potential forms of that plant. Epigenesis provides the mysterious missing ingredient. "Depending on how, when, and where you plant a seed, a limitless variety of forms can arise.... The concrete forms are emergent characteristics that arise out of a germinal state and develop in the interplay between the plant's plasticity and the environment. In particular surroundings the potential of the plant is evoked, but what appears is only one manifestation of the myriad ways in which this plant could develop."[14]

There is also the matter of the bulk and quantity of "extra" data packed into the dense coils of creature DNA. We in fact have little idea how much of this stuff there is and what it represents—in terms of either what coded it (gave it information) originally or what atavistic and primordial structuring it still remains capable of (and under what conditions any of that can be activated). We don't know how it got turned off and how its latency is maintained in a suspended state. If the dormant "hereditary junk" in our cell vaults is actually still amino-acid capable and tappable in mysterious ways as forms and functions ("invented" and then used and reused a long time ago by other bionts—even by prehistoric animals, fungi, and plants), then a virtually limitless variety of expressions can occur in one genome. Even if it is agreed that genetic transcription and activation are themselves highly regimented and almost always restricted to immediate precedent, the mere possibility that other forms, meanings, and phenotypic characteristics can enter from "elsewhere" under special conditions opens the system to a variety of novel and unpredictable manifestations of life.

HEREDITY IS NOT FIXED and rigid traits but "the capacity to develop out of an undetermined state."[15] A general symmetry of petals and florescences and an orange hue may originate in the mechanisms of DNA and RNA, but not the *kind* of dandelion we get. Likewise, we cannot know the physical and psychological idiosyncrasies of any person from his or her genes. Nothing fixed is inherited—"that is, no thing is inherited,"[16] because any characteristic (color, size, form) is subject to its context, and there is always the possibility that an embryo might not even germinate beyond a blastula, hence all "traits" would vanish. That is why genes are not real things, at least not in the way chromosomes are. They do not really exist in and of themselves.

The ingenious configurations of organisms may originate from genes and require genes for their ontogenesis, but the cell-making and shape-morphing functions of genes could not themselves have arisen without prior organisms. So what are genes? Where do they come from? What makes an organelle genetic?

Cytoplasmic Inheritance

"Development starts from a more or less spherical egg," writes biologist C. H. Waddington, "and from this there develops an animal that is anything but spherical. . . . One cannot account for this by any theory which confines itself to chemical statements, such as that genes control the synthesis of particular proteins. Somehow or other we must find how to bring into the story the physical forces which are necessary to push the material about into the appropriate places and mould it into the correct shapes."[17]

Most closed biological fields begin with simple symmetry. Some initial ripple disturbs the pattern and sets a gradient flowing along one trajectory or another. Chemical components (steroids, hormones, protein receptors, and the like) synthesized from an early-acting set of genes persist in either the cytoplasm or the extracellular medium, and these function as a kind of separate memory, calling separately upon the genes for their own maintenance and later differential expressions. Cells march to the nuclear drum, but they also sustain and interpolate its signals in intracellular structures and tissues.

Cytoplasmic memory is a property quite different from and more elusive than genomic imprinting. It is a synergistic set of operations performed by protoplasmic sequences without constant prompting from a script. Morphologies not only elicit structures from nuclear memory; they induce them from themselves. Higher-level cytoplasmic forms recur to a micron's breadth as meticulously and unerringly as genes reputedly replicate themselves, generation after generation. In fact, the forms are *more* rigorous and coordinated than the indirect transmission, translation, and expression of genetic data would seem to equip them for. That precision is apparently epigenetic, a property of emergent fields that do not even require genes for their signification, yet proleptically anticipate DNA's requirements.

In a medium of tightly-packed cells, chemical gradients provide their own configurational momentum, turning initial asymmetries into complex shapes and func-

tions. Beyond nuclear governance, organizing activity in the cytoplasm of protozoa and the meristems of plants leads to cilia and leaf spirals, respectively.

THERE ARE THUS TWO REIGNING MODELS or metaphors of biological form. One is
absolute genetic determinism, the genes as everything, complexity as merely an
elaborate machine run by genes. The other is intrinsic, inchmeal, epigenetic coalescence.

If viewed from an informational basis, genetic programs *are* truly capable of
making just about anything at all out of an interplay of nucleic acids, proteins, lipids,
polysaccharides, minerals, and a few metals. The kingdoms of protists, fungi, plants,
and animals that flourish on Earth are but a modest demonstration of DNA's performance. Given enough time, thousands of hypothetical chromosomal monkeys
typing on thousands of amino-acid typewriters would produce not only this universe but innumerable others. But that doesn't mean that genes *do* invent plants
and animals. Alone, they are vines without lattices.

In models based on epigenetic control, the genes are not true innovators or
inventors; they use forms already provided by nature; and, with gradual tweaking,
they contrive and assemble myriad species. By this argument something other than
genes must have entered the system in order, for instance, to turn "worm stuff" into
"human stuff," to achieve greater complexity from the same scale of information.
That "thing" must hold and transmit form too.

Yet the genes cannot be completely dethroned, for, as antecedently forming tissue nexuses modify nucleotide expression (both phylogenetically once and ontogenetically again and again), genetic authority persists, supplying new amino acids.
Most advocates of epigenesis admit that tissue shapes are somehow by-products of
the cumulative affiliation and outcome of genes.

The epigenetic challenge to genetic determinism must then be that, if chromosomes lie at the roots of epigenesis, their "prime mover" status cannot solve the
problem of emergent form. Genes do not provide the attributes of life forms, at
least not in any linear projective fashion. It only looks that way after the fact.

If epigenesis is not simply fancy "gene-esis," what it is it? What is holding up
the system besides genes? We hardly dare to probe too closely for fear either that
there are only genes wearing the emperor's new clothes or that genes can't pull it
off and Humpty-Dumpty will come tumbling down out of thin air (us with him).

Bacterial Chemotaxis

A similar genetically ambiguous situation occurs among bacteria responding to
attractants by changing their tumbling frequency and progressing toward one another

in chemotaxis. This microbial event would appeared to be regulated by prokaryote DNA and its protein receptors and tumbling-control proteins. However, stable characteristics arise not only from chemotactic network governances but a "network architecture" transcending the mere biomolecular components of the system. Changes in gene expression notwithstanding, the individual bacterial phenotypes defend kinetic parameters in a manner more resembling wave phenomena and shear force than quantitative biogenetic modulators. Explaining this behavior would require a systemic shift from pure reductionist genetic algebra and induction to complex qualitative differentiation. In fact, chemotactic bacteria in their networks are as thermodynamic as they are biochemical. Biologist Richard Strohman concludes: "Genetics (molecular biology) and dynamics are irreducibly complementary. Clearly, we still have a long way to go in the exploration of this new interdisciplinary paradigm in which genetic information is seen as the essential source of functional networks which generate robust behaviors that are irreducible to the genetic agents which provided for their origins. Genetics without dynamics cannot bridge the gap between inheritance and phenotype, between genes and function, between mutations and disease or, finally, between inheritance and evolution. In all cases it is the dynamics of context-dependent developmental process that must be understood together with the genetics."[18] Life is not mere genetic output, but a semistable gruel arising in fact from the same agglutinative physicochemical forces that made genes, then merging with them and their agenda.

Something profoundly extrabiological imparts form for free. Biomorphology behaves like the inexorable outcome of the multiple frequencies of connectivity implicit in all combinatorial systems (molecular and atomic ones as well).

Morphodynamic Factors

Mechanical and physical forces are embryogenically inevitable, given their ubiquity and universality (on the one hand) and the relative isolation and conservatism of genes (on the other). As noted in the previous chapter, most embryogenic processes have dynamic counterparts. Simultaneously geophysical and genetic, tissues "arise inescapably from properties of cell aggregates considered as physical matter."[19] Ooplasmic fluctuations stem from density differences and sedimentation. Epiboly expresses differential adhesion and layer submersion. Invagination transposes strain in a manner not unlike diastrophism. Epithelial delamination is a mica-like splitting of cellular sheets along parallel planes. Stripes and insect tagmata are congealed reaction-diffusion couplings. Parallel interpenetrating profusions (microfingers) are congealed convective events in the context of gravity and surface tension.

Biological forms are cohesive and fluid; they *look* more like interactions of matter in physical environments than incremental assemblages of bits of nucleic data and random mutational constructions. Lions prowl, falcons dive, and their intestines digest their kill more like rivers and clouds than binary mosaics. Developmental morphologies would thus appear to be both cellular and epicellular, coming into being not only within emergent cell-stuff but exterior to it—twin trajectories working their way into the interlocking spirals and cycles of biological patterning.

One can get a sense of this process by a simple experiment: dripping olive oil into a pan of water. Little oil-slick bubbles form—ideal cell templates. By gently oscillating the pan and producing wave motions at different pulses, you can generate transient organelle and organ motifs in the oil. If you keep at it long enough, you may even make a "kidney," a four-chambered "heart," and a passable "brain."

Homoplasy

While organic evolution gives rise to a startling multiplicity of forms, it tends to reuse similar morphologies in different contexts, as though tissue masses and genes working in concert with each other through embryogenesis are exploring physical mechanisms already built into the cellular configurations of primitive organisms, likewise into early embryos and the primordia of organs.

Without any shared sight-possessing ancestor and long after their divergence, mollusks and vertebrates independently developed near-identical ocular organs (see Chapter 18). The segmentation of vertebrates is totally different from that of arthropods, but they reflect similar embryogenic mechanisms. The occurrence of analogous and congeneric anatomical features in deviating bloodlines that do not share a common ancestor bearing the trait in question is known as homoplasy. A more localized instance is the reduction of salamander hind digits from five to four in three totally independent lineages.[20]

Convergent evolution with recurrence of forms across phyla has two possible explanations: it may indicate, as neo-Darwinians prefer, adaptation to similar ecological niches, or it may simply result from tissue exploration in domains of delimited formal possibilities. Though both exigencies probably bear to differing degrees, selection along divergent pedigrees from a vocabulary of morphologies would take biodynamic priority over a unique random adaptation each time. There are too many instances of convergent tissue formations in vastly different organisms for the common choice to be fortuitous in every instance, or fortuity probably had a reserve of latent primordial structures on which to draw.[21]

Such a reserve is also a possible explanation for the resemblance of the organs of vertebrates to whole invertebrate animals (see Chapters 14 and 19).

The Interaction of Genetic and Nongenetic Factors

Gravity

In the end neither genetic nor physicodynamic activities by themselves can explain embryogenesis or phylogeny. Despite our century-long infatuation with reified genes, molecular machineries clearly honor physical forces too. Experiments as early as the 1940s showed that, after fertilization of normally oriented frog eggs, dense cortical cytoplasm, immiscible with deeper cytoplasm, slips approximately thirty degrees to one side under the influence of gravity. At the same time, the sperm entry point slides ventrally. Thus, a mechanical vector amplifies a preexisting chemical boundary. Even in rotated eggs, gravity reestablishes orientation between cortical and deep cytoplasm. In some species of amphibians it is the very rotation of the cortex relative to the cytoplasm that discloses (and probably helps arrange) the organizing site of pigmented cytoplasm, the gray crescent. No doubt gravitational bias combines, both before and after genetic input, with viscosity, textural strain patterns, and chemical waves to orient and differentiate embryogenic systems, providing a spindle and fulcrum for molecular factors to position themselves.

Gravity is the force that is always already there. "[B]ecause gravity may suffice to drive cytoplasmic reorganization under all but the most unusual circumstances, it could have been the phylogenetically original determinant of cortical rotation and axis specification. A specific microtubule-based force-generating mechanism may have subsequently been selected on the basis of its ability to enhance the dependability of an event originally driven by a generic physical process."[22]

Morphogenetic gradients throughout early embryogenesis "are exquisitely sensitive to the presence of gravitational fields, which can influence the pattern of chemical waves attained at steady-state. Because of the chemical complexity of eggs, and of developing systems in general, it would be expected that multiple generic [physicomechanical] effects could contribute (along with locally acting genetically specified molecular interactions) to bringing about specific morphological outcomes."[23]

Genetic Capture

Although the inside environment of an organelle or emerging cell gradually becomes less susceptible to external, dynamic rearrangement as it is integrated into tissue, it can escape neither the universe nor its own history. Biomorphology (like timespace) is all of a piece. There is no real distinction between the physics of genes and the physics of bubbles. There is also no dynamic gap between the formation of cells in primeval tidepools and the present ontogenesis of cells and blastulas within embryos; there is merely a skein of condensation, synopsizing, cataloging, and refinement. In the first throes of Precambrian evolution, the insides of emerging cells were "bubbles" that became consolidated by primitive genetic molecules. Gradually they turned more machinelike, their physics and chemistry originating and supervising local effects. Some indigenous mechanical factors were overridden.

However, with the amassing of cells in multicellular aggregates, physical properties began to throw their convective and gravitational weight around anew. Living bodies became tiny moons floating in Luna's tides. Still, genetic evolution did not terminate with spongification or vermiformation even as physical influence did not cease with the advent of genetic macromolecules; there were no abrupt jumps either way. Mechanical factors continued to torque and marshal the products of chromosomes; genetic factors continued to mold whole organisms. As transitions and feedback between them occurred in nonlinear phases, the genes captured large mechanically derived forms as well as their own delicate nucleic designs. An earlier intracellular regime, already imprisoned by genetics, gradually extended its dominion out into the multicellular realm which simultaneously refracted back into the cell's chamber. Strain, adhesion, buoyancy, convection, and reaction-diffusion transmitted their products, in code, directly through the cell nucleus where they were encrypted again and again (by divergent mutations over epochs), reindexed, and serialized into more complex, holographic living versions of what they once were.

By themselves the genes are not capable of such originality and artistry. Only preanimate configurations rustling outside the nucleus became (after capture and conversion) myriad structures conducted from within the nucleus.

BUT HOW COULD wild natural shapes be copied and converted into tiny, rigid, amino-acid codes? The only answer is that the genes—opportunistic, ever changing, mutating, garbling, providing novel chinks and nodes—eventually found a way to trap and seize the attractive forms roiling in their midst. Equally the children of wind, sun, and rain, chromosomal units captured the ripples and convective flows imprinted by gravity and chemical inhomogeneity in protoplasmic pools. They internalized and sublimated them into new forms of biochemical integration.[24]

Biologist Stuart Newman surmises "that genetically specified molecular mechanisms have evolved to reinforce ... inherent tendencies, and to limit or specify the conditions for their occurrence. The evolution of mechanisms indifferent to, or in opposition to these forces, while formally possible, would probably have occurred less frequently."[25] The system tended to resonate around features it already contained, to synergize them. The genes bevelled toward mirroring and copying the interfaces, strains, flows, and bubblelike chambers in which they were already immersed. A progressive overtake of physical forces by genes is perhaps the only way we can explain how conservative, menial, and interned DNA could get ahold of such a startling and marvellous array of topologies and architectures.

The morphogenetic cannot escape the morphodynamic, and the morphodynamic (apparently) must become morphogenetic under scrupulous conditions of element distribution and climate (not on Venus or Neptune, but the way things are on Earth). Still one does not want to place too provincial a scope on a process about which we know little at large, having but a single example—DNA.

Sources of Information and Stability of Form

The mystery of development lies somewhere in the conceptual gap we impose between genetic sources of information and thermodynamic forms. Since gravity, convection, interfacial tension, phase separation, buoyancy, reaction-diffusion, electromagnetic fields, and molecular adhesion all precede genetic structures, we must assume that they played substantial roles in creating chromosomes and providing them with ontological pathways. Genes are tiny, infectious, self-replicating ripples that once attached to other protein ripples and, developing as hollow sheets trussed by tubes, replicated their basis too. Physicomechanical factors are implicated in morphogenetic space from the outset; they can never be rendered obsolete by nucleic sophistication; and they continue to layer themselves in developmental processes at new levels of complexity and organismic depth. Bionts are assembled in sequences of generic forces which, at their deepest and most nonlinear tier, turn into genetic mechanisms and molecular machines.

Creatures, far from being digital robots or megabytes, are resonance waves, whirlpools, crystals, and clouds—pagan events sculpted exquisitely by genes, themselves produced by more ancient meteorological phenomena concentrated in feedback loops in minute spaces. Physical and mechanical forces provided the repertoires that led to biological form. These ragged systems were then sharpened, refined, and modulated over time by genetic specification under mutational revision. They became living machines.

Seen from a different angle, perhaps proteins mysteriously came first; then their circus was tamed by feedback from generic forces, reinforcing and augmenting discontinuities and rough boundaries, and turning them into metabolic homeostases.

As universal forces act upon tissue, both in its primordial form phylogenetically and in its contemporary form ontogenetically, the molecular residues of biogenetic events exploit the formal and dynamic possibilities presented. "Changing patterns of gene expression during development can drive morphogenesis and pattern formation by making tissues responsive to fresh generic effects. Genetic change during evolution can act to conserve and reinforce these morphogenetic tendencies, or in rare instances, set phylogeny on a new path by establishing susceptibility of the embryo or its tissues to different generic forces. Such generic-genetic interactions will not give rise to all conceivable forms and patterns that may be constructed from living cells and biological macromolecules. They may nonetheless provide a concrete account of why organisms achieve the particular variety of forms with which we are so familiar."[26] The DNA ripples get extrinsic motifs to organize.

In fact, as noted in the previous chapter, genetic mechanisms do not so much create form as "limit and constrain pathways that have been set by generic physical effects, a reversal of the usual attribution of all morphological novelty to random genetic change...."[27] Through natural selection and genetic regulatory feedback, the initially loose relationship between genes and extrinsic forms would develop into not only intrinsic but failsafe redundant mechanisms. These could not continue to be mischievously perturbed by physical-mechanical forces or they would lose their integrity. A balance between genetic and physical mechanisms eventually led to a phylogenetic stasis appropriating not only the morphogenetic effects of molding forces but also those of mutations. The creation of such a jurisdiction marked the advent of biology.

Genetic Redundancy

One of the keys to genetic capture of physical mechanisms is informational redundancy. Since genes were not mandated by prior facsimile, they mimicked and carried over configurations in fragmented, repetitive, and overlapping stages. Sometimes they grasped a large chunk of a partial form, other times an infinitesimal but necessary piece. Purely morphological processes must be reconstituted unceasingly in genetic and then biodynamic space. In order for these events to be woven together successfully, organismically sustained, and passed on hereditarily, each captured puzzle bit must ineluctably be a part of some other unit coded elsewere in another way. Like the blind men of the Sufi parable, the protogenetic molecules felt different (and partially coinciding) parts of the emerging elephant, thus reported that it was shaped differently; working together, though, they assembled a complete pachyderm in a different dimension of time and space.

Genetic capture of native form was not just a straightforward matter of DNA being a great artist right from its inception, able to photostat and replicate the physical events and sequences in its midst in single gulps. It takes at least twenty-five different genes to specify the segmentation of the *Drosophila* (fruit fly) body plan; there is enormous redundancy as well as overdetermination in each factor involved. Yet this is the most likely way in which an already existing chemicomechanical segmental tendency could have been captured. Insect segmentation could hardly have been accomplished incrementally gene by gene, for how would the intermediate phases of creatures have survived their "transitional" compartmentalization?

Recoding

However they begin, feedback loops between morphogens and mechanical forces consume free energy while maintaining spatial nonuniformity. Transcriptional factors and gene promoters interact complexly with already-ingrained physical components. Backed by an emerging and elaborate genetic machinery, gradients then become (or provide the basis for) organic structures. Such structures may be temporary, periodic, or constant, but in all three states they are codable, thus inheritable. Genes define themselves by imposing their unique activity within a membrane.

After initial recruitment of new forms, a period of mutation and evolution (with DNA rearrangement) was necessary to salvage the functional aspect of forms partially and irregularly incorporated into genetic space. As organelles such as microtubules and intermediate filaments evolved with their unique and stable cytoarchitectures, then larger forms could be encompassed (and inherited) in pliable intracellular matrices that nonetheless resisted full mechanical deformation. These could then be retranslated to DNA along with information for creating new, more or less identical generations of organelles. Natural selection would reinforce and improve the more successful (though still imperfect) renditions of compartments, lumens, segments, multilayering, passageways, folds, stripes, and the like. Over time cell clusters representing these would graduate into the primordia of fascia, glands, kidneys, intestines, hearts, muscles, and limbs. They would become heritable not as "abstract genomic representation[s]"[28] but outcomes of generic effects in tissue masses subjected also to genetic regulation. In the same fashion as it propelled simple metazoans into invertebrates, generic-genetic interaction could project reinforceable subsets (organs) within those invertebrates, assembling them in larger, coupled templates for vertebrates.

ENTITIES SIMPLER TO CODE and regurgitate (like striping in the worm-like ancestor of the insects) would have been favored evolutionarily, selected less because they furnished any functional advantage and more because they were codable by genetic circuitry that itself was loosely and randomly assembled. For instance, it is possible that reaction-diffusion mechanisms provided, first of all, assimilatible and flexible ratios of scale between a chemical domain and a later tissue domain, and secondly, a simple set of genes and morphogens that could be manufactured and replicated easily. Stripes and segments would reinforce themselves and even supply (along with their own continuing morphodynamics) redundancy of patterning. It would be harder to squeeze spirals and spots and plaid (with their extra orthogonal gradients and promoters) into gene circuitries. Similarly, "eggs that were both generically *and* genetically determined to take on a spherical shape would be much more likely to maintain this shape than those formed only by generic forces."[29]

The process of recoding and preserving these generic-genetic templates would be mostly covert, buried at deep levels of DNA by historical episodes of disruption and rearrangement. Yet the deeper and more latent the forms became, the more stable they were—the more available at multiple levels for later ecological niches and selective regimes to seize and further adapt. Because they were already templated, they would be less incremental, hence less maladaptive to inherit. Organisms could draw on ancient genetic space for whole preexisting, dynamically cohesive templates rendered over millions of millennia by physical and genetic vectors; they wouldn't have to try out each of the genes' arbitrary renderings. Eliminating a 99.99% failure rate obviously hastened evolution.

Phenotypic Change without Genetic Change

With so much covert depth of information, many morphological shifts likely occurred without substantial new mutations—for instance, hundreds of distinct varieties of cichlid fish in the East African Lake Victoria alone, representing 200,000 years with minimal genetic change. The phenomenon of metamorphosis in the transition of tadpoles to frogs, caterpillars to butterflies, larval to adult sea urchins, clearly demonstrates the variety of bodies and lifestyles that can be stored in a single genotype.[30] Phenotypes are not rigidly templated by genotypes; organisms arise from interactions between sets of genes and shifting environments.

Early in evolution, generic forces worked on limited systems of genes and their products to produce a great variety of "raw material of the evolution of form."[31] So outside energies perturbed cellular space, and cellular space then captured outside energies and not only transformed but coded and stored them, to be drawn on at later times when phylogenetic systems were more stable, neither requiring nor susceptible to physical forces. It was much more effective to have the application be molecularly precise and originating from within; i.e., vectors of storms and strain patterns became vectors of proteins. Ripples became segments between organs, tubules, vertebrae.

The dynamic fluidity of this process would have been evident early in evolution before genetically stabilizing mechanisms established themselves. As mechanical and thermodynamic functions participated together, gradual organismic change (caused by random gene drift and mutations) would be disrupted at those radical moments when the genes seized an event outside the system. Climate changes would have had especially dramatic effects on nascent organisms. Since generically derived tissue structures were already buried in the most ancient repositories of the genes, in crises they came flying out in unexpected ways, giving rise to combinations of atavistic and hypermodern creatures. Evolution moved in bursts and florescences, in echinoderms and mollusks, in ferns and flowers.

Later there was a greater need for adapting to these changes, for morphological stasis in place of volatility. This is when genes took on a role of canalizing, stabilizing, autoregulating, and reinforcing the physical templates of body-plans through molecular mechanisms, locking in what was already there (and fooling biologists into thinking they were the innovators). After millennia of natural selection, new plans would remain substantially dormant, though preserved within the genotypic lineage. Success would be maintaining stable phenotypes in changing environments.

With long-term stabilization and many layers of chromosomal redundancy, genetic changes may have little or no effect on phenotypes. Cichlids in Lake Tanganyika show six times the genetic variation of those in Lake Victoria, yet without perceptible morphological expression.[32]

The Relationship between Morphodynamic and Genetic Factors in Embryogenesis

Nowadays genetics and physical forces restage their ancient shadow dance, chiselling the modern embryo, reimposing the outcomes of "small, viscoelastic, chemically active parcels of matter ... in each generation, when fertilization and cleavage give rise to a new multicellular aggregate."[33] As each embryo assembles itself, disequilibria are rediscovered and deployed anew (as once upon a time phylogenetically) in overall development. In such a manner, a rivulet or microfinger—over generations of compression, interfusion, cybernetic packaging, holography, and reprecipitation—becomes a lung or phallus.[34]

While the outcomes of prehistoric forces are locked deep in the genes, actual interfacial forces from the viscoelastic elements of the immediate gastrula (and then neurula) mold the products created by those genes (of course, morphogenetically active tissues *are also* hereditary events). The genetic-physical homeostasis is stable enough to trap and securely imbed lineages of past mechanical events (most of them from the early days of life on Earth) while, from embryo to offspring embryo, organizing present tissue forces rigorously and congruently. There will always be a slight vacillation between originary physical factors distributed by genes and contemporary dynamics enacted spontaneously (morphodynamically) by tissues as they are spun wet and afresh. Yet embryonic templates coordinate potential discrepancies,

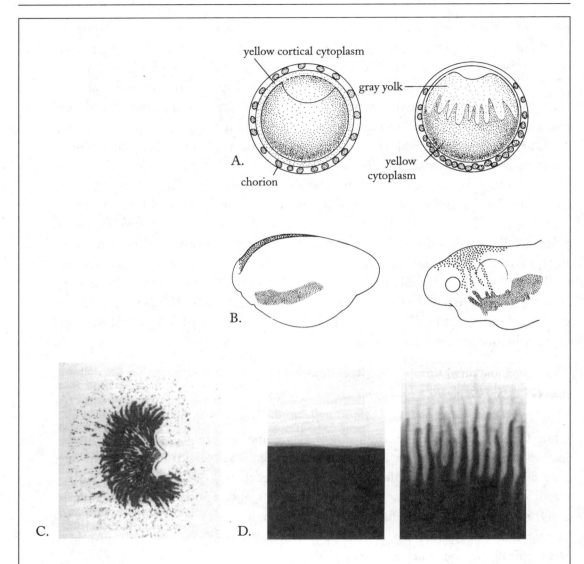

FIGURE 12E. Microfinger patterns in living and nonliving systems. A. Cytoplasmic rearrangement in a tunicate egg. (Left) Before fertilization, inner gray yolky cytoplasm is surrounded by a peripheral layer of yellow cytoplasm. (Right) By five minutes after fertilization the yellow cytoplasm has streamed to the vegetal pole, exposing the gray yolk. Microfingers of yellow cytoplasm continue to flow vegetally. B. Normal and ectopic (positionally abnormal) cranial neural-crest migration in the axolotl embryo. (Left) The right neural ridge of the head has been stained (coarse stippling). The left ridge has been excised, stained, and implanted horizontally lower down on the same side (fine stippling). (Right) Ectomesoderm (coarse stippling) from the right side is migrating down the left side, as it does normally, meeting streams of cells that migrated from the graft in the dorsal direction.

following identical pathways pretty much unerringly in the creation of each new member of a species. Slackness (or anachronicity) in generic-genetic interplay would not lead to discrete organisms (such as we have).

Where are the primeval mechanical forces in modern ontogenesis?

The organic system is reciprocal and homeostatic at levels ranging from organelles to tissues. Mechanical forces incite chemical gradients, while chemical inhomogeneities honor deep mechanical vectors. By one etiology, physical factors may remain dormant after ovulation (or even fertilization) until cells in a blastula respond to genetic morphogens; then they are swept into morphogenetic activity in the context of local signalling. Conversely, the production of morphogens (signalling) may require gravity, convection, and interfacial strain to sponsor their tasks insofar as these dynamic topologies tell the genome within the cell where it is and what biochemical activity is required. Again, past mechanical events have become purely genetic; they have otherwise disappeared. Present events straddle genetic and physical vectors.

When someone asks, "Where are the primeval mechanical forces in modern ontogenesis?"—the answer is: they are no longer manifest in the same ways. They were vividly determinative in raw protoplasm millions of years ago; then they were captured and metamorphosed by the genes. Now development exhibits none of their unruly splashes, swift bubbles, or reaction-diffusion stripes; its gastrulas and neurulas are coordinated by tight gene-tissue networks. Still, the templates of those forms originated in freewheeling environmental dynamics.

Darwinians have no trouble explaining even drastic modifications of existing forms (from moles to rabbits or worms to moths), but the origin of form itself gives them nightmares. Now we see that forms may arise as physically driven, self-organizing processes in nature, phenomena that accrue in external environments and continue to interact with aspects of those environments via their cell-forms. As

---

Figure 12E. *continued*

C. Autoradiographic image of a slice of rat kidney perfused with a triturated gaseous compound. The filled collecting ducts, which are between thirty and fifty microns wide, are in a microfinger arrangement. D. Time evolution of structured flows in a polymer system containing dextran and polyvinylpyrrolidone (PVP). (Left) Initial preparation. (Right) After forty minutes. The microfingers are on the order of five hundred microns in width.

From Stuart A. Newman and Wayne D. Comper, "Generic physical mechanisms of morphogenesis and pattern formation" *(Development, #110, 1990).*

these complex networks generate shapes and products, the genes hop on board and snare them, reinforcing and stabilizing morphologies and templates that already exist for nongenetic reasons.[33]

Plans of Species versus Plans of Phyla

Punctuated Equilibrium

An interplay between generic and genetic factors leading to evolutionary change fits the "punctuated equilibrium" model assigned to the fossil record by most modern palaeontologists. Within homeostases, potential changes build up in the gene pool, leaving no fossils, and then explode (on some morphogenetical or environmental signal) in a variety of new species. Ancient rock strata appear to show aeons of "morphological stasis punctuated by episodes of rapid structural integration."[36] Because of the relatively small amount of morphological information initially captured by genes and stored in chromosomes, minor mutations could tip the dynamic balance, with wide-ranging and novel anatomical effects, far exceeding the demands of natural selection. Wriggling whorl shapes, hardly respectable bodies yet, more like something a satellite probe might illuminate under moon ice near Jupiter or Saturn, fluctuated among uncertainty states, possible futures. Then, after phylogenetic stabilization, single mutations had to work in a more Mendelian fashion, mandating small phenotypic changes in color, size, form, etc. Ontogenies became narrower and more canalized, deepening from the level of families and genera to the level of species and varieties, hence (in more modern epochs) yielding different kinds of butterflies and songbirds, flowering plants and small mammals. Genes have interpolated and delimited random forms into species.

The Genetic Basis of Competence

In 1935 German embryologist O. E. Schotte excised a clump of cells from the bottom of a frog embryo, cells that would have become skin, and grafted it onto the prospective mouth of a salamander blastula. Ordinarily frogs have toothless, horny jaws with suckers on either side; salamanders have teeth and paired swimming stabilizers. The frog cells in Schotte's salamander developed a toothless jaw flanked by suckers. Directed by their salamander context to become a mouth, they responded, but in terms of their frog genes.

Competence now always has a genetic basis; it cannot transcend its lineage to supply parts on demand for exotic new structures. In the larva of a housefly, if one imaginal disk is transplanted to the site of another, the grafted disk will emerge from metamorphosis as a structure congruous with its origin regardless of its location.

This cell memory is heritable through generations of proliferating cells. If leg cells of a chick are grafted into the tip of a wing bud, they are able to recognize their distal location, but they will develop into toes not wing digits. If a chick somite is replaced by its quail equivalent early in development, quail muscle cells will sprout among chicken ones in the puzzled bird's wings. Their essential nature has been established irreversibly by their early developmental history.

Tissues grafted onto a foreign species will not develop organs idiosyncratic to the host. If salamander ectoderm is transplanted onto the site of the ventral suckers of a frog embryo, it will develop the balancer organs it would have developed in the equivalent site on a salamander (it matters little if the ectoderm is removed from a region distant from the prospective balancers; it will form these organs if it has not lost its competence through maturation). Likewise, a frog sucker will develop epidermally on a salamander if competent frog ectoderm is transplanted to the balancer site.

One embryologist explained: "It is as though the transplant says, 'I recognize my new position, but I must respond in my own way.'"[37]

This suggests a possible relationship between genetic and dynamic heritage. Genes provide the hereditary specifics of organs—"the differences between attributes of systems"[38]—while fields determine the overall form-principles of the organism—"the integral subjects which carry and display those attributes."[39]

We follow how embryogenic activities are transmitted from parent to offspring. The epigenetic plan, though, is a mystery, for it seemingly arises through genetic activity, yet is independent of it. In truth, "ontogeny is not a gradual revelation of a plan stored in the genome, as there is no such 'plan' in the genome, but only a vast amount of specializing and distinguishing detail."[40] What pulls this detail together is the mysteriously originating morphogenetic field—in fact, the hierarchy of fields—in the emerging organism. It is a series of "equilibrated spheres of coordinated action" into which the program of proteins flows.

THUS, WE KNOW what makes a beetle different from a butterfly rather than what makes them both insects (or animals). We may safely assign color and shape of fins or musical ability to genes, but not a basic piscine or human gestalt. Though these have their genetic basis, it cannot be located in any congery of traits. It was here before we were; its agency is seamless and old.

That is, each vertebrate has distinctive specializing characteristics; yet there is another instrumentality that confers its overall creature identity—a template underlying the subphylum. This plan has more of a phylogenetic than ontogenetic etiology; its provenance is lost in the tangles of its own miniaturization and condensation over time. We can no more uncover the origins and blueprints of phyla than we can unwind DNA helices all the way back through their cybernetic packing and genealogy to simple linear spirochetes (or whatever they come from).

Planetary forces induce phylum templates.

Forms for plants and animals resided everywhere, terrestrially and extraterrestrially, long before life on Earth. It is possible that the more comprehensive body-plans on this world—those of plants, fungi, worms, insects, fish, primates, etc.—have substantial origins in ecophysical effects. Meanwhile narrower arrays of anatomical features—such as those distinguishing a pig from a possum, an ant from a beetle, an oak from an elm, a horsetail from a clubmoss, etc.—represent the later "superimposition of genetic modifiers such as homeobox-containing nuclear proteins ... [i.e.,] as nonuniformly distributed in developing limb buds.... refin[ing] the rough pattern by influencing local patterns of chondrogenic gene expressions."[41]

Unable to shape structures globally (because organisms are exponentially beyond their scale), genes and nucleic acids act locally and incrementally, "conserving existing successful body-plans rather than causing species to undertake the first steps leading to major structural rearrangements"[42] Planetary forces are uniquely capable of inducing and transforming phylum templates on worlds because not only are tides, etc., inexhaustible and discrete at the same time but they wantonly exceed the ordinal assemblage and feedback loops of genes and proteins at any one juncture. A very long time ago wind and water provided emergent gene networks with surpluses of potential morphology for tissue; likely most of this was programmed and locked into codes when morphogenetic grids were relatively pliable—an account drawn upon gradually over millennia. Most of the primary kingdom and phylum templates (fungus, moss, vascular plant, mollusk, arthropod, vertebrate, etc.), no matter how and when they were first expressed in the archaeological record, suggest great age and genotypic latency. Their guiding tissue matrices are nucleicly far older than their ultimate embodiments (their fossils); hence, they leave no seams in the modern embryo for biotechnologists to juggle.

Kingdoms represent the most fundamental templates on the present Earth. Even their cell types are exclusive; they make unique structures and ecologies. Plants differ from animals in growing and spreading outward, unfurling their macromolecular cadences in roots and branches, petals and carpels. Though intricate in their own right, they are ineradicably linear, as they exteriorize themselves into earth and air, and they are limited by their rigid cell walls from exploring as great a range of morphodynamic possibilities. By contrast, sheets of zoological cells, many of them lacking walls, curl and embrangle inwardly as well as by arborescence. Thus, many unique kinds of nesting and interpolation are possible in animal tissues, leading to deeper orders of internal structure and metabolism and complexities of behavior.

Over billions of years, sun became leaves, air became birds, water became fish, and lightning bursts became mind.

Summary of Morphodynamic-Genetic Relationships

"Genes do not impart higher order upon orderless milieus by ordainment."

Without metabolic and reproductive restrictions the universe explodes galactically and sends ripples and stones hurtling through space. It erupts in hot springs; it culls tides through estuaries. Natural effects do not have to obey transitional adaptive states. Invading nascent biological systems on planets and moons, cosmic forces respond nonlinearly to "changes in control variables. Thus, a small change in density of an ooplasmic determinant could lead to large changes in its spatial distribution. A minor alteration in interfacial tension between two tissue compartments could strikingly change their relative configurations.... In each of these cases, profound alterations in morphology, reproducible from generation to generation, would ensue, virtually at one stroke. If the resulting variants proved successful in establishing and populating new niches, eons of genetic evolution could follow, stabilizing and reinforcing the new outcome. The alternative [genetic] model, i.e., major morphological evolution by increments, would be analogous to bridging a chasm of indeterminate breadth."[43]

However they originated, plans of kingdoms and phyla, organized in the gene-protein nexus, owe their consecution to deep mathematical and topological relationships and invisible geometrical grids combining environmental and intercellular influences within biological fields, overriding but integrating the minutiae of local incentives. In addition, morphogenesis is always overdetermined and redundant, compiling cornucopias of potential states within fragile, metastable forms. Enormous amounts of information are buried within transitory structures, each assembled by a single playback of proteins.

FROM OUR VANTAGE POINT, order always came first; there is no beginning to it, only a prior organization, and then one prior to that, and one prior to that ... seamlessly flowing back through a series of architectural fields to an initial set of episodes and equations we cannot begin to imagine. There form began and was synopsized and encoded; from there (and from other, later thresholds) form emerges anew in embryogenic translation. It is almost as though the ultimate symmetry of a class of organisms already infiltrates the pathways to that symmetry; an intimation of higher order entices that very configuration into itself. The genes play a critical role, but they are dependent on information entering the post-transcriptive network from somewhere else.

"Genes, highly organized in themselves, do not impart higher order upon orderless milieu[s] by ordainment," declared experimental embryologist Paul Weiss, "but ... they themselves are part and parcel of an ordered system, in which they are enclosed and with the patterned dynamics of which they interact. The organization of this supra-genic system, the organism, does not even originate in our time by 'spontaneous generation'; it has been ever present since the primordial living systems, passed down in uninterrupted continuity from generation to generation through the organic matrix in which the genome is encased."[44] At the age of seventy-one in 1969, he lyricized: "We encounter here the phenomenon of emergence of singularities in a dynamic system—unique points or planes—comparable, for instance, to nodal points in a vibrating string."[45]

Like the particles of quantum mechanics, the nodes of biological fields operate in an acausal, atemporal unity. To mechanists interested in living systems only as the ultimate machines, this is a disappointing regression toward vitalism.

Genes are mainly conduits and holdfasts in this process.

The Darwinian-Weismannian version of traits and genes, emphasizing the random play of aberrations and valorizing freaks, is no longer the cat's only meow. "[T]he alternative view holds that the various types of organisms that populate the biosphere are the virtually inevitable formations of living matter, much as the elements of the periodic table are inevitable formations of subatomic particles."[46]

As we have seen, bionts are complex multicellular realizations of the innate attributes of gels, flows, crystals, and plasmas inculcated in protoplasm at the earliest phases of its evolution. First they were turned inside-out and copied by primitive genes; then they were flipped outside-in as gastrulas. They resemble origami toys that, folded into a wad, then (with a puff of air) inflate into fullblown aviaries and rainbow-colored fish. Creatures do not organize, reproduce, and speciate by chance aberrations but "a range of organic possibilities to which any evolved genetic 'programs' must necessarily conform."[47]

This is one reason why a substantially common gene pool gives rise to such divergent creatures. The pathway from a sponge to a clam or tunicate may include many data points (codons) and elaborate and novel nucleotide arrays (mutations), but these alone are neither sufficient nor (in all likelihood) inauguratory. Outside forces (as vast as the gravity and heat of the Sun, as taut as the stress planes and adhesions of plasma surfaces) continually shaped and potentiated the raw protein nexus, providing shifting field states and geometries for macromolecules to investigate and redeploy. Once metastable entities were "corrupted," their stable elements imposed a new order from a cryptic series of possible arrangements. Creatures thickened, tangled, twisted, changed size, and sublimated; novel metabolisms and hydraulics occurred.

Lost and latent DNA codons may become activated as complex resonances and feedback loops within organisms create fresh dynamics of activation (by means presently unknown to us). This is more brazen than just "inheritance of acquired characteristics"—already an act barred from modern biology; it is instantaneous inheritance of new (ancient) protein motifs and biological fields, shifts of tissue moiré patterns. Whether these are then inheritable (in whole or in part, linearly or nonlinearly) by progeny is a whole other question.

The genes (such as they are) are mainly the conduits and holdfasts in this process. They ensure that entities evolving from disturbances are efficiently composed, tightly organized, and rigorously maintained (either that or, famously, they perish).

Genes are mere props for some designs. A degree of organization would arise even in their absence. In this way they do resemble traffic lights more than blueprints.

Undoubtedly genes also introduce formal elements (substances, shapes, and contiguities), but chromosomes are not the creators (at least not the singular and prime inventors) of functional biology. They are too provincial, too remote from the arena of action. The metamorphosis from the template of a *Caenorhabditis elegans* worm to that of a chimpanzee was accomplished with almost three-quarters the same gene base because those genes were given denser and more intricate materials (in succession) in which to deliver their wares. The loci of conveyance changed, and some of the raw materials also changed, but mostly the dialect and gross national product of the polity evolved. Pouring sugar and spice into something that looks like an octopus or pig is a lot different from dispatching those same ingredients into a configuration organized like a comb jelly.

The genes have a political agenda.

The notion that the genes control everything is not ideologically neutral. It has at its heart (and well underway) the hegemony of human space by economics and algebra. It justifies turning over our whole future to biotechnologists and corporations. It heralds not only the death of God but the death of humanity. Ads are placed shamelessly on every surface, for life is just a shuffle of competing products and amusements.

There are now genes for "happiness," "melancholy," "novelty," "creativity," and "sexual orientation." Genetic reductionism herds individuals away from their own mystery and toward the rule of the algorithm and the serial number. The search for any other source of biological form is an attempt to reclaim spirit, freedom, and unknown destiny. It is also legitimate because the genes are neither omnipotent nor peremptory.

The universe has become complicated many times over.

The shapes we see repeated in plant and animal bodies (curves, spirals, spheres and semi-spheres, bubbles, insides and outsides, insides inside insides and outsides, insides outside outsides and insides, striations, ripples, anastomoses, etc.) are all events which exist in nature prior to life but in simpler forms. The genes—or, more properly, the entire genetic apparatus and regime—transmutes them by layers into other combinations of shapes by internalizing them and improvising them in many different structures, always under the edict of function, survival, and reproducibility. One of the prime acts of multiple internalization and miniaturization (leading to compound forms, anatomical metonymies, synecdoches, and the like) is the placement of raw morphological information into geneticizing (i.e., symbolicizing and algebraicizing) series that have their own internal logic and exigencies for input and output.

Coding within them means assigning something that can be made to act like a number and a boundary, or at least part of a number and a fractional boundary, to something which is wild and random, and then using that code to project and transmit it into another, often more complex configuration. These systems exist, at least calculably, at the levels of DNA, chromosomes, amino acids, and proteins—each of the tiers different from the next, each reorganizing information according to its own algebraic rules different from the rules preceding and succeeding it. And this is merely the part that is calculable, that we can (more or less) see. The rest is hidden in not only evolving systems but systems which have been evolving erratically so long that much of their etiology has become miniaturized and multiply submerged in codifications beyond visibility or decipherment, even in the genotype.

That is, the interplay of action and time have buried what we call "links" and "hypertext" in operations that seem to occur for different reasons or (usually) no reasons at all. The very nature of coding—for lack of a better term to describe the radical and extreme process by which one system steals another and imbeds it in itself through rules and mnemonic devices it invents along the way—is that it seems to transcend the linearity of time and reverse cause and effect. While deforming its original materials so much they cannot be recognized (and certainly cannot be morphologically traced), it also startlingly keeps returning to them, at least partly because there are relatively few formal and aesthetic patterns to draw on in nature, and likely also because every system is in some way entrained (oriented around) the information that exists at its roots, that gives rise to it.

This is the way in which a feather is a series of rivulets around a stick in a stream and a primitive kidney is a reaction-diffusion bubble. A snake is water moving on land. They are all first hieroglyphicized and packaged for a long journey; they are then decrypted and spun out multidimensionally in surprise embodiments. Lines, boundaries, and spirals disappear into the black hole of genetic deconstruction and reappear in utterly new embodiments and costumes resembling creative distortions and condensations of their original forms. Meteorology becomes biology and linguistics—and again in the totemism, petroglyphs, and ceremonies of tribes incubated over aeons in primordial cellular marl and emerging (at least on Earth) during the late Ice Ages. Thus does the universe become complicated many times over.

Splashes, gradients, and sunspots do incarnate.

The world of living form, from spiralling cones to echinoderm origami, from heliozoan spires to fractalling gyri, from gemmed rose carpals to anastomosing scales oozing feathers and plumes and jointed carapaces worming and barnacling, is all accomplished from one pot of genes on one cyclonic world—not because of genes' implacably rigorous mimicry (though impossible without it); not as their transmission of information into form, but their translation of form into information, back into form. Genes do not create life by sewing together randomly occurring information packets. The nascent fibers find compact shapes in their midst—bubbles, mottles, lineations, laminae, pellicles—and simulate them in their own totally different idiolect. In the process they rewrite them, not in their original textile, but in proteins. Deep, deep originary shapes are ubiquitous and cosmogonic. They arise with the physical universe and recur at its various levels; in fact, they mingle to compose them. As ceaseless variations of molecular bonds are shuffled through polyhedra of gene kaleidoscopes, protein molecules assemble favored polymers at varying scales, textures, densities, and proportions, and in varying states of functional integration.

Pattern lineages are arranged in layers and contexts provided by kinetics and landscapes: amoeba, fungus, anemone, crab, spider, shark, wart hog, penguin, eel ... it is all the same thing—i.e., the same stuff, the same algebra and geometry, the same nuclei and nucleic acids—and yet, at this level of phenomenology and meaning, it is equally all different; the events it generates mean utterly different things to the creatures that experience them.

THERE CAN BE NO FORM without genetic underpinning, no plants or animals made by magic out of thin air and molecular debris, but the essentialities of plants and animals originate as much in currents of thin air and debris as in genetic components.

Despite our present idolatry of cybernetics and information (as opposed to form), biomorphology plays a central role in both phylogenesis and ontogenesis. It is a great polar bear hidden in a vortex of genetic ants—the most obvious, yet the most invisible, feature of the landscape. Splashes, gradients, sunspots, dust devils, fog, hail, cyclones, stalactites, quartz crystals, and intersecting streams do incarnate. It takes them a long time, numberless nonlinear equations, and a perpetual series of miniaturizing and interiorizing fissions gastrulating into more and more infinitesimal and densely textured space.

They are *still* incarnating—and with them the spirit of the invisible universe.

The genes are merely their shepherds and scribes.

ANY OTHER SUBTLE AND PARAPHYSICAL ENTITIES, hyperdimensional forces, vital energies, or radical and divine acts implicated in this process will be left, for now, to the imagination of the reader.

What we don't see took billions of years to manifest.

EMBRYOGENESIS IS A CUMULATIVE INTERDEPENDENT PROGRESSION; any chain of amino acids may express itself phenotypically along diverse gradients with their own thresholds and pirouettes. A mutation affecting one gene generates a ripple of changes throughout an organism as proteins from other genes behave differently in order to maintain the functional unity of the field across space and time.

The fact that extraneous substances can trigger the same responses as endemic inducers does not make development hopelessly nondiscriminatory; after all, embryogenesis and evolution are finalized only in terms of a chronological relationship between genes and the creatures that emerged from them, individual by individual and species by species.

A developmental process that has undergone such turbulence in the currents and storms of the planet has survived by incorporating alternate equations for stable configurations. What appears to be nondiscrimination is actually the inherent complexity of hierarchical centers and overlapping realms of influence. Those landmarks that were used historically became inductors embryologically. But nothing required that they be exclusive or one-way maps. In fact, it was better that they be permutable—flexible and interchangeable—given the turbulent working conditions and lack of an architect. Morphogenesis is a "spatial-temporal order that arises from periodic wave propagation over an excitable continuum."[48]

Embryologists have centrifuged fertilized eggs, removed sections, pricked them with needles to make multiple activation points, and poisoned them with chemicals. In a large number of trials the embryos reestablish their symmetries and polarities and develop normally. It would seem that something so fragile, at the beginning of such a long and delicate assembly, could be easily and fatally disrupted, but the zygote is remarkably adaptable. Even when thwarted, organic fields reintegrate. It might take four or five waves of tissue movement to accomplish what one would have fashioned initially, each libration "correcting" a single out-of-place factor, but the primary theme prevails, riding out, as it were, the distortions without losing the melody. Though artificial interference falsifies histories and triggers established sequences out of order, morphogenetic patterning resonates toward development rather than nondiscrimination.

In some cases a wounded fetus may incorporate the disruption, becoming an anomalous creature—a freak. A tadpole embryo centrifuged perpendicular and then parallel to the animal-vegetal axis forms as Siamese twins joined ventrally at the gut. They have two mouths and two throats but only one midgut and anus.[49] Only in a surprisingly small number of cases are the effects totally lethal.

Laboratory reenactments introduce the kinds of jumbles that evolving tissue had to deal with in much more desperate and enduring situations probably trillions of times during evolution, but because the trials are not identical to game-day crises, they lead to partial resolutions or monstrosities.

The key events in evolution were likely carried out by Michael Jordans of the cell world—"in the zone," 6.6 seconds showing on the clock. "A lot of players and coaches can look at film afterward and point their finger at the exact moment when a game slipped away, but Jordan could tell instantly, even as it was happening.... It was ... as if he were in the game playing and yet sitting there studying it and completely distanced from it ... as if [he] had already lived through it."[50]

Just as we cannot locate the genetic basis of phyla, we cannot uncover the source of form; we can but manipulate its principles in labs and see their quantal states of syntax and alternating homeostases.

We weren't at the game. Yet we can tell, after the fact, who won. And we can do that much only because form already exists—in us and in the tissue of experimental bionts.

Although scientists try to provoke the kind of ingenuity with which crises were solved in the primordial seas, their experiments are no longer dealing with raw evolving tissue matrices but highly specified state-of-the-art outputs after millennia of experimentation in the wild. This is a different thing in a way we cannot fathom, for although the laws of nature may not change, mysterious disjunctions lie between epochs. The miracles of life occurred without audience by spells that we cannot excavate or reconstruct:

"He held the ball, faked a move to the basket, and then, at the last minute, when he finally jumped, fell back slightly, giving himself almost perfect separation from the defensive player ... defying gravity ... willing the ball through the basket."[51]

If it can happen in our artificial games and staged dramas, it can happen in the long and deep pond of nature. Events can fall into "the zone," and suddenly, uncannily, skip all the seemingly necessary steps up to the next plateau of order.

The baffling and disappointing arrays of partial organizations that now appear in laboratory dishes speak to the experimental effects of condensation, dissociation, and discontinuity applied anachronistically to myriad-level opaquenesses that were not originally hypothetical and resolved themselves only in terms of singular, actual transformations.

What we don't see took billions of years to manifest, far more time than we have to reconstruct it. A few million more years in a random soup under stellar bombardment might well sort most of these experiments out. In any case, we are too late and too slow; we don't see it *and won't see it forever.*

The use of sequential waves to restore disrupted cohesion suggests that evolutionarily there were always alternate pathways to equivalent life forms and that, when organized genetically, protoplasm is resilient and inventive, flowing, pulsing, reversing direction, and using its own viscosity to fashion membranes and tubes through which it diverges and propagates. As genes are altered by mutations and random sorting, the gene pool is stirred, resulting in "an effective search through the potential space of morphogenetic trajectories, an exploration of the possible forms in some of which the living state can be expressed as robust and viable species in suitable habitats."[52] Because the genes continue to recode only in terms of a closed amino-acid field, old motifs inevitably return at new levels of structure and scale, reflecting homoplasy. The ciliated tentacles of the simple entoprocts so suggest bryozoan lophophores that these creatures have been jointly named "moss animals," despite the fact that the former is more allied to rotifers and the latter to clams and snails.

Mutations combine existing elements in unforeseeable ways. The webbed feet of the duck, the talons of the eagle, and the pig's hooves are novel variants of similar traits that come to express utterly different personalities and lifestyles. As we have discussed earlier, there is latent potential for radically divergent species in any organism.

In the primordial laboratory of the Earth—the only place where it counted—combinations of multiple crisscrossing pathways and complex repetitious (even tautological) hierarchies set novel creatures waddling through the deep with enough syntax in reserve to later fabricate organs for hunting and procreation and, later, for shore and air colonization. A worm was deformed into a mayfly, its sister into a lobster. Fishes became frogs and turtles. These descendant creatures breathed vapor with their protein "deformities." Chromosome alterations may be random, but proteins are deeply structured.

Not only must a new morphology inhabit its tissue, but the creature must be able to grasp its wholeness and translate it into meaningful actions. Even the duck-billed platypus was an organized whole, not a collection of separate intentions (a reality sometimes missed by the nineteenth-century naturalists who discovered it); its unity was expressed in its single personality and ecological integrity.

Intelligence arose initially because the protein field has the capacity for sorting and storing information and, under conditions of bulbous inductions of the ectoderm, integrating surplus neural cells into coherent activities. But it spread from mere swimming to philosophy because it encountered object relations; it was cajoled and confirmed. It saw itself reflected everywhere else; it had someone to speak to. This pulled it along in daring vaudeville acts like a juggler controlling more and more balls in the air. Because neuralized animals used their enhanced sensoria to occupy untapped niches on the planet, it was possible for more of their kind to proliferate, and other diverse branches—including ours—to follow.

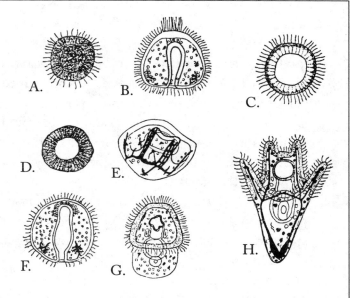

FIGURE 12G. Effects of chemical alteration of surrounding medium upon sea-urchin development. A. Without OH, ciliated solid blastula; B. KOH has been added; C. Normal blastula; D. Blastula in potassium-free medium; E. Reared in a K-free medium and replaced in normal sea-water; F. Raised in a medium devoid of magnesium; G. Pluteus with three-parted gut, mouth, and coelomic sacs, but neither skeleton nor arms; reared without $CaCO_3$ or $CaSO_4$; H. Normal pluteus.

From William E. Kellicott, *A Textbook of General Embryology* (New York: Henry Holt & Company, 1913).

Systems Theory

THE ORIGINAL RENEGADE PARADIGM to emerge in place of either pure mechanism or vitalism was a structural organicism first proposed in the 1920s as "general systems theory" by Ludwig von Bertalanffy. Since (at least on Earth) biology maintains hierarchies of delicate structure—information systems built of amino acids and metabolizing proteins—organicists proposed a self-organizing chemistry: regulation by wholeness, organization by prior organization.

The basic laws of matter of course must operate without exception at each level of organization (atoms, molecules, cells, tissues, organisms), but levels are stacked so that each also has its own discrete set of corollaries not applicable to the other levels. A system at any level possesses emergent properties that are characteristic of the system as a whole but not locatable in any of its parts. Each system is stabilized by feedback between levels (but also destabilized so that new forms arise from their dynamic equilibrium). Higher levels of organization express expanding complexity of information, refinement of the system, and extension of its range.

If "general systems theory" sounds like embryogenesis, it is no accident. Bertalanffy was a colleague of Weiss, and he realized that the embryo (i.e., the developing organism) is the link between simple physicochemical reactions and complex social and symbolic systems. There is certainly no other candidate, metaphorical or otherwise.

The difference between a cybernetic "systems" machine and a mill or automobile is that the former transcends the prescribed cause-and-effect relations of its parts to impose unpredictable patterns that have no starting point or causal seam. On and off, go and stop, cause and effect cannot be reduced to the properties of either constituent or extrinsic elements, DNA molecules or inducing proteins. It is all of these but none of them individually.

Weiss described the living organism as "the exact antithesis of a classical machine."[53] A biological field is "a rather circumscribed complex of relatively bounded phenomena, which, within those bounds, retains a relatively stationary pattern of structure in space or of sequential configuration in time despite a high degree of variability in the details of distribution and interrelations among its constituent units of lower order."[54]

Driesch's sea-urchin eggs had restored entire functioning systems out of parts; they had regulated themselves. Needham translated this remarkable and seemingly inherent capacity of nature into his own biological Marxism:

"... every level of organization has its own regularities and principles, not reducible to those appropriate to lower levels of organization, nor applicable to higher levels, but at the same time in no way inscrutable or immune from scientific analysis and comprehension. . . .

"From ultimate physical particle to atom, from atom to molecule, from molecule to colloidal aggregate, from aggregate to living cell, from cell to organ, from organ to body, from animal body to social organization, the series of organizational levels is complete. Nothing but energy (as we now call matter and motion) and the [degrees] of organization (or the stabilized dialectical syntheses) at different levels have been required for the building of our world."[55]

Weiss himself concluded: "The patterning is inherent and primary; . . . what is more, human mind can perceive it only because it is itself part and parcel of that order."[56]

From there it is but a short step to neo-Platonic metaphysics, time travel, and Carl Jung's archetypes of the collective unconscious.

Epigenetic Landscapes

IN THE 1960s, C. H. WADDINGTON proposed an individuation field spreading through multidimensional space. The fertilized egg, according to Waddington, in its transformation into an organism crosses an "epigenetic landscape." This domain, invisible to us in the usual sense, is an unfolding space-time wave. Tissues becoming organs are its ripples. Their fields, self-organizing and mutually integrating, emerge as though time were running backward and their wholeness assured in advance. In fact, time *is* running backward: phylogenetic time is running backward into ontogenetic time which is running forward. This is how coordinated tissue emerges so mysteriously in our midst and, seemingly, without mechanical cause. This is how structure escapes genetic authority and puts its own stamp on matter. Frogs that did not yet exist were already inducing their templates in prior genetic fields. Life was summoning itself backward from Silurian time into Precambrian time.

The physicochemical identity of any specific embryo is a manifestation of both a prior and a *subsequent* whole.

Under this theory organismic events are integrated by evolutionary paths or process chains Waddington called chreods (based on the Greek for "necessary" and "path"). Chreods are not configurations of tissue, or cells, or even molecules; they are globs of space and time with complex attractor surfaces that bring displaced points back into line so that biological stability is maintained. Thus, molecules and cells are compelled into patterns by transdimensional topologies. Such a model requires a violation of the current laws of physics along with a presumption that some aspect of a future event can "trickle" back and influence, however faintly, an event yet to happen of which it is a result![57] So do completed organs guide raw cells into their grids.

Chreods can be represented as valleys canalizing traits by their naturally sloping sides. For "events" to escape they would have to get over the ridge into the next chreod, but the tendency would be for them to roll back toward their original field. This tug appears as a resonance, or moiré, of tissue patternings.

The individuating power of chreods is vividly illustrated by insect larvae which obliterate their sculpture and gist in order to create entirely new animals with their own plans. By this paradigm the butterfly/caterpillar system contains two different chreods that are activated from different points in space-time. In the passage from one to another the brain, hindguts, and heart alone survive; the rest of the morphology is broken down into molecules and transmogrified.

Chreodes

IN 1981, RUPERT SHELDRAKE, a British biologist, fused general systems theory with Waddington's chreods in a universal cosmology of matter and mind. Calling his activating entity a "chreode," Sheldrake postulated a morphogenetic field transcending all other forces of nature, able to influence events by a previously unknown energy which he dubbed "morphic resonance" and defined as an asympathetic coinciding of vibrating systems.[58] This transcendental force does not involve mass or energy and need not act according to the laws of thermodynamics or even quantum physics. "This process," Sheldrake later explained, "the energetic flux of the universe, underlies time, change, and becoming.... Matter is now thought of as energy bound within fields—the quantum matter fields and the fields of molecules and so on. I think there are many of these organizing fields that I call the morphic fields, and they exist at all levels of complexity."[59]

Morphic fields imbue the transitions between levels of complexity, disrupting prior form and introducing new principles of organization. The chreodes are not chemical substances or molecules but, like chreods, centers of resonance radiating shapes in multidimensional frameworks and crossing space and time instantaneously.

Behavior learned by one population of monkeys is immediately transmitted to other monkeys elsewhere in the world. Moths, snakes, and humans likewise transmit activating shapes globally and across generations.

Chreode "telepathy" and telekinesis are not only properties of living things. When a previously unknown crystal pattern occurs in a factory it may soon begin turning up at other plants. The present "illusion" is that seed particles are blown planetwide through the atmosphere from the original crystals, but, according to Sheldrake, it is the crystals' new chreode resonating across dimensions.

In this fashion the chreode of "bee-ness" travels in no time at all between generations to reach newborn bees with scripts of their activity, and a turtle chreode trains recently hatched babies to rush into the sea (and later provides them with maps to find their way back to their birth sites to lay their own eggs). Other chreodes explain the highly refined social patterns of termites at birth as well as the migratory routes of swallows inherited from generation to generation, the territoriality of cats, and the mastery of language by infants. Sheldrake elaborates in an interview:

"I am suggesting that this memory that each species has is not stored in the genes, which is the conventional view, but rather is drawn upon directly by the process I call morphic resonance. An organism tunes in to similar organisms in the past.... Memory is not about space; it's about relationship to time—and it's cumulative. The whole idea of morphic resonance is that the past is potentially present everywhere, and that you gain access to it by resonating with it. And out of that grows the future...."[60]

The genes therein function as an abacus for tracking order and indexing events already certain to occur.

SHELDRAKE INVOKES MORPHIC RESONANCE to explain simultaneous inventions and discoveries (among them, nitrogen in 1772, oxygen in 1774, the telegraph in 1837, photography in 1839, the planet Neptune in 1845, the telephone in 1876, the phonograph in 1877, etc., each by at least two individuals with no contact).[61] Just as ancient ants transmit to modern ants through chreodes, human society is in touch with ancestral and future stages of itself (a theme dramatized in the movie *The Philadelphia Experiment* in which the "inventor" of an entire technology was accidentally blasted back in time from a future in which that technology already existed, allowing him to "invent" it). We are now transmitting chreodes to both our ancestors and descendants.

Nature, a British scientific journal, called Sheldrake's book, *A New Science of Life,* "the best candidate for burning there has been for many years."[62] A more favorable British magazine, *New Scientist,* offered a prize for anyone who could devise an experiment that would at least suitably test Sheldrake's hypothesis.

MORPHIC FIELDS ARE A USEFUL CONCEPT in that they have the range and complexity necessary to explain the plans of phyla and even embryogenesis itself. They transcend and encompass biological fields in such a way that the origin of life from inanimate matter and the induction of genetic form by prior gene-directed form are no longer etiological problems. But a morphic field is merely a metaphor, an intellectual construct like a gene. It is not even as "real" as a gene, for it lacks material basis in anything as concrete and textured as a chromosome. It is unfortunately also a countercultural placebo. It serves an epiphanic rather than a scientific function. It represents an ideological attack on science's sterility.

The goal of Sheldrake and other sacralizing biologists is to reintroduce spirit and meaning into the sterile mechanisms now proposed as the sole basis for life, yet without violating their deep algebraic structures and with attention to salvaging and reframing the technical language and conceptual rigor by which they are contrived. As the terminologies of physical science and transpersonal psychology are blended, the dehumanized biological field is replaced by "a field of living energy which is self-conscious and has a primary thrust toward growth, maintenance, restoration, and the development of optimum wellness for ever increasing self-knowledge."[63]

This is an admirable venture, a worthwhile antidote for our oppressive reign of quantity—for explaining the complex organization of primordial life, for getting genetics and epigenesis together.

But chreodes are not new. All that is novel are the attempts to present them as scientific rationales and experimentally testable hypotheses. The West has known chreods and chreodes by other names for millennia. They are the talismans and stellar signets of sixteenth-century natural magic and appear full-blown in the third century A.D. writings of Plotinus:

"Particular entities thus attain their Magnitude through being drawn out by the power of the Existents which mirror themselves and make space for themselves in them. And no violence is required to draw them into all the diversity of Shapes and Kinds because the phenomenal All exists by Matter's essential all-receptivity and because each several Idea, moreover, draws Matter its own way by the power stored within itself, the power it holds from the Intellectual Realm."[64]

When science presumes to concretize archetypes and spirits, it loses its value as science. Once we are discussing chreodes we might as well invoke prana, ch'i, and Navaho sand-paintings; these at least come from sacred lineages and are refined by millennia of rigor. A chreode is finally just a packaged sound bite, of no value to thermodynamics. If the event it describes "exists," it will only be in some immaculate way having nothing to do with chreodes.

The thread is utterly and completely unknown.

IN THE END, MOST UNIFIED BIOLOGICAL-FIELD MODELS, including those yoked to quantum mechanics and space-time acrobatics, are wild-goose chases, caught up in valorizations of tropes and sound bites, blind to how nature makes aeonic changes and establishes order among systems subject to entropy.

Genes clearly are not alone, nor are they first. Something right on the heels of genes, or that genes are on the heels of, braid together code and codable stuff, then sort them through quasi-random events that, while selecting the fittest of them, are overridden by another kind of order—systemic, hyperdimensional, or both. Those forms that emerge from this broth continue to be braided into new, more intricate forms.

The invention and construction of organisms has occurred on a scale of unimaginable aeons, during which mechanisms of intricate and minute assemblage have sealed themselves *in* one another over and over, packing morphologies so tightly across protein layers that cause and effect are camouflaged within ever-emergent complicacies of microstructure. This happens in such a way that a raw (yet perfect) synopsis of hyperlinks among discrete ancestral organisms emerges in almost the same way each time ontogenetically, while seamlessly adding novel and inextricable twists of structure. Forms become concealed so deeply within one another that chronological specification, to all intents and purposes, vanishes, offering a deceptively guileless surface of genetically-indexed physicochemical effects—so deceptive that biologists are led to believe they can recreate (or hypothetically reconstruct) the equivalents of etiological phenomena. Their so-called historical events in a molecular environment are actually technocratic metaphors, fables, and blind catechisms.

The true causal factors—even if they were physicochemical once—have been so infinitesimally and microtexturally woven among themselves and one another that they provide no hope of hypothetical reconstruction in a modern syntax. Their unknown avenues of incipient structure (and function) manifest seamlessly, not as the protoplasmic computers by which we represent them but as simulcra impersonating events in a conventional universe—laboratory ghosts. No matter what they look like now, their native domains and originary principles we don't begin to know.

The great mystery is that atoms, molecules, and cells collaborate perfectly on organisms, episode after embryogenic episode, ignoring (first of all) that genes cannot and do not provide a plan (do not even name a kingdom, let alone a species); secondly, that DNA code is vacuous without a hierarchy of inexplicably antecedent structure; and, third, that the entirety of molecular energy for embodiment must be summoned, catalyzed, and deployed anew each ontogenesis. Scientifically inclined humans almost smugly expect to gestate fresh humans (their children) by ordinary laws of matter, but in fact the guarantee they are counting on exists not on the surface (where it deceptively and guilelessly seems to happen), but deep in a thing compressed and condensed within another thing, packed and compacted within *another* thing, reminiaturized and enclosed again and again within an almost bottomless, timeless sequence of ova and metacellular grids, one within another within another (not so much *within* as flowing right through the very substance of), all the way back to mere simple chemical reactions.

Genes exist only insofar as cell chemistry gives them context. Yet the inside of a cell is permeated by the insides of tissue layers it emerges within—tissue layers that are permeated by and inseparable from an organism. The organism is assembled out of and metabolized by an environment. The environment, while microbiologically invading and altering cells and tissues, flows across topographies and semi-permeable boundaries of a planet. The planet congeals out of galactic dust showered by cosmic debris. All of these indeterminate substances and signals crisscross at multiple levels of structure and function. In fact, a gene requires the whole universe to express itself, but that universe requires genes in order to transmit interstices and delicacies of animate form.

PARENTS ARE COUNTING ON NOTHING, on a chimera of logic and law. The thread that holds embryogenesis together is utterly and completely unknown. Its serial (or nonserial) nature is unknown. How it "knew" to fashion prior networks for its informational helices is unknown. How it inveigled itself into physical form is unknown. How it invents and corrals its aesthetics and phenomenology is unknown. How it keeps assembling more and more complex living structures without an authorizable genetic plan is unknown.

We have not begun to approach the real mystery. Declarations of "holism" and telekinetic resonance do not convert physical equations into biological systems, and "general systems theory" itself is a sister to "probability theory" and does not explain the elevation of blithering nonsense to elegant design. Something is missing between the inanimate drawing of the cards and the emergence of the game complete with rule books and players.

Chaos, Fractals, and Deep Structure

The Statistical Basis of Life

THE TRADITIONAL VIEW OF BIOLOGICAL SYSTEMS is that they represent a kind of special landscape by comparison with the tumult of oceans. A planet with life forms—episodes of prey and predator (wormlet and fishlet), and ultimately tribes, languages, and civilizations—is presumed to be qualitatively different from present models of Jupiter or Saturn which, although they comprise their own complexities, are gargantuan chaoses without discrete, meaningful events by Earth's standard. Their flux is incomprehensible: small eddies within bigger eddies (running up and down a ladder from eddies many times the size of Eurasia to wee invisible ripplings), all incessantly consuming energy and moving randomly.

Reconstructing an imaginary trail of events leading from such chaotic flux to the first life forms is well beyond human calculation. However, the advent of computers encouraged the fantasy that if we could discover every variable in a system (every atom and its temperature, or similarly, every factor influencing a marketplace), we could predict the outcome of that system at each moment. This naively optimistic viewpoint made it hypothetically possible, from knowing the universe's initial state and the laws that govern its change, to predict its history (and even its biologies and sociologies) till the end of time.

From the 1950s into the 1960s evolutionary biologists isolated supposed constituents of the Earth's primeval brew and ran them through speeded-up computer time, setting creatures aswim in cybernetic oceans. Getting the first animate droplet to coalesce in ocean water subject to the Second Law of Thermodynamics—which requires that all structures (including any incipient life form) inevitably degrade and disperse—yielded plots worthy of Gothic novels with molecules as characters.

Sesquimillennial epochs and fortuity alone could not explain vast, deeply organized realms of protoplasm. No model ever created satisfactory terrestrial organisms, though many gave the illusion they could within the time frame of the emergence of life (as we reconstruct it from fossils) after the postulated origin of the Earth.

MORE THAN MIDWAY through the twentieth century most astronomers and biologists considered the stars vacant, our planet an implausible garden requiring its own unique explanation. After all, it was amazing enough that such unlikely things as living creatures emerged even once. No matter how vast the universe, it was considered highly improbable that any bionts could have syncretized through matter under more inhospitable strictures likely prevailing elsewhere.

Since the late 1960s this bias has turned around; we expect the universe to be teeming with life forms and civilizations—other planets quite congenial for their own indigenously adapted bionts. We presume that, one way or another, chaos organizes into metabolizing structures. Life is now considered natural and ubiquitous, quick to arise on planets warmed to a moderate clime by proximate stars. Biology is an assumed outcome of the most probable chemical reactions befalling the most ordinary atoms in the universe. Fossils in Martian meteorites and redatings of the first microorganisms and wormlings on Earth (pushing them dramatically close to the formation of the planet) suggest that, for reasons we don't fully understand, biology is an ordinary proclivity of the cosmos. Once initiated, it is considered capable of building on itself, of diversifying by genetic refinement and natural selection alone.

The Power of Infinitesimal Events

BIOLOGICAL SPECULATION TOOK A DIFFERENT COURSE in the 1960s as an offshoot of the inquiry into highly complex and marginally predictable systems—fluctuations of animal populations, semi-periodic flooding of rivers, and irregular movements of commodities markets. Silicon chips were the great equalizer, for they provided a single context for the study of stock markets, cells, and galaxies. On a primitive computer at the Massachusetts Institute of Technology in Boston, meteorologist Edward Lorenz modelled the movement of clouds and winds across the globe. In another cubicle in Westchester County, New York, on more advanced machines at IBM, Benoit Mandelbrot translated variations of cotton prices and personal incomes into equations. As deviant as the ingredients of these systems were from one another (operating at totally different scales of abstraction), they shared a hidden mathematical structure and periodicity, and of course they were transcribed into the same algebraic language. At their core, they generated startlingly similar

equations which seemed even to fulfill the Newtonian promise that the world proceeded "along a deterministic path, rule-bound like the planets, predictable like
eclipses and tides,"[1] long after, in fact, that promise had been dashed by quantum
physics. Just as randomness had intruded where it was least expected (at the heart
of the material universe, the atom), now order reemerged where it made least sense
(across boundaries of helter-skelter events).

To Lorenz the goal of understanding weather patterns seemed modest and reasonable within the existing bounds of science. A computer could condense months
and years into hours. If one programmed enough variables (the temperature, speed,
and pressure of wind and water, and the gradients of regional landscapes) into silicon, then eventually, from the cumulative thrust of calculation upon calculation,
future atmospheric conditions would be fully displayed.

Yet the interaction of factors behind even a fractional degree of temperature is
simultaneously so manifold and so infinitesimal as to defy capture. One can never
know all the agencies in a simple system of wind, fog, clouds, rain, etc., because
there is always one more middleman ... and one more ... and so on. The components providing a planet's weather juggle (and are juggled by) instability and turbulence at every point along the ground, in the air, and from the ionosphere and
beyond. No number of satellites in orbit feeding computers on the ground will ever
record the role of every draft and rivulet, branch and wagging tail, in creating heat
and motion. Even the wobbles of Venus and Mars and the flutter of a butterfly's
wings in Tokyo infinitesimally alter barometric pressure in New York.

Outside an idealized Newtonian, Laplacian universe with its orderly, deterministic probabilities, real systems of incalculable components fluctuate wildly and
capriciously. Lorenz stumbled onto this condition when, rounding off his decimals
(.506127 to .506), he innocently perturbed artificial weather patterns into El Niños.
He had prejudged that extremely tiny changes would be negligible, absorbed in the
general thunder and thud. This was not true. Elements that might not amount to
a draft across a room in a day or two eventually coalesced into tornados. A trifling
puff of air had the same ultimate consequences as a hurricane.

Lorenz's problem turned out to be not only the linear one of knowing the kinetic
state of every atom at every moment (or enough to make a sound prediction at the
level of the system under analysis); it was the overall nonlinear dynamics themselves underlying systems. These dynamics everywhere render the positions and
fates of individual atoms—already unknowable—secondary and irrelevant. Computer analysis showed deep systems to be changeable not just along simple linear
paths but in bizarre ways in which aspects of them become unexpectedly entangled
in one another, producing entirely unpredictable landscapes.

THE UNIVERSE IS MADE UP of uncountable events at depths beyond thousandths, millionths, billionths, trillionths, or even quadrillionths, any of which might (most often) dissipate entirely at a fractional level or (on rare occasions) build into an avalanche transforming an entire domain. Outcomes thus depend on octillions of minute, disjunctive, and irregular features that can neither be predicted nor controlled. In a crude but deceptively complex example, the spinning buckets of a waterwheel do not simply move faster with more rapid flow, for filling with water also slows them down and produces chaotic effects, including unaccountable reversals of direction. Patterns negate and reinforce one another in exquisitely countervailing ways.

If we were to assemble an exact replica of the Earth down to the position and motion of every molecule at any moment before life began, we still could not model what then transpired because of incalculable variables driving the system. A clump of oceanic foam might stay together or be fissioned by a wave, its pieces travelling ultimately to opposite ends of the planet; yet no amount of prior information could enable one to predict which subclump went where, the exact pattern of fissuring, or the fate of all *its own* individual subclumps.

Just as each snowflake presents twisting dendritic boundaries to the air as it falls, picking up water molecules, spiking out in a way that records wind direction, humidity, and gravitational resistance in all the zones it passes through, so each aboriginal coacervate and zooid was a replica of the history it had experienced, until it began to maintain its own structure genetically (and even then its future replications were subject to transformation through random mutations).

We face similar dilemmas in daily life. Does choosing paper over plastic at the supermarket checkout line aid or harm the environment—and by whose parameters of evaluation? Does taking a car to be scrutinized at the local mechanic before a trip make an accident more or less likely? Perhaps the time spent at the garage, though providing for a safer vehicle, leads to a later start and a collision with a drunk driver. At any moment one is subject to vigintillions of unknowable factors operating at different levels of widely dispersed systems.

When scientists are explaining the effects of colliding billiard balls in Milwaukee, they do not take into account whirlpools on Jupiter or the falling of a leaflike entity on a planet circling a sun in Andromeda. These influences are considered too small to fuss about. Yet what is mostly true in the case of billiard balls and autumn leaves turns out to be less true when applied to systems like weather—or that epiphenomenon of weather, life. The falling of leaves—or what passes for leaves—on other planets may have negligible sway on mundane events (for all we know), but episodes that might have once been considered almost as absurdly peripheral as these "leaves" do have effects when their totality is considered over time and within the dynamics of turbulence.

Perhaps a single twirl of foam in a tide is vitiated in the ocean (that is, neutralized by incalculable other twirls), but constant twirlings of foam ultimately produce patterns that diverge radically. Each wobble and variation may lie out beyond the novemdecillionth decimal point, but if it is sustained in some way and reinforced by repetition, it will contribute irrevocably to the system's outcome.

At each moment, oddities—inexplicably ordered patterns—must occur, and no one can ever know what would have happened otherwise, i.e., in the same system lacking any one minute event. "The computation is so vast that it can only take place in real time, the very time of . . . lived experience; and the universe itself, in its entirety, is the only computer big enough to crunch all the numbers."[2]

Clearly we cannot fit this Univac in the house. But merely knowing about it spurs fantasies of the marvels we might perform on somewhat smaller models.

We need all the computational power we can inveigle, for we are dealing with environments that have not only infinite freedom but infinite dimensions of enactment. The emergence of biological organization is the advent of not just complexity but *infinite, infinitesimal* complexity.

Unexpected Sources of Order

FROM OUR VANTAGE WE SEE ONLY EVENT UPON EVENT, ad infinitum, without mercy, without relief, without clarification or intelligence, extraterrestrially to the maelstroms of Jupiter and the most distant galaxies, inside to our bloodstream and individual cells. Yet there is apparently an unknown and inherent principle of order arising relentlessly from chaos and complexity themselves.

What laws govern the formation and persistence of Jupiter's red spot (several times larger than the whole Earth) or similar abiding cyclones on Saturn, Uranus, and Neptune? "You see this large-scale spot, happy as a clam amid the small-scale chaotic flow, and the chaotic flow is soaking up energy like a sponge. . . ," says one scientist. "You see these little tiny filamentary structures in a background sea of chaos."[3]

The red spot is not inherently different from the erratic tumult around it. Yet it is self-organized, sustained, and regulated by that same twisting, tumbling, agitated medium—chaos, yes, but stable chaos. In truth, order and disorder arise in the same environments. Dissipation occurs in one part of turbulence; in another part a vortex emerges. Chaos and coherence emanate together, inseparably so.

Jovian meteorological patterns resemble Lorenz's weather systems and the mathematics that arose from Benoit Mandelbrot's analysis of cotton-price variations.

Although each particular market blip was circumstantial and unpredictable, the equations for daily changes and monthly transitions replicated one another exactly, ignoring even chance and scale. The curve maintained its basic shape for sixty years despite the intrusion of World Wars I and II. Organized bands and cyclonic spots likewise emerge on all the gas giants in this solar system.

Infinitesimal and haphazard factors that make predictions impossible in a Newtonian universe somehow coalesce into their own determinate order. Despite the madcap destructiveness and dissolution of their parts, their sums (in complex and vast environments) are cohesive and elegant.

Amid arbitrary and chaotic factors, Mandelbrot seemed to have found an inexplicable governing principle, a phenomenon of scaling "with a life of its own—a signature."[4] Chaos had proven to be a mask over another, profound mathematical order. This is what salvages us from the horror of infinite spaces and infinitesimally infinite depths, what redeems the bottomless hell of detail and fractionality.

Randomness represents the true profundity of the universe, its creative harmony. Nature apparently sustains itself and evolves not despite chaos but because of it.

Tipping Points

TIPPING POINTS—a statistical phenomenon discovered during the 1990s—shed new light on the infrastructure of all cause and effect. Scientists had long known that infectious diseases do not surge and wane according to linear rules. Up to a point new infections enter populations at steady, non-epidemic rates. Then something intervenes—a seasonal ceremony, a civil war filling refugee camps; greater numbers of people come in contact with one another; more cases occur. Initially the increase is solely quantitative. But when the tipping point is reached, a very small number of additional infections explodes into an epidemic.

The same pattern that epidemiologists observed with flus social scientists and ethologists saw in zoological systems. When an animal population is pushed below a certain level, the species is set on a course toward extinction (no matter how many buffalo or eagles are later born). Yet sometimes one or two additional breeding adults, a number seemingly too small to change the course of history, provides the critical mass not only to salvage the species but to cause it to flourish.

In either case, a few new flu victims (or rhinoceri) make a stupendous, unlikely difference.

Criminologists recognized that the types of phenomena they were studying also maintained statistical levels above which dramatic increases of anti-social behavior occurred and below which equally sudden drops ensued. Regardless of powerful individual motives behind each episode, homicides, robberies, and rapes seemed to follow statistical patterns. In each one-time good neighborhood, deterioration was gradual for a very long time. Yet, as signs of danger multiplied, store-owners and families fled, sparking the flight of even more families. Suddenly, seemingly overnight, the area had become a full-fledged slum.

In the mid 1990s, police and social scientists began to pay more attention to seemingly superficial signatures of deterioration—graffiti, peeling paint, broken windows. In city after city they found that the mere repair of windows and removal of tags had effects far beyond what was predictable from the cosmetic events. When neighborhoods were restored (even marginally), more people moved in, boosting confidence and attracting additional people.

Tipping points were used to explain the sudden drop in crime in New York City through the mid 1990s—more than fifty percent of homicides and burglaries in some areas.

Thresholds no doubt played a similarly major role in evolving populations. Characteristics built up very slowly in gene pools until suddenly they spread like wildfire, either from a few additional mutations, an influx of outsiders bearing a particular gene, or some other, imponderable factor (like skewed distribution of conjugating pairs for a few generations). Once a new equilibrium was established, even though it had been catalyzed initially by a tiny blip, it maintained itself in the population and changed the demography and biology. Land-dwelling organs, bipedalism, and even consciousness probably arose once in such fashion.

According to one policeman in the Seventy-Fifth Precinct of East New York, "there was a time when it wasn't uncommon to hear rapid fire, 'like you would hear somewhere in the jungle in Vietnam....' It is possible [now] to see signs of everyday life that would have been unthinkable in the early nineties. There are ... ordinary people on the streets at dusk—small children riding their bicycles, old people on benches and stoops, people coming out of subways alone."[5]

But this is also a description of the Earth just a few billion years after its separation from the Sun.

What is the Relationship between Chaos and Morphodynamics/ Morphogenetics/Natural Selection?

"The central force on a planet is nothing but a gravitational attraction toward the sun."

The inquiry into the embryo and its agency—into the form, source, and nature of life—began, in the lineage of Western philosophy, with Aristotle or, more accurately, the pre-Socratics and their own unknown progenitors, perhaps along the cusp of the final millennium B.C. Democritus inherited a Stone Age animism from which he opined, "By convention there is sweet, by convention there is bitter, by convention hot and cold, by convention color; but in reality there are only atoms and the void."[6] He grasped the componential and transitory nature of form. When Heraclitus warned, "The fairest universe is but a heap of rubbish piled up at random,"[7] he foreshadowed the algebraic yardstick by which all cosmic edifices would one day be measured.

A little over a century later, Aristotle addressed phenomena in a more systematic, empirical fashion; he understood that something dynamic happens within an egg when it differentiates, when complexity arises from homogeneity. He did not insist that a cavalcade of interlocking shapes must lie hidden, preformed, in each embryo; instead he deemed that form and complexity develop intrinsically and reciprocally. Depicting nature as made up of four prime constituents (earth, air, fire, and water), he proposed that all mundane events consist solely of these elements in variegated interaction over time (i.e., bubbling, steaming, blending, leaving residues). While transcending the superstitious scholia of his culture, he could hardly ascertain the true origin of elements or their kinetic pathways, although he knew that such things must exist temporally and perform with inviolable regularity.

He conjectured the heavens to be made of separate, simpler substance. Composed of quintessence *(quinta essentia)*, the celestial orbs revolved in endless perfect circles around the universe's center, the Earth.

Throughout the Middle Ages the goal of Christian philosophy (and by default, science) was to assay how much of the terrestrial sphere was spawned by independent, atomistic processes (including ones under divine orchestration) and how much of it was imprinted by nonmaterial, supernatural templates. Having mislaid the texts of Greek science, Western Europe (by comparison with the Islamic south) stuck mostly to theological assertion for over 1500 years. In the thirteenth century Thomas Aquinas rediscovered the classics and merged Christian and Greek cosmologies in a set of icons and canons, transforming Aristotle into a priest.

It took another three hundred years for the first astronomer-physicists—William Gilbert, Nicolaus Copernicus, Tycho Brahe, Francis Bacon, Johannes Kepler—to confront Aristotle on his own terms and open the doors of celestial mechanics. By shattering the dichotomy between earth and sky, microcosm and macrocosm, Isaac Newton educed universal laws of mechanics and laid the groundwork for a modern science (thermodynamics) describing the passage of mechanical energy and heat within and between systems.

In 1686 in *Philosophiae Naturalis Principia Mathematica* Newton wrote: "I derive from the celestial phenomena the forces of gravity with which bodies tend to the sun and the several planets. Then from these forces, by other propositions which are also mathematical, I deduce the motions of the planets, the comets, the moon, and the sea."[8]

In a Newtonian universe not only do all stars, planets, and moons affect one another gravitationally, but "all objects in the world attract one another with a gravitational force like that existing between a falling stone and the earth; consequently the central force on a planet is nothing but a gravitational attraction toward the sun."[9] Gravity as prime mover became an axiom from which no subsequent science could depart without risking a return to either vitalism or theology.

Newtonian Mechanics and Life's Complexity

After Newton, naturalists gradually embraced the cosmic machinery and attempted to cram all phenomena, plants and animals among them, into its hydraulics and gears. According to the reigning teleomechanist paradigm of the nineteenth century, life is not a unique force, emergent or otherwise; it is simply the sum effect yielded by a singular organization of physical and chemical materials. Animation requires no exogenous spark and does not violate the law of conservation of energy.

In the late 1830s German physiologists Matthias Schleiden and Theodor Schwann attempted to explain the fissioning and proliferating of cells by purely physical laws, likewise their differentiation and assemblage into functionally organized domains. In order to gain residency in the Newtonian kingdom, models of cells had to carry out concrete (if unidentified) chemical reactions along mandated pathways, as energy and mass were constrained, condensed, and rechannelled; thence snails and dragonflies would emerge from mineral-like chrysalides.

For Darwin, a generation later, to unlock the riddle of ontogenesis and speciation (egg and chicken), he had to apply, both as dogma and cultural datum, a facsimile of the quantification of matter and energy codified by Newton. Newtonian physics underlay Darwinian biology, not immediately and explicitly but inherently and profoundly, for Darwin had to get energy, heat, matter, and form into his system without introducing vitalistic effects.

Darwin moved seamlessly from observations of nature to a sense of provident order to a body of laws — a journey set down in his incidental writings and the revolutionary book published in November, 1859 — *On the Origin of Species by Means of Natural Selection*. What Darwin saw on the islands visited by the *HMS Beagle* — an exquisite interplay of metabolisms among environments — disclosed to him the shaping engines behind the whorls of nature, not their origination (of course) — though he speculated what that might have been — but their activity late in the game, after likely millions of years ("... animals on separate islands ought to become different if kept long enough apart, with slightly differ[ent] circumstances . . . Galapagos tortoises, mocking birds, Falkland fox, Chiloe fox, English and Irish hare . . . Aegyptian cats and dogs, ibis. . . . As we thus believe species vary, in changing climate we ought to find representative species; this we do in South America closely approaching"[10]). From these adaptive lineages he construed a cosmic, transplanetary biology corresponding to Newton's universal physics.

Darwin could not reduce life to Newtonian equations or anything resembling them, in part because the factors are too many and too deeply imbedded, and in part because he did not have access to the miniature realm inside the gene where the denominators rest. He also lacked any method of systematizing highly complex, highly irregular systems (like seas and atmospheres) in the way Newton and Galileo could quantify orbits of planets and moons and the descent of apples. Thus, he made guesses and committed blunders which now exile many of his writings outside the theology to which he gave his name. He was far too involved in an economical and intelligent view of nature to write biology's ultimate callous equations.

When Weismann disproved, for all intents and purposes, the inheritance of acquired traits, he marginalized the unriddling of adaptation as somatic information flowing through pangenetic pathways. This opened a since-widening gap between Darwin himself and Darwinism. The latter is now a blind faith, giving rise to various neo-Darwinisms in a manner in which there can be no neo-Newtonianisms. Scientists accept the Newtonian sector as a quasi-demarcated zone in a vast universe; that is, Newtonian laws do not apply in the usual manner to subatomic particles. Yet, ironically, the same ilk of researchers have continued to impose an etiquette and rigor on life that atoms, by demonstration of Max Planck and Werner Heisenberg, evade. It is as though once things have become molecules, they are required to honor Newton's coda (even when they are parts of organisms with mysterious and disjunctive attributes).

The entire neo-Darwinian inquiry into the relationship between morphodynamic and morphogenetic factors is an attempt to square the exquisite shapes of matter and energy in biological systems with the original landscapes of planets before biology; not only to extrapolate the possible assemblage of genetic apparatuses by Newtonian equations—an enterprise all but impossible—but to describe how those apparatuses continue to appropriate form from a "heap of rubbish piled up at random" (i.e., to rehabilitate natural selection).

Our world garden, fecund with peonies, roses, passionflowers, and the like—even *sans* animals—must be explained in the context of wind and water, gravity and mass. What has somehow been captured and conserved are shapes, resonances, symmetries. Their umbels, cymes, and pigmented pinwheels are now all about us and in us, atomic essences frozen into skins, as the universe probes its own nature the only way it knows, the only way it can. Even Darwin realized that this life dream is somehow different from and more than all the sums and analyses of how it came to be:

"The weather is quite delicious," he told his wife in a letter from Moor Park in April of 1858. "Yesterday, after writing to you, I strolled a little beyond the glade for an hour and a half, and enjoyed myself—the fresh, yet dark-green of the grand Scotch firs, the brown of the catkins of the old birches, with their white stems, and a fringe of green from the larches made an excessively pretty view. At last I fell asleep on the grass, and awoke with a chorus of birds singing around me, and squirrels running up the trees, and it was as pleasant and rural a scene as ever I saw, and I did not care one penny how any of the beasts or birds had been formed."[11]

Probability

Dynamics gives rules for mathematical descriptions of the instantaneous state of any physical system as well as formulas for determining the past and future states of that system. Newton and Galileo were dynamicists of their era, concerned with a study of change, rate of change (velocity), and rate of change over time (acceleration). Darwin meant to be a dynamicist too.

But does life conform to dynamic principles? Schleiden and Schwann assumed so, but they were limited to deductions from gross aggregations of organic chemicals under unrealistically regularized conditions. Later in their century James Clerk Maxwell attempted to save Newtonian appearances by subjecting the more inextricable jumbles of nature to a novel set of laws. First, he proposed mélanges in which incalculably large numbers of elements interact in apparently random fashions; then, although he could not provide tidy equations for every element and event among their aggregates, he showed that their collective pathways could nonetheless be predicted with great accuracy merely by summarizing the individual entities and motions and averaging their total.

Maxwell introduced differential equations as corollaries under Newtonian physics that predicted outcomes for the collisions of millions upon millions of molecules of gases jostling randomly with one another. This algebra provided a means for congealing organic molecules into life forms, tepid slicks into robust microbes. Scientists have since averaged out trillions of complex but still gravity-driven, hypothetical inanimate phenomena and derived the bodies (and minds) of creatures subject at every stage of their development to the laws of thermodynamics. In theory at least, there is a Newtonian formula for life. The birth of the probability universe, however, did not so much salvage Newton and Darwin as translate both of them into another dimension.

Three main obstacles still stand in the way of a purely Newtonian biophysics. First, genetic networks are themselves complex beyond tracking or summarizing. Genes are not separate objects like beads on a rosary or molecules of gas; their arrays and trajectories do not average out. Second, chance plays at least as significant a role as natural selection. Thus, matter develops self-organizing properties that do not have accessible dynamic solutions or even proximal causes. Third, mind is irreducible.

Biologists and biomechanists in the lineage of Darwin, when attempting to quantify and systematize the origin of life and emergence of species, have repeatedly encountered nature's labyrinths and its confounding capacity to manifest order and subtlety out of chaos. Conventional dynamics cannot deal with the biosphere's quantum leaps in information and organization much as Newtonian physics cannot track the uncertainty states of electrons and neutrinos.

Phase-Space and Strange Attractors

Phase-spaces provide a terminology for discussing the trajectories of dynamical systems. Geometrical models describe the states of an object, of congeries of objects, or of an aggregation of averaged objects (like Maxwellian gases) in terms of the number of variables (degrees of freedom) that define them. Maxwell's differential equations were used to follow an object or aggregate through phase-space. Two variables—momentum and direction—were considered, each in three dimensions, giving rise to six dimensions. In a dynamical model of the universe, all systems travel through six-dimensional phase-space, ultimately stabilizing in regimes resistant to subsequent perturbation. The states in which they settle are visualized as attractors. A pendulum stops swinging because the drag of friction has put it at equilibrium, i.e., subjected it to a point attractor; in its present situation no further motion is possible.

Most systems are too busy to resolve as point attractors. A few stabilize in closed, periodic loops, i.e., limit-cycle attractors. The majority of systems in the universe are more complex than that; yet classical dynamics allows only point and limit-cycle attractors; otherwise, there is no attractor at all, only damnable chaos, Jovian disorder, perhaps the transient mirage of a false attractor. The mathematical idealization does not match the physical display.

When computers finally allowed the rapid modelling of phase-space within highly chaotic systems over long periods of time, scientists discovered a strange sort of inexplicable order (described earlier in this chapter). Massive, billowing ensembles, beyond classical dynamics, form patterns. These have no classical physical explanation but are presumed to be governed by a special region of phase-space—strange attractors (now called Lorenz attractors). Their exotic dances, spinning bands of order and tranquility out of bedlam, include cells arising in the dawn-time oceans of the Earth.

In 1985 the Institute for the Study of Complex Systems was established in Santa Fe, New Mexico, to forge credible models for the evolution of cell types and other discontinuous phase-transition events. There physicists, economists, biologists, and assorted interdisciplinarians developed new metaphors and domains of strange attractors to augment classical dynamics and natural selection. The title of a 1993 book by resident developmental geneticist Stuart A. Kauffman, *The Origins of Order: Self-Organization and Selection in Evolution*, betrays the extent to which neo-Darwinism had become fused with chaos dynamics.

Life is a bounded hurricane within a membrane.

In order to come alive, natural nonliving systems somehow drift out of thermodynamic equilibrium; they then maintain themselves in trigger-sensitive, stably unstable states, capable of incorporating nonlinear trajectories. Through something like this skein of events they self-organize.

Life probably emerged from systems that long ago explored the esoteric phase-space between order and chaos. In a narrow wash on the brink of chaos, on the edge of order, relatively simple elements interacted under native dynamics in ways that spilled out complex, semistable curd—a viscid burlap with emergent global properties. Protocell bubbles of preliving muck induced chemical reactions adding to their bulk, increasing their stability, and accreting other bubbles, their stuff ultimately corroding and reknitting into denser bubbles.

In chaotic preanimate systems, self-organizing and energy-dissipating vectors merged creatively. Natural selection must have favored energy utilization and transfer side by side with physical stability. Thus, organized chemistry within primitive envelopes became not only thermodynamic but incipiently energetic, metabolic. It is unclear if thermodynamics laid this groundwork along intricately girdled pathways or whether energy was organized in some other fashion for which thermodynamics provided molds and deepening channels. Somewhere along this path through the sacred labyrinth, what was becoming genetics met what was being organized by emergent properties of order within chaos.

Original cell gumbo incorporated density gradients, coupling buoyancy, interfacial tension, thermal diffusion, and adhesive differences (viscous forces), eventually using the richness of chaos to manufacture and sustain autocatalyzing hubs. Energy-and-matter flows with differential properties exploited their own products far more efficiently than did their chemical competitors. Autocatalytic cycles developed true membranes and became structures; structures interacted with one another and meshed into deeper, finer structures; thermodynamic potential increased. The result was a kinetic pathway, ultimately a cell.

Such an etiology somehow underlies the system of life, though retroactively life cannot be reduced to it.

BUBBLES, AS WE HAVE SEEN, are life forms in the making. Waterspouts and tornados are kinetic pathways, raw metabolisms. The red spot on Jupiter is not only "happy as a clam"; it is the forerunner of a clam. Life is a bounded hurricane within a membrane, i.e., an energy flux with restraints. Yet the demarcation between what is alive and what is not is blatant. Beyond the strange attractors that might have coaxed its predecessors into protoplasm, wriggly, gelatinous seaweed now arises totally separate of the rocky gneiss on which it rests. It maintains ontologically novel properties.

Once autocatalytic systems became ecologies, informational macromolecules in their midst established and stabilized homeostatic, metabolic pathways, allowing external environments to be read and responded to creatively. Resources drawn selectively into autocatalytic systems were recycled; energy flows were subtilized. As informational macromolecules became more intricate and iterative, ecologies turned into ontogenies. Organisms cut themselves out of the environment, i.e., self-selected as integrated ecosystems. Phylogeny represents an unfolding of ontogenetic, ecological events through phase-space.

"There is a grandeur in this view of life with its several powers," Darwin writes at the conclusion of his opus, "having been originally breathed by the Creator into a few forms or into one; and that, while this planet has gone circling on according to the fixed law of gravity, from so simple a beginning endless forms most beautiful and most wonderful have been, and are being evolved."[12]

The Twisted, Tangled, and Intertwined

SCIENTISTS NOW KNOW THAT THERE IS NEVER PURE DISORDER. Wherever randomness seems to rule, a kind of nonrepetitive order is emerging—just as wherever order seems to exist, disintegration is already underway.

Stated differently, disorder has in it an exquisite order, a sequencing of near repetitions suggesting a pattern. Chaos may be preliminarily described as "a kind of order without periodicity" or "the complicated, aperiodic attracting orbits of certain (usually low-dimensional) dynamical systems."[13] A stable chaotic system maintains persistent, periodic irregularity. It is not unstable but metastable.

The old linear mathematics of Darwinian biology and general systems theory says little about a nature that is prevailingly nonlinear, a universe of differential geometry. Yet chaos has a fine structure. Without extraneous thermodynamics or inputs of energy it doubles and bifurcates. Whether inanimate or animate, it migrates from spatial homogeneity and anarchy to patterning. Hidden fluctuation points and sites of unexplained bifurcation improvise almost vegetal symmetry and aesthetics.

"Clouds are not spheres.... Mountains are not cones. Lightning does not travel in a straight line. The new geometry measures a universe that is rough not rounded, scabrous, not smooth. It is a geometry of the pitted, pocked, and broken up, the twisted, tangled, and intertwined. The understanding of nature's complexity awaited a suspicion that the complexity was not just random, just an accident. It required a faith that the interesting feature of a lightning bolt's path, for example, was not its direction, but rather the distribution of zigs and zags.... The pits and tangles are more than blemishes distorting the classic shades of Euclidean geometry. They are often the keys to the essence of a thing."[14]

Protean life forms did in fact have far more—and more subtle—resources to draw on than are obvious from the laws of physics and biochemistry. And, in addition, these zags and tangles were not semi-miraculous events at the dawn of time—mere fortuities—but are part of the average arsenal of any breeze or puddle.

The paradox is that we are looking at randomness; yet randomness only exists with relation to any one event. Where a particular piece of sea foam or gust of wind travels is in itself dependent on many variables—likewise whether its energy is degraded or synergized—but the overall pattern maintains a chaotic dynamism.

The origin of cohesive biological fields may well reflect the way in which order develops out of aspects of disorder. The unceasing patterning of disorder, while not foreshadowing or guaranteeing life, intimates that oceans throughout the universe, unpredictable in a Newtonian sense, are probably subject to a kind of patterned bedlam that favors membrane-enclosed forms and their morphogenesis. Chaos embodies preadapted "oil slicks" that precede biology, that precede even crystallography, and yet are apparently abundant at the core of any primordial ocean on any planet.

The Intricate Structure of Disorder

WHEREAS THE PROPONENTS OF SYSTEMS THEORY have declared that the features of constituent entities are fundamentally transfigured in passage from the realms of atoms to those of molecules, cells, and organisms, respectively, chaoticians insist that there is a dynamics of the whole that does not change from quarks to galaxies, from glaciers to birds, or from genes to stock markets. It is almost as though the randomness and "everything-everywhere" quality of matter tires of its unruly romp and organizes by default.

But the mess is so bad it cannot be sifted and arranged on its own level at its own scale. While ripping itself into pieces of pieces of pieces forever, disorder concedes higgledy-piggledy and kicks tidiness up the ladder to the next dimension within which the prior bedlam vanishes or becomes purely componential and proportionate. Chaos at one level of random activity generates exquisite order and organized behavior at another. Complexity births complexity. There is no other place for chaos to go, and there is no other place for order to come from. Molecules become cells, and cells become organisms. Of course, there are discrete and novel properties introduced at different levels of organization, but *the intricate structure of disorder itself does not change and this alone is what transmits phenomena across scales.* Systems evolve not because of stable destabilized field states succeeding one another from subatomic particles to organisms, but because the energy transmitting and sustaining them is at every level aperiodic, intermittent, diffeomorphic, and structured. The universe is not a muddle; it does not lose its way in its own depths.

ONCE UPON A TIME, genes were the only hardcore game in town. Life was, by far, more complex than genes, but it was not more complex than their hypothetical algebra (which was not hypothetical in the face of Earth's extant biology). Now there is another game: complexity itself as an organizing principle. Life could still arise from random chemical cohesions generating unique temporal forms, with gene-orchestrated preservation and projection of those forms—but, at the same time, the random configurations themselves, the patterns they made, and (later) the sequential elements of their chromosomes were substantially and decisively boosted into existence by an inherent tendency of combinatorial systems to thicken and complexify. Physical and dynamic systems are configured not just by gravity, adhesivity, and convection, but by complexity itself, by a random proclivity to order and pattern. Genes may peerlessly encode prerequisites of form and transmit structured elements to each next generation, but another intrinsic morphology, arising automatically out of chaos, supplies endlessly novel, intricate designs for the emergent properties of all connective systems—meteorological to embryogenic.

Where formerly the two main models for life were genetic and epigenetic, now the center-stage combatants square off as genetic determinism and complexity theory. But these are not really in competition. Life needs the randomness and order of both in order to happen, for the causal mandate of genetics must preserve the random generic order of complexity, and the chaos of complexity must provide elemental properties for the establishment of order. In the end, genetics and complexity must merge; at the same time, we must not forget, they are not biology, they are metaphors for aspects of biology. Life itself ignores their paradigmicity, grabbing hold of something that is inseparably both yet neither of them, while exploding into being.

Fractals: Repeating Scales of Irregularity

FRACTAL NUMBERS (representing partial dimensions) provide a method for calculating the irregular regularity of things. If a contour cannot be measured in a pure linear sense, it can still be defined in terms of brokenness, or fractionality—its fractal dimensionality. Depending on the persistence of the measurer, a coastline can extend longer and longer seemingly without limits. Initially a surveyor can map just the major bays, coves, and inlets, then the little ones within those, then the ones within those, down to sub-sub-coves and sub-sub-inlets, all of them having coves and bays within themselves *ad infinitum.* At each new measurement, the length of the coastline will increase, although always within the geometric finitude of the actual physical coastline.

Furthermore, the patterns have a hidden recursive nature. Coastlines, though irregular, sustain signature irregularities at each level of finitude. A map of the coast of a bay in Maine will resemble a map of the whole Maine coast, but so will a map of an inch or less of Maine beach. As the scale of irregularity changes, from a half mile of a stream to a river of hundreds of miles (in one direction) or a trickle of a few millimeters (in the other), the gross pattern of the irregularity is self-similar and symmetrical. Great fruiting clusters branch into both smaller and larger clumps of fruits.

Temporally, likewise, patterns of order and disorder precede and follow one another such that the pattern of disorder each second contains patterns of order precisely resembling those also occurring each minute, hour, millisecond, and millennium of a series within which it is an interval. Phenomena themselves do not know or care at what part of a scale they exist or how long their forms have been sustained. Randomness with iteration across spatial and temporal scales *is* complexity.

The universe apparently extends itself by sustaining creative tumult across scale. Galaxies are thunderclouds massing and dissipating. A jet of ink from a squid is a nebula. Just as Jupiter is an opal fish writ large, in truth an orchid is galactic turbulence miniaturized. Without being told its scale in advance, we may find it difficult to gauge if a cloud is five thousand feet from our plane or a hundred feet away—likewise if we are looking at a photograph of a planet, an onyx crystal, or a membrane under a microscope. The mechanics may be vastly different across this range, but the relationship between chaos and complexity is fundamentally the same.

Fractal Uses of Space and Information

THERE IS A REASON the real world is scabrous, pitted, turbulent, and irregular. Despite D'Arcy Thompson, simple geometric objects are poor ways to fill up space because they are too smooth. Fractals reveal how things fuse, branch, and splinter; their objects crowd one another with their replicating roughnesses, squeezing, as it were, coastlines into coastlines. This is critical in the human body where tremendous amounts of functional tissue and information must self-construct and maintain themselves within limited topologies. "Blood is expensive and space is at a premium. The fractal structure nature has devised works so efficiently that, in most tissue, no cell is ever more than three or four cells away from a blood vessel. Yet the vessels and blood take up little space, no more than about five percent of the body."[15] The intestines are equally fractal (see Chapter 19).

Phylogenesis apparently chose this route to complexification. Biological networks began in closed sets of polymers that had the "capacity to catalyze one another's production."[16] Systems developed "properties of self-maintenance and self-replication ... in the structure of autocatalytic set(s)."[17] As these closed sets became pressure cookers for sustained metabolism, they continued to explore and catalyze themselves and one another, interacting, replicating, and combining to generate additional sets.

A mammal was a worm fractally imposed again and again upon itself in a way that maintained the creature's structural integrity by removing weight at some points

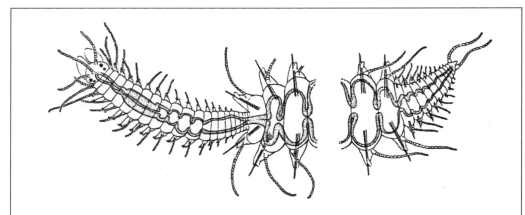

FIGURE 13A. Anterior and posterior regeneration from an original six-segment piece of a syllid.
From N. J. Berrill, *Growth, Development, and Pattern* (New York: W. H. Freeman & Company, 1965).

to accommodate density and organ depth at others. Interfacing from scale to scale by their roughnesses, surfaces filled gaps and surrounded bumps. Compact, dense matrices grew around and through one another.

With its inherent self-similarity, DNA does not have to specify every bronchus, bronchiole, and alveolus, every fiber of every neural network. All it has to do is spawn a reduplicative, self-similar paradigm of bifurcation and fractal arrangements, a panoply of iterative forms. Though never absolutely identical and operating at varying scales, these express traits of the same chromosomes. The complexity of tissue is the result of relatively simple instructions interpolating themselves in endlessly chaotic patterns, in the contexts of gravity and woof.

The limited amount of information in the spore of a fern would seem to curtail its intricacy. Yet a fern is complicated because of fractal repetition, using single alleles to build multiple levels of organization. Starting with meristem cell multiplication, "an excitable cytoplasm bifurcates to an initial gradient that is stabilized by plastic deformations of the tip, whose growth then makes possible the appearance of the next mode of a pattern sequence, an annulus."[18] Later tip flattening and bifurcation break the concentric symmetry of the annulus to recreate the whorl at a new scale. Bifurcation leads to morphogenesis, and morphogenesis stimulates further bifurcation; branching and spirality are transposed back and forth across smaller and larger scales. The moving boundary leaves behind congruently deformed and reproportioned shapes. "Ordered complexity ... emerges through a self-stabilizing cascade of symmetry-breaking bifurcations that have an intrinsically hierarchical property, finer spatial detail emerging within already established structure, as whorls arise from tips and fine branchings occur in growing laterals."[19] Genes define the "parameters that specify what morphogenetic trajectory [each] zygote will follow."[20] Morphospace is gametogenesis projected outward. The evolution of species is a radiation of successive patterns out of digressing morphogenetic fields.

Structures assembled using tensegrities and generic forces require only elliptical contributions from their genes. What is most important phylogenetically is that quite different structures can be transformed into one another by mutations of single genes. Inhibiting and unmasking one another's expressions, certain genes in plants, for instance, may operate as coefficients, turning on and off the production of leaves, sepals, petals, stamens, and carpels, using fields of dynamic tensions in the meristem.[21] Plants would thus represent diverging composites of concentric spiralling forces grounded in different primes. Leaves are their zero state, propagated when there is no other genetic interference. The contributions of other combinations of genes then lead to the synthesis of separate organs. Concentric leaf whorls and reproductive florescences are equivalent interpretations of one pattern with very different functional implications.[22]

Oscillatory Patterns

THE ROOTS AND BRANCHES OF THE BOTANICAL KINGDOM remanifest in the trees of the animal body—neural, vascular, and pulmonary. We see similar organization in the urinary collecting network, the biliary duct of the liver, and the His-Purkinje network of fibers carrying electrical current to the contracting muscles of the heart.

A heart embodies patterned order within the chaotic medium of its fibers. "It is a self-exciting system, designed to operate in a particular dynamic mode—the familiar thump-thump-thump of the repeating contraction wave pumping blood around the body,"[23] much like concentric patterns of chemical mixtures or slime mold aggregations but with greater velocity and rhythmic frequency. A heart retains the resonant periodicity and oscillation of the amoebas and organelles at the center of its field. Flapping flags, dripping faucets, rattling mufflers, and differentiating cells are all likewise systems of deterministic chaos. Other periodicities catalyze the conversion of sugar into energy by yeast.

A fibrillating heart may show little pathology in its separate parts, but the rhythm of the whole is uncoordinated. Infarcts within heart tissue, however tiny, change the normal excitation pattern when a wave has to negotiate a path around the obstacle. "The cardiac pump is thrown out of gear and the last of its vital energy is dissipated in a violent and prolonged turmoil of fruitless activity in the ventricular wall."[24] The thump-thump-thump disintegrates back into the chaos it comprises. A jolt-generating defibrillation device can sometimes be used to attract rhythmicity back.

It is possible that emergent biologies, in defending against disruption by nature's ceaseless noise, exploited nonlinearity. In the face of wildly disruptive factors, they couldn't afford perfect symmetries. As proteins conduct energy, the heart maintains its pulsations and the nervous system transmits swarms of sensations to the brain. Randomness challenges at every portal, but it is dispelled by nonlinear algebra. Disorder is resonated back into order along multiple interchangeable pathways. The wonder of cardiac and neural systems is not their sudden dysfunctions but the sustained reliability of their normal beats, their creative nonlinearity.

Still other oscillations spark the activity of the immune system with its millions of separate processes and the medley of molecules orchestrating odors in the olfactory bulbs of cats. The electrical flow through the thalamus and cerebral cortex is also a periodic pattern. Alzheimer's Disease may in fact be one consequence of brain-wave destabilization.

FROM THE STANDPOINT of the science of chaos, natural selection is a misnomer. Although it may appear as though species exterminate one another through fierce, genocidal competition, in fact what is happening is that forms with different emergent properties supplant one another in sequences of dynamic stabilization. Symmetries are busted and replaced by other systems of increasing complexity. Competition may serve as a kind of coarse filter for the selection of randomly shuffled, gene-based configurations, but even inanimate systems stabilize in series of refractory forms.

Genes do not have to be very precise (as we saw in the two preceding chapters). They cannot in fact be very precise, for they have too many states to test and too many disruptive challenges to overcome. The thousands of genes even in a single cell have so many potential patterns at their disposal that it would take billions of times the age of the universe for them to sort and choose from among the options. This is "what it means for a system to be dynamically complex."[25] The genes juggle and sift random generic themes, morphing them through one another in irregular approximations. They are scions and engines of the chaos that formed them. Perfect ontogenetic transistors, they make shape because they are shape.

"How can they cope with such overwhelming complexity? The answer . . . is: they get order for free."[26] The fundamental chaos of dynamic nonlinear systems leads to totally unexpected organisms, unpredictable in every way from either the genes or the environment. Dynamic patterns seize renderings up and stabilize them until the world is a riot of shape, color, comedy, and tragedy. This is not only who we are; it is the complication we intuit.

In the old physics, thermodynamics ruled like a Biblical Jehovah; now something is emerging out of nothing.

Chaos and Zen

CHAOS IS WHAT MAKES THINGS REAL. It is not senseless motion against infinite darkness. It is a cauldron of novelty, surprise, and increasing awareness—an invitation to a dada, abstract expressionist universe, a universe with a sense of humor.

This is the way nature must organize in the context of an unimaginable vastness of terrain and mass within a limited array of physical laws. What looks like imbalance and topsy-turvydom is actually a subtle and potentiated form of balance, using everything, in fact using the vastness of resources to overcome the anarchy of the void.

From protoplasmic globs propagating waves out of their rear and crawling across muck, to gossamer brains transmitting pulses—what has transpired is not so much a change in the underlying equations as a refinement of a method of pitting already pitted surfaces and subsurfaces in order to penetrate and activate them at deeper, more infinitesimal degrees of profundity, packing information into miniaturizations, storing organization within structure, structure in organization, creating immense "systems liberated to randomly explore their every dynamical possibility."[27]

From termite colonies to large-scale ecologies to solar systems (and back), order now inhabits the precarious boundary of chaos. Both the order and the chaos are temporary and dynamic. Forms that stray into chaos bounce back out from their own rhythmic properties. Forms that become too orderly crackle with chaotic disruptions restoring their creative potential.

The artificially demarcated, unnaturally binary world that we cling to (with its whole numbers and "actual" adjectives and nouns) is replaced by a "fuzzy logic" in which everything is enmeshed in and becoming (sooner or later) everything else.

Somehow each thing got entangled in its own mechanism and was forced to become a different thing. Matter entered and was snared by the exigencies of its own existence. It was forced to invent, to yield a series of equations.

"Why is it that the silhouette of a storm-bent leafless tree against an evening sky in winter is perceived as beautiful...? Our feeling for beauty is inspired by the harmonious arrangement of order and disorder as it occurs in natural objects—in clouds, trees, mountain ranges, or snow crystals. The shapes of all these are dynamical processes jelled into physical forms...."[28]

A practitioner sitting zazen no longer sees a bevy of discordant events but a single congruity:

"To live in the realm of Buddha nature means to die as a small being, moment after moment. When we lose our balance we die, but at the same time we also develop ourselves, we grow. Whatever we see is changing, losing its balance. The reason everything looks beautiful is because it is out of balance, but its background is always in perfect harmony. This is how everything exists in the realm of Buddha nature, losing its balance against a background of perfect balance. So if you see things without realizing the background of Buddha nature, everything appears to be in the form of suffering. But if you understand the background of existence, you realize that suffering is itself how we live, and how we extend our life."[29]

Nature has no alternative; it produces tragic realms of ceaseless, fragmented events against which it generates a near invisible but eternal stability.

What is the ontological status of forms not lodged in chemistry?

IN THE FOURTH CENTURY B.C.E. Greek philosopher Plato made a distinction between eternal, ideal forms and actual things of experience. Eternal forms exist solely in a higher realm, essentially the mind of God; they are imparted only secondarily to matter. Although this pronouncement and its corollaries were issued a long time ago, they still rule much of progressive Western cosmology, partly through their force of persuasion and partly because they are refracted through countless world-views negating them. "The safest general characterization of the whole Western philosophical tradition," declared Alfred North Whitehead, "is that it consists in a series of footnotes to Plato."[30]

Plato's student Aristotle differed from his teacher in his supposition that forms have no existence independent of things in the world in which they manifest. The enterprise of progressive Western science henceforth was to discover and describe the proximate conditions activating objects and events in nature and, using logical and rational proofs, to make their relationships clear to the human mind. No value remained in sifting endlessly among atoms and predispositions of substances for their ultimate meaning and source. Those lay, if anywhere, outside the system, outside time. They could be intuited through their worldly relationships. Physics and dynamics thus replaced Platonism and its perennial archetypes.

Aristotle characterized causes as fourfold: material, efficient, formal, and final. However, with the ratification of Darwinian theory over the last two hundred years, final and formal causes were banished from speculative science (ostensibly for good).

Final causes were dismissed as incompatible with thermodynamics and natural selection. Nature does not operate in pursuit of a goal or to achieve a result; it has no excuse for biological activity. Life occurs for the most mundane reason: it cannot *not* happen.

Formal causes succumbed when physicists realized that there are no prior patterns outside nature to mold matter, not in higher dimensions, not in the mind of God.

What were left were Aristotle's material and efficient causes, both imbedded in direct activity in the world. These do not express themselves through transcendent forms or teleologies; they provide mere rules of engagement for things as they are.

Analysis of material and efficient causes led quickly to arithmetic exemplification because only numbers precisely capture the constituents of substance and trajectories of motion. Isaac Newton's equations became the constitution of the universe. D'Arcy Thompson later applied them to emerging shapes of simple cells.

Since there are now only material and efficient causes, things in nature must arise solely from happenstance events; life forms can be no more than concatenations of mathematically circumscribed vectors. This is long familiar ground. However, the discovery of nonlinear dynamics in complex systems has obliged a re-opening of pre-Darwinian inquiry into the status of ideal form in biomorphology, foreshadowing a retreat from purely mechanical causation to some indefinable class of eternal objects (though not yet to final causation, the ultimate bugaboo of biological determinism).

Of course post-modern scientists believe that chaos principles are the antithesis of Platonic forms, especially since they are based not only in biochemical homeostases but even more rigorous cybernetic numbers. Probabilistic and componential to their very bones, these numbers plunge beneath algebraic membranes and geometric properties of cells into the fractal, pre-biological domain of chaos. But they also give rise to extra-biological elements operating at the level of determination; thus they provide matrices that are for all intent and purposes non-material. Their cover story is that they represent mere numerical calculation. Yet chaos principles have introduced new terms, rescuing unstably mutating species of animals by metastable sets of patterns with emergent properties, thereby bringing eternal and atemporal aspects into the causation of living systems. The fractal "complexity" universe suggests that formal and final causes may live, though in a non-Aristotelian, non-Euclidean domain.

In the opening decade of the twentieth century, mathematician David Hilbert proposed locating all science under one roof, one set of geometrical equations and logical axiom-sets and their corollaries. This initially promising enterprise was successfully challenged some forty years later by Kurt Gödel who showed, in a famous proof, that any formal system for ordinary arithmetic must be either inconsistent or incomplete, thereby dashing the hopes of modern formalism to evade the ontological problem of the nature of mathematics.

An axiomatic system must employ concepts and procedures from outside the system to maintain its consistency. Yet if it does, the system cannot be derived from its axioms. And if it doesn't, the system has other irreparable defects. So calculi cannot be neutral players in this world, and we are hoisted onto our own petard, our essential contradiction, at the precise point at which abstract number intersects actual nature. Platonism thus remains a kind of default ontology that few people anymore defend outright with much happiness but everyone is stuck with.

Mathematical forms—1, 2, 3, 3.14, 3.1416, etc.—are *our* things, not nature's. Their ontological status has hardly been resolved in the domain of mathematics itself, so how can they be transferred, whole hog and nonideologically, to biomorphology?

How can we apply them to the origin of life and mindedness without dragging in the baggage of their abstract and unresolved footing in the universe at large?

Numbers are symbols. Mathematics began as mere marks on paper; these were not real things, but rules of procedure—literally calculi, so the order generated by equations (including cyberneticized ones) is not *a priori* concomitant to the order of nature. It is true that matching numbers up with empirical data gave us science in its modern incarnation, but this was always our convention, not a requirement of nature. We lapse into the trance of mathematical/physical congruence because, at least since the turn of the twentieth century, we have fueled technology almost solely by calculations.

With the discovery of unexpected mathematical patterns in living and symbolic systems (and in complex structures in general), the unresolved status of numbers and calculi is now back on the table. Morphological patterns emerge in a coherent sequence (phase shifts) where specific thresholds are crossed but where the sequence remains invariant over several heterogeneous material substances or domains. Something remains constant, and something is determining its patterns, something that is not explained by the material conditions in which the patterns emerge—after all, the same patterns emerge under radically different material conditions. What are we to make of the essential role played by such patterns in governing heterogeneous material contexts? If numbers are too anthropomorphic and biased to be their compass, what measure or constant do they respect? Is there a better universal language to inscribe on our greeting plaques flown out of the Solar System?

The metaphysical relevance of any theory of life can no longer be assessed apart from recognition of our current uncertainty regarding the ontological status of mathematical patterns themselves. At the same time, something clearly goes beyond our temporal application of number because the red spot on Jupiter is a real thing, not a computer model; it was a pattern in a storm no doubt not only long before Archimedes but long before the first eukaryote cells formed.

"The quest for absolute truth is subverted by the very act of writing it depends on."[31]

THE PHENOMENOLOGICAL PREMISE is that emergent form is *not* caught up in time, in flux and radical impermanence, and, though immersed up to its eyeballs in the turbulent nature of things, imposes atemporal and elegant patterns relentlessly and ineradicably. We have seen that throughout this chapter. At the same time, the powerful abstract forms emerging from chaos are historical; like Plato's original geometric solids, they arrive in human minds at a moment in time, altering everything they contact, including representations of themselves. They are simultaneously here and there. This is the paradox philosopher Jacques Derrida addresses in his introduction to Edmund Husserl's *Origin of Geometry*.

All geometries, including those of language and number, have ineluctable historicity. While recognizing the temporal aspect of these idealities, Husserl nonetheless felt they were never entirely reducible to historical or psychological factors. They existed independently of them—in sheer contradiction. The wagon wheel may be obvious, but it is not *a priori* given, certainly not to those cultures that didn't discover it. Language is likewise given only after the fact, and only to its speakers. Emergent form is both temporal and atemporal, Aristotelian and Platonic.

Derrida has pushed not only the purport of this discussion but all discourse into a situation where it cannot be taken for granted at the level of its intended meaning and cannot be salvaged from its half lives even by an identification of most of its culturally determined subtexts. For instance, he requires that we take into account the meaning of the very act of writing, the sounds and inscriptions that support it historically, the biology of its execution (our neurons keyboarding or scribing), and the shifting milieus of the consciousness that utters and the consciousness that receives and interprets text.

The past is fundamentally unintelligible; in behalf of this contention, Derrida cites "the silence of prehistoric arcana and buried civilizations . . . the entombment of lost intentions . . . the illegibility of the lapidary inscription."[32] To this we might add the indecipherability of fossils, the mirage of Golgi bodies viewed through a microscope, the illusion of ribosomes, the fuzz of DNA, microtubules, gastrulation, and the like. The meanings of these are undermined by the very attempts to excavate their absolute meaning. In pretending to signify them, we inculcate only their vestigial shells into our own predetermined logos of things. The more we labor their textualization, the more meaningless they become. However, as they lose sense, they regain a different sort of meaning—more tenacious, less circumscribed; more us, less them (who "de" be anyway?). The same is true of the words and histories behind such unstably complex rubrics as viruses, cells, genes, proteins, etc. They become less real as things and more real as signs, semes.

There is also "the surface of the page, the expanse of parchment or any other receptive surface."[33] These letters printed in dye on bleached pulp in folios speak as much to the nature of the printing industry and the origin and standardization of alphabets and grammar as they do to such extraneous matters as embryos or cells.

The discussion of biology cannot ever be separate of the act of writing and the being of us. However, this superimposition/fissure cannot be assuaged by mere assertion of it glossed onto what is otherwise a manifesto (as to say: animals are making this text, so it can never be more than animal mutterings in the face of crisp biological facts); it must constantly redefine each act at every instant and heterogenous aspect of its coming to be in the context of what it is trying (and failing) to say, while admitting both the attempt and its inevitable failure (as well as the failure of the failure and its admission)—and even this falls short. We can "neither overcome being nor make it intelligible."[34] But we make closer orbits when maintaining the tension of unintelligibility than when immodestly proclaiming the prefabricated slogans of intellectual life. Likewise, we are better off deconstructing numbers as we build the edifice of civilization from them. We should still build, but not heedlessly and not without stumbling (Marx Brothers clowns) through our own absurd gaps.

Number, name, character, ideogram, letter, phoneme, syllable collide. "E-numeration, like de-nomination, makes and unmakes, joins and dismembers, in one and the same blow, both number and name, delimiting them with borders that ceaselessly accost the borderless, the supernumerary, the surname." They sponsor "overproduction—and surplus-value—without which no (trade) mark ever gets registered."[35] Hence, systems float like meteoric debris about an invisible asteroid, more or less substantialized, more or less rewarded. Genes, cells, organisms bow listlessly to textbook charades. Chreodes and clones are among the ample dust that comes cheaply enough when so many stones grind one another. Numbers and emergent properties are made of much harder rock. Still, we can never see the full surface created by number and name, not the least because it is broken and fractured, exposed where invisible, camouflaged where flagrant, distorted while configured, superfluous where conscripted. It is not what it appears to be, and yet of course it is everything that it appears to be (what else could it be?)—and then some.

THIS BOOK, *Embryogenesis*, has an earlier version. After having its print run through an OCR (optical character recognition) scanner, I have written over the original dozens of times, incorporating changes as they occurred to me, addressing the responses of others to words I wrote fifteen years earlier and then to the various revisions and supplements I have made since. There is no actual book, only an urgency in a literary medium to approximate a felt truth that is also a paradox. I have seen virtually none of this stuff in action. My text is not even operating at the level of concreteness that cells have for real biologists doing experiments. I am writing ethnography, not science. I am an intruder (who last took biology in ninth grade in 1959), though I have dreamed often since then of 1960 and '61 laboratory tanks of strange crablike phyla and almost-luminescent multicolored worms tangled around graduate-level sea-plants of indeterminate morphology, awaiting my truancy.

Where did the rapt innocence of taxonomy go?

Now I am chanting an ode to the current historical claim that we are structures of cells—no more than cells, no more than a song—a wounded cry to be made whole by shouting enough of the masquerade of facts that envelops us. Sometimes the flow of language seems deeply ingenuous and moves me. Other times it seems verbose, derivative, and utterly boring and I would rather sit in the sun or go to a movie. All of these things must attend (as static) the true text the reader never receives (the true text must also include what I think but don't say, say but don't think, and what will be transposed and forfeited by translation into other languages or morphophonemic mutations through time). The reader will initially struggle along those lines I have carefully carved out, tracks that my editorial fiat ploughs him or her into. Yet he will not read those either. He will read his own rendition of the true text, altered and enriched by his distractions, flights of fancy, deteriorating memories, and the weather and/or movies playing in his town. And we will still be heaps of cells (and words) (or not).

It will never get better than this, more accurate than this, but given the hollow of interstellar space and roar of sun-stars, this is pretty good, pretty close to meaning. It is what has sustained us through our history, and that is what number systems are the archon of, how they have become the golden boy of "philosophing." Just because we now pay attention to holes and subtexts and static more than meaning doesn't mean that the holes are not in something—something substantial, something worth keeping. Yet to ignore the vacancy is to trick ourselves into believing we are in the marketplace of ideas rather than where we are, is to confine us to a tyranny of systematic truths and trademarks which are also neuroses against the onslaught of time and the triumph of death.

In order to write about "things" (including the so-called emergent properties of complex systems), I am having to create a highly devious narrative and inveigle myself and the reader into it. Citing this becomes all the more crucial when addressing the status of chaos, which exists at the level of the unresolved status of my text (and text itself) rather than at the level of semantic and mechanical determination (which well suits avowals of cells and tissues and similar material occurrences). In truth, the book is about the impossibility of our situation—semantically, biologically, epistomologically, simultaneously. It exists in words because words have created the crisis, the trap.

Giving Derrida the last say: "The geometry of this text's grid has the means, within itself, of extending and complicating itself beyond measure, of its own accord, taking its place, each time, within a set that comprehends it, situates it, and regularly goes beyond its bounds after first being reflected in it. The history of the text's geometries is a history of irrefutable reinscriptions and generalizations."[36]

Of course it is never the last say.

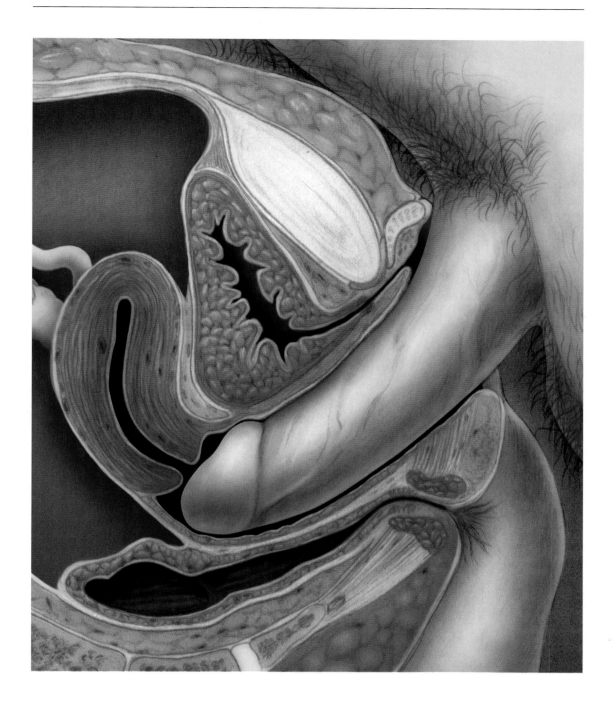

Figure 17. Intercourse. Illustration by Phoebe Gloeckner.

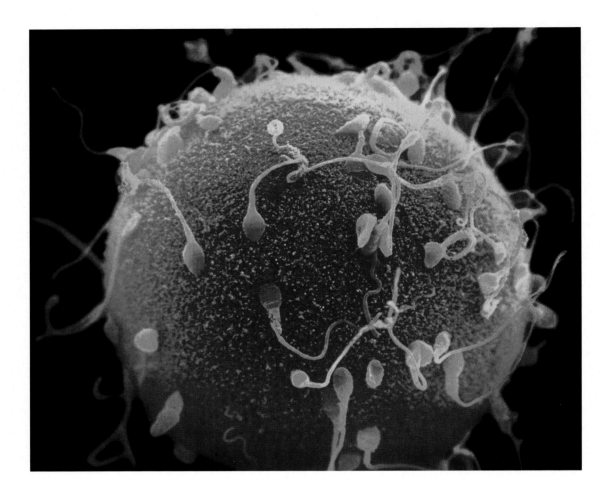

Figure 18. Color-enhanced scanning electron microscope image of human sperm fertilizing an egg. Magnification = 345x at 35mm. © David M. Phillips/The Population Council. Courtesy of Photo Researchers, Inc.

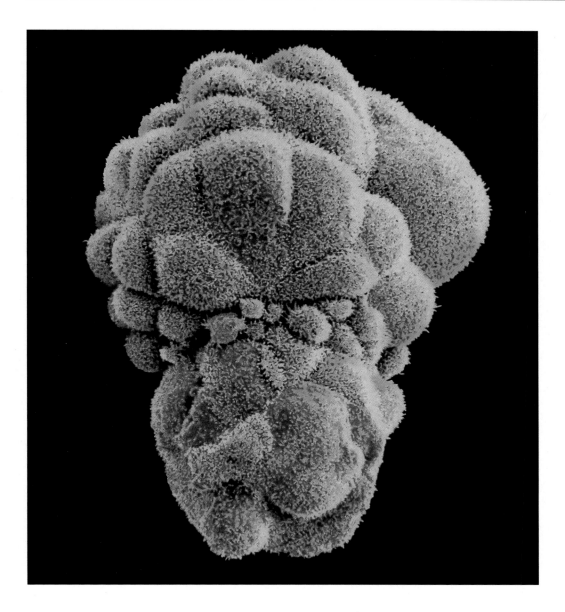

FIGURE 19. Colored scanning electron micrograph of a human embryo at the blastocyst stage, six days after fertilization. It has fully hatched from the *zona pellucida* (not seen), the protein shell that originally surrounded the unfertilized egg. The blastocyst is a hollow ball of cells (blastomeres) with a fluid center. Most of these embryonic cells will form the placenta and membranes around the embryo; only a small group (the inner mass) will form the embryo proper. At this stage, the blastocyst is in the uterus and is ready to implant on the endometrial wall of the womb. Magnification: x600 at 6 x 7 cm. size. © Dr. Yorgos Nikas/ Science Photo Library. Courtesy of Photo Researchers, Inc.

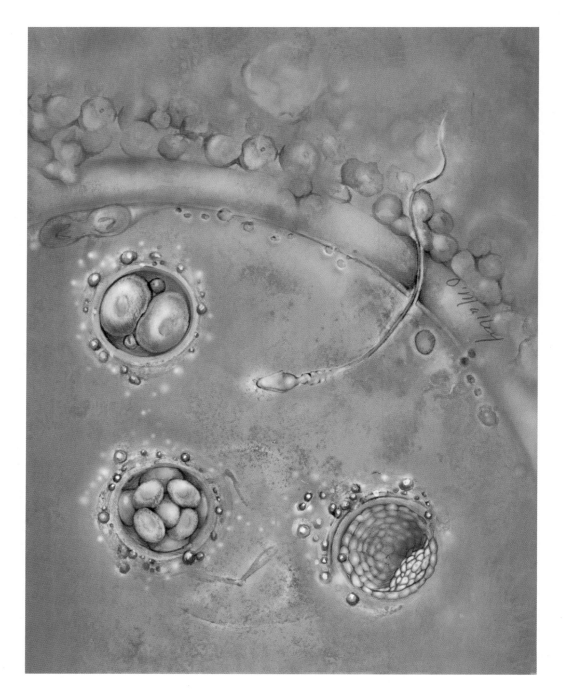

FIGURE 20. Fertilization, cleavage, and hatching blastocyst. Illustration by Jillian O'Malley.

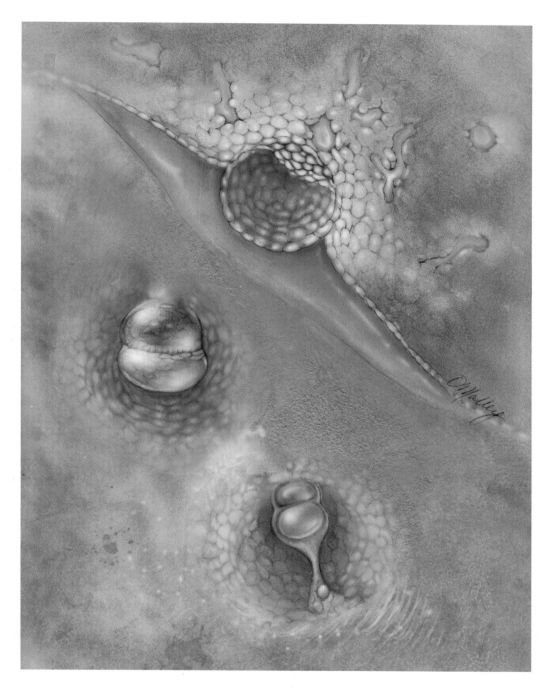

FIGURE 21. Implantation (blastocyst is made up of epiblast and hypoblast). Formation of amnion and yolk sac. Illustration by Jillian O'Malley.

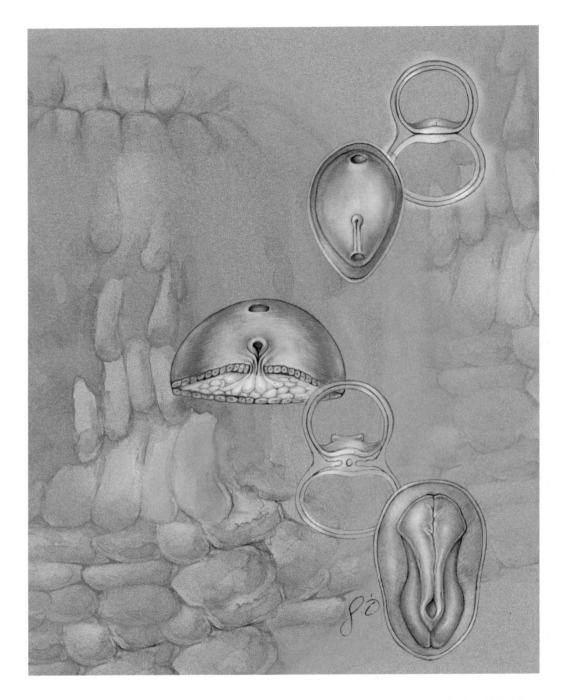

FIGURE 22. Gastrulation (formation of trilaminar germ disc). Illustration by Jillian O'Malley.

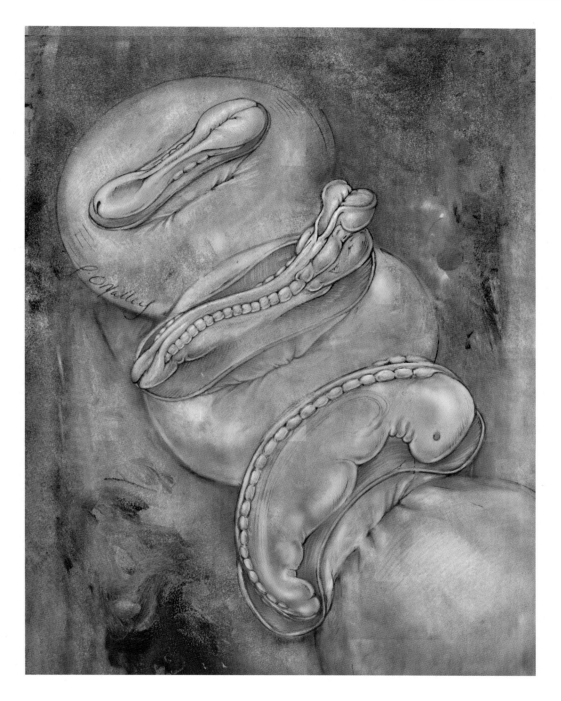

FIGURE 23. Folding (neurulation). Closing neuropores. Illustration by Jillian O'Malley.

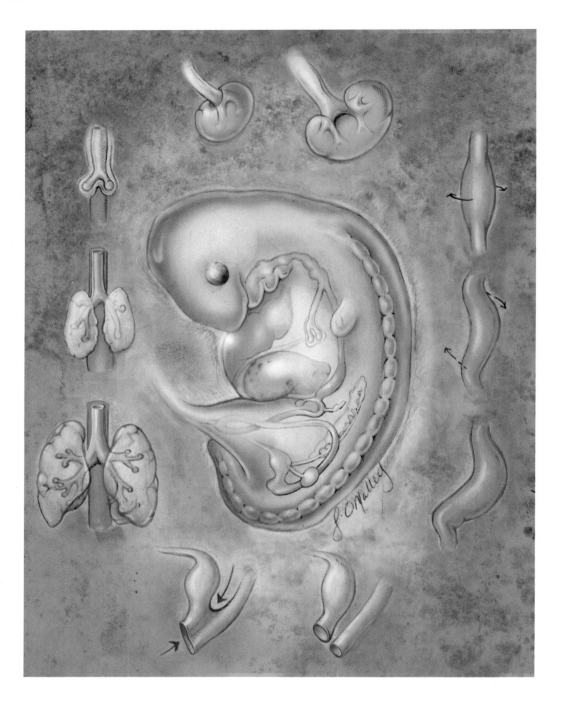

FIGURE 24. Organs. Formation of kidneys, stomach, urogenital tract, and lungs. Illustration by Jillian O'Malley.

14

Ontogeny and Phylogeny

The Web of Similitude

LONG BEFORE OUR PEDIGREE IN THE ANIMAL WORLD WAS DISCOVERED, a proverb held that our embryos reenacted the phylogenetic history of our species. From at least as far back as Aristotle, the temporary gill slits and fish-like arteries in the developmental stages of mammals seemed to disclose the appearance of an actual aquatic creature. So striking was this appearance that the presumption of metamorphosis was not subjected to rigorous scrutiny until surprisingly late in the scientific revolution.

Contemporary with the discovery that humans had evolved in the ocean was the corollary that our fetuses *were* fish—not for any functional reason but in order to validate the "piscean phase" in the womb.

The sympathetic imagination imposes signs and symbols throughout nature, creating a spiral ruled by similitude. Plants, for instance, were long assumed to carry signatures depicting the organs for which they were medicinally propitious (walnuts the brain, hepatica the liver). Stones bore emanations from affiliated stars. Marks and discolorations everywhere had potential divinatory and diagnostic value. Nature was a divine (or hermetic) puzzle, bristling with clues as to its ultimate cosmic, evolutionary meaning and littered with scrapings and scraps of its intelligent design. There were no wasted or circumstantial anatomies. "Chiromancy," explained Paracelsus, "is a science which not only inspects the hands of men, and from their lines and wrinkles makes its judgment, but, moreover, it also considers all herbs, woods, flints, earths, and rivers—in a word, whatever has lines, veins, and wrinkles."[1]

In his seminal text (translated into English as *The Order of Things*) Michel Foucault explained the thrall in which resemblance held the Western world:

"Up to the end of the sixteenth century, resemblance played a constructive role in the knowledge of Western culture. It was resemblance that largely guided exegesis and the interpretation of texts; it was resemblance that organized the play of symbols, made possible knowledge of things visible and invisible, and controlled the art of representing them. The universe was folded in upon itself...."[2] It had indelible residues of analogy, reversibility, and polyvalency, "furrowed in every direction."[3]

Ontogeny recapitulates phylogeny.

IN A VIRTUAL ENCYCLOPEDIA OF WRITINGS from the 1860s to the turn of the century, Ernst Haeckel codified the striking resemblances between the human embryo and more primitive life forms as a "biogenetic law." To Haeckel it appeared that the same essential principle continued to push each embryo through stages of development as once impelled its lineage of ancestors through progressively more complex organisms. Recapitulation was not simply a pattern of visible resemblances or an *aemulatio* refracting from a distance—it was the sole driving force behind embryogenesis. When Haeckel pronounced, "Ontogeny recapitulates phylogeny," he meant: "Phylogeny is the mechanical cause of ontogeny."[4]

If Haeckel's "phylogenetic force" were real, it would have to convey itself through an actual agency (like gravity)—a mode of embryogenic magnetism pushing traits back through stages of development, condensing some and excising others in the elapse of ontogeny. The recent organs of more progressive animals would be added always at the end of the sequence. There was no other possible trajectory.

Ongoing terminal addition of traits would re-route each embryo back through a series of its actual ancestors' developmental anatomies. Such a summarization would explain why, for instance, a chicken begins as a tiny worm and then suddenly sprouts wings. Its template *was* a worm, so its manifestation must reenact precisely that same worm.

The terminal addition of traits may have been proposed as an irrevocable law of nature; yet it remained, even in Haeckel's time, an unresolved paradox. Nothing else in the universe unfolds backward in linear retro-progressive chains. The evolution and development of terrestrial species was a strange exception to the majority of nature to fall under its own inexplicable rubric.

Haeckel buttressed his axiomatic criterion by identifying many contemporary embryonic creatures in the lineages of other contemporary embryos, i.e., fish and newts in human fetal development. However, his peer, biologist Louis Agassiz, perceived that stages of ontogenesis could signify only *extinct* animals—the actual species in the lineage of the embryo in question rather than divergent lines of survivors also extant today. The "recapitulated" animals, represented concretely only in fossils, would never be found among fauna on the Earth.

By limiting recapitulation to what would become a genetic domain, Agassiz unintentionally tarnished Haeckel's principium. Mammals do not recapitulate worms and fish as we know them today. They recapitulate only their own quite different and more primitive "worms" and "fish."

However, recapitulation of extinct ancestors, while primarily a metaphor, contains more than a smidgen of biological truth. Although twentieth-century scientists have uniformly rejected and spoofed Haeckel's law (and even Agassiz's improved version), they contradictorily accept that, in some fashion, ontogeny recapitulates phylogeny. They have little choice. There is no other place for ontogeny to come from. A creature could not possibly invent its own entire genetic blueprint in a single generation, so it must inherit its mode of becoming, step by step from its ancestors—and not just one of them or certain key ones but the full unbroken lineage of all of them. It requires millions of generations of creatures, each incorporating its ancestors' prior development (as summarized in genes), to birth even the simplest modern organism.

DURING HAECKEL'S ERA the renowned comparative embryologist Lorenz Oken published a sweeping "Naturphilosophe" describing correspondences and homologies throughout the animal kingdom. In his system the higher animals pass through the permanent stages of the lower animals as they add on organs. Homoplastic repetitions of morphologies were interpreted as linearly evolving linkages underlying every embryonic structure. Haeckel later adopted this escalating hierarchy as dogma.

According to Oken (and Haeckel) the intestinal organs of the mammals represent the infusorians at one level and the rats and beavers at another. The vascular organs signify clams and sloths, but also snails and herbivorous marsupials, and finally squids and carnivorous marsupials. The stomach, said Oken, was once the simple vesicle of an infusorian, which became doubled in the albumen and shell of the corals, vascularized in the headless clam, and infused with a blood system, liver, and ovarium by the bivalved mollusks. Our muscular heart, testicles, and penis mark our transit through the snails. "The whole animal kingdom," concluded Oken, "is none other than the representation of the several activities or organs of Man; naught else than Man disintegrated."[5]

But how did Man disintegrate before he existed? The anachronism of this pantheon did not dissuade Oken because he was transfixed by its powerful and blatant resemblances.

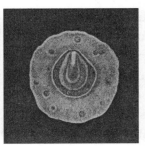

FIGURE 14A. Franz Bauer's nineteenth-century drawings of the development of the chick.

From Arthur William Meyer, *The Rise of Embryology* (Stanford University Press, 1939).

Oken's successor, the medical anatomist Etienne Serres, classified aborted fetuses by the stage of development at which they were arrested; for instance, a headless fetus was a clam (actually a brachiopod) on the basis of its apparent cutaneous respiration. The human embryo, according to Serres, must pass through every major class of animal, including the insects (in which its limbs first sprouted) and the birds (in which it initially sucked air into its lungs). These fetuses not only *looked like* clams and insects; by being aborted before final development, they *were* them. Nature was tidy and economical, wasting no structure or species in its final goal of equipping a human animal.

Once again, biologists had stumbled upon a partial truth—a similitude of morphology and metabolism we today assign to selective gene expression and tissue dynamics but which, in the culture of *aemulatio,* meant synopses and signatures of whole species.

Totemism and Taxonomy

RECAPITULATION IN TRUTH fuses two historically distinct systems of belief—a prescientific cosmology in which the species of animals were considered eternal entities (along with lakes, rivers, Sun, Moon, stars, etc.), and a Darwinian view of speciation through evolution. The totemic aspect stretches back into the Stone Age when the first philosophers perceived creatures as spirits—manifestations of numinous forces. Animals ostensibly received their permanent characteristics from a cosmic, supernal dimension and imparted their seeds to lineages and clans of human beings.

In South American aboriginal creation myths, tobacco orginated from buried jaguar-woman, wild pigs from lustful humans, bats from excrement, and toads from burning sperm—not by alchemy, sodomy, or transmigration but through progressions of totemic classes.[6] To the South American aborigine, the crocodile and opossum are sacred, inalterable entities; they could not have more primitive ancestors. The jaguar could hardly be a transitional visitor to the Earth, for he is the perennial source of fire and the custodian of the cooking hearth.[7]

French anthropologist Claude Lévi-Strauss proposed that tribal peoples *think* in plant and animal categories and that these categories are the basis of their social and religious philosophy. "How animals and men diverged from a joint stock that was neither one nor the other (and) how the black-nosed kangaroo got his black nose and the porcupine his quills"[8] are not legends but irreducible elements that explain not only the zoology and etiology of the Australian desert but also the origin of Aboriginal tribes, clans, and languages; as well as the rationales behind exogamy, sister-exchange, circumcision, and why men and women must die.

Long before Aristotle constructed the rudimentary categories of taxonomy for emerging Western civilization, there were "pagan" tribes in the Aegean too. For these Stone-Age philosophers, each beast, flower, and planet signified an etiology. The specific designations were not necessarily inherited intact by Renaissance scientists, but the underlying totemism was. Animals entered biological taxonomy as quintessential forms, and they remained immutable until recapitulation imposed a physiognomic alchemy on them.

For the eighteenth-century preformationists, evolution could only be the unravelling of archetypal creatures wound into primal germs at the beginning of time. If birds and mammals "evolved" from jellyfish, then their seeds must have already been encapsulated within the sex cells of the medusa, requiring only the maturation of intervening seeds (species) to emerge full-blown from the chrysalis. This cocoon-like unbraiding would be the phylogenetic force behind ontogeny, a spinning top whose motion was zodiacal. At the same time, German philosopher Johann Wolfgang von Goethe "maintained that the archetypal plant not only enabled him to recognize every plant as a specific expression of this principle, but also to imagine nonexistent plants that, given the required conditions, could exist."[9]

Once Darwin showed that species originate in time, eternal templates for plants and animals became superfluous. The Kantian species are the last flicker of a neo-Platonic system. Yet not until the twentieth century did it become utterly clear that animals (*and* stars) are ceaselessly changing fields and do not *mean* anything in and of themselves. "The different types of organisms are just arbitrary groupings of continually changing populations into convenient categories such as plants and animals, animals with and without backbones, animals with and without a placenta for bearing young internally. . . . These categories are a result of the history of adaptive response to changing environments and the accidents of heredity that confer better survival capacities on some rather than others."[10]

Haeckel was dealing in information theory and deep structure, not natural science. His real ontology was cybernetic and syntactic rather than mechanical. Because he preceded structuralism, though, he ordained his integers in a primitive, sterile phase. He made them zoological facts instead of subtextual signs, and he wrote a grim natural history rather than a heraldic bestiary or morphological dictionary.

The first creatures must be included
within all creatures descended from them.

PHYLOGENY ASSEMBLES CREATURES out of prior creatures much as ontogeny does. This is a given. But each step in ontogeny does not recapitulate its whole comprehensive phylogeny. Like phylogeny it condenses and excises while embodying. Organisms edit and abridge the long history that preceded them.

Even as it abbreviates and elides, evolution cannot make giant leaps. If it tried, it would become teratology. Only small changes—usually infinitesimal ones—can be incorporated in a single generation. If these are sustained beneficially, additional changes may be synergized from them.

Creation cannot obliterate stages from its blueprint; it can only add (or subtract by condensing), using what already exists as the basis for the new. The universe has no way to make life except to return to what is left of its original commutation each time and follow an elliptical path back to the present. In order to arrive at each human being, fetalization must embody the same general procedure and design by which it became human in the first place, including the quantum leaps by which it historically condensed itself through gastrulation and organ formation—for the same factors of gene expression and tissue dynamics continue to prevail.

The steps, even from parent to offspring, are never, however, precisely identical. Minor variations collect and tip into unpredictable configurations. Yet, somehow, despite cataclysmic changes over aeons, every stage and configuration is accounted for (sometimes by being supplanted, i.e., by the very kinetics of replacement). Every mutation or morphogenetic change is inserted in the biological field of an existing organism, to appear, perhaps only latently, in its offspring and eventually to develop and fructify in its offspring's offspring.

Since one cannot enlarge on a prior motif without disturbing and then includ-ing it, the first creatures must be included within all creatures descended from them. Although they have been replaced, protein by protein and organ by organ, their replacement has been solely in terms of their original presence and configuration, so their "erasure" as well as their reality lies at the basis of all subsequent organisms in their line. The gaps could not be too large because each of them had to inhabit a body that was coherent enough to survive. The genetic message may no longer include even a trace of some of them, but it would not be the same message if they had not participated in it.

They are the rocks in the current.

CELLS KNOW HOW to make the animals in their direct lineage, using laws of induc-tion, collage, and splicing, only because the cells preceding them made roughly those same animals—and then only because individual units of meaning (amino acids) were selectively and serially preserved and potentiated in chromosomes.

Chromosomes inscribe their own ancient initial informational states, plus all of the later "mutations," aberrant particle distributions, whorls, viscosities, microfin-gers, reaction-diffusion states, phase separations, and sequence transpositions that conferred on them their singular codes (including, of course, ones from coacervate and protist times, before they were chromosomes). There is nothing in ontogeny that does not express a historical event, i.e., the morphodynamic resonance of a mutation or a series of mutations (multiplicities) in tissue states. The discrimina-tion of matter and the development of texture occur only locus by locus, concor-dantly, cumulatively, and through discrete, field-bound waves of feedback. Thus, ontogeny recapitulates a welter of phylogenetic, morphodynamic, homoplastic events in jumbled sequence, not any one lineage of phylogeny.

If the plan of an organism is assembled layer by layer over millennia, ontogeny is a discrete program for reassembling the layers accumulated up to a given gener-ation. It is the temporary finis of a performance, a dance that codes, miniaturizes, and incarnates as it unfolds. Phylogeny is but the ceaselessly factored sum of bil-lions upon billions of separate mutations and the differential production of off-spring. It is not a force. It is an improvisation somaticized in progress each new generation through ontogeny.

LEG RUDIMENTS ROUGHEN in a legless creature; gill slits expand into primitive lungs. Collectively and over many generations, these mutations bring whole new cate-gories of experience—species—into the world. Though, paradoxically, no offspring can ever be of a different species from its parents, a gradual course of transition can produce as exotic a result as grandfather moles and great-grandfather worms for apes such as us.

Whales and dolphins, like all other mammals, inherit gill slits from extinct fishes and reexpress them early in ontogeny. The vestigial gills of mammals, though, are uniformly reabsorbed and do not become functional in adults. Therefore, even though they are mammals ancestrally and pelagic ecologically, whales and dolphins cannot use these simple filaments for breathing, as would make total sense from a design standpoint (or if ecology could directly inform anatomy in Lamarckian fashion).

Their permanent return to the deep notwithstanding, sea mammals must develop a respiratory apparatus secondarily from the clumsy air-breathing organs of the land animals and carry these around underwater, breaching periodically at great energetic expense and risk in order to fill them — despite the fact that their lineage has not walked on the earth since a small wolf-like hippopotamus diverged from the ancient forerunner of an elephant fifty million years ago, filled its lungs, and went fishing. That a bulky secondary land-breathing system takes precedent over a compact water-breathing system in a water-dwelling mammal is a demonstration of how only fragments of stages recur in ontogeny, not entire animals. Once gill slits became vestigial, they were mere ornaments, fossils of an archaic lifestyle. They could not be pressed into service despite their clear superiority and efficiency over lungs.

If an extinct human race leaves any genes on this planet, creatures a billion years from now may well contain us within them. Even if lines of our descendants lose intelligence, they will experience that loss as a gradual descent (gene by gene, creature by creature) over an invisible precipice.

All systems flow from homogeneity to heterogeneity.

IN 1828 KARL ERNST VON BAER offered an explanation for the resemblance of embryos to ancestors that was to be far more compatible with Darwinian science than Haeckel's recapitulationism, but it was overlooked (for the most part) until the decline of Haeckelism. According to von Baer, ancestral features persist only because they were once the general organic configurations from which the specific traits common to any line of descendants developed. Prior stages of organization are always the raw material for subsequent differentiation. If they were completely eliminated, the embryo would have no history; there would be nothing from which new organs could emerge. But they are *not* the regressed and condensed replicas of adult animals; they simply follow a lineal trail of tissue assembly.

Early embryonic stages of vertebrates somewhat resemble invertebrate embryos because the majority of invertebrates have not departed as significantly from the last equipotential state. They remain, from an adult vertebrate perspective, generalized embryos in an oceanic womb.

We might say, simply, that ancient features return in the life of the embryo only because they were never eliminated; they exemplify no force, no law of development. They represent (in Darwin's words) "a community of descent" prior to their divergence by mutation.

For Haeckel the egg simply marched through its programmed stages, its memory traces of ancestral beings. For von Baer the egg was a germinal mass produced during early phases of evolution. It could no more regress than any mature animal could. It too was a "terminal adult," but at a different station in its development.

Von Baer, a vitalist, and Darwin, by comparison a materialist, provided the two poles for modern evolutionary theory. Whereas Darwin decoded the phylogenetic mechanism for speciation and survival, von Baer discerned the cosmic pattern of differentiation. His mechanism applies equally to galaxies, solar systems, oceans, and primeval cells. Systems of energy and matter flow from homogeneity to heterogeneity, from primal density to microstructure:

"1. The general features of a large group of animals appear earlier in the embryo than the special features.

"2. Less general characters are developed from the most general, and so forth, until finally the most specialized appear.

"3. Each embryo of a given species, instead of passing through the stages of other animals, departs more and more from them.

"4. Fundamentally therefore, the embryo of a higher animal is never like a lower animal, but only like its embryo."[11]

Psychological and Cultural Recapitulation

THOUGH DARWIN HAD LITTLE SYMPATHY for Haeckel's writings, recapitulation advertised Darwinism to the general public as well as to the emerging practitioners of social science. The chain of atavistic ancestors within us became a psychosomatic correlate to the descent of species without. It informed the way educated nineteenth-century people viewed the seemingly primitive tribes of the Indies and the nonsensical minds of young children.

A symbolic version of the biogenetic law has served ever since as a yardstick for human and cultural development. Its uses in the twentieth century have stretched from Rudolf Steiner's cosmic evolution to Jean Piaget's theory of child development. For Freud recapitulation was a necessary biological correlative to the unconscious levels of psyche (and an explanation for spontaneous regression). For James

Frazer and Lucien Lévy-Bruhl, Haeckel's paradigm became an anthropology, a source of biogenetic precepts whereby they could inscribe animism and totemism in the genes of primitive peoples and then recapitulate them as fantasies and superstitions in the developmental stages of children in civilized cultures.

The Mechanics of Recapitulation

WHATEVER OUR OPINIONS of the psychological and cultural embellishments of recapitulationism, we must remember that Haeckel intended his succession of ancestors as anatomical fact, not a metaphor for stages of consciousness or cultures (though he did not shy from wide-ranging applications either). Recapitulation was more than just classificatory totemism or archetypal biology; it was the single law of evolution, the one concept unifying the life sciences into a field. The human fetus is actually a worm, a clam, and a fish as it develops, and this is *why* it develops. Gastrulation occurs in each embryo only because of an ancient somatic track which invaginated the primeval blastaea (to use Haeckel's name).

"If we now want to explain the phylogenetic origin of the gastraea (repeated, according to the biogenetic law, by the gastrula) on the basis of this ontogenetic process," Haeckel cautioned, "we must imagine that the single-layered cell-community of the sphaerical planaea began to take in food preferentially at one part of its surface. Natural selection would gradually build a pit-shaped depression at this nutritive spot on the spherical surface. The pit, originally quite flat, would grow deeper and deeper in the course of time. The functions of taking in and digesting food would be confined to the cells lining this pit.... This earliest histological differentiation had, as a consequence, the separation of two different kinds of cells—nutritive cells in the pit and locomotory cells on the outer surface."[12]

Ontogeny was an inheritable resonance, a tendency toward motions acquired in phylogenesis. Germinal "atoms" were recorded in the nervous systems, transmitted through hierarchies of tissues to the germ plasm, and ultimately imbedded in genital ridges. Thus, it was eventually the Lamarckians with their theory of acquired biological attributes who kept Haeckel's law alive into the twentieth century. According to their interpretation, every characteristic acquired by an organism represents a new necessity and is transmitted to its progeny; these progressive traits were added at the end of ontogenesis specifically because they were developed by mature animals during their lifetimes and transmitted from the phenotype to the genes; dynamically, there was no other route for acquired traits to follow except to go to the end of the germinal line.

Contradictions to Recapitulation

Traits are aggregated randomly.

Recapitulation was so compelling an image that its influence far outweighed its validity. From early in the nineteenth century, there were already persuasive alternative explanations for the resemblance of the stages of the embryo to "lower animals," and, throughout the sorting of knowledge during that century, most scientists saw that pure recapitulation was contradicted by abundant discrepancies from all phyla. Yet they stared right through these overt obstacles.

No whole adult ancestor is ever really recapitulated in ontogeny, as one would expect it to be if new traits were simply added at the end of a developmental sequence. Characteristics emerge in unique dynamic configurations throughout the differentiation of embryos of each species.

Furthermore, single organs and resemblances to organs are not appearances of whole animals. Gills alone do not make an embryo a fish, and limb buds certainly do not make it a fly. It is impossible in all but the most metaphorical sense to assign a jellyfish or flatworm stage to the embryos of advanced phyla. There may be morphodynamic remnants of such relative stages in the genes and tissues of reptiles and mammals, but the physiology of a three-layered gastrula is already more complex and less specialized than a jellyfish or sea cucumber, and it has no potential for anything as exotic as a stinging polyp.

In truth, creatures form dynamically and epigenetically, not linearly, from inherited blueprints. They represent complex resolutions of levels of mutational disruption such that the elements underlying their assemblage are substantially juggled at certain critical evolutionary junctures. At such points radical divergences lead to new bionts. Those of them that fail to rearrange their hereditary elements successfully (in fact, most) perish. The rest juggle the various possible forms and anatomies, phase by dynamic phase, through their own peremptory assemblage in such a way that they come up with a viable plan for both embryogenesis and maturation. Any appearances of other plants and animals during these phases may as likely be circumstantial as lineal. And they are never recapitulational in the sense of an impelling biogenetic force.

Traits emerge asynchronously.

Organs that seem to be accelerated in relationship to other organs eventually fall out of synchrony, miscegenating ancestral forms. It would also appear that some organs are accelerated faster than others: the human heart and brain both appear far earlier in the embryo than they would in a purely parallel recapitulation of the phylogenetic sequence.

Apparent recapitulations are fetal adaptations.

Many embryonic organs are quite obviously uterine metamorphoses; the placenta, for instance, does not occur in any adult ancestor. What appears to be recapitulation may simply be a series of adaptive changes in each embryo, resembling phylogenesis only because the watery environment in which the zygote spawns (especially as a free-swimming larva) resembles the primal ocean. The sequential bodies of each embryo betray not its ancestors but ancestral ways of surviving at every crisis of development, most of which occurred larvally or fetally—plus mutations and random divergences.

The relationship of the salamander to the frog is a clear indication of the way in which fetal (or larval) adaptations can be water-based and recapitulational at the same time. No one doubts that the salamander loosely represents the ancient forerunner of the frog (and likely other amphibians too), just as no one doubts that, in the millennial ontogenesis of the frog, the salamander must have secondarily embellished the nuances of its aquatic adaptation. It is not impossible that, in the process of taking its position ontogenetically, the salamander totally replaced an ancestral amphibian with itself. It would then be a constellation of fetal mutations reorganized into a new creature masquerading as an ancestor because of its niche.

How many "salamanders," ancients and aliases, lie disguised in the embryos of modern creatures?

JUST AS MACROEVOLUTION TAKES PLACE in the Earth's various watery and surface abodes, microevolution (of tiny, rapidly "evolving" organisms—i.e., ontogeny) takes place within the watery membranes of creatures—their own and the extraembryonic tissues of any eggshell or uterus within which they gestate. But microevolution itself has the same two, slightly disjunctive meanings: one diachronic (macrophylogenetic) and the other synchronic (microphylogenetic and ontogenetic at the same time). Microevolution proceeds from the dynamics of its own chromosomes, membranes, and biological fields, which themselves synoptically and incompletely recapitulate series of ancestral microevolutions. We might say that the embryo recapitulates not its adult phylogeny—the extinct organisms in its lineage—but *its own concealed* fetal phylogenetic history—that is, the abbreviated and condensed phylogenetic adaptations of all of its forerunners. The terminal and intermediate elements of its fetus represent fetal stages in more primitive primate and generalized mammalian wombs, respectively; the formative elements of its morula even more deeply compress the lifestyles of these ancient preuterine habitats.

Embryos are kinetic, metamorphosing creatures with their own histories and modes of adaptation. They must struggle as creatures in perilous habitats until their stages become superfluous, are leap-frogged by other stages, or condensed; there cannot be temporary nonfunctioning stepping-stones or shortcuts to more complex organisms. That in itself is a reason why species must gradually elide stages of their history they no longer need. Sequences of larval and/or adult creatures, once necessarily inherited, become duplicative and burdensome. Otherwise, as noted earlier in this book, embryogenesis would take too long. At the absurd extreme, ontogeny would have to repeat its entire phylogeny in order to arrive at a life form; thus each creature would require the whole history of the biosphere since the first cell just to germinate.

Ontogeny does not recapitulate phylogeny.

ONTOGENY CANNOT RECAPITULATE PHYLOGENY because the stuffs inside the cell nucleus and the egg both are dynamic, fluid, nonlinear, and metachronological. They do not transmit information serially, nor do they unfold in merely three dimensions. Ontogeny works in all possible thermodynamic phase-states, balancing on the membrane between chaos and order, but insofar as the relationship between ontogeny and phylogeny is basically recapitulative, the embryo must disperse the separate chronologies of its hierarchically stacked systems among one another—fetal systems, adult systems; primeval systems, evolving systems; nucleic systems, amino-acid systems, protein systems; morphodynamic systems, morphogenetic systems—to assemble a new system, a new chronology, a new dynamic folding, based on a simultaneous and urgent consideration of all prior events in the solo climate and chemistry of their habitat. And this is not only ontogeny; it is phylogeny, going back to the original tidepool and primitive membrane dynamics leading to the first tissues and cells. It is an active relationship between heredity and dynamics that potentiates new functions and new forms.

Palingenesis, Cenogenesis, and Heterochronic Displacement

HAECKEL AND HIS FOLLOWERS had ready explanations for apparent exceptions. "All of ontogeny falls into two main parts," Haeckel wrote, "first *palingenesis* or 'epitomized history,' and second, *cenogenesis* or 'falsified history.' The first is the true ontogenetic epitome or short recapitulation of previous phyletic history;

the second is exactly the opposite: a new, foreign ingredient, a falsification or concealment of the epitome of phylogeny."[13]

Cenogenesis, according to Haeckel, could be charged to a number of factors: uterine modifications; the interference of yolk cells in the differentiation of the blastula and gastrula; the displacement of cells from one layer to another, for instance, the migration of the gonads from one of the primary germ layers to the mesoderm; and disjunction in the developmental timing of organs in relation to one another. Additionally, fetal condensation would blur sequences and relationships by crowding aeons into hours.

"The development of each organ is entirely and exclusively dependent upon phylogeny," declared one of Haeckel's defenders. "But we must not expect that all the stages evolving together in a phylogenetic series will appear *at the same time* in the ontogeny of descendants because the development of each organ follows its own specific rate."[14] Haeckel called such displacement "heterochrony," but he considered it only a distortion of palingenesis, his "true natural history."

The issue of timing is critical and, once the science of genetics became sophisticated enough to deal with fractal levels of gene expression, synchrony and heterochrony actually discredited rather than affirmed recapitulation. If ontogeny were purely chronological and serial, olden animals might be distinguishable as phases, but complexity is never a matter of rigid terminal addition; once a new criterion is implicit in a system it is integrated backward and forward disjunctively and anachronistically through every stage of development. Its presence shuffles and recomposes the creature's entire sequence of inductive hierarchies. A new biological gestalt emerges—and with it, new embryonic phases.

In addition to all of these exceptions, the early embryonic features of some ancestral animals seemed to occur in the adult forms of their descendants. Phylogenesis was "reversed."

Nineteenth-century recapitulationists were confounded by the axolotl, a Central American amphibian that retained larval features in its clearly adult stages. It was a salamander that gave birth only to salamanders; the frog had been eliminated from its cycle. Sexually mature larvae of comb-jellies and starfish were also glimpsed during the nineteenth century. Yet Haeckel's "law" stuck and these fetalized creatures were regarded as anomalies of one sort or another. In fact, Haeckel's followers confabulated an ingenious explanation: The youthful features of these animals were merely senile second childhoods caused by an overacceleration of development pushing new traits back into the most persistent juvenile traits, hence producing degenerate forms in violation of progressive evolution.

So powerful was the attraction of recapitulation that it survived even an absolute refutation of its mechanism. Its supporters were convinced enough of its correctness that they perceived a bizarre series of overlapping epicycles instead of the simple linear reversal that was self-evident. Recapitulation (to them) was so obviously and blatantly the way of all flesh that every dynamic mutational series reflected it, either directly or in camouflage. After all, why should nature be devious and mysterious in its complex application of its own treasured axiom!

The most notable instance of larval retention was also the one closest to home. The human being seemed suspiciously to have retained the traits of juvenile apes through sexual maturity: a flat face, hairlessness, small teeth, and a brain abnormally large in relation to the rest of his skull.

The Tissue Mechanics of Development

FROM THE POINT OF VIEW OF MODERN EMBRYOLOGY, Haeckel's law is a sterile abstraction, a similitude that does not explain how energy is transferred from system to living system or how an embryo unravels into an organism. Even during the heyday of recapitulationism, embryologists were beginning to examine the actual thermodynamic development of creatures. Using elastic sheets to represent visceral layers and rubber tubes for the brain and gut, the anatomist William His was able to imitate ontogenetic processes. He split tubes, bent them back on themselves, and stretched them with remarkable resemblance to various embryonic stages. Haeckel was appalled not because His tried to demonstrate a possible mechanical basis for tissue structure but because he showed only *the immediate physical cause* of morphologies and ignored phylogeny, the so-called deep cause. Haeckel, not realizing how quantitative research would rule the coming century, considered His an engineer, a hack, not a scientist concerned with ultimate causation. Yet the Kantian theoretician could never demonstrate that phylogeny was the cause for ontogeny at *any* level and, as scientists came to ask *how* rather than *why,* he was passed over … though the shadow of recapitulation continues to haunt us for other reasons, as we shall see.

THE LANDMARK CONTEMPORARY WORK on this subject is palaeontologist Stephen Jay Gould's *Ontogeny and Phylogeny,* published in 1977. Gould reveals on the opening page that the "topic has fascinated me ever since the New York City public schools taught me Haeckel's doctrine, that ontogeny recapitulates phylogeny, years after it had been abandoned by science."[15] Probably they continued to teach it because it *is* fascinating—and inexplicably compelling.

Gould's book is not only an account of the role that the putative relationship between ontogeny and phylogeny plays in the history of biology but also a scientific analysis of the actual (i.e., mechanical) links between embryos and their ancestors.

As a Darwinian palaeontologist (and a modernist), Gould summarily dismisses the psychological and anthropological parallels to recapitulation (including Freud's explanation of primary development). It is needless to challenge him on this issue, for his major interest is in biogenetic timing, not sociology. Ultimately, he banishes universal Haeckelism from research biology to psychology and metaphysics, which he then dismisses as folklore and superstition rather than science.

In the arena of pure science Gould compares Haeckel's version of the ontogeny of the liver with that offered by the early experimental embryologists Wilhelm Roux and Hans Driesch. According to Roux, "the multipolar differentiation of the liver cells . . . causes the transformation of these cells from the tubular to the framework type"[16]; i.e., organogenesis lies in the cellular differentiation of basal and secreting surfaces. But, he notes, Haeckel would have searched for an ancestor that had a tubular liver in its adult state. He would have missed the physical basis of morphogenesis.

The Replacement of Terminal Addition by Genetic Space

WE NOW PRESUME THAT GENES CONTROL both absolute morphology and rates of growth through their coding of proteins and enzymes. Novel forms have arisen in two ways only: "by the introduction of new features or by the displacement of features already present."[17] Insofar as all mutations occur at discrete points in history they are expressed finally in ontogeny (either as a divergent creature or a lethal defect).

Before a mutation, two variant creatures must share a common ancestor; they are one lineage without any foreshadowing of a division. When a mutagenic event alters a genetic codon, a new group emerges from the individual bearing the change, and its members (if they survive) continue to embody the pattern as if it had been inherited from time immemorial.

We must not regard such transformations as rare or abnormal for, in fact, all tissues and organisms arose once by random nucleotide alteration. Fundamental changes with far-reaching multidimensional effects cleave back to the root of the biological field; they are not just lineal terminal additions.

The late-nineteenth-century rediscovery of the work of Gregor Mendel provided mechanical and mathematical principles for Darwin's "origin of species." Twentieth-century microbiology has since located its physical basis at the heart of the cellular nucleus. For the orthodox, chromosomes provide the whole kit and caboodle of both ontogeny and phylogeny: the elements of structure (proteins), the context for development (tissue), the algebra of morphology (purines and pyrimidines), and the syntax of change (mutations).

As Darwin himself intuited, evolutionism and pure recapitulationism are actually in contradiction. Animals are not completed prototypes; they are transitory motifs in a current. Speciation occurs only because systems cannot be frozen; in the random churn of nature they express solely and quantally the energy stored in them by the chance coherence of prior patterns.

It may look as though cows are born of cows, hornets of hornets, and human beings of human beings, but they are each born of cells. They are created anew and metamorphically from something which is not them. And the sole blueprint for the process is the cumulative genetic record back to the first cell.

SINCE THE LATE 1970s it has been recognized that DNA is organized in coding sequences that continually are shuffled in countless combinations to generate actual proteins. These functional sequences, repeated in different combinations, make up only a portion of the DNA in most complex organisms. As much as thirty percent of the nucleotides comprise long noncoding sequences that produce no proteins, yet have a role in regulating the expression of the active sequences, enhancing some and repressing others. They also ameliorate potentially lethal recombinations of vital units of meaning by their distancing intervals.

Through genetic recombination and transposition, many active sequences (exons) are interspersed and rearranged, with some genes duplicated and reduplicated. The consequence is ceaselessly evolving novel morphologies.

The expressions of genes are also altered by exposure to different combinations of nonessential sequences and noncoding introns. The surplus nucleic material has a role in protecting and regulating the productive message. Introns, lying on either side of active genes, contribute to variation by providing multiple protein sites for recombining exons.

As long noncoding series are interfiled with comparatively brief coding sequences, DNA segments must continue to be shifted, excised, exchanged, and reintegrated; bionts continue to morph. Recent genetic analysis of different primates has shown that a surprisingly large number of replications of two transposable DNA sequences appear to have engulfed the basic mammalian chromosomes in their detour to bipedalism, enhanced eyesight, cerebralization, etc. Enhancing the expressions of some mutations and dampening the disruptive effects of others, mobile modules frame potential form states of organisms as they travel through temporal and eco-dynamic space.

Steven Shaviro brings this algebraic index up to date with a semeiological flourish:

"When we look at the molecular-genetic basis of life, all we can find are differences and singularities: multiple variations, competing alleles, aberrant particle distributions, unforeseeable sequence transpositions. These multiplicities never add up to anything like a distinct species identity. Postmodern biology deals not with fixed entities and types, but with recurring patterns and statistical changes in large populations, whether these be populations of genes or populations of organisms.... Look at the mutations and transpositions haunting any genome, or observe the behavioral quirks of the cockroaches invading your apartment. You will find what [Gilles] Deleuze and [Felix] Guattari call 'molecular, intensive multiplicities, composed of particles that do not divide without entering another multiplicity and that constantly construct and dismantle themselves in the course of their communications, as they cross over into each other at, beyond, or before a certain threshold.'"[18]

MUTATIONS IN MOST GENES (as noted) lead to early death of the embryo. Mutations altering protein synthesis in specialized cells lead to malfunction of affected organs (birth defects) and occasional prodigies (when defects become assets with different functions, i.e., swim bladders as rudimentary lungs). Mutations in control genes alter the overall body plan and give rise to dramatic new motifs; most of these sideshows quickly perish. Over long periods of geological time a few become successful new creatures with totally unique functions, and these bear futuristic, virginal lineages of bodies and behavior.

As we have seen in preceding chapters, many mutations and transpositions of genes do not merely alter protein expression on a part by part basis; they create whole new classes of structure from identical starting points. Insects are apparently mélanges of discrete segments, each with their own set of homeotic selector genes. As these genes synthesize proteins, the cells receive general positional values — rough addresses. When a mutation relocates a patch of cells, the units discover they are somewhere else and assemble a different structure.

The split structure of some genes and their selective activation and suppression by distant enhancers in and of itself potentiates significant genetic shuffling. Cataclysmic bursts of mutational transpositions lead to simultaneous changes in multiple properties of an organism. While the chances for lethal expression are much greater than for functional cohesion, the simultaneous shifts of two or more transposable elements in an organismic plan open the possibility for innovative traits to coalesce in a hybrid function, increasing the talents of the biont and enhancing its opportunities for survival. Clusters of plural mutations may have given rise to the innumerable genera of worms, insects, and crustaceans by modular gradations of displacement, as if the chimeric hodgepodges of a child's Lego set took life.

Natural Selection and Unpredictable Speciation

EVER-DYNAMIC CLIMATES, TOPOGRAPHIES, ECOSYSTEMS, latent caches of DNA, and morphodynamic-morphogenetic interplay have potentiated obscure mutants of long-established species, so new kinds of creatures teem into changing environments. These transitions have often been cataclysmic, such as an alteration in the balance of oxygen and carbon dioxide, shifts in gravity, floods, earthquakes, glaciers, volcanoes, asteroids, tidal waves, meteorite showers, radiation storms from distant stars (shuffling the genetic alphabet), and, of course, land bridges thrusting up between continents once separated by water, likewise channels cut between seas. Any of these might cause the sudden arrival of predators like sharks or wolves, or the decline of whole orders like the dinosaurs. They might also debut new lush meadows and maiden streams chocked with nutrients. It is little wonder that ubiquitous birds and butterflies, marsupials and marmosets have arisen amidst chaos at the beginnings of new epochs.

The living fount is unpredictable, reflecting complexity and emergent properties rather than strategic assignments. It rushes into unlikely habitats while ignoring other obvious ones. Its lotteries are bizzare, especially considering that far simpler life histories could have been assembled but were not. Pond flukes must reach the intestines of songbirds in order to breed, and they accomplish this only by infesting certain snails in a manner that makes the snail's feelers look like "brightly colored caterpillars."[19] Cuckoos lay their eggs in other birds' nests (and these embarrassingly huge intruders apparently appeal to the species that must nurse them, so their kind flourishes even though they retain no nest-building skills of their own). Salps form hermaphroditic chains in alternate generations. One clam fashions an artificial lure from its brood pouch and outer skin, and an angler fish baits itself by somehow having a dorsal fin modified and attached to the tip of its snout. Both decoys are so perfect that they bear the precise dorsal and anal "fins" and "tails" of the tiny fish they are imitating.[20]

Generations of bee-hawk moths have been spared from predators only by their inherited resemblance to the stinging bumblebee, a creature to which they have no near kinship. Likewise, swallowtail butterflies have spread through Africa and Madagascar mimicking the foul-tasting danaids. From region to region they match them by species, varying from black and orange to black and creamy-yellow to dappled black, white, and orange, while imitating danaid flight.

Innumerable races of flies are avoided by birds and reptiles only because they have developed mimicry of stinging insects—a few hairs to suggest the thick fur of the bee, antennae, or, in the case of the bee-fly, smoky-brown translucent wings with their leading edge darkened by a single vein. Of course, there cannot be too many mimics before the lesson is lost; if harmless varieties outnumber noxious ones, then both are attacked with a reasonable percentage of success until the former are exterminated and the threat is restored.

The filament of life on Earth composed of discreet modular protein matrices extends everywhere, from the surface of Arctic ice to the liquid crust, thousands of Fahrenheit degrees. Even volcanic vents on the ocean floor house microbes under cubic units of pressure no three-dimensional organism should be able to bear.

Some worms dwell in the mummified ectoderm of other animals, squeezing just enough liquid out of their dessicated cells. Wheat seeds taken from sacred urns in the tomb of Tutankhamen have sprouted after more than three thousand years in the dark, while beetle eggs from the pharoah's bandages, upon exposure to light and humidity, hatched. A chameleon, no more than half an inch long, dwells on a leaf, then another. It waits for a single raindrop that will drench it, and trundles off. A newborn wallaby, lacking back legs, crawls through "miles" of fur into its mother's pouch. Australian ants feed their larvae their own unhatched eggs. The bandicoot carries a pouch full of glistening children, each weighing less than a thousandth of an ounce. Some species of birds bury their eggs in sand; their offspring dig their way out. The seahorse squeezes its young from its body, two or three miniatures of itself at a time, fragile transparent embryos that latch onto underwater topography. Every grain of soil is an ecosystem more denseley populated by spiders, mites, and microbes than the five boroughs of New York.

"Let it be borne in mind," wrote Darwin, "how infinitely complex and close-fitting are the mutual relations of all organic beings to each other and to their physical conditions of life; and consequently what infinitely various diversities of structure might be of use to each being under the changing conditions of life."[21]

From porcupines and echidnas to sea anemones and Venus flytraps, from hippos and foxes to lichen and bacteria, life is a single papyrus. If all the living creatures that ever inhabited the Earth appeared before us, we would behold a panorama of partial folds and stumps, misshapen rudiments, half-formed wings, irregularly twisted shells, creatures barely able to move, quasi-amphibians drowning in their own breath—all of them surviving at least a few generations.

The gaps between existing types of plants and animals are filled by life forms that became extinct. The present plant and animal landscape has been whittled from a huge block of raw material as exquisitely as a frieze of soldiers and courtesans out of marble. Bionts seem discrete only because the debris has been cleared.

Evolution is blind and amoral. Because it is unconscious it can be neither blood-thirsty nor nurturing except by circumstance. After following a trail of species for a billion years the life current can suddenly abandon it, leaving only a fossil of shells or wings in sandstone. It can also suddenly adopt a strategy it seemingly rejected in another lineage in a prior millennium. Note the seeming backward (land to sea) transition of Pakicetus, the "wolf" who chose life in the water while retaining its land mammalian limbs and ears. Its successor a million years later, Ambulocetus, had developed aquatic ears while retaining the hands and feet of a generalized mole. Another three million years passed before Rodhocetus swam with a flexible back-bone. Its descendant after six million years, Dorudon, was a primitive whale with the teeth of a wolf. (Of course, all of these time frames mark fossils rather than living creatures.) Now whales travel the oceans as if natural regents of the deep, though they are, in truth, awkward regressions of landed mammals.

Nature is extrinsic, idiosyncratic, and karmic in the sense that its experiments are incarnated and live their destinies. ("They offer you a body forever. To shit forever."[22]—William Burroughs.) Ants, worms, and birds have no choice except to embrace their anatomy right down to each tiniest synapse of phenomenology ... spinning threads, preening, stalking, becoming fat and sessile, living inside another creature. Some critters are left eating their own children or laying an egg every half-second or secreting spiral matrices many times their size. That is literally their price of being born, their fate. Once the hornet and pollen are linked in the maelstrom, they lock, and their sentence is written again and again in the gnosis of the world as fresh as a spring breeze and purple clover, each time as if it never happened before. Chaos becomes complexity. Molecules become seeds. Accidental and remote connections turn into semi-conscious acts. Sustained by energy and eros, they cannot be exterminated as long as their cycles continue and sex cells are exchanged. This is how life in nature is supposed to be, the only way it can be.

What does *not* happen is the recapitulation of olden life forms in the embryos of later ones. All resemblances are combinations of fetal adaptations, convergences, homoplasy, achronological gene displacements, the fractal nature of gene expression, plus some scant thread of actual retained genetic history.

Recapitulation and Paedomorphosis

IN AUGUST OF 1971 Julian Huxley told Gould that Haeckel's law of recapitulation is "a vague adumbration of the truth,"[23] and Gould concludes that this truth must be the importance of temporal displacement of genes in evolution. Certain mutations cause particular cells to behave in the way that their daughters or grand-daughters would; other mutations cause selected cells to act like normal cells at the stages of their parent or grandparent. The result is that some portion of the genome may be repeated many times, retarding development; or another part may accelerate and skip intermediate stages, leading to mature states being attained prematurely. Heterochronic mutations affect the relationship between cellular division and differentiation and thus reset the tempo of biochemical development. They also disclose a mysterious clock in the embryo, a timer that is not calibrated by each mitotic division.

In a situation in which generic/genetic coupling has cached multiple and even antithetical pathways and motifs deep within most genotypes, changes in order of gene expression and timing could liberate radically divergent morphologies and body plans.

For Gould (unlike Haeckel) heterochrony more precisely describes the dissociation of traits from one another *in either direction* during development. The acceleration of some traits relative to others leads to recapitulation of older ones; the retardation of some traits (relative to others) leads to paedomorphism of juvenile ones. Gould's point (in keeping with the ethos of modern biology) is that these displacements represent neither progressive nor regressive evolution and reflect no preference for either juvenile adults or recapitulated ancestors in the formation of higher phyla; they are simply different strategies of survival made possible by heterochronic mutations.

Gould reminds us that as early as 1918 the geneticist Richard Goldschmidt, in his work on geographic variation in gypsy moth populations, intuited the existence of "rate genes"—genes which caused large differences in patterns of pigmentation from small changes in developmental timing. Goldschmidt wrote:

"The mutant gene produces its effect . . . by changing the rates of partial processes of development. These might be rates of growth or differentiation, rates of production of stuffs necessary for differentiation, rates of reactions leading to definite physical or chemical situations at definite times of development—rates of those processes which are responsible for segregating the embryonic potencies at definite times."[24]

Without "rate genes" phylogeny could occur only by abrupt introductions of new material (with an accompanying dishevelment of each living system). It would be a hit-and-miss process—mostly misses—because tissue is rarely able to organize a radical change from a single locus. Heterochrony, however, allows creatures to use their existing complexity and organization as the blueprint for multidirectionally diverging variations; one level of complexity can turn into another. The evolution of dense, pliable, triploblastic creatures opened an immense range of somatic paths, new morphologies desynchronized and displaced at varying scales from single ancestral creatures. For a bevy of new species to be realized, genetic expressions had to be transmuted in a multitude of indiscriminate directions, over generations and in unique ecospheres—all from reciprocal loci reorganized temporally in relationship to one another. The conservative aspects of heterochronic gene expression (as well as its invariable doubling back on prior forms) give the insidiously misleading appearance of recapitulation.

The first coelomate worms were highly specialized mutants, but some of their embryos, through further mutations, retarded linear "worm" aspects and potentiated sexually mature juveniles with radically different loci, including templates for billions of kinds of insects, spiders, and crustaceans, all heterochronically tweaked from an amorphous, less organized source with lots of cell potential and tissue mass at its disposal. In addition, "insects usually manage to adapt to changed environmental circumstances a lot faster than we do, thanks to their greater propensity to generate mutations, and their far higher rate of genetic recombination over the course of much shorter reproductive cycles."[25]

As noted numerous times, most new "organs" initiating classes and phyla probably began as pathologies and were lethal in all but a few inheritors of them, in which they became functional by way of the fortuitous discovery of niches occupiable soley through a "deformity."

Mutations resulting in recapitulation or paedomorphosis can enter the embryogenic motif at any stage, so there will always be two kinds of heterochronic potential in a genotype: one, continuing a strategy of specialization by retaining adult forms of ancestral animals, condensing them, and surpassing them by terminal modification; and, two, radiating from their partial development short of full ancestral maturation and adapting to a variety of microenvironments through different expressions of their retarded genetic potential.

Without heterochrony highly complex organisms would become dead ends. However, heterochrony allows for the rearrangement and reorganization of whole sequences of development by slight displacements of single genes—a deep-structure paradigm rather than a linear progression. Such changes become patterned and functional because of the highly variable nature of gene expression and the role of context in determining and coordinating any expression. As we have seen, a gene activating tiers of organization engenders multiple outcomes. It doesn't just insert raw material at one level with one consequence.

RECAPITULATION AND PAEDOMORPHOSIS can occur in quite opposite contexts with different evolutionary meanings. If some aspect of somatic development is retarded while embryogenesis proceeds at the ancestral rate, then the adult is juvenilized by neoteny. If the embryo becomes sexually mature precociously, i.e., while still a "child," then paedomorphosis has occurred by progenesis.

In a relatively untenanted environment with abundant resources, some aphids apparently spawn wingless forms which mature rapidly by progenesis. Since there is ample vegetation to feed on, extra energy need not be consumed in sensorimotor mobility; they just sit in place, eating. Other progenetic forms become parasites, developing their endodermal organs so rapidly that they end up as little more than gonads and stomach. If their hosts are abundant, all the rest of their physiology can be dispensed with in the "haste" to produce offspring and fill the new environment. Or, in the language of neo-Darwinism, such activity fills the environment with paedomorphs faster than with more mature and endowed species.

Dwarfism may also be adaptive, especially where tiny creatures enter otherwise uncrowded niches; for instance, parasites in the organs of clams and fish. It is possible that whole phyla of small metazoan creatures such as roundworms and rotifers have progenetic origins. The actual line leading to the land vertebrates could have originated from tunicate tadpoles with short larval phases. Although progenesis seems primarily to be a strategy for abundance of offspring at the expense of complexity of tissue, genetically plastic paedomorphs (like these tadpoles) might have developed entire diverse lineages if transferred by chance to suitable environments.

According to Gould, neoteny is a more promising mode for the emergence of higher taxa, for it preserves the morphological plasticity of unspecialized juvenile forms. Whereas progenetic paedomorphs may lose evolutionary potential, the more conservative neotenous paedomorphs, when the development of crucial organs is retarded along with maturation, usually gain potential.

The two different types of paedomorphosis arise in insects: metathetely (or neoteny) when an increase in the amounts of juvenile hormone causes childhood features in adults; and prothetely (progenesis) when the juvenile hormone is suppressed and adult traits appear larvally, often at a premature molt (relative to ancestral forms), with subsequent molts suppressed. Activated by opposite biogenetic mechanisms, progenesis and neoteny result in different modes of adaptation despite their expression in similar appearances. It is no wonder that Haeckel and his generation were confused. Ontogeny seems to recapitulate phylogeny when actually it is tracking multiple levels of structure fluctuating within homeostases of membrane-trapped energy. In truth, it is *seeming* to do lots of things, but recapitulation alone captured an atavistic imagination.

IF A MUTATION CAUSES an ancestral trait to be displaced backward, then recapitulation occurs by acceleration, the classic Haeckelian mode. If full somatic development continues at an ancestral rate while maturation is delayed, i.e., if only gonadal development is retarded, then another kind of recapitulation can occur, and this is called hypermorphosis and often leads to larger, more differentiated organs like the antlers of elk and giant mollusk shells. When ontogeny elaborates way past its prior termination point, it can also lead to immense creatures like dinosaurs and whales in relatively few generations. Once the scale of growth and maturation is tipped, animals can shoot from one size range to another until the biophysics of tissue imposes its own limitation.

Neoteny is more common than hypermorphosis or progenesis in the situation of a relatively favorable but bounded environment with a harsh and perilous surrounding terrain—classically, small ponds in arid regions without predators themselves but impinged on by predators. Not only would there be little advantage to population growth by rapid maturity (or enlarged bodies), but there would be no incentive to colonize the outlying region. Thus axolotls and other such paedomorphs do not even mature sufficiently to live their ancestral lives. Their development slows down, so they retain larval anatomy. Axolotls never become fully amphibious; they are able to stay in the water and reproduce without having to brave the shoreline environment and its carnivores. Such neoteny may be ephemeral and, in some species, it can be counteracted by experimental doses of thyroid—the animals then mature. Most neotenous paedomorphs, however, have developed hereditary resistance to metamorphosing hormones and do not respond to treatment. The juvenile state is their permanent adult state.

Human Evolution by Acceleration and Retardation

THE INTUITION THAT ADVANCED HUMAN DEVELOPMENT was paedomorphic rather than recapitulationary and accelerated was disturbing to many Eurocentric nineteenth-century anthropologists. If juvenilization was the desirable characteristic for advanced status, then it was clear that the Mongoloid races were more deeply fetalized in most respects and thus capable of the greatest development. But then recapitulation seemed to favor the African races with respect to other traits. The implicit contradictions ran deeply enough that the human being was gradually conceived of as a simultaneously retarded and accelerated animal. To a certain degree this is accurate, for the expressions of mutations locate in groups of tissues, not universally; and whereas many key human traits may be paedomorphic, others are more likely recapitulationary. The growth of the brain may be either.

In general, advanced mammals have evolved through retarded development, smaller litters, and long gestations. Most simple mammals are born with nearly full survival skills. Humans have become secondarily altricial, apparently because their immense brain expansion has outstripped the capacity of the birth canal. A brain which matures among a diversity of external stimuli also has certain neuropsychological advantages (see Chapter 21, pages 563–566).

Gould argues that man and woman are paedomorphic not because of any one juvenilized trait but because an overall retardation of development changed the selective matrix in which all aspects of human morphology were environmentally and culturally selected over time. The primates were already retarded in relation to the rest of the mammals, so the hominids merely continued the paedomorphic trend.

The implication is that, if individual human beings were somehow allowed to continue developing indefinitely, they would slowly become more simian, like Aldous Huxley's Fifth Earl of Gonister in *After Many a Summer Dies the Swan*, who, by his 201st birthday, from using an extract derived from the intestinal flora of carp, had turned into a hairy, inarticulate, muscle-bound ape.[26] As it is, only retarded development allows us our already unnaturally long life span by primate standards, and it is to be presumed that if our rate of maturation could be slowed even more by heterochronic mutations, we would become more childlike, i.e., more human. We would also be brainier and live longer without degenerating.

Men and women remain embryogenic even after they leave the womb: Witness the late eruption of their teeth, their bodily growth through adolescence, and, most notably, the postnatal expansion and convolution of cerebral tissue—indispensable aspects of the human condition. The longer fetal development rates are retained through adolescence, the more biological fields get to translate latent possibilities into actual configurations.

Recapitulation and progenesis still occur with regard to certain human traits, even in the overall context of neoteny. Genes and mutations have no loyalty to purity of heterochronic mechanism. Gould cites as recapitulationary the "early fusion of the sternebrae to produce a sternum; the pronounced bend of the spinal column at the lumbo-sacral border; the fusion of the centrale with the naviculare; and several aspects of pelvic shape."[27] Progenetic traits include relative loss of pigment and body hair, orthognathy, labia majora in women, loss of brow ridges and cranial crests, general thinness of the skull bones, long neck, thin nails, eye orbits under the cranial cavity, and reduced teeth. It is the underlying trend which is neotenous, i.e., extension of the life span, persistence of cranial sutures, secondary altricial dependence, and the general lengthening of the time of body growth.

THE COLLECTIVITY OF HETEROCHRONIC MUTATIONS throughout our evolution is what has led to our departure from the simian line more than any excision and replacement of genes—that is, displacements of existing elements rather than brand new traits. Astonishingly, ninety-nine percent of our genes are identical to those of the apes (yet we probably could not even breed with one of our hominid forerunners if we were ever to discover a tribe of these creatures hiding from us in remote caves). Shifting biological fields have organized the same basic genotype into a radically different animal, drawing exponentially greater complexity from commutations in morphodynamic patterning and timing. How would a visitor from another solar system explain our remarkable capacity to reconstruct this planet when he might initially classify us as "a third species of chimpanzee"?[28]

In 1926, biologist Louis Bolk wrote: "I would say that man, in his bodily development, is a primate fetus that has become sexually mature."[29]

The effect of heterochrony was profound and irrevocable. Hominoid, and then hominid, populations became demographically distinct, and, in the context of culture and language, callow men and women became domestic, educable, and symbol-possessed and possessing. These creatures suddenly leapt the seemingly uncrossable chasm separating nature from culture, and timelessness from time.

This is apparently how the universe invents itself and sires new meanings out of prosaic themes. Cycles occurring at one scale repeat at another, and another, not only dislodging but reinforcing motif shifts, redistributing and transcending chaos, in moiré-like waves. Neither ontogeny nor phylogeny can escape the series that binds them to each other, but this is almost syllogistic, for they *are* each other, separated by intervals of immensity and the human mirage of time.

As ancient primate and mammalian motifs fell into latency and were sublimated, new forms arose, psychically as well as physically. With our combination of recapitulation and fetalization, mental phenomena of different orders no doubt existed simultaneously. These linger as unconscious phases in strata of our minds, but in the fossil record we see them as successive species, Australopithecus, Pithecanthropus, Cro-Magnon, followed by the various tribes and races of humanity. Gradually, the physical and the psychological came together, and retarded and accelerated features merged in a creature which obliterated their antitheses.

Biotechnology

The Great White Hope

THE GENETIC AND EMBRYOLOGICAL EXPERIMENTS of the first three quarters of the twentieth century have led, in its last quarter, to the swift and steep rise of their application, biotechnology—along with vocal camps of its supporters and detractors. Optimistic futurists laud biotechnology's insights into the mechanics of cells as well as its potential to devise new medicines for hereditary and life-threatening diseases, plus hardier crops to squeeze into the six million square miles of diminishing arable land to which the human race now seems permanently restricted.

Pessimists augur long-term perils from artificially mutated viruses as well as plants and animals with untested chromosomal capacity. Social and philosophical critics of technology (in general) point to its inherent limitations and shallow conception of nature; they also presume the heralded benefits of forthcoming sci-fi product lines are self-servingly exaggerated.

A glowing cover article in the January 1999 issue of *Time* greets the final year of this millennium with this proclamation: "Ring farewell to the century of physics, the one in which we split the atom and turned silicon into computing power. It's time to ring in the century of biotechnology. Just as the discovery of the electron in 1897 was a seminal event for the 20th century, the seeds for the 21st century were spawned in 1953, when James Watson blurted out to Francis Crick how four nucleic acids could pair to form the self-copying mode of a DNA molecule. Now we're just a few years away from one of the most important breakthroughs of all time: deciphering the human genome, the 100,000 genes encoded by 3 billion chemical pairs in our DNA."[1]

At the unravelling of the nucleic code, exponential leaps in medicine and agriculture wait in the wings, for scientists will gain proximal access to the protein factory of nature. To understand and control life at this level will elevate humans if not to the level of gods, at least of that of second-tier makers. All animals-yet-to-be-born and future generations of us, take heed! There is a new spinner among the Fates, one with a progressive utilitarian agenda and a reputation for careless expediency.

Biotechnology portends the most radical transformation of life on Earth, if not since the eukaryote cell, then since the evolution of the hominids, for it juggles parameters of proteomes (entire protein complements of genomes) instantaneously without having to wage the sluggish, directionless paths of mutation and natural selection. Thus, it threatens to take us anywhere—anywhere at all—from every imaginable utopia to every dreaded apocalypse . . . from every "brave new world" to every "planet of the apes." In a time of skyrocketing human biomass and maxed-out food production, it has been deemed "the single most promising approach to feeding a growing world population while reducing damage to the environment."[2] It is material science's best claim ever to transmutation and magic, perhaps the last "great white hope" of the West. Biotechnology critic Jeremy Rifkin has baptized the genetic alteration of living machines as our transition "from the age of pyrotechnology to the age of biotechnology"[3]—the cellular equivalent of harnessing fire.

Keep in mind that biotechnology requires neither a unique axiom nor a previously unknown energy. It is not yet magic; it is not a whole new paradigm. It follows from prior technologies of the industrial era, ones involving mineral and plant identification, extraction, storage, manipulation, and transformation. Biotechnologists do the same kinds of things that other engineers do with resources; they sever and mine (using extremely tiny scalpels on infinitesimal objects); they redirect kinetics into antientropic machines even as petrol into cars or water across dams. What makes biotechnology unique is that it taps and redirects energy contained in membrane systems. In place of manufacturing plastics out of molecules or alloying metals into machines, it harvests and alloys cells—it is the practical industry of gene identification, segregation, storage, and manipulation.

I AM IN NO POSITION to evaluate the claims of biotechnology; I am not sure any of my contemporaries truly are, either. At times I think that the keys to life will remain forever a secret and that tampering with DNA will yield only gaudy disappointments ringed with unforeseeable disasters. At other times I see no reason why the genetic basis of heredity should not be decipherable and the human genome unmasked as straightforwardly as were the molecule and the atom.

In a remarkably short time already, the surface riddles of the material world have melted before humanity's onslaught like grade-school puzzles (the core mysteries, beyond simple materialism, remain another matter). What other course is there for the rampaging locomotive of science, what other worthy challenges for the generations that will inherit the spoils of the illustrious, miracle-rife twentieth century (besides—it goes without saying—reversing the imminent demise of the Earth's ecosphere)? And what reason is there not to grant technologists the ability to rearrange the integers of life and concoct new traits and creatures much as they have rearranged molecules of matter to fashion toy soldiers and transmit quanta of electrons across power grids and telephonic wires?

As the same time, what reason is there not to suspect that we will encounter at the threshold of the gene, as at the threshold of the electron and neutrino, the elusive enigma of form and the inalienable paradox of mind, matter, and energy?

ADHERENTS ON ALL SIDES of the biotech debate invariably distort opposing positions. Anyone who does not admit that technology has unimaginable power to transform our planet to its marrow need only look at what has been done already to an indigenous landscape: roads, pyramids, downtowns, skyscrapers, factories; trains, oil rigs, fiber optics, cell phones, web sites. A mere half century after a glider initiated air transport by floating one hundred and twenty feet to the cheers of two brothers and their friends, thousands of jets a day soar over oceans and mountain ranges, hauling loads of people and their luggage from site to site across the Earth's continents. In twenty years computers have evolved from dinosaurs to a global internet. Who would question the debut of marvellous artificial species, warehouses of cells to replace damaged organs, customized crops, cyborgs, and designer babies?

Yet anyone who does not recognize the limitations and risks of profit-driven industries should inspect the devastations wrought by the automobile and other petroleum-based devices in less than a century, or assess the environmental and sociopolitical consequences of splitting the atom. Biotechnology may well someday provide replacement hearts, superior eyes, cures for cancer and AIDS, and exponentially increase agricultural production; it may even "improve" the human genome—but it will not solve the existential crisis of life on Earth or illuminate our existences. At worst, it may spawn monsters and tyrannical institutions that will set our ecology and spiritual growth back by centuries, or even eradicate biology on Earth.

Anyone who thinks that biotechnology can be halted by fiat or ethical contrition should remove the rose-colored goggles. Remember the beleaguered attempts at nonproliferation during the first decades of the atomic bomb. Look at the results of treaties against biological weapons—signed with fanfare and routinely ignored. Now these various bombs and delivery systems march in an unbroken column from Israel to North Korea. The jinni never goes back in the bottle—never. In an epoch of venture capital, spiritual apostasy, fundamentalist terrorism, and scientific vanity and virtuosity, tabooing acts in the old tribal sense is futile. DNA overseer James Watson speaks with the collective bravado of mankind when he advises: "Never postpone experiments that have clearly defined future benefits for fears that can't be quantified."[4] This has surely been the motto of the century.

Gene Splicing

As noted, genes are very hard to "find"; they are not concrete subcellular entities but dynamic ripples of DNA molecules, letters of an alphabet like the ones in which these words are being commuted. While DNA ribbons can be routinely shattered into fragments, the detached pieces are far from legible templates made up of coherent nucleotide sequences; they are alphabet soup in which gobbledygook occurs much more readily than sense. Even to begin to identify the sources of traits, biologists have to isolate chromosomes with cogency.

Their best strategy has been to hoodwink intracellular entities into writing their messages in our domain and then replicating their subsets for predesignated assignments. After all, they are already in Rome and speak the language, while we have no way of insinuating ourselves among the Lilliputians.

By the late 1960s purified enzymes from bacteria were enlisted for precision gene whittling. These "restriction nucleases" detach sequentially prescribed lengths of DNA, known as restriction fragments. Nuclease actions also endow some of the fragments with tiny cohesive tails at either end, complementary base pairs serendipitously suited for linking double-helical DNA fragments from different creatures to each other. The resulting hybrids can then be mass-produced in the chromosome of a bacterial virus (see below).

When part of one gene is fused to a different gene, novel proteins emerge—with often radically divergent results in terms of functional properties, amounts of polymers synthesized, and even cell types in which the proteins are produced. Yet biotechnology is more than prodigal feats with chemical utensils in subcellular realms; it requires nurturing and interpreting the products of its experiments, and devising incrementally subsequent experiments. Otherwise, it would be little more than spooling a biological kaleidoscope and watching patterns rise and tumble from nucleic dialings.

The earliest dramatic breakthrough in biotechnology occurred in 1973 when, after some thirty years of labors in their laboratories, Stanley Cohen of Stanford University and Herbert Boyer of the University of California at Berkeley succeeded in combining two isolated patches of genetic material from organisms unrelated to each other. Their scalpel was a restriction enzyme. It was first applied to nucleic material in such a way as to split DNA molecules from a donor; then a like enzyme was used to snip a piece of genetic material from the body of a plasmid, a short strand of independently replicating bacterial DNA much like a virus. The two segments were hitched and bonded at their adhesive ends. The hybrid plasmid was grafted into a bacterium; the zooid absorbed it, reproduced it, and (if it still exists) will reproduce its DNA again and again, hypothetically forever.

An altered portion of a gene's nucleotide sequence can also be synthesized and then combined with a predesignated strand of DNA containing a nucleotide sequence from a genome in which a redesign of traits is sought. The consequences of this activity materialize only when a tampered-with gene is inserted in a live organism and variants emerge. Though this is the premier feat of recombinant-gene technology, its execution does not mean that traits can be supplanted or modified with the same ease as chromosome fragments; in fact, as we know, there is no linear space in which to interchange one purported gene and its traits for another. Furthermore, only in simple yeasts is it possible to substitute engineered nucleic materials for their endogenous counterparts; in complex mammals, there is no way to supervise the biochemical integration of mutated DNA into new chromosomes.

Just as embryologists cannot trace the complete and final effects of single genes through epigenetic fields, so biotechnicians cannot predict the full outcomes of their own transpositions. Since living systems have intrinsic metabolism and motility and the capacity to grow and diverge, they perturb all altered DNA into complex, meta-stable forms. Researchers do not know where a piece of foreign code will be integrated into an existing plan or how its integration will be expressed phenotypically, so they proceed by trial and error, making guesses about trajectories to gain results like glowing tobacco, tomatoes that resist freezes, cows that give more milk (using the recombinant bovine growth hormone rBGH), and bacteria that mine copper ore by eating salts. While microbiologists "do sometimes succeed in isolating a single, crisp gene with a single known function," more often they "get no further than marking off fragmentary stretches of DNA that may be thousands of bases in length."[5] Finding the real genetic information in these is one level of challenge; tracking how it is deployed phenotypically is another.

The industrious enzymes of the cell bind the alien fragments floating their way in long tandem arrays, then toss them into randomly selected genetic locales. Thus, altered DNA injected into a cow or mouse egg may or may not have observable effects on the ensuing animal, may or may not be traceable chromosomally, and may or may not end up in germ cells to be passed on to the offspring cows and mice of future generations. The experiment also may or may not be repeatable.

Even when hereditary material is transplanted with initially linear, single-locus results, the true consequences of repositioning DNA must await the passage of generations, even as it had to in the initial instance of prokaryote splicing and subsequent mutations and adaptations, or in the millennial divarication of magnolias, phloxes, millipedes, mollusks, etc. We simply do not know all the factors of epigenesis and environmental dynamics that we are ruffling.

Technological time cannot rival deep phylogenetic time.

Transgenic Creatures

OUR SPECIES' ORIGINAL "BIOTECHNOLOGIES" were plant breeding, animal husbandry, and synthesizing foodstuffs, medicines, and other raw products from diverse life forms. These ancient crafts dealt with phenotypes, entire organisms. Though hybridization at its core entails manipulating genotypes, it does not require handling their imperceptible genes—even Neolithic farmers could bring together phenotypic vectors (herd animals and garden plants) with hereditary consequences. They could breed mules and corn, but not unicorns and basilisks.

Present-day biotechnology, by contrast, deals with genotypes (chromosomes and genes), its warrant extending from the cardinal discovery that species do not harbor self-characteristic DNA. Though the fur of a raccoon may differ radically from the petal of a rose in chemistry and texture, the nucleic codes in which they are written are identical and can be merged. A giraffe and a human (however alien to each other) are scripted in the same amino acids; so are a squirrel and a trout. Chromosomal sections from any of these can be cut out and spliced into any other, with the result that they will be recognized intracellularly as universal DNA rather than squirrel, rose, or any species, and will make good protein in the context of its new locale. Thus, presumptive squirrel stuff can become rose stuff.

Scientists were no longer restricted to experiments with organisms as integrities of traits. Instead, they could go directly to the fount of common genetic material in each and, siphoning and recombining pieces of it, cross all imaginable mating boundaries. Initial attempts were based on practical and humanitarian considerations. When Factor VII (clotting) genes from humans were inserted in bacteria or yeast, the much more rapidly dividing host cells amplified and "manufactured" voluminous amounts of the inserted section (for medical use) as if it were their own. As different in practice as this is from a field botanist collecting plant specimens and brewing pharmaceutical compounds, it is not so different in concept as it might seem, for the biomolecular outcome is the guiding goal.

WHEN HUMAN GROWTH-HORMONE GENES were relocated in mouse embryos in 1983, the mice grew twice as rapidly and to almost double the size of ordinary mice. Furthermore, their offspring inherited this condition. The growth genes were no longer human but mouse, as if a particular tribe of mice had developed this variant of DNA by a mutation. In a sense, it *was* a mutation but from a human rather than a natural source.

In 1984 scientists bred a sheep-goat chimera (a "geep") by combining embryonic cells from a goat and a sheep.

Transgenically altered tomatoes grow denser and less pulpy and mature longer on the vine without cellular decomposition.

Translocated scraps of DNA have spawned fruits with natural pesticides (hence, not requiring haphazardly sprayed toxins), beans and grains chocked with protein (reducing the worldwide call for meat), potatoes with less water and more starch, caffeine-less coffee beans, and sugar-diminished strawberries. (A concerned Prince Charles of Wales proclaimed in a newspaper editorial that "transferring genes between utterly unrelated species—fish to tomatoes, for instance—'takes us into realms that belong to God, and to God alone.'"[6])

Transgenic bacteria churn out human insulin and other biological products necessary for metabolic processes. Lines of "natural" vaccines are manufactured and packaged in much the same way. Soil microbes transgenically altered are strewn in the fields and naturally drawn up into the vascular systems of plants. The botanical genomes then synthesize proteins stitched in by alien DNA as well as their own. Seeds for fruits and vegetables bearing these mutated chromosomes can be dispatched throughout Africa and Asia and harvested in even the most remote gardens and fields such that citizens and tribespeople are immunized without having to import or store pharmaceuticals.

In an act of resplendent showmanship in 1986, biologists transposed firefly genes into tobacco plants (crossing kingdoms from the zoological to the botanical) to produce tobacco leaves that glowed when "watered" with luciferin, a light-emitting chemical. Blue roses are a future target.

Processes emanating from these experiments might one day stock entire industrial farms with technograins and herds of genetically identical sirloined cattle, a far cry from the ranches of the frontier, and not necessarily a fate to aspire to.

Less benignly, The United States Department of Agriculture (USDA) has invented a technology, underwritten by public money, to strip seeds of their capacity to propagate. The process is being patented worldwide on behalf of Monsanto, through a subsidiary (Delta and Pine Land Company). Monsanto is preparing to splice this "Terminator" gene into its transgenically enhanced high-yield crop seeds, rendering the fruits of their harvest sterile. This means that poor farmers through-out the planet will have to purchase fresh seeds from a multinational corporation after every crop. It also means that companies are willing to risk letting sterilizing mutants loose in the biosphere in order to defend their patents and profits.

For all the careful planning by Delta and Pine Land and Monsanto, there is the possibility that genetically altered organisms will share their suicidal genes with other species. When farmers sow Terminator seeds already treated with tetracy-cline, the recombinase will have acted, leaving the toxin coding sequence next to the seed-specific promoter, tripped to fire at the end of the next embryogeny. Of course, as incapacitated seeds grow into plants and manufacture pollen, every grain will bear a ready-to-act toxin gene. If a Terminator crop germinates near a field planted with a wild variety, then sterile pollen may well be transferred by insects or blown by wind to the adjacent field. Any eggs fertilized by the poisoned pollen will now bear and transmit one toxin gene.

THE CANINES ABOARD poet Edward Dorn's "Tan Am" flight out of Lima, Peru—animals with human DNA who can do everything we can—enact the deadly para-dox of a cornucopia we invite:

"Several bipeds turned their heads/and squeezed their Newsweeks in discom-fort./'Jesus—transgenic dogs,' one of them muttered,/'why didn't they take the Air-bus!'//Odin ran his tongue over his impressive teeth/and observed: from the minute that species/stood up and walked the planet was doomed.//But seriously, we all know now/what the man meant when he said/'You ain't seen nothin' yet!'—/it's when the genome comes home to roost:/Protein Chaingangs dressed up to look like scientists/in white coats and droopy socks and dumb hair./More crooks in the banks than in the prison system."[7]

Cloning Organisms

To CLONE A PIECE OF DNA—to alter it, reinsert it, and obtain functional and inheritable results—represents one operation with its own obstacles and scales of difficulty. To clone an entire genome is another transaction altogether. Of course, nature clones genomes all the time; this is its fundamental way of reproducing and maintaining life. Most of our cells replicate themselves again and again by mito-sis, cloning in the zygote to make the blastula, cloning in legion to pack the morula with macromeres and micromeres, cloning in every man and woman to keep their bodies vital and alive with new protoplasm. Without our cells cloning themselves inexorably, we would wither in a matter of days.

Various species of plants, bacteria, protozoa, jellyfish, worms, and some other animals which have totally lost the ability to reproduce sexually, generate gametes from cells of their adult bodies; hence, clone their offspring. In species utilizing sexual reproduction, fertilized ova may also randomly clone themselves, resulting in twins, triplets, quadruplets, quintuplets, sextuplets, etc.

The technology of cloning genomes is far more easily carried out using old-fashioned embryo twinning than it is by grafting biological material from adult organisms into ova. For the former, all that is necessary, as Hans Spemann found out seventy-five years ago, is to split a blastula (or morula) at four, eight, sixteen, or thirty-two cells (or thereabouts, depending upon species of subject and caliber of tools) into two or more clumps. In the case of mammals—as opposed, for instance, to sea urchins—the separate clumps must be implanted in a uterine environment for further development. Successful implementation of this cycle leads to the birth of genetically identical animals. It is also possible to insert foreign genes into imma-ture clumps of germ cells and produce transgenic organisms, creatures with genes from sources extraneous to them, usually a plant or animal with which they would never have mated.

Cloning is possible only with the cells just a few divisions removed from their zygote stage. Afterwards they gradually begin to specialize, to shut down their bat-teries of genes. Once the DNA is programmed in expression loops on their chro-mosomes, their chromatin cannot go back in time and reset itself. The cells lose their capacity to function as zygotes.

Successful cloning of adult cells, as in the conception of Dolly the sheep or Cumulina the mouse, requires tricking fully determined nuclei into regaining the capacity to express their original DNA complement (see Chapter 8, pages 139–140). Ian Wilmut and his team in Scotland began their lamb cloning using early morula cells; then later fetal cells; and finally, only after developing successful retrodiffer-entiating techniques, preserved adult cells from a deceased ewe. The initial prob-lem was in synchronizing the tempo of cell division between donor and recipient cells. Cells fissioning at different rates led to nonfunctional embryos, so the donors for Dolly needed to be kept as undifferentiated as possible. Wilmut's contribution was to place them in a nutrient-deprived solution, thereby starving them and pre-venting cell division altogether. Keeping cultured cells in such a resting state made their nuclei more pliable. No one had done this before.

A nucleus from one of the old ewe mammary-gland cells was extracted by suc-tion with a micropipette (thinner than a hair) and inserted into an unfertilized egg prepared for its guest by the prior removal of its own nucleus. A surge of electric-ity then substituted for acrosomal enzymatic action, perforating the membranes and fusing two cells into a single organism.

(Dolly turned out to be a rather playful and spoiled sheep. She bleated in delight at the approach of her admiring visitors and kicked over her food bowl each time she didn't like the menu.)

When Ryuzo Yanagimachi cloned scampering mice (instead of sheep) in 1998, he introduced a gap—one to six hours—between inserting the foreign nuclear material into an enucleated egg and then activating the egg, and he used a chemical solvent instead of an electric shock. These refinements resulted in a five times greater success rate.

Later in the same year, the journal *Science* released the results of experiments carried out by a group of biologists under Yukio Tsunoda at Kinki University in Nara, Japan. This ambitious team cloned beef cows using cumulus cells and cells from the linings of Fallopian tubes gathered from entrails at a local slaughterhouse. They injected cumulus cells into 99 enucleated eggs, Fallopian-tube cells into 150 more. Forty-seven of the cumulus and 94 of the Fallopian-tube cells began to develop. Of these, 38 total eggs became full-fledged embryos; ten survived to be transferred to surrogate mothers; a remarkable eighty percent of the latter were born as calves. This was a dramatic improvement over Wilmut's 400 eggs yielding 29 embryos and only one lamb.

CLONING HAS A SINGULAR meaning in animal husbandry but takes on manifold subtexts when applied to human beings. Replicating human genotypes is fraught with social and psychological perils. What would it be like to enter a world in which we could view our precise genome at all its different stages of maturation? As children, we would meet ourselves as teenagers, middle-aged men and women, crones and geezers, while they would be looking back concurrently at their own child and adolescent bodies. How would it feel to be the clone of someone deemed a genius, and have to live up to his achievements—or to be hatched from the cell of a genocidal dictator later overthrown (after cloning himself)?

It is questionable whether people want to view their pasts and futures marching around among them. This would steal some of the novelty and surprise of their own lives from them.

Clones may suffer in another way—by inheriting cells with Hayflick-limit life spans reduced by the number of divisions of their progenitors. Dolly was born an adult, at least at a cellular level, with a correspondingly shortened life-span.

Without intervening meiosis, cells also miss the sorting and splicing-out of chromosomal errors that occur during crossing-over.

A subtler dilemma would arise in trying to establish the personal identity of a somatic-cell-grafted entity. A clone is not the source-cell person over again. Experiences are untransferably unique to each individual, not each genome. When a second or third individual is cloned from a genome, she does not inherit events in the lives of her predecessor any more than she inherits her knowledge of Mediaeval history or tattoos. Grieving parents of some future society, attempting to "bring back" their child killed in a supersonic automobile crash through cloning one of his cells rescued from the funeral parlor, will find that they do not reincarnate the dead child but a twin, a totally new being without the first one's personality or emotions, likely without the same skills and interests, and certainly without the memories of the dead child. What is the use of begetting one life by exact reference to another that it is unable to access? Such a child would inevitably be plagued by unfulfillable expectations, the kinds of burdens that are projected even onto normally born siblings of tragically deceased children.

IF HUMAN BEINGS were to reproduce by splitting into exact twins, we would experience a curious identity problem. Which of our fission products (if either) would we become? Might each of the twins have our thought patterns as well as our nuclear material? If not, how would a clone without our memory know who or what it was? Or how would "twins" that began with one body individuate their separate existences as mother and daughter?

Frank Herbert asks these questions in the latter volumes of his "Dune" science-fiction sequence. The warrior Duncan Idaho is reembodied thousands of times from a single patch of skin preserved from the corpse of the first Duncan slain in battle. Though his remembrance-tracks up to death are preserved in the "hard drives" of each successive clone, they all become independent personalities with their own idiosyncrasies, ignorant of one another's inner lives and later memories.[8]

However prescribed by the force of a previous generation, new cells are rebellious. Herbert has given a hypothetical answer to a hypothetical problem, but beneath it lies the old dilemma of who we are, if anyone, when we awake (again) at the beginning of time.

Genetically altered plants and animals are not adulterated.

DESPITE THEIR OTHER LACKS, recombinant DNA, gene splicing, and cloning are not artificial procedures imposed on nature; they are versions of trademark biokinetics the planet has carried out randomly and experimentally since the advent of the biosphere. Humans are merely attempting to impose direction and utility on the promiscuous spread of DNA, to exert self-interested regulation over certain limited aspects of an ongoing, boundaryless, selective process.

The results of biotechnological activities are not robots, cyborgs, or artificial organisms in any sense. There are no additives or toxins in the milk of a transgenic cow or a Bt potato. The entities issuing from biotechnology are more natural than any metal or plastic, any margarine or synthesized drug; for cloned and transplanted cells and their tissues behave as organically as so-called unaltered cells and tissues (which are far more deeply infringed upon on an ongoing basis than genetic engineers could ever manage, though by nature not scientists).

Biotechnology fiddles with very small bits of cells' germinal order; then it places its jimmyings back into the native embryogenic process to do what it will. Far more than ninety-nine percent of all the stuff inside the nuclei of transgenic plants and animals is nature's ordinary assemblings; a tiny portion represents the human introduction of a message into the blueprint. The playback is totally uncooked.

Humans are in fact latecomers to the game. Viruses have been transferring plasmids between organisms for millions of years, so genes have been promiscuously strewn throughout the biosphere just about forever. Bionts might be specific, but DNA remains global and indiscriminate. It is no wonder that we share so many nucleic units with worms and mice. Each of us is an old-fashioned neurally agitated alimentary tract with a mouth and anus. We wriggle the same wriggles and fire the same ATP.

All creatures are transgenic; all creatures are spliced and altered.

Environments are nested seamlessly in one another.

NO IMPERMEABLE MEMBRANE segregates an organism from its environment, or, in the other direction, from the environments inside its cells. These flow together, permeate one another, and are altered and maintained by one another (see Chapter 11, "Morphogenesis," pages 238–240). When my son Robin was a graduate student at the University of California at Santa Cruz, he was excited to study plants and animals in environments, though much less enthusiastic about dissecting them. His teacher, renowned biologist Todd Newberry, observed his bias and told him, "The distinction between outside the organism and inside the organism is the most arbitrary of boundaries."

As noted, we are already engaged in biotechnology when we breed plants and animals; we are engaged in it again when we fertilize fields and feed antibiotics to herds (including humans). In all these cases we manipulate protoplasm. Because genes dwell in cells, cells in organisms, and organisms in environments, the effects of genetic engineering can only be interpolated in nestings of preexisting dynamic systems: "Every organism is continuously going beyond a mere object relation to its surroundings. What was outside is now inside—not inside as in a drawer filled with things, but, rather, inside as incorporation, as unification."[9]

The organism in the ecosphere is much like the cell in the organism, altered in subtle ways by all the various mechanics and chemistries that contact it. In the case of primates, we may add social influences to those of morphogens and environments.

Just as the extant membranes and tissues of a biont direct the expressions of its genes, and as the membranes and tissues are influenced by their macro- and microenvironmental contexts as well as the foodstuffs they imbibe, so genetically intruded-upon embryos interact with their own displaced biological fields and the ecology around them and are subtly transformed back and forth by all. Let into the wild, they will change the world around them.

So-called killer bees swarm north, disrupting insect and flower co-ecologies. "Alien" crabs and snails deposited accidentally on the outsides of ocean-crossing vessels threaten remote ecospheres by predation. Weeds like Queen Anne's lace and dandelions, once transported across the Atlantic with pilgrims' grains, become integral parts of the New World forests. Guerrillas travel from Cuba to Angola, religious warriors from Yemen to Afghanistan, mercenaries from Belgium to Swaziland. These are all transcellular, even transcontinental, biological events.

GENES ARE NO MORE CONCRETE, independent units of heredity than orchids or periwinkles are independent florescences sprouting in a void. They are equally transhumants, weeds. Chromosome patches cannot be switched usefully between organisms as if genes were widgets any more than a horse can be happily placed in a lake, a cactus in a marsh, or a grocer in the outback. "The relation between seed and forest is similar to the mutual dependency between DNA and its host organism."[10] Genetic engineering likewise cannot barter in ready-made traits like phosphorescence, ice resistance, or growth. Instead, it supplies nucleic material; the traits are then created by conditions within and about the recipient embryo (i.e., as events in fluctuating micro- and macro-ecospheres).

Yet the very words "engineering," "cutting," "splicing," "reading information," and "manufacturing protein" suggest targets and substantiality. "Such language is on the one hand mechanomorphic *(cutting, machinery, manufacture)* and on the other hand anthropomorphic *(information, code, expression)*. The combination of the two makes it sound as though one could actually see and understand all that is going on.... If you visit a genetic engineering laboratory, you may be disappointed to find that none of the processes described above can be observed. This is not because the cutting is being hidden, but because it is occurring in thought."[11] Genes do not give commands; they do something that is not even a metaphorical version of a command. As Martin Heidegger noted, "The essence of technology is nothing whatsoever technological."[12]

Genes as Units of Meaning

PEOPLE DON'T REALIZE THE FULL IMPLICATION of a belief system based on total genetic determinism. In such a worldview every animate thing must arise ultimately from the information content of what we call genes—life, mind, society, environment, philosophy—with no other originator, no other *efficio*, no other arranger of molecules. We must go back to the sortings of genes and genes alone for how things are.

Yet there is so incredibly much information that defies and exceeds the limited repertoires of any version of actual genes.

Try playing a game in which players take turns naming things that perhaps don't come from genes; then other players imagine where else these things might come from—a cat waiting by a mousehole, ability to compose a symphony, musical taste, criminal behavior, a droll sense of humor, charity, thatch houses, the word "gene," etc.

Well, where do these come from?

Biotechnology is based in theory on the concept of something like genes as unique pilots of form. Where such gene things derive form and organismal meaning is at best an enigma because inherent in the whole concept of a gene is the belief that nucleotides and mRNA arose by chance interactions of molecules, thus carry and transmit protein-based form arbitrarily and fortuitously.

Genetic engineers then reorganize the originators—in practice to change a limited range of specified things, but in theory to change everything, including meaning. After all, if there is no source of meaning other than genes, any living or symbolic meaning at all can be supplanted by a change in codons.

Change genes and you change reality.

This fails to take into account the etymologizing thread that orients any cultural system of meaning. "Genes" do not precede meaning; they are an outcome of a very mature linguistic as well as physical inquiry (and at more than one level of logic and technology simultaneously). Genes reflect deep layers of grammatology, figuration, and reinscription. They are only artificially routed back through the cumulative act of being written into an inflated role of originator of a biologistic system of meaning. Genes may come first in epochal chronology, but that chronology is an origin myth, not a true imprint in a cellular landscape (see likewise the imponderable turtle who preceded the creation of the world in Chapter 28, page 728).

Genes are our symbolic writing of units of our own meaning, traces (i.e., mor-phophonemic strings) which were in existence for tens of thousands of years (at least) before the term "gene." So we have woven ourselves into a web of our making and displaced our meaning by a tautological contrivance. As long as "life" continues to arise (and lie dormant) in its own domain, the deeds of biotechnologists will fall well short of the prospect of genes as protean originators. The limits of technology will demonstrate ultimately, if they have not already, what so-called "genes" are not and what, by default, they are—i.e., properties, utility functions, place markers, traffic lights, rheostats, etc.

Where the rest of all and everything comes from is a mystery.

Levels Too Deep To Be Deciphered

NONETHELESS, THE ACTIVITIES OF BIOTECHNOLOGY are publicized as con-crete—transferring phosphorescence to tobacco, girth to mice. Scientists *seem* to be dealing in traits as literal expressions of genes. Apparently some genes yield consistent biochemical characteristics, even in widely differing species; most genes, however, are far more field-oriented and do not translate into hard traits. Some express both perceptible traits and generalized field characteristics. This means that any use of genetic material, transgenic or otherwise, that results in some consistent, tangible change also has other effects, many of which are not immediately expressed or apparent. If this weren't the case, then worms and humans could not be written from fundamentally similiar codes. From a growth and form standpoint, if you have one creature produced from componential units and a very different creature produced from a somewhat varying assortment of both the same and different componential units, those units (especially insofar as they represent code rather than traits) can be manipulated only with effects throughout the whole organism, like ripples from a rock in a pond. The reason that biotechnologists can't turn wormlike bionts into the equivalents of land mammals is that they are (fortunately) not allotting enough time and space—generations of resorting and environmental selection—for the full range of factors they are potentiating to be realized. Yet this remains the latent time bomb of transgenic splicing.

The use of gene maps to link nucleotide sequences to cellular and organismic effects (including physiological and pathological outcomes) must always be subject to the qualification that genes (as well as proteins) are not merely componential, cumulative, and expressive (or not) but may be pleiotropic (determining more than one characteristic), complexly interdependent in their functions and morphological significations, Lamarckian in their intercellular dynamics (with covalent marking of DNA and chromatin by cellular fields emerging from that very nucleic material), and epistatic (nonreciprocally suppressive of each other's expressions). Proteins are also additive in a manner that is genomically complex and subject to factors of multiple redundancy, mutual interaction, and multidimensional interconnection. There are no set, synoptic gene-protein syllabi.

As perturbed factors are passed upward in the hierarchical organization of a functional creature, epigenetic regulation takes over in an unplottable manner, imposing merely "the most proximal of a hierarchy of constraints extending outward from DNA structure to the cell boundary and beyond."[13] In fact, there may be as many unknown as known factors, and they may extend further into the universe and deep evolutionary web of species than we imagine.

No MATTER WHAT SPIN we put on them, the rules nucleic expression and epigenesis obey "are extragenomic and are most likely to be found not in molecular mechanisms per se but in their integration into complex gene networks and, more peripherally, into their connectedness with regulatory networks (metabolic and other) of cellular dimensions"[14]; i.e., membranes and environments.

What we splice into these systems will sink beneath the various interior and exterior surfaces, engage ancient complexities, and become interpolated in manners identical to and yet different from anything introduced randomly during evolution. As long as we believe natural selection is merely arbitrary, we will not take the dangers seriously. After all, what more damage could we do than nature, blind and bumbling? Yet it is possible that most root biological templates originated in primordial times when organisms were primitive and more pliable; modern species are the delicate result of millions of years of mechanical and genetic equilibration. Thus we end up tampering with meaning at levels that are too deep to be deciphered or penetrated productively; likewise we place our inflated goals at shallow levels. Or we do both at once because, worst of all, we do not have a clue as to what we are really doing.

Even where genes seemingly prescribe hard features now, they did not initiate them change by change along incremental pathways (biotechnology-style) or by mere linear adventitious shuffles. Instead, genes inculcated themselves into independent epigenetic processes of both animate and inanimate origin, thereafter serving as a casement to stabilize discrete organisms. The so-called end-products are not rebuses of interchangeable traits but liquid topologies with genes latching on in all different ways with a variety of meanings and consequences, not unlike neural synapses. Attempts to remodel these as if they were "genetic houses" totally misunderstand how genes and organisms came into their current arrangements.

The forms we manipulate are simultaneously archaic and semistable, having become "more themselves" both despite and because of mutations (see page 269 *et seq.*). Introducing purely intellectual changes at incalculable levels of homeostasis and depth while ignoring the rudders of genome resilience, we may well undermine not only species integrity but anatomical dignity and personal identity.

THE FAILED PROGENY OF TRANSGENIC EXPERIMENTS demonstrate most poignantly the fragility and unruliness of engineered genes in unpredictable environments. For instance, in an attempt to make male mice out of females (reported in *Nature*, May 1991), ninety-three mice offspring yielded five mice with transgenic material. Two of these were normal masculine mice, showing no effects of the transplanted DNA. Two genotypic females bore the transgenic material but had no male characteristics. One transgenic mouse was produced with very small, sterile testicles (yet that one of the ninety-three appeared on the journal's cover with the caption "Making a Male Mouse," disguising the fact that ninety-two other mice were also "made").

At around the same time, an experiment intended to grow wooly hair on mice from sheep DNA bred no curly mice and only one with periodic baldness and broken hairs. Another experiment transplanted human iron-binding protein (lactoferrin) into cows as the first stage in developing a method for manufacturing pharmaceuticals in transgenic cattle and harvesting them from their milk. In this trial, 981 out of 1,154 eggs injected with human DNA survived, but only 129 embryos were successfully transferred into the oviducts of cows. From twenty-one pregnant cows, nineteen calves were born, one transgenic, one with transgenic DNA only in the placenta. "Moreover, a rearrangement had occurred involving a deletion of part of the [foreign] DNA construct."[15] Thus, other contrary vectors were already eliding the intrusive syntax.

The USDA's implantation of human growth genes in pigs yielded gigantic bug-eyed animals with severe muscle and joint damage, rendering their normal behavior impossible. The biological engineers had added a second switch to their program that should have tripped growth hormone only when the animals were fed zinc. However, it failed to express itself.

INSOFAR AS AN ORGANISM is not just a panoply of traits but a living part of the ecosphere in which it arises, its genes are no more existential than its habitat. The poor overweight pigs lived in constant pain without any excuse for existing. The experimenters noted that "they could be penned in such a way that they would no longer need to carry out most of their behaviors"[16]; hence, ostensibly they were spared the indignity of their plight. But is a pig only a model like an Oldsmobile or Toyota, currying new designs? Is a cow a milk-generating machine without a lifestyle? Will the protein produced by such a compromised beast be of the same vitality and quality as that molecularized by a healthy animal grazing in a field? What reason is there for a tobacco plant to glow, a fly to have legs on the sites of its nonexistent antennae, a mouse to grow wool? The ultimate livelihood and fates of all of these flimsy creatures will be determined by their prosperity and durability in environments, not by experiments alone, not by the parameters of chemical companies—for these creatures emerged from environments and *are* environments.

Technologists are not creating systems; they are perturbing systems someone else made, systems they could not make from scratch with raw materials. Everything real about these artificially mutated bionts reflects their genesis in deep evolutionary history and ecology; everything gimcracky about them represents the shallowness of their manipulated gelding. Ultimately, genetic engineering runs the risk of twiddling kite strings without any awareness of winds or power-lines or even the shape and size of the kites. And these are not your usual toys. They are dragon kites that change size and shape and alter all waves and particles they encounter.

Scientists may pretend that life can be subjugated to mechanism, but protoplasm and DNA have properties that defy the very meaning of technology. "The plasticity whereby an organism selectively incorporates aspects of its environment, internalizing them and entering into a nonobject-like relation with them, is an essential characteristic of life. This plasticity is a prerequisite for all genetic research and genetic manipulation. Without an egg's ability to take DNA into itself, no genetic manipulation could succeed."[17]

Epigenesis is at least as important as genetics, for the program in the genome is in no way isomorphic with the emerging organism. Yet the sole burden of concreteness has been misplaced onto DNA rather than the egg's (and planet's) wondrous succession of differentiating robes.

Likewise, in an obsession with number and product, we have missed the roles of faith, prayer, magic, fun, romance, zydeco, zen, *Chi Gung*, doo-wop, night trains, and everything else that doesn't fit a Puritanical model.

Genomic Medicine

MEDICINE IS THE AREA in which biotechnology augurs the most promise and also the most serious immediate disruption of human existence. In placing genetic science in the best possible light *Time* reminds its readers: "Before this century, medicine consisted mainly of amputation saws, morphine and crude remedies that were about as effective as bloodletting. The flu epidemic of 1918 killed as many people (more than 20 million) in just a few months as were killed in four years of World War I. Since then, antibiotics and vaccines have allowed us to vanquish entire classes of diseases. As a result, life expectancy in the U.S. has jumped from about 47 years at the beginning of the century to 76 now. . . . The next medical revolution will . . . conquer cancer, grow new blood vessels in the heart, block the growth of blood vessels in tumors, create new organs from stem cells and perhaps even reset the primeval genetic coding that causes cells to age."[18]

This is futurism at its most glowing and irresistible—a visitation of Earth by beneficent and gifted overlords in the form of our own descendants. If the prognosis is correct (more or less), pharmacy will soon be conducted at the level of DNA rather than tissue. Doctors and computer technicians will select medicines by genetic profile and customize them to match not only general pathologies (as now) but the precise susceptibility revealed by the nucleic code underlying a patient's biological template. The hypothetical nature of the gene will become secondary to the predictive value of genetic analysis and the therapeutic successes of genetic manipulation.

From the standpoint of genetic medicine, traditional pharmacy (which was considered ultramodern and progressive only yesterday) is "like shooting a quiver of arrows into the air and then running around to see what they hit."[19] The discovery of penicillin in 1928 by Alexander Fleming represented its epitome, a fluke of wildly good luck. When pathologies are orchestrated at different levels by dozens of separate genes (as high blood pressure is), prescribing once had to be a matter of guesswork and trial and error. Now doctors can select a medicine on the basis of a DNA map of the patient compared to equivalent maps of other patients for whom the therapeutic results of a variety of medicines are known. They can aim directly at targets with arrows cyberneticized for their tasks.

For the alleviation of depression and anxiety, Prozac and its cognates have been targeted at the serotonin receptors of the brain. Tagamet and Zantac have been customized to mitigate acid indigestion in the stomach. Hundreds more biological mechanisms suitable for selective therapy have since been identified. Dr. Wayne Grody, head of the DNA diagnostic lab at the UCLA Medical Center, foresees "a new paradigm—genomic medicine—with tests and ultimately treatment for every disease linked to the human genome."[20]

In the future oncologists may be able to set a small sample of malignant cells on the glass bed of a computer chip, run the chip through a program, and be told which mutant genes are involved in the cancer. Based on the tumor's genetic profile, a sequence of medicines will be selected to impede further growth and metastasis. This would reduce cancer to the level of a flu.

It has been a long time (forever, in fact) since anyone knocked on the door of a nucleus and tried to interest it in something other than the ciphering of its own stuff. Now scientists are attempting to impress the great polymer itself into *their* cybernetic tasks. Future microchips may even incorporate DNA in their grids. If its activities were placed at the service of human goals rather than its own pagan, untamed plan, the double helix could become the ultimate computational device on the planet (see page 92). Linked to silicon hardware, it could be used to analyze simultaneously the complex interactions of genes and proteins.

In one model involving potential AIDS treatments, sequences of tens of thousands of genes are downloaded in rows of anti-codons onto chips. These hieroglyphic braids string themselves across microns of a glass plane, silicon mirrors of phrases in a genetic alphabet, insignias of nucleotides in a noncellular medium. A solution is then derived from the blood of a patient with a virulent HIV infection—his immune system actively generating RNA molecules to assemble appropriate proteins while pumping out millions of cells to try to neutralize the attack. After RNA is extracted, split into sections, and tagged fluorescently section by section, the solution is sluiced onto the chip. As the RNA finds its DNA complements on the glass, fluorescent tags mark the matches. A computer then identifies the "hits" and prints out a register of what genes were expressed in the HIV infection. A subsequent comparison of immune responses from different patients who are more or less successful at fending off the virus will then provide information as to which genes offer the most effective "cures."

Somatic Gene Therapy

WHEN RESEARCHERS DEVELOP GENETIC MEDICINES, they leave the matter of getting recombinant stuff into actual cells to the recognized experts in that task—viruses. These take to cells like prairie dogs to dirt. However, first they must have their own genes removed or altered in such a way that they cannot spread disease. Then, ideally, salutory genes are spliced into what is left of their genetic material. The new carrier virus is mixed with human cells to make a medicine. Because viruses cannot carry the sorts of large, complex genes that would be effective in many conditions, somatic gene therapies are limited in use. Still, a whole new pharmacopoeia has been compiled from the biological products of altered genes.

According to Inder Verma of the Salk Institute in La Jolla, California, by placing "beneficial genes into the cells of patients ... and consequently the protein that [they] encode ... 'you either eliminate the defect, ameliorate the defect, slow down the progression of the disease or in some way interfere with the disease.'"[21]

Varieties of recombinant DNA have led to affordable remedies for patients with the rare adenosine deaminase (ADA) deficiency as well as for those with cystic fibrosis and other diseases. Not all sufferers are benefitted equally (or even at all), but enough show improvement to lure huge amounts of venture capital into companies like Geron and Genentech.

An incorrect gene in those born with ADA deficiency renders the T cells of their immune system incapable of synthesizing the essential ADA enzyme. The cells die, leaving "bubble boys" and "bubble girls," children who must stay in sealed environments to keep from being infected with bugs their bodies cannot fight off. While periodic injections of ADA protected by a chemical sheath allow temporary survival outside quarantine, infusions are required weekly at a present cost of $60,000 a year. As an alternative, billions of faulty T cells taken from one "bubble girl" have been subjected to defanged leukemia viruses spliced with human ADA genes. The viruses invaded the cells, consorted with their DNA, and transferred functional ADA genes. After altered cells were reinjected in one source patient, her ADA level went up to 25 percent of normal, sufficient for immune protection while playing point guard in basketball. Insofar as that patient's own blood marrow cannot synthesize the necessary cells, somatic gene therapy is not a cure; regular reinjections are required. However, introduced genes have significantly mitigated the hereditary disease and reduced the cost of treatment.[22]

Selected patients with angina too severe for bypass surgery come to St. Elizabeth Medical Center in Boston for somatic gene therapy. A medicine is injected directly into their hearts through a tiny slit in their chests, a solution containing billions of clones of part of a human gene (VEG-F) that incites proliferation of blood vessels. Without understanding the mechanism of cell penetration, doctors have nonetheless succeeded in actuating biological effects using naked VEG-F DNA (no viral assistance!) to penetrate cells and code fresh nucleic material. Though the raw DNA shuts down in a few weeks, the proteins it stimulates spread to legions of contiguous untreated cells with exponentially therapeutic effects. In one trial, sixteen heart patients improved after treatment, with six able to return, pain-free, to their normal lives.[23]

GTI-Novartis of Gaithersburg, Maryland, has developed a unique gene therapy for brain tumors that demonstrates the algebraic structuring of the ciphering system and the circuitous logic necessary to infiltrate its labyrinth of meanings. The Novartis carrier is a retrovirus (an RNA virus that infests only cells in mitosis); its spliced-in package is a herpes virus. This hybrid is transmitted into the brain. Since brain cells do not divide, they are not affected by the retrovirus. However, the dividing tumor cells are quickly invaded, and the herpes gene is smuggled inside them. Then the herpes drug ganciclovir is dispensed to the patients with the goal of making "the tumor cells commit suicide."[24] Incredibly, this ruse has worked.

Other promising somatic gene therapies have not yet been tested on humans. In one dramatic animal trial, DNA molecules for an insulin-growth-factor protein were packaged in the shell of a "safe" virus and injected into the skeletal muscles of mice with "fountain of youth" results: aged and atrophied muscles were enlarged and restored to adolescence and vigor.

As EXPECTED, adding ostensibly efficacious genes to cells and removing faulty ones also affects other wide-ranging aspects of the morphogenetic field, imparting new functions to cells and inciting a plethora of unpredictable and not always happy results (including cancer and heart disease). Viruses used in gene therapies have spurred debilitating inflammations, and both adenoviruses and retroviruses may have deleterious immunological side-effects. Also sometimes new genes are not expressed at all, or express themselves and do not produce the expected proteins, or produce the proteins without therapeutic results. Sometimes they start off efficaciously; then their performance vitiates, as if they weren't there, or were recognized for the party-crashers they are.

Other gene therapies are flawed because they overproduce the protein needed. Normal insulin genes for diabetics, for instance, trigger a dangerous surplus of insulin. One resolution may be to splice a biochemical rheostat into the mutated gene, rendering it inactive except in the context of another substance, which must be ingested separately. Pills could be administered to regulate the expression and shutting down of the gene and its proteins (as was attempted with the oversized pigs and their zinc). However, evolution is difficult to replicate on short notice.

Germ-Line Therapy

LARGE NUMBERS OF EXPERIMENTERS (and investors) now believe it might be more lucrative to predict diseases and correct them in the embryo, using repaired or transgenic genes or artificially synthesized strands of DNA inserted into human germ cells (sperm, eggs, or early blastulas), rather than waiting until traits are inculcated in an organism. This more controversial embryogenic method is called "germ-line therapy"*; it is presently not far enough along to be applied in humans, plus the ethics and legality of experimenting with our genomes are problematic. Gene manipulation would not be of much use to people who are already sick.

---

*The term "germ-line alteration" would be more accurate and less evangelical; after all what does "therapy" mean with regard to someone who does not yet exist as a person?

Germ-line alteration also risks disturbing deep biological equilibria, with complex and uncertain long-term effects. It has little use beyond the questionable one of species enhancement, yet tampers with the lives of the unborn without their consent.

In an early exploration of germ-line potential in the mid 1990s, researchers from Case Western Reserve Medical School in Cleveland designed human artificial chromosomes (HACs) in cultured cells—chromosomes that behaved normally, replicating when the cell did. This theoretically allowed genetic engineers to write their own progams and insert them into morula cells. The "therapeutic" genes would have control switches (rheostats) so that they transcribed only in the presence of a specified chemical, ideally one not ordinarily synthesized in humans (for instance, ecdysone, an insect hormone). Then a tripping substance would be fed to a child or adult at the proper time (depending on the target disease), activating the gene and its propitious effects.

A BREAKTHROUGH WITH POSSIBLE APPLICATIONS combining aspects of both somatic and germ-line methods was reported in November of 1998 independently by two cadres of researchers, one at Johns Hopkins, their rival at the University of Wisconsin in Madison. Both laboratory teams were apparently able to isolate and culture stem cells from early human embryonic tissue, the Wisconsin ones from blastocysts of roughly 140 cells (donated as excess by couples attempting in-vitro fertilization), the Maryland ones from aborted fetal tissue. These cells, as yet unprogrammed and undetermined, are able to differentiate into every and any one of the 210 types of cells in the human body, something they do automatically and selectively during fetal development.

From its cache of embryonic stem cells, the Wisconsin group was able to induce separate cultures into bone, muscle, gut, blood, and nerve tissue, suggesting the possible future synthesis of fresh tissue for patients with untreatable and degenerative conditions: heart disease, diabetes, Alzheimer's. Once integrated into the body's fields and induced by surrounding structures, these artificially differentiated stem cells could become new heart muscle, insulin-manufacturing pancreatic cells, or neurons to replace damaged cerebral tissue.

In 1999 researchers at the National Institutes of Health grew two types of nerve cells—astrocytes and olgiodendrocytes—from stem cells of embryonic mice. Transplanted into rats with a genetic malady impeding formation of nerve-insulating myelin, they stimulated growth of the critical neural sheaths. Such a therapy could be adapted to human neurological diseases, including Parkinson's and multiple sclerosis (in which the body attacks its own myelin). Other modes of treatment could ultimately be developed for brain and spinal-cord repair.

It may also be possible to reset body cells to a pristine state and thus derive the equivalents of stem cells for people from their own mature tissue. Each person would be a repository of his or her new organs for transplants or undeveloped tissue to replace diseased or aged tissue.

On the premise that aging is caused by the shortening of the telomeres at the tips of the chromosomes during each cell mitosis, researchers have also attempted to rejuvenate cells by a reconstitution of these zones, a sequence successfully ventured in 1998 by scientists at Geron Corporation of Menlo Park, California, who achieved twenty divisions past the Hayflick limit.

If some of these techniques are developed to their potential, the urban landscape could be infiltrated by a whole new commerce: stores providing miracle "stem cell" drugs, home renewal kits, and organs on demand. Tissue could be extracted, stem cells generated, and fresh body-parts (arms, eyes, kidneys) bred and ice-packed to a surgeon.

The Human Genome Project

AT THE HEART OF MEDICAL BIOTECHNOLOGY is a $3 billion attempt to map the human genome. Enfranchised under the auspices of the National Institutes of Health (NIH) in 1989, the National Center for Human Genome Research was placed under the initial directorship of James Watson. The explicit goal of this project is to "sequence the entire 3-billion-letter human genome with high precision as a prelude to figuring out eventually what protein each gene produces and for what purpose."[25] This is about as labor-intensive as scratching a line in the surface of the ground, anthill to tree to ravine, up and down buildings and trees and meadows, all the way from Cape Cod to Puget Sound.

To "read" a piece of the genome scientists must work at the scale of molecular processes of cells, isolating the protein-coding aspects of chromosomes. A gene-bearing fragment of DNA is severed from a chromosome; cloned millions of times; sorted chemically by nucleotides; and the different nucleotides, separated by electric charge, are fed into a gene sequencer which, in essence, coopts their linguistic capacity, inducing them to write their message in colored dyes rather than proteins, a pattern that can later be read by a laser like a bar code.

Since the advent of the Genome Project many rival groups have pioneered short-cuts through the painstaking process, either ignoring geneless regions of DNA (comprising the bulk of it) and focusing on the 0.6% related to major disease-causing malfunctions or speeding up the process (with attendant errors) by using RNA and highly automated gene sequencers. In the late 1980s at NIH, biologist Craig Venter directed cells themselves to locate genes (as is their wont). Without laborious cutting and customizing, he duplicated strands of RNA protein-assembling code directly in bacteria. Then "the bugs [were] ripped open and their DNA [was] run through a gene-sequencing machine."[26]

Watson dismissed Venter's shortcut as the equivalent of a monkey hammering away on a typewriter, producing mostly gibberish, and he bolted the Genome Project for the laboratory at Cold Spring Harbor, New York; Venter departed too, using venture capital to start his Institute for Genomic Research. "If this is the book of life," grumbled new director Francis Collins, "we should not be satisfied with a lot of mistakes or holes."[27]

In any case, the Human Genome Project has splintered into a number of different grail quests under different auspices, and the results and their applications are as uncertain as volatile. The most insidious shadow that now hangs over the future of genome-mapping is the possibility that codes will be patented and privatized, leading to organ blueprints and medicines controlled by closed cabals and self-serving individuals. At one level this would deny DNA-based treatments to those who couldn't afford them; in an even darker prophecy it could create the most impregnable upper class and aristocracy of all time—a society in which the privileged and wealthy were able to extend their lives indefinitely while harvesting the biological and genetic products of the underclass (even as they presently harvest Third World botanical and mineral resources).

Someone who claims to own life (and has the weaponry and militia to back it up) can impose an absolute slavery on the disenfranchised masses, manufacturing them at his whim and using them as he wishes (because his lineage has held the patent on their source codes for generations). The familiar and deadly combination of tyrannical government and militarization of science now threatens annexations and concentration camps far more antipathetic and implacable than Oaxaca, Kosovo, or Sri Lanka. Though genetic formulas may create feeling-depleted warrior monsters and weakened, servile lackeys, biological fields have their own impenetrable integrity and will mutiny in unexpected ways. The grass-roots rebellions of the twenty-second century may sound more like today's science-fiction stories about chattel colonies on moons of Jupiter than Mao's great march or the liberation of Zimbabwe.

Will mapping the human genome unlock the secrets of diseases and their cures?

WITH HUMAN GENOME RESEARCH NOW PARADING as a kind of internal Hubble Telescope, more and more DNA continues to be mapped and interpreted.

Through early January 1999 "some 7% of the human genome has been sequenced in encyclopedic detail."[28] Yet there are major hurdles to its application medically, for (as we have seen) genes are not traits (or anything else). What is being manipulated are nucleotide chains. Their code can express itself in innumerable different ways at different levels of structure, different stages of development in the same biont, different bionts, and it can even be translated from biont to biont. "[A]lthough detailed genetic maps for a variety of cellular structures [are being] established, the nature of the processes being perturbed by gene manipulation remains a black box."[29]

EVEN IN CASES OF DISEASES triggered seemingly by single mutations in amino-acid sequences (monogenic causality), the connection between a gene and an undesirable trait evades consistent explanation. Only one out of 574 amino acids differs in a comparison of sickle-cell hemoglobin to normal blood; that single deviant amino acid is the demonstrable culprit, the lineal cause of sickle-cell anemia. Yet the phenotypic expression of the disease, in terms of time of its onset and severity, differs radically from individual to individual. A DNA-certified sickle-cell mutant may show no sign of anemia even in her fifties, whereas another with the same amino-acid discrepancy may die from it in childhood.

Similarly, the gene related to cystic fibrosis has been shown to bear over 350 separate mutations with outcomes totally inconsistent with any mutation-to-trait isomorphy. Many healthy individuals have the exact mutations of those with fibrosis.

In cases of polygenic causality, an undiagnosable interplay of genes, experiences, and environmental episodes contributes to most chronic diseases and physiological functions. Hundreds of different genes participate in coronary artery disease. Likely thousands of genes collaborate in other conditions. "A complex disease like colon cancer is now acknowledged to include not only large-scale mutation but also profound changes in patterns of gene expression. Genetic instability in the forms of loss of heterozygosity [separate alleles for a single trait] and aneuploidy [excess or deficiency of chromosomes] also complicate the simple single or even multiple gene mutation theories of cancer.... Considering in addition the classical but mostly unrecognized uncertainties inherent in widespread epistasis and pleiotropy, the present emphasis on dominant gene effects and on single-gene or protein-based diagnosis and therapy for common human diseases must be seen as unrealistic."[30]

It is finally true that the 1999 terrain of biotechnology is neither deep enough nor wide enough. The trip from Cape Cod to Puget Sound will eventually be accomplished, mound by mound and declivity by declivity, but the map still may be hopelessly lacking in relevant detail because epigenetic texture does not originate in the dimension travelled. The attributes of cells and organisms overwhelm their raw genomic database. Only by ignoring these boundaries can biotechnologists (and journalists) be unmitigatedly optimistic. Their enthusiasm does not take into account all the operant factors, for they limit themselves in advance to those vectors they want to regulate, defining them in such a way as to oblige them to seem material and sequential, thus imposing an abstraction onto an idea and making the quasi-concrete even less concrete by pretending to convert it into linearities with hard outcomes.

Biotechnicians have marketed their experiments effectively, conveying tabloid images of transgenic mutants and somatic-gene miracle potions. But they have not solved the epigenetic problem yet, and they are not masters of the gene. (It is probably just as well they don't know enough because if they do and this is as good as it gets, we are in worse shape than the doomsayers even imagine.)

One thoughtful observer notes:

"The entire public justification for the Human Genome Project is the promise that some day, in the admittedly distant future, diseases will be cured or prevented. Skeptics who point out that we do not yet have a single case of prevention or cure arising from a knowledge of DNA sequences are answered by the observations that 'these things take time....' But such vague waves of the hand miss the central scientific issue. The prevention or cure of metabolic and developmental disorders depends on a detailed knowledge of the mechanisms operating in cells and tissues above the level of genes, and there is no relevant information about those mechanisms in DNA sequences. In fact, if I know the DNA sequence of a gene I have no hint about the function of a protein specified by that gene, or how it enters into an organism's biology.

"What is involved here is the difference between explanation and intervention. Many disorders can be *explained* by the failure of the organism to make a normal protein, a failure that is the consequence of a gene mutation. But *intervention* requires that the normal protein be provided at the right place in the right cells, at the right time and in the right amount, or else that an alternative way be found to provide normal cellular function. What is worse, it might even be necessary to keep the abnormal protein away from the cells at critical moments. None of these objectives is served by knowing the DNA sequences of the defective gene. Explanations of phenomena can be given at many levels, some of which can lead to successful manipulation of the world and some not.... An easy conflation of explanations at the correct causal level may serve a propagandistic purpose in the struggle for public support, but it is not the way to concrete progress."[31]

The apparent goal is relatively simple: to introduce adequate "correct" nucleic material into a sufficient number of cells, then to get it to transcribe, and finally to keep it active long enough to bring about any change, ideally our desired change. Of course, insofar as we are imposing coarse linearities on a system with much subtler and deeper syntax, all outcomes are chancy. Yet we are looking for those single crisp gene-protein-attribute loci.

A template which took several billion years to assemble has anisotropic, interdependent facets of incalculable depth. Condensed into an extraordinarily tiny space, body-making elapses in an absurdly short time. To crack open this kernel and insert new meanings would require a translation between remote scales of meaning. The compression of time and space is so great in organisms and the coordination of parts, both genotypic and phenotypic, so minute and precise that there is no landscape in which parts are not annealed and amalgamated in one another.

At this stage of things the particular genes or combinations of genes underlying the vast preponderance of even the simplest attributes and conditions of human beings have not been pinpointed or denominated, so we cannot engineer their nucleotides. The kernel remains for all intents and purposes impenetrable.

Drawbacks of Genetically Engineering the Genome

NINETY PERCENT OF PREGNANT WOMEN in the United States have blood samples taken in search of proteins betraying spina bifida (a hole in the fetus' spinal cord), neural-tube defects, and Down's syndrome. Fetal cells can also be extracted from the amniotic fluid or placenta, and these may show chromosomal flaws, for instance the extra Down's-syndrome chromosome, or enzyme errors such as an insufficient level of hex-A expressed in the fatal metabolic condition Tay-Sachs disease (common among American Jews from Eastern Europe). DNA tests also exist for various forms of mental retardation, susceptibility to breast cancer, cystic fibrosis, Huntington's disease, Duchenne muscular dystrophy, and deterioration of the nervous system, including the brainstem, spinal cord, and peripheral nerves. More complex (but not insoluble) configurations may underlie diabetes, stroke, cancer, Alzheimer's, depression, etc. Could addiction, criminal proclivity, and anti-establishment behavior be far behind?

In fact, we do not even have to wait until a defective fetus needs to be aborted (with concomitant medical and social problems). Sperm from the father can be mixed with eggs from the mother in a Petri dish—a process known as in-vitro fertilization. Genetic tests can be conducted on fertilized ova at the sixteen-cell stage, with only promising blastulas implanted in the mother's tissue. A child is launched without sexual intercourse and at reduced risk of birth defects.

Once we become this involved in determining our children, we are changing not only the human genome but destiny itself, and we are altering the meanings of our own lives, moving from the level of simply experiencing the world into micro-management of an uncontrollable universe. Some interference is no doubt desirable, as a multitude of obvious plagues and other deadly ills have been redeemed through the arts of civilization, but we clearly do not know where to draw the line.

THERE IS ALSO the problem of identifying and circumscribing an appropriate target. Given the complex, interdependent conditions under which species evolved, and the resourceful tendency of biological fields to use whatever they find present in novel ways, we cannot foresee what other characteristics we may be eliminating in attempting to excise pathologies alone. What about flukey creative talents, musical and mathematical abilities, odd ways of seeing the world, future unorthodox geniuses like Bach or Picasso, Melville or Samuel Beckett, potential breakthroughs in wind and solar energy, unborn diplomats who might prevent world wars, avatars and shamans—will some of them be eradicated too along with otherwise deleterious mutations? They will if the guilty genes have hidden roles elsewhere in initiating novel qualities in organisms. Will we discard as well the cheerful, good-natured Down's syndrome children who bring their own standards of humanity into existence despite (by our measurement) their physical and mental limitations?

Do we actually prefer *our* world, our Las Vegases and Disneylands, to the unruly and unpredictable planet that flew off the solar whorl? Will we choose zoos, malls, and laboratories over the African savanna and the Brazilian rainforest? Will we take a run at restricting our experiences to arenas ruled by phobias, Price Clubs, and compulsions and reject the vast and astonishing domains deeded us by an unknown god?

When we attempt to incarnate our ideas rather than "nature's"—to make nature over in our image—we also exterminate radical and unconventional but as yet unformulated attributes of our genome. We shackle the wild hand of creation.

Drawbacks of Even Deciphering the Genome

IN THE CASE OF ALREADY INCARNATED MEN AND WOMEN the potential suffering caused by the identification of individual genetic risks may well outweigh the benefits. The problem is that too little information is given to patients, in a form that is both terrifying and medically useless. For instance, one gene, identified in 1995 as BRCA2, comprises 10,254 nucleotides. Those women lacking just nucleotide 6,174 have dramatically increased susceptibility to breast cancer. Yet there is no therapy to replace the missing nucleotide; there may never be such a therapy; and in fully formed adults replacement would be like closing the barn after the horse has escaped. Possessing this gene puts a woman's breast-cancer likelihood into a range of from sixty to ninety percent (depending on other factors) and ovarian-cancer probability at about twenty percent. However, no calculation guarantees developing either form of cancer and, given the purely statistical basis of the danger, no one can explain why some people carrying the mutation stay free of malignancies while living to old age. Additionally, the only form of prevention currently available is surgical removal of ovaries and breasts, which reduces the risk, as one woman's doctor explains to her, "by ninety percent but not to zero."[32]

What does it mean for a healthy person to have to sustain such a dire prognostication—especially when not every carrier develops the disease seemingly potentiated in their genes?

"Like the twists and turns of the gene-bearing DNA molecule ... the message is fundamentally problematic. Genetic testing tells us things about ourselves we may not want to know."[33] It is not unlike going to a palmist to have the lines on your hands read to see how long your life will be. Only, of course, the lines are inside your body and only *they* can see and read them.

When chromosomal diagnosis is misread as medical diagnosis, the individual is frozen in the hands of the genetic Fates. This is, literally, hexing. What a horrible sentence, in some ways equivalent to a fatal disease itself!

AFTER HIS JEOPARDIZED PATIENT HAS LEFT, the doctor responsible for the BRCA2 discussion (above) stares at the photographs mounted on his desk—his wedding day, his daughter Emmy on a swing "in a moment of fearless glee."[34] Given his own family's history of breast cancer, he wonders:

"What terrible aberrations hid in the fabric of her DNA, waiting for age and hormones and the myriad triggers of the environment to unleash them? Would the effort to unravel DNA condemn Emmy to the twilit terror that Karen had just entered? Perhaps it was best for all of us to remain ignorant, so that life could progress naturally, without the burden of deadly prophecies. It sometimes seemed as though the decoding of our genome would cause a fundamental change in how we perceive time—as if we would come to ponder not the infinite time of an expanding universe but the sharply limited span of our existence. Like Karen, all of us will face the choice of learning our probabilities of illness. In addition to those in BRCA1 and 2, genetic mutations that predispose people to Alzheimer's disease, colon cancer, Huntington's disease, endocrine tumors, and melanoma have been identified. The list will grow until it encompasses all our potential pathologies. We might try to shrug off the knowledge or run from it, but when we had quiet moments during the day or woke in the middle of the night we would be forced to accept it as our constant companion, because we could see its features in our very being."[35]

We would have allowed an implacable demon into our dialogue with ourselves, banished spirit from flesh, and turned our destinies into machines.

We would in fact forfeit all our quiet moments without any care for their measure in preventing disease and giving life meaning. Our human identity would then be condemned to its most minimal denominator.

It is a curse identified at the very beginning of Western civilization and put into words by that old sixth-century B.C. master, Heraclitus: "Man is estranged from that which is most familiar." We are more and more (since then) estranged from intimacy with our own bodies. The universe of biological data, while pretending to disclose our being to us, is contradictorily the extreme of the unfamiliar.

In truth, we emerge out of a vast field of DNA possibilities, even individually, and what is conferred hereditarily is only a flux of physical and mental traits. Perhaps latent DNA, unused for generations, is still accessible to organisms under just the right combination of cell signalling and metacellular spells. Beings might unconsciously and expeditiously change their history and presumed fate. They might "cure" at least some of their defects. Shamans, yogis, Reiki masters, faith healers, and other energy remitters might undo the curse and literally rewrite the fatal message in DNA by sequences too complex to imagine or enumerate.

This is not to suggest Lamarckian intrusion in genotypic transmission so much as to open the door to a different, more post-modern blasphemy—that all organisms are nests of pliant characteristics, interchangeable layers of morphologies and functions, all from single genotypic sources. Though the thread of germ plasm may not be capable of alteration in constructive, ideological ways by transmission of biofunctional data during the lifetimes of organisms (for this is not how nature works), the organization of amino acids and proteins may be fluid and subject to "cognitive" and vibrational perturbations.

The idea of the genotype itself as an utterly fixed (and therefore) doomed unit is likely naive and antiquated—an artifact of twentieth-century biocybernetic puritanism. Bionts may inherit not absolute blueprints but guiding ledgers of multiple designs—perhaps a major motif with latent motifs shepherded under its aegis. The phenotype wouldn't have to wait a generation (or generations of incremental adjustments) to get new proteins and tissue shapes; it would merely have to summon the appropriate resonance (or biomutative mantra) within itself.

The illusion pushed by modern molecular biology is that all traits can be tracked and quantified, and then we can be reduced to them. Maybe we can, but that is a political-industrial surmise rather than our guaranteed future. We might also be radical, scalar, and ultimately irreducible phenomena. We might be magic.

Who's in Charge?

As the above narratives foretell, it would seem as though humans are about to take over the mechanisms of life and, from their performance with other fragile ecosystems, that would be our worst nightmare. We are not the stewards or mages we need. And yet again, perhaps *we* are the result of extraworldly biotechnicians tweaking ape DNA, shifting the small number of nucleotides that lie between ancient primates five million years ago and *Homo sapiens.* It is unlikely that aliens concocted DNA from raw molecules (unless it was a very, very long time ago) ... and then who invented them?

There are even Silicon Valley executives who believe that our entire cybernetic revolution, including fiber optics and biotechnology, has been back-engineered from materials discovered aboard a crashed UFO at Roswell, New Mexico, in 1947 and secreted to Bell Laboratories and IBM on orders from then U. S. President Harry Truman. Similar conspiracy theories identify sites of other interplanetary techno-debris, secret "Majestic" branches of the American government, and treaties with aliens, obtaining trinkets of their technology in exchange for permission to kidnap humans and conduct biological experiments on them. It is hard to know whether the purported extraterrestrials are behaving more or less cruelly than we will at their stage of development (or conducting themselves just about the same).

Any way we slice it, there is no exit. Twentieth-century citizens are cyborgs in a robot-serviced illusion, riding in dream vehicles toward a void.

Our consolation must come from the fact that all things arise in the mind of the universe, which is neither mind nor matter but a great hieroglyphic wave of atomisms. From the Gnostic world-view we exist only as figments in the Divine Mind, of which our own minds are reflections, so if we read any version of nature's alphabet and meddle with its tech, we are messing only with our own handiwork, rewriting our generative sequencings at another dimensional level.

I would probably prefer unconscious renderings borne nondiscursively through nature, and the schools and flocks they give rise to, to our regulative mentality, but then I don't have a say.

And I might be wrong.

Part Three

Organs

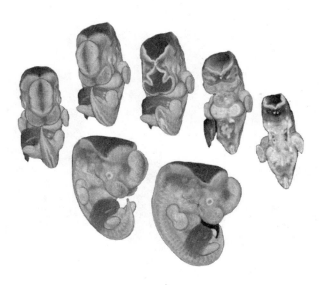

Magnetic resonance microscopy of developing
human fetus; image by Dr. Bradley R. Smith.

The Origin of the Nervous System

The Universality of Neural Activity

SENSATIONS OCCUR IN FREE-LIVING CELLS even without nerves. Paramecia respond to light, find food, reverse direction after collisions. The coordination of their cilia "mimics" neuromuscular comprehension, as they orient and shift about in their environment. Amoebas have been observed tracking prey and maneuvering to escape as they are being engulfed by other amoebas.

Sentience and life apparently share a site of origin. It makes no sense anatomically; yet it seems ontologically true. Awareness itself is an emergent property, irreducible to its substratum.

The coalescence of single cells into multicellularisms, simple into complex tissue, membranes into organs, and organs into creatures, is also the only way we can understand their subsequent interpolation into herds, societies, and civilizations. Emergent properties continue to arise from interfusing components, working their way up a ladder of complex morphology.

IT IS UNCLEAR HOW PROTOZOANS heed impulses and reciprocate. Their most likely "neural" organelles are various subcellular fibers; yet, even if these conduct excitation, they cannot be the precursors of true nerve fibers because the transmission of information in metazoa involves *cell-to-cell* electrochemical properties of membranes.

Conductive nervous systems likely emerged from the ocean with multicellularity as one of the basic aspects coordinating cells and maintaining creature unity. As metazoans added layers of girth, they became dependent on integration of tissue down to its finest yarn. Although all cells maintain a standing charge, a mutation in one lineage caused its progeny to specialize in the passage of current along their

FIGURE 16A. Embryo of the skate, *Raia binoculata,* removed from shells, the younger ones bearing external gills.

From Emil Witschi, *Development of Vertebrates* (Philadelphia: W. B. Saunders & Company, 1956).

membranes. Like the fiber optics of a miniature telecommunications grid, these distended tubes came to generate and conduct indivisibilities in a protoplasmic flow; its collective buzz translated cellular into multicellular agency. The membranous conductivity of discrete protists became massive blankets of them impregnated by sensate strands. With the emergence of full-fledged neural mazes and relay nodules, motile expressions of lifestyle and, ultimately, apprehension of existence took hold.

BENEATH THE MOVEMENTS OF creatures is a jig of particles. Young puppies prancing frenetically and lapping at their mother's teats both contain and replicate darting protists. Flocks of swallows wind-surfing above meadows are bacteria-packed globules endowed with heads and wings; they flutter like zooids in sea currents.

It takes billions of paramecia ancestors to make up one deer; yet that deer is a hewn clump of dependent cells. It employs some of their bodies to feed itself, some of them to hold itself up, some of them to cover its chassis. Great bundles of others trigger reactiveness to stalk and bolt; still others feel branches along themselves, sniff blossoms and dank leaves, and transmit these raw sparks along networks of creatures in dense concordance into a bulbous hive of cells conscripted long ago to the manufacture of thought.

A landscape of raccoons/rabbits/pond is the same panorama a microscope reveals in a droplet *from* the pond—Earth, life, protein, DNA, metabolism. From electrons that charge atoms in molecules, to synapsing cells, to the pirouette of a ballet dancer integrated in her cerebral cortex, life is little more than water, coagulated, compressed, pulsating.

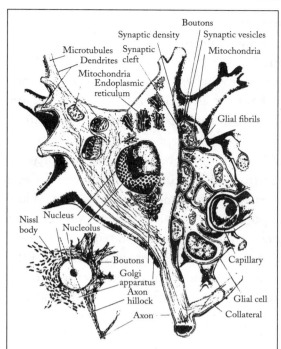

FIGURE 16B. Neuron cell body, with a capillary and supporting glial cells.

From Deane Juhan, *Job's Body* (Barrytown, New York: Station Hill Press, 1987).

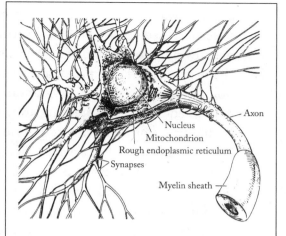

FIGURE 16C. A detail of a typical synapse. The impulse is passed on to the next cell when the seminal vesicles release a neurotransmitter substance into the synaptic cleft, which excites the membrane of the next cell (the postsynaptic membrane).

From Deane Juhan, *Job's Body* (Barrytown, New York: Station Hill Press, 1987).

Cells are electrochemical nodes.

THE ORCHESTRATION OF NERVE CELLS into synapses and ganglia (aggregations of nerve-cell bodies) poses a substantial riddle for development. All other cells operate as points; they each have a position and an inherent role. They interact with other cells only mechanically and chemically; nerve cells not only extend organelles across great distances—they somehow engage in a manner than integrates them in a full ontological system.

The sensory aspect of cells originates solely in their bioelectrical potential, the effect of differing environments inside and

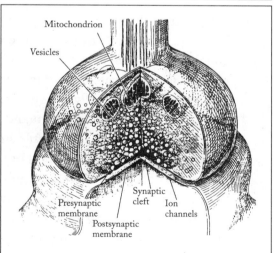

FIGURE 16D. Terminal fibers from axons synapsing to a cell body and its dendrites.

From Deane Juhan, *Job's Body* (Barrytown, New York: Station Hill Press, 1987).

outside their membranes. There may be thirty times as much potassium in the semi-sealed cell medium as in the fluid surrounding it; meanwhile the intracellular sodium may dwindle to one-tenth of its concentration just without. Although the actual chemistry is more complex than this one disproportion, a sodium-potassium equation rests at the heart of the standing charge of any cell. Sodium and potassium tend to equilibrate by draining away from their respective areas of concentration. Living tissue prevents this equilibrium; in fact, the disproportion of internal and external milieus is the very basis of a cell—without imbalance there is nullity.

Cells are relentless in defense of electrochemical identity. Their membrane's molecular shield is able to discriminate between very similar positively charged ions of sodium and potassium, and to bind potassium selectively. Negatively charged proteins, which are too large to negotiate the pores, hold potassium ions in the cytoplasm, while negative chloride ions slip back into the environs in their place. Of course such vigilance takes a great deal of work; a cell must use energy from the conversion of glucose and other foodstuffs to keep pumping out sodium and protecting its boundary differential.

With each cell negative inside and positive on the outside, the potential difference across the membrane is in the range of seventy to one hundred millivolts, a very small amount of latency but conducted across a space only a millionth of a centimeter in thickness. Cell charge is thus a powerful hundred thousand volts per centimeter.

EXCITABILITY IS (LITERALLY) THE CURRENCY of nervous systems. In cells and plants it simmers as simple tropisms; in neural tissue it spreads as pulses of polarization and depolarization; in muscles it synchronizes by expansion and contraction. While paramecia are activated by contact, temperature, light, and chemical gradients, and respond by differentially wiggling their cilia, mammals dispatch stimuli through protracted fields of protoplasm and react by tensing and releasing muscles. They are both alive; they are both tactile and tactical—but in the transition between their domains the molecular becomes organismic.

Set in motion by external stimuli upon tissue, depolarization waves break down the electrical potential of distended membranes in individual multicellular creatures. As a wave passes along a membrane, sodium and potassium suddenly change their positions—sodium flooding through the membrane, potassium vacating. The cell then must oxidize glucose immediately to restore the imbalance, which is depolarized by the next wave. As exhausting as it sounds, every nervous impulse passes only by means of a fluctuation in the resting potential of the membrane. There is no other way of packaging information—no less belabored way and no way that rises to the greater complexity and eminence of human thought. Waves of electrical currents, lasting from 0.3–10 milliseconds each, may occur from dozens to hundreds of times per second in a given neuron. These generate pulses of polarization and depolarization, and the pulses alone conduct excitation across tissue.

Axons

PACKED WITH MICROTUBULES AND INTERMEDIATE FILAMENTS, neuralized cells (neurons) take on their specialization by elongating embryogenically as compressive force is shifted from the organelles themselves to the extracellular matrix. As the balance of pressure deviates, the structure of the cytoskeleton and the geometric parameters transmitted across the cell surface are altered in ways that reprogram cell chemistry and subsequent gene activation—leading to elongated membranes with electrical properties. Billions of these make up sensation and thought.

The splay that transmits depolarization currents—impulses—away from the cell body to another neuron, or to a gland or muscle, is the cell's axon. This long-branched electrified filament is generic and universal from Coelenterates to primates.

Simple invertebrates have naked axons. In more advanced creatures, axon extensions are usually surrounded by nutritive cells without nervous properties (tissues called glia in the central nervous system and Schwann's cells elsewhere).

Synapses

AXONS TRANSMIT THEIR IMPULSES THROUGH SYNAPSES, a name bestowed in 1897 by Charles Sherrington (from the Greek for "clasp"—to express a dynamic interchange). Synapses are linked pathways between neurons, points at which depolarization transits one cell to its neighbors across the tangency of their membranes . . . and so on in a chain. A synapse turns an electrical signal into chemical information, dumping neurotransmitters onto the next station in a maze of ascending pathways.

Physically the passage of information through synapses involves complex protein interactions. How, after all, do electrons and inanimate juice transmogrify into memories and ideas? Although ancient nerve cells may have secreted very simple hormone-like substances, their descendants eventually patented pharmaceutical pathways in the context of depolarization waves, shooting their drugs (sophisticated neurotransmitters manufactured mostly in glands and the brain) across the gap junctions between cells—nascent synapses.

How were such atomistic tidings recorded, packaged, stored, made coherent, and accessed? Short of spying into twenty-fourth century science, we can only guess. The cytoskeletons of neurons are made up of regalia of microtubules linked by protein bridges. On an even finer scale microtubules comprise tubelets, welded columns of paired twin molecules (dimers) arranged in alternate configurations based on their bioelectrical polarizations. In this pygmy universe, patterns apparently flow in fields among dimers and along microtubules and their protein bridges. In the even more miniature realm within dimer columns, information must exist in quantum-mechanical states. Ordered threads of water and amino-acid chains fluctuate in sequences of unknown vintage, preserving and organizing data in vast atomic and subatomic fields that gain depth and capacity from their algebraic uncertainty states. Trans-dimensional packets gather, hold, and transmit quantum-coherent oscillations which, fluttering through the dimers, imprint the microtubules with permutabilities and kinetic codes. This doesn't explain pulsation or thought—or even come close—but it suggests the infinitesimally dense, disjunctive subterrain that underlies the biochemistry of neural activity. Thought is neither information nor cinema. Cells—in fact, by the millions—can be removed from the brain or killed without loss of memory or image.

A synapse is not a concrete object like a wire or buckle; it is a recognition, from our vantage, of the quantal nature of axonal firing. The internal cell cytoplasm does not simply flow from neuron to neuron (as was believed earlier in this century); instead there is a potential-gauging valve between them. If the firing at a particular synapse crosses the threshold of excitability there, the message passes through it and the next cell fires. Stimuli below the threshold simply do not "exist."

Cyberspace, virtual reality, and video images are likewise built by black and white bits. In fact, we have deeded to the whole pantheon of artificial systems our own all-or-none binary response.

It is hard to believe we must go through so many firings, use so much glucose, and transfer so much sodium and potassium even to think the smallest impulse. It is equally beyond imagination that neurons can transfer such sophisticated information, such sustained images and emotions simply by saying yes or no. But this is apparently how things are, how *being* arises from *nothingness*. The cybernetic revolution confirms the power of the binary network; even Da Vinci's Mona Lisa examined under a microscope reveals only the presence and absence of dots and the presence and absence of degrees of color and texture.

If this makes reality seem flimsy indeed, there is a compensatory factor often overlooked—without synapses we would have a dull, nonprobabilistic nervous system, streams of information coming together unilaterally, data indiscriminately converging. Evolutionarily this did not happen. Neurons give a clue as to how (in the universe at large) disorder becomes order, and chaos complexity. Synapses do not stream; they fire. And when they fire, the next neuron has a choice (through the synapse) of whether to pass on the impulse or not. An axon may also have a presynaptic potential accumulated from earlier episodes; at a given moment it considers all the synapses impinging on it and fires, or not.

Thoughts and ideas are built anew from discontinuous sequences. All is novelty: there are no prepackaged events or concepts in the biological realm. Notions must be reduced first to single all-or-none responses and then reassembled in ganglia.

Images are beginningless and nowhere. Masquerading as the means by which we have escaped the ocean of oblivion, they are instead the buoys that keep us from drowning in a vast, anonymous chemical transaction. Neuralization *limits* sensory charge, or at least must inhibit the bulk of it in order to translate the rest into shapes and events.

Dendrites

THE OLDEST SYNAPSES OCCURRED DIRECTLY on the bodies of adjacent cells but, as creatures became stringier and more complicated, another nerve process developed for integrating synaptic input—the dendrite. Dendrites were apparently an evolutionary response to the increasing complexity of data flowing into single neurons. They are branched thorny extensions of cell bodies (somas), often indistinguishable from them.

The axon of a cell is a long smooth flap usually insulated in myelin fat. The dendritic spines are short, irregular, and occur at each synapse upon each individual neuron. Among invertebrates and in vertebrate spinal ganglia, dendrites are interwoven with axonal terminations, but vertebrate dendrites are fully segregated structures developing on somas.

Since synapses in many creatures occur without dendrites, these briery processes are assumed to be refinements of the primordial nervous system; they subtilize the passage of information. The dendrite is subsequent to the axon, not only phylogenetically but ontogenetically insofar as it differentiates rather late in embryonic life (or, among mammals, after birth).

Neurological devices record axons as sharp spikes, bursts of excitation; the synaptic regulation of the dendrites generates slow, smooth brain waves characteristic of normal mammalian activity. (Of course, it is not known precisely what electroencephalograms measure, but the proposed dichotomy is suggestive: Mind is first a scatter of distinct firings, a raw excitability; then a modulated flow of bits.)

Neurons

As COLLECTORS OF INFORMATION about the environment, neurons tend to spread colonially through the body, gathering data about light, smell, taste, gravity, movement, etc., always instantaneous (within the limits of the physics of transmission). In collaboration with skin, muscles, blood, and with the networks of which they are part, the roots and branches of the "axon dendron tree" (so named in the poet Robert Kelly's extended pun) help induce tissues and organs: "cells/sea cells/pretty/she sells/but what/to buy/so precious/what she-//cells/(xs & ys)/he-cells/start/in yolk sac/thence/migrate/ . . . thru all/her (his) body. . . . "[1]

And elsewhere he summons our mysterious embodiment and knowing: "Is it positive?/Are you positive?/Are you negative?/Are you neutral?/Do you hear me?/Are you neural?/Do you feel me?"[2]

As COMPACT STORAGE NODES and integrators of sensory information, neurons also tangle and intertwine in ganglia. These centralizing clumps transpose waves of sensate data into coherent phenomena. They are primitive brains, forerunners of brains.

The vegetal quality of the nervous system is apparent. Even before their integration into formal ganglia, neurons sprout jungles, tufts, tassels, and taproots, bloated cells climbing one another like ivy on a grapevine, or branching through each other like overgrown shrubs and panicles. Other neurons interdigitate like rosette fingers and dig into the soma with claws and cups of petals. Varieties of impulses flow into their dendrites, often many thousands on a single process. The collective foliage around it alternately excites or inhibits a neuron.

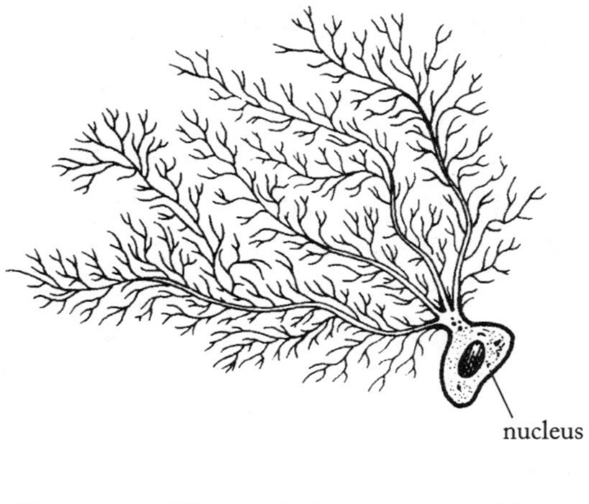

Neuralization is a prerequisite of somaticization, also a result of it. After all, without sensation permeating it, protoplasm is useless—numb blubber. Thus, as creatures dilate and thicken, nerves and tissues induce each other, historically in evolution and ontogenetically in each organism. Neurons are the signal-standards coordinating mesodermally spreading organs, thereby avoiding the undifferentiated

FIGURE 16E. Nerve cell. Illustration by Harry S. Robins.

mucilage of the jellyfish and the seamless tubing of the ribbon worm. Axon-packed muscles give animals range and independence; neuralization projects movement through space.

Darwinian Considerations

FOR THE LAST CENTURY, since Darwin's *On the Origin of Species* convinced us that all living creatures must have arisen from chance mutations and blind selection of the fittest, we have assumed that intelligence on Earth evolved because lineages of creatures with rudimentary brains were uniquely successful at finding food, eluding hazards, and procreating. There is no other intrinsic justification for "mind." If moss animals, sponges, and jellyfish could have seized enough of the biomass and energy on this planet, waves would have washed to shores and trees fallen in forests unheard for eternity.

Neurons made canny cell clusters—animals with novel skills; these survived, mutated, flourished, and gave birth to even more intelligent offspring. According to biologists, brains were the outcome of an accidentally originated "strategy" pursued to its inevitable conclusion.

Yet sensory and ganglionic structures seem more volatile than that; they leap into being in the Cenozoic, seemingly overshooting the requirements of their niches. The excessive and luxurious expansion of nervous systems (in seeming defiance of the minimum required for successful competition and survival of the fittest) has always given pure Darwinians their strongest challenge. If they recognize them at all, biologists claim the metaphysical aspects of mind must serve survival in some mysterious way. Else why would nature sponsor cerebral ganglia so heavily?

There is no explicit answer to this question—not even a likely candidate—but possible partial explanations (starting here) lie imbedded in many different ontological discussions throughout the remainder of this book.

The Phenomenology of Consciousness

WE EXIST NOT BECAUSE OF AXONS and dendrites but despite them (i.e., despite their seeming circumscription). Consciousness does not so much prove neural wiring as belie it. In fact, most neurological activity never makes it into consciousness. According to one researcher (Benjamin Libet), unconscious neurons—rather than an ego—initiate all action. When his human subjects were wired with electrodes, signals could always be machine-detected in the brain a good half-second before a conscious decision moved a muscle. The desire or intention to act therefore does not precede the brain's activity; it merely displaces it into an illusion of free will.[3] This is no doubt why skilled martial artists can discern an opponent's intention an instant before he or she actually commences to strike (even before the flick of a nervous eyelid), though it is unclear how they could train such an ability in the face of the "user illusion." We act without real awareness of our motivations, and the mind makes sense of it all, reconstructing an ongoing narrative very much after the fact. At least so the reductionist argument informs us.

From the condition of aborted fetuses scientists know that by thirty-two weeks in utero the nervous system is ready "to transmit signals back and forth throughout a complex mass of unnumbered cells, signals which miraculously arrive at all the right muscles, glands, and organs. How these electrochemical signals are ultimately transformed into meaningful messages, ideas, decisions or memories cannot be explained in physical terms alone."[4]

The contradiction implicit in the very fact of sentience—the existential riddle—is the irreducible incalculability of phenomena, both inside and outside the nervous system, on which sentience is based. There is no absolute, scrutable nexus leading threshold by threshold to Shakespearean flights and fancies—or even to quack-quack and waddle in an ordinary duck. These are "qualia," inexplicable subjective enhancements of the quantitatively measurable attributes of stimuli. If axons provide the conduits of sensations, something in the brain ineffably weaves them into luminous underwater geographies and groves of orange trees swaying in the breeze. And this is not a "user illusion."

Perhaps cells and tissues invent the phenomenological realm by their sheer interactional complexity, or perhaps they discover an ideal reality that underlies nature. Though scientists classically seek rules in the physical linkages themselves, there is no ground state for biological meaning and its metachronological domain.

The gestalt of a butterfly is the activity churned up by a discrete crystalline heap; it is spatially errant and roves in a way that confounds any physics and mathematics of its nervous system. Behavior has spatial, rhythmic, and hierarchical qualities which transcend the composition and anatomy through which it is delivered. Totally without warning, neurons reflect the unreflected. They contemplate splintery, mongrel events as numinous effervescences, single bubbles arising from fizz. A bright red cherry to peck at. A tawny flash of cheetah peril. A comely mare. Cold pings of rain.

Perception is thus not a sum of neural patterns but "the paradoxical phenomenon which renders being accessible to us."[5] Existence occurs finally not as nerve impulses *per se* but from the dialectic of organisms in milieus, living assemblages waking to themselves. These relations break out of all behavioristic loops, cannot be predicted in advance, and—by comparison to the information they provide (i.e., the immense gravity of both galactic and philosophical systems)—weigh virtually nothing. They are the universe's closest approximation to antientropy, a counter-force to its mindless mass. French philosopher Maurice Merleau-Ponty writes:

"The ambivalence of time and space at the level of perceptual consciousness reminds [us] of the mixed notions by means of which modern physics goes beyond the abstract simplicity of classical time and space. It should not be concluded from this that forms *already* exist in a physical universe and serve as an ontological foundation for perceptual structures. The truth is that science, on the basis of certain privileged perceptual structures, has sought to construct the image of an absolute physical world, of a physical *reality,* of which these structures would no longer be anything but the manifestations."[6]

Cells cannot explain their own subjectivity or meaning; thus, ". . . the universe of naturalism has not been able to become self-enclosed, and . . . perception is not an event of nature."[7]

It is almost as though neurons are a by-product of perception rather than its agency. Despite biology's materialistic fundamentalism, no agenda arises from the fact of synapses and ganglia. We act solely because we find ourselves in a world. The rest, including explanations, is collateral, superfluous. "We must take care of things simply because they exist."[8]

We will explore the mind-matter paradox more fully in Chapter 17, "The Evolution of Intelligence"; Chapter 21, "Mind"; and Chapter 27, "Spiritual Embryogenesis."

Coelenterate Nervous Systems

IMPRESSIONS OF EVERYDAY LIFE mesmerize us into thinking that time flows evenly and is irreversible, an illusion initially confirmed by Newtonian thermodynamics. However, this is a biological prejudice, built into tissue during evolution and presumed by the first physicists. If creatures could not simplify the flow from their senses, order it, and respond to it, they would perish. The processing of neural information creates a narrative time-line mirage and an illusory present moment. This is not only the basis of civilization but the source of each simple animal urgency. *Now!* In zen meditation, a student learns through her breath that the universe wants her to exist solely in the singular biology of the moment.

Although there are many creatures with few and simple sense organs, the Coelenterates are considered closest to the ancestral line in which multicellular systems first developed. The first terrestrial clock was probably jellyfish-like pulsing through an epithelium, marking moment-to-moment quantal sensations.

In most simple metazoans, awareness is an incipient property of ganglia, but the jellyfish/anemone nerve net is disperse enough that the number of neurons per zone of protoplasm is for all intents and purposes unvarying. These are ganglia-less creatures. Coelenterate mazes form series of crisscrossing paths duplicating and quadruplicating one another's links, making creatures a uniform repetitive pulse of being. In the same epithelia food-sensing cells make up their own separate nerve net.

Excitation spills through these nets in all directions with barely any discrimination between the electrical flow of information and concomitant neuromuscular response, leading to (perhaps) a vague spackle of cognizance. The life beat of the organism is the throbbing of its bell, modulated solely by mechanical contact with the environment, including constant hydrostatic pressure. Cuts in the nets do not disturb the animal; the same information continues to flow through remaining pathways.

Global pulses, punctuated by sudden bursts of electrical activity during contraction or when exposed to light, suggest the universal autonomic consciousness at the base of all living systems, i.e., the unconsciousness.

THE NERVOUS SYSTEMS of today's jellyfish are simple not because they have been reduced but because they are inherited intact from primitive creatures who once upon a time gave rise to the other metazoan phyla. Jellyfish have retained the primeval lifestyle and biology of their ancestors. Only medusoid forms have true sense organs—photoreceptor cups with lens-like cuticular masses and equilibrium-measuring organs called statocysts. These sensory pits and vesicles make contact with the animals' nerve nets, exciting generalized responses to light and currents.

Jellyfish-like creatures feed, defecate, and contort and sway with seeming intention. Perhaps ganglionic forerunners exist in nerve rings close to the margin of their bells and in touch with plexuses in the sub- and ex-umbrellas.

In the related phylum of comb-jellies, specialized sensory structures of apparently mixed modalities form under the comb-plate rows. Upon stimulation they change their rhythm of ciliary beat and spread luminescence through the combs.

The Origin of a Central Nervous System

THE COELENTERATE NERVOUS SYSTEM is anatomically fundamental. From primitive comb-jellies to the music of Bach and the formulae of Einstein, there are only homeostases of neurons buttressed by quantum leaps in their numbers and relationships. In human embryogenesis likewise, neurons multiply and gather in nodes, condensing evolution.

Similar elaborations have occurred not only along the vertebrate line but through-out invertebrate phyla; they generally gravitate toward central nervous systems with anterior brains—pulses coordinated in ganglia and hierarchicalized, then returned selectively through effector fibers to muscle tissue for swimming, crawling, and ingesting. The cerebral ganglion may have initially been no more than one of the specialized fluid-filled sacs that function as organs of balance in invertebrates (a statocyst); it could also have begun as an aggregation of receptors along a margin of protoplasm. Whatever its origin, it gradually became the organ of animal iden-tity, taking charge of regional relays. It was the gravitational core to subsystems of nerves and ganglia, the parabolic locus for the advent of mind.

Central nervous systems differ from nerve nets in that they favor determina-tion and discretion. In the process of hierarchicalizing flow and mustering ganglia, they create a labyrinthine world, conferring capacity for spontaneous action, regional expression, and grace. Even sponges and comb-jellies are felicitous; and earth-worms, grasshoppers, and crabs have their own peculiar elegance. Hierarchical con-trol was such a successful mode of animal survival that it was differentially favored in all lines, so advanced nervous systems likely evolved independently from nerve nets countless times. The brain probably lies at the congress of innumerable adap-tive trajectories.

Despite centralization and cephalization in most phyla, reduced systems with a paucity of sense organs—and a tendency toward decentralization—also mani-fest throughout the invertebrates, even in some of the most advanced phyla. The Echinoderms, close relatives of the Chordates, have no brains to speak of and their radial mesh of nerve cords resembles in many ways a jellyfish net. The few sense organs they retain are of simple construction. But note the coordination when a starfish chases and captures prey or rights itself radially. Echinoderms are not "jel-lyfish." In contrast to their primitive receptors, they have an advanced central ner-vous system with a circum-oral neuromotor ring and linearly arranged bundles that conduct and return excitation. Evolution is neither progressive nor linear, and crea-tures can lose ontogenetically traits that their ancestors accrued phylogenetically.

Obviously, relative complexity of nervous system is only partially a consequence of phyletic position; the other factor is niche and habit of life. Reduced systems are common in sessile forms, parasites, and to some degree, sedentary animals in all phyla. Sense organs are often lost when they are not used, their signal elements and assorted genetic seeds falling back into unused sections and silent fragments of the genotype, then perhaps (untold generations later) all the way back into the undif-ferentiated alphabet soup and molecular chaos from which all biological form orig-inates. Seemingly hard-won centralization returns to regional ganglia. Clams, chitons, bryozoans, and sea squirts all show what is apparently secondary reduction of their nervous systems as consequences of immobile lifestyles. Their more lively larval forms demonstrate the purely collateral nature of the loss.

Although there is a way in which nature favors sensation and cerebralization, there is another, more entropic pathway along which nature not only doesn't care but would just as soon slough off the cost of mind and individuation; we inherit aspects of both.

THE PATH TO NEURALIZATION and ganglion-formation is highly versatile, perhaps because of intrinsic neural potential. Neural-crest cells and neurons invade expanding layers of tissue, penetrating protoplasm at every embryogenic opportunity. This gives new lineages myriad sensory opportunities—bases for fresh themes.

Apparently the forty or so basic body-plans that have survived (from among the hundreds or more that were tried in Precambrian and Palaeozoic epochs) are all modular variations on a motif that is organizing cohesion, mechanical elegance, and relative ease of reproducibility. Although neuralization *per se* may not be the sole yardstick by which plans are meta-environmentally "judged," it is the dominant epistomological element holding together phylogenesis. It and metabolism may be the universal traits that life forms on all physicochemical worlds share and by which they will recognize their mutual "biology." Whether an invisible eschatology biases the maze of modular cell pathways toward bioelectric information and sensory systems—whether the elemental properties of substances have a vitalistic predisposition toward individuation and expression—are ultimate mysteries of nature.

EACH NERVOUS SYSTEM has been molded and defined by ancient events. Even as systems change in capacity and configuration, they incorporate prior habits and modes of behavior within their novel networks. Instinct and responsiveness are inherited and amalgamated from creature to creature, species to species. Amino acids and proteins are constellated in the new ways, and axons and synapses translate them into blocks of pristine behavior, mortised timelessly in embryogenically arising layers of flesh. The topology of rudimentary neural grids becomes a series of phenomenologies woven into a thick web of competing, collaborating life forms.

An ant is what it is, a snail too, each of them idiosyncratically and as witness to the mystery. Where tangible nerves contact the abyss, fully equipped creatures of knowledge shuffle into the world and carry out the labors of their kind.

The Evolution of Intelligence

Nervous Systems in Worms

Although intelligence occurs in an enormous diversity of creatures, there appear to be only two major lineages on Earth representing distinctly different orders of mind. One path leads to Mollusks, Echinoderms, and Chordates, hence to land vertebrates. The other one ushered in Annelids and Arthropods, and the great insect and spider societies.

Both genealogies appear to have diverged from the same primitive ancestor whose closest present-day counterpart dwells among the Platyhelminthes (flatworms). Species in both lineages eschewed the seas for the continents. While the Chordate line apparently postponed intelligence and social order until it had accumulated enough neurons to escape the trap of pure genetic praxis, the insect race developed its mind inside a rigid, segmented shell in which its descendants continue to dwell.

A flatworm is a diffuse ball laddered lengthwise—an anemone in a tube. A Coelenterate-stage nerve net persists, but it is wired and augmented by two long cords trailing into a cerebral knot. Flatworms gain peripheral apprehension in a number of submuscular nerve plexuses, including pharyngeal, genital, and visceral ones, all connected to the rudimentary brain. On either side of this cerebral crown, taste and light receptors spark, while additional neurons differentiate across the scant body, recording light, "odors," and tactilities.

Within flatworms' neural concentrations, nerve processes have begun to separate from their nerve-cell bodies, forming dense tissue known as neuropile, with glial cells filling the spaces between neurons. This is the ground state of all central

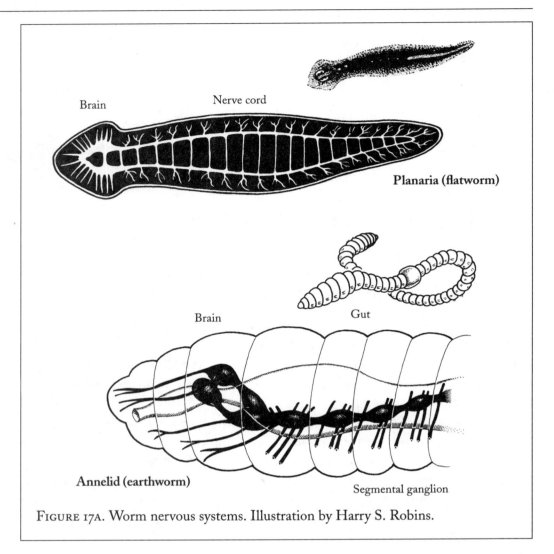

Brain

Nerve cord

Planaria (flatworm)

Brain

Gut

Annelid (earthworm)

Segmental ganglion

FIGURE 17A. Worm nervous systems. Illustration by Harry S. Robins.

nervous systems and brains.

The diffuse nerve nets and simple plexuses of the polyps are summarized and then superseded historically by hierarchies of neurons leading to denser regional grids. Ganglia are control centers for sets of organs. In the most sentient creatures they themselves are overridden by one cerebral ganglion, but, as we shall see, such centralization is relatively weak throughout the invertebrates, and in trauma the separate control centers take over, even in animals with brains.

Regional independence persists in the peripheral and autonomic regions of our own nervous system where stomach and intestines digest food like independent self-sufficient creatures, and lungs and heart pulsate intelligently beyond cerebral regnancy. The fact that humans afflicted with a pathology that causes them to be born brainless can carry out enough basic metabolism to survive for a time shows the degree of decentralization we still embody.

The fragments of the worm go on crawling and the penis of the mantis keeps copulating after excision. The arm of the octopus swims about like an independent organism after amputation, for it has its own "brain."

ANNELIDS AND ARTHROPODS share a body plan based on the formation of successive sections in segments. Individual segments, called metameres, arise embryogenically from the iteration of mesodermal divisions. Subsequent segments bud off existing ones to create a reinforced length of neuromusculature. Metamerism was likely introduced by a series of fractalizing and homeotic mutations in these creatures' flatworm-like ancestor, and then diverged like wildfire. Integrated segmentation provided speedy, serpentine creatures designed around central cavities. Some of their descendants remained vermiform, while others developed more exotic designs and topologies and became insects, crustaceans, and spiders (see Figure 13A, p. 314).

Metamerism is repetitive but not superficial. As high-school biology students learn, the divisions of earthworms go through grooves in their mesoderm to internal organs. Each metamere contains all three layers of body tissue and three pairs of nerves from its larval segment. The embryonic coelom forms when cells on either side of the gastrulating worm pull apart, the resulting segmental cavities coalescing until the wall of mesoderm is hollowed out (except for the mesentery of the gut and thin sections of tissue between segments).

PRIMITIVE OCEAN-DWELLING ANNELIDS discharge their sperm and eggs into the sea, and fertilized zygotes develop into ciliated gastrulas with apical sense organs (called trochospheres). The metameres bud from their anal regions, pushing formed segments forward. This bilateral symmetry overrides the larval sphere and establishes a grid all segmented worms will adopt (including Frank Herbert's gigantic sandworms on the imaginary planet Dune).[1]

The sections of worms are integrated mesodermally. As suckers (called setae) make points of contact with external objects, neurons fire and muscles contract in response to stimuli. They then must be restored to their precontraction lengths by antagonistic muscles. Pulling is turned into pushing as one set of muscles uses its torque to operate another; a hydraulic torso bulges backward in forward movement.

Nervous Systems in Arthropods

Chitin

An exoskeleton of chitin distinguishes Arthropods from all worms. Chitin differentiates from the epidermal layer and lodges within it as an intercellular matrix of nonliving cells—long unbranched proteins wrapped in a waterproof lipoprotein shield around a primitive pulsating pump, a heart with ostia. The bearers of such armor include insects and their allies.

Retention of moisture within chitin provided an internal milieu, a spacesuit of their own cells, for ancestral Arthropods to matriculate from the sea—archaeologically the first major group of animals to do so.

The Arthropod outer skeleton does not remain part of its body; therefore as the animal grows, it must excrete a new cuticle while molting the prior rigidified one, removing calcium from it and returning it to the blood. The entire ectodermal epithelium is then renewed, hardening around the enhanced soft dimensions of its inhabitant. These metamorphoses are fired by similar hormones throughout the phylum—from crustaceans to insects—and occur several times during each creature's life.

Metameres

Crustaceans begin as nauplius larvae with three pairs of jointed limbs. As they molt, appendages are added at their rear (similarly to Annelids). The nauplius becomes the head, and the rest of the Crustacean swells from budded sections.

Metameric development can lead to sheer unrefined bulk. In the Decapod order (which includes shrimp and crabs) the nauplius may be suppressed, hypermorphically gastrulating a more developed and mature zoea larva, a creature with legs, chelae, and abdominal segments that, in the signal instance, iterate into a fifty-pound lobster.

Among Annelids serial organs repeat from segment to segment with regional differentiation for respiration, excretion, and copulation. In Arthropods, however, the different metameres specialize and take over the incipient functions of organs. The three major regions (tagmata) become the head, the thorax, and the abdomen, respectively. Embryonically, the insect head consists of six segments, the thorax three, and the abdomen seven.

In spiders the prosoma (corresponding to the insect head) takes seed in the embryo as separate segments which become fused in adult forms so that its metameres superficially vanish. The head, appendages, and eight legs are borne on the prosoma; the twelve segments of the rear opisthosoma also become a single tagma.

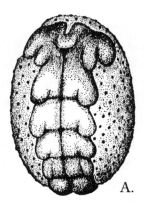
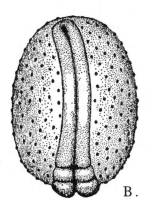
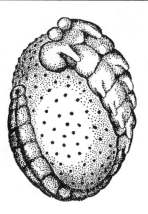

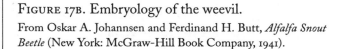

Third-day embryo A. Head and first two thoracic segments;
B. Second and third thoracic segments and abdomen.

FIGURE 17B. Embryology of the weevil.

From Oskar A. Johannsen and Ferdinand H. Butt, *Alfalfa Snout Beetle* (New York: McGraw-Hill Book Company, 1941).

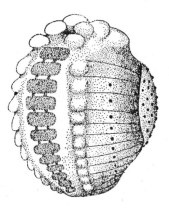

Fourth-day embryo.

Eighth-day embryo.

Tenth-day embryo.

The fossils of ancestral trilobite forms show all three tagmata, so compacting is clearly a modern feature, only partially developed in scorpions and other primitive arachnids.

Limbs and Exoskeletons

The limbs of the oldest land Arthropods are efficiently arranged in a series of levers around a fulcrum. Although their exoskeleton is petrified throughout, it is flexible at its joints. The trunk moves forward without swaying, and the angles between the sections of the limbs change continuously, keeping the same distance between the midline of the animal and its point of force against the environment. A backstroke thus propels the body. The identical mechanical pattern is used in crawling, running, digging, jumping, climbing threads, and running upside-down.

Multilegged, wingless insects like millipedes and centipedes likely descended from walking worms with paired limbs. Other creatures evolved from these, i.e., extinct worm-bugs and spiders with long limbs and articulations closer to their centers of gravity.

The mild "psychology" of the worm, preserved in bug larvae, is transformed in insect adults, often in a bellicose and inflexible direction. The molding of worms into insects probably occurred heterochronically from mutations lengthening embryogenesis and using the increased period to organize proteins and secrete hormones for shells and complex, though stringent, social reflexes. While a respectable worm manifests ontogenetically in insect larvae, insects are utterly different chreodes. When living parts (like the walking legs of a spider) link together plates of armor, animal and machine seem almost interchangeable.

Flapping of insect wings comes from opposed contractions of muscles deforming the thoracic segments that bear them. The two sets of muscles work in rhythmic alteration, changing their angles during both upstroke and downstroke and twisting the wing so that its leading edge moves down during downstroke to give thrust as well as lift. Indirect flight muscles do not always contract in synchrony with their motor axons. Instead, impulses may set the muscles in excited synaptic states from which thirty or more contractions occur. Dragonflies make only one stroke per nerve impulse, but hive bees can beat their wings 250 times per second, and some midges can flap a thousand times in a second—a real olympian feat.

The Origin of Insect Nervous Systems

The most likely contemporary vestige of an ancient worm brain occurs in the trochosphere larva of aquatic Annelids. This lump takes shape as a peristomial segment around a mouth from which palps, antennae, proboscis, jaws, and eyes swell and carve themselves out. The concentration of nerve rings under its apical sense organ develops into a rudimentary brain, while the organ itself becomes the prostomial segment.

A different, primitively bilobed "brain" arises in the miracidium larva of the flatworm fluke. Two medullary cords run out of a ganglion to the rear of the animal, each radiating branches to the margins of the body and periodically crisscrossing nerves like the ties on a ladder. A replica of this crane will occur in all instances of higher neural development, though it will be variously reconstructed and reoriented.

Medullary cords in simple flatworms and ribbon worms sprout as direct outgrowths of the "brain." Equivalent cords in Annelid larvae arise, quite separately, from a pair of ectodermal ridges. Only later do these unite with the "brain" through circumesophageal fibers. This would appear to be the more primitive situation; yet it occurs in the more advanced creature.

In Arthropods the cerebral ganglion originates as an anterior continuation of neural ridges of ectodermal material at either side of the stomodeum (primitive mouth). Other ectodermal cells migrating inward contribute to peripheral ganglia.

The central ganglion is divided into three parts. A protocerebrum of separate neuropile masses includes optic lobes and association neurons for self-initiated action. A unique visual apparatus emerges first as a series of neuropile clumps joined to each other by tracks between the retina and protocerebrum.

Insects' distinctive compound optica were celebrated by Goethe in his poem about a dying fly which unwittingly continues to imbibe her poison:

"The numbness spreads, she barely feels a thing;/yet on she sips, and even as she does,/death covers with a cloud her thousand eyes."[2]

The deuterocerebrum forms the neuropile for the first antennae; the tritocerebrum contains the nerves to the anterior alimentary canal, the upper lips, and, in some crustaceans, the association center for the second antennae. A subesophageal ganglion controls chewing while maintaining tonic excitation of the other ganglia.

The Behavior of Insects

INSECTS ATTAINED "INTELLIGENCE" with very few neurons—virtually no associational ones—so they must adhere religiously to ancestral rules. Their wiring is strict and inflexible. They have limited memories and intentions; they can learn virtually nothing; background and foreground to them are one. Their functions imprinted into their tissues, they act automatically, without mind, at least by human standards. A bee trapped in a room and shown the way out—even a hundred times—cannot learn it, and will die there eventually, batting itself against a photoattractive window.

"Every insect is a singularity without identity," writes Steven Shaviro. "The fringe biologist Donald I. Williamson even goes so far as to argue that larval stages are remnants of symbiotic mergers between formerly independent organisms.... The body of an insect ... is perpetually 'other than itself.'"[3]

Everything about the insect world suggests not only the irrelevance but the nonexistence of the individual. Ants in a colony do not have enough neurons each to achieve consciousness, but they themselves are like neurons, or nerve plexuses. Sluggish members are impelled into their chores by active comrades moving to sites adjacent to them—their activity, however, will soon cease if they are not stimulated again. Worker bees resemble cells swarming in embryogenesis, creating the metaphorical cells of their hive, each symmetrical polygon the same as the previous one, as in tissue. They are "elements driven by chaotic dynamics ... mobile cellular automata"[4] obeying rules. A hive is a superorganism, sharing hormones, breeding specialized cells, arising in a cloud like the cloud from which it came. No wonder Eugène Marais spoke of "the soul of the white ant" and described an invisible influence emanating from the queen into the "minds" of the other termites.[5]

GENERALIZED KINESES THROUGHOUT ARTHROPOD BODIES synapse in a variety of modes: jumping twitches (in locusts and fleas), wing-beating (in hornets and termites), and feeding, mating, attacking, and marching in columns (in ants). Kineses provoke motion in random directions based on intensity of stimuli. As a rock is turned, each bug hiding under it rushes about until it finds the dark again, where it is quiet.

Taxes are another kind of genetic pathway, initiating movement directly toward or away from some stimulus. The blowfly larva rotates its head alternately right and left while swinging its body out of the direction of stronger light. Ants maintain a regular gradient in relation to the sun, a "mindless" trigonometry. These animals are "compulsion energized," their instincts welded to their behavior.

Crickets give out and respond to a repeated trill. Male phosphorescent beetles emit light patterns; females acknowledge these. Moths spew moth perfume, attracting other moths.

The Minds of Insects

INSECT BEHAVIOR, THOUGH CELL-LIKE, can synergize in complex episodes. For instance, some ants build nests in tree trunks, leaving an ingeniously tiny hole as an entranceway, large enough for a single ant. "Their community includes a not very numerous caste whose entire mission in life is to act as doorkeepers. They have enlarged heads, flattened in front, that fit exactly in the entrance hole so that they can function as live plugs. Morever, the texture and color of the head, as far as it is visible from the outside, is such that it can hardly be distinguished from the surrounding bark. A doorkeeper will sit for hours in the entrance hole. She admits only members of her community demanding entrance by taps with their antennae, and these only if she can also recognize their smell. Should the hole be slightly larger than the head of the doorkeeper, the ants use a substance like papier-mâché to narrow the entrance until it fits the head exactly. When the opening happens to be exceptionally large, several doorkeepers may block it jointly."[6]

This resembles both Buckingham Palace and ringaleevio—but insects play more deadly games too. Maurice Maeterlinck describes the mayhem as a bee-hive is cleansed of its male overpopulation: "Each [drone] is assailed by three or four envoys of justice; and these vigorously proceed to cut off his wings, saw through

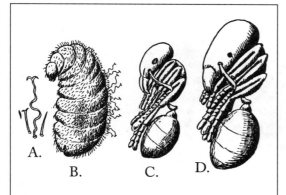

FIGURE 17C. Immature stages of the brown house ant. A. Enlarged view of hairs on larva; B. Larva; C. Worker pupa; D. Soldier pupa.

From S. H. Skaife, *The Study of Ants* (London: Longmans, 1961).

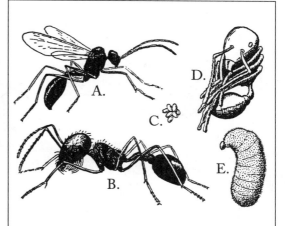

FIGURE 17D. The bearded ant. A. Male; B. Worker; C. Larva; D. Pupa; E. Eggs.

From S. H. Skaife, *The Study of Ants* (London: Longmans, 1961).

the petiole that connects the abdomen with the thorax, amputate the feverish antennae, and seek an opening between the rings of his cuirass through which to pass their sword.

"No defense is attempted by the enormous, but unarmed, creatures; they try to escape, or oppose their mere bulk to the blows that rain down upon them. Forced onto their back, with their relentless enemies clinging doggedly to them, they will use their powerful claws to shift them from side to side; or, turning on themselves, they will drag the whole group round and round in wild circles, which exhaustion soon brings to an end. And, in a very brief space, their appearance becomes so deplorable that pity, never far from justice in the depths of our heart, quickly returns, and would seek forgiveness, though vainly, of the stern workers who recognize only nature's harsh and profound laws. The wings of the wretched creatures are torn, their antennae bitten, the segments of their legs wrenched off, and their magnificent eyes, mirrors once of the exuberant flowers, flashing back the blue light and the innocent pride of summer, now, softened by suffering, reflect only the anguish and distress of their end. Some succumb to their wounds, and are at once borne away to distant cemeteries by two or three of their executioners. Others, whose injuries are less, succeed in sheltering themselves in some corner where they lie, all huddled together, surrounded by an inexorable guard, until they perish of want. Many will reach the door, and escape into space, dragging their adversaries with them; but, toward evening, impelled by hunger and cold, they return in crowds to

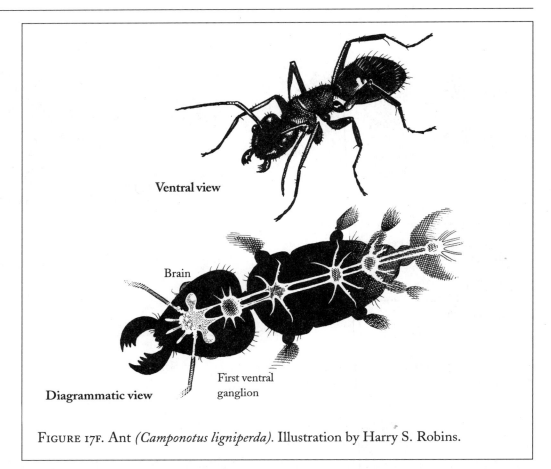

Ventral view

Brain

Diagrammatic view

First ventral
ganglion

FIGURE 17F. Ant *(Camponotus ligniperda)*. Illustration by Harry S. Robins.

the entrance of the hive to beg for shelter. But then they encounter another piti-less guard. The next morning, before setting forth on their journey, the workers will clear the threshold strewn with the corpses of the useless giants...."[7]

It would have been hard to foresee such incipient compulsion and aggression among worms; Arthropod gastrulation produced completely novel classes of crea-tures with revolutionary ontology. Bees run "nurseries" and "factories" that look to us like sci-fi laboratories; they excitedly communicate the whereabouts of nectar. A "queen" bee soars amidst drones in her mating "dance." A single "husband" from among them fertilizes her; she kills him, tearing his genital out of his abdomen and carrying it off with the sperm.

"The insect brings with him something that does not seem to belong to the cus-toms, the morale, the psychology of our globe," adds Maeterlinck. "One would say that it comes from another planet, more monstrous, more dynamic, more insen-sate, more atrocious, more infernal than ours."[8]

Beyond human parameters lie unknown phenomenologies. Yet, when we look at bees we are gazing into light, not darkness, for it takes mud a long time to become as fiery and sweet as a hive.

TERMITES SWARM, FALL TO THE GROUND, chew off one another's wings, and then couple. After a period of "chastity," the mates begin intercourse. Attracted to a secretion at the female's rear, the male keeps her permanent company. As she lays a fertile egg every two seconds, the pair breed thousands of workers who gather food and care for them and for their young. The workers also dispense hormones that hatch additional workers and "soldiers" who repair the nest, scout for danger, and drive off enemies.

The king and queen perch together in their termitary, she swollen and fat and laying her eggs continuously. Every now and then he squeezes under her body to mate. A stream of workers licks up their secretions and carries off their eggs.

Ants "fight wars," "build cities," and "manage selective breeding programs."

Robert Kelly writes:
"... here is an alien being doing alien things
wordless timeless neither beautiful nor ugly
as readily on a planet of Toliman as here

while from our need to identify with the ant
& render its institutions such homage
as it may be to call them by our names

we may observe (here is the lesson) how lonely we are."[9]

THERE IS NO INSECT MIND in our lineage, yet we might wonder if we are not sometimes more like them than like ourselves. Our cities and factories are great termitaries. Our battles (with their elaborate trenches) are as grim and pointless as attacking columns of ants. Our genocides and ethnic cleansings resemble hive massacres.

"We can kill individual insects, as spiders do," Steven Shaviro reminds; "but we can't for all that extricate ourselves from the insect continuum that marks life on this planet. The selectional forces that modulate insect bodies and behaviors are also restlessly at work in our own brains, shaping our neurons and even our thoughts. Does such an idea revolt you? The problem might be that we can't read insect expressions: we don't know what they are thinking, or even if they are thinking. But this is nothing but an unwarranted vertebrate prejudice; after all, 'insects are naturally expressionless, since they wear their skeletons on the outside.'"[10]

We may have escaped the cul de sac of genetically imposed economies and behavior, but we have reimposed it socially and politically, instituting *"traditions* and *norms of critical reflection,* the better to police our identities and prevent our minds and bodies from going astray. Education, after all, is just a subtler and more sadistically refined mode of operant conditioning than the one provided by direct genetic programming.... Our mammalian talents for memory and self-reflection serve largely to oppress us with the dead weight of the past."[11]

We may not be insects, but we *are* terrestrials. Though we do not secrete shells, we have constricted ourselves in other sorts of extracellular matrices. Like bees and termites we are social beasts with royal hierarchies.

Nineteenth-century humanitarians tried to get us (and God) out of this dilemma by proposing (romantically) that insects experience their lives as collective rather than personal. The single Bee imbibes hive eros. The slaughter of citizens does not diminish consciousness and, though the maimed workers emanate pain, their suffering is not individuated. (Try telling that to the Nez Perce or Cherokee). Shaviro asks:

"What has changed in this picture in the last hundred years? Only one thing. We have come to understand that such alien splendor is precisely what defines the cruelty and beauty of our world."[12]

Participating in the same DNA that we do, sharing the same axons and dendrites, insects are part of consciousness and psychic life in general; they are the philosophical kingpins among earthworms, crabs, and spiders, the highest form on a separate branch of intelligence.

Perhaps among galaxies of hydrogen-silicon, architectures of their kind rule giant planets (from where science-fiction writers imagine their starship invasions of the future Earth). Then we will know for sure if we are "us" or we are "them."

Nervous Systems of Mollusks

THE ENTIRE INTERMEDIATE RANGE of the evolution of consciousness occurs within the Mollusk phylum alone. Chitons—small intertidal-zone mollusks with colored shells—are little more advanced than flatworms, whereas octopi are as sentient as vertebrate fish. This gradient of intelligence is accomplished through one underlying morphology—six ganglia usually around a gut, each ganglion paired and cross-connected by long commissures. Constellated as rinds of cells with central cores of fibrous nerve processes, the ganglia respectively form neural centers for head organs, visceral mass, pedal musculature, ctenidia (gills), mantle, and radula

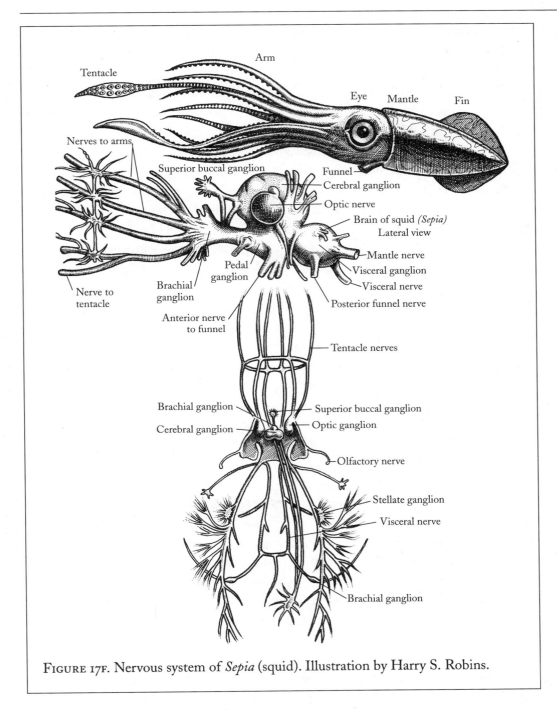

FIGURE 17F. Nervous system of *Sepia* (squid). Illustration by Harry S. Robins.

(toothed tongue). Cerebral ganglia send connectives to and receive them from the other neuropile centers, activating sense organs, muscles, gills, genitals, and, in cephalopods, tentacles.

In chitons, cerebral ganglia are barely more than medullary strands. The whole creatures are leathery sacks of viscera with tactile, tasting, and balancing receptors strewn over their ectoderm.

Gastropods (snails) show increased cephalization through a subesophageal fusion of their ganglia, but they are basically sluggish "worms" lacking quick reflexes or learning capacity. Their skeletons transmit force only as retractor muscles contact remote antagonists, so their functional level is ciliate and Annelid. Intelligence manifests in their fine awareness of locales and their classically Molluscan ability to sort through particles and separate them one from another, the edible from the inedible. Tactile sorting surfaces occur not only on the feeding organs but in the guts of many mollusks.

The idiosyncratic intelligence of the phylum is dramatically realized in the most advanced class of all invertebrates, the Cephalopods. Octopi and their kin have many well-developed sense organs: statocysts that measure direction, speed, position, pressure, angle of acceleration, and sound; olfactory pits; chemoreceptors and mechanotactile sensillae on their arms and suckers; and visual organs with corneas, iris diaphragms, lenses, and retinas (that are not inverted). Supported by massive optic lobes, these eyes coalesce, in some classes, to cover a full third of the body surface.

The Mollusk nervous system is the result of a longstanding and deep-seated phylogenetic pattern—the twisting of internal organs into loops—concretized in the familiar spiral shells of Gastropods. In many dextrally coiling Gastropods the right-hand member of paired organs in the mantle-cavity degenerates as the fetus rebalances body with shell.

Twisting rearranges nerves and viscera and makes for the unique lifestyle of this phylum. Although the ecological benefits of torsion are not obvious, it may be of survival value to the larval form—reorientation of organs allowing head and velum to be withdrawn quickly in life-threatening situations.

Torsion is the result of series of related mutations in different classes. While primitive limpets are merely conical, most Gastropods are not only coiled but helicoid as well. Such extreme spirals do not endow other classes, yet there is still a Molluscan proclivity toward displacing internal organs and realigning neural commissures.

THROUGH THE LONG mutational and epigenetic tunnel of Cephalopod evolution— as the two main Molluscan axes of symmetry have converged—the head-foot region and the mantle cavity came into contact in the anterior portion of the animal. In squids and octopi central nervous ganglia, distinct in the other classes, were transposed into one another in a mass around the esophagus, forming the only true invertebrate brain. Even on top of this, Cephalopods developed complex new ganglia, structures with no homologs elsewhere among Mollusks. These merge with the esophageal brain and subesophageal neural centers to compose a thick fleshy organ. In effect, the overlays of decentralized Molluscan ganglia became lobes of a large and amorphous octopus brain.

Neural clusters refine the talents and gestures implicit among Cephalopods, e.g., coordination of head tentacles and suckers, integration of the mouth organs, and storage of memories. The subesophageal region, subject to only marginal hierarchical dominion from superior lobes to which it is linked, pilots the chromatophores, mantle, ink sac, gut organs, arms, and initiates jet propulsion and the general movement of the mantle. Stellate ganglia along the inside of the mantle run hierarchies of giant neurons to effector muscles; these synchronize movements of the mantle wall associated with the jet and contract both circular and radial muscle fibers. Octopi and cuttlefish locomote at high speeds in all four directions, spin horizontally, hover, and perform an exquisite repertoire of maneuvers.

With up to four million classifying cells in their optic lobes, Cephalopods can distinguish and remember irregular geometries. In experiments in which crabs are fed to captive octopi in association with subtly divergent shapes (one leading to shock and the other innocuous), the animals discriminate quickly and learn the consequences of varying figures.

The eight tentacles for which the animal is named are in continuous motion, their own axial cords provocatively reminiscent of vertebrate spinal columns. Rich supplies of nerves run from these medullary bundles to the skin and muscles of the arms and to the suckers, as though eight animals danced together and collaborated. One author describes these organs as "rather like the fingers of a blind person sorting through the contents of a jumbled drawer."[13]

At least three Cephalopod cranial regions have developed densities of cells with no known sensory or effector connections. These are likely areas of association and learning, silent realms corresponding (in their own way) to the cerebral cortex of human beings.

In the eyes of the octopus we find no mammalian empathy, but rather billions of generations of jellyfish, worms, and clams come to apperception of their existence in the world ocean. Long before vertebrates and insects multiplied, the ancestors of squids were the unchallenged seers of this planet.

The Emergence of a Central Ganglion

WE THINK OF THE BRAIN AS OUR CENTER OF IDENTITY. During the executions of the French Revolution a morbid curiosity led some officials to pick up newly guillotined heads and address them. It was claimed that many of them knew their situation and tried to answer. But self/brain is a vertebrate obsession. Among invertebrates brains are less centers of identity and decision-making than back-up and amplifying computers, experimental add-ons to an already functional product. In just about all invertebrates the cerebral ganglia can be cut or ablated with minimal effect on the creature's activities.

In octopi secondary ganglia and motor centers are especially strong. The central brain, in fact, never becomes aware of the different weights or relative positions of objects held in the various arms. In this respect the tentacles are more "intelligent" than the creature itself.

Removal of the entire brain of a flatworm slows locomotion and makes it harder for the animal to find food. Instead of turning around after a number of unsuccessful encounters, it will persist at an obstacle. After excision of their brains, most insects continue to feed and move about, though the loss of cerebral-visual proprioception leads to hyperactivity. Even in Cephalopods, where there is clear cerebral centralization, the most convoluted region can be removed without any noticeable effect on the animal. It continues to learn most mazes and tasks. However, it takes in information more slowly and cannot do anything which goes against well-established behavior.

In all phyla the higher cerebral lobes steal their consciousness from existing lobes or ganglia. If the lobes grow large enough and incorporate enough functions, they subordinate organs, but they never assume all autonomic functions.

A more intelligent brain requires a rearrangement of the basic invertebrate body plan.

THE ANATOMIES OF ALL INVERTEBRATES—Annelids and Arthropods as well as Mollusks and Echinoderms—necessitate esophageal brains (if there is to be cerebralization at all); this is the topology through which ganglia interpolated themselves, defining the lifestyles of these phyla. In the beginning it was not a bad plan: brains grew larger; animal behavior became more complex. Yet brains cannot expand indefinitely in any organism at the expense of its guts or they would so condense the animal's esophagus as to make digestion impossible. Spiders are almost such "mistakes": in order to suck food through their bellies they must reduce their prey to a fine liquid. Additional neural limitations are imposed by the shells and chitinous exoskeletons of crustaceans, insects, and snails. When these are secondarily reduced, as among Cephalopods (abdicating skeletal function and becoming mere bits of cellophane-like cartilage), the brain finds room to expand and centralize circumesophageally; but it still cannot intrude upon the gut.

Cerebralization among invertebrates appears to be pushing at many species, its inklings embodied in ganglia and neuropile. Yet it is realized only in a Chordate group of obscure bottom-crawlers. Having struggled through generations of frustration among worms, insects, and mollusks, immanent consciousness turned to an unprepossessing side branch of all these creatures and "tried" a new design that would require aeons for its realization.

Of course, this is our anthropomorphization, but it dramatizes how the hypothetical trend toward higher consciousness proceeded to dead ends along separate invertebrate paths. While Arthropod societies covered the land, swarming over meadows and forests; while squids and prehistoric octopi jetted through oceans, the philosophical brain lay dormant and disguised in an unsegmented worm on the ocean floor.

The Nervous Systems of Acorn Worms and Tunicates

THE LINEAGE OF THE VERTEBRATES diverged from Protostomate phyla at roughly a Coelenterate stage. Thus, all vertebrate internal organs developed according to unique Deuterostomate patterns. In Protostomia the blastopore of the embryonic archenteron usually becomes the mouth or is divided into mouth and anus, whereas in Deuterostomia the blastopore becomes the anal opening and a mouth forms secondarily from a new perforation of the body wall. Annelids, Arthropods, and even Mollusks are Protostomate; Echinoderms and Chordates branched off from primordial ancestors of Deuterostomates through a long-extinct creature resembling starfish and sea-squirt larvae.

The Deuterostomate brain of vertebrates embodies itself differently from any other cerebral ganglion. Here alone the nervous system proceeds from an introversion of ectodermal tissue to form a hollow tube. The central nervous system *per se* is neither frontal nor oriented around the gut cavity; it is phylogenetically (and thus ontogenetically) dorsal. This peculiar orientation arose, probably for incidental reasons, in simple vermiforms who neither exploited it nor had neurally advanced descendants for thousands of generations. The critical Gastropod-like reorientation had little or nothing to do with intelligence initially; it was just another spiral in a world conducive to coiling.

Present-day Chordate acorn worms are very similar in appearance to more common Protostomate worms but have reduced nervous systems even by Annelid standards. They themselves are not ancestral to the vertebrates but retain features which were shared with our common ancestor, including a hollow dorsal nerve cord and a notochord-like structure in their proboscis. Caudal to these, an acorn worm is pure ribbon worm. However, some of its ancestors apparently swam into unexploited niches where they became subject to slightly different selective pressures from the ancestors of other worms; likely they were transformed by mutations. In any case, they developed a "malignancy"—a true notochord.

This supple rod of supportive tissue, the partial forerunner of the vertebrate spinal column and organizer of vertebrate organ development, in the lower Chordates is a cylindrical sheath of fibers enclosing a core of cells with vacuolated cytoplasm and serving as a brace, a fulcrum for swimming movements. Tissues on either side of it contract alternately as the tail wiggles from side to side. Such a lever exists contemporarily in the lancelet, but also, more significantly, in the larval forms of sea squirts (tunicates). As adults, these creatures become motionless lumps in "tunics" (hence, their name), with upper oral openings drawing water through sheets of glandular mucus into buccal cavities. Filter feeders, sea squirts are upright sacks with enormous pharynxes interlaced with elaborate basket-like gill slits—but their larvae betray that this adaptation came secondarily from embryos which were more Annelid than Molluscan, hence carrying (by human standards) more evolutionary potential.

Another series of mutations must have rescued our vertebrate progenitors from becoming another race of larval notochord-bearing forms whose awkward nascent spines were degraded and absorbed in development. Rearranged chromosomes working through protein fields would not have had to invent a spinal organ in these creatures so much as fail to suppress spiny mesodermal protrusions. Embryonic eel-like elaborations could then proceed randomly over millennia.

Adult tunicates have a single small cerebral ganglion with modest sense receptors and plexuses of nerves on their body walls. There is even some evidence of a jellyfish-like nerve net between their siphons. Yet these primitive features occur in the context of very advanced structures like gills, a frontal nerve collar, and neuropile clusters.

The true advanced feature, from our standpoint, does not even occur in the adult. It is the hollow central nervous tube of the free-swimming tadpole and its propulsive tail which uses a classic functional notochord as its axial skeleton. Dorsal to the notochord is a nerve cord, enlarged anteriorly into a light receptor and an organ responsive to tilting. As the nerve cord extends into the tail it loses its neural character. A non-neural vestige of it remains in the adult form behind the neural gland.

In metameric animals, larvally derived ganglia fuse and neural structures repeat along the segmented girth of adults. In Echinoderms and tunicates this does not happen; the neural features of the larvae are isolated, ignored by the anatomy of the mature animal, and ultimately deactivated. The ancestral jellyfish calls the creature back from the piscean revolution. Sluggish Molluscan/Coelenterate styles of creatures take the place of the promising starfish bipinnaria and the sea-squirt tadpole. An incipient mobile and neural way of life is abandoned—a collective act of evolution that may appear retrogressive to us but which birthed a variety of species that continue to participate happily in the autonomic currents of ocean life. The tunicates, brittle stars, and sand dollars fetally "threw off" the tyranny and tension of a backbone and nerve cord and inhabited the rich nurseries of the deep. They never had to enter the imperiled worlds of the squid or hive bee. But one of their lines chose (unconsciously) to continue splashing about, to brave such witchcraft. And we are, for better or worse, their will and testament.

Metamorphosis of the Starfish Embryo

DURING GASTRULATION, STARFISH ARE MOLDED as advanced Deuterostomes. A blastocoel opens, and the blastopore becomes the anus; the stomodeum breaks through later. This is the heritage of the bilaterally symmetrical ancestor common to all Echinoderms, long extinct. Those starfish that have returned to bilateral symmetry have done so only secondarily through radially symmetrical larvae. Developmentally they cannot skip the radial phase, so they go from bilateral to radial back to bilateral symmetries, recapitulating an indecisive phylogenetic history. Most

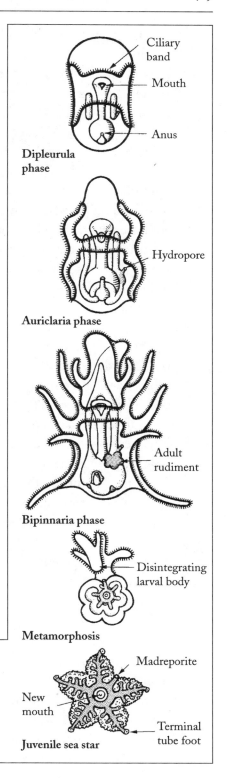

FIGURE 17G. The metamorphosis of a sea star from bilaterally symmetrical dipleurula through bipinnaria phases to a radially symmetrical juvenile sea star.
From Vicki Pearse, John Pearse, Mildred Buchsbaum, Ralph Buchsbaum, *Living Invertebrates* (Palo Alto: Blackwell Scientific Publications, 1987).

mature starfish become radial animals with elaborate arms beating tube feet, and, in some species, feeding tentacles. All these organs arise almost exclusively from the left anterior region of the ciliated, bilaterally symmetrical brachiolaria stage of the starfish larva and are nonchordate specializations. The rest of the neurally advanced, bottom-attaching protofish embryo is absorbed into the "star." Metamorphosis proceeds in a direction reversing that of insect grubs and winged adults. In one line of starfish embryos bilateralism was preserved and enhanced paedomorphically. This "fishiness" apparently led to the Chordate body-plan with its unique strategy of neuralization—a plan we inherit at our anatomical core, a chrysallis we are built upon. The original ocean currents that oscillated and longitudinalized the first multicellular animals and provided a wavy, hydrodynamic body-plan for primitive fishes are preserved in the elongated, vertebral torsos of land mammals. We are sea-made creatures, carrying the shape of the deep in our spinally induced embryogenic template. Each of our organs maintains its own aqueous subset within the greater musculoskeletal ocean.

A FULLY CENTRALIZED BRAIN evaginating at the anterior pole of the neural tube is the Chordate and vertebrate hallmark. The central nervous system is part of this tube; it emerges embryonically with the notochord and neural groove and is imbedded in it end to end. Neither a ring nor a hierarchy of ganglia, the tube becomes surrounded by skeletal tissue produced by the somites; muscles and nerves are then attached to the backbone at its vertebrae. Afferent nerves shoot out to the sense organs; efferent nerves impregnate the muscles and limbs.

When the head of the neural tube expands to become a brain it has no point of contention with skeleton or gut, only with the structural requirements of the spinal column and the rest of the facial skeleton.

The basic vertebrate brain expresses its grandeur initially in three modest frontal swellings, ontogenetically as well as phylogenetically. Sense organs forming in concert with them, they emerge gradually as full subsidiary lobes. The forebrain develops with an olfactory lobe, the midbrain with an optic lobe, and the hindbrain as an otic organ comprising balance, vibration, and general equilibrium.

The hindbrain is the lingering compass of the invertebrate realm and of primitive vertebrates; it coordinates functions critical to their ocean abode. The forebrain evolved in early vertebrates (fish) as a scenting ganglion. However, it continued to expand beyond its olfactory base to become the cerebrum, and in the higher primates (in their embryos as in their ancestors) it has swelled out over all the other convoluted structures, engulfing prior strata of intelligence and lobes in its own. It is in this cerebral cortex that mind lays claim to a replica of the cosmos.

Is intelligence more than neural quantification?

THERE IS NOTHING ABOUT THE BRAIN viewed from outside that suggests an experience of mind and identity, although in a cybernetic age we are lulled into believing that microcircuits and chips can be packed into any shape, size, or space to generate a full repertoire of images and information. We conclude (by default) that we exist as an epiphenomenon of meat. That is, the billions of cells which flow toward (and away from) the brain fire our collective mental apparition within its lobes.

Although there is no clear point in the history of creatures when sum and concentration of neurons cross the hypothetical threshold and ignite mind, we presume that it is the sheer number of sense cells and the critical juxtapositions of their organization in grids and hierarchies that generate our thoughts and awareness. What could never be explained qualitatively is given a solely quantitative justification. The volume and relationship of neurons becomes not just a decipherment of mind but a declaration of reality.

This machine synthesis of intelligence is then applied back to nature; the more circuits, the more "mind." Chimpanzees remain forever subhuman because they do not have quite enough wiring; wolves have even less, frogs less than wolves. The worm is presumed not even to know it exists. The jellyfish and Porifera, for all intents and purposes, *do not exist.*

Yet if we are a mere quantitative degree removed from a sponge, we are not removed at all.

Computers assail our other flank. Toward the end of the twentieth century enthusiastic software engineers have engendered an unexamined assumption among them that continued improvements in microchip technology will soon yield machines far more intelligent than humans. According to Gordon Moore, chairman of Intel during the mid-1960s, every two years new computers provide twice as much power and capacity at about the same cost. One futurist software inventor, Ray Kurzweil, calculates that capacity has already doubled thirty-two times since the first computers and will continue to double until around the year 2020 when he presumes we will reach the theoretical limits of the physics of silicon. At that point, though, he expects we will have discovered a superior molecular technology and will continue increasing cybernetic capacity exponentially.[14]

At a certain point, humans will become antiquated and superfluous. Given that electronic circuits possess (or will possess) more "capacity, speed, and reliability"[15] than neurons, Kurzweil asserts that it will become expedient to have one's brain and nervous system scanned and downloaded into silicon (or its successor). Mapping the more than ten billion neurons of a human cortex and then replicating them in molecular holographs would be a way to contrive a cybernetic replica of someone's consciousness. "[L]iving in this slow, wet, messy hardware of our own neurons may be sentimentally appealing, like living in an old shack with a view of the ocean,"[16] but it is vulnerable, mortal, and proscribed. Using nanotechnology—the process of "building objects atom by atom and molecule by molecule"[17]—we will make ourselves over in an indestructible, even immortal form, capable of limitless algorithmic pondering and assaying. Mind would be back-engineered from a holographic blueprint of the neural biology of the brain. This, of course, assumes that mind is coterminous with cortical physiology and that a reconstruction of mere cerebral circuits will ignite the epiphenomenon of sentience, the "user illusion," and the subjective sense of identity. Such a lugubrious process is probably quite different from nature randomly integrating simple nanobiologies into more complex ones by trading in organic carbon-based material. Yet software enthusiasts are undeterred.

"We will be able," continues Kurzweil, "to reconstruct any or all of our bodily organs and systems, and do so at the cellular level. . . . We will then be able to grow stronger, more capable organs by redesigning the cells that constitute them and building them with far more versatile and durable materials."[18]

If your "body" happens to be destroyed in a plane crash or by some other accident, then a replacement copy of your database can reconstruct you. Additionally, as software, you can be run in a variety of hardware, choosing your unit according to present needs or desires. You can be strong, beautiful, sexy, intelligent, swift, mechanically handy, etc. In fact, everybody can get to be everybody else, trying out their machine bodies. Virtual pleasure will replace present unreliable sources of happiness. In virtual sex, the precisely appropriate circuit of a computer brain can be stimulated to highly refined partialities without even another "body" or person present. "Virtual sex will provide sensations that are more intense and pleasurable than conventional sex, as well as physical experiences that currently do not exist."[19]

At such a point we and the computers we build will converge. "What, after all, is the difference between a human who has upgraded her body and brain using new nanotechnology, and . . . a robot who has gained an intelligence and sensuality surpassing her human creators?"[20]

ACCORDING TO ANOTHER FUTURIST, Hans Moravec, "machines are the next evolutionary step, with organic tissue but a blink in the eye of cosmic history. Once intelligence is created by natural selection it will be only a matter of time (a very short one by cosmic standards) before the products of intelligence outshine their creators, finally displacing them altogether."[21] By this prognosis, just as multicellular entities trumped cells, replicating themselves into a vast, diverse biomass, so artificial noncellular entities will soon supplant the plant, animal, and human populations with machines. Animals are merely nature's transient vessels for the cyberneticization of intelligence. Like bionts (but in a more syllogistic and foolproof fashion) cyberonts can be programmed to handle all conceivable environmental occurrences (and to improvise effectively when something novel intrudes). They are perfect organisms portentously assembled by imperfect cellular creatures. Once free of us, they will run their own industries, rebuilding themselves autonomously out of minerals they mine down to the Earth's core, providing "foodstuffs" for their own metabolisms. They will be everything life is—and more.

Future cyberonts may include minute nanobots—machines approaching molecule size. Using the fuzzy logic of decision averaging while acting along a gray scale of options rather than chip by binary chip, these superminiaturized computers (literally "dwarf robots," after "nano-," representing one-billionth of a unit) will be able to operate inside both organisms and environments. Set loose by ambitious technologists, they may run amok, carrying out millions of computations in a second, as they clamber and flutter about on microbial insectlike limbs. With the capacity to utilize and alter matter on an atomic scale, they will be able to replicate themselves indefinitely on a surrounding matrix of almost any raw material. Cloning at intrinsic nanoscale speed, they will produce one offspring every five seconds, in effect doubling their population in that time. There will be no soil left for photosynthesis, no habitable water. Unless these organisms gobble up the entire planet, they will become immortal and replace all biological entities. Visitors from outside the Solar System will arrive only to find a huge robot-serviced factory, its makers long vanished.

On the same drawing board are foglets: "tiny, cell-sized robots, each more computationally powerful than the human brain, that are equipped with minute gripping arms that enable them to join together into diverse physical structures. At ease the foglets are just a loose swarm of suspended particles in air, but when you press a button they execute a program for forming themselves into an object of your choosing."[22]

Thus humans could orchestrate an entire new layer of nanomorphogenesis, using the robotic equivalent of cells and membranes. Upon command the foggy swarm would arrange itself into a house or an exhaustive three-dimensional mosaic of a favored environment; it could carry out machine photosynthesis, produce artificial nutrients, and provide virtual friends and vacations for its owner. Foglets would eventually reengineer this rugged watery and rocky clime orbiting a sun into a technocrat's dream of a smoothly run asteroid, reconstituting the Earth more efficiently by processing photons in great artificial silicon-chlorophyll machineries rather than botanically.

Of course this is just a technocratic fantasy, an extension of commodization into the molecular realm. It cannot hold a candle to the Earth's real nanotechnology, carried out by the cells of bacteria, plants, and animals. Lynn Margulis and Dorion Sagan write:

"Bacteria have mastered nanotechnology; already miniaturized, they have control of specific molecules about which human engineers dream. Far more complex than any computer or robot, the common bacterium perceives and swims toward its food. Choosing and approaching destinations, bacteria propel themselves by flagella, corkscrew-shaped spinning protein filaments attached to living motors in the membranes of their cells. Complete with rings, tiny bearings, and rotors, they are called 'proton motors' and spin at about 15,000 rpm."[23]

IN AN ERA IN WHICH SCIENCE already preserves life by grafting organs, and virtual reality taunts the primacy of the natural world, it has become increasingly difficult for us to distinguish our seemingly innate cellular intelligence from the manufactured mentations of electrodes. If copper and silicon can duplicate every network and bundle of nerves and ganglia, when does machine become animal; when does virtual become real? As computers approach and (eventually) surpass the computational capacity of the human brain, there is a tendency to view them as sentient entities, beings with all the prerequisites of mindedness. If robots can be loaded with enough synapses to strategize chess moves, prove theorems, manipulate symbols, store complex data, analyze corporate profits, and speak the rudiments of emotions, when do we decide that they are "thinking" as opposed to mimicking thought? If a machine behaves quasi-consciously—carries out human operations and movements—does that mean it has actually become conscious? Or is it simply a symbol-manipulating machine using *our* minds?

How much prosthetics can be added to a nervous system while still maintaining its human identity? What is human identity anyway—cells or epiphenomena? Can a brain wired to a cyborg speak truly for the being whose "memories" it carries? How can programs of the same person's brain and nervous system be downloaded into different hardware units and have the same identity?

If humans ever escape their biological "wetware" and acquire "immortality" by successfully relocating the circuits of their identity in computers, would these new cybernoids be "people" or mindless robotics capable only of juggling the external residues and symbols of terminated intelligences? Would the thread of individual sentience from their point-of-origin brains continue or evaporate? Does embryogenesis under a morphogenetic template implant some kernel that can never be kindled in an artificial assemblage? Is organic tissue the only way by which nature can negotiate consciousness, i.e., engender a qualis, a subjective experience of being? If so, then why?

John Searle considers it ludicrous to believe that a computer can actually understand chess. Writing of IBM's Deep Blue, the unit that defeated champion Gary Kasparov, he concludes:

"The computer has a bunch of meaningless symbols that the programmers use to represent the positions of the pieces on the board. It has a bunch of equally meaningless symbols that the programmers use to represent options for possible moves. The computer does not know that the symbols represent chess pieces and chess moves, because it does not know anything.... [W]hat was it thinking about? Certainly not about chess.... The symbols in the computer mean nothing at all to the computer. They mean something to us because we have built and programmed the computer so that it can manipulate symbols in a way that is meaningful to us."[24]

This is simulated cognition, not artificial intelligence or existential being.

IN SOME UTTERLY MYSTERIOUS MANNER (that, even so, does not explain consciousness), synapses forge their own logic, their own rationale, and the extension and exponentialization of their fabric out of their own synapsing. The primitive "nerve netting" of neurons feeds on itself and its own circuits to invent mind. Mind is the sole outcome of contagious synapsing and has no other apparent antecedent or agency.

If centuries from now computers still cannot synthesize qualia, even with every nerve of our body and brain replicated holographically in their hardware, what element will they be lacking? Where can they look for the singular emergent property, the missing link? What, literally, turns mind on?

How can we know if machines have minds without knowing what gives us *our* minds?

In what fabric of materiality does the "usness" of us reside? How do we become "real" while a world external to us simultaneously becomes real?

How do spheroids of primeval gas, propelled at terrific speeds around larger, often burning orbs, themselves maintained along tracks of hyperdimensional gravity, become theaters for self-aware characters portraying the hunt, romance, and other archetypes?

Where is the mind in matter? Where is self among synapses and ganglia, in these slender textiles spun out of mere cosmic junk?

That which exists through itself is called meaning.

IRONICALLY, THE MORE WE SEEK ANSWERS, the more we find there is no question. The more conscious we become, the more we perceive the imminent disintegration of consciousness.

In koan practice, trained meditation on nothingness (over many years) can lead to a state of cognition "beyond mind"—the enlightenment recognized by zen monks. The biology of the evolution of consciousness, though skeptical of satori, tends to validate this path. That ephemeral sensations of axons and dendrites collaborate in something as cohesive, egoic, divisible, and profound as mind is the greatest paradox in the universe.

In absence of honoring their thoughts beyond mind, humans in Western civilization become trapped by cultural demarcations and tech idioms, limiting existence not to what it is (which remains unknown) but various materialistic definitions of having a body or the epiphenomenon of a life. Experience is negated by appearances of its façade. Death-obsessed paradigms of the laboratory and biomedical establishment insist that we are obsolescent machinery requiring constant recommoditization, natural or artificial—ultimately hopeless. Yet, being made of cells should be a miracle and an opportunity, not a limitation or a death sentence.

Behind the veil of personality, behind hormones and neurons is naught but cells dividing, replicating their own ancestral patterns. This is not trivial nothingness but ultimate nothingness. If we are nothing but cells, fissioning and replacing one another in transitional motifs, then we do not exist (even though we do). Our dreams are nothing, our hopes nothing, our divertissements nothing. Our identity, which rests not even upon cytoplasm, atoms, synapses—or anything—is energy and form, arising through substance, transferring itself like clouds in moist air, a geyser in a lake.[25]

We get conceived, get born, then *are*—an owl, a goat, a cricket, a sea anemone, a twenty-first-century man or woman. A fish has the mind-totality of a fish—a bird, of a bird: feathers, waves, flutter, light.... "... feed me because I cry louder .../because I am alive and make noise/because I can crack the cheap bowl of your sky with my shriek...."[26]

Where there is no negotiation, there can be no diminishment. Cells teem out from eternity to put their stamp on matter. Meditate on this long enough and, in becoming nothing (while still awake and aware), you become everything. Your essentiality, never born, cannot die.

Neurulation and the Human Brain

The Neuralization of Tissue

FOR DAYS AFTER THE MERGER OF SPERM AND EGG, stem cells within the blastocyst are totipotent and pluripotent; they have the capacity to become any kind of tissue in the body, and they can reproduce essentially forever.

Through early embryogenesis cells within the same germ layer remain equipotent. Until they are demarcated from their neighbors they share genetic potential.

There is no such thing as pure uncontaminated sentience or (within living beings) totally unconscious flesh or bone. All nonsentient tissue is potentially neural (suppressing its axon-making capability), and all neural stuff remains partially collagenized and epidermalized (supporting girth and structure). Tissue that will form ears and eyes could also become hair or tusk.

Neither deep gut nor connective-muscle tissue—but ectoderm, the delicate surface stratum—is the source of mind. This makes sense, for it alone is the Ur protoplasmic stuff and our electro-permeable boundary with phenomena. While skin is unsensitized nerves, brain and sense organs are polarized epidermis, cells which could become fur, scales, or teeth if their tubules and filaments were not induced by nonsensory tissue—by the notochord and surrounding mesodermal epithelia.

WITHIN A YOUNG TADPOLE GASTRULA, the prospective lateral mesoderm, notochord, and somites are arranged around the coelom; a germinal nervous system covers the dorsal hemisphere, lying atop the somites and notochord to the rear and above the endodermal lining of the foregut. As the blastopore closes, the neural plate identifies itself by segregating from the rest of the ectoderm. A thickening, dorsally moving epithelium, its cells that are to become neuralized elongate while cells of

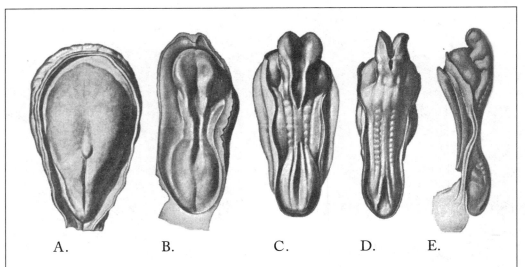

A. B. C. D. E.

FIGURE 18A. Developmental stages of the human neural groove and tube. A. Pre-somite embryo, with neural plate and primitive streak; B. At three somites, with deep neural groove; C. At seven somites, with closing of tube beginning midway; D. At ten somites, with closure extending into brain region; E. At nineteen somites, with closure complete except for neuropores.

From Leslie Brainerd Arey, *Developmental Anatomy: A Textbook and Laboratory Manual of Embryology* (Philadelphia: W. B. Saunders & Company, 1946).

the future stratified epidermis (of the skin) remain flat. The edges of the neural layer thicken further and fold above the plate along its length. As they ascend, the region of cells between them collapses into a groove, the entire plate contracts transversely, and the folds touch at their midline and fuse to form a tube with a frontal opening, the neuropore.

One can recreate this morphogenesis by different mechanics with a lump of clay. First flatten it; then squeeze its free edges until they become lips. At the same time make the piece smaller by pressing its substance toward the center. As the lips come into contact, round their surface over the gap to enclose the neural tube.

Ontogeny *must* recapitulate phylogeny here. Neurulation of the embryo is concomitant with elongation. As cells flow in crisscrossing battalions toward their destinies (resembling also whirlpools), the animal stretches to fill a spine it suddenly and historically embodies: an Echinoderm sprout has metamorphosed into a proto-vertebrate; a neurula has infested a gastrula.

The Phylogeny of the Neurula

THE SPINAL CORD OF PRIMITIVE SEA VERTEBRATES persists subordinate to the brains of reptiles, birds, and mammals—a signature retained ontogenetically. The trunk of quadrupedal and bipedal mammals is first the fulcrum of a fish.

Small burrowing fish-like lancelets are generally regarded as the living Chordates closest to the vertebrate line. Filter feeders throughout the world's oceans, these virtually headless swimmers bear flexible notochords with muscle units on either side (see the previous chapter). Their central neural pathways and cerebral ganglion, fed by afferent and efferent fibers, are located in a tube running the length of their body. Afferent nerves conduct impulses from peripheral receptors into the central nervous system, and efferent ones send commands from the central nervous system to organs, triggering secretions of glands and contractions of muscles.

This simple tube and brain stem represent the whole Chordate brain.

Early in the development of the human embryo, an archive of this structure forms—a cylinder with a slight bow in it—but, as the brain bulges anteriorly (an event totally foreign to lancelets), it develops a number of deeper bends. The so-called cephalic flexure forms at the spot where the forebrain curls downward in front of the midbrain. This hump, which obtrudes at the end of the first month after conception, is so pronounced that the organ is almost bent in half. Its dramatic early appearance in ontogeny reflects a millennial departure for our lineage— perhaps a series of changes coinciding with mutational spikes. From that crossroads, complexity and convolution will pack one another fractally to a depth and profundity in no way prefigured by insect or octopus ganglia.

ALONG THE VERTEBRATE LINE the spinal column gradually enclosed the neural tube. Gray matter (pink in living creatures) was wrapped in white myelinated fibers carrying sensory information about touch, temperature, and muscle kinesthesia, and transmitting instructions coordinating the arms, legs, shoulders, and neck. Occasional exposed butterfly-shaped sections along the spine's length betrayed rich nerve complexes flowing off to the body's peripheries.

Where myelin historically covered the neural tube, no further expansion of nervous processes was possible. However, a fresh zone of neuralized (gray) matter swelled out in the one place it could find an opening—over the head of the tube— establishing the palaeocortex, forerunner of the brain.

This cerebral hemisphere probably originated in fishes as fibers of olfactory bulbs, no more than amplifiers of smell. Inherited by long-vanished mammals from

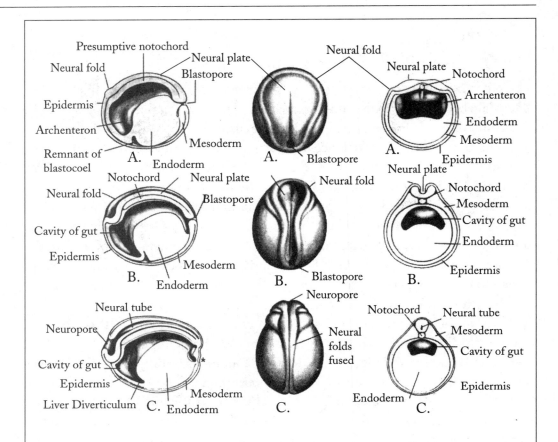

FIGURE 18B. Neurulation in a frog embryo. The drawings in the middle are the whole dorsal view. The drawings on the left show the right halves of embryos cut in the median plane. The drawings on the right show the anterior halves of embryos cut transversely. A. Very early neurula; B. Middle neurula; C. Late neurula with neural tube almost completely closed, the blastopore closed, and an asterisk marking the spot at which the anal opening will break through.

From B. I. Balinsky, *An Introduction to Embryology*, 5th edition (Philadelphia: Saunders College Publishing, 1981).

ancestors they shared with amphibians and reptiles, these bulbs are recapitulated in the human fetus after gastrulation when the front of the neural tube—the brain stem—thickens in sections that will become the rhombencephalon, mesencephalon, and prosencephalon, i.e., the forebrain, midbrain, and hindbrain, respectively.

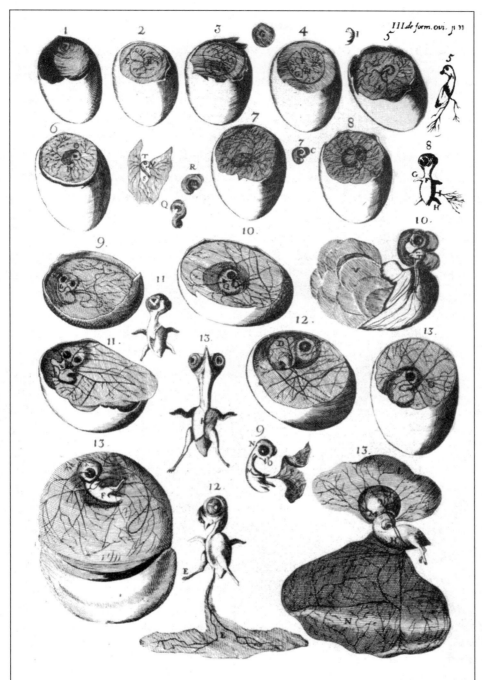

FIGURE 18C. Plate from *De formatione ovi* of Hieronymus Fabricus ab Aquapendente (1687), illustrating development of chick.

From Arthur William Meyer, *The Rise of Embryology* (Stanford University Press, 1939).

Differentiation of the Human Nervous System

In human neurulation, the ectodermal surface of the neural plate gives rise to the central nervous system—the spinal cord and brain (this picks up the description in Chapter 10, overlapping slightly). After an elongation of the notochordal process induces extension and thickening of the plate, on the eighteenth day after conception the entire plate invaginates and a groove pushes up along its main axis, a hollow mesodermal injection between ectoderm and endoderm and open to amniotic fluid above. Secondary folds press on either side like buttes adjoining a plain. Driven by cresting cells, the crimps of tissue reach over their plain like outstretched arms and close to form the neural tube. Their fusion, originating at a central point in the neutral groove, becomes a zipper sliding simultaneously cranially and caudally.

The neural tube also thrusts upward, sustained against gravity incrementally by the nascent spine and general skeleton. Cells of the old hydrosphere sag into valleys along which axons will run. Virgin foliage springs up.

In the mesoderm the chorda (notochord remnant) is trailed by cubelike pegs on either side, bilateral ridges of somites in forty-four pairs, forerunners of bony skeleton, muscles, and dermatome. After inducing the emergence of the vertebral column, the remnant becomes encased in it and persists as the *nucleus pulposa* of the intervertebral discs. Chordal lengthening induces foregut and hindgut within the endoderm, the former rolling cranially, the latter caudally. This tissue underlies future intestine, itself the genatrix of organs such as lungs and liver.

The already-fat anterior of the neural tube continues to plunge downward, thickening and lengthening. It will become brain and brain cavity; its narrower, sinking tail will develop as spinal cord.

Once closed, the neural tube drops beneath the dorsal surface of the ectoderm, not yet skin. Its cresting (neural crest) cells fill a zone between the tube and the ectoderm. There they pair into spinal ganglia,

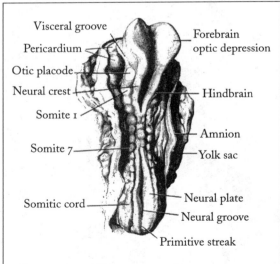

Visceral groove
Pericardium
Otic placode
Neural crest
Somite 1
Somite 7
Somitic cord

Forebrain
optic depression
Hindbrain
Amnion
Yolk sac
Neural plate
Neural groove
Primitive streak

FIGURE 18D. Human neurula at seven somites, dorsal aspect with upper and lower neuropores.

From Emil Witschi, *Development of Vertebrates* (Philadelphia: W. B. Saunders & Company, 1956).

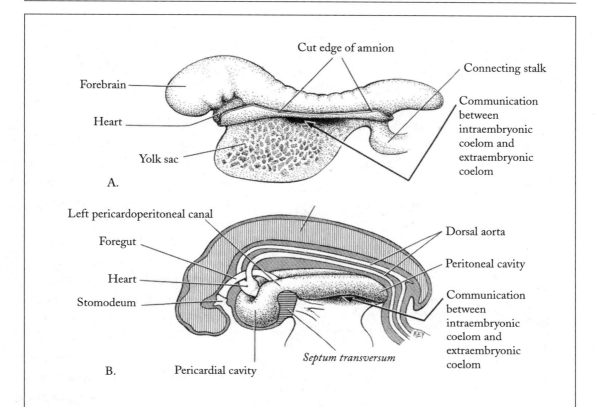

Cut edge of amnion

Connecting stalk

Forebrain

Communication
between
intraembryonic
coelom and
extraembryonic
coelom

Heart

Yolk sac

A.

Left pericardoperitoneal canal

Foregut

Dorsal aorta

Peritoneal cavity

Heart

Stomodeum

Communication
between
intraembryonic
coelom and
extraembryonic
coelom

B.   Pericardial cavity

Septum transversum

FIGURE 18E. Effect of head fold on the intraembryonic coelom. A. Embryo at 25 days beginning to fold; the forebrain is large; the heart is ventrally located; B. After the passage of another day or two, the embryo has folded, reorienting the intraembryonic coelom.
From Keith L. Moore, *The Developing Human: Clinically Oriented Embryology* (Philadelphia: Saunders College Publishing, 1977).

segregating on either side of the tube. As the tube identifies itself completely free of overlying epidermis, the separated outer layer knits along its edges to cover the back of the embryo as skin.

The pace of the neural tube's extension surpasses the growth of the more ancient embryonic disk—a near catastrophic disjunction causing "multiple doubling-up in deep folds in the cranial and middle part of the . . . tube, forming what will become the central part of the brain."[1] The series of pouches at its fore end incubate the front brain cortex, midbrain, and hindbrain. "The cranial end . . . now develops a bilateral bulge which first grows outwards and then downwards, forwards, upwards, backwards, downwards, and forwards again, ultimately covering the central part of the brain as the two cerebral hemispheres."[2]

Along the length of the tube and its branches, cell bodies dispatch axons, fibers,

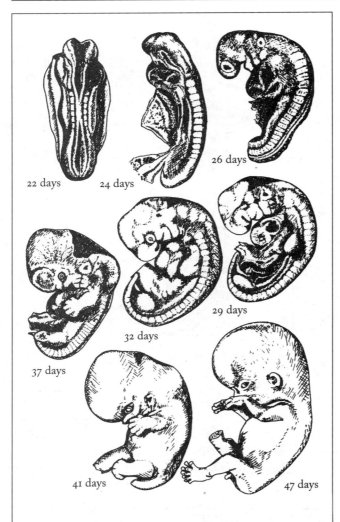

22 days　　24 days

26 days

29 days

32 days

37 days

41 days　　　　　47 days

FIGURE 18F. Stages of development of the human embryo.

From B. I. Balinsky, *An Introduction to Embryology*, 5th edition (Philadelphia: Saunders College Publishing, 1981).

and nerves—threads from which the spinal cord will be braided. Twelve pairs of cranial nerves sprout therein, then thirty-one dyads of spinal nerves. Rudimentary sense organs begin to poke out of corresponding ectoderm. Incorporating brain, spinal cord, and nerves, the neural tube establishes a permanent unity between the neuromusculature and cerebral cortex.

THE ABRUPT TRANSLATION from a radial "jellyfish" to a "larval worm" does not portend that an identically dramatic departure occurred during phylogenesis, for (as we know from other examples) epochal histories have been both synopsized and subordinated to fetal adaptations. It does suggest that this specific divergence in body form, gradual and intermittent in the ancient oceans, took on distinct life meanings and ecological consequences leading to a quantum leap and leaving a developmental gap between the embryonic disk and the neural tube. Retroactively embracing the profundity of this gap, the vertebrate embryo summarizes and then bridges its interrupted evolutionary history.

Once upon a time, functions were in fact passed hierarchically upward to superior lobes which swelled over and encapsulated prior ones. In order for this to happen, rates of tissue growth for the invading neural tube *had to outstrip* the old embryonic disk, perhaps spawning generations of excessively cranialized eels. Other mutations conferred neuroepithelial expansion, crinkling, and intorsion. Nexuses of cells that had reached their seeming endpoints as centers of identity in fishes and newts became mere relay stations for subsequent cerebral hemispheres, their sensory and proprioceptive data translated upward into lobes that themselves were later superseded by the cerebral cortex.

Over millions of years the visage of the Earth changed from the stare of a lamprey to the Renaissance squires of Albrecht Dürer.

The Medulla and the Pons

THE CONTEMPORARY HUMAN BRAIN STEM begins in the medulla oblongata, a vertebrate expansion of the spinal cord as it enters the skull. The medulla coordinates our basic Chordate existence: heartbeat, breath, digestion—the mainstays also of the lancelet's swimming-based habitat. Bundles of nerve fibers converging here are combined in networks and conveyed to higher lobes.

In more cerebral animals, a cervical flexure develops between the medulla and spinal cord, and a pontine flexure twists the opposite way, thinning out the roof of the hindbrain. This bend becomes distinct in humans during the middle of the second month after conception, at which time the midbrain is most enlarged in relation to historically subsequent lobes. The pons develops its own consolidated bands of connecting fibers between the cerebral and cerebellar cortices and the spinal cord.

Some neurologists believe that a zone buried in the lower pons is the locus of paradoxical sleep. In this state (by contrast with slow-wave sleep) the muscles relax, the closed eyes scan excitedly, heartbeat and blood pressure decrease, and dreams arise deeply and often. Apparently two chemical transmitters—natural mammalian hallucinogens—are involved: serotonin during slow-wave sleep and noradrenalin during paradoxical sleep. These "induce" aspects of the mind much as other polypeptides induce gene expression in the cell, removing constraints on latent material and substituting one psychoactive landscape for another.

When the vertebrate cortex quaffs the dream ambrosia, its standard functions vanish; and something else, mysterious by waking standards, takes their place. Forgotten experiences, neural chatter, glimmers of vestigial functions, emotional traces of organs, and assorted environmental and cosmic rhythms all contribute charge to dream formation in the pons and brain stem. As Freud divined: there is no time in the unconscious. There is also no dimensionality or scale.

"The mind can make/Substance," wrote Lord Byron, "and people planets of its own/With beings brighter than have been, and give/A breath in forms which can outlive all flesh.//. . . . A slumbering thought, is capable of years;/And curdles a long life into one hour."[3]

From the beginning, the brain had to mediate between alert and subliminal

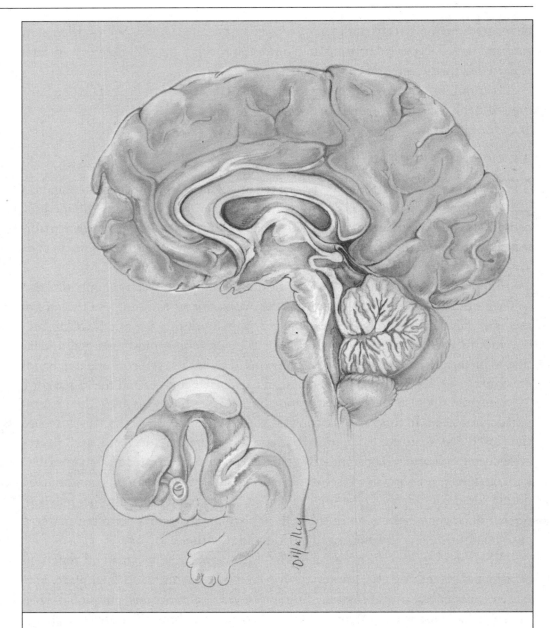

FIGURE 18G. The developing brain. Illustration by Jillian O'Malley.

phases of mind, the pons a living relic of the outcome. Without dreams we might hallucinate continuously or never be quite awake, never fully aware of the brain's simulation of a planet with star-filled skies. Our ancestors could not have attained a waking state if unable to dream while asleep, so forced to conjure while wayfaring.

Surviving creatures all perfected some form of hibernation or paradoxical sleep, with phases of recurrent, fluttery eye movement; the rest have long since sleep-walked off this plane.

The Cerebellum

THE OTHER MAJOR ORGAN OF THE HINDBRAIN is the cerebellum. Buried in folds with only about a sixth of its surface lying open inside the skull, it is one of the most convoluted regions on Earth. Formed from a series of expanding bulges in the brain stem, the cerebellum receives connections from many points in the body, including muscles, joints, sense organs, and the cortex itself. Because it sends back only a third of the number of fibers it receives, we assume that the cerebellum forms interneuron links between afferent and efferent paths, condensing and coding multiple sets of impulses in reflex arcs, and remitting them as single messages.

Although not particularly an organ of higher consciousness or creative endeavor, the cerebellum integrates the movements of the motor system, fine-tuning muscular activity and adjusting equilibrium. A calligrapher inscribing lines of a character with a pen, an Eskimo artist carving scenes of animals and flowers in a pebble, an astronomer aligning a telescope to a remote galaxy, a baseball hitter timing a ninety-five-mile-an-hour fastball are all calibrated in the cerebellum. Without this organ, people would not be able to gauge distances and would undershoot or overshoot what they reached to. The trembling of older people betrays a breakdown of cerebellar function.

The cerebellum originates at the back of the brain in animals travelling (or once travelling) on legs. Birds have an especially enlarged cerebellum, no doubt for maintaining balance during flight and for alighting.

The cerebellum comprises lobes from three different archaeogenetic eras. The small archicerebellum at the lower rear (just above the medulla) has fibers connecting to the chambers of the ears. The next oldest palaeocerebellum is an anterior lobe that processes sensory data from the limbs. The intermediate neocerebellum, the actual posterior lobe, developed last and is the source of subtle limb movements and timing.

Early Sensory Functions of the Vertebrate Brain

THE MIDBRAIN EVOLVED IN BONY FISHES, probably 450 million years ago. With their vertebrate axis of motility and sharpened eyes, these carnivores established themselves throughout Ordovician and Silurian oceans. Their midbrain was functionally their forebrain, a ruling lobe, incorporating newly enhanced powers of sight.

Eyes may have begun as simple pigmentation spots in Chordates epochs earlier, but the neural tube ultimately elicited full retinal surfaces with image-forming sites. Eye spots of lancelets deepened in their descendants into pits with light-recording cells at their bottoms. These gradually came to meet the stalk of the optic nerve in the brain stem and culminated in the two swollen lobes of the optic tectum which dominate the midbrain of fishes.

Sense pigments in protozoans and anemones recall the origin of ocular receptivity in roughened ectoderm. Full-fledged eyes are the later result of "the natural-state changes of morphogenetic fields: calcium-cytoskeleton dynamics, localized cell growth and deformation, bucklings of cell sheets, and directed cell movements over surfaces."[4] At their basis are light-sensitive bacteria to whom the Sun revealed itself. Ideal probes, swift photons bounce unswervingly off solid objects.

A rough goggle of translucent epiderm shielding a cavitation of axon-packed neurons is already an optically excitable organ, a primitive imaging cup. From there actual eyes with lenses have evolved at least three times independently: in the insects, the Cephalopods, and the vertebrate line leading to the mammals. An inherent capacity of raw ectoderm, sight induces image-forming cells and storage chambers for visual remembrances and logic. Inevitable quantum-dynamic states, eyes are as natural and robust as flowers, for "... there is a large range of parameter values in morphogenetic space that can result in a functional visual system."[5] On a sunlit planet, tissue will eventually "see," if not by one neuroanatomy, then by another.

ALTHOUGH FISHES HAVE CEREBRAL GANGLIA rudimentarily like ours, they retain an invertebrate separation of functions—distinct "brains" for olfactory, optic, auditory, and visceral functions, and for the surface of their epidermis. Their cerebellum is undeveloped; there is no cerebral cortex. From the era of the fishes the prospective mammalian optic center has been shifted to the forebrain, the tectum reduced to four small swellings—the colliculi. Signifying the brain of fishes ontogenetically, these lineaments in humans are formed from the same plate of neuroblasts as our afferent nerves; a superior pair relays visual impulses, an inferior pair auditory reflexes.

A new mode of intelligence arose first among those fish with genotypic potential for amphibious life. Their forebrain split into a central diencephalon with the vesicles of a telencephalon on either side of it, an event loosely recapitulated in each descendant embryo. The diencephalon then took over as the brain.

OUTSIDE THEIR WATERY ancestral home the prehistoric forerunners of newts experienced echoes, aromas, and far-flung landscapes. They "invaded" land, and the land invaded and shaped them.

In modern mammals the diencephalon is primarily involved in conducting nutrients and oxygen to the nerve processes of the brain. The dominant fish hemisphere, during amphibian and reptile ascension it came more to govern hormones.

This olden aeon is reincarnated in the human embryo when neuroblasts in great numbers accumulate in the lateral walls of the emerging diencephalon and protrude into the underlying brain cavity as the thalamus, hypothalamus, and epithalamus.

The thalamus continues to swell until the third ventricle of the brain is squeezed to a mere slit. In reptiles this organ coordinates all incoming afferent pathways, a role retained among mammals where it serves as the last ascending station for messages below the cerebral cortex. It also harbors the initial optic synapses outside the retina. But it is not just a passive electrical line; it fuses and coordinates impulses. Sensory modalities which are disperse prior to entering the thalamus leave as coherent gestalts, in which state they pass into the cortex for mentalization. The thalamus is perhaps where "emotional, time-oriented appetites live."[6]

Awareness and consciousness already exist in the thalamus of noncortical animals. Human babies born without a cortex function typically at first. Still brainlike enough in the cortex's absence to organize rudimentary behavior, the thalamus reclaims its former role among amphibians and reptiles and reenacts it as the upper lobe of debilitated infants in which its reigning function would normally have been superseded.

The Neuroendocrine System

THE EMERGENCE AND REFINEMENT of nervous systems require delicate chemical regulation. Nerves without hormones would be pure fire; they would scorch their own organisms. Glands modulate neurons and participate with them in sustaining complex psychological states and versatile behavioral ranges. In a sense, nerves and glands represent two different types of information conveyance, the former compact and decisive, the latter diffuse and pervading.

The relationship between endocrine cells and neurons is evolutionarily fundamental and critical to the cohesive organization and functioning of complex multicellular animals.

IN ANY LIVING ORGANISM (as noted in Chapter 11, "Morphogenesis"), cells continually send each other messages across varying distances. These transmittances, however they are written and interpreted upon reception, are packaged in similar classes of molecules and dispatched to correlative sorts of protein receptors elsewhere in the body. The contents of the packets are usually favorite molecular instructions such as: "Make more of this"; "Stop making that"; etc. These bits of communication differ from one another mainly in the discrimination of their recitals and celerity of their delivery. Sensations and stimuli are thereby synchronized reciprocally with the manufacture of chemical substances that have metabolic, psychosomatic, morphogenetic, and hormone-prompting effects.

Each cell in a living system must carry out a crucial and sensitive function, so it usually has multiple receptors that can receive and activate a range of signals from the bloodstream, some modulating or even reversing others. Cells respond only to those molecules for which their receptors are specified, and they take action only in terms of their position in tissue and state of specialization.

ENDOCRINE CELLS ORIGINATING in independent glands routinely secrete hormones into the extracellular (interstitial) fluid, from where they diffuse into capillaries and enter the bloodstream in highly diluted form; these molecules then travel throughout the body, delivering their signals (over the course of minutes) to complementary cells in tissues which have been phylogenetically prepared to receive them. Endocrine signalling is a relatively slow mode of transmission.

For example, an increase of glucose in the blood may originate from pituitary command, but it must be modulated by cells in the pancreas which, alerted to the oversupply by neuroendocrine transmissions, release stored insulin (a protein formed by two amino-acid chains, one comprising twenty-one units, the other thirty). The sudden increase of this substance in the bloodstream triggers fat and muscle cells to take up more glucose. Intracellular vesicles bearing membrane-bound transport proteins are propelled by exocytosis to the plasma membrane. As they engulf their glucose prey, sugar is absorbed and removed from the blood; insulin production wanes. The glucose-carrier vesicles are later restored to the intracellular pool by receptor-mediated endocytosis.

Cycles of this sort are performed with mostly flawless *élan* by interactive organs throughout the body-mind.

CELLS ALSO EXUDE short-lived chemicals that act only on other cells in their immediate vicinity, within about a one-millimeter circumference; afterwards, these molecules degrade or become inactive, and they do not enter the main bloodstream in any significant quantity. This process is known as paracrine signalling.

Synaptic signalling (as described in Chapter 16) occurs solely within the nervous

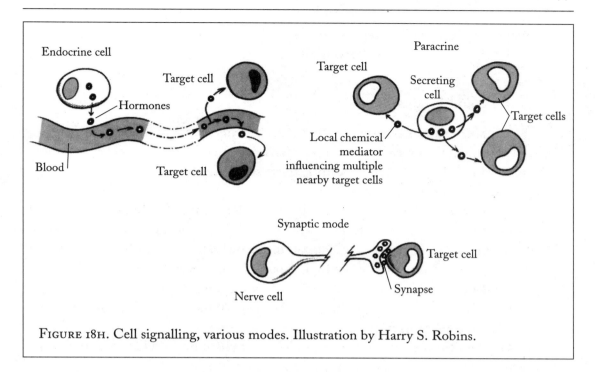

FIGURE 18H. Cell signalling, various modes. Illustration by Harry S. Robins.

system, usually at ranges of fifty nanometers (billionths of a meter). Neurotransmitters are secreted from chemical synapses across a synaptic cleft to the adjacent post-synaptic target cell, which receives the message by binding the cueing molecule and carrying out the activity mandated by its presence. This mode of conveyance is extremely rapid, impulses jetting along nerve processes at speeds approaching one hundred meters per second. The synaptic mode is also more precise in that it ignores intervening cells that have receptors for the same neurotransmitter, delivering instructions only to its next target cell in less than a millisecond—a state that is loyally relayed from cell to cell without corruption of text.

After synaptic signalling, enzymes and transport proteins clean the synaptic cleft of the debris of transmission, keeping the conveyance prompt, precise, and brief.

As THE NEUROENDOCRINE SYSTEM integrates itself at multiple levels of anatomy and reverberates throughout the billions of cells of a higher organism, foodstuffs are digested, wastes are eliminated, protein is selectively manufactured and trimmed, delicate balances are maintained, and moods and emotions germinate. The kinesthesia of all of these events gives rise to higher-level signals, kindling complex mentations (both conscious and unconscious) which initiate additional feedback loops throughout the viscera. In humans a critical threshold is crossed, as luminous cephalic bursts try to answer in some fashion the exigencies, appetites, and paradoxes posed by the deep, swift-flowing network itself.

The Pineal Gland

FROM MID-DORSAL DIENCEPHALON SITES where embryonic neural folds come together, the pineal body and parietal organ later evaginate. Ocular in amphibians and reptiles and their ancestors, these structures have been transformed into glandular organs in mammals.

The pineal, a tiny organ in the shape of a pine-cone (hence, its name), originates in the fifth week of human life as a blind sac and branches off the diencephalon, a thinning stalk of it maintaining a connection. It finally locates near the roof of the third ventricle of the brain, roughly where a third cyclopean eye might sprout. Some prehistoric amphibians and reptiles in fact grew a pineal eye in the backs of their heads. Though the pineal body is not the human third eye, it is (perhaps) its rudiment in the brain.

Over a lifetime this ocular organ loses its glistening eyelike singularity and becomes fibrous and coated with calcium scales.

Responsive to sensory information from the optic nerves, the pineal cannot itself see but responds to light in a variety of other ways. As a gland, it synthesizes a spectrum of hormones, including melatonin which contributes to the inauguration of puberty in girls and regulates their menstrual cycles as women. The pineal increases melatonin manufacture each dusk as sunlight diminishes, then correspondingly retards production during the day. Thus the gland serves as one of the body's internal clocks. For this reason herbal melatonin has been widely used as a nontoxic sleeping pill.

Producing serotonin and dopamine as well as melatonin, the pineal helps regulate other glands in the body and influences regions of the contiguous brain. By generating a slight magnetic field itself (like a tiny moon) and responding to changes in the Earth's magnetic field, this mysterious organ also attunes our circadian cycle and other biorhythms.

The psychospiritual power of the pineal, long intuited among indigenous peoples (for instance, native Americans and Australians), is heeded by painting sacred symbols or anchoring feathers at the spot on the forehead corresponding to the cone's interior location. In esoteric and occult circles the pineal eye is regarded as a bridge between consciousness and unconsciousness, objective and subjective psyche. An unmade eye that "sees" without light, it is a remnant of the animal mind in the human brain—the part of our anatomy corresponding to lead in alchemy—

which may be converted by meditation into spirit, or gold. Psychics who discern auras and other invisible vibrations are presumed to be tuning transdimensional rays through submerged pineal eyes. This radiation is then transferred to the brain, which interprets it in ghostly shapes and images.

The Hypothalamus and Pituitary Gland

THE ANCIENT HYPOTHALAMUS CRADLING THE BRAIN and bearing the pituitary gland (see below) is the anatomical and functional site for the bridging of nervous and endocrine systems. It regulates the internal milieu—coordinating information about body temperature and blood pressure, instigating panting and shivering to alter the flow of blood to different regions, measuring water balance and fullness, and sending out signals of thirst and hunger. The hypothalamus also conducts states upward that we interpret as erotic—stimulation of this region causes monkeys to begin mating.

In other experiments conducted in the mid-fifties a so-called pleasure center was discovered in the hypothalamus, a region in themselves which captive rats chose to stimulate by electrodes. Even when parched or starved the animals preferred this "happy zone" to food or drink, and activated its button until they passed out from exhaustion. Electrical stimulation of other areas of the rat hypothalamus have produced, ambiguously, rage and docility, terror and total absence of fear; so it would appear that it is a mediator rather than an originator of emotions.

WHERE THE ECTODERM of the primitive mouth cavity fuses with the neuroectoderm of the diencephalon floor the pituitary is induced. Anatomically part of the hypothalamus, which lies just above it, the pituitary is a chief regulator of endocrine (hormonal) activity in vertebrates. Functionally a gland, this bilobed organ is fused from two totally different embryogenic layers of tissue (endocrine and nervous) separated by fibrous lamina. The pituitary comprises cells that have simultaneous neural and endocrine qualities. No bigger collectively than a pea, its anterior and posterior aspects protrude from the undersurface of the brain at the end of a twiglike stalk, taking sanctuary inside the saddle of the sphenoid bone like a tiny skull within a skull. The larger kidney-shaped anterior portion derived from ectoderm of the cheek cavity bears the small round posterior node, a nodule of the embryonic brain, in a snug concavity.

The pituitary is the endocrine lieutenant of the brain. Axons flow from the hypothalamus through the pituitary stalk into the organ itself where they synapse in legion. The gland thereby serves as a control station for the biochemical equilibrium of the organism and the coordination of its many functions by specialized cells. It

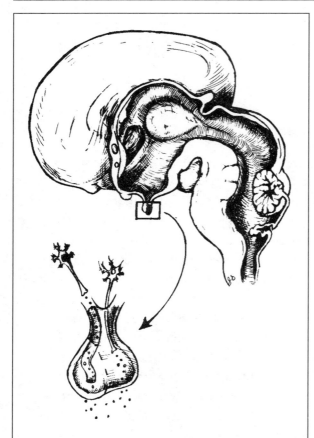

FIGURE 181. Sagittal view of diencephalon with magnified view of anterior and posterior pituitary. Illustration by Jillian O'Malley.

has a similar relationship to the body as the nucleus does to the cell, transmitting code for morphogenetic events. As other elements in the brain stimulate them, hypothalamic neuroendocrine cells secrete peptide hormones into the pituitary stalk. From there cells deliver them either into the gland itself or directly through the main bloodstream to protein receptors throughout the body. Those that end up in the gland do not travel outside it but induce the manufacture and secretion of other hormones which the pituitary then releases into the blood; these secondary proteins then carry esoteric information potentiated in the hypothalamus. (See also the description of glucose and insulin on page 440.)

The anterior pituitary specializes in basic growth hormones which accelerate the transport of amino acids from digested proteins out of the blood into cells where they form new tissue (especially connective viscera); it is primarily involved with the development of the body's frame, its musculoskeletal elements, and the brain itself; pituitary malfunction can result in maladies ranging from incomplete teeth and sparse hair to giantism (excess hormone) or dwarfism (a deficiency of stimulant). Growth hormones also assist in breaking down fats and slowing the catabolism of glucose (hence, keeping it in the blood and increasing blood-sugar levels).

As noted, the pituitary discharges many hormones which regulate the synthesis of other hormones; products of anterior pituitary stimulation include estrogen, thyroid, adrenal cortisol, melanin granules for skin cells, breast-developing prolactin, follicle-stimulating hormone (FSH), and luteinizing hormone (LH). (For a description of the uses of FSH and LH, see the account of the development of sex organs in the next chapter, page 508.)

The posterior pituitary may induce the hormones that initiate and control hibernation in animals and later wake them from their long sleeps. It also discharges sexual, neuromuscular, and metabolic hormones. Its antidiuretic potion helps reclaim water from the tubules of the kidneys into the blood, hence reducing urine flow. The other major posterior-pituitary hormone, oxytocin, is secreted by women at the end of pregnancy, stimulating contractions of the smooth muscles of the uterus and initiating the onset of labor, then sustaining it. Later it stimulates the breasts' glandular cells to trickle milk into their ducts.

In Indo-European esoteric traditions the pituitary has been considered the female element of the brain, the Hindu Radha or Christian Virgin Mary, by contrast with the masculine pineal body. "In the Egyptian mythos, Isis in her aspect as the pituitary body conjures Ra, the supreme deity of the sun [the pineal gland] to disclose his sacred name.... When stimulated by the disciplines of occult philosophy, the pituitary body begins to glow with a faint roseate hue. Little rippling rings of light emanate from it to gradually fade out a short distance from the gland itself. If the stimulation be continued, the emanating rings about the gland grow stronger and a distinct pulse beat is apparent in the flow of the forces. The emanations are not equally distributed, the circles gradually elongating into elliptics, with the body of the gland at the small end. The elliptic extends back from the gland on the side adjacent to the third ventricle and reaches out in graceful parabolas to the pineal gland. As the stream of force becomes more powerful, the luminosity lights the interior of the ventricles, approaching ever closer to the slumbering eye of Shiva. At last tinging the form of the gland itself with a golden red light, it gently coaxes the pineal gland into animation. Under the benign warmth and radiance of the pituitary fire, the 'divine eye' thrills, flickers, and finally opens."[7] The material and terrestrial permute into the spiritual and cosmic.

In the words of Madame Blavatsky, "The arc of the Pituitary Gland mounts upward more and more toward the Pineal Gland, until finally the current, striking it, just as when the electric current strikes some solid object, the dormant organ is awakened and set all aglowing with the Akasic Fire. This is the psychophysiological illustration of two organs on the physical plane, which are the concrete symbols of and represent, respectively, the metaphysical concepts called Manas and Buddhi.... Once the sixth sense [pituitary] has awakened the seventh [pineal], the light which radiates from it illuminates the fields of infinitude; for a brief space of time, man becomes omniscient; the Past and the Future, Space and Time, disappear and become for him the present."[8]

This is an organ functioning not as biological substrate but epiphanic avatar. How much of the body represents occult anatomy and distributes esoteric and supernatural forces will never be revealed by cell biology and conventional embryology.

It is a matter of astrology, sympathetic vibration, microcosmic signaturing, and ch'i flow. Somehow either a series of synchronicities imprints emerging protoplasm, or the organs themselves are as much the outcome of invisible archetypal whorls as morphogenesis.

The Reptile Brain

AN ESPECIALLY CONVOLUTED CLUSTER OF CELLS sends axons from the spinal cord and the thalamus into the cerebellum; this is the inferior olive—it synapses the proprioceptive data of the shoulder girdle and neck. An accessory olive, singularly developed in vertebrates that swim, integrates the wriggle and power reflexes of the trunk and tail muscles.

Phylogenetically, all speculative mesodermal organs had to attract neurons in order to develop functions and thrive. The olivary nuclei are ontogenetically the products of neuroblasts migrating ventrally from the same neural plate as the colliculi of the midbrain and the afferent nerve fibers.

THE BASAL GANGLIA (including the globus pallidus) form to the front and sides of the thalamus; they are the processing center for the discharge of thalamic responses to the cortex. Fibers from this zone branch out to the thalamus, hypothalamus, and the cortex, as well as to the brain stem and reticular formation. Instinctual activity is probably coordinated in this center, including mating, nesting, and territoriality—reptiles and birds have sophisticated basal ganglia.

This is the area of the brain referred to as the R-complex by biologist Paul MacLean; its ambitions remain substantially reptilian—aggressive, ritual, prowling for food, guarding fiercely its status. Within our own higher lobes we inherit, nearly intact, the ganglion of a crocodile, enforcing "its own intelligence," says MacLean, "its own sense of time and space, and its own memory"[9]: pack hunting, hoarding, and boastful exhibitionism. In an experimental test of this supposition MacLean cut into the globus pallidus of an unfortunate squirrel monkey and thereby stopped its ritual displays.

When human beings perform reptile-like acts, it is not (by MacLean's premises) that they are reliving dinosaur memories but that olden brains are continuing to dispatch instinctual signal configurations to higher lobes. The basal ganglia express themselves through our social classes, armies, fashions, and compulsive ordering. The Sioux Sun Dance, with its prayersticks, feathers, beads, and chants; the Saint Patrick's Day Parade, with its flags, floats, and marching bands, are (in MacLean's philosophy) reptile pageants embroidered by gaudy symbols of the cortex. Basal ganglia are capable of such multidimensional extravaganzas when the higher lobes are put at their service. The crowd standing as one and roaring at a great catch for a touchdown reenacts corroborees of prehistoric primates. This expression of cruciality and triumph is truly millennial and cannot be explained through the rationales of the cortex.

"The reptilian brain is filled with ancestral law and ancestral memories," adds MacLean, "and is faithful in doing what its ancestors say. . . . It is not a very good brain for facing up to new situations. It is as though it were neurosis-bound to an ancestral superego."[10]

The Limbic System

THE NEXT LAYER OF PSYCHOANATOMY is the limbic system, which is made up of a variety of structures, including the amygdala, hippocampus, and part of the cortex and olfactory bulb. The amygdala is a bulge imbedded in the temporal lobe of the cortex—a relay center for afferent messages from the motor cortex, olfactory lobes, reticular formation, and other proprioceptive areas. So many sensory signals converge on single cells of the amygdaloid nuclei that it is impossible to guess what impulses and behavior are condensed and packaged there.

The hippocampus is morphologically (and etymologically) a seahorse-shaped section of gray matter folded into the cortex and connected to the greater brain by fiber bundles. Aboriginally, it was the limbic brain, with its origin in the dominant archicortex of early mammals. In humans the hippocampus is our relic shrew or fox brain, a chamber of short-term memory, more concrete and linear than the cortical zones where selected long-term memories are stored (it appears that, long ago, short-term and long-term memories actually overlapped so that the first mammals experienced an eternity of being). The hippocampus contains enigmatic "counting cells" that regularly tap out rhythms of four or five numbers or are activated only when a discrete number of stimuli have occurred. It also includes novelty-recording cells, cells that are silent except when a new stimulus amuses; then they fire once, but not again even when the stimulus is repeated. Such "idea formation" in actual neurons betrays the anatomical basis of some aspects of cognition and temporality.

The limbic system is MacLean's old mammalian brain, not by itself but through its interactions with the hypothalamus and the autonomic nervous system. It is a coordinating center for homeostases of viscera and glands, hence, for primitive emotions, ancient drives, and obscure passions. Fear, anger, and desire are all embodied in their passage through the limbic nodes. We may romanticize the more subtle and personal aspects of these emotions, their cortical refinement, but their essences are chemico-electrical, without premeditation or meaning, as sudden as epileptic fits which arise with inexplicably intense moods on the borders of the limbic system. We usually sublimate, subtilize, or transform our prehuman outbursts, but we cannot purge or inoculate them from their animal quintessence.

According to MacLean: "Affective feelings provide the connecting bridge between our internal and external worlds and, perhaps more than any other form of psychic information, assure us of the reality of ourselves and the world around us. The limbic system contributes to a sense of personal identity integrating internally and externally derived experiences."[11] MacLean somewhat disingenuously ascribes the horror of Nazi Germany to a sudden and irresistible eruption of the limbic system, all the more powerful because the participants did not experience its archaeozoic roots, only its patriotic cortex symbols.

IN THE END we can assert little more than that all levels of the brain embody rudimentary phenomenologies, some of them ancient, some of them newly arising but always in the context of ancient ones. In the emotionalized centers of reptilian higher consciousness and cerebral hierarchies of ascending mammals, it is impossible to know when and how each layer of mind arose, or what emotional and neurological factors in the lives of these creatures served to induce new structures for the brains of their descendants.

The Peripheral Nervous System

WITHIN THE EMBRYONIC BRAIN STEM AND SPINAL CORD, other clusters of neuroepithelial cells proliferate. Some differentiate as neuroblasts, some as glioblasts, and others line the central canal. The lateral walls of this canal thicken irregularly, producing a *sulcus limitans,* a long shallow groove between two lamina—the alar plate in which the afferent functions of the spinal cord originate, and the basal plate which is the source of efferent spinal functions. The cerebellum develops late in embryogenesis from symmetrical bulges of the alar plates which protrude into the ventricle of the forebrain. Wherever limbs form peripherally, there is corresponding development in nerve centers on the spinal cord supplying them.

The peripheral nervous system consists of spinal and cranial nerves connecting the central nervous system with skin and muscles; it is the sum of nervous connections not located in the spinal cord or the brain. The formation of nerves joining the spinal cord to organs and limbs of the body requires neural-crest cells that originate outside the entire structure of the neural plate and subsequent neural tube—primordial ectoderm that separated itself from both the epidermis and neural plate during the formation of the tube and then travelled as mesenchyme. These zooids stream into spaces between the epidermis and mesoderm, between the neural tube and the somites (and even through the somites), and among the rudiments of organs. Some locate behind and within the eyes to become cartilage and ciliary muscle; some contribute to the skull and teeth; some become sheaths of nerves and membranes around the brain and spinal cord (meninges). Others migrate from above the neural tube and cluster segmentally in groups along the spinal cord; these later become spinal ganglia and participate in the formation of the peripheral and autonomic nervous systems. Others become pigment cells (except in the pigmented retina). Still other neural-crest cells are induced into specialized states by regionally developing tissue and incorporated in the design of organs.

Mind incarnates in organs not only as a field potential of individual tissue layers but as a concrete feature of the neural crest's migration. While the nervous system is still being formed, these partially neuralized cells disperse throughout the body and viscera, laying the basis for connective and skeletal tissue associated with neuromuscular complexes. If impregnation could have been achieved without such an extensive odyssey of mesenchyme, then surely muscles and bones would have induced sensations locally in neural tissue. Obviously, comprehensive intelligence requires deep, episodic linkages. Common migratory origins bring distal organs into coordination in stages of sequential neuralization.

NEURAL CONNECTIONS between the spinal cord and peripheral organs are partially induced by the organs themselves. Nerve processes originating in spinal ganglia creep outward until they contact a limb or a gland (at the same time they travel back to the spinal cord), so the transmission of messages from the central nervous system to organs (and back) goes through spinal ganglia. Some of these axons grow incredibly long; nerves will detour around obstacles or leave their customary paths to intercept a limb that has been transplanted. The attraction between organs and axons is so generalized that processes may be drawn into almost any proximal tissue mass, even an irrelevantly transplanted limb bud—an eye grafted onto a torso. Neuromotor awakening follows only when there is a match between nerves and their terminal organs.

The longer neural processes growing out to the skin are afferent. They converge with efferent fibers travelling out of the ventral columns of the spinal cord. Together they form sensory-motor nerves and, as mixed afferent-efferent processes, branch out to different regions of the body. As noted earlier, the afferent fibers bring sensory information from throughout the skin, viscera, and other organs, including proprioceptive messages from muscles, tendons, and joints. The efferent fibers carry impulses back to limbs, muscles, viscera, and glands.

A SEPARATE, AUTONOMIC BRANCH of the peripheral nervous system regulates a series of subliminal activities including blood pressure, salivation, digestion, body temperature, cardiac rate, respiratory rate, dilation of the pupils of the eyes, blood sugar amount, urine excretion (through the kidneys, ureter, and bladder), and erection of the penis—coordinating these functions in a homeostasis of nerves, glands, smooth muscles, and internal organs. It also provides the muscle tone that allows us to sit up and move about.

The autonomic nervous system itself comprises two complementary bifurcations: one sympathetic, the other parasympathetic. In general, those activities stimulated by one are inhibited by the other. The sympathetic accelerates the heartbeat and invigorates the lungs by dilating their bronchi. The parasympathetic sedates these processes, but on the other hand stimulates peristalsis and gastric secretion in the digestive tract and arouses vegetative functions. Between them the systems maintain an internal balance regulated through organs such as the hypothalamus and medulla.

Even before spinal nerves start their migrations, sympathetic ganglia sprout in neural-crest cells flowing dorsally across the neural tube. After connecting in pairs the ganglia amass a twin longitudinal nerve cord which synapses with the spinal column alongside it.

The parasympathetic system originates more obscurely in different clusters of neural-crest cells, some of which migrate as far as the mesencephalon. For a long time the parasympathetic was believed to be a separate autonomic system arising directly in the neural tube or regional mesoderm. Its ganglia do not form chains but are located individually next to glands and muscles. Its paired longitudinal columns lie alongside the spinal cord, with one ganglion each in the visceral efferent branch of the cord and another per column extending to a muscle or gland. A sacral section regulates the lower colon, bladder, urinary and anal sphincters, and genitals, and a cranial one sends branches of axons out to the head and face, even to the lens of the eye where the autonomic system works through ciliary muscles in focusing images.

Sympathetic and parasympathetic functions are not just limited to nerve reflexes; they are sustained neuroendocrine states initiated by hormones and ganglia. They embody the visceral component of the emotions and transmit it to the spinal cord. Although primarily unconscious, the autonomic nervous system is directly responsive to tension, anxiety, desire, fear, and other emotional states (some of which it participates in generating).

THE CENTRAL AND PERIPHERAL nervous systems are a single emanation, medulla and nerve bundles inducing each other right up into the midbrain. The efferent cords of this supersystem wrap around and through the body like an aura, its afferent fibers spiralling into a knot at the brain's center.

The Reticular Formation

ANOTHER UNCONSCIOUS SYSTEM—even more disperse and shadowy—the reticular formation converges cranially as a diffuse core of gray matter, generalized and undifferentiated, running from the medulla and midbrain right up through the thalamus. Fed by fibers from the cerebellum, the colliculi of the midbrain, the hippocampus, and other chambers through which it passes, the reticular formation sends inputs to the thalamus and the cortex itself.

The cells of the reticular formation make up as many as ninety-eight distinct clusters. Its ascending influence is described as a wakefulness, a readiness, a general "take note of." When the reticular formation is stimulated, animals react more alertly. The descending influence of the reticular formation both facilitates and inhibits motor activity (with the lower medulla being the most inhibitory). Lesions in the reticular formation will cause a victimized cat to go to sleep, and it is impossible to arouse the animal no matter how disruptive and discordant the attempt. At best, it will briefly stir.

Apparently without this faint background stimulation we remain unaware of phenomena. Our link to reality is a tuning within the old Chordate brain stem, a wake-up call that tells us anything at all is worth our attention—not only heavy objects bumping into us but the end of a night's sleep from which we startle back to landscape. A change in reticular rhythm alerts us that a mirage created by brain waves has been superseded by another mirage, of external vistas.

The Emergence of the Cerebral Cortex

THROUGH THE EVOLUTION OF MAMMALS the control center has continued to be translated upward into the neocortex, its lobes induced by olfactory rudiments so that its hemispheres "hemorrhaged" out over the rest of the brain. The ascension of the telencephalon is relived ontogenetically: nerve cells migrate up from the other lobes—the more intelligent the animal, the more abundant and encompassing their sheets. By the third fetal month they dwarf and almost cover the diencephalon in a separate layer. In another month the cerebral cortex spreads over the cerebellum, which has also begun to expand. At this point it is smooth.

Suddenly its texture changes, folding and developing irregularities, fractally increasing effective neural surface. The fully developed human cortex is three to four millimeters thick and enfolds most of the external surface of the brain.

EPOCHAL BURSTS OF NEURONS fill the first six months of fetal life, and then subside. On the average, twenty thousand neurons are hatched every second before birth, recapitulating and bridging whole geological epochs. Histogenesis becomes noogenesis.

The general (though not universal) belief among biologists is that these cells cannot divide, so each of us is born with all the nerves we will have. Though 10^4 of them will disintegrate before the person dies, that is still only 3 percent of the total. Glial cells, however, continue to proliferate after birth, their lipid sheets lining the infant's gray matter.

The most crucial postnatal episode in the nervous system is a sudden florescence of dendrites and the inculcation of synapses through the expanding tissue of the cortex. Axons continue to creep outward, their stretched neurofibrils responding to transmitter proteins of synapses. By three months after birth the cortex has begun to convolute more deeply, its cells interacting along thickening new pathways. Nerves from the lower chambers of the brain penetrate the upper lobe, and more and more regions are annexed under cortical control.

THIS MAMMALIAN CORTICAL ORGAN captures the highest functions of the other lobes—coordination from the cerebellum, visual integration from the midbrain, memory from the hippocampus, and creature identity from the limbic system. This is, of course, a misstatement: the cortex does not take these functions away; it incorporates aspects of them, and in so doing, transforms them collectively into a new gestalt. As the cortex redefines all activity at its level, it grows in another way—with deep, silent regions of gray matter, association areas that have no direct motor outputs, no historic functions in mammalian and reptilian lineages. Gray patches which do not project outside primarily interact with one another, processing information, creating imaginal reality. Most of the brain is not in fact linked to anything other than sections of itself. Each cortical zone in us associated with a sensory modality is also surrounded by an orbit of cells specializing in the integration of symbols and ideas. The neurons in these regions do not receive information from external sensors; all they can gauge are representations of the world and symbols generated in other parts of the brain. It is no wonder that a complex self-referential system develops and establishes its own subjective reality. The brain is a sorting device of unknown energies as much as an organizer and regent of sensation.

The Cerebrospinal Hydraulic System

As the human cerebral hemispheres expand like balloons filling with air, they eclipse, one by one, the diencephalon, midbrain, and hindbrain, and collide and spread horizontally, their medial surfaces coming together and flattening, trapping mesenchyme in between. During the sixth week of development, a distinct swelling originates in the floor of both cerebral hemispheres. Bearing this enlarged structure (the *corpus striatum*), the floors of the hemispheres grow less rapidly, which causes the hemispheres themselves to curve. This in turn molds anterior, posterior, and inferior horns within the lateral ventricles of each, remnants of the old central cavity of the Chordate brain.

Meanwhile, loose mesenchyme around the neural tube gathers into a primitive meninx, its innermost cells derived from the neural crest. This will become the three meninges of the cerebrospinal system. While the outer layer of the meninx thickens into *dura mater*—a viscid, unmalleable tissue fused to the internal aspect of the skull—the inner layer remains thin and pliant; it will develop into the *pia mater* and the arachnoid membrane, which will fill with cerebrospinal fluid. The vertical aspect of mesenchyme trapped in the fissure of the cerebral hemispheres becomes a membrane separating them—the *falx cerebri*.

The *dura mater* (or dural membrane) comprises, vertically, the *falx cerebri* and the *falx cerebelli* (intervening between the cerebellar hemispheres); and, bilaterally, the horizontal sheets of the *tentorium cerebelli*, which partition the cerebrum and the cerebellum from each other. The *dura mater* is the hydraulic holding bladder for the cerebrospinal fluid, hence its pressure valve.

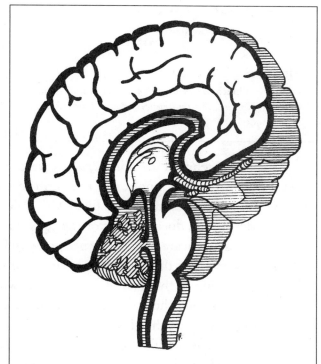

FIGURE 18J. The neural pump: cortex, midbrain, brain stem.

From Stanley Keleman, *Emotional Anatomy: The Structure of Experience* (Berkeley: Center Press, 1985).

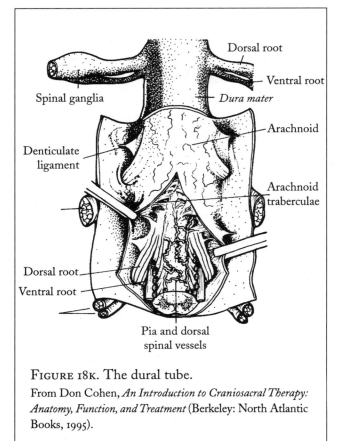

Spinal ganglia

Dorsal root

Ventral root

Dura mater

Arachnoid

Denticulate ligament

Arachnoid trabeculae

Dorsal root

Ventral root

Pia and dorsal spinal vessels

FIGURE 18K. The dural tube.

From Don Cohen, *An Introduction to Craniosacral Therapy: Anatomy, Function, and Treatment* (Berkeley: North Atlantic Books, 1995).

"The two layers of the dural membrane are tightly attached except where venous sinuses are formed. The outer layer is attached to the inner surface of the bones which form the cranial vault.

"At the sinuses the dura separates away from itself and from the bone. [This gap] affords space for the collection of blood and then adheres to the dura from the opposite sides of the sinus to form either a falx or the tentorium. It is this endosteal [membranous] contribution of dural membrane to cranial vault bone which enables [an osteopathic physician] to use these bones of the cranial vault to diagnose and treat the intracranial membranes. The dural membrane forms the functional, if not the strict morphological boundary of the hydraulic system."[12] (For a further discussion, see Chapter 24, "Healing.")

The arachnoid membrane, a soft, vascularized component, is insulated by subdural and subarachnoid spaces from the *dura mater* external to it and *pia mater* within. As these cavities are filled with fluid, the arachnoid membrane floats independently and does not mirror the convolutions of the brain.

Itself a delicate membrane packed with blood vessels and conveying blood, the *pia mater* winds through the labyrinths of the brain and spinal cord, parallelling and adhering to them and to nerve roots which it enwraps.

By this time the medial wall of the cerebral hemispheres has thinned along a groove known as the choroid fissure. Originally continuous with the roof of the third ventricle, the fissure migrates to its medial wall. With expansion of the hemisphere restricted rearward, its caudal pole twists down and forward, carrying the ventricle and choroid fissure with it and forging the inferior horn noted above. Invaginated by the vascular *pia mater* during the third month of pregnancy, the medial wall shapes the choroid plexus and extracts cerebrospinal fluid from surrounding blood along the lateral and third ventricles of the brain.

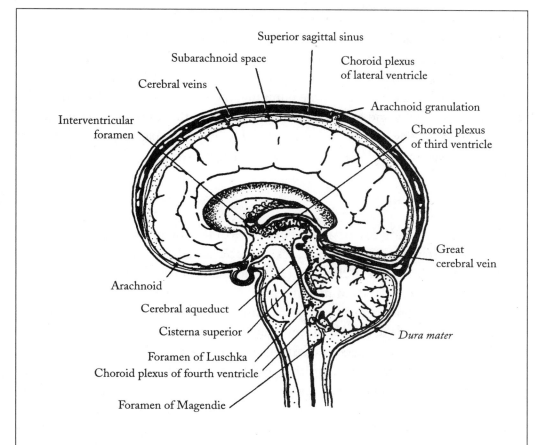

Superior sagittal sinus

Subarachnoid space

Choroid plexus
of lateral ventricle

Cerebral veins

Arachnoid granulation

Interventricular
foramen

Choroid plexus
of third ventricle

Great
cerebral vein

Arachnoid

Cerebral aqueduct

Cisterna superior

Dura mater

Foramen of Luschka

Choroid plexus of fourth ventricle

Foramen of Magendie

FIGURE 18L. Aqueducts, foramens, plexuses, ventricles, and other structures involved in cerebrospinal-fluid flow.

From Don Cohen, *An Introduction to Craniosacral Therapy: Anatomy, Function, and Treatment* (Berkeley: North Atlantic Books, 1995).

It is the folding of the *pia mater's* blood vessels into the brain's ventricles that secretes components of the choroid plexus' blood (lacking red and white cells) as cerebrospinal fluid. Only certain of the blood's constituents are allowed through the walls of the plexus; large molecules and those with undesired electrical charges are excluded. The arachnoid villi, a thin layer emanating from the arachnoid epithelium and the endothelium of the sinus, then returns the blood to the venous system.

The cerebrospinal fluid (CSF) is sprayed with some force as if out the head of delicate jets. It pours over the brain, the brain stem, down the spinal cord, even apparently seeping into other aspects of the neuromuscular system. For instance, evidence of CSF has been found among collagen helices.

The outflow and return of cerebrospinal fluid, pumped by the brain's pulsations, establishes the cohesion, vitality, and ceaseless transitional quality of neural life. This craniosacral motion is a "rhythmic, mobile activity which persists throughout life ... in man, other primates, canines, felines, and probably all or most other vertebrates.... [Its normal rate] in humans is between 6 and 12 cycles per minute."[13]

During the extension phase of the pumping, the entire body, including the skeleton, rotates externally and broadens ever so slightly. In the immediately subsequent flexion phase the body narrows, rotating internally. Thus, the passage of fluid through the nervous system has global effects which for the most part go unnoticed.

Eyes

AT AN EARLY STAGE OF HUMAN NEURULATION the forebrain develops evaginations of its own—two lateral sacs, optic vesicles from which eyes will form (see the description earlier in this chapter). These clumps are induced from the substratum by adjoining islands of mesenchyme. Mushroom heads on stalks swell laterally while their connections to the forebrain shrink. The vesicles now induce the adjacent ectoderm of the head into lens placodes. A deep telescoping follows. From the center outward each placode collapses, creating a lens pit; the boundaries of the placodes surge forward into each other and fuse as lens vesicles. The lens at this point is a canopy of epithelial cells, one cell-layer thick, around a central cavity. Those cells facing backward toward the retina elongate into fibers, synthesize crystallin proteins, and deposit them in their cytoplasm. As their nuclei are degraded and protein synthesis terminates, a thick refractile body forms, composed of long, lifeless cells wedged against one another. Although very little of the adult version comprises exact molecules that were deposited in the embryo, the rear of the lens retains some of its original embryonic components with no later turnover of those contents.

The optic vesicles also invaginate; they become double-layered optic cups. The outward-growing cups will house the lenses. Their outermost layer bears a pigment epithelium, which forms continuous with the pigment epithelium of the ciliary body. The lens epithelium looks out into the world, a thin layer of low cuboidal cells proliferating rearward. The lens grows as they boost their production of crystallines and differentiate into fibers (at a slackening rate throughout the life of the organism). Variations of the refractile index between the earliest embryonic crystallins and subsequent ones enable the eye to self-correct the types of optical aberrations that form in more homogeneous lenses made of glass.

The inner layer of the optic cup differentiates as a thick neural zone and, induced by the lens, teems with neuroblasts. This becomes the embryonic image-forming surface, the retina. Oddly the neurons that conduct retinal information into the brain lie external to the eye's light receptors, so luminosity and color must pass through them first to form the image they then receive and relay.

The cells of the retina further particularize into photoreceptor cells, rods and cones, and bipolar and ganglion cells. The axons of the ganglion cells travel into the inner wall of the optic stalk, making it a single long nerve. The cones form images in daylight; they are receptors of colors. The rods pick up shapes in dim light. They each comprise their own distinctive networks of protein with visual pigment. Though these cells do not divide, their photosensitive protein molecules are regularly replaced. Old membrane layers of the rods are cannibalized by cells of the pigment epithelium and digested, as new layers flow steadily outward from a site near the nucleus.

The ciliary body is mostly a forward projection of the non-neural part of the retina, whereas its muscle is derived from mesenchyme beside the optic cup. The eyelids develop from folds of ectoderm bearing cores of mesenchyme.

The lens vesicles meanwhile separate completely from the ectoderm and sink into the optic cups. The contractile membrane, the iris, is induced from the lens-covering rim of each optic cup; its connective tissue is mesenchymal. The dilator and sphincter muscles of the iris differentiate out of neuroectoderm on the optic cup.

The long cell columns of the lens are induced by vascular mesenchyme; the lens itself encourages the ectoderm over it to become the cornea, which refracts light through it onto the retina. The vascular layer of the eye, the choroid, and the delicate membrane beneath the cornea, the sclera, both braid from mesenchyme surrounding the optic cup. The sclera is a direct extension of the dura of the spinal cord and brain.

The lens-ciliary muscle system, developed in darkness before birth, controls the focus of images on the retina, while the iris expands and contracts the lens at the pupil to accommodate decreased and increased amounts of light. Lying directly in front of the lens, the circular fibers of the iris can reduce the pupillary membrane to a minute pinhole when external brightness would otherwise make image formation impossible.

By imitating these principles we have invented cameras and light-magnifying instruments, a chronology Sherrington reverses to show us how miraculous it is:

"If a craftsman sought to construct an optical camera, let us say for photography, he would turn for his materials to wood and metal and glass. He would not expect to have to provide the actual motor adjusting the focal length or the size of the aperture admitting light. He would leave the motor power out. If told to relinquish wood and metal and glass and to use instead some albumen, salt and water, he certainly would not proceed even to begin. Yet this is what that little pin's-head bud of multiplying cells, the starting embryo, proceeds to do. And in a number of weeks it will have all ready. I call it a bud, but it is a system separate from that of

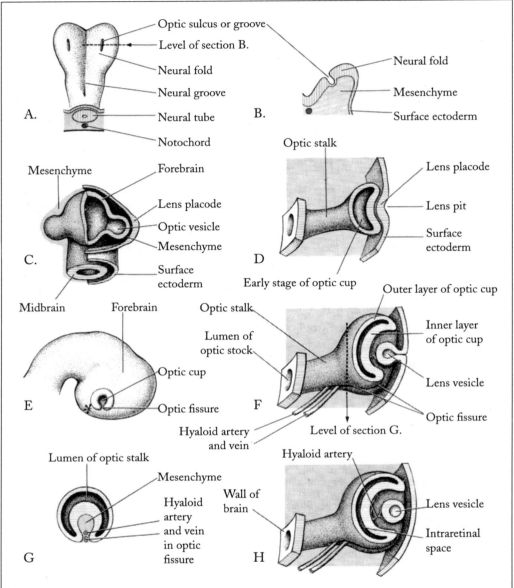

FIGURE 18M. Early eye development. A. Dorsal view of cranial end of embryo at about 22 days, showing the first indication of eye development; B. Transverse section through an optic sulcus; C. Forebrain with mesoderm and surface ectoderm at 28 days; D., F., and H. Development of optic cup and lens vesicle; E. Lateral view of the brain at 32 days, optic cup manifesting externally; G. Transverse section through the optic stalk, showing optic fissure and its contents.

From Keith L. Moore, *The Developing Human: Clinically Oriented Embryology* (Philadelphia: Saunders College Publishing, 1977).

its parent, although feeding itself on juices from its mother. And the eye it is going to make will be made out of those juices. Its whole self is at its setting out not one ten-thousandth part the size of the eyeball it sets about to produce. Indeed it will make two eyeballs built and finished to one standard so that the mind can read their two pictures together as one. The magic in those juices goes by the chemical names, protein, sugar, fat, salts, water. Of them 80% is water."[14]

The Cerebral Hemispheres

THE CORTEX IS DIVIDED DOWN ITS MIDDLE by a longitudinal sulcus. The hemispheres emerge as mirrors of one another, reminiscent of twins forming from the same blastula. We have two brains: a left one and a right one. The left hemisphere of the brain connects to the right side of the body, usually the dominant one; it is supposedly engaged in rational and analytical thought and is thus empirical and critical. The right hemisphere is said to be more intuitive and creative, identifying patterns and forming images; it connects to the left side of the body.

The twin hemispheres of the brain are by no means exclusive in what they do; they have homologous topographies and, to a large degree, duplicate each other's functions. There are eyes and ears in both, albeit connected to opposite halves of the body.

Andrew Weil calls the the left and right brain "symbolic designations of the two phases of mind."[15] He laments that certain schools of creative training now prescribe binding the right arm so that the intuitive hemisphere is forced to develop. This kind of "New Age" scientism is a false literalization of a far subtler and more complex dichotomy.

CEREBRAL HEMISPHERES ARE CONNECTED by the *corpus callosum*, a long band of fibers like the zone joining the hippocampus to the neocortex. Made entirely of white matter, this commissure is bent double in front and curved ventrally. Material continues to be added to it as the cortex expands. The basic line of communication between the hemispheres, the *corpus callosum* fuses their dual realities. When it is cut, one side of the brain does not see objects presented to the other side. This isolation of landscapes is a natural occurrence in animals who do not have a band of connecting fibers.

Not all regions of the two hemispheres are homologous. The speech center appears in the dominant hemisphere of the brain only, which is, in most people, the left region (coordinating the right side of the body). Thus, when the *corpus callosum* is damaged, a person may not be able to name an object presented to the left eye (for instance, a spoon) but still may use it properly. In the portion of the nondominant hemisphere corresponding to the speech center no function has yet been discovered.

A blood arterial system runs over the surface of the brain and through its interior, diffusing oxygen and glucose, and carrying away waste. Although the brain is only 2% of the body, it consumes 20% of its oxygen in adults (50% in infants). Collective cerebralization in the biosphere involves a massive expenditure of the Earth's resources.

The moment this torrent of blood from the heart ceases or is poisoned, consciousness ends. Our mind is as fragile as breath and as physical as fog over a lake.

The Cerebral Lobes

THE CEREBRUM IS ALSO DIVIDED into four highly convoluted lobes which arise (at least in part) from their own internal hydraulics and become separated from one another by fissures. Cutting connections between these regions in monkeys, brain scientists have deduced approximate functions for each of the zones. The parietal zone, the most developed at birth, coordinates motor input and output and the senses. The occipital lobe is a region of visual reception. The frontal and temporal lobes are associated with speech, learning, memory, and symbol formation. Numerous discrete processes originate here—representation of objects, naming, simple classifying, higher orders of classification, and philosophical abstraction from classes. Just for the evolution of primitive speech, all of these symbolic functions must be coordinated with raw sensory data and motor control of facial, thoracic, and other muscles.

The developmental sequence of neuron proliferation, convolution, fissuring, segregation, and coordination is a cumbersome choreography for phylogeny to impart to ontogeny, and an ontogenetic marvel to abbreviate and reconstruct it without losing congruency or function en route to mindedness.

Consciousness

OUR KINESTHESIA AND PROPRIOCEPTION incorporate all the internal connections of the neuralized layers and lacunae within us; the coalescences of quantal sensations from cells and subcellular topologies; the interactions of uncountable surfaces, subsurfaces, and milieus (shallow and deep); and the density and hollow of unneuralized tissue. Signals from fluids, membranes, viscera, and zones of the body fission and fuse. We may act as though mind were cerebralized only, but separate of organelles and organs, a brain is an abstraction without living experience.

"We are not flesh with a spirit or genetic code dwelling in us," declares Stanley Keleman. "We are an event that sustains a particular life style. We are not a machine with a mind or with a spirit. We are a complex biological process that has many realms of living and experiencing ... a layered, ecological environment of ancient and modern lives...."[16]

With every breath we foster textures of spleen and gall bladder, tension contours of sphenoid and sphincter of Oddi, tight spiral rotations of lungs and kidneys, vibrations of bones and ligaments within viscera, numbnesses of epithelia no longer accessible. Each realm of tissues develops its own proprioception and, at the same time, contributes to the collective proprioception of the organism—the movement and texture of blood, lymph, and cerebrospinal fluid; the density, granularity, and placement of liver, stomach, and lungs; the photoreceptivity of the retina; the structure and leverage of bones and cartilage. "Muscle gives rise to sensations of rhythm, containment, holding, releasing, shortening and lengthening. Bone introduces sensations of compression and pulling. The intestines produce sensations of swelling, fullness, and emptying. The uterus, like the heart, is an empty space surrounded by dense, rhythmical tissue. The abdomen is a central cavity containing fluids and organs surrounded by bone and muscle. The lungs and heart are organs which are contained by a rigid wall. Thus hollow, soft, and dense tissue produce different sensations and feelings."[17]

As layers and organs dynamically fuse and interact, the overall proprioception of existence deepens and the complexion of the organism changes.

"The pump of the internal viscera, and the neural hormonal pump ... [create] the pressure that organizes body spaces to maintain their structural integrity. This pressure also reflects an internal state and generates the feelings that we recognize as ourselves.... There is a dialogue of sensations from hollows to solids, from liquid chambers of the brain to densely packed muscle cells. This overall relationship generates a basic tissue state that forms a continuous pattern of consciousness."[18]

For all its range and power the brain never contacts actual objects or things, never tastes real food, never hears real songs. Everything is conveyed to it as signals, code. We never imbibe perfume or feel the body of a lover. What we experience are millions of axon relays and synapses from throughout our physiology, pings of chemicals and pressures against sensors which neurons translate into electrochemical charges and deliver to the cerebral cortex which interprets and reassembles them into images and events we honor as the real McCoy. Passions are expressed, as bodies seamlessly change chemistry and sink into imageries as deep as dreams.

It is amazing that proxy works as well as it does, that we believe it and accept it as completely as we do. The real seems, in fact, truly real, and we are moved to respond deeply and unconditionally to it. In truth, we wander from trance to trance, self-hypnotized into a drama of events.

A neural ghost continues to revisit us. A child after suffering brain damage suddenly considers his parents impostors. A man with Tourette's Syndrome walks down the street telling strangers to "fuck off." After a stroke a woman laughs uncontrollably. Neurologist Oliver Sachs made a legendary tale out of a patient who one day mistook his wife for a hat.

People whose arm or leg has been amputated often feel the presence of their phantom limb, sometimes experiencing jolts of pain in it. These organs existed in the brain so long that homunculi of them continue to glow there even in their absence.

Yet such homunculi are all we know of each other or even ourselves.

WE LIVE AS WELL in a larger sensorium that passes through our boundary. Within moist tissue we feel the mass and electromagnetism of our own planet; the equilibrium of Earth, Moon, and Sun; and, to some degree, the minute but profound gravity of the universe through the weft of the Solar System. We perceive subtle pressures of underground water, vibrations of invisible radiations, and a diurnal-nocturnal polarity imbedded in the circadian rhythms of our bodies. We also inherit instinctual complexes hard-wired into our cortex—senses of number, courtship, sexual jealousy, mate-selection, child-rearing, appreciation for inequality in social interaction ... fears of dangers long past, love for things no longer incarnate. To one degree or another these are phenomenologies from crises of survival in ancestral pre-primate environments.

And then we have possible "extrasensory" senses, still evolving or vestigial.

In inexplicable episodes a person may suddenly speak a language he never learned or shoot images from his "mind" directly onto photographic film. The literature of parapsychology is replete with paradigm-shattering feats, events that seem to violate laws of thermodynamics, suggesting intelligence without a body and transfer of information at speeds greater than the physical limit of light. Such phenomena are doubted or vehemently denied by most scientists, but until their apparitions are explained, all laws of mind or matter are "patent pending." We do not know in truth what either mind or matter are, what separates their domains, or how they marry in the cortex.

More "normal" talents corroborate the depth and multiplicity of the human brain: the concertos of Bach, the relativity theory of Einstein, the lines of perspective in Leonardo da Vinci's *Last Supper,* and the discursive myth cycles of South American Indians. The incredible memory and calculating capacities of so-called "idiot savants" attest likewise to both the complexity and mystery of the brain.

When people memorize an entire page of a telephone directory at a glance or routinely multiply seven-figure numbers by each other with flawless results and (accounting for leap years and changes in the calendar) instantly name the day of the week on which an event occurred centuries earlier, it becomes clear that the sheer unexplored depth of the neural complex outstrips even those remarkable aspects that are ordinarily manifested.

Redundancy of Cerebral Tissue

THE CELLS OF THE CEREBRUM are neither as predetermined nor regionally distinct as brain mapping seems to suggest. Experiments have shown something else—brain tissue (echoing the blastocyst) is neurally equipotent before it is induced and regionalized by interfaces and pathways. After injuries and strokes it regains embryogenic flexibility and can take on new functions. Abilities learned in one region of the cortex can be transferred to another intact, even as memories are passed from one molecular cache to another throughout a lifetime.

Our enormous redundancy of tissue shows up in many ways. Long-time victims of epilepsy learn to skip damaged portions of the brain and retain their memories—from current events to virtuoso violin skills and how to solve algebraic equations. If the impaired tissue is removed, there is not even noticeable diminishment of knowledge or change in personality. During the 1950s Israeli physical therapist Moshe Feldenkrais made use of this capacity of the brain, developing methods of restoring the former activities of stroke victims by teaching them to re-route enterprises through different relays, to arrive at old behavior by new neural paths.

The duplication of cerebral data and skills has become an implicit justification for treatment of mental disorders by electric shock and lobotomy. If disturbing associations and emotions can be ablated without notable loss of memory and function, then phobias and depression might, in a sense, be mechanically extracted from personalities. The only noticeable side effect from shock treatment is a subjective one—less profundity of character, more stereotyped and flatter emotions. It is as though the intention were to push a human being a notch down the evolutionary ladder, not far enough to become a different animal but enough to no longer participate in the more painful anomalies of the human condition.

Psychotropic Drugs

THE OVERALL NEUROENDOCRINE COMPLEX of biochemistry, phenomenology, memory, early development, trauma, and social interaction is insolubly entangled. At times it would seem that the old Greek humors totally dominate the phantom of mind. We wander in mazes of our own mysterious motivations and behavior, apparently instigated or at least sustained by unconscious events. Mothers suddenly take a gun in the middle of the night and hold it to the heads of their beloved children when they are asleep, slaying them in the paranoid delusion that they are protecting them. Other persons become convinced that they are Napoleon, or Jesus of Nazareth, or that a movie star is in love with them, or that they are being followed by aliens or government spies. More ordinarily, people contend with outbursts of self-destructive rage, forfeiting their jobs and families in petty incidents with virtually no compensatory gain. Others sink into incurable depressions despite fulfilling lives. Still others become phobically attached to familiar surroundings or terrified of certain situations like being in elevators, finding oneself too close to a snake or spider, or looking down from heights.

Once upon a time these were considered real events, existential crises on the road of life and death. Shamans, priests, and psychologists tried to heal them and the meanings they expressed. Then they were either karmic echoes or spirits. Now they are mere hereditary or hormonal flaws.

The growing tendency to practice psychotherapy as a branch of pharmacy demonstrates the degree to which our epoch allies disingenuously with the physical side of the mind-body paradox. The argument of neurotransmitter determinism suspiciously parallels that of genetic determinism. When professionals encounter what they view as dysfunctions of behavior, they inevitably devolve to viewing them as chemical lesions, presuming that, at the bottom of this affair, there is nothing more than atoms, molecules, and chemical reactions, mimicking existential reality. "Clinical psychopharmacology ... threatens to virtually replace a psychology of experienced self and affect with what amounts to a psychological equivalent of the reductionistic sociobiological position in which we are encouraged to believe that the vicissitudes of neural transmitters are the only dimension of relevance."[19] If our existence is solely a chemical event, its deepest meanings can be adjusted, reversed, or excised by pharmaceutical substances.

For instance, what were once diagnosed by post-Freudian analysts as behavioral pathologies, neurotic effects of childhood traumas, and developmental narcissisms are now considered hereditary "deficiencies in hardwiring and neuronal control mechanisms ... [or] temperamental hyperactivity of a hardwired, serotoenergetically based shame system,"[20] sometimes inducing agoraphobic compensations that evaporate after psychopharmacological treatment of the primary disorder.

Idealists counter that the psyche has an independent existence transcending any apparent cerebral locus or molecular tropisms; its core cannot be reached by drugs.

To antidote materialist orthodoxy with a view that mind must rule over and impress its own nonphysical stamp on body is an extension of dualism, a mirror image of reductionist materialism. In truth, biochemical events generate neural ones, which elicit phenomenological ones. Neurotransmitters are morphogens, and morphogens induce neurotransmitters (along with other psychosomaticized peptides and organelles). Mind and matter, thoughts and cells are coeval vibrations within a gradient of dynamic form. They embody each other. Most mental disorders are hybrid genetic, hormonal, neural, linguistic maladies. They come into being only when disparate and myriad flows of sensation cannot be organized in a way that allows satisfying social behavior and/or a sense of well-being and ego stability.

There *are* atoms and molecules at the basis of all thoughts and emotional behavior—neurotransmitters and inhibitors—and these can become locked with neuromusculature in self-destructive cycles reinforced by the autonomic nervous system and long-term habitual feedback loops.

Depending on their specific hormonal/emotional expressions, some cycles can be arrested temporarily by antipsychotic, antimanic, or antidepressant medications such as imipramine, Valium, Prozac, Neurontin, Zoloft, and the like. Their roles in either blocking neurotransmitters or augmenting the body's serotonin output may not be that different from the roles of enzymes in suppressing or encouraging the transcription of particular genes. Once again expression and its negation at a subcellular level are hopelessly entangled with form and meaning at a cognitive level. No discrete thread or substance generates reality.

Psychotropics are stop-gaps more than whole-system resolutions, for life (at any stage) supersedes reduction to *any* of its components.

Mind and body remain irreconcilable. Psyche is physically tethered yet multidimensional; mind is neuromuscular yet transpersonal and collective. Tragedy and comedy (like anxiety and depression) are based in chemistry but also epiphenomenal. This is why Oedipus' and Hamlet's dilemmas cannot be solved by a mere chemical dosage, and why "Hector . . . in Ilium, far below,/ . . . fought, and saw it not—but there it stood!"[21]

Suppression of Consciousness

IN AN AGE OF BIOMECHANICAL DETERMINISM we have elevated the cerebral cortex to the explanation for ourselves. An abundance of neurons alone seems to have made us more complex, more reflective, and more compassionate than animals, though we are still quite lacking in these attributes even by our own standards. So vivid is our image of higher intelligence as convoluted cortices that we invent higher beings, science-fiction creatures from our future or from higher dimensions and other worlds, who have swollen crania, new lobes bulging out of their neocortices, and thus are more intelligent and civilized than we.

Whether our problems could be solved with additional folds of neurons is impossible to know. I doubt it. We have developed our culture and technology from the Stone Age without accessory neurons (and perhaps with somewhat fewer than Neanderthal and Cro-Magnon). We have no reason to believe that an increase in the mass and complexity of the cortex would lead to intelligence beyond ours. Much of our brain is already unused. Perhaps becoming conscious in the way we have is a plateau we cannot transcend along the same lines.

We should not fantasize that conventional mutations could give us powers that would civilize us further or prevent our self-destruction; genes are not issue-oriented. Additional signals might have driven us mad.

One of the main roles of ganglia in achieving personality has been that of limiting consciousness. Like gene expression, intelligence seems to be a process of selecting aspects of universality, then cultivating them by channelling them into contexts from which they derive meanings. Our mind is a kinesthetic whole, created equally by excision, suppression, amnesia, and cognizance; a great portion of its energy goes to inhibit not to increase consciousness.

While modes of survival dependent on cerebral function require neuralization and substantial reserves of tissue (hence, repetitive structure), apparently the self cannot allow simultaneous expression of too many of these complexes or its coherence evaporates into senselessness.

WE TRANSCEND THE RIGIDITIES of insects and dinosaurs; yet we carry out many of their dread and mindless missions. In fact, we raise their mindlessness to the level of mind where we embroider it with the phantasms of civilization.

We have translated sensations into signs, signs into symbols, and symbols into artifacts in a mode unlike any previously on this world. We have made over the planet from the inside of our brain. But we have not made over the laws of nature, so we remain epiphenomena, tied to a physical evolution we cannot transcend.

Organogenesis

Our organs have their own animal identities.

WE ARE ALL MICROENVIRONMENTS. The viscera of our bodies—wrapped in membranes, suspended by folds and fibers—float on branches of state-of-the-art coral in a Precambrian sea. Though it is sheer folklore to believe that vertebrate organs are invertebrate animals (as some early recapitulationists did), these tissues clearly retain rudimentary invertebrate function and sentience. After all, they are programmed by the histories of their predecessors, and their development in embryogenesis is an interpolation of systems they already comprise. Our cells specialize into organs because their forerunners carried out the similar, successive metamorphoses in ancestral creatures in our lineage. None of our internal milieus could exist without a substratum of templates going back to polyps and amoebas.

"The web between our fingers, the membranous *dura mater* and esophagus, the suspiciously protozoan curve of our brains and viscera that lie pulsating in water, are vestiges of ancient worlds here before we were,"[1] notes movement teacher Emilie Conrad. Traces of departed animals have woven together in federations of their offspring, their depleted genomes infiltrating one another and overlapping, forming new animals.

Lymphocytes lurch and sprawl through our fluids as if free-living placozoans, engulfing invaders. Neural-crest cells and mesenchyme migrate through skeleton, cartilage, fluids, and nerves to individuate organs. Formative heart gel within our gastrula reverberates with throbbing heliozoans. Our lungs are colonies of pseudophoronids and quasi-bryozoans; our genitals, pudgy tunicate crabs. Our gut embodies ctenophores; our intestines wind in ribbon worms and entoprocts, their stalks of villi swaying in dense digestive currents, feeding as the pedicellarias of sea urchins do.

Imbued with nerves and blood vessels, lined with muscle and skin, permeated by fibers connecting one to another, our urinary bladder, pancreas, kidneys, and spleen are clams and chitons, swallowing and metabolizing. As they convert phosphates, we discharge their sludge in caterpillar-like sensations.

The simultaneous fashioning of the many once-independent digestive, respiratory, excretory, and reproductive subanimals from mere series of creases and folds in the gastrula, as well as their integration with one another and with neuromuscular and circulatory tissue and fluids, is a feat of embryogenic fusion and function.

As we have seen repeatedly, relatively terse amounts of hardcore data and genetic architecture deviate, in the context of mutually inductive fields, into radically errant organs. Urinary bladder and brain are "failed" hearts—bags of folds in different positions in organismic fields. Teeth, tusks, and bones are hardened concrescences; the sphenoid behind the face is a small, elongated pelvis; the skull a second, aborted body. Hair, glands, skin, nails, and even eyes are variants of ectodermal buds plunging downward and contacting mesoderm. The fine detail, fractal texturing, protein-secreting specialties, and metabolic collaborations of organs are a result of the exquisite and iterative subtlety of their fields and subfields rather than exhaustive amounts of initial heterogenetic detail. The Earth has devised one program—one wheel—for zoological assemblage and diversity.

"We are a process of millions of years of an open-ended experiment," adds Conrad. "Our forms have been designed and redesigned, unendingly adaptive and innovative. Chemical codes alone determine whether we will have a snout or a nose."[2]

Resemblances and homologies thus express historic, histological events—familial lineages and functional topologies—the unique phylogenesis of each organ obscured by veils of ectoderm and fasciae that hold them together and ganglia, blood fluids, and other tissues that fuse and obliterate the meanings of their independent origins.

Cell Differentiation

A UNIVERSAL BLASTULA ORGANIZES ITSELF into regions by redefining the contexts of its separate cells. This process gains momentum through gastrulation, designing a basic body-mold for each species.

At the basis of all elaboration are autonomous zooids. These stem cells have no differentiated function except to reproduce more of their kind, but they carry enormous potential detail and patterning locked within their DNA packaging.

Inherent competence is not so much a matter of the genetic make-up of a cell as it is of the activating of its nucleic component through inductions. As long as a cell retains its full complement of DNA somewhere in its nuclear maze, it has the theoretical capacity to be reprogammed back to totipotency—something that does not occur either randomly or often. Competence is mostly lost as cells continue through development, each induction further narrowing their potential. There is no pan-biological starting point for this diminishment of competence; in some species capacity is fundamentally reduced in each cell at the two-cell stage.

ALTHOUGH INDIVIDUAL CELLS within a particular region remain equipotent to a greater or lesser degree (and, if transplanted, can form radically different structures in accordance with their new locales), they otherwise become fated by position: the contexts of locally emerging fields of influence and a larger field—the phylum and order—of the creature itself.

During blastulation and gastrulation some cells may become predetermined for a particular class of tissue while retaining generalized potential within that class. Epidermal stem cells manufacture fresh layers of skin; muscle satellites fission into replacement skeletal muscle; and spermatogonia yield generations upon generations of spermatozoa.

IN ORDER TO form tissues, cells respond to genomic regulation in two ways: they proliferate by simple mitosis, or they differentiate from their parent cell. Proliferative cell cycles lead to aggregations of the same kind of cell. Some of those may divide henceforth without limit, at least during the lifetime of the organism; others may become terminal.

Conversely, quantal cycles particularize cells. Such cells may then continue to fission in their new state, or they may also lose that ability (like red blood cells, primordial lens cells, and bone) and spend the rest of their existence performing a specialized function. The relationships between these two cycles are ancient and evolutionarily deep. They are propagated by zooid-organelle interactions, ramified by generic forces, selected through environmental/metabolic nexuses, and preserved and altered by chromosomal indexing and mutations. They gradually become organs.

Tissue and Organ Differentiation

THE HUMAN NEURULA EMERGES, a swelling dervish with sphincters marking its entrances and exits and pumping liquids and molecules to fuel its metabolism. "From this pouched tube," somatic therapist Stanley Keleman tells us, "will [later] develop the various compartments—head, chest, abdomen-pelvis. At the pelvic end, that area where end products are transformed, the genitals, anus, bladder and legs develop. At the other end will form the mouth and entrance for the major senses as well as the breathing tube. In the middle will begin pouches of transformation and inner circulation—the heart, abdomen, viscera. Rings of separation between the pouches develop into diaphragms, separators, sphincters."[3]

The original, asymmetrically triplicate layers of tissue (outer, inner, and middle) participate mutually in the induction of organs in one another and, in many cases, fuse to form joint organs.

Ectoderm generates the central and peripheral nervous systems, most of the epidermis with its hair and nails, some glands (including the mammary and pituitary), and the enamel of the teeth.

Endoderm is primarily gastrointestinal; it forms the epithelial lining of the main esophageal and respiratory tracts, tonsils, liver, pancreas, part of the bladder, parts of the ears and tympanic cavities, and various glands (including the thyroid, parathyroid, and thymus).

Mesoderm thickens the body and holds it together with connective tissue, cartilage, bone, muscles, heart, blood and lymph, kidneys, gonads, spleen, and various membranes lining body cavities. The muscle and connective-tissue layers of the gut and its derivatives (and of many otherwise ectodermal organs) are mesodermal. The intraembryonic coelom, arising as islands within the lateral mesoderm, grows into a horseshoe-shaped cavity and splits the mesoderm into two cosmogonic zones—a somatic layer extraembryonically continuous with the amnion and a splanchnic (visceral) layer which extends over the yolk sac. The latter provides the nonendodermal component of the organs of the digestive tract.

In small animals (like houseflies) in which the surface of the body rivals its volume, the exoskeleton coordinates development and serves as a positional compass for the orientation of organs. However, in vertebrates, internal connective tissues not only supply the supporting architectural framework but orient and guide overall pattern formation. The human plan is basically mesodermal, following a prototype set by the lamprey and shark.

REGARDLESS OF THEIR LOCATION, the body's tissues share certain requirements. They all rely on the extracellular matrix secreted by fibroblasts to maintain their structural framework. They are all permeated with blood vessels insulated with endothelial cells to supply nutrients and evacuate waste matter. Most of them are innervated by axons cloaked in Schwann cells. Their melanocytes contribute pigments for protection and decoration. Tissues also house macrophages to clear their debris, dead cells, and excess matrix, and lymphocytes to guard against invasion. Fresh sanitizers continue to infiltrate exogenous tissues throughout adult life.

Except for cells carrying out an intrinsic local function, the rest are migrants that have arrived from other zones of tissue during embryogenesis and continue to maintain some indigenous qualities throughout a lifetime abroad, ultimately passing their collective cell memory onto their progeny. The motley consistency of most tissues is a result of an intricate admixture of cell types, weaving lives and functions to form them. Bodily organs are not "things," but curdled nodes in process.

Ectodermal and Ecto-Mesodermal Organs

Skin

The proliferating surface cells of the embryo spread and thicken to form an outer coating of simple squamous epithelium—the periderm. The cells of this temporary fetal skin are continually transformed and exfoliated through synthesis of a fibrous, sulphur-rich protein called keratin. As they harden, new strata push up into their places from a basal cell layer beneath. The desquamated cells seal the periderm with curdled oil *(vernix caseosa)*.

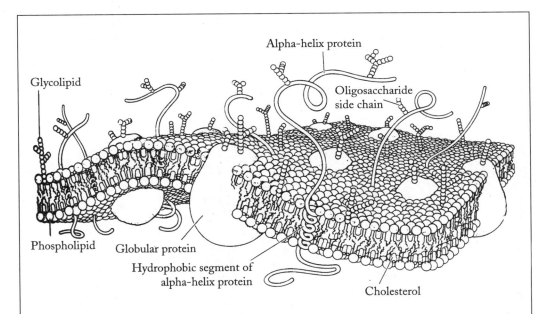

FIGURE 19A. The skin barrier's selectivity is achieved by the embedding of large molecules in the phospholipid bilayer. These large molecules penetrate both layers, their shapes and chemical affinities facilitating the passage of some substances across the membrane, and blocking others.

From Deane Juhan, *Job's Body* (Barrytown, New York: Station Hill Press, 1987).

The basal layer also develops regional downgrowths that will become the ridges and grooves of palms, fingers, soles, and toes.

The neural crest, engaged elsewhere in the manufacture of material for the peripheral nervous system, dispatches herds of melanoblasts into the epidermal region where they differentiate into melanin-synthesizing granules (melanocytes). The degree of pigment in these cells will determine darkness of skin, the most common marker of caste among humans. At birth, their chromatic distinctions are minor, but exposure to light stimulates melanin production, fully signifying this socially inflated tag.

As they fuse with deeper mesodermal material (dermis), epidermal cells knit a vast organ, the skin.

Dermis derives from mesenchyme of two sources: a thinning lateral column continuous with the yolk sac and amnion, and dermatome from the somites. It uniquely controls the type of structures that assemble in the epidermis (fur, glands, claws, etc.), conferring their character and patterning.

The adult epidermis itself remains an epithelium of many layers of keratin-synthesizing cells. Its outermost layer is made up of mostly dead squames stacked in interlocking hexagonal columns. Having forfeited their organelles, they stick together in flattened scales replete with keratin and buttressed by tough, cross-linked intra-

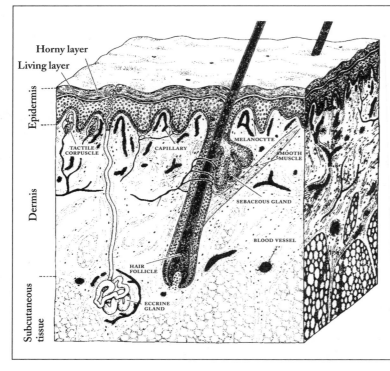

FIGURE 19B. Cross-section of skin. Connective tissue fibers in the dermis bond outer layers to deeper subcutaneous layers.

From Deane Juhan, *Job's Body* (Barrytown, New York: Station Hill Press, 1987).

cellular protein (involucrin). Beneath these are several layers of prickle cells anchored to keratin filaments by their desmosomes. Below them lies a mitotic basal-cell layer—the immortal stem cells issuing constant fresh skin.

As core members divide and percolate outward, their advancing peripheral layers are recruited into the prickle-cell layer. Loss of contact with the basal lamina triggers their terminal differentiation into skin. Gradually they start to relinquish organelles and nuclei and submit to keratinization. Their old age is spent as squames. In death, one textbook reminds us, they "finally flake off from the surface of the skin ([to] become a main constituent of household dust)."[4]

Hair and Nails

The epidermis gives rise to a host of keratinized codicils, including hairs, feathers, claws, nails, and scales. Epiderm is the source likewise of nodes becoming glands. The mechanics of these little organs begin as globalized instructions for cell-matrices and

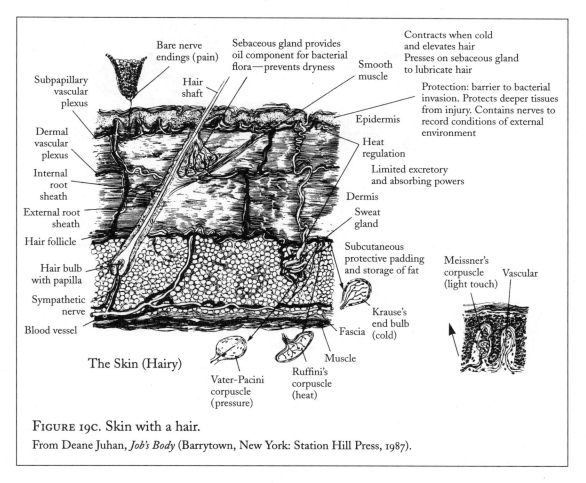

FIGURE 19C. Skin with a hair.

From Deane Juhan, *Job's Body* (Barrytown, New York: Station Hill Press, 1987).

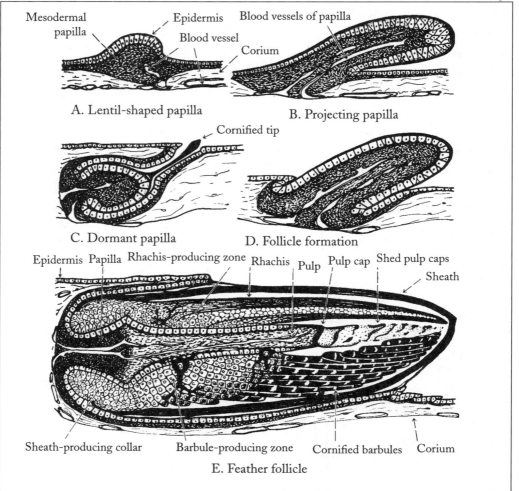

A. Lentil-shaped papilla

B. Projecting papilla

C. Dormant papilla

D. Follicle formation

E. Feather follicle

FIGURE 19D. Stages of feather development. A. to C. Early stages; D. Dormant papilla of sparrow fetus; E. Sagittal section through mature feather germ (enlarged more in width than length), formation of rhachis on outer side, barbs and barbules on inner side. From Emil Witschi, *Development of Vertebrates* (Philadelphia: W. B. Saunders & Company, 1956).

cell-to-cell adhesions. Fibroblasts exerting traction on their own collagen tend to agglomerate at the sites of future appendages.

Hair follicles originate in hard epidermal declinations into the dermal layer, the deepest part of each bud becoming a club-shaped bulb invaginated by a small mesenchymal papilla—the germinal matrix of a single hair. Cells in the bulb proliferate until they bulge outward into the shaft, a keratinized bristle breaking through the epiderm, jutting up above the skin. Additional melanoblasts migrate into the bulb until melanin production is taken up by the hair-forming cells themselves.

Eyelashes evolve similarly. Feathers originate in tracts of feather rudiments that blanket the back, wings, and upper legs of birds.

Fields on the dorsal tips of the limb digits (fingers first, then toes) induce folds of epidermis which bulge outward, specialize in keratin, and harden into nails.

A superficial layer of epidermis surrounding each of the thickening sites degenerates to its base where a remnant cuticle persists.

Similar processes in other families and genera yield hooves and horns.

Glands

Deep-lying organs extracting, storing, and chemically altering secretions from blood, skin, and other sources are called glands (from their general appearance, after the ancient Indo-European root for "acorn"). Sebaceous glands germinate from budding at the sides of developing epithelial roots of the hair follicles. As they sprout along adjacent connective tissue to form alveoli and ducts, their central alveolar cells disintegrate, leaving an oily secretion (sebum) which extrudes into the hair follicles onto the skin to meld with the cheeselike oil of the periderm and cover the fetal surface.

Sweat glands, like hair follicles, begin as proliferative downgrowths from epidermis into dermis. The elongating crown of the bud then coils to form the eventual secretory portion of the gland; the organ's epithelial attachment to the epiderm gives rise to a duct, its central cells gradually degenerating into a lumen, or passageway. Sweat secreted by a single epithelial layer at the bottom of the tube travels upward along an excretory duct (itself a mere two layers thick) to the surface of the skin.

Mammary glands comprise branching networks of excretory ducts submerged in connective tissue. They begin much in the manner of sweat glands, but each initial downward-wedging bud soon fissions into secondary buddings which differentiate into lactiferous ducts. Adjoining mesenchyme supplies fat and fibrous connective tissue. Because of their widely varying size and shapes, their prominent thoracic position, and their role as a primary erogenous zone, these culturally adorned semes of female identity absorb an astonishing variety of laudatory, derogatory, and ambivalent appellations in every culture.

Later, epidermis at each gland's tip will collapse to form a mammary pit; out of this a nipple will erupt by proliferation of mesenchyme. Subsequent development of mammary ducts and fibers lapses until puberty when breasts (in female genomes) fall under the influence of the ovaries' secretion of estrogen and progesterone and complete their development as lactiferous bosoms. With the dispatch of female hormones into the bloodstream, the duct cells proliferate, their terminal portions branching into outpocketings (alveoli) with secretory cells stimulated into milk production by post-natal hormones.

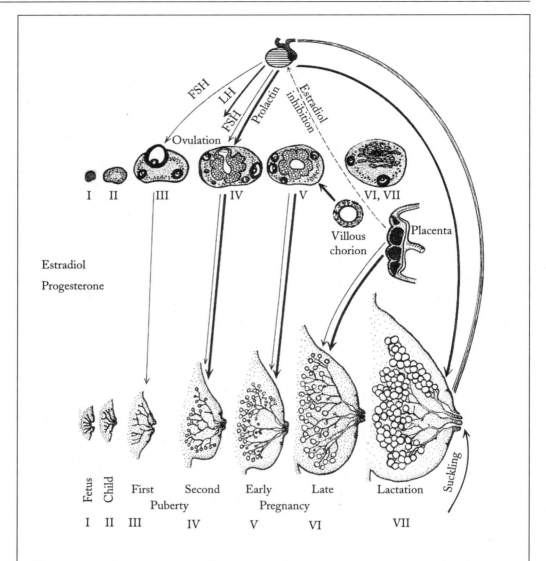

Estradiol

Progesterone

FIGURE 19E. Representation of hormonal relationships in mammalian development and lactation. From Emil Witschi, *Development of Vertebrates* (Philadelphia: W. B. Saunders & Company, 1956).

Teeth

During our long evolution through vertebrate, mammalian, and primate phases our teeth were subject to shifting ecological and dietary pressures. Initially lumpy granules, they matured as nested gradients of overlapping hardened epithelia. With mutations flowing through elaborate tissue fields, relatively simple initial structures diversified and deepened into true bony gems.

Enamel is ectodermally derived within the oral cavity; other dental tissues begin as mesenchyme and differentiate stage by stage in layers.

Tooth buds originate in the surface epithelium where they are invaginated by a region of condensed mesenchyme, the dental papillae. Only those adjacent to the inner portion of the enamel epithelium are specifically prompted as odontoblasts—cells that will make and deposit dental pulp and pre-dentin (the forerunner of calcified dentin). Conversely, the inner portion of the epithelium is elicited by the dentin of odontoblasts into enamel-producing ameloblasts. This dually-induced alloying, of enamel on the outside and dentin on the inside, begins at the cusp of each tooth and continues downward to its root. The root's ensuing development is orchestrated by a folded sheath of enamel epithelium invading mesenchyme.

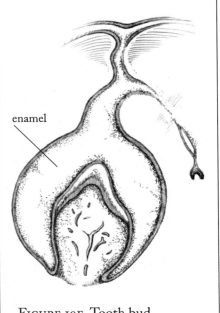

FIGURE 19F. Tooth bud. Illustration by Jillian O'Malley.

The teeth will eventually excite bone formation and become covered by bone except at their crowns.

It should be noted that dental organs formed embryonically are deciduous, pushing up through the oral mucosa after birth, usually within two years. After they are shed, their roots are absorbed by bone-related cells.

Permanent teeth will develop in a virtually identical manner, erupting sometime during the time span from the sixth year through adolescence.

Endodermal, Meso-Endodermal, and Ecto-Endodermal Organs

The Central Gut with its Openings

In gradated sectors from front to rear the unfurling endodermal tube is induced regionally by surrounding mesoderm into a foregut, midgut, and hindgut. Whereas the neural tube was originally open to amniotic fluid, the gut is a closed, looping double cul-de-sac. It will later perforate cranially and caudally to open nostrils and mouth, anus, and urethra.

At its forward margin, endodermal tissue contacts the ectodermal layer to form the oral plate of the mouth; to the rear its margin branches into the anal primordium.

During neurulation the head of the gut forges a number of its own outpocketings: the primordia of the lungs, liver, gall bladder, and pancreas—all still closed protrusions. As foregut consorts with splanchnic matter, a rich topography of jointly induced endodermal and mesodermal structures emerges.

The hindgut incorporates the allantois into its invagination. After developing a common urinary and rectal passageway (the cloaca), this structure will split, as a septum of tissue drops between them like the curtain of a theater, into bladder and rectum (see below).

Anteriorly, the foregut will bear the pharynx and thyroid gland; posteriorly, the stomach and duodenum. Remember, all organs are "chakras"—energetic gradations of eddying streams of cellular material, evanescent configurations of organic stuff in transformation.

Pharynx, Thyroid, Parathyroids, and Thymus

The pharynx reembodies a series of histories from both sea and land vertebrates.

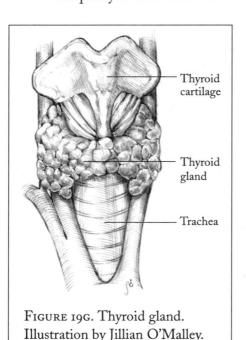

FIGURE 19G. Thyroid gland. Illustration by Jillian O'Malley.

Thyroid cartilage

Thyroid gland

Trachea

As it materializes, it induces a row of paired pouches. At their slits endoderm swells outward to contact ectoderm, eliciting new membranes. In humans the first pouch comes to encompass the middle ear bones as it gives rise to the tympanic cavity—widening into the pharynx through the eustachian tube. The endoderm of the second pouch fuses with mesenchyme around it, becoming tonsillar crypts (the remaining mesenchyme spawns lymph nodules). The third and fourth pouches give rise to the formative masses of the thymus and parathyroid glands, respectively.

The four small parathyroid glands (in superior and inferior pairs) ride on the back of the thyroid gland and secrete a hormone that stimulates osteoclasts (bone-resorbing cells) to intensify their erosion of the hard bony matrix and thereby liberate calcium and increase its concentration in the blood (see next chapter). The precise regulation of calcium is also, however, critical to the operation of every cell in the body. Too much of this molecule will occasion hallucinations and ultimately paralyze the heart's beating; too little will cause cells to become so overactive they bombard the muscles with impulses causing them to spasm.

Initially the pharynx rests atop the cardiac region which bulges into it—an etiology with psychosomatic implications. To have our "heart in our mouth" is to return to a state of aboriginal fright.

THE THYMUS TAKES SHAPE as dense clusters of endoderm are broken apart by travelling mesenchyme cells which will supply the gland's blood vessels, connective tissue, and perhaps even some of its repertoire of small lymphocytes (whose company also includes reprogrammed stem cells migrating out of the yolk sac). Thymic corpuscles arise as compacted evaginations of endodermal epithelium (see the next chapter for discussion of lymphocytes).

Beginning as a ventral pocket in the floor of the pharynx, this horseshoe-shaped organ is later closed off and displaced to the rear—a tiny, soft, pyramidal body that regulates growth up to puberty, retrogressing thereafter. There is a pseudoscientific belief that criminal behavior is stimulated by overactive thymus glands that maintain their vigor after adolescence.

Thymus injected in a tadpole keeps it from metamorphosing into a frog, rendering it a tadpole for life. The thymus also seems to have a role in the formation of shells in birds and reptiles and possibly the evolution of the human ovum.

THE TISSUE FOR THE THYROID is unrelated to the pharyngeal pouches forming the thymus and parathyroid glands. Arising as a single evagination of endoderm along the pharynx floor between the first and second segments, the thyroid expands posteriorly into the mesobranchial region, obliterating all hints of its point of origin except in a few species of sharks. Meanwhile it differentiates into a mound of tiny closed epithelial follicles linked in vascularized connective tissue. Finally locating against the laryngeal cartilage of the neck

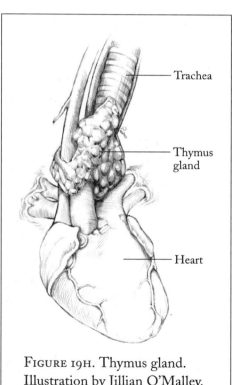

FIGURE 19H. Thymus gland. Illustration by Jillian O'Malley.

"in what Plato calls the isthmus between the body and the head, the thyroid is the mediator between the emotions and the thoughts, and the common denominator of animal and intellectual life."[5]

A primitive vertebrate nodule, the thyroid has ancestral forms in most proto-chordates—the first creatures with gill slits. In these fishlings, it secreted mucus to engulf food particles, dispatching these into the pharynx—an anatomical function completely vestigial in mammals. Now synthesizing its mucus as a hormone, the thyroid has become an iodine-processing metabolic feedback unit, uniquely absorbing dark gray, poisonous halogens from the environment through the blood capillaries and building them into a useful hormone (thyroxin). Caching its elixir in its colloidal follicles, the thyroid liberates it into the bloodstream in discrete amounts, thereby speeding up the release of energy from digestion of foods and regulating the biochemistry of all of our trillions of cells.

Mouth, Tongue, and Palate

An embryonic mouth (stomodeum) originates as a slight cratering in the pharyngeal ectoderm. The increasing depression is separated from the pharynx by an oropharyngeal membrane (combining ectodermal and endodermal layers). The collapse of the membrane three and a half weeks after conception links the primitive digestive tract to the amniotic cavity.

The tongue originates from a median triangular swelling of mesenchyme in the floor of the pharynx. Two round nodules emerge on either side of it, and the three structures fuse. Mesoderm from the branchial arches provides connective tissue, blood vessels, and some muscle fibers; additional muscle is supplied from the primordial myotomes of the occipital somites (see the next chapter).

The epithelium of the tongue derives from pharyngeal endoderm. Its taste buds (which transmit sweet, sour, salty, and bitter sensations to the brain) begin as vallate and foliate papillae induced by terminal branches of glossopharyngeal nerves (themselves originating as neural-crest cells laying pathways connecting facial and masticatory regions to the central and autonomic nervous systems). Each taste bud has only about fifty elongated cells arranged like the staves of a barrel; a small taste pore contacts the exterior and serves as a conduit for molecules to be sampled. Taste-tranducer cells pass this flavorful information to nerve fibers permeating the bud, which relay it swiftly to the brain. Evolutionarily this function developed more for vulnerable mole-like creatures to spit out life-threatening poisons in time than epicureans to savor wild berries and Italian cuisine.

The nerves are the aggressors here; they seek taste information and induce appropriate molecule-sensing buds. If the nerves are severed, the buds vanish but, when

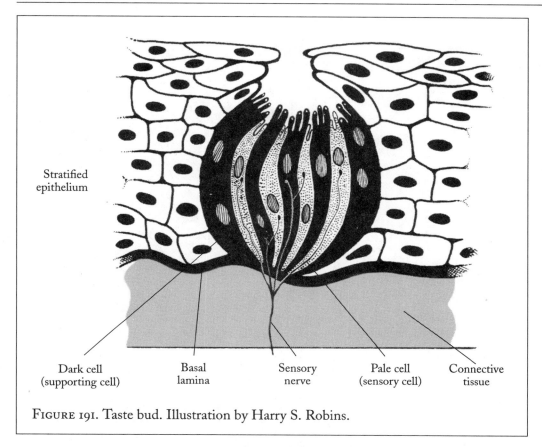

Stratified
epithelium

| Dark cell | Basal | Sensory | Pale cell | Connective |
| (supporting cell) | lamina | nerve | (sensory cell) | tissue |

FIGURE 191. Taste bud. Illustration by Harry S. Robins.

the nerves regenerate, they induce new epithelial cells into functional taste buds.

The processes of the palate, composed of mesoderm in contact with ectoderm, arise from the jaw and project downward on the sides of the tongue, which descends as the jaw develops. Lateral palatine processes then flip to a horizonal position and fuse with each other and with the forward palate and a downward growth (the nasal septum). Subsequently, bony membrane molds the hard palate.

Trachea, Throat, and Larynx

Combining as the floor of the brain and the roof of the mouth, the tongue and sphenoid-ethmoid palate are a single complex packed with blood and hormonal and cranial nerves. In concert with the nasopharynx muscles, the glotus, the hyoid, sternohyoid, and omohyoid bones, and the clavicle muscles, this structure functions as a muscular membranous partition (a diaphragm) regulating the flow of pressure into the trachea pouch and the lungs and helping to maintain upright posture.

In the lining of the ventral wall of the pharynx, internal folds fuse to make the tube of the larynx in which the cords of the voice-box develop; these are attached

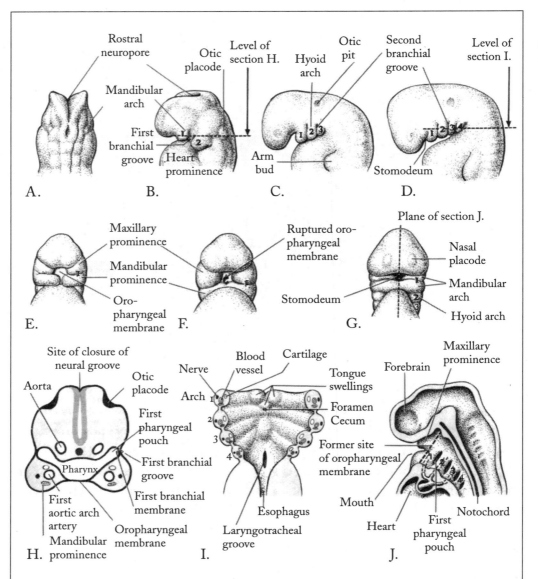

FIGURE 19J. Development of human branchial apparatus. A. Dorsal view of cranial part of early embryo; B. to D. Lateral views showing development of branchial arches; E. to G. Facial views, illustrating relationship of first arch to stomodeum (primitive mouth); H. Transverse section through the cranial region; I. Horizontal section through the cranial region, showing branchial arch components and floor of the primitive pharynx; J. Sagittal section of upper region, showing openings of pharyngeal pouches in the lateral wall of the primitive pharynx.

From Keith L. Moore, *The Developing Human: Clinically Oriented Embryology* (Philadelphia: Saunders College Publishing, 1977).

to the respiratory tract by folds of mucous membrane and articulate with the trachea, the windpipe.

The lobes around the neck, throat, and hyoid, and the tongue, jaws, and teeth sculpt a channel that is used secondarily by mammals for storing and releasing traumatized or otherwise fixated energies in styles of breathing, frowning, orating, singing, howling, gnashing, etc. As experiences reshape still-malleable adult tissue, impressionable organs may rigidify neurotically and bind one another, with functional and psychosomatic consequences (see Chapter 24, "Healing").

Facial Sense Organs: Ears and Nose

In anterior spots where the ectoderm is neuralized, mesoderm or endoderm (or both) contact it; there capsules are induced, and large-scale sense receptors pop out, nostrils, orbits, and ears.

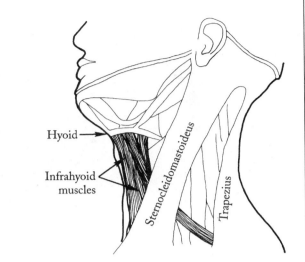

FIGURE 19K. The neck muscles and fasciae which attach to the thoracic inlet from above (the fascial component of the neck is continuous with that of the thorax). Note the oblique courses of the muscles and their fasciae, exerting influences on both the structure and mobility of the region. This area is highly complex anatomically with high potential for fascial restriction, CSF reduction, osseous dysfunction, and free mobility of many other viscera throughout the body.

From John E. Upledger and Jon D. Vredevoogd, *Craniosacral Therapy* (Seattle: Eastland Press, 1983).

A pair of regional thickenings (placodes) sprout in front of the neural plate, appropriating material from the neural fold; these become two olfactory bags. Nasal sacs then work their way into the developing brain.

Similar placodes of ectoderm and neuroectoderm invaginate into otic pits on the sides of the hindbrain and form vesicles (otocysts) which then detach from the epidermis and twist about to make the labyrinth of the inner ear. Some epithelial cells flatten to become membranes; others elongate in patches of sensory ectoderm. The cochlea of the inner ear develops from an expanding diverticulum of each otic vesicle after it coalesces from its invaginated placode.

The ear is a riverine cave, auricular hillocks curling from proliferating mesenchyme of the first and second branchial arches around the margins of the first branchial groove. The ear's characteristic labyrinths occur because its expansion is

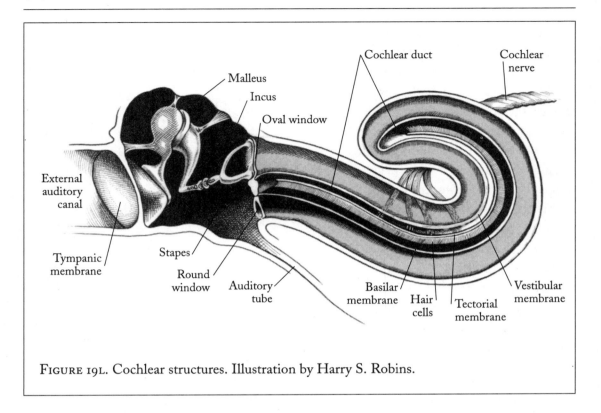

FIGURE 19L. Cochlear structures. Illustration by Harry S. Robins.

cramped in some areas while swelling in others. The vesicle itself, induced by meso-dermal contact with ectoderm, then induces surrounding mesenchyme to produce a spiral of cartilage.

As EACH OTOCYST loses contact with surface ectoderm, its dorsal portion (the utri-cle) develops a hollow diverticulum that elongates into an endolymphatic duct and sac—a primitive sound chamber. Meanwhile the otocyst's ventral section (the sac-cule) forms three smaller diverticula; their central portions fuse, atrophy, and van-ish, while their peripheral aspects compose semicircular ducts with their own subtle vibration-processing characteristics. These remain attached to the utricle as they are enveloped within the bony labyrinth's canals.

Dilations (ampullae) at either extremity of each ear's semicircular canal develop vibration-sensitive nerve endings. The utricle and saccule, however, induce slightly different sensory structures—maculae; these gelatinous organs are imbedded with otoliths and bear their own neural fibers and receptor hair cells. The otoliths within each macula are small calcareous particles that calibrate gravity (they have a similar function in the statocysts of invertebrates). *Maculae utriculli* and *sacculi* use a com-bination of otoliths, hairs, and fluid to register the position of the head relative to

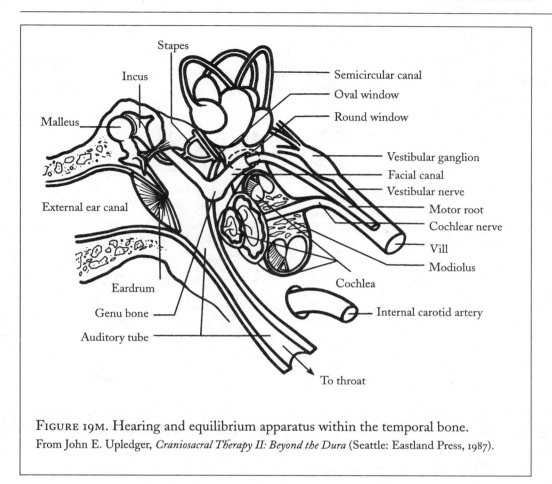

FIGURE 19M. Hearing and equilibrium apparatus within the temporal bone. From John E. Upledger, *Craniosacral Therapy II: Beyond the Dura* (Seattle: Eastland Press, 1987).

gravity (see the discussion of the vestibular mechanism below).

The tubular ventral portion of the otocyst continues to grow inward, creating an expanding diverticulum (the cochlear duct) that coils into the cochlea of each inner ear. Inducing its own rich lining of auditory neurons, the diverticulum cuts a spiral conduit in the shape of a snail shell with a bony screwlike core. As the connection of the saccule to the cochlea becomes physicodynamically restricted (like a river bend), it forms a narrow *ductus reuniens*. A cochlear subregion consisting of hair cells and generating action potentials in response to sound waves—the organ of Corti—materializes from cells in the cochlear-duct wall. This minute structure is filled with sensory hair cells bearing hairlike microvilli and embedded in a gelatinous ledge; its body is penetrated by nerve processes (synaptical terminals) from ganglion cells of the eighth cranial nerve as they migrate along the coils of the cochlea. Afferent fibers from sensory neurons in the cochlear ganglion combine to form a cochlear nerve which later joins the vestibular nerve from the otoliths and runs to the brain.

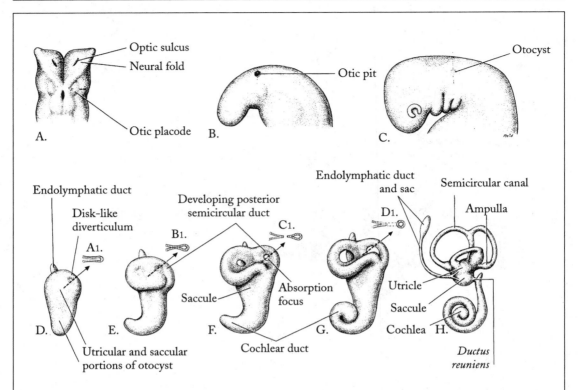

FIGURE 19N. Early development of the inner ear. A. The region at twenty-two days of development, showing the otic placodes; B.-C. Development of the otocysts; D.-H. Development of the otocyst into the membranous labyrinth from the fifth to eighth week after conception; D1 to H1. Diagrammatic sketches of the development of the semicircular duct. From Keith L. Moore, *The Developing Human: Clinically Oriented Embryology* (Philadelphia: Saunders College Publishing, 1977).

Mesenchyme around the otocyst differentiates into a cartilaginous capsule. As the underlying labyrinth expands, vacuoles in the otic capsule fuse to form a cavity (perilymphatic space) filled with perilymph. Within this reservoir the membranous labyrinth is suspended in fluid. Meanwhile the cartilaginous aspect of the capsule ossifies as the bony aspect of the inner ear's labyrinth.

SOUND WAVES GATHERED by the outer auricle strike the middle ear at the tympanic membrane—a complicated structure woven by all three primary cell layers: ectoderm from the dorsal edge of the first branchial groove, endoderm from an expansion of the first pharyngeal pouch (the entire tympanic zone develops within an expansion of the recess of this pouch), and late-arriving mesenchyme from the first and second branchial arches.

The tympanic membrane's vibration initiates vibrations in the three delicate ossicles of the middle ear, the malleus, incus, and stapes—extensions of the temporal bone of the cranium. These interlocking tuning forks are formed embryogenically from branchial-arch cartilage, then encased by the endodermal epithelium of the distal portion of the pouch's recess (in the tympanic cavity).

The bones' vibrations oscillate the membrane of the oval window of the cochlea, producing waves in the perilymph of the cochlea and causing the basilar membrane to vibrate. Sound is detected by stimulated hair cells in the organ of Corti, and these induce action potentials in the cochlear nerve. The nerve transmits those impulses along axons to the cochlear nucleus in the brain stem where they are repackaged and dispatched to the inferior colliculus of the midbrain; from there they shoot through the thalamus into the auditory cortex of the cerebrum.

The inner ear is a vortex of protoplasm, which separates waves of sound in the air into notes on membranes and passes them into the brain as a replica of sound (which I recognize as shouts of children and a tapping of keys on this machine).

How WE HEAR is the ultimate "Rube Goldberg" machine. If the steps of its sequence are considered separately, the coherent packaging and transmission of sound waves into meaning is absurd but, as with all other morphogenetic assemblages and homeostases, the whole transcends its parts, both mechanically and existentially, to produce a confident functioning organ system undeterred by its many interlinked and complicated components and routings.

The passages of the ear are also more than just aural canals. Cochlear nerve fibers meeting vestibular nerve fibers from the maculae of the vestibule form the vestibulocochlear nerve (cranial nerve VIII), which carries joint sound-and-balance messages to the brain. The vestibular mechanism in the ears' inner recesses (vestibule) functions as the body-mind's kinesthetic compass and stabilizing bar. It receives information from proprioceptors, interceptors, and kinesthetic receptors throughout the body, enabling us to locate ourselves in space—telling us where we are, how we are oriented, and which way we are moving. As the head adjusts, the canals of the ear register gradations and shifts of velocity and transitions in time.

Thorax, Diaphragm, and Lungs

The viscera of lower invertebrates all lie in the same cavity, but the coelom of mammals is divided into two sections—the thorax, which includes the chambers of the heart and lungs, and the abdomen, which holds the whole digestive tract, including liver and kidneys, and reproductive organs. These are separated by the body's major diaphragm—a sheet of tendon whose muscles contract and expand in breathing. A

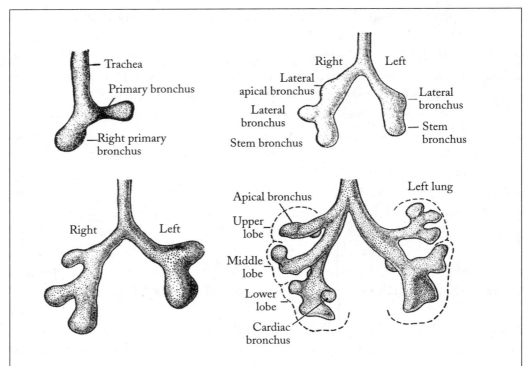

FIGURE 190. Development of the chief bronchi of the human lung in ventral view. From Leslie Brainerd Arey, *Developmental Anatomy: A Textbook and Laboratory Manual of Embryology* (Philadelphia: W. B. Saunders & Company, 1946).

continuous partial vacuum sucks air down the windpipe into the lungs and then expels it by relaxing. Birds, with no diaphragm, must use the muscles of their rib cage and their wings for breathing. Reptiles and amphibians pump air in by their throat muscles.

Connected to the pharynx through the trachea, the lungs expand into large spongy sacs. Actually, they are concretions of multiple branched tubes ending in minute air-bags. The initial lung bud sprouts at the caudal end of the laryngotracheal tube and splits in two, a bipolar expression of one set of instructions.

As an evagination of the alimentary canal, the lung cavity is as endodermal as gut. Digestion of food and air are differentiated aspects of the same blueprint; they occur in one endodermalized tube that oxidizes nutrients. The mouth is a gateway to both esophagus and trachea. The thorax breathes air while the abdomen admits food. Two worms—one of gut, one of breath, both Chordate—lodge in each other like Siamese twins.

When lung buds contact surrounding mesenchyme they differentiate into bronchi and limbs. Lungs then project bronchial trees—vast arborescences sprouting down-

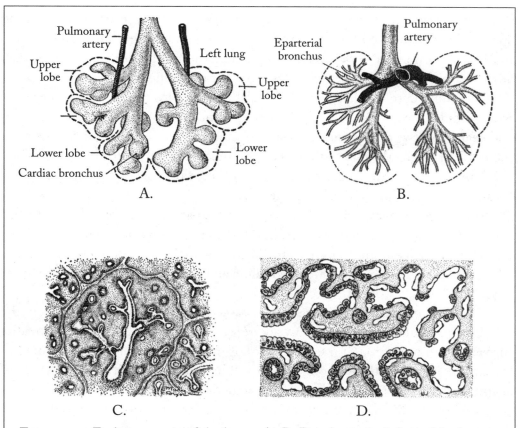

FIGURE 19P. Embryogenesis of the lungs. A.-B. Developmental plan of the human lungs, in ventral view. C.-D. Cross-sections of the human lung; C. Developing lobules, at four months; D. Loss of epithelium in the terminal air passages, at eight months.
From Leslie Brainerd Arey, *Developmental Anatomy: A Textbook and Laboratory Manual of Embryology* (Philadelphia: W. B. Saunders & Company, 1946).

ward, forking dichotomously to the sides and back. The unbranched section, suffused by cartilage from raw mesenchyme, becomes the trachea.

In lower vertebrates the lungs are mere folded sacs at the end of the bronchi, but in mammals the bronchi diverge outward, forming as many as ten multiple stems in the right lung and nine in the left. Air sacs (alveoli) then develop on the terminal branches so that the total fractal area exposed to air is as large as a tennis court.

When this system expands into the thorax, surrounding mesenchyme supplies connective tissue and envelops the lungs in membranous bags (pleurae), which lubricate them with a coating of fluid. The bronchi grow into the body wall of the thorax and, as the embryo folds transversely, come to lie adjacent to the heart.

THE LUNGS ARE MORE than bags for oxygen exchange; their in-and-out pumping spreads oxygen and other subtle currents and energies through the organs, inspiring image-formation. Psyche is an epiphenomenon of breath, represented as a winged internal goddess. Keleman describes her rhythmic dance:

"The continuum of inhalation and exhalation is like a wave. The breath increases in amplitude, rises to a crest, and then gently wanes. We inspire, the wave emerges and peaks, then we gently expire, pause, and inhale again. If we are excited the wave increases in pitch. When we are relaxed we breathe deeply into the belly. When life demands we breathe vigorously, we recruit more of ourselves by extending our breath into the abdomen, neck, and head. . . .

"Breathing is a pump with a total organismic expansion and contraction of 18–22 cycles per minute. [It] goes from head to toe as a pervasive and constant activity. It can be compared to the pattern of the heart as a total spiral contraction, unwinding, filling; a separate but synchronized filling and emptying of the upper and lower chambers."[6]

All serious meditation, yoga, rebirthing, ascension, and shamanism begin in the mind of the breath. Internal alchemy and *chi gung* take their elixir from the breath's furnace—not just through the mechanics of transpiration but in its discharge of sensation and prana throughout the body. The lungs transmit the medicines of acupuncture, chiropractic, and massage via meridians to viscera.

Every outbreath is a small death for the organism, the dissociation of its ego. Every time we expire we must trust that we will inspire and restore our selves.

But we never know for sure.

That unconscious fear is the source of some of the anxiety and rigidity in our personalities; it manifests as shallow breathing, subconscious but stubborn compulsions to prevent the dissolution and restructuring that occur in full breath. The body/mind tries to force reality to conform to its limited ego-sense, but the universe wants to move on, through the lungs and heart into new realities, fresh surges of cells.

IN MODERN FISHES an organ sprouts similarly out of the endodermal wall of the alimentary canal, the swim bladder, which allows the animals to rise and sink by filling with air and emptying. This hydrostatic sac may have originated as a primitive double-sac in the thorax of extinct fishes, a structure which also gave rise to the lungs of the land vertebrates. Thus was an original use converted into a different one when a new creode coopted it. The ability to breathe atmospheric oxygen through this sac allowed the lungfish prototypes of amphibians to cross land during droughts in a desperate search for new pools of water. Such journeys were consumated over generations, as most of these adventurous "fish" became dehydrated and died en route. Apparently, chance mutations transformed a few of their descendants into part-time terrestrial animals with rudimentary lung sacs. They laid their eggs in a familiar medium, water, and their descendants sired amphibians, reptiles, birds, and mammals.

Woven into the daily world, the common metamorphosis of tadpole to frog is a vivid signature of the path our forebears took from water to land. It is both anatomical and ecological.

The mammalian embryo reenacts (at least partially) this stage of development; its alveolar ducts are small immature bulges siphoning amniotic fluid which keeps them half inflated. At birth this fluid is suddenly expelled, some of it into arteries and veins and a good deal of it through the mouth and nose from the pressure of the birth canal on the thorax. The lungs then respire only because the alveolar capillary membrane is thin enough to allow gas exchange — a state which originates ontogenetically while still "underwater."

A millennial change of habitat is accomplished in an instant: The history of fishes, amphibians, and mammals is recapitulated in a singular cataclysmic event — a water-

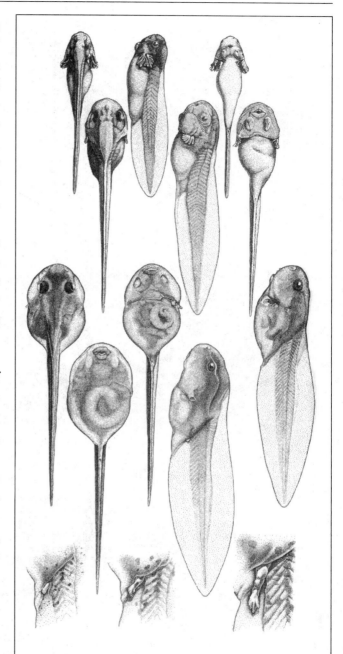

FIGURE 19Q. Standard stages of early development of frog *(Rana pipiens)*.

From Leslie Brainerd Arey, *Developmental Anatomy: A Textbook and Laboratory Manual of Embryology* (Philadelphia: W. B. Saunders & Company, 1946).

breathing eel turned into an air-respiring newt. Among humans the imperative to breathe at birth is a source of both primal trauma and lifelong difficulties inhaling and exhaling fully:

"The infant is not given an opportunity to make a transition. . . . We breathed one way in the womb and, because of the premature cutting of the umbilical cord, we are forced to learn to breathe outside the womb instantly, and in a do-or-die situation.

"Air striking the lungs for the first time results in unbelievable searing pain. Yet the infant must breathe; there is no alternative, the cord has been cut.

"Breathing then becomes subconsciously associated with the pain, fear, and panic of the first breath. This results in perpetual anxiety and feelings of urgency. In order to keep this suppressed, we learn to breathe in a very shallow manner."[7]

Since the 1980s, mothers trying to avoid such shock have delivered their babies directly into tubs of warm water, then left their umbilical cords attached for hours afterward to make the sea-to-land transition more gradual.

Stomach and Intestines

As the pharynx expands during the first seven weeks after conception, the esophagus elongates within the foregut. Mesenchyme and neural-crest cells provide it with material for smooth muscle and a visceral plexus. The creation, molding, and positioning of abdominal organs embryonically is a complex reenactment of phylo-

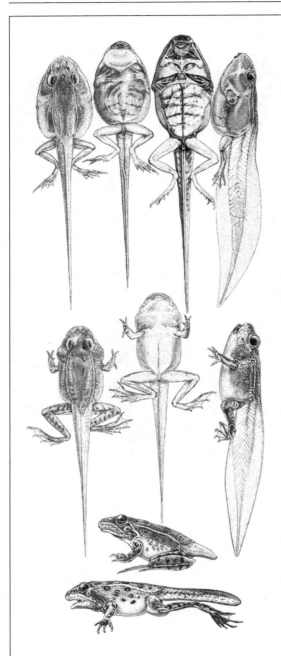

FIGURE 19R. Standard stages of late development of frog *(Rana pipiens)*.

From Leslie Brainerd Arey, *Developmental Anatomy: A Textbook and Laboratory Manual of Embryology* (Philadelphia: W. B. Saunders & Company, 1946).

genetic brinkmanship. Many asynchronous, separately derived plans are interpolated seamlessly, reflecting both ancient chronology and fetal exigency. Maintaining function while assembling layers of fractalized structure yields an embryogenic path simultaneously rigorous and meandering.

The caudal part of the foregut dilates as the stomach, its dorsal tube expanding more rapidly than its ventral border and imposing a gradual curvature. The stomach swells into the abdominal cavity, rotating ninety degrees clockwise on its longitudinal axis so that its original left side lies toward the belly and its original right toward the back. It is suspended from the wall of the cavity by a large mesentery (an abdominal membrane fold) that develops coalescing gaps in its surface and becomes the lesser peritoneal sac.

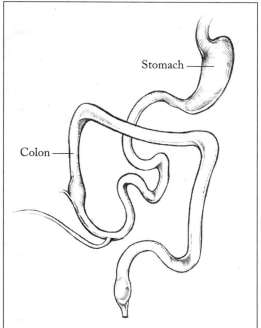

FIGURE 19S. Stomach and partial intestines. Illustration by Jillian O'Malley.

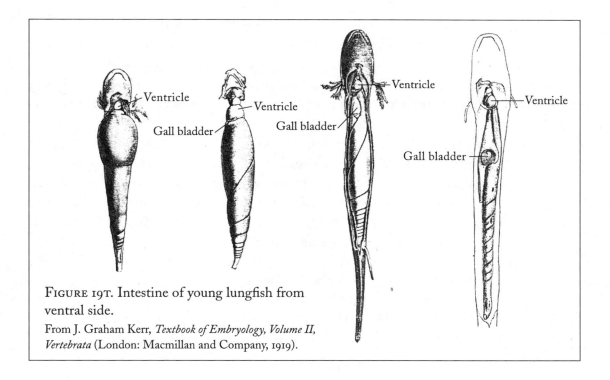

FIGURE 19T. Intestine of young lungfish from ventral side.

From J. Graham Kerr, *Textbook of Embryology, Volume II, Vertebrata* (London: Macmillan and Company, 1919).

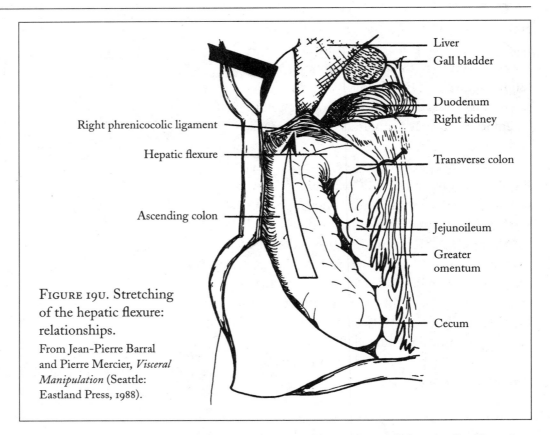

Liver
Gall bladder
Duodenum
Right kidney
Transverse colon
Jejunoileum
Greater omentum
Cecum

Right phrenicocolic ligament
Hepatic flexure
Ascending colon

FIGURE 19U. Stretching of the hepatic flexure: relationships.
From Jean-Pierre Barral and Pierre Mercier, *Visceral Manipulation* (Seattle: Eastland Press, 1988).

In mature organisms the stomach begins the process of chemically digesting food received from the esophagus. The smooth fibers of its pyloric sphincter detain the ingested contents—chewed plants and animal (and occasionally mineral) matter—for about three hours and then release them, transformed, through the opening of the pylorus into the small intestine.

Initially suspended from the dorsal abdominal wall by a short mesentery, the midgut twists and coils, as its cranial corridor grows faster than its caudal bow and competing organs annex much of the space of the coelom. Then the cells connecting it to the yolk sac gradually break down, leaving only an umbilical stalk. As they elongate more rapidly than the body itself, the intestines are projected spirally into the umbilical cord, herniated from the abdomen by the expanding liver and kidneys. Only by the tenth week after conception do they return inside the body, encouraged by both the slower growth of the competing organs and the enlargement of the abdominal cavity. Then they are gradually rotated almost completely around, filling the coelom, as the liver and kidneys continue to decrease in proportion to the overall enlargement of the cavity. Pushing their mesenteries against the cavity wall, the intestines fix themselves in place. Gyration of the stomach pulls along the duodenum and causes it and the pancreas to fall to the right, pressed similarly against the dorsal abdominal wall.

WINDING IN LOOPS loosely attached by mesenteries to the abdominal wall, the small intestine (actually twenty feet long) becomes a conduit through which the digestion and absorption of foodstuffs are consumated. Its fore portion, the C-shaped duodenum, curves around the top of the pancreas in a ventrally projecting loop from the end of the foregut to the beginning of the midgut; it carries out most of our chemical digestion. The inaugural duodenal segment receives semi-digested food (chyme) from the stomach; its mid section, permeated with duct openings, is inundated with pancreatic digestive juice and bile from the liver.

The mid and distal portions of the small intestine, the jejunum and the ileum, float freely in a series of loops attached to the posterior abdominal cavity. The large intestine extends from the ileum to the anal perforation at the end of the alimentary canal.

The lining of the small intestine continues to differentiate and induce subtle embodiments—a classic display of fractal geometry. Its folds (plicae) develop thousands of mucus projections (villi); the epithelial cells of the villi bristle with microvilli; the microvilli project their own bumpy protrusions, and so on. Once the process of sprouting and nesting begins, it torques into its own cavernous domain—an ingenuity of soft visceral packing. Pursuing Zeno's paradox (of the arrow that never reaches its destination), the intestinal wall continues to cram and twist into space that is almost but not quite full. Like Mandelbrot sets swirling into limitlessly pliant subsets, the total surface that the plicae, villi, microvilli, micro-microvilli, etc., offer to contact material passing through the small intestine is functionally limitless.

Each phallus-shaped villus emerges in an epithelial lining around a lymphatic tube. The tube, wrapped in arteries and veins, absorbs lipids and fats from the chyme. Between the villi lie deep crypts that plunge into underlying connective tissue. At their base, intestinal stem cells are generated and slide upward along the plane of the epithelial sheet, ultimately borne to the surfaces of the villi; they are finally shed from the villi tips.

Deep at the core of our vulnerability, carpeted with more than a hundred million neurons, the intestines perform as virtually a separate nervous system, a "second brain." While carrying out digestive functions, they express longing and ire and implement a belief system of sorts, all in the absence of signals from either the cerebral cortex or the spinal cord. The intestines' concerns are predominantly gut matters, inarticulate yet profound and lusty. They grumble, roar, chortle, sigh, fart, and hunger, often to the chagrin of the ego-centered brain.

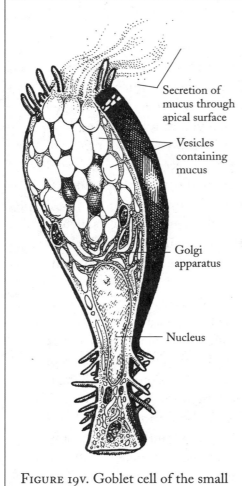

FIGURE 19v. Goblet cell of the small intestine. Illustration by Harry S. Robins.

Secretion of mucus through apical surface

Vesicles containing mucus

Golgi apparatus

Nucleus

These winding sausages are, in truth, their own great growling dragon, submerged in an excitable medium, imprisoned and tamed and ceaselessly fed, but struggling to impose their inexhaustibly nostalgic memories through peristaltic waves of satiety and sorrow. They generate a lifelong zero-state of mild orgasm. Though their bursts of release are not unlike those of the genitals, we do not romanticize them; they are neither stylish nor social. In fact, intestinal sensations and acts are, for the most part, deeply private. When these organs fail to churn in normal fashion, we become irritable and attempt to reinitiate their native spasms, often more avidly than we seek any other satisfaction.

At the same time, the intestines emit hallucinogenic bursts of jubilation and nausea, ravenousness and vacancy, vestiges of former and prehistoric embodiments. We feel many different kinds of waves of empathy and compassion from our guts. We consult them unconsciously at moments of decision, for the bowel is a great reader of character and hidden intentions (so-called "gut feeling"). Hypersensitive, it gives rise to its own "mental diseases" and nervous disorders—gastrointestinal maladies, many of which embody spasmodic, psychosomatic compensations for stress and fear.

The intestines also (astonishingly) produce more than ninety-five percent of the neurotransmitter serotonin, a critical mood regulator in the brain and no doubt a donor to the temperament of the bowels as well.

AS THE INSIDE OF THE GUT was being formed and the organs rotated into place, the embryo was forced to integrate hourly turmoil of its own reorganizing shape. The original twisting became a repository, throughout biological life, for intrauterine longing and intense feelings, often of torment and rage. The suppression of such passions represents a belated attempt to prevent the structuring of the soma, a declaration of ego identity against the autonomic floods of life.

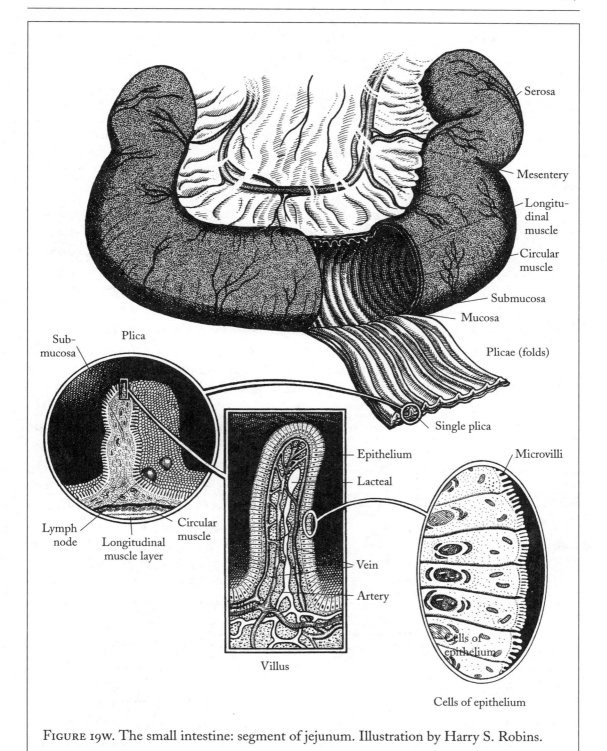

Serosa

Mesentery

Longitu-
dinal
muscle

Circular
muscle

Submucosa

Mucosa

Plicae (folds)

Plica

Sub-
mucosa

Single plica

Lymph
node

Circular
muscle

Longitudinal
muscle layer

Epithelium

Lacteal

Microvilli

Vein

Artery

Cells of
epithelium

Villus

Cells of epithelium

FIGURE 19W. The small intestine: segment of jejunum. Illustration by Harry S. Robins.

"The conflict is going on in the abdomen ... the biological conflict," Stanley Keleman says of a client. "We can see the actual struggle in the abdomen as it pulls in and lets out."

He asks him to squat down and hold his hands, rocking him slightly on his heels: "... breathe ... let your head back ... breathe into your belly."[8]

This frozen energetic component of organ creation and bodily experience must be assisted and released for normal, healthy functioning.

Liver

The liver is the body's most massive gland. Its major loyalty is to a larger group of glands in immediate communication with the gut, all of which are involved with different aspects of digestion, carbohydrate and lipid metabolism, and excretion of waste. The liver takes nutrients digested in the gut and transferred through the blood and prepares them for their participation in specialized functions of other cells throughout the body.

Embryogenically, the liver diverticulum spreads to the front of the alimentary canal, its forewall billowing into folds which enclose its small internal cavity. The tiny posterior opening that remains joins to the duodenum as the bile duct. Pleats of the diverticulum ultimately break up into strands of cells with blood vessels and sinuses.

Originating close to where a major vein adjoins the wall of the primitive gut, the liver derives its cells (hepatocytes) from the gut epithelium. The association of gut epithelium with mesenchymal tissue induces clusters of specialized cells in the organ (much as different characterstics are jointly induced in the lungs, thyroid, pituitary, etc.).

As the hepatocytes fold in sheets facing blood-rich crevices (sinusoids), their surface layer of flattened endothelial cells exchanges metabolites with the blood, receiving rich scarlet fluid directly from the intestinal tract through the portal vein. The hepatocytes then produce and degrade a number of chemicals,

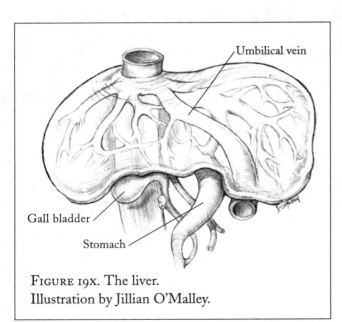

FIGURE 19X. The liver.
Illustration by Jillian O'Malley.

Umbilical vein

Gall bladder

Stomach

caching nutrients for future dispersal and returning altered products to the blood, including the bulk of the blood plasma's protein.

Hepatic cells remove from the blood substances that are poisonous to cells, intercepting them between the intestines and the heart and converting them into urea, which is passed on to the kidneys via the bloodstream. They also synthesize emulsifying bile from blood (useful for digesting fats). They secrete it out through ducts and narrow channels, the canaliculi, into the gut lumen along with metabolic debris and exhausted blood cells.

Because these hepatic cells proliferate so rapidly, the liver fills most of the abdominal cavity during early embryogenesis. While the right lobe grows faster, the left lobe develops a partial split, giving the organ its characteristic leafy shape. The liver appropriates resources so voraciously that at ten weeks it represents ten percent of the human fetus by weight, and its participation in embryonic blood-making gives it a bright red color (the Sumerians located the seat of consciousness here rather than the brain). The adult liver dominates its fellow organs like a sun with planets.

The stem cells of this organ remain primally mitotic and continue to renew themselves through the lifetime of the organism. As a neutralizer of venoms, the liver must be self-regenerative. Certain herbs help to flush out the toxicities of modern civilization by stimulating the production of bile. The Amerindian New World desert chaparral and turkey rhubarb root work indirectly through the bowel and smooth muscles of the digestive system, respectively, and the Chinese *shan zhi-zi* (gardenia from Shan Province) directly milks the biliary tree.[9]

In esoteric lore the liver is considered the bodily counterpart of the cerebellum—a storehouse and clearing chamber. It is also a seat of both desire and melancholy.

Gall Bladder, Pancreas, and Spleen

Another sheet of endoderm, a division of the diverticulum, wraps around the space beneath the mesoderm as the rudiment of the gall bladder. Located under the right lobe of the liver, this organ stores bile and articulates with the bile duct. Stimulated by hormones from the intestinal mucosa, the gall bladder releases alkaline liquid drained out of the liver into the duodenum.

The pancreas initially buds dorsally and ventrally from endodermal cells at the tail end of the foregut. A small later-forming protrusion develops near the bile duct, and when the duodenum (pushed by the stomach) rotates to the right, it is carried dorsally with earlier-forming elements. The major portion of the pancreas is constituted by the original dorsal budding, whereas the pancreatic duct arises from the ventral bud. Connective tissue originates mesenchymally as in the other gut organs. The pancreas is particularly friable, healing slowly when injured.

Loose mesodermal material apparently induces the distinctive secretory cells of this organ with their rough endoplasmic reticulum and swollen Golgi bodies. As a large gland lying to the rear of the stomach, the pancreas shoots alkaline digestive enzymes into the duodenum and spurts insulin into the bloodstream. Pancreatic fluid, which enters the small intestine at the site of the bile duct, includes sodium bicarbonate which neutralizes the hydrochloric acid of gastric juice.

The blood-filtering spleen, comprising dense, pulplike lymphocytes, is the source of hematopoeitic production until late in the life of the fetus (and a lifelong synthesizer of lymphocytes and monocytes). In concert with lymphatic glands and bone marrow, it manufactures and renews the cellular components of the blood and removes spent elements. It forms directly from masses of primordial mesenchymal stuff between the dorsal mesenterial layers. As the stomach rotates, the left portion of the mesogastrial organ is brought into contact with the peritoneum over the left kidney, where it fuses, coming to lie to the downward left of the diaphragm. A mixture of lymphatic nodules, tissue, and venous sinusoids with fibrous partitions, this mushroom-shaped oblong mass is about five inches long, four inches wide, and only about an inch-and-a-half thick in adults.

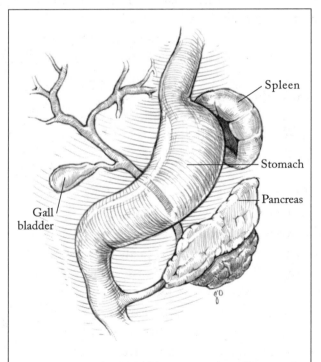

FIGURE 19Y. Gall bladder, stomach, spleen, and pancreas. Illustration by Jillian O'Malley.

Considered the governor of both mirth and sorrow, the spleen is classified as a subsidiary of the liver, its aide-de-camp which executes its incompleted tasks. This gland is also considered "the outer vestment of an invisible organ intimately concerned with the distribution of the solar force throughout the parts of the body."[10] In the flood of vital force through its cells the spleen turns pale rose and radiates luminosity through the nerves of the being, playing the role with them that electricity plays in a telegraph system. In the abdomen of a medium, the spleen is also the etheric site for the oozing of ectoplasm in the materialization of disincarnate entities.

The splenic artery is one of the earliest-forming aspects of the circulatory system and is often palpated by visceral therapists through layers of other, later viscera and muscles in order to guide the whole psychosomatic system in a reorganization of itself.

Hindgut, Bladder, Kidneys, and Adrenal Glands

The hindgut is made up of the colon, rectum, the upper portion of the anal canal, and part of the bladder and urethra. Its expanded rear portion, the cloacal membrane, is accordioned inward by a sheet of mesenchyme. As these folds converge, they partition the cloaca into the rectum and upper anal canal and, in front of these, the urogenital sinus. The hindgut fuses with the anal pit; endoderm meets ectoderm at the anus, the opening at the end of the alimentary canal.

Urinary and genital systems arise primarily from nephrotomes, a long streak of partitioned mesoderm that lies alongside and lateral to the somites. The rapid growth and transverse folding of the embryo cause this intermediate section of mesoderm to become detached and migrate ventrally, linking to the allantois.

Packed with tiny tubules, themselves interfused with clusters of fine blood vessels (the glomeruli), the kidneys are pure excretory organs, eliminating toxic substances and adjusting fluid balances. A thick basal lamina in each glomerulus functions at the molecular level as a filter, supervising the transit of macromolecules from the blood into newly forming urine. The high pressure of glomerular capillaries is necessary for deriving and then purifying the blood's waste products in this manner. Nitrogenous debris from protein breakdown, filtered from plasma passing through the tubules of the nephron unit, is ultimately discharged into the bladder, while serviceable components are reabsorbed by the blood. In lower vertebrates the ducts open directly into the cloaca, but in higher ones a separate pair of tubes, the ureters, develop from the end of the so-called Wolffian ducts and connect to the kidneys. These drain into a common tube.

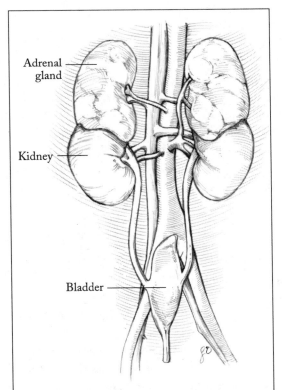

FIGURE 19Z. Adrenals, kidneys, and bladder. Illustration by Jillian O'Malley.

THE UROGENITAL SECTOR of the mammalian excretory system is divided into three masses, each arising from discrete sections of nephrotomes. A frontal pronephros is reproduced only recapitulationally in the advanced vertebrates and is nonfunctional, its cervical cells enacting an archaic tendency to flow into the Wolffian duct formed by their fused nephrons. In amphibians a pronephros coalesces from mesodermal thickenings along the second, third, and fourth somites. A functional kidney in the larval bony fishes, the pronephros is no doubt the original excretory organ of our lineage which induces subsequent renal sections through mutations and natural selection—a lost history reenacted ontogenetically.

While it may have some temporary embryonic function, the actual mesonephros is nonfunctional in mammals (though providing a grid for urogenital convergence). Formed out of mesenchymal tissue (from the dissolution of the nephrotomes), the mesonephros separates in clumps that subsequently develop lumina. Each one grows into an S-shaped tubule which extends laterally until it reaches the common (Wolf-

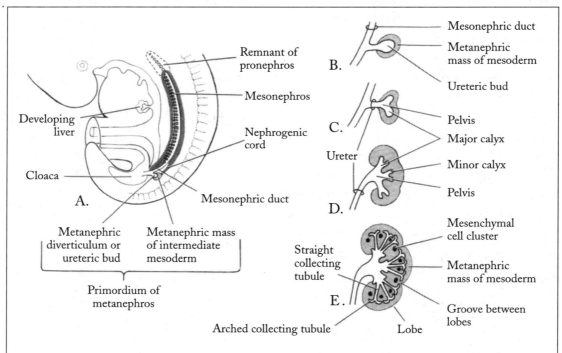

FIGURE 19α. A sketch of a five-week embryo showing the primordium of the metanephros. A. Lateral view at five weeks after fertilization; B. to E. successive stages of development of ureteric bud into ureter, pelvis, calyces, and collecting tubules from fifth to eighth week.
From Keith L. Moore, *The Developing Human: Clinically Oriented Embryology* (Philadelphia: Saunders College Publishing, 1977).

fian) duct. The medial end of each is flattened and pushed inward by glomeruli so that a cup forms. This phylogenetically secondary kidney joins the primary duct that induces it. In mammals with very loose connections between maternal and fetal tissues (like pigs), the mesonephros operates as an excretory organ prenatally.

The permanent kidneys of mammals develop from a posterior section of nephrogenic mesenchyme—the metanephros. Adjacent to the cloaca a node germinates on the mesonephric duct and protracts rearward. This single ureteric bud uniquely induces sets of collecting tubules, for the metanephros is not a duplication of older kidneys. As the bud contacts and grows into metanephric mesoderm, its cranial end expands into calyces, and these branch dichotomously to form lineages of collecting tubules. Mesenchyme provides glomeruli at their tips. Stalks formed by the ureteric bud hollow into ducts, extensions of the bud connecting the kidneys to the urinary bladder—these are the ureters. As the embryo's body grows caudal to the kidneys they are shifted from the pelvic region to the abdomen.

The muscular urinary bladder is a membranous storage sac capable of bloating and swelling outward. While its epithelium is endodermal, its muscle layers and lining are imported from adjacent splanchnic mesenchyme. As the organ expands embryonically, its dorsal wall annexes caudal sectors of the mesonephric ducts. These ducts are later absorbed, their epithelium supplanted by an endodermal equivalent from the urogenital sinus.

Meanwhile, the curved cavity of the endodermal cloaca is divided by a membrane into a dorsal rectum and ventral urogenital sinus. The rectum is continuous with the alimentary canal, and the sinus expands into the mesonephric ducts to form the later epithelium of the urinary bladder. As the ducts are absorbed, urine drains out of the collecting tubules of both mature kidneys down left and right ureters into the bladder. Narrow tubes lined with mucous membranes and a dense, muscular wall, the ureters are as peristaltic as the intestines, propelling their contents by visceral waves. A quarter of an inch wide and a foot long, they are urinary pipes.

As the canal from the bladder initially penetrates the allantoic stalk, urine blends with amniotic fluid drunk by the fetus. This extraembryonic organ, the urachus, later disintegrates.

The sinus behind the bladder narrows into the urethra, the eventual urinary duct; it derives its muscle from contact with mesenchyme. As the kidneys produce traction in their cranial migration, the ureters of males open laterally and cranially to form sperms' ejaculatory ducts. The rudiments of these degenerate in females.

THE TRIANGULAR ADRENAL GLANDS sit atop the kidneys "like little cocked hats"[11] and, through the early fetal period, rival them in size. They ultimately settle at about one and one-half inches of length, a slightly narrower width, and half an ounce of weight. The adrenals materialize complexly, their original gonadally derived cortex wrapped in cells from the coelomic epithelium, their core supplied by migrating neural-crest cells. A series of self-enveloping differentiated structures, this glandular mass is a source of sudden bursts of energy, no doubt critical in the ancient hunt. Its main product is epinephrine, a hormone that functions much like a neurotransmitter, with wide-ranging receptors and psychosomatic effects throughout the body. While increasing cardiac output, epinephrine also breaks down stores of glycogen into glucose. Discovery of the adrenal role helped reinforce a Darwinian, anthropocentric mythos:

"During the long evolutionary rise to power of the human race the adrenals were man's bulwark in the survival of the fittest.... Our jealousies, hates, fears, struggles for wealth, power, position, lusts, and our superstitions all call upon the reserve supply of adrenal secretion—the fighting or energizing secretion—until the glands are exhausted...."[12]

In society, adrenalization can mimic emotional authenticity and neuromuscular function, recalling the prehistoric struggle of life and death, but such states distort personality and cause seemingly intense moods which burn out meaninglessly.

Adrenal exhaustion is an epidemic of modern civilization.

Male and Female Genitals

Germinal Epithelium and Gonadal Cortex

In humans, male and female gonads appear first as outgrowths of the coelomic epithelium on the midline of the urogenital ridge (a region called the germinal epithelium because it was once thought to be the source of primordial germ cells). Soon after the body's primary tissue masses have been defined, the primordial germ cells uniquely arise outside of all three of the layers, becoming visible along extraembryonic endoderm in scattered bands marking routes of migration from where they were birthed to where they will eventually roost. Conditional to species, they may first show at a number of assorted endodermal or mesodermal extraembryonic sites (in salamanders, for instance, they materialize in closer association with median mesoderm). Expressing royal lineage and destiny, they maintain berths aloof from the mortal creature being assembled.

Direct descendants of free-living zooids, gametes are large, spherical, and contain vesicular nuclei. No doubt some of the sensation surging up through layers of adult tissue and coalescing ultimately in fluids and images is a projection of their own latent urgency to be released. They are primitive autonomies revelling to individuate, which is what makes them "germinal" and perhaps even gives seduction its gamy charge. The gametes transmit their breach from the rest of the body through eros into flesh and consciousness.

Even as the germ cells are formed, they are vigorously immigrating toward their epithelium by amoeboid crawling and in bloodflow (depending on species). Populating the epithelial cortex and medulla, the gametes imbed as it thickens and swells into the coelom.

Although ultimately fused into common organs, germinal cells and the genital in which they become implanted have radically different sources. The structural aspect of the gonads derives from intermediate mesoderm situated on either side between the dorsal aorta and the primitive kidney strip. This genital ridge gains mass partly from its increasing numbers of chubby nonmesodermal recruits.

Within the medulla of the ridge, mesenchyme collects around the primordial germ cells in strands as sex cords, a sheath of tissues interfused with blood vessels and nerves — the primordia of both testes and ovaries.

Only after the gametes migrate from the wall of the yolk sac to the site of gonads do they differentiate into oogonia and spermatogonia and become incorporated into genital development. Prior to that they are ambisexual and diploid.

THE DIVISION INTO male and female is a secondary induction of tissues around sex cells, assembling organs for their storage, maturation, delivery, and development into blastulas. In land vertebrates most reproductive tissue is appropriated from discarded renal masses used in fabricating the more primitive excretory systems of water-dwelling ancestors. As an antecedent phylogenetic function irrigated their mesoderm by budding, bifurcating inductions, the fetal excretory river continued to branch ontogenetically deeper and deeper in caudal cloacal extensions. More sophisticatedly designed urinary tributaries arose embryogenetically, replacing older renal networks, which lay vacant and ready for some other use. The emerging mammalian reproductive system colonized and redesigned old nephrotome plumbing, carving a tribe of new sex organs from its parts while eroticizing aspects of its vestigial anatomy. Abandoned urinary channels and aqueducts became passageways for sperms and eggs (see below).

If it were not for the bisexualization of species, reproductive activity could occur easily within the body-plan and mechanics of a single gender and phenotype. Yet, as evolution segregated germ cells into morphologically distinct sperms and eggs, contiguous mesodermal tissue differentiated into male and female organs (see Chapter 22, "The Origin of Sexuality and Gender"). This primeval epochal event occurred prior to organ formation for reasons probably (in part) having to do with chromosome authentication and repair, using exogenous templates—males for females, females for males—to copy undamaged and novel codons (see Chapter 4, "The First Beings"). Whether or not this is a reason (and we can only guess), the development of male and female sex types represents the eminent dichotomization of phenotypes among humans (overshadowing size, coloring, race, etc.). Males and females remain identical in almost all their organs and chromosomes. Yet a minor codon variation leads to quite different reproductive tissues, body types, personalities, tastes, styles of behavior and dress, and (most notably) roles in the genesis of new embryos.

The engenderment of the male system is in many ways more intricate and complex, for it requires secondary induction of tissues tending naturally toward the manufacture of female organs. Sex-specific hormonal induction is equally compulsory for breasts, labia, oviducts, vaginas, and the like; however, the fact that the undifferentiated genital material bears more female than male characteristics indicates that "woman" is the base state of the underlying body system.

Genetic Basis of Gender

In embryos with a so-called XX chromosome, the gonadal cortex will become the ovary and the mesenchymal medulla will not develop beyond a thin epithelial layer.

In the XY (male) complex the medulla hollows out into seminiferous tubules and develops as a testis; the gonadal cortex regresses. The Y chromosome transcribes proteins which, in the general morphogenetic field, have the effect of inducing the medulla of the indifferent gonad, so the sex cords differentiate as tubules. In the absence of such induction in females, an ovary forms, swelling with the expansion of a cortex and germ cells within. Those nearest the surface of the ovary's cortex become the primary oocytes.

A TDF (testosterone-determining-factor) gene on the Y chromosome initiates the gendering process. If, during the first meiotic division, this factor is translocated from the Y to the X of a developing sperm, the person inheriting the Y chromosome that lacks the TDF gene will be female, even with an XY complex. Conversely, a person receiving the X chromosome with the translocated TDF gene will be an XX male (see Chapter 25 for transsexual variations).

Ovary and Testis

Even though sexual fate is established genetically at the time of fertilization, sex organs must be induced and developed, so until the seventh week of human embryonic life the anatomy of male and female are identical: one androgynous organ. This raw genital embryogenically precedes the male and female organs in which it will have its singular biofunctional expressions.

In female development, as the medulla withers, the primordial germ cells lodging in the cortex continue to increase its thickness. Meanwhile, mesenchymal cells from the inner cortical surface split into clusters about the germ cells, as these become primary oocytes. Other germ cells lounge nearer to the surface of the cortex, a reserve of new eggs from puberty until menopause.

The phallic portion of the indifferent gonad will ultimately elongate into a penis under the influence of male hormones (androgens) or differentiate into a female clitoris.

THE MALE TESTIS and spermatic cord (later comprising testicles inside a scrotum) originate within the mid-groin area of the body cavity, attached to the floor of the primitive pelvis by a rigid ligament (the *gubernaculus testis*). They develop in complete independence from the penis (which has an embryonic cohesion with the anus). As testicular seminiferous tubules forfeit their connections with the germinal epithelium, they become enclosed in a fibrous capsule. Supported by their own abdominal fold, the developing testes slide away from the mesonephros in sections (lobules), each comprising a long, narrow, coiled tubule.

In the ovary, germ-bearing cortical epithelium splinters into primordial follicles—single oogonia surrounded by flattened cortices. The ovary also pulls away from the retreating mesonephros and is suspended by a mesentery.

The testes of the XY fetus manufacture hormones (including testosterone) which stimulate the mesonephric ducts of the vestigial middle kidney; these become the male genital tract, the *vas deferens*. They have already disarticulated from the germinal epithelium within a thick band of membranous fibers, the *tunica albuginea*, the eventual lining of the scrotal sac.

With the elongation of the fetus, the *gubernaculus testis* pulls the organ down through inguinal canals (formed from the lining and muscular fascia of the

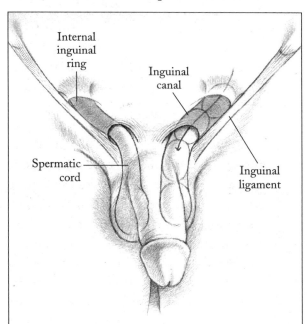

FIGURE 19β. Pathway of descent of testicle prior to birth. Internal inguinal ring is exit point of testicle from abdominal canal.
From R. Louis Schultz, *Out in the Open: The Complete Male Pelvis* (Berkeley: North Atlantic Books, 1999).

abdominal cavity), over the pubic bone, into the initial scrotal swellings. During the seventh or eighth month the testes migrate into their pouchlike scrotal sac and remain there, suspended by a thin cord of tissue comprising blood vessels and nerves as well as the sperm duct. The testicular walls contain a combination of epithelial Sertoli cells from sex-cord tissue and germinal spermatogonia.

The ovaries of the female descend through the abdominal cavity in similar fashion to the testis of the male.

ENDOCRINE ACTIVITY UNDERLIES ALL of these morphogenetic episodes. FSH induces both the ovarian follicles and seminiferous tubules and then helps differentiate them separately into female and male sex organs; it also stimulates the expulsion of ripe ova during ovulation. Pituitary luteinizing hormone (LH) participates with FSH in its stimulation of follicle cells to mature and (in females) to secrete estrogen and initiate ovulation. In the male, LH stimulates interstitial cells in the testes to differentiate and secrete the male sex hormone testosterone.

Penis, Clitoris, Scrotum, and Labia

Within the first month of a human embryo's life, a prominence develops at the front of the cloacal membrane; swellings and folds form on either side of it—the clitoral Ur-phallus. "It is only at the five-week mark that those fetuses destined to be males endure the spontaneous abracadabra that transmogrifies their clitorises into penises."[13]

Androgens secreted by the testes accelerate the growth of the clitoral mound, and it elongates as a penis, pulling the urogenital folds forward so that they mold the lateral walls of a urethral groove underneath it. Lined with endoderm from the urogenital sinus, the folds fuse from rear to tip, shaping the penile urethra. Columns of lightly spongy tissue arising from mesenchyme surround the urethra (the *corpus spongiosum* in front, bearing the penile urethra with the glans at its tip, and in back, the highly vascularized fascia of the *corpora cavernosa*). The phallus is composed mostly of this soft, porous, erectile tissue.

Ectoderm grows back over the penis' tip, leaving a cellular strand which splits and meets the urethral groove. As the groove zips closed, its external orifice is pushed to the tip, the glans differentiating around it. A tiny mound of ectoderm on the margin of the glans pushes inward as a plate, or cellular cord, which, upon cleaving, joins the urethral groove within the body of the penis and provides an external orifice at the glans tip. A foreskin grows over the glans, and the labioscrotal swellings creep toward each other, meet, and coalesce into the scrotum. The tissues of this sac become continuous with those of the penis.

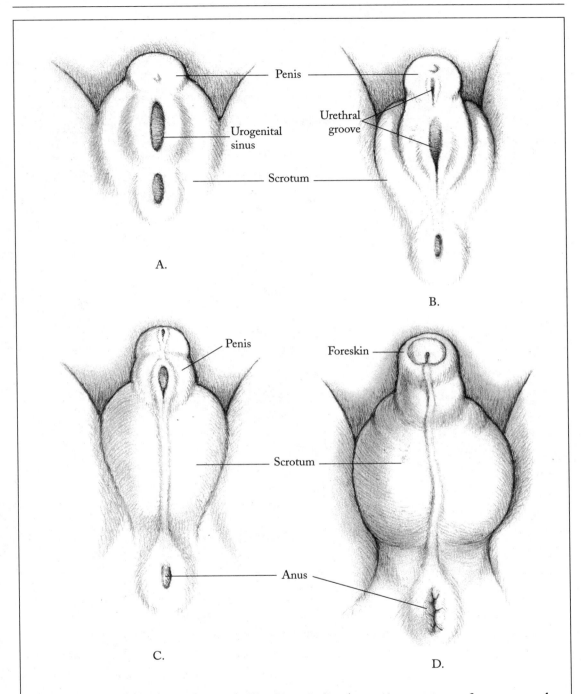

FIGURE 19γ. Stages in development of male genitalia. Approximate stages of pregnancy: A. 7 weeks; B. 10 weeks; C. 12 weeks; D. Near term. From R. Louis Schultz, *Out in the Open: The Complete Male Pelvis* (Berkeley: North Atlantic Books, 1999).

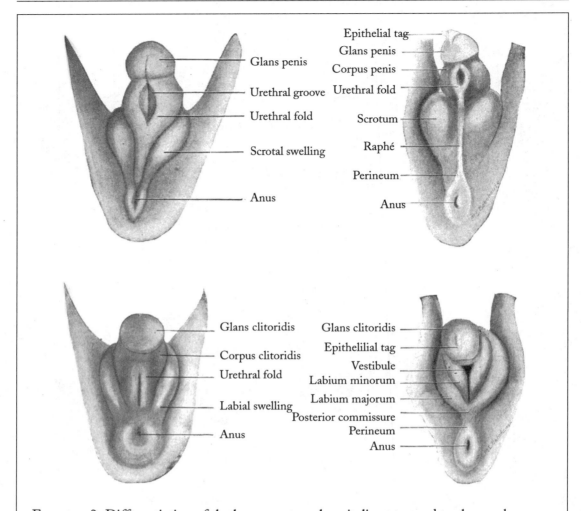

FIGURE 198. Differentiation of the human external genitalia at ten and twelve weeks.
From Leslie Brainerd Arey, *Developmental Anatomy: A Textbook and Laboratory Manual of Embryology* (Philadelphia: W. B. Saunders & Company, 1946).

The penis is a thick ejaculatory flower rooted one to three inches deep in fatty tissue over the pubic bone above the perineum and anus. Bearing a fusion of germinal epithelium and testis in its scrotal aspect and meso-endodermal tissue in its phallic head, it is a double appendage—a snake embodying the swift journey of the sperm to the ova, toting an ancient pouch of alchemical seeds at its base.

The penis is mesodermally integrated with the rest of the body, its fascia continuous with the perineal fascia and the muscles of the abdomen and spine—in fact all the way to the fascia of the respiratory diaphragm (see next chapter).

EVEN LACKING ANDROGENS, a morphologically similar though less developed genital, composed likewise of erectile tissue (the clitoris) germinates in the female at the vaginal opening. Waxing swiftly at first, its pace gradually subsides until the completed organ is smaller than the penis and without its fusion of urogenital folds except at the very front of the anus.

The labioscrotal swellings expressed as the scrotum in the male remain unfused in the female as the labia majora. Extended folds of fat and glands, covered with hair externally, the labia majora bear rings (or folds) of labia minora cupped more deeply and compactly within. Together these create an adduction zone, quite different phenomenologically from anything in the male. The clitoris locates at an equivalent site to the penis in the female—the anterior juncture of the labia minora. A urethral orifice exits between the clitoris and the vagina, the latter usually covered by a membrane, the hymen. A small area below the labia majora and above the anus forms the female perineum.

Kidney ducts are converted to sperm ducts in the male.

In males, mesonephric buds germinate as seminal glands, while outgrowths of the urethra in surrounding mesenchyme produce a doughnut-shaped prostate gland around themselves. The gland's thin, milky secretion ultimately envelops the sperms, capacitating them as seminal fluid during their passage out of the body through a complicated series of ducts and

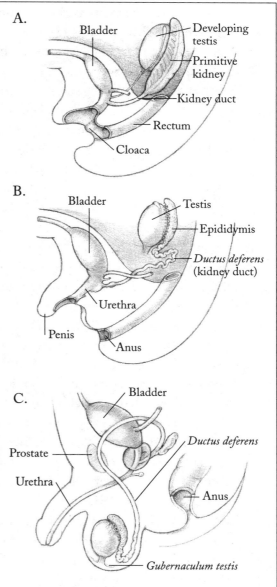

FIGURE 19ε. Development of the male reproductive system. A. Development of the testes in relation to the primitive kidney; B. Degeneration of the kidney with the duct system remaining; C. Descent of the testes into the scrotum.
From R. Louis Schultz, *Out in the Open: The Complete Male Pelvis* (Berkeley: North Atlantic Books, 1999).

glands. The journey from the scrotum commences along a tightly curled, hooked tube, the epididymis; this eventually meets the *vas deferens*. As the *deferens* canal transits the posterior margin of the bladder, a duct bearing seminal fluid meets it to form an ejaculatory pipe. This passes through the hole in the prostate into the urethra, through the inguinal canal and penis, culminating at the external urethral orifice, the exit point for both sperm and urine.

The mesonephros develops renally (as the mesonephric duct) and reproductively (as the sperm duct). Evolutionarily the Wolffian duct comes to play two roles simultaneously, as the ureter and sperm duct.

There is very little space in the body, especially at critical nodes where organs, crammed together by hydraulics, are connected by culverts. Metabolic processes seize what tissues and lumina they find in their path, regardless of architectural elegance. Reproductive machinery is actually trapped within the male and female digestive systems by a series of historical events resulting in urinary-genital convergence. Fecal, urinary, and reproductive materials must pass through the same cloaca in order to reach the outside world. Because sexual organs border on the nephric ducts, renal channels in males are in essence coopted for the passage of sperm to the cloaca. Only in the second month of pregnancy is this common channel subdivided by membranes and folds, the rectal path thereby sealed from the bladder and urogenital tract. The latter continue to share a conduit through the male urethra. Neuromuscular and psychosomatic elements of excretory and sexual activity merge here, leading to male and female postures and stances, both fluid and repressed, aggressive and vulnerable.

The sheer proximity of the gonad to the renal organ leads to this secondary partitioning of the primary nephric duct and its exploitation for transporting eggs and sperm to the cloaca. Basically the primitive kidney, the mesonephros, degenerates in the context of the development (lower in the abdominal cavity) of the metanephros, but its vestigial ducts (as described earlier) remain available for adoption by the testes in the delivery of the sperm. The excretory/reproductive channel is conveniently borrowed (taken phylogenetically) from lingering mesonephric tubules that become ductules and commingle urinary and genital delivery systems through the mesonephric duct.

Potential female ducts and glands in the male are suppressed by AMH (antimüllerian duct hormone secreted by Sertoli cells of the developing testis) and remain histological remnants through the lifetime of the organism. Unless so inhibited, they develop autonomously to form a female reproductive tract.

The urogenital confluence, anatomically inevitable, carries mixed libidinal contents and highly charged, hybrid meanings that oscillate between the elimination of tissue debris and the discharge or attraction of gametes (see Chapter 25). Sexual represssion can carry over into urinary or fecal dysfunctions, including constipation, incontinence, and flatulence. A scrambling of anal, genital, and peeing sensations may likewise elicit sexual aberrations, including sadomasochistic acts that implicate the excretory system with its peristaltic functions in the eros of ejaculation and orgasm.

Urogenital and rectal neuromusculatures are too crowded embryologically not to overlap and contend.

Vagina and Uterus

The female manifests the complementary aspect of this biphasal system. Her hormones—FSH (follicle-stimulating hormone from pituitary secretions), estrogen, and progesterone among them—suppress the mesonephros and stimulate the differentiation of the Müllerian duct into oviducts, uterus, uterine tube, and vagina. All of her mesodermal sex organs are different versions of faux kidneys, the paramesonephric network originating from invaginations of the coelomic epithelium parallel to the unused male ducts (the Wolffian duct is used in the embryo to carry urine from the mesonephros to the cloaca). While individual invaginations fuse and run caudally into the uterovaginal canal, the unfused cranial chambers become uterine or Fallopian tubes. It is through these oviducts that the mature ovum will pass out of its connective-tissue capsule (the Graafian follicle) to lodge ultimately in the uterus as the zygote.

Like the delivery of sperms, the transport of eggs has evolved radically in the direction of urogenital convergence in the vertebrate line. Lampreys have no oviducts, shedding eggs as mesodermal musculature squeezes them through a body wall into a urogenital sinus exiting behind the anus. Amphibians manufacture oviducts by accreting cells from their coelomic lining. These pile into a backward-forming ridge along the kidney line. At its anterior end a funnel breaks into the ridge, transforming it into a tube. Posterior amassing of cells draws the tube back into a position to empty ultimately into the cloaca.

In mammals, as the oviducts spread rearward, they cross over the ventral section of the developing kidney ducts forming ureters and join at the midline to the rear of the urethral aspect of the urogenital sinus. In many species, these paramesonephric ducts evacuate directly into the cloaca. In humans the urogenital sinus is shortened along an anteroposterior axis and lengthened dorsoventrally, becoming a slit between the labia minora. As vaginal and urethral openings push closer to the surface of the body, the vaginal orifice eclipses the urethral outlet.

SITUATED IN THE PELVIC CAVITY, the uterus is induced from a paramesonephric tissue mass into a thickly muscular organ (the size of a pear), a small cavity persisting inside it. The uterus has an upper body and a lower neck (cervix) opening through the cervical canal into the vagina. After impregnation of its egg, this organ will expand to many times its size in order to encompass both an embryo and a considerable amount of fluid, its ultimate range extending to the upper margin of the abdominal cavity.

The body of the uterus meets the oviducts just below its most bulging prominence (the fundus). At its other end, twin bulbs from the urogenital sinus intercept the end of the cervical canal and form a solid vaginal plate; its median cells disintegrate, leaving the lumen of the vagina. While the oviducts are travelling backward, meeting each other and becoming tubular, the prospective posterior vagina persists as a solid mass of cells combining smooth muscle with mucous membrane. Its boundary with the uterus is indistinct and epithelial. Later it will differentiate as a four-inch-long dilatable tube.

Female hormones also catalyze mammary-gland maturation (see earlier in this chapter).

Male and Female

Men and women have unique chemistries and morphologies, endemic internal spaces, shapes, hormones, layers of tissue, tubes, and muscles, albeit built from the same core components. Thus, they feel the same tissues molded in very different ways, defining a proprioception of gender both separately and in relation to one another.

Male and female nervous systems and brains develop their own phenomenologies and emotional ranges. "Hormonal floods of testosterone or estrogen organize us toward male and female behavior: the erection of the penis, the release of the scent of estrus from the female, the bringing back of the pelvis to open the vaginal and uterine tubes. These predispositions have begun during the early weeks of life embryologically. Hormonal releases of androgen and estrogen give a feeling, a behavioral thrust to gender. Blood carries the secretions that give the generative organs their specific tubal motility, sending the sensations that state 'I am a male' or 'I am a female.' These liquid floods are the flushes and waves of desire. The blood arousal, this cellular passion galvanizes gender. The fluids of the ductless glands are carried to all the organs where identity is based."[14]

The external male phallus is a spherical, tubal zone of sensation, a vehicle whereby general bodily imagery is translated into sexual feelings and activity. Under the correct signal the parasympathetic fibers of the autonomic nervous system incite the small arteries to dilate, causing blood to flow into the sinuses of vascularized tissue. The *corpus spongiosum* and *corpora cavernosa* swell; pulsations draw the seminal fluid bearing its germ cells down the *vas deferens* and ejaculatory duct, to be expelled in ejaculatory spasms through the urethral orifice. These spasms give rise to additional sensation, causing a general bodily orgasm to spread through the viscera and skin, resolving the tensions that led to them. This includes "involuntary rhythmic contraction of the anal sphincter, increased breathing rate, increased heart rate, and elevation of blood pressure. There is a tingling sensation throughout the body."[15]

Afterwards the sympathetic fibers constrict the arteries, blood in the sinuses flows back into the veins, and the penis becomes flaccid, an ordinary lump of flesh or organ, like an externalized kidney or inside-out sphincter.

This entire event is involuntary and relaxed. Though the penis is not a muscle to be flexed, there are muscular elements associated with it that may give males sensations of intentional control and mastery, leading to macho sexual personae. At the base of the ischial tuberosities a pair of ischiocavernosa muscles extend into the base of the *corpora cavernosa,* and these may augment or help sustain erection, but they do not cause it. A bulbospongiosus muscle arising from the perineum knits into the base of the *corpus spongiosum* and helps expel the final spurts of urine from the urethra after the bladder is empty (and can stop the urine flow midstream in an embarrassing circumstance where unplanned restraint becomes judicious); it possibly contributes some of its urogenital grip to an erection.

One way or another, all of these nerve and blood innervations and incremental neuromuscular effects contribute to the sensation of male discharge.

THE CLITORIS AND VAGINA likewise undergo throes of release which are transferred throughout the body. Less thrusting and more complexly transmitted through tissues and membranes,

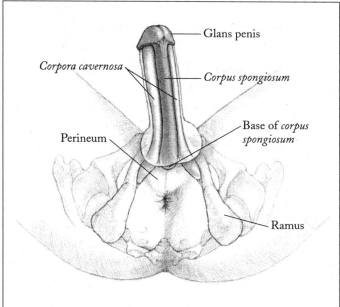

FIGURE 19ζ. Structure of the penis.
From R. Louis Schultz, *Out in the Open: The Complete Male Pelvis* (Berkeley: North Atlantic Books, 1999).

these spasms transmit a prism of subtle sensations through which the depths of female identity—labia, vagina, breasts, uterus, and even ova—are also savored.

"The hollow spaces of the vagina and uterus generate feelings different [from] those produced by the pulsating tubes of the penis. One gives rise to a filled throbbing assertion, the other to a milking, filling feeling. The male feels the imperative to thrust, to expel, to fill; the woman feels the imperative to reach and be filled."[16]

The nerves in the penis and vagina are connected not only to autonomic sexual functions but to myriad symbolic representations and expressions of eros throughout the neuroendocrine system and in the cerebral cortex as well.

The pulsations of the erect penis and clitoris are akin to the peristalsis of intestines and the expanding and contracting waves of heartbeat and blood vessels. Intrinsic cellular movements fall subject to localized anatomies. Sexuality involves deep trances (not unlike the calms of sunning animals and hypnagogic sleep, though rhythmically agitated), leading to more fluid tissue states, culminating in spasms and exchange of active, embryogenically capacitated cells.

These organs sow, collect, and organize many sensations not associated with sexuality *per se,* having more to do with empathy, overall kinesthesia, central-channel (Conception Vessel) circuitry, somatic core, neural dispersion, ch'i energy, and personality formation. In fact, this is true to a greater or lesser degree for all organs. However, in an overly genitalized culture such as ours (in which sexuality is valorized and commoditized), information from genital sensations is often lost or artificially eroticized. The result is a diminishment of texture and spaciousness as well as an escalation of compulsion and arid ritual.

Clitorises, penises, vaginas, and other sex organs incite consciousness as well as offspring. When our insides speak through our organs and fluids, and impart sensations of filling and emptying, reaching and grasping, penetrating and sucking, male and female are embodied. They bestow on each other polar vortices intrinsic to the whole of nature and the destiny of incarnation. In searching together for their individual identities, they collaborate across their gap of tissues in fathomless, transpersonal acts.

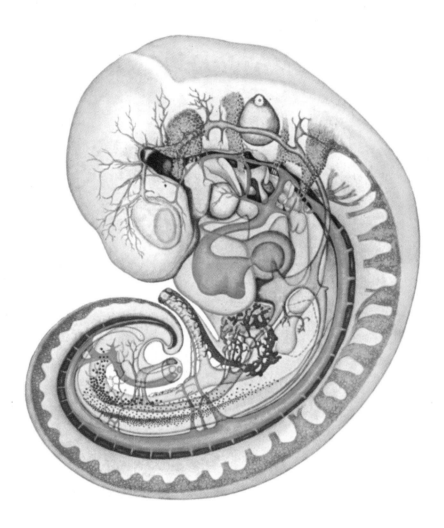

FIGURE 25. Human embryo of 32 somites, 4½ weeks old.
From Emil Witschi, *Development of Vertebrates* (Philadelphia: W. B. Saunders & Company, 1956).

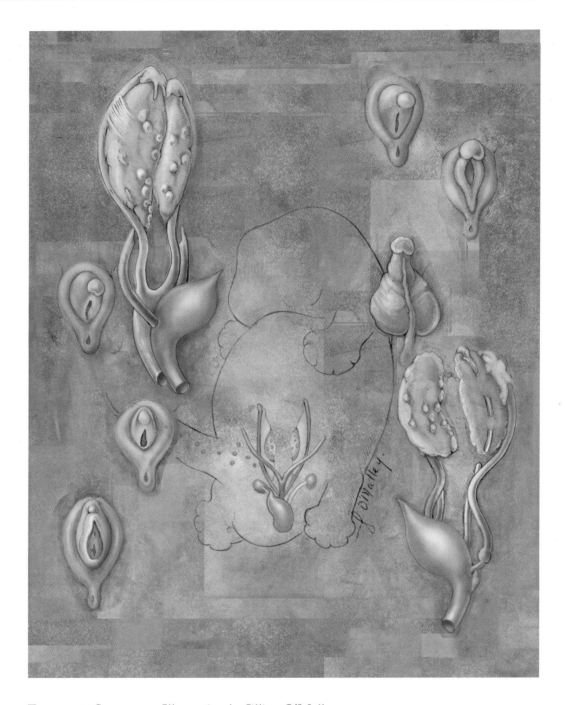

FIGURE 26. Sex organs. Illustration by Jillian O'Malley.

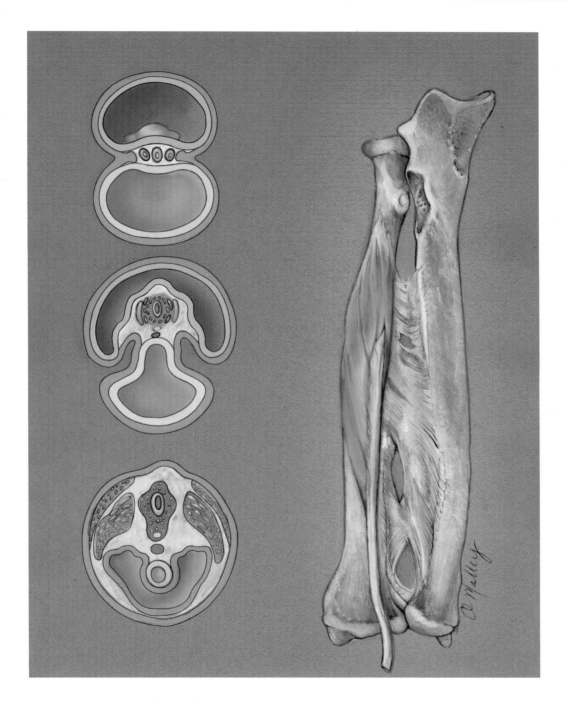

Figure 27. Bone, muscle, tendon, and fascia. Cross section of embryo showing somitomeres dividing into: sclerotomes (vertical column), dermatomes (skin), myotomes (muscle). Illustration by Jillian O'Malley.

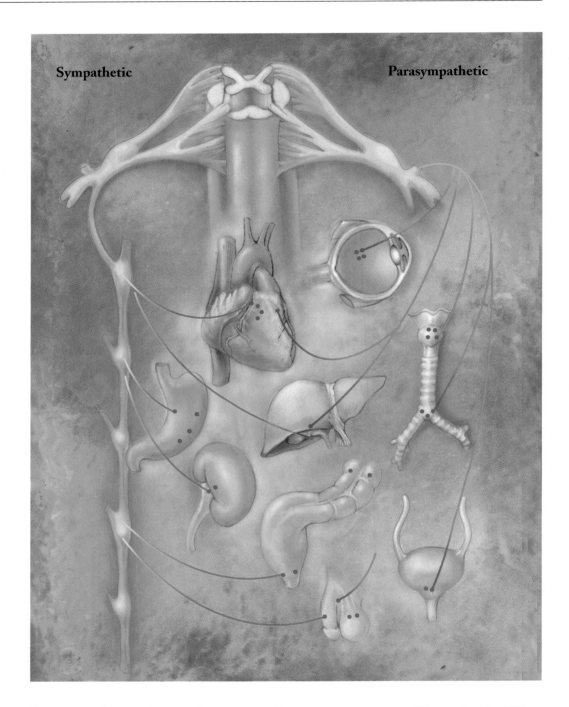

FIGURE 28. Sympathetic and parasympathetic nervous systems. Illustration by Jillian O'Malley.

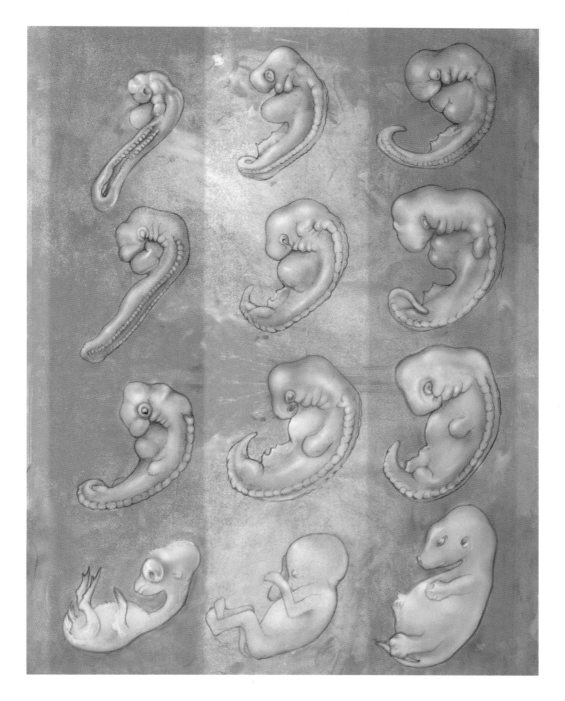

FIGURE 29. Neurulation and organogenesis of chicken, human, and pig. Illustration by Jillian O'Malley.

FIGURE 30.

Mandelbrot Set with Feather Filling.

The Mandelbrot set M is a fractal form discovered by Benoit Mandelbrot. The Mandelbrot set lies inside a disk of radius two around the origin of the plane, with the "stinger" just touching the edge of this disk. Its border is so highly irregular that it is thought of as having a dimensionality greater than one. In this image, the interior of the Mandelbrot set has been filled with a "feather" pattern relating to certain periodicities in the computation of the set.

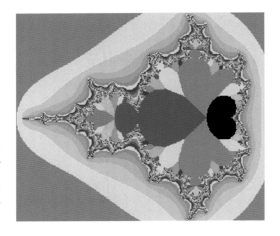

There are two key lessons to glean from the Mandelbrot set. First of all, a very simple computational process can produce an organically rich-looking structure. Secondly, although such a computational process is conceptually simple, it may take an exceedingly long time to carry it out.

FIGURE 31. Roadkill Cubic Julia Set.

Related to the idea of the Mandelbrot set are the Julia sets, named after the mathematician Gaston Julia. For every point c in the plane we can find a Julia set J_c, which will be some shape in the plane. To decide if a test point p lies inside a quadratic J_c, we define a sequence $Z_c(n,p)$ by $Z_c(0,p) = p$ and $Z_c(n+1,p) = Z(n+1,p) = Z(n,p)2 + c$. If the sizes of the $Z_c(n,p)$ grow larger than 2, then p is outside the Julia set; otherwise it is on

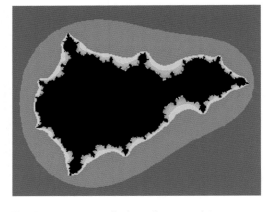

the border or interior of the set. (This "Roadkill" Julia set is actually based on a cubic process which we will discuss in the caption to the "Shmoo" image to follow.)

Some J_c are connected forms like squashed disks. Others thin into dendritic forms like snow crystals, while a third kind breaks into clouds of dust. It turns out that a J_c is connected if and only if the origin point O of the plane lies inside it. If you have a mathematical bent of mind, you can see that this means that the Mandelbrot set is a kind of catalogue of the quadratic Julia sets; that is, a point c lies inside the Mandelbrot set if and only if its corresponding J_c is connected and disk-like.

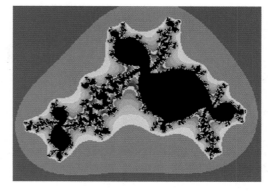

FIGURE 32. Shmoo Cubic Julia Set.
The "Roadkill" and "Shmoo" Julia sets shown here are based on a cubic rather than on a quadratic iteration method. We use two complex numbers c and k to specify a cubic Julia set Jck. Given a point p, we decide whether or not p is inside the set by examining the growth of the sequence Zck(n,p) given by Zc(0,p) = p and Zc(n+1,p) = Z(n+1,p) = Z(n,p)3 + k*Z(n,p) + c. If Zck(n,p) grows arbitrarily large, then p is outside of Jck; otherwise it's on the border or interior of Jck.

The cubic Julia sets come in four kinds. As well as the disk-like, dendritic, and dust-like forms, they also can have a nodular form, like the root system of a potato or a peanut plant, such as the "Shmoo" shape shown here.

A lesson to draw here is that by adding more complexity to our formulas we can get a broader spectrum of possible shapes, with less obvious kinds of symmetries.

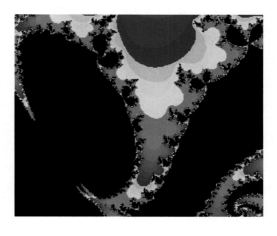

FIGURE 33.
Martian Landscape in the Rudy Set.
The quadratic Mandelbrot set M can be defined as the set of all c for which the quadratic Jc is connected. By analogy, for each k, one can define a cubic Mandelbrot set Mk as, roughly, the set of c such that the cubic Julia set Jck is connected. The Mk themselves come in shapes similar to that of the cubic Julia sets: some of them are connected disks, some nodular, some dendritic, and some dust-like. Abstracted to yet one level more, the so-called Rudy set is, roughly, the set of k for which the cubic Mandelbrot set Mk is disk-like.

The image here shows a zoomed-in view of a detail of the Rudy set. The complexity of this set's definition is such that the structures one finds inside it come as a complete surprise.

Illustrations and captions by Rudy Rucker. These four images were generated by a program called James Gleick's Chaos. The Software can be downloaded for free from Rudy Rucker's website: www.mathcs.sjsu.edu/faculty/rucker.

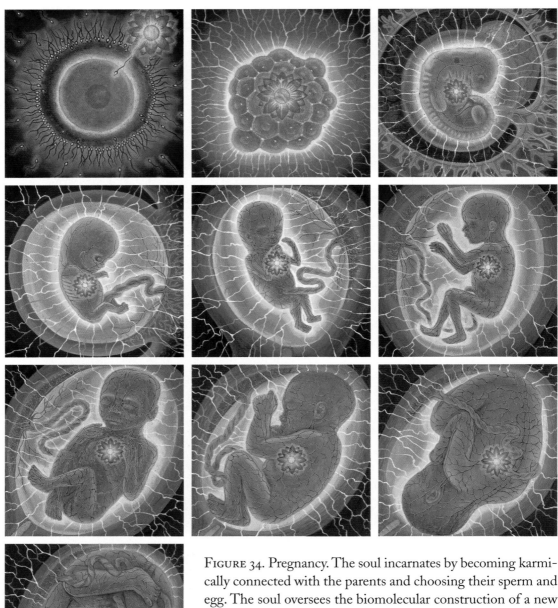

FIGURE 34. Pregnancy. The soul incarnates by becoming karmically connected with the parents and choosing their sperm and
egg. The soul oversees the biomolecular construction of a new
body, and barring damaging influences, the new body will be
nearly perfect. The embryological bloom of creation, starting as
a single-celled zygote, miraculously unfolds into trillions of cells
working harmoniously in the various systems of the body. This
is a time of radical transformation for the incarnating soul, as
well as for the new parents. This painting was done while Allyson
was pregnant with our daughter, Zena Lotus Grey. Illustration
and caption by Alex Grey.

The Musculoskeletal and Hematopoietic Systems

The Stability of the Body

EVEN THIS LATE IN THE BOOK, organismic unity is a startling occurrence. Individual cells move about, extending lamellipodia, rearranging microtubules, responding to cortical tension, dissolving their own membranes—and still we stand tall and look one another in the eyes through masks of dermal designs. Despite the vagueness of so many independently linked organs rustling inside us, we negotiate a muddle of competing sensations into a wholeness.

Lugging twenty-four-hour-a-day protein factories as if they were neutral wetsuits, conducting our business in extra-ectodermal garments, sashes, and girdles, we are somehow believable to ourselves. The mouth speaks the presumed truths of hormones, enzymes, neurons. Under neuroendocrine injunction, the body-mind propagates meanings and engenders symbols and lifestyles. Finding outlets of expression in the body wall, sense organs, genitals, and limbs, internal waves and pulsations radiate the simultaneously neuromuscular and hormonal basis of character.

Eyes are where they should be, person after person, set in an anterior cranium, flanking the nose from above, subtending the forehead, despite different races, nationalities, families, and genetic quirks. No two people are the same; yet we recognize humanity instantly. Even in crowds of strangers, only a rare man or woman elicits a second glance; they are all variants of people we know.

Our torsos come off a mitotic assembly line, replicated in different proportions and scales, packed onto subway cars, strapped seat by seat across planes, matched in football and wrestling matches, convened in assemblies to make hominoid rules.

Notwithstanding the multiple complexities of systems making up the body, it does not collapse or warp; it does not unravel or deteriorate. It maintains local and global symmetries. Cells do not jumble or lose their way. They not only proliferate and differentiate in suitable relative quantities, but they hold strict relative positions. They travel along precisely induced paths, adhere on chemical and mechanical bases, differentiate and selectively quarantine their components with epithelia of differential permeability, and are restrained from invading one another's territories by contact inhibitions. As they spread and interfuse, soft parts swell and interpolate, fill their niches, stitch wounds, synthesize the constituents of metabolism, and evacuate debris.

ALTHOUGH EVERY TISSUE SURFACE and structure in the body contributes to cohesion, the connective-tissue framework is the mold—the soft, elastic grid—for the preservation of a global vertebrate structure during cellular turnover of its modules. Connective tissue interpenetrates and imbues bone and cartilage in the skeleton, dermis, tendons, muscular ligaments, envelopes of muscles, blood vessels, and neurons, as well as intervening fasciae that connect and tie these modules together. As we shall see below, the cells of connective tissue all arise from the activity of mesenchymal fibroblasts migrating from the lateral-plate mesoderm adjacent to the somites of the embryo and becoming ingrafted in the collagenous extracellular matrices they secrete everywhere they go.

Collagen

COLLAGENS COMPRISE A FAMILY OF FIBROUS PROTEINS, the fourteen or so members of which contribute to tightly strung, extracellular matrices in tendons, ligaments, skin, the webbing of muscles, and bones. Secreted by connective-tissue cells (the afore-mentioned mesenchyme-originating fibroblasts), collagens are the most plentitudinous proteins in mammals, comprising at least twenty-five percent of the sum mass.

The collagen molecule is a left-handed polypeptide helix with three helices braided into a right-handed superhelix. Held together by hydrogen bonds, the helices (comprising polypeptide chains of about a thousand amino acids each) buckle into a variety of crystalline architectures able to disassemble and fabricate textural units. It is the combination of the ringed structure of the amino acid proline and the tight spacing of glycine (the smallest of the amino acids) that gives collagen its stable left-handed torque and tight packing. The chains themselves are assembled on membrane-bound ribosomes implanted in the endoplasmic reticulum, where they interthread with indigenously secreted proteins and exogenous

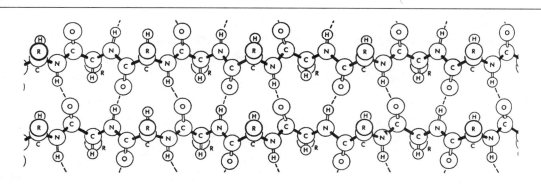

FIGURE 20A. Separate molecular chains linked with collagen-like hydrogen bonds. This is how single tropocollagen molecules tie together to form fibrils, and how fibrils stick together to form structures. Exposed surfaces of collagen structures also have hydrogen and oxygen atoms sticking out, and this is how structures meant to be separate become "glued" together. From Deane Juhan, *Job's Body* (Barrytown, New York: Station Hill Press, 1987).

amino acids. The plasticity of this construction underlies all cartilaginous and skeletal self-healing in adults.

As procollagen molecules are secreted outside the plasma membrane, they assemble into cable-like fibrils, several millimeters long and a few billionths of a meter in diameter. Propeptides prevent fibrils from accumulating among intracellular spaces where they would clog or choke the cells; yet once these regulative polymers have catalyzed the formation and delivery of fibrils, they vacate the collagens, for if they stayed, they would render them brittle.

Covalent cross-links between lysine side chains within the collagen molecules further strengthen them. These bonds proliferate in tissues where protractile strength is a requirement—the Achilles tendon, for instance.

Fibroblasts express their tensility and tangledness in the collagen medium they are secreting, slinking over it and yanking it into spatial configurations, making sheets, strands, and yarn. Its moist components contract in globules; its dense aspects spread in mesh. Each of these qualities is passed upward to the muscles, ligaments, and bones of the body. Even as their handiwork emerges in great sheets, fibroblasts continue to tug and fuss at it like tennis netting, refining and subtilizing it.

Puppets, sailboats, crossbows, stretch fabrics, tents, and string games all emulate aspects of musculoskeletal systems.

As WE FEEL our basic wholeness and connectivity, we appreciate the fineness of the embroidery and its dense neuralization. Connective tissue provides our springy oneness, the girth of our electrical wiring—our stretches, yawns, stress patterns, and synchronized, purposeful movements. Across the extracellular matrix, collagen in its manifold forms links together the intelligences and activities of cells and weaves their morphology into a fine three-dimensional textile unlike any cloth sewn on a mere linear shuttle. Without the musculoskeletal system other organs have no fixed locales or contexts. Furthermore, the components of the connective system are ontogenetically and phylogenetically seminal, inducing the rudiments of many other tissue plexuses. They are a gene-transcription compass, retaining and confederating our underlying worm, sponge, fish, and monkey shapes.

Fibroblasts, cartilage cells, and bone cells are individually sources of connective material. As they secrete collagenous extracellular matrix, they collaborate with their close cousins, fat cells and smooth muscle cells, to establish the contextual framework of the body as well as to repair tissue and heal wounds. Fibroblasts are the least determined and most versatile of these cells, for they can become cartilage- or bone-generating cells under subsequent induction by the extracellular matrix. Cell shape and tightness of anchorage also govern gene expression—strong adherence stimulates rapid mitosis.

Fibrils induce one another and, once spun, tend toward multiplication by self-assemblage. The number and variety of them in mammals testify to the might of duplicating DNA and its iterative patterns of amino acids. Fractal output bastes a small animal into a larger one and a relatively buoyant, wormlike insect or tunicate into a dense visceral quadruped.

In the extracellular matrix also dwells a complex of elastic fibers of similar amino-acid make-up to collagen (abundant in proline and glycine, cross-linked by lysine residues) but with some key differences. The elastin polypeptide backbones flop in random coils without folds. This combination of cross-suturing and willy-nilly spiralling produces a network as resilient as a cluster of rubber bands. As elastin and collagen fibers interweave with each other, they confer both loose and tight qualities on tissue—its ability to stretch but also its cohesiveness and resilience. Living matter keeps its shape and does not easily tear.

Using noncollagen macromolecules to guide and orient their collagen, cells regulate the particular uses and properties of the protein in their matrix. As noted, underlying cross-linking can vary, leading to different overall properties in aggregations of cartilage, muscle, fascia, bones, etc. In the cornea of the eye, arranged in parallel layers, collagen is transparent; in muscles its striated fibers surround individual cells, according ductility to their strands.

Common Bases of the Musculoskeletal System

THE GENESIS OF CARTILAGE, BONES, MUSCLES, and fasciae enacts a continuum whereby migrating cells participate with local tissue in synthesizing specialized structures with unique protein products and distinctive matrices. These organs then repeat and spread on their own like crystals or vines, encompassing one another and choreographing the body. Centered in the mesodermal layer, they integrate deep endodermal material with fluid ectodermal components, rigging and leveraging the organism with halyards, spars, buntlines, jibs, shrouds, etc.

As noted, skeleton and muscle both originate in somites, aggregates of mesodermal cells distributed in postgastrulation movements about a central cavity. As embryogenesis proceeds, they flatten, their cavity constricting to a shallow slit. There they become arranged in linings: a thick inner sheet and a very thin outer sheet. The lower half of the inner sheet, known as sclerotome, breaks up into mesenchyme; its cells migrate into the spaces around the notochord and spinal cord, enveloping them while differentiating into cartilage. In areas of prospective cartilage the mesenchyme condenses and its components multiply, accumulating in nodules ranging from ten to several hundred cells.

Chondrogenesis (cartilage production) depends upon the secretions of bunched-together cells in a small cavity inside the extracellular matrix. Their fabric is synthesized from collagen and other complex molecules consisting of chains of sulfated polysaccharides affixed to a protein core. Cartilage then expands by swelling, as cells in the matrix continue to secrete more collagenous tissue around themselves. A layer of connective tissue enveloping the cartilage performs like a corset, restraining the expansion and conferring a succession of dynamic shapes. This sheath, the perichondrium, is also the source of precursor cells for new chondrocytes.

Sclerotome is the primordium of cells for the vertebrae and ribs. Dermatome provides the dermis of the trunk. Myotome yields tissue for almost all the muscles of the body, including the segmental ones of the trunk.

Bone

A BONY SKELETON IS THE DEFINING TAXON of the vertebrate phylum. Invertebrates are lumps of flesh supported by their own jellies or by rigid exoskeletons (shells). Bone provides body for tissue, leverage for muscles, and shielding for viscera. A skeleton is primarily a bony layer deposited on a scaffold of cartilage. The embryo is kept in proportion because the scaffold also extends as soft tissues expand

and pass through stages en route to final shapes. The design is malleably scaled up without distortion or visceral-muscular interference. This coordination is crucial in going from a tiny neurula to a full-grown adult.

Bone consists of approximately equal volumes of inorganic crystals reinforced by collagen fibers. The much denser crystals are minute individual clusters of calcium, phosphate, and hydroxyl ions fused to the fibers, and contributing the bulk of bone's weight.

The especially tough collagen fibrils of bone are arranged in layers like plywood, units crisscrossing from layer to layer to form a leathery substratum. Collagen is long-lived compared to most cellular proteins; the individual molecules survive for ten years or more before they degenerate and are replaced.

Cartilage and bone are different media of cells inset in solid matrices. Cartilage permeates a flexible matrix that expands by tumescence, bone a rigid one that expands only by apposition. But bone is still a pliable, tensile organ.

Whereas dermal bone forms from the same layer as skin without a cartilaginous precursor (though induced in the opposite direction and with greater metamorphosis), replacement bone precipitates out of preexisting cartilage by calcification. After individual cells obey a signal to increase in size, they calcify and die; a thin layer of their crystal is deposited around a shaft. While some adjacent mesenchymal cells break up into hemapoietic marrow, others become osteoblasts and deposit bone matrix upon already calcified spicules under the fibrous surface membrane of the hardening cartilage.

The skull is an exception, laid down as dermal bone without any cartilaginous precursor.

Like bone, dentine (ivory) consists of collagen with crystals deposited extracellularly. Cells that lay down grids become imbedded in the bone they form but not in the ivory.

Dense yet radiative bony organs are deposited less as mortar than snowflake spikes in tough, elastic sinews. When, under stress, a crack spreads into the collagenous matrix, the bone will be slightly deformed but usually will not break.

Bone also arises directly in mesenchyme by the differentiation of cells into osteoblasts, which then manufacture calcifying substance for spicules. Once embedded in hard matrix, the descendant osteocytes continue to secrete tiny quantities of additional matrix, though they lose the capacity to divide. Bony spicules gradually thicken like crystals precipitating around a mineral spring, consolidating in compact plates between which the mesenchyme, remaining spongy, differentiates into marrow.

Within the marrow, hemapoietic stem cells give rise to monocytes which, after travelling the bloodstream to sites of bone resorption, merge with one another to constitute large multinucleate osteoclasts (resorption cells). These eat away at the matrix, forming deep tunnels in the bony matrix. While blood capillaries populate the centers of these tunnels, osteoblasts attach themselves to their walls. As the osteoblasts continue to secrete matrix, the tunnel narrows into a snug canal around the blood vessels. The osteoblasts become osteocytes.

While the tunnelling in some areas of bone is being filled in, osteoclasts are perforating adjacent areas, riddling both new bone and prior systems so that the overall gemology is constantly recrystallized into layered patterns suggesting overlapping concentric rings of fossil tree trunks growing together. Animate cells constantly degrade old bone and deposit new matrix.

Osteoclasts also erode embryonic cartilage in the manufacture of fetal bone. Invading the hollow cavities of mineralized cartilage, they degrade the matrix and pave the way for osteoblasts to set down bone matrix in their wake.

Bones do not lie in direct contact with each other; joints and sockets couple them. Between long bones, mesenchyme separates peripherally into ligaments and vacates centrally, leaving behind tiny seams—the rudiments of knees and elbows. Bone joint surfaces have liquid or semi-liquid interiors.

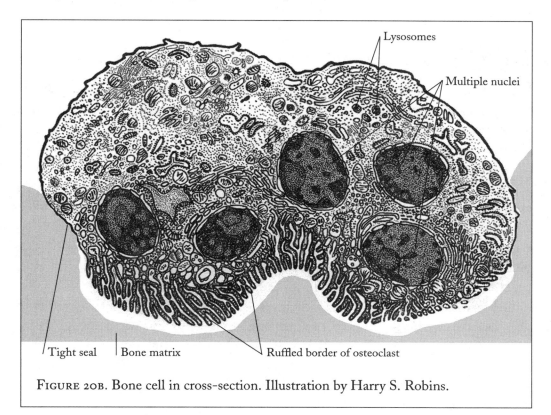

Lysosomes

Multiple nuclei

Tight seal Bone matrix Ruffled border of osteoclast

FIGURE 20B. Bone cell in cross-section. Illustration by Harry S. Robins.

Stored embryogenic cartilage in the connective tissue around the bones is capable of repairing breaks. Reclaiming its primal function, it orients itself as the first stage for a fresh calcium matrix.

Continuously molded by resorption of material and expanding at its open edges, bone is fifteen percent living protoplasm covered by a skin, the periosteum. A full network of arteries, veins, and nerves works its way through minute channels to reach the innermost bony cells.

"Bones are living tubes, inner honeycombs sheathed by dense, compact cells. With this arrangement the body is able to withstand tremendous pressure, compression, and tension. Bones have a rich nerve supply on their surface, and thus can feel pain. All muscles attach to and move bones. The skeletal frame gives the tubes support yet is moveable, so we are not just robots stuck in space...."[1]

The tensile strength of bone is fifteen thousand pounds and its compressive strength about twenty-five thousand pounds per square inch, thus it has great elasticity and can resist blows and return to its original shape after distortion. Because of the buoyancy of the skeleton the t'ai chi master advises his students never to strike with just a single fist: "In push-hands the hands are not needed. The whole body is a hand and the hand is not a hand."[2]

Phylogeny of the Skeleton and Skull

THE ORIGINAL SUPPORTING SYSTEM of ancient Chordates was their notochord, a skeletal forerunner which is retained by lampreys, sharks, and rays in their adult states and which exists vestigially at the nucleus of the vertebrate skeleton as the primary inducer of cartilage in the axial system of the early embryo. An ontogenetic fish frame is still needed to make an ape. The migrating sclerotome of the higher vertebrates surrounds this ancient grid, giving rise to the intervertebral discs and centra of the vertebrae. Each vertebral section develops from two adjacent sclerotomes. It is at this point, ontogenetically, that the notochord degenerates. Sclerotome travelling dorsally covers the old neural tube, providing the material for the neural arch. A subsidiary section moving ventrolaterally into the body wall forms costal processes which will later be induced across the thorax as ribs.

As terrestrial creatures evolved, their cranial plates became capsules of individual small bones over their heads, ossified skin lining a mouth cavity. The head of a fish is continuous with its body; the skull of a salamander is barely distinguishable from its vertebrae. Gradually, through selective mutations, a rough composite of skin, cartilage, and bones developed around the old Chordate region where the gut, the seat of the nervous system, and the primary sense organs reside. In dinosaurs it was the armor which sheltered the core of the nervous system and sense organs, and buttressed the great jaws (after all, these reptiles were scaled-up bony snares activated by "snapping neurons").

The modern skull is molded and carved from mesenchyme around an emerging brain. Traditionally its sectors are classified as neurocranium (brain case) and viscerocranium (jaw skeleton), but embryogenically it has three separately originating zones: the chondocranium (primarily capsules surrounding sense organs), splanchnocranium (visceral skeleton including gill arches and jaws), and dermatocranium (surface skeleton that hardens from dermis and becomes primarily the flat skeleton of the dome of the skull).

Overall, the skull is a highly complex crystal of crystals, its massive and tiny bones and webs of tissue fitting together irregularly like three-dimensional puzzle pieces with pliant gaps. Several separate sections of cartilage fuse to mold the skull's floor and sides; subsequent ossification of mesenchyme above the brain creates a cranial vault, with fibrous areas of dense connective tissue between the skull's flat bones (sutures).

The sense organs of vertebrates are fortified with additionally induced cartilaginous capsules—otic ones around the developing ears, nasal ones around the sacs of the nose, and orbital ones supporting the eyeballs. Other sections of cartilage meld in the vicinity of the forebrain, their upper edges gradually becoming wedged between the brain and the rudiments of the eyes and the nose. While neural-crest cells migrate into their frontal zone anterior to the eye sac, their posterior section remains pure cartilage.

The skull is internally mobile and active, carrying out its own cycle of pulsation and undulation. Its sutures allow the cranium to compress in the birth canal and later to accommodate postnatal enlargement of the brain. The cranial vault expands rapidly during the first year of life and continues to grow until some time during the seventh year.

During osteopathic manipulations, sutures can be tractioned apart and compressed therapeutically by hands placed strategically at different positions on the skull. The crystal remains both embryogenic and axial to the body's musculoskeletal vectors; it is a physicodynamic healing node.

Ontogeny of the Musculoskeletal System

IN THE FOURTH WEEK OF THE HUMAN EMBRYO'S EXISTENCE, branchial arches protrude on either side of the future head and neck, little angular mounds fusing intricately from all three tissue layers and accommodating neural-crest cells that

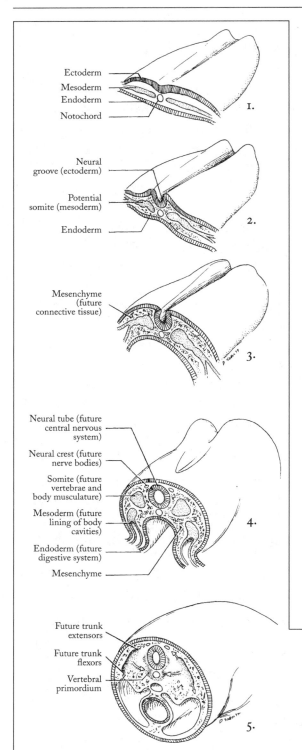

Ectoderm
Mesoderm
Endoderm
Notochord

1.

Neural groove (ectoderm)

Potential somite (mesoderm)

Endoderm

2.

Mesenchyme (future connective tissue)

3.

Neural tube (future central nervous system)

Neural crest (future nerve bodies)

Somite (future vertebrae and body musculature)

Mesoderm (future lining of body cavities)

Endoderm (future digestive system)

Mesenchyme

4.

Future trunk extensors

Future trunk flexors

Vertebral primordium

5.

swarm into their mesodermal cores. They will become structural, muscular, ligamentous, and neural and will participate in bone, nerves, arteries, cartilage, etc. Their mesodermal portion transforms into muscle; from their neural-crest cells are derived connective tissues of the lower face and neck. Their neural components—trigeminal, facial, glossopharyngeal, and laryngeal-vagus—travel directly from regions of the skull into the brain.

The first arch splits into a greater mandibular prominence and lesser maxillary process—lower and upper jaws plus muscles for mastication. The second provides the structure of the hyoid bone and surrounding neck; its muscles sustain facial expressions. The dorsal ends of both of the first two arch cartilages contribute to the bones of the middle ear. The intermediate portions regress, leaving their perichondria (outer fibrous membranes) to form the sphenomandibular and stylohyoid ligaments, critical to the muscular and social activity of the face and neck.

Four more arches emerge caudally—two full-blown ones and two rudimentary. The third arch cartilage ossifies into part of the hyoid bone and its horn (cornu); the fourth, fifth, and sixth provide the mater-

FIGURE 20C. Stages of development of major human tissue systems during the third and early fourth week after fertilization.
From R. Louis Schultz and Rosemary Feitis, *The Endless Web: Fascial Anatomy and Physical Reality* (Berkeley, California: North Atlantic Books, 1996).

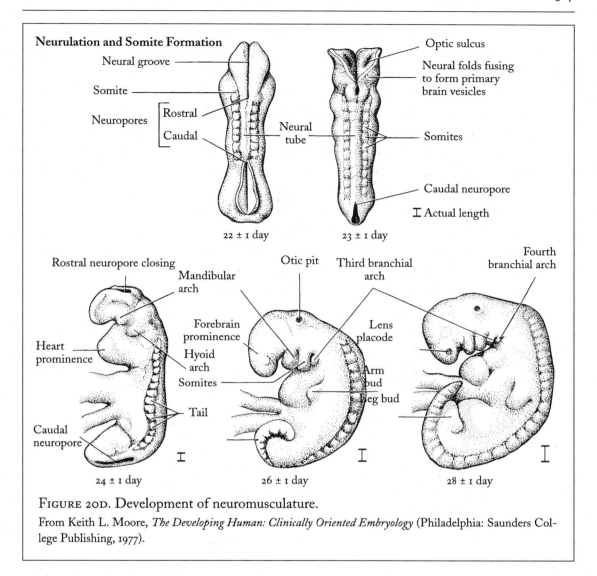

Neurulation and Somite Formation

FIGURE 20D. Development of neuromusculature.
From Keith L. Moore, *The Developing Human: Clinically Oriented Embryology* (Philadelphia: Saunders College Publishing, 1977).

ial for pharyngeal and laryngeal cartilages, the largest of which is known as the Adam's apple.

As the branchial grooves are effaced, the neck becomes a continuous skeleton.

MEANWHILE, THE OUTER WALL of the somites (the dermamyotome) generates a connective layer of tissue in the skin (see Chapter 19). In vertebrates the upper part of this (the myotome) supplies raw material for the somatic muscles of the vertebrae and back.

Some cells arising in the somites are induced as precursors of skeletal muscle cells and migrate into proximate mesenchyme. After a period of intense proliferation, these myoblasts elongate, cluster in bundles, and meld to form myotubes, multinucleate tube-cells assembled on a repeating hexagonal grid. Transformed by intense DNA activation, the cytoplasm of these tubes is saturated with myofibrils, each one containing millions of actin and myosin filaments. Actin-binding struts join myosin molecules in a symmetrical lattice, transforming them into cross-striated muscle fibers. Once this process is triggered it continues as tissue self-assembly without further DNA synthesis. The myoblasts proliferate, stretch out in parallel bundles, and then fuse. Whole muscular grids form from multiplication and adhesion of their components.

Because the nuclei of these fused cells will never again clone their DNA, muscle cells cannot divide; the human fetus has all such cells it will ever have. However, myoblasts can expand (putting on bulk). They can also be fused together (increasing their length) and, as body-builders appreciate, their contractile myofibrils can be stimulated by activity to multiply and dilate. Within the basal lamina, a few tiny stem myoblasts linger in inactive states, to be recruited if a muscle is severely damaged, for these satellite cells are uniquely capable of generating virgin fiber.

Through organogenesis the myotomes continue to realign themselves as longitudinal columns, each a muscle unit separated from the next by a layer of connective tissue. As the organism develops, the segments spread down between skin and lateral mesodermal plate. Lower vertebrates retain primitive segmentation in their mature forms, but the segments are obliterated secondarily in vertebrates adapted to land.

THE FIBERS OF OUR MUSCLES (and those of other animals) align in a variegated pattern, with each longitudinal fiber matched to a fiber crossing it. These bundles of finely tuned twine are interfused with nerves which fire their synchronized potentials electrochemically. Neuromusculature means "neuralized armature," a proprioceptive dominion moving and coordinating viscera as well as bones. As each prospective muscle divides, the developing spinal nerve splits and projects a branch into a myotome.

Most myoblasts migrate, with a rich neural accompaniment, away from their somites. This multidimensional network is induced mutually by its own components and adjoining tissues. The extensor muscles of the neck, the vertebral column, and the loins organize from cells dispatched by the dorsal segment of myogenic epithelium. Ventral myotomes meanwhile protract in sections—thoracic for lateral and ventral flexor muscles, and inferior for the pelvic diaphragm and striated muscles of the anus and sex organs. Occipital myotomes generate myoblasts for the tongue, whereas myoblasts from the branchial arches travel to the sites of masticating muscles, facial-expression muscles, and the connective tissue of the pharynx and larynx. Ocular muscles arise from other mesenchymal cells. Even ectodermal mesenchyme differentiates into sets of muscles—for the iris, the mammary, and the sweat glands.

FIGURE 20E. Muscle cell. Illustration by Harry S. Robins.

Striated muscle of the heart and smooth muscles of the gut originate separately in a visceral layer of mesoderm in contact with endoderm. The latter not only propel food through the digestive tract but participate in the emotional life of the intestines by shooting up erect hairs in recognition of cold and fear.

Unstriated myoepithelial cells are derived from ectoderm and activate epithelia, dilating the irises of the eyes and extruding saliva, sweat, and milk from their respective glands.

Functionally muscles operate as waves—long, slow, peristaltic ones resisting gravity and supporting posture and mobility; short, fast ones reacting quickly to situations and retreating or striking out. Pulsations at their core flow outward through tissues and translate into feelings, goals, and actions.

The central neuromuscular modalities are expansion and contraction, pulsation and fluidity.

The Fascial Web

BETWEEN THE FIRST AND SECOND MONTHS of pregnancy in humans, muscle is both induced and tugged within the general directional pull of connective tissue. Expanding and contracting easily, spongy muscle differentiates into its final mature form along templates drawing it in taffylike contour lines. Likewise, cells in the process of becoming muscle induce traction and friction in the adjacent and less differentiated potential fascia. Because of this interlocked chronology, bones, ligaments, tendons, muscle, and myofascial aspects of connective tissue have a continuous, generalized interrelationship among one another throughout the body. Remaining connected in a layered fabric, they form a mechanical whole. As noted in the previous chapter, activity of the penis (and clitoris) even affects (and is affected by) the respiratory diaphragm along a fascial trajectory. Connections among the eye muscles, gut, and limbs—sustaining profound psychosomatic states—are equally neurofascially generated.

Histologically, "the fascial wrapping of mature muscle is not a true wrapping. It is better described as an area of greater concentration of connective tissue. There is no beginning or end to these structures. Ligaments and tendons do not really attach to the bone—they are continuous with the periosteum ([the] fibrous covering of the bone), which in turn is continuous with the next tendon or bone. . . . It is more accurate to say that tendon goes through muscle than that the muscle lies within the tendon."[3]

Mesodermal differentiation occurs along queues of strain that are established when cartilage forming pathways for bone thrusts through the reservoir of connective tissue. Fiber materializes in accordance with these stress patterns, which themselves stimulate increased directional pull and the formation of more fiber. A single biological field expresses itself, as a deeply situated mesodermal nucleus catenates indivisibly to its furthest branches and nodes in bone, muscle, and fascia. Potential tendons or ligaments already have potential muscle developing within them. The connective tissue of that muscle muffles tendonous propensities and mutates into fascia. Separate but connected blankets of tissue spread from strands stretching between and among layers. They all come off a single spool, their texture and character transmuting and transmitting by layer and context—blends of specialized chemistry, sheer force, iteration, and tensegrity.

The fascia emerges finally as a continuous, fractal envelope of lubricated tissue enwrapping the whole inner body and keeping "our livers from falling out, our lungs and heart from exploding, our intestines from falling down into the bottom of our pelvises, and [enveloping] each and every structure of our body. The tiniest nerve has its own fascial sheath or envelope, as does the largest bone. About half of the muscular attachments of the body are to fascia, so that muscle tone or state of contraction has a lot to do with how tight or loose the fascial sheaths and envelopes are in certain areas of the body at any time. . . .

"Fascia has been described in various ways. It has been called the body stock-

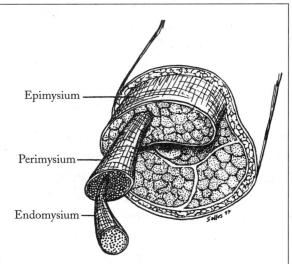

FIGURE 20F. Cross-section of the arm showing how muscle tissue is embedded within its connective-tissue wrapping.

From R. Louis Schultz and Rosemary Feitis, *The Endless Web: Fascial Anatomy and Physical Reality* (Berkeley, California: North Atlantic Books, 1996).

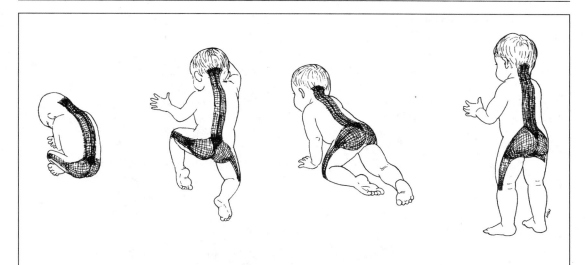

FIGURE 20G. Movement of fascial sheets: before birth, creeping, crawling, and standing. From R. Louis Schultz and Rosemary Feitis, *The Endless Web: Fascial Anatomy and Physical Reality* (Berkeley, California: North Atlantic Books, 1996).

ing under the skin which helps to hold us together. It has been described as tubes within tubes within tubes. It also has been viewed as a series of lamina which cohere, separate into envelopes, and cohere again."[4]

The distinctions among muscles, ligaments, cartilage, fascia, and even bone are transient ones, for the system is a single work of evolutionary art.

Limbs

THE MUSCLES OF THE LIMBS are derived from mesenchyme surrounding developing bones. The myoblasts of the skeletal muscles differentiate from the myotome of the somites, with contributions from mesenchyme, as in the branchial arches. Through the movements of gastrulation, some vertebrate somites become specialized as the precursors of skeletal muscle cells, and they migrate to sites of the limbs. At this stage, without expressing specialized contractile proteins, they look like other non-somite cells in the limb buds. Arms and legs appear first as small mounds of tissue sheathed in ectoderm. Anatomically, they are fins. The unshaped nodules go through many metamorphoses: they swell, protrude, narrow, and gracefully etch digits and sensory tips. A limb bud containing inductive cartilage rudiments, its morphogenetic elongation draws nerves and vessels out along its same track, right to fingertips.

At each site where a limb will form, a layer of lateral-plate mesoderm is induced

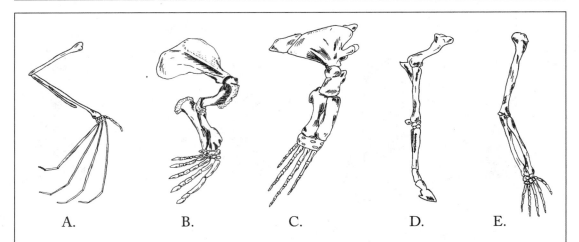

FIGURE 20H. Homologous structures: the right pentadactyl limb of five different mammals. A. The wing of a bat; B. The paddle of a seal; C. The fin of a blue whale; D. The leg of a horse; E. The arm of a human (illustration by Adrienne Smuckler).
From Willis W. Harman and Elisabet Sahtouris, *Biology Revisioned* (Berkeley: North Atlantic Books, 1998).

to thicken. Although the mesoderm remains a continuous sheet, individual cells break their connections to it, accumulate outside its margins, and attach themselves to the inner surface of ectoderm. This activates the inductive field of the limb, which proceeds outward as cells multiply within the disc. In humans, arm buds germinate first opposite caudal cervical segments; a few days later leg buds amass opposite lumbar and upper sacral vertebrae.

A limb is raw equipotential tissue. When half of it is destroyed, the other half will still develop as an arm or leg. A split limb bud prevented from re-fusing will form two complete limbs. And two buds experimentally fused will merge into one normal limb, larger at first but gradually returning to scale. However, destruction of neural-crest or somite tissue in the early embryo before the formation of limb buds usually leads to muscle-less limbs with normal skeleton, nerves, skin, and tendons (these degenerate in the absence of muscle stimulation).

Only the latent genome knows where a limb will form and which one will become an arm, which terminate in a foot. The forelimb bud and hindlimb bud comprise the same types of tissues with different spatial orientations. Positional values alone give them disparate developmental trajectories and shapes. A corresponding polarity of fields causes the leathered wings of the goose to diverge from its webbed claws.

DRUGS USED EARLY in pregnancy (during primary limb formation) may inhibit one or more of these inducing factors. Limbs may wither partway through formation, fail to form hand and foot plates or digits, or perhaps not form at all (thalidomide impeded mainly the proximal region of a limb).

Locally the ectodermal ridge exerts an inductive influence on limb mesenchyme, drawing it out and gradually rippling its field distally into hands and feet—plates with marginal digits (between phalanges, the cells deteriorate). Retinoic acid in varying concentrations apparently plays a role in diffusing limb-bud gradients; this protein acts on cell receptors similar to those for steroid and thyroid hormones, binding specific DNA sequences.

As limbs elongate, cartilage that will become bone creeps outward, crystallizing along their loci; myoblasts gather, line each extremity, and separate into dorsal and ventral (extensor and flexor) segments. Ultimately, arms and legs torque in opposite directions. The upper limbs rotate ninety degrees laterally so that elbows point backward, and the lower limbs turn medially at almost the same angle so that knees point forward. Spinal nerves migrate along both dorsal and ventral surfaces of these buds as they expand.

In the uterine world, limbs are useless appendages but, immediately after birth, the infant will grasp blindly with his or her ungainly pathways, expressing inarticulate yearnings and fending off unknown dangers.

From a primordial Chordate body—a Palaeozoic lump on the ocean bottom—limbs shoot forth like rays, histogenic icons of our extension into cosmos.

The Heart

ALL CREATURES CIRCULATE INTERNAL FLUIDS and disperse oxygen to their organs. The jellied waves of the primordial ocean are regenerated in the internal fluxes of plants and animals.

"The first 'hearts' seem to have been nothing but faint waves of peristaltic motion (like the waves that nudge food through intestines), which gradually became localized and developed into swellings with a pulse. As circulation was mostly open and unconfined by blood vessels (as it still is in clams, shrimps, insects, etc.), heart action was more comparable to gently stirring soup with a spoon than to anything that could be called pumping—which may explain why the squid needs three hearts, the grasshopper six and the earthworm ten. And even when the heart evolved its valves with completely channeled blood flow, it still awaited a future history extending from the single-loop circulation of fish to the loop with a side (lung) branch of amphibians and finally to the now well-perfected double-loop circulation of mammals, which uses a two-chambered heart to pump blood first to the lungs to absorb oxygen, then to the whole body to distribute it."[5]

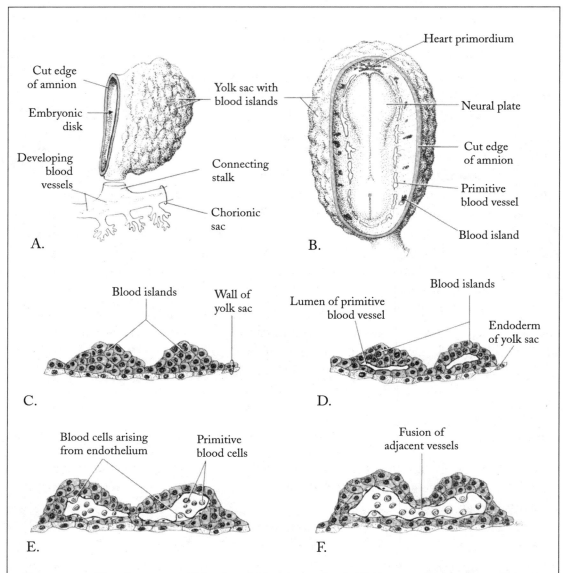

FIGURE 201. Development of blood and blood vessels. A. Yolk sac and a portion of chorionic sac at about eighteen days; B. Dorsal view showing the embryo exposed by removing the amnion; C. to F. Progressive stages of development of blood and blood vessels.
From Keith L. Moore, *The Developing Human: Clinically Oriented Embryology* (Philadelphia: Saunders College Publishing, 1977).

The mammalian heart is a mesodermal constituent composed of three separately arising structures: an inner lining (the endocardium) which is continuous with the endothelial lining of the blood vessels; a heart muscle (the myocardium);

and a very tough cellular membrane (the pericardium) around the myocardium.

This organ originates ontogenetically in a sheet of mesoderm advancing forward from the blastopore and travelling between ectoderm and endoderm. The dorsal section is swiftest moving; the ventral slowest—at first little more than a pulsating ripple.

The heart is the only organ that does not emanate within the embryonic disc. Situated in front and ahead of it, it becomes incorporated in the body only as the embryo extends cranially and captures it. At five weeks the heart is a huge bulge barely held inside a little creature. In this form it is an autonomous animal, in some ways more creature-like than the embryo itself; its rhythmically pulsating cohorts, the blood islands, maintain protozoan independence likewise.

By the completion of the neurula phase, both the dorsal and dorsolateral segments of the mesodermal mantle have contacted the head region, creating an open forward space with a broad anterior base while engulfing oral and pharyngeal zones. The rear of this space will be filled gradually by formative cardiac mesoderm. These early phases of heart-making are endodermally induced.

IN HUMAN CARDIOGENESIS a pair of elongated mesenchymal strands develop lumens, drift together, and fuse into an endocardial tube. A gelatinous connective tissue, a cardiac jelly, collects around it. Surrounding mesenchyme forms a mantle which will give rise to myocardium and epicardium. At this stage the organ has the fluttery-edged, transparent look of an underwater plant.

After neurulation the free edges of mantle begin to converge and thicken ventrally in presentiment of an organ. The ventral cells flow freely like mesenchyme; then they converge in a longitudinal strand which becomes a tube with a lumen, i.e., the heart cavity.

A cylindrical lump gradually lengthens and begins to dilate in spots and thicken in others, forming sacs, arches, and a ventricle, including the aortic sac and arches, the atrium, and the sinus venosus (the mature pacemaker into which thread the umbilical, vitelline, and common cardinal veins of the chorion, yolk sac, and embryo). Because some regions swell faster than others a loop develops, the bulbus cordis on one side of it and the ventricle on the other—twin lobes of an S-shaped organ.

As a head fold unfurls in the expanding embryo, heart and pericardial cavity fall in front of the foregut and sink behind the membrane of the pharynx. The tube of the heart lengthens and bends; the organ is submerged in the dorsal wall of its cavity, suspended by a fold.

The internal stuff of the heart condenses; atrium and ventricle begin to be partitioned. Veins and arteries interfuse tissue, some of them developing as evaginations

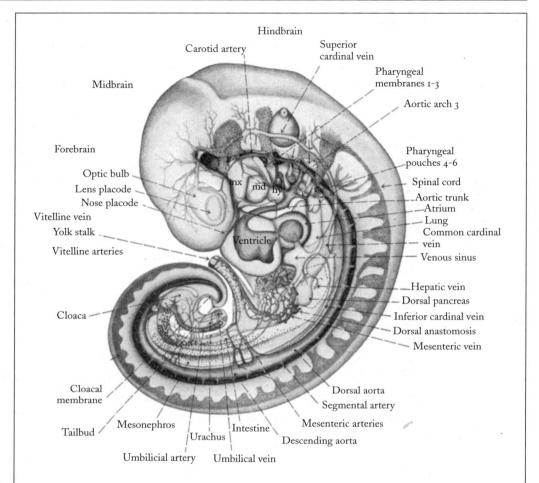

FIGURE 20J. Reconstruction of a human embryo of thirty-two somites. Black dots on lower gut, mesentery, and mesonephros represent loci of primordial germ cells in migration. Mx and my are maxillary and mandibular processes of the first visceral arch; hy is hyoid arch.

From Emil Witschi, *Development of Vertebrates* (Philadelphia: W. B. Saunders & Company, 1956).

of the atrium wall. At the same time, ridges of tissue become hollowed out into valves controlling the flow of blood into the atrium; when the ventricle contracts, blood is pumped back out. The muscle layers of the atrium and ventricle knit together until they squeeze peristaltically, at first activating an ebb and flow between themselves and the embryo but eventually using coordinated contractions to establish a unidirectional stream.

A mutant pump which began in ancestors we share with the lungfish is now a life-support and fluid-draining mechanism in all large, mobile vertebrates.

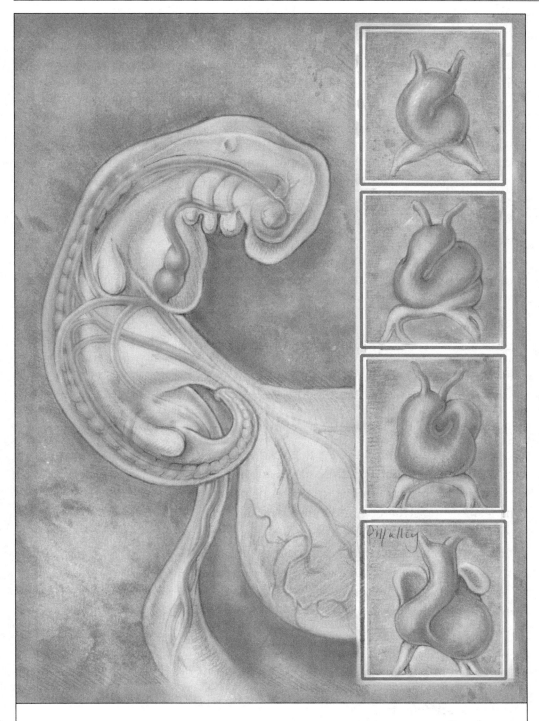

FIGURE 20K. Developing heart and vascular system. Illustration by Jillian O'Malley.

P RIOR TO ITS INTERNAL ASSEMBLY, half the heart could be transplanted and still develop as a whole organ. Once the chambers and valves begin to form and muscle fibers penetrate ventricular walls, cardiac induction occurs primarily between fields within the organ itself. With the subsequent migration of the developing heart from the pharyngeal region to the thorax, a mature system takes shape, separated into a right half carrying blood to the lungs and a left half receiving it from the lungs through pulmonary veins and sending it out through the body. Embryo and uterus develop in mutually pulsating tandem.

Heartbeat differs from muscular activity in its ceaseless pump. Though it may vary in pace and breadth, its metronome is the register of life, breath, and identity. Isolated cells removed from an embryonic heart will pulsate in place with their own separate rhythms but, if they touch, their cadences quickly synchronize.

With the heart driving vital fluids through the body, the vagus nerve connects it to both the esophagus and the dome of the diaphragm. As breathing and blood-pump combine, nerve impulses ignite and stabilize with the ceaseless hum and glow of ordinary reality.

Even though they do not divide, heart muscle cells regularly synthesize RNA and protein, changing shape and size in response to blood pressure and the load on the organ itself. The heart is also a giant gland, producing hormones that affect almost every organ in the body and helping regulate the limbic system.

B Y THE JUNGIAN CONCEIT that the gods of old Greece have been reduced to organs and diseases in the modern body, the heart is the daemon of love and war. This is not a sentimentality but an intimation of an embryogenic connection, an emotional constitution emerging in tissue that itself is being invented and shaped, developing meaning and structure on the fly. Demythologized and embodied, the gods are no less capricious and omnipotent than they were on Olympus, for power now comes not from an ethereal realm but intractable squads of cells. Therapist Robert Sardello describes the aboriginal connection between heart and brain:

"In the human embryo they are not at all distinct. In the first twenty days after fertilization of the ovum, what is to become the heart lies nestled around what is to become the brain, as if the heart is its crowning glory. As the brain emerges, the heart submerges. I suspect the heart forever remembers this intimacy: The heart attacks when the brain thinks that it no longer needs the throbbing rhythm of the life below, when it thinks it can be more productive without the interference of emotion, sentiment, feelings."[6]

The Body's Fluids

The Source of Emotional Life

Bearing microscopic jungles of sprites and basilisks, our insides truly resemble the Palaeozoic waterways from which they arose. "We are moving water brought to land," writes Emilie Conrad, "and our relationship with our planet is maintained by the resonance of our fluid systems with all fluid systems. Blood, rivers, oceans, cerebrospinal fluid [CSF], all are in a state of resonance, a unity without boundary. We are the flowing expression of a divine and complex intelligence that has formed us for a purpose we may never know."[7]

Tides of antibodies, hormones, blood, and CSF implement unconscious agendas. Organs and fluids generate senses of buoyancy, space, orientation, aspiration, sweetness, balance, release, and playfulness. Joy, sadness, fury, jealousy, and grief likewise have roots in aqueous substrata. These sensations are not only dramatic and recreational; they are the language of tissue.

Hematopoietic Stem Cells

Hematopoietic stem cells originate in the liver embryonically. In adults they develop in the bone marrow, associating with fat and connective-tissue cells in a thick mesh of collagen and extracellular components. These pluripotent zooids give rise to a variety of committed-progenitor blood creatures, all of whom (despite their different pedigrees) share a common ancestor. Since then, classes of their descendants have been induced into terminally differentiated states with radically divergent functions.

The hematopoietic stem cells that generate blood yield a number of other quantal derivatives — a tribe of descendant protists including general leukocytes, monocytes, a specialized line of lymphocytes, and huge megakaryocytes which make their home in the bone marrow from where they squeeze new platelets through the endothelial lining of sinuses into the bloodstream. The cells in cerebrospinal fluid are another hematopoietic by-product (described more fully in Chapter 18 in the section on the hydraulic mechanism of the brain and in Chapter 24 in the sections on somaticization and craniosacral therapy).

As NOTED IN CHAPTER 10 ("Gastrulation"), the entire cardiovascular system emerges as mesenchymal cells swarming together in blood islands — angioblasts. Induced in the context of yolk, their "purpose" is vegetal — to bring food and oxygen to multiplying cells. The virtual lack of yolk in the mammalian oocyte makes immediate circulation a necessity, for the embryo must obtain nutrients and oxygen to survive its own matriculation. The heart-pumping-blood mechanism is thus the first living system to become functional in a mammalian embryo.

A transport network is assembled in place by angioblasts forming epithelia around cavities and linking up in chains. These separate channels extend by budding and fusing with exterior vessels.

Red Blood Cells

The liquid part of blood is essentially salt water. Its "solid part consists almost entirely of coin-shaped red cells that are remarkably elastic, so flexible they can elongate and fold up and sneak through a capillary of barely half their own diameter."[8] This sea is warmed by muscle contractions and the metabolism of digestion and oxidation. Cold-blooded lizards must be heated by the sun before they dart about (we concede that solar "hit" in order to remain active at night and in winter).

Red blood cells, erythrocytes, are out and away the dominant hematopoietic derivative. Erythropoiesis—red-blood-cell formation—is a regional specialization of blood epithelium induced by an enzyme. Erythrocytes transport oxygen and carbon dioxide through the bloodstream. Upon a decrease in oxygen or a shortage of red blood cells, erythropoietin synthesized in the kidney is secreted into the bloodstream. Erythroid progenitors then matriculate in the bone marrow. These stem cells are potentiated to synthesize and store the metalloprotein hemoglobin, which actively captures and releases gases. Hemoglobin has the basic molecular structure of chlorophyll, but with four atoms of iron in place of chlorophyll's central magnesium atom. This explains blood's strong ferrous attraction, and also suggests the far distant aeon when plants and animals were a single creature.

Blood cells originate in the same epithelium of the yolk sac as blood vessels. Under enzymatic influence they lose their nuclei and give up reproductive capacity, becoming committed erythrocytes. About the sixth week after conception, erythropoiesis migrates to embryonic mesenchyme; thereafter it wanders from side to side in mesoderm, from liver and spleen to lymph nodes, ultimately taking up residency in the bone marrow, where blood cells originate thereafter.

The molecular, hydraulic relationship between the heart and blood recapitulates, at least metaphorically, the gravitational bond between the Sun and its planets. Once sovereign entities, blood cells are captured and dispatched into vessels to transport

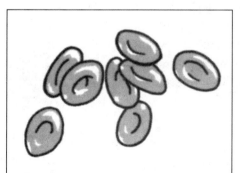

FIGURE 20L. Red blood cell. Illustration by Harry S. Robins.

oxygen throughout the body and remove unwanted gases (such as carbon dioxide). As biconcave discs, these hematic zooids are biochemically and physiologically specialized—with all other functions discarded in order to improve their geometry for oxygen transport and traversing microcirculation networks.

BLOOD VESSELS ULTIMATELY permeate the entire adult body. They are "attracted" to vascularize any part of the soma in need of nutrition and oxygen, even transplanted tissues. Spreading by budding and fusion, vessels branch outward into long, constant riverine tubes crossing vast topographies and detouring around obstacles. Since the system is supple and flexible, it is irrigated in part by the amount and direction of flow itself, but, at the same time, it is not initially dependent on a heart, emerging and trickling locally even in its absence.

Those branches that receive the heaviest flow become arteries; those that receive too little deteriorate. Blood networks proliferate in regions between mesodermal and endodermal viscera, especially in the heart and the principality of the kidneys, where erythropoietic hormones are synthesized.

Anteriorly and posteriorly around the heart, blood vessels assemble the paired continuation of the heart tube. Induced endodermally, two posterior vessels become vitelline veins which collect blood from the surface of the gut; an anterior pair become ventral aortae below the endodermal pouches of the pharynx (the aortic arches lie *between* the endodermal pouches).

In the embryo blood vessels also arterially connect ventral to dorsal aorta in six pairs of aortic arches. Although most of these arches dissolve during human development, the fourth and sixth pair continue to supply blood to the back and the lungs, respectively.

Leukocytes

White blood cells migrate into tissues and help combat infections and digest debris. These leukocytes are subdivided into three main groups: granulocytes, monocytes, and lymphocytes. The lysosome-rich granulocytes include neutrophils that phagocytose bacteria, eosinophils that attack large parasites and neutralize allergic responses, and basophils that secrete histamine and serotonin to reduce inflammations. The monocytes mature into macrophages, many of which fuse with neutrophils to form phagosomes, vesicles that ingest microorganisms. Lymphocytes manufacture antibodies and slay cells infected with viruses (see below). The induction of each type of blood cell in the bone marrow is signalled (often chemotactically) by molecules produced throughout the organism by a variety of organelles and other cells (including white blood cells, histamine-secreting cells, platelets, and nerve endings), and by the protein products of inflammation and antibody-antigen reactions.

During inflammatory episodes, signalling molecules travel in the blood to the bone marrow and, upon arrival, induce it to synthesize extra leukocytes. Other molecules simultaneously prepare the disturbed region for an army of oncoming white blood cells, loosening its endothelium and making its cell surfaces adhesive enough to capture passing leukocytes like bugs on flypaper.

The Immune System

Lymphoid progenitors (primary lymphocytes) migrate in the bloodstream to the thymus gland among the branchial pouches (see Chapter 19). There, under local induction, they propagate as thymocytes. With prototypical surface topographies, separate clusters develop capacities to customize single kinds of antibody molecules. In this mature form they are called "T cells" and their immunity is termed "cell-bound" because they carry antigen-combining sites right at their surfaces and directly neutralize foreign bodies. Substances alien to an individual organism are called antigens for their antibody-provoking capacity. T cells regularly examine other cells for changes in their surfaces and attack molecules of unfamiliar shape—a vigilance which prevents potential tumors from developing and also causes the frequent rejection of grafts and transplants. They aggressively surround these interlopers, corroding them on the spot.

Our response to infections and immunizations (humoral immunity) arises from plasma cells of different lineage. Another type of lymphocyte (termed the "B cell") develops somewhat mysteriously, an antigenic reaction independent of the thymus. This reaction is multidimensional and requires T cells as helpers as well as nonimmunocompetent macrophage cells which initiate contact with invaders.

LYMPHATIC VESSELS FORM the same way that blood vessels do; they are part of the venous system and may even arise as capillary extensions of its epithelium. Like the fetally hematopoietic spleen, the lymph nodes are assembled by invasions and aggregations of mesenchymal cells in the context of local cavities and mesenteries. Six lymph sacs materialize along the median axis of the body—first, two jugular; then, two iliac; then, a pair in the abdominal region. Vessels grow out from lymph sacs and follow the main veins. Where mesenchyme encounters these sacs they dissolve into separate channels which become the lymph sinuses. Some of the mesenchyme then differentiates into the capsule and connective tissue of each emerging lymph node. Squadrons of lymphocytes spew out henceforth.

The denizens of the lymphatic system emerge very early in ontogenesis, replicating the guard cells of eukaryote colonies. The first zooids originate in the yolk sac, though their generation is eventually taken inside the fetus and continues in bone marrow after birth in concert with the formation of red blood cells (of which lymphopoiesis in the spleen is a variant). Even stem cells from the yolk sac must enter the thymus to become immunocompetent; they migrate out of it subsequently as full lymphocytes. Other lymphocytes, in apparent contradiction of their lineage, develop directly from mesenchyme.

Clearly, complicated phylogenetic histories have been displaced into organs where their protagonists are now linked in labyrinthine migratory cycles within the body, performed anew each embryogenesis. These vision-quests (through the kiva of the thymus) also transform journeying zooids into molecular shamans entrusted with ancestral secrets.

Immunity is as old as animal life, for without this mechanism the elaborate circulation of fluids and other substances would spread any local infection throughout the body. Plants differ: they maintain diseases locally and wither in sections; they have the ability to branch out again from almost any part; they do not have immune systems.

THE MAMMALIAN IMMUNE SYSTEM is highly complex, for it must recognize the many-faceted chemical identity of a substance at a molecular level. Mature lymphocytes are apparently able to read all the intricate three-dimensional topologies of antigenic polypeptide chains—their peculiar ruts, extensions, and electronic and spatial configurations. It seems unlikely that thousands of distinct antibodies arose independently in the evolution of animals, nor does such an explanation for immunal capacity account for its spontaneous ability to destroy new toxins—for instance, industrial poisons and imported viruses. Immunocompetence appears to be an inheritable proficiency of uncommitted lymphocytes which develop an appropriate neutralization only when exotic antigens appear. We do not know how such information is stored and utilized in the immune system, though it would appear that raw DNA itself would have to be tapped in a variety of contexts to generate matching lymphocytes for each novel antigen. Apparently there is a constant and a variable region in the gene. In the variable region, nucleotides are shipped out; each developing lymphocyte makes a unique antibody, thereby introducing contemporary sequences into a traditional formula.

If extinct animals of the Earth are stored in our immune system (or in another subcellular library), as was suggested initially at the end of Chapter 6 ("The Genetic Code"), these "creatures" perhaps could tell us (in the language of biological specification) all they know, providing a set of references for immune reactions.

A creature must learn to discriminate its identity from others', its body from its environs. If grafts from another animal are made soon enough after birth, the organism may accept them (as well as all subsequent grafts from the same animal). However, the learning period must come to an end if the creature is to survive the lifelong onslaught of microbes.

A related process blocks maternal T cells from rejecting the fetus. In a certain sense, maternity is a form of death to the mother, who temporarily gives up part of her biological identity to her progeny and is "willing" to preserve its life at the expense of her systemic exclusivity.

We keep a separate record of what-is-not-us alongside our genetic identity. We understand poisons because we are potentially toxic; we bear the agents of our own extinction treacherously close to a benign embryogenesis that they shadow and sometimes grotesquely and atavistically mime. We survive by suppressing not only billions of potential cancers but trillions more inherited messages that are morphogenically "tempted" to use our matrix to embody themselves.

Placental Trophoblast as an Anti-Cancer Vaccine

BELIEVING THAT "tumors in the body originate from small clusters of embryonic cells which remain unchanged during the organism's development" and that "oncogenesis is a blocked ontogenesis,"[9] the late-twentieth-century Russian scientist Valentin Ivanovich Govallo developed what he called "the conditional equivalent of an anti-cancer vaccine" by "immunization with placental extract which contains a broad spectrum of trophoblast antigens."[10] The clinical success Govallo claimed—the recipient of major skepticism outside the old Soviet Union—he attributed to the "ability of the trophoblast to destroy cancerous cells ... and to reverse the development of malignant tumors through immunization with cells of embryonic origin."[11]

The primitive streak is, in a sense, a cancer, but not malignant because it has the intrinsic capacity to harness malignization into "normal" morphogenesis. Normal growth, conversely, is a highly regulated tumor.

Without lingering embryonic (gastrulation) cells, Govallo believes, cancer would not occur at all, for cancer is ontogenesis gone wrong. Whereas "embryo antigens ... are necessary for the maintenance of malignization, ... activation of oncogenes does not contribute to the emergence of tumors, but does contribute to their progression."[12] While the trophoblast is itself a partial malignization (with properties of unrestrained multiplication and metastasis), its potential tumorhood is suppressed by its concomitant antibody cells—a therapeutic metabolism that weakens in the maturing organism.

The injection of placental trophoblast into adults is meant to restore immuno-logical recognition by flooding any vestigial malignization with antisuppressory antibodies.

The wonder is not that people develop cancers but that instead of turning into heaps of cells, we assemble normally most of the time. Each living bug on the wind-shield and cat on the porch—not to mention the mobs in their autos on the free-ways—are a testimomy to cell malignancy regulated with exhaustive rigor into tight, functional organisms covered with skin and tuned by muscles and nerves.

Twin potentials of a single morphogenetic capacity (one complexly develop-mental, the other crude and lethal) apparently share a razor's edge in the embry-onic body. The deep layers of multiply redundant code inside each organism must constantly correct renegade and malignizing sequences, heading off tumors. It usu-ally takes many separate failures to override this utility and produce a carcinoma.

Autoimmune Diseases

OPPOSITE KINDS OF PATHOLOGIES beset immune systems. Tumors are collective failures of the lymphocytes to respond to a malignancy and attack it; the immune system does not recognize the exotic nature of the invader or the seriousness of its threat. In autoimmune maladies the cells confuse the self with the "other" and attack their own body. In lupus, cells of the genome combat *themselves* as if outsiders. They destroy DNA, with consequent damage to blood vessels and the kidneys.

These lesions represent crises of biological identity. Whereas so-called psycho-logical (and even psychosomatic) ailments reflect conflicts in the formation of per-sonality, immunal disorders point toward unresolved ambiguities in the fact of cellular identity (which sits in fragile equilibrium).

A popular theory propounds that autoimmune diseases originate in infections: a pathogen—perhaps a mycoplasma, hard to detect or identify—slips into an organism and hides out. One day the dormant invader wakes up and begins repli-cating itself; the immune system responds by attacking those cells which harbor the pathogen, causing an autoimmune flare-up. This results in diseases like lupus, rheumatoid arthritis, and multiple sclerosis.

WE ARE ALL DNA, but we are all also aliens, antigens, carnivores, and prey. The oppor-tunists of nature threaten us from the moment of our crystallization until they get to devour our meats at the end. In fact, at the nether end of this high-flying century, the biggest treat awaiting an appropriate mutant carnivore (probably bacterial or viral) in the dwindling jungles of Indonesia or Africa is the burgeoning human biomass.

Mind

A T THE MOMENT LIFE BEGINS, MIND BEGINS TOO. Even in sponges, where cells are barely associated, matter surprises itself. "How?" it asks. "How am I here?" We see a hint of reconnaissance in the feints of crabs and snails, in the dull eyes of fishes. Predators prowl, slash, and feed, but they bide a glimmer of ego; a doubt; a brief, plaintive dissociation between self and act. Lion and zebra confront and become each other at the kill, two sad eyes apiece in their mammalian skulls. "It is ludicrous that I should be this," all animals bray, howl, squawk, "*I*, guardian of the universe, source of possibility." The bear whines restlessly; the lynx paws at the sky. The walrus carts his immense blubber across rocks. Each creature is a combination of something and nothing, a question that tries to ask itself before succumbing to a heap of feathers. Cats and dogs look to us for the answer—us who have separated them from their own world. Monkeys in the foliage stare down at Indians painting themselves for the ceremony.

The Evolution of Primates

Tree Shrews

The order of primates was emerging from a branch of mammals almost seventy million years ago (at the end of the Cretaceous and the beginning of the Palaeocene era). Their lineage descended from the same reptile-like forager that marsupials, carnivores, and ungulates also claim as progenitor. In this long-ago time nothing resembling us dwelled on the Earth; rodent-like fauna were our closest kin.

While the poor-sighted ancestors of moles tunnelled into the mantle of the Earth and became underground predators, the forerunners of otter shrews and marsh shrews took to the water. Primitive anteaters, aardvarks, and other insectivores

FIGURE 21A. Young lesser tree shrew.
From W. E. Le Gros Clark, *The Antecedents of Man: An Introduction to the Evolution of the Primates* (New York: Harper and Row, 1963).

coevolved with prey, anatomies whittled by feedback from their specialized diets. Some groups of shrews migrated into trees; from these evolved flying chiroptera (bats), tree shrews—plus the extinct ancestors of the primate line.

Tree shrews are the living primates most closely related to the hypothetical progenitor of our order. These foragers have few specializations; yet the adaptation of their grandmothers to life in trees—though hardly original for small carnivores—had global ramifications in this one instance. Through a circuitous course beginning with arboreal evolution, their descendants garnered disparate mutations into an efficient mode of survival and, more significantly, an unforeseen realm of intelligence. These meek, unprepossessing mammals—not the more formidible ancestors of jackals, lions, and caribou—were alone preadapted to symbol-creation.

IF ONE ANATOMICAL CHANGE preceded and catalyzed others, it may have been the differentiation of limbs into two distinct pairs. Fore and hind (and their joints and digits) were polarized early in the history of the primates, most likely in response to locomotion through trees. In jumping, an animal is propelled by its hindlimbs working as a lever; and in climbing, it pulls itself through the branches with forelimbs. But primate limbs are more than specialized props and levers for support and motion; almost acrobatically loose, they are attached mobilely to a spine, its girdles, and each other. The distal joints of the fibula and radius also rotate, providing orbits for hands and feet to explore at a variety of angles. Even the bones of the forearm (the radius and ulna) are distinct, flexible structures. Comparatively, the skeletons of horses and weasels are fixed, machine-like chassis. These creatures were not going to work on assembly lines, paint the ceilings of chapels, or (for that matter) lift a pebble.

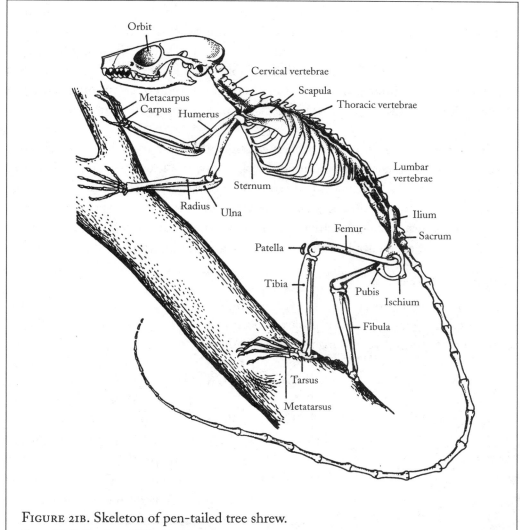

FIGURE 21B. Skeleton of pen-tailed tree shrew.

From W. E. Le Gros Clark, *The Antecedents of Man: An Introduction to the Evolution of the Primates* (New York: Harper and Row, 1963).

Primate mobility is most restricted (at least among contemporary members) in the tree shrews, for they specialized strictly in a rodent direction counter to the generalized arboreality of their order. The fore- and hindlimbs of these animals are moderately differentiated from each other; their shoulder girdle is somewhat more rotatory than that of terrestrially oriented mice and anteaters. Shrews are alert and agile climbers with sharper vision and more cerebral cortex than any insectivores. Once upon a time their ancestors reverted back to a safe domain of nimbleness and claw.

After all, by pure anatomical standards, the early primates were ill-equipped—a genetically handicapped order founded by lame cousins of the mice and moles. Crooked limbs denied them the natural fleetness of their mammalian heritage; indeterminateness of their teeth and claws robbed them of natural weapons; and (later, in higher primates) reduction of their snout and concomitant atrophy of olfactory lobes of their brains would cost them their exquisite mammalian tracking ability.

The direct human lineage represented, first, the least specialized vertebrates, second, the least specialized mammals, and finally, the least specialized primates and anthropoids. Our precursors had few physical tools or athletic skills. If we imagine the various ecological niches being deeded out as kingdoms, the creature who waited the longest and most patiently was ceded the most limited domain of all—one that did not even entail an explicit somatic gift but instead had to be imagined.

Lemurs and Tarsiers

Back in the Eocene, some fifty to sixty million years ago (after the tree shrews had abandoned the generalized but progressive line, claiming their squirrelly niche), a second major group diverged from simple aboriginal primates; the present-day genera most closely resembling them are lemurs and lorises (tarsiers are their close relatives). By comparison with shrews these prosimians are largish, fully tree-adapted, and retain few insectivore features. Their fore- and hindlimbs are differentiated for climbing, thrust, and support, and their clavicles and neck-jaw muscles are oriented so that they can turn their heads and look every which way. Lemurs also sit up. Their center of gravity has shifted backward juxtapositionally to their hind musculature, a tail specializing as a rudder. Without this rearward shift, such bulky animals could not distribute their force properly; their leaps would displace into spins. The lemur face is also less prognathous and bestial by our standards, its features flatter and "cuter."

From neckless, snouty fishes to stalky, cranial hominids virtually no bones have been added, even within the head, so existing skeleton had

FIGURE 21C. Ring-tailed lemur.
From W. E. Le Gros Clark, *The Antecedents of Man: An Introduction to the Evolution of the Primates* (New York: Harper and Row, 1963).

to be constantly reshaped, realigned, and reoriented—braincase molded from muzzle. In the facial region alone, mutations had to direct and recalibrate layers of mesoderm and neural-crest squadrons, redistribute reservoirs of mesenchyme, tilt angles of cartilaginous and fascial grids, rotate axes of osteoblast migration, reprogram branchial arches and trigeminal and facial neurons, reinvent linkages of sphenomandibular and stylohyoid ligaments, etc. Later, on the road to men and women, similar morphogeneses had to occur in the rest of the body, notably the pelvic region.

Whereas the jaw of the average mammal is little more than another claw, the lemur jawbone is significantly atrophied and deflected toward a position below the braincase and at an angle to it. With prompt snapping less a necessity, the powerful forces of compression in the skull can be applied more toward crunching vegetation.

Changes in the proportions of the head opened skeletal space for an expansion of the cranium and its contents.

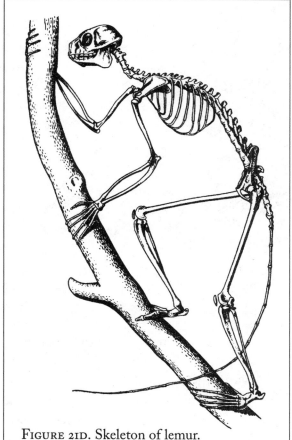

FIGURE 21D. Skeleton of lemur.
From W. E. Le Gros Clark, *The Antecedents of Man: An Introduction to the Evolution of the Primates* (New York: Harper and Row, 1963).

As a NOVEL primate lifestyle became more distinct in the forest, one transition initiated another, disparate trends were fused, and anatomy was restructured allometrically. Lines continued to diverge from the generalized stock. The ancestors closest to tree shrews and their kin went pretty much unnoticed, but lemurs put on quite a show in the branches, establishing the legitimacy of primate style. They might not have had the winged grace of birds or the swift prowl of the cheetah, but they demonstrated something quite distinct and ingenious, worth taking note of in the bestiary of its time.

Tree navigators must be able to sight and interpret three-dimensionality; thus, for generations, as shrew ancestors turned into prosimians, assorted mutations enhancing visual receptors were likely reinforced by the success of their bearers.

FIGURE 21E. Tarsier.
(Renaissance drawing, source unknown).

Fibers from corresponding optic areas of the brain were induced into the new tissue. Even in lower primates, the nuclear elements of the thalamus—particularly those supporting sight—began differentiating quite early. The cerebellum became fissured, packing enough "memory bits" to coordinate a greater variety of activities. But the biggest change of all occurred in the cerebral cortex and its sensory-motor projections to the limbs. This corticospinal web kept swelling and crinkling until awkward ancestral lemurs were reinvented as coordinated trapeze artists.

CONSCIOUSNESS BEGINS TO REVEAL its distinctive topology in the lemur cerebral cortex, a separation of lobes commencing in the temporal and occipital areas. Yet with frontal and parietal areas lacking sulci, overall the ancestors of prosimians were likely not as cerebral as the ancestors of monkeys and apes. With their special rod retinas, many proto-lemur species were no doubt adapted to night foraging.

Tarsiers are small, wide-eyed prosimians with freakishly enlarged tarsal bones in their feet (for leaping through trees). Their gigantic eyes feed a huge visual cortex and a correspondingly enlarged occipital lobe. This gives them a startling humanoid appearance. Despite the fact that tarsier brains are otherwise primitive and lack the convolutions of the lemurs, some zoologists have derived the human line directly from this leprechaun (rather than from monkeys or apes) on the basis of its upright posture and ocular advancement. (Perhaps even more psychosymbolically, the French linguist Jean-Pierre Brisset proposes the direct descent of humans from frogs on the basis of their jealousies and battles and the tactility of their language and orgasmic screams!)[1]

Monkeys and Apes

The first anthropoids, the ancestors of monkeys, apes, and humans, diverged from the common ancestor they shared with lemurs and tarsiers some thirty to forty million years ago during the Oligocene era. The cerebral cortices of these animals were already deeply convoluted. In living ape and human brains, secondary sulci have overgrown and obscured simian layers underneath them. Even the most primitive modern monkey brain outweighs by three times a comparable prosimian brain.

The anthropoid brain also changed in qualitative ways. A whole new series of nerve tracts trails directly from the cortex to the spine. These fibers (called pyramidal) were apparently induced in the collaborative evolutionary induction of the spine and the cortex, and they superseded slower, more diffuse extrapyramidal relays between the brain stem and the motor neurons. Many cortical fibers also skipped their older internuncial relays and ran directly to motor neurons—the greatest percentage of these to the distal muscles of hands and feet but also some to proximal trunk muscles. There was a concurrent development of sensory filaments that record and discriminate muscle contractions. Information from these proprioceptors and from the tactile neurons of the skin was transmitted in quick, discrete quanta— at least by comparison with the disperse peripheral pathways of the lower mammals.

The dual eyes of these classes of animals have moved toward one another, creating an overlapping stereoscopic field for leaping and veering from point to point. Heavy wingless mammals navigating perilously above the ground must interpret irregular patterns of branches and zigzagging images along gravity's delicate slide. Birds do not require abstract algebra, for their wings lift them into open space; but the flight of arboreal primates changes rapidly and unpredictably in the geography of trees. Our internal organs may be aquatic, but the lobes of

FIGURE 21F. Gibbon with long brachiating arms.
From W. E. Le Gros Clark, *The Antecedents of Man: An Introduction to the Evolution of the Primates* (New York: Harper and Row, 1963).

FIGURE 21G. Capuchin monkey.
From W. E. Le Gros Clark, *The Antecedents of Man: An Introduction to the Evolution of the Primates* (New York: Harper and Row, 1963).

our brain are holograms of receptors hurtling through branches. We still use these synapses to find our way through abstractions and to measure the distances between stars and to track the "jumps" of subatomic particles. It was in treetops that the relationship between space and time became ontological.

THE FORELIMBS OF THE EARLY ANTHROPOIDS were prehensile, useful for clinging, gathering food, examining objects, and hanging from branches. The sensor pads of the manus and pes (hands and feet) gradually grew tender-tipped receptors reminiscent of octopus tentacles. Some lines of apes developed enough pliancy to oppose their thumbs to their other fingers, in preadaptation to tool making and counting. ("The hand," declared Martin Heidegger, "is infinitely different from all grasping organs—paws, claws, or fangs—different by an abyss of essence."[2])

FIGURE 21H. Male gorilla.
From W. E. Le Gros Clark, *The Antecedents of Man: An Introduction to the Evolution of the Primates* (New York: Harper and Row, 1963).

The arm structures of these anthropoids were modified for brachiation, that is, for suspending the weight of their bodies and swinging through the trees. This performance required muscular support in the thorax and freedom at the shoulder joint. The frontal clavicle guided the limb in a wide arc, and the dorsal bone of the shoulder girdle, the scapula, was able to move back and forth across the thorax—forward for pushing, and backward for pulling and climbing. The scapula also rotated vertebrally to permit the fullest elevation of the arms.

Meanwhile the arms developed freedom to rotate at the joints of the radius and ulna, and the hands became flexible. Mutations lengthening phalanges and reducing thumbs went much further in apes than in hominids, as required for brachiation. The thorax of apes broadened by comparison with that of monkeys and flattened from front to back rather than laterally, which gave support to the trunk.

ALTHOUGH THE LINE of extinct hominids leading to our species seems more similar to apes than monkeys (at least in fossil chronologies), we did not, in truth, evolve from either of them. Our

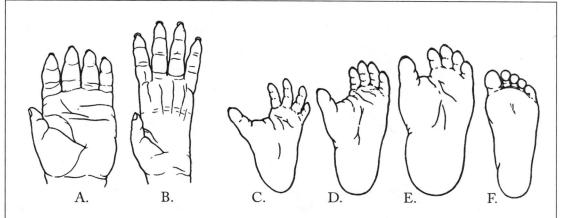

FIGURE 211. Comparison of hands of forest gorilla and orangutan, and comparison of feet of chimpanzee, forest gorilla, mountain gorilla, and *Homo sapiens*. A. Hand of a forest gorilla; B. Hand of an orangutan; C. Foot of a chimpanzee; D. Foot of a forest gorilla; E. Foot of a mountain gorilla; F. Foot of *Homo sapiens*.

From W. E. Le Gros Clark, *The Antecedents of Man: An Introduction to the Evolution of the Primates* (New York: Harper and Row, 1963).

ancestor most likely diverged from the line leading to the apes sometime between ten and twenty million years ago at the end of the Miocene or during the early Pleistocene era (though we cannot exclude the possibility that hominids branched off earlier and that we share our last ancestor with other anthropoids at a point before even the discursion of apes and monkeys).

The apes today are represented by four genera: gorillas, chimpanzees, orangutans, and gibbons. We resemble these more than any monkeys in the following key characteristics: the dimensions and configuration of our brain, our general skeletal construction, our internal organs, the development of square cusped teeth at the back of the jaw (the molars), and our atrophied tail.

In size and in number and configuration of cortical folds the brain of the ape lies, anatomically, somewhere between the monkey and the human brain. The putative first humans, the Australopithecines whose fossils indicate they appeared in Southern Africa some two to four

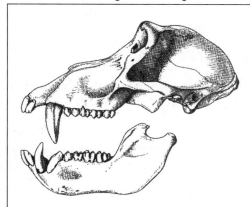

FIGURE 21J. Skull of male baboon with large canine teeth.

From W. E. Le Gros Clark, *The Antecedents of Man: An Introduction to the Evolution of the Primates* (New York: Harper and Row, 1963).

million years ago, had brains only slightly larger than those of chimpanzees and gorillas (at least so far as archaeologists can discern from measurements of their fossil skulls). This means that individual chimpanzees and gorillas may be brainier than some of the first men; yet they probably lack subtleties of neural configuration not evident in mere skulls. Or perhaps culture is a gradual collective insight that modern apes might yet claim in rudimentary form if there were not other hominids now blocking the way.

Hominids

The early hominids were semibrachiators, their arms having already been freed for tool use before they left the forest. Their stout humerus bones were designed for weight-bearing, and their spine had stiffened as a vertical rod (a fossil Australopithecine scapula falls, morphologically, somewhere between orangutan and human). Our ancestors could not have been full brachiators or their bone structure would have become too specialized for the range of limb movements they needed subsequently.

Though the mutant apemen were unknowingly preadapted to bipedal existence on the plains, in no way would they have been tempted to leave their protected arbors for such a barren ecosphere. The actual human progenitors were probably marginal tree apes, outcasts from a larger simian community. A number of factors would have forced single bands onto the savannah, for example, overpopulation, natural disasters, and contraction of native forests (for which there is evidence in Africa at the time hominids emerged). Among the earliest fossils that remain classified as hominid are those of the Indian and Kenyan creature Ramapithecus, known only from jawbones and teeth. He is not an ape, and he is not a human, so he is considered an extinct primate near the line leading to *Homo sapiens*.

Once these awkward "yetis" colonized the open savannah their unused quirks and talents—semi-upright structure, head and limb mobility, acute vision, and incipient intelligence—took on radically new ecological meanings, enhancing their opportunity of survival in a neighborhood across which fierce beasts tracked and swift prey hid. They were uniquely hampered, for semi-brachiators are not great gallopers, and sensitive phalanges are ineffective weapons.

Outside the woodlands the array of selective factors working against hominization (as we retroactively define our special case) would have been somewhat neutralized and, in a few cases (bipedalism, for instance), immediately counteracted, but the creative potential of a new phenotype would have taken generations to blossom into a full-blown lifestyle. During that time Ramapithecus and his kin likely survived by imaginativeness. They might just as easily have died out—no doubt many of their communities did—until one finally gave rise to the ancestors of a successful band of precocious hominids, perhaps in a protected meadow with abundant fruit bushes and small prey.

Sanctuary over a number of generations safeguarded the genetic and behavioral gambles necessary for full hominization.

The Invention of Tools

AT FIRST HIS FATE MUST HAVE SEEMED LIKE EXILE to Ramapithecus (and his equally mythical—or real—successors). He had lost the asylum of the grove and was stalked everywhere by hungry, talented adversaries. In truth, he was hamstrung by the very features of incipient bipedalism that preadapted him to the realm of symbols and tools.

The ability to grasp objects—stones, sticks, bones—was a turning point in the saga of hominid survival. Probably millions of times before culture developed, anthropoids spontaneously found, used, and then abandoned weapons and tools in Pliocene forests. Modern apes maneuver props to get to food placed out of their reach and, in the wild, have been observed manipulating sticks to dig termites out of a mound. But the full potential of these objects eludes them. They do not create or apprehend artifacts for imagined future tasks.

Armed with sticks and bones, capable of conceiving applications of objects through time, early hominids were a fair match for the ancestors of wolves, lions, and pigs. Hurling spears and stones was one of the first deeds of our lineage; the baseball pitcher and football quarterback remain warrior-heroes of our clan.

Archaeological evidence of the hunt dominates early Australopithecine sites— bashed-in skulls of antelopes and other game animals, skeletal parts which surely yielded raw material for future weapon design. These warriors were barely, if at all, more intelligent than apes; they half-limped, half-crawled by modern standards of bipedalism. But they had entered the imaginal realm.

The human brain apparently contains preadapted hardwiring for response by language, for thought by strings of symbols. A modern child learns to speak as innately as she learns to walk. But this is because languages have already been invented. In the beginning of speech, deep phonemic syntax probably came about more gradually. Yet it emerged everywhere as a single, identical structure. Linguistic frames as remote from one another as Finnish, Nahuatl, Mohawk, Aranda, Hebrew, Serbo-Croation, and Welsh all share a word-forming, sentence-catenating grid—a secret operational skeleton. The underlying principles common to language itself are what children learn in mastering the sounds and grammar of any one particular language.

Likewise, adults must regain that embryonic flexibility of structure to translate their thoughts into a different subgrid.

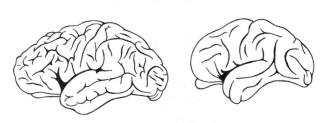

FIGURE 21K. Comparison of brains of gorilla (right) and *Homo sapiens* (left).

From W. E. Le Gros Clark, *The Antecedents of Man: An Introduction to the Evolution of the Primates* (New York: Harper and Row, 1963).

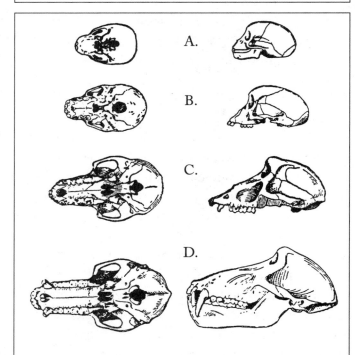

FIGURE 21L. Heterochronic growth of face relative to the cranium in the baboon. A. New-born; B. Juvenile; C. Adult female; D. Adult male.

From Paul Weiss, *Principles of Development: A Text in Experimental Embryology* (New York: Henry Holt & Company, 1939).

TOOL-MAKING REQUIRED not just a chance eureka but a committed continuum of intelligence, a sustained tradition (generation to generation) of artisans and masons. As extrabodily prostheses were substituted for teeth, prehuman incisors and canines became smaller and the hominid jawbone shrank too, perhaps because random mutations reducing their sizes were no longer maladaptive. Bone space was fortuitously available for further cranialization.

Meanwhile mutations expanding the cranium and elaborating its contents were clearly propitious, providing sulci for the phenomenology of hunting. With their arms freed to bear rudimentary spears and knives, symbol-wielding, bipedal creatures went to battle as tactical bands rather than sylvan galoots.

Tentative tool-making must have arisen thousands of times independently among different groups. The first tools were probably stripped branches, graspable rocks. The oldest extant remnants of aboriginal men and women are pebbles and bits of river gravel chipped by Australopithecines, likely used as hand knives, chisels, and scrapers. Eventually their descendants began to sculpt more refined implements.

Our devotion to inventing technologies misleads us. Artifacts are dangerous, and new ones subsume generations in being imagined, flickering in and out of recognition before finally taking hold and spreading like wildfire. The more radical the technology, the longer the phases of its introduction. Whereas the radio and computer took little more than decades to engulf their predecessors, electricity required centuries, the harpoon and the wheel thousands of years. The original stone tools, language-making, and controlled fire—the forerunners of all machines—emerged from the technological void only on a scale of tens of thousands, perhaps hundreds of thousands, of years.

ANIMAL PHENOMENOLOGIES, first submerged in cells, are then cloaked in neural matrices, then filtrated back through discursive layers of tissue. No wonder they are (in the process) sublimated and displaced—into often ambivalent and unfinished acts.

There is an implicit resistance to an act even as naive as rock-chipping, for to imagine one's self cutting a preconceived shape in stone is to become conscious, and to entertain, as well, the borderline spirits that attend consciousness. To make a tool is to make its user. No way to avoid staring into the mirror, this was the beginning of *I*, the remote apperception that *I* and *It* are different. Grasped between the opposable digits of a curious animal paw, an altered pebble reflected selfhood with a concreteness and purity all later inquiries into the origin of being fail to recover.

The first tools may have been crude, unidirectional cuts in the borders of small rocks, but they were images, even more luminous and powerful than those etched millennia later on the walls of the Lascaux caves, for they were prior and seminal to them.

Skeletal Adaptations in Hominids

AUSTRALOPITHECUS WAS EVENTUALLY SUCCEEDED by the more cerebral fire-using creature Pithecanthropus, a far-flung race including Heidelberg man in Europe, Sinanthropus in ancient Cathay, and Java man in Southeast Asia (aliases from our world for their fossil remains). Fully bipedal, Pithecanthropus is traditionally considered the first member of our genus: *Homo erectus.*

With the perfection of bipedal locomotion, various mutations and shifts in biological fields occurred, including ones affecting early differentiation of somites, chondrification of the mesenchymal primordia of the limb buds and other skeletal elements, sclerotome migration, myoblast aggregation, and general histogenesis of cartilage and bone in relation to each other and to the changing topologies of viscera and neurons. Cells, flowing at minutely deviating angles, reproportioned and repositioned soft structures that then ossified with new load lines and fulcra.

The human pelvis expanded in three dimensions as well as in breadth; flexor and extensor muscles were attached further out from their pivot points. In general the blade of the ilium shortened, widened, and became curved; it now lies behind the point of articulation of the femur bone instead of to its side. The bending outward of the ilium created a new point of attachment for the muscles of the floor of the pelvic basin, but this curve has not proceeded as far in women, who require a pelvic opening for the birth canal. The sacrum also shifted its orientation to receive more of the weight distributed through the iliac spine. Muscles from the thorax to the ilium now lift the trunk over the spine at each step and transmit the center of gravity over the foot. This potentiates bipedal locomotion, with the weight on one leg at a time. Meanwhile the human foot has narrowed, distributing its muscles along finer load lines. Grounded in the tripod of the big toe, small toe, and heel, this appendage has lost its grasping tactile phalanges (which shortened and weakened), but it became an innervated volar pad with sensitive terminal receptors on its digits — a base for bipedal movement.

The Emergence of Culture

DRIVEN FROM WINTER TO WINTER, firepit to cave, Pithecanthropus enjoyed short cloudberry summers in the North, long rainy autumns in the South. Our families roamed the epochs we call the Pleistocene—glaciations hundreds of thousands of years long, stalled by brief temperate interludes (of which the Holocene of our recorded history may be only the most recent). Harsh climates no doubt encouraged dependence on campfires, weapons, gods. The ice packs are determiners of evolution in a major way, though we don't know whether these titans will succumb to global warming or march again upon the continents, scraping clouds.

Pithecanthropus crafted thousands of generations of handaxes of lava, quartz, quartzite, and flint. His so-named early Chellean tools were flaked around the edges with strokes in alternate directions so that two faces crossed in a staggered margin. These scarred bifaces "migrated" in the packs of their makers across Eurasia (where they are now found beneath the archaeologies of ancient civilizations). By the second interglacial (some 300,000 years ago) they had initiated much more sophisticated styles.

Prior to the third interglacial, Acheulian handaxes were roughened initially by flaking, then finished with a second sharper edge by percussion from a bone or wooden baton. In the later Pleistocene, one hundred thousand years ago, Neanderthal arrived, a cerebral omnivore with state-of-the-art Stone Age skills (although, according to 1997 DNA studies of his remains, apparently not a direct human ancestor). He augmented his side-scrapers and retouched chert blades with fine edges. He also learned how to prepare large cores from which dozens of separate tools could be manufactured, each with the single stroke of a hammerstone or baton. Whether or not Neanderthal Woman is in our direct lineage, her Palaeolithic culture is representative of hominid society at that time.

DURING THE FIFTY THOUSAND YEARS prior to the first writing tablets and villages, tools mirrored the complexity of the cerebral cortex. Cro-Magnon's "kit" included awls of antler, points, burins, concave saws, needles, fishhooks, willow-leaf blades, arrowheads, lunar counting devices in bone, small statues of either mythical beings or actual people, and pigments of crushed oxides and plants. Sacred caves from Spain to Queensland and Argentina announced (at different stages in history) the birth of the supernatural. Muscular elephants, bison, kangaroos, and horses—strikingly realistic and totemistic at the same time—were tinted onto stone. These are pure and iconic in a way that a gold-leafed Christ from Byzantium or bull of Picasso grandiloquently ordain. The hunt was a ceremony, game animals numinous beings.

Writing about a subsequent era, Erich Neumann captures an aspect of how the twilit realm of these Pleistocene artists must have appeared:

"In the early phase of consciousness, the numinosity of the archetype . . . exceeds man's power of representation, so much so that at first no form can be given to it. And when later the primordial archetype takes form in the imagination of man, its representations are often monstrous and inhuman. This is the phase of chimerical creatures composed of different animals or of animal and man—the griffins, sphinxes, harpies, for example—and also of such monstrosities as phallic and bearded mothers. It is only when consciousness learns to look at phenomena from a certain distance, to react more subtly, to differentiate and distinguish, that the mixture of symbols prevailing in the primordial archetype separates into the groups of symbols characteristic of a single archetype or of a group of related archetypes."[3]

In the initiating aeons of the human epoch, distinctions between human and animal, self and other, and culture and nature were ambivalent and unresolved. By the time nineteenth-century anthropologists discovered equivalent systems of cosmology and ritual among so-called primitive tribes, their sigils were already parabolic, tens of thousands of years thick.

The usual explanation for animism and totemism is that ancient men and women used mythologies and clan symbols to explain and alleviate the violent, weird world all about them, its displays of lightning and thunder, its comets and disappearing moons, its civets and hyenas. However, it was not nature that threatened early humankind—nature was concrete and reassuring; it was the devious structure of their own consciousness, the relentless pulse of mind, the excruciating peril of social life, the existential fact of being that baffled and frightened them. "The mistake," opined anthropologist Claude Lévi-Strauss, "... was to think that natural phenomena are *what* myths seek to explain, when they are rather the *medium through which* myths try to explain facts which are themselves not of a natural but a logical order."[4] I.e., those of gnosis—family, eros, animals, gods, stars, burial, ghosts.

Totemism, like language itself, is rooted in the design of the nervous system, the phenomena of nature, and the organization of social life. Through its practice by indigenous philosophers and shamans, it became the basis of mores, taboos, customs, and institutions thereafter.

[Or, in the words of poet Gerrit Lansing: "wha you say, 'Nay-cher'?... I said gNature, 'birth,' prae-gnant/from (g)nasci, to be born. (I no say, 'Gno'....")][5]

Neural Darwinism

DESPITE THE LONG-STANDING BELIEF of computer scientists that artificial intelligence is an inevitable consequence of computer advances, new paradigms of social psychology suggest that mind does not arise in isolation. It is contextual and cultural only; thus, no amount of silicon circuitry can even approach the most rudimentary human thought, let alone the phenomenology of consciousness. As we saw in Chapters 16 and 17 (on the origin of the nervous system), though density of ganglia and complexity of neurons are prerequisites for the higher domains of consciousness (and unconsciousness), paradoxically, cells do not explain mind.

Compared to syntaxes of language and philosophy of problem-solving, neuron gradients and modular maps of the brain are absurdly simple. Neural hardware likely contains no more than five percent of the textile of thought and none of the brain's emergent properties; "those properties [are] not inherent in any of the individual components but aris[e] solely from a particular level of interactive complexity that has the potential for a truly novel 'jump' in capacity."[6] The human genome probably encodes basic perceptual systems, their organizing principles, and a simple set of programming instructions such as avoidance of certain odors, tastes, sensations, etc., and attraction to other ones. Axons extend and differentiate in cell bodies, as more "memory" and operating capacity become necessary. The remainder of our elaborate pantheon of knowledge is filled in "nonmaterially" by contexts of real events.

We are truly ghosts in machines. Even individual birds in a flock are probably dealing with no more inherited data than "flap and stay three feet away from each other"; the rest is conjured by the act of group flight; the synergy of wind, air, light, and one another's existence. Schooling fish function as complex parallel processing systems "without the necessity for any single locus of executive command."[7] Just as the interactions of cells through morphogens, hormones, and other polypeptides congeal epiphenomenally into an organism, the interactions among the individual fish *are* "mind." Human artifacts and symbols likewise permute into culture. All are emergent properties of chaotic systems that impose order on their nexuses (and have the fruits of that order then imposed back on them).

Intelligent machines fall short. They lack a malleable substratum of cells and a collective social modality. They are apparently, by the strict order of what they are and what nature is, an "emergent property" dead-end. And it is not because they are inanimate—most of the astrophysical universe is inanimate—it is because they are rigidly structured *not* to become chaotic or to use chaos as an organizing principle; and when they succumb to chaos, they have no capacity to use it creatively in their present form. They must first be returned to molecules.

Mind is not simply neuronal quota or output. The cells of the different nervous systems and organizing ganglia of humans surpassed all hardware requirements for thought and speech long before anything was thought or spoken. They constellated in networks and ganglia for evolutionary and morphogenetic reasons having little or (more likely) *nothing* to do with the deep analytical and representational properties of intellect. Yet, once they were federated and indexed, the fact of their being alive and inherently complex led to novelty and complexity, which generated more novelty and even greater complexity.

Psyche arose from virgin alchemical spume already pregnant with her wiles— preadapted crystals. This evolutionary process is reenacted in each human fetus.

Without social experience to help with processing their entanglement of neurons, it is doubtful humans could have achieved full intelligence. Psychiatrist H. Robert Bagwell summarizes the role of social life in the development of integrative processing:

"Our primate ancestry was intensely group oriented, and much of our integrative processing design seems to reflect that group history. We are designed to process our feelings sitting around the campfire talking about them.... The rhythm of Pleistocene life was probably to have experiences, to cope, and then to return to the home circle and talk about it and let strong feelings have their play. Plains Indians would raid, then return to tell, or dance, their story.... And, in fact, from our understanding of the brain mechanisms involved, in a certain sense those feelings don't really exist outside of such a process, as we actively assemble them in dialogue. Experiences leave complicated states of physiological arousal, but without necessary coalescence into the kind of specific mental states we describe as feelings. And, without such coalescence, certain kinds of higher-level integrative process cannot be accomplished. In our basic design we are not self-contained emotional processing units. We evolved as co-dependent group members. . . ."[8]

Each modular human unit is a holographic representation of the culture seen from one angle. Each memory bank is not only personal and solitary but communal. That is how the group maintains the individudal when he or she is independent or separated.

"Single humans [have] developed the psychological capacity to tolerate the anxiety of separation from the group and go off alone for various periods of time, to explore, seek food, etc. It seems likely that such enormously adaptive capacity was made possible by the coalescence of memory traces of the group, and our capacity to have that memory 'playing' as a kind of background to our more focused mental efforts of the moment. That formed the internal security system, the world of remembered others who are brought to life in this background narrative. It is the animated version of our representational world."[9]

"Alienated" means "out of integrative context" and, in its current epidemic form in the West, it is a symptom of the breakdown of society.

OUTSIDE OF CULTURE, the raw tissue material of thought apparently does not arrive at mind—or anything resembling it. Like multicellularity and organic form, intelligence and language represent not a network or plexus but an emergent property of the dynamic behavior of prior complex systems. As such, they are a direct outcome of the life of the hunting band, the primeval family, and the tribe, unavailable to other primates not so much because they lack the neurons (though they likely do) but because they lack the cultural patterns and contexts underlying the birth of symbols.

Yes, "wild children" reared by wolves or lions have cognitions of a sort, but they are nothing like the inner worlds of the children of bands, tribes, and civilizations. Also, we must not discount the roles—and possible roles—of maternal empathy during ontogenesis, global telepathy, unknown neural parameters developed since the advent of language, and Lamarckian inheritance in conferring rudimentary thought even on children raised outside culture.

Feral children of symbol-bearing adults are quite different from the first hominid children. The latter had only phylogenesis and the immanence of blank tissue to educate them, so they required the synergy and social context of the group. Ontogenesis gave them nothing. They had to invent language, invent meaning, invent themselves.

The richness and excitement we experience in being alive go well beyond our biological heritage; they encompass moieties and metaphors into which we were taken at birth, empathies and longings into which we are initiated. We inherit the hopes, dreams, and disappointments of the elders even as we silently incarnate their phenotypes and take up their endeavors. From our first breath we are never alone, never vacant, never sovereign, never exempt.

UNACKNOWLEDGED THOUGHT GOES ON in the background of all our dialogues with ourselves, a ceaselessly reassuring babble and debate stitching our identities and ratifying our existences, individual and collective. Conversations between parts of ourselves become conversations with other individuals; conversations with other individuals improvise the terms for our self-dialogues. Even private and personal ideas—the most secret, idiosyncratic fantasies—arise as collective text somewhere between the group and the individual. Almost no one perceives this promiscuous activity, for materials pass seamlessly back and forth between signifiers.

It is even doubtful whether "inside the mind" is a separate locale from "outside the mind" insofar as a thought in language, a feeling integrated, only exist in strings of collaborative symbols that are exchanged as freely as viruses transfer DNA back and forth. Fashions of behavior, slang and vernacular expressions, insights and ways of understanding, scientific formulations and religious ideologies emerge in single minds solely from group reciprocality. (Our ancestors may have awakened into a wilderness ruled by chattering and cawing. We emerge in a mall drenched with metonymies and replicas, a landscape completely signified.)

The Freudian unconscious dwells somewhere within this schema of overlapping personal and collective mentation and, though its contents may incorporate extrasensory, archetypal, and paraphysical elements as well as the symbol-less activity of organelles, cells, and organs, its representations, including its visions and nightmares, are all furbished and blazoned in culturally recognized stuff and portrayed by figurines and sigils everyone else knows or (if they don't) can imagine and endow with their own personal meanings.

WITHOUT A RIGOROUS SOCIAL REALM, human beings have no meaning; they are like bees which, if removed from a field in Portugal and placed in one in New Zealand, would be no wilder. Society rescues humans from the dimensionless void of hunting, eating, mating, giving birth, sleeping, dying they otherwise inherit as animals. Social life even recreates domains of sheer nature—Dreamtime rocks and caves, mythic rivers and seas, constellations of the night sky. As creatures conceived storms and stars, porcupines and parrots—things separate in their own innateness—nature penetrated society a second time as classifications, signs.

"What is this world," asks Lévi-Strauss, "unless it is that to which social life ceaselessly bends itself in a never wholly successful attempt to construct and reconstruct an approximate image of it, that world of reciprocity which the laws of kinship and marriage, in their own sphere of interest, laboriously derive from relationships which are otherwise condemned to remain either sterile or immoderate?"[10]

BY ITSELF MIND is madness and blind rage—ungoverned instincts illuminated at best with demonic apperceptions. Between animal and cultural experiences lie great neuroses and much trauma, especially among long-extinct primates struggling to bridge the gap. Beneath our dense forest of symbols, a gaping wound remains—the portal through which Freud and Jacques Lacan viewed our slips into hidden crises and unfinished meanings. Human existence is not only a work in progress but a paradox hanging on a precipice. When a society runs amok these days we see flashes of the incognizant mayhem and cruelty endemic to the primal horde.

Like organisms, societies are built up in layers, which is why democracy cannot be instantly imposed in Third World countries, and overthrown military dictatorships in Africa often crumble into their mirror images. Similarly, without intervening stages of cell connection and emergent properties, a mammal cannot be fashioned from a fish.

Hunting bands become tribes; tribes amalgamate into chiefdoms; chiefdoms give rise to agricultural principalities and kingdoms (as in Sumeria and Mexico); kingdoms become feudal duchies; archduchies become nation-states and conquer continents of tribes, bands, and chiefdoms. From the standpoint of political and hierarchical strata, British, Spanish, Dutch, and Portuguese colonialism in the New World and Africa resembled triploblastic creatures descending upon protozoa and sponges. If the United States of America collapses into total anarchy, restoring constitutional government would be just as difficult as creating it from scratch in Somalia or Liberia.

The holocaust of Nazi Germany was all the more disturbing, requiring hundred of times the scholarly inquiry of the Khmer Rouge regime in Cambodia and the Rwandan *interawahme* put together, because it occurred in an already complex civilization.

The Gift

AN INEVITABLE OUTCOME of hunting and mutual aid, society was a biological and psychological necessity.

As noted, one of the absolutes of primate evolution was the extension of the prenatal life of the embryo (a general mammalian trend). A long gestation allows additional neuralization in a protected uterus. However, the pelvic girdle can accommodate only so large a cranium, and cerebralization *in utero* beyond that point would be lethal for both mother and child. In human development the cortex must continue its expansion outside the womb, so infants are born unfinished and must be cared for in their early years.

Neuralization requires society, and society requires neurons — cerebral ganglia. Together they lead to the parthenogenesis of social institutions out of cell life — and out of the inchoate debris of animal phenomenologies. Whether it truly happened that way, we imagine that the responsibilities of guardianship and education of the young led to full-blown tribal societies with sexual division of labor and complex family structure (mothers and fathers, aunts and uncles, lineals and collaterals, nobles and stinkards, etc.). Migratory bands of men and women, clans of interrelated families, founded landmark communities, farm villages. All of their main streets, churches, and meeting sites have long since been obliterated and buried beneath what are now Africa, Europe, India, China.

As new beings attained manhood and womanhood they were resurrected through the sodalities and rites of their groups. So powerful has this symbolic realm become that today it has all but replaced apperception of the physical planet.

MARRIAGE, THE NUCLEAR FAMILY, and the division of labor between sexes lie at the basis of full hominization. Incest taboo was one of the first belief systems — a combination of mysterious customs with a precedential injunction. "Marry out, or die out,"[11] declared Lévi-Strauss in justifying a primal superstition among human societies. Marriage replaced incest by becoming it, i.e., by turning a forbidden endogamous act into a commoditized exogamous one. Perhaps this was how truces between Pleistocene bands were negotiated — by kinship.

A prior French anthropologist, Marcel Mauss, proposed that the initial act of culture was a gift; the first gift was the spouse, a marriage partner from an outlaw tribe. The peace offering to an enemy became, after the fact, dowry between kinfolk. Writing from the unexamined male perspective of his time, Mauss described the process this way:

"Food, women, children, possessions, charms, land, labour, services, religious offices, rank—everything is stuff to be given away and repaid. In perpetual interchange of what we may call spiritual matter, comprising men and things, these elements pass and repass between clans and individuals, ranks, sexes, and generations."[12]

The gift seems to have unconscious forerunners in the fish passed between swallows, the insect wrapped by the spider as its nuptial libation. Even the nests of songbirds and sticklebacks are tokens of a heedless generosity.

"No marriage," adds Lévi-Strauss, "can ... be isolated from all the other marriages, past or future, which have occurred or which will occur within the group. Each marriage is the end of a movement which, as soon as this point has been reached, should be reversed and develop in a new direction.... Since marriage is the condition upon which reciprocity is realized, it follows that marriage constantly ventures the existence of reciprocity."[13]

Marriage served an inductive function in ameliorating borders and coalescing and enlarging cultural systems. Neighboring groups soon consisted of sisters, brothers-in-law, nephews, patrilateral cross-cousins, *orang samandos*. They could not be sworn enemies.

Societies coalesced even as cells once did, from a network of signals, positions, and meanings.

Arising powerfully through neuroendocrine substrata, eros restrained the barbarian and the beast, keeping skirmishes and cannibalism to a minimum. Beneficial genes flowed from society to society, from Africa to Europe, to Asia and Southeast Asia and back, later to the Americas and Australia. Through intermarriage humanity became a single family.

The Death of Pure Instinct

DNA THAT SPREAD DURING THE PLEISTOCENE reflected the "humanizing" of culture. With the reduction of the muzzle and corresponding enlargement of the cranium, the center of gravity of the human head retreated almost to its point of pivot upon the spine. This meant that the nuchal muscles, which attach at the back of the skull and support the head, gradually atrophied and, by the late Pleistocene, the bony crest at which they attached on either side of the back of the skull had been almost eliminated.

Community life required cortical control of the limbic system and basal ganglia. Excessive boastfulness, untamed passions, inconsolable fury, xenophobia, territoriality, and homicidal jealousy had to be sublimated through ritual and play.

They still occurred (in fact, they could not be prevented), but they became politics and pageant.

Society is born through an uneasy alliance of opposing impulses and the death of pure instinct. Without taboos on incest, rape, cannibalism, murder, and hoarding of subsistence goods, the entire tribal enterprise would have collapsed, aborning, into a primal horde. According to Freudian dogma, civilization is the collective projection of a taboo, expressed individually through the emergence of the super-ego in each personality. The pure libidinal energy of the animal world is displaced and ritualized. Repression of unbridled lust and absolute desire is essential for "long-term cooperative, reciprocal relationships."[14] Modes of self-deception and intentional ambivalence are critical, culturally-reinforced elements in driving passions underground. Every aspect of raw instinct and desire had to be reflected, challenged, and totemized.

Our species pays a heavy price for consciousness, but perhaps not too high a one considering the alternative.

As ADRENAL AND SEXUAL FUNCTIONS were suppressed cortically, facial features which supported their emotional outbursts subtilized. Biological evolution became psychosocial evolution, and a new gestalt shifted the human inductive field into something beyond all its prior hominoid states. Through generations the ferocious and mute look of the prognathous beast was replaced by the more dreamy, civilized gaze of a being impregnated with neurons, entertaining a stream of culturally derived events.

The human neck drew even longer, the face even flatter. Bone and musculature receded from regions of eye-to-eye contact and speech. The countenance became a zone of communication and contact, with degrees of sensitivity and nuance. Its masks marked neuroglandular episodes that left no archaeological trace among the bones of the face.

The human body shed its claws and coat of hair. At the same time, abundant sweat glands developed in layers of skin—their apparent function, to moisten the epiderm and sensitize tactile receptors. Our "nakedness" may be an indirect result of diurnal hunts which required long-distance marches on hot afternoons, but it had psychosexual consequences, as did the perfumes of the sweat glands. It is no surprise that hair loss and fatty glands coincide in a creature who is becoming intimate and romantic to him/herself. Men and women were able to touch directly, to experience their samenesses and differences nakedly, to seduce as well as mate. Ares and Aphrodite emerged from the shadows of Olympus.

Agriculture, Medicine, and Story

THERE WAS NOW A PLACE IN CULTURE, i.e., in nature, for an animal who composed songs and mapped lunar cycles. There was a niche for a hominid who collected and classified—a role for a story-teller, healer, and herbalist. Society became a great memory bank; the individual biont was replaced by the superorganic tribe.

Cro-Magnon was to "invent" agriculture by the good fortune that everywhere he disturbed the climax forest, weeds and herbs sprang up—amaranths and sunflowers. By making beads and sand-paintings out of wild seeds, he crossbred and hybridized them. His garbage heaps were bait for animals he would later tame, animals which hastened their subjugation by domesticating themselves. The dumps also became gardens when rotting roots and berries gave rise to new crops and medicines. Gradually his sons and daughters discerned the meaning of fertility and made farming and ranching their rites. The descendants of hunters stayed put and became villagers.

The Ice Man of northern Italy, preserved as a mummy in a glacier at his death 5,300 years ago, was carrying leather thongs threaded to two walnut-sized spheres bearing a preparation of the woody fruit of the tree fungus *Piptoporus betulinus*. Originally thought to be tinder for starting fires, the ingredients turned out to be a medicine—their laxative oils toxic to parasitic worms whose eggs were found in Ice Man's intestines.

This ancestral European was either a doctor or the patient of a well-trained tribal physician. The skill behind his diagnosis and treatment suggests a medical system going back millennia. Two thousand years later an Egyptian papyrus gave a prescription for a parasitic purge called "aaa," an herbal brew of steeped pomegranate bark and beer (previously considered the oldest extant Western pharmacy).[15]

Genesis tells a story of bedouins, shepherds, and farmers, calling out their clan origins:

"And this is the lineage of the sons of Noah, Shem, Ham, and Japheth. Sons were born to them after the Flood. The sons of Japheth: Gomer and Magog and Madai and Javan and Tubal and Meshech and Tiras. And the sons of Gomer: Ashkenaz and Riphath and Torgamah. And the sons of Javan: Elishah and Tarshish, the Kittites and the Dodanites. From these the Sea Peoples branched out ... each with his own tongue, according to their clans in their nations."[16]

Aranda, Tikopia, Ainu, Eskimo, Yahgan, Ndembu, Zuni, and Zulu chant parallel lineages, matrilineal and patrilineal: Wuningi, Nolingi, Kandingi, Bagali, Malan; Kafumbu, Nyamakayi, Kami, Machamba, Wadyang'amafu....

The awakening of the luminous pillars of culture is a dream (likely it once proceeded in a mycelial trance or hallucination). The moment when man and woman became conscious is as ineffable as the flicker between being and not-being in ourselves. It is so revolutionary and transforming it is objectified as a ray beamed from an extraterrestrial intelligence in Arthur Clarke's *2001: A Space Odyssey* and Robert Anton Wilson's *Cosmic Trigger*. Whether as true divinities, cosmic avatars, plant spirits, or projections of our own psyche, wise ones preceded us and oversaw our transformation by proxy in the cerebral cortex. The collective unconscious of the entire bestial world spoke.

It is no wonder that psychosomatic links have been mysterious and imponderable from Ice Age shamans to radiologists.

The disjunction between society and nature has never been completed. This far into history, nature has been unable to reestablish pure unconsciousness, but we have not been able to extinguish, by cities, machines, newsprint, or great chemical fires, the dormancy that constitutes most of this planet still.

How do ganglia get to know where they are?

THE BASIS OF MIND is far more enigmatic than it even appears—and, from Parmenides and the early Greek naturalists to Martin Heidegger and Jacques Derrida, it is quite enigmatic enough. The architectural plans of Frank Lloyd Wright and symphonies of Gustav Mahler are clearly mental constructions, phenomenologies, but they have equal bases in animal carapaces and autonomic dances of worms.

Somehow an organizing principle, antecedent to mind, to even the most rudimentary expression of mind ("ganglion"), inspissates matter. The mind knows to "think," to individuate its remarkable situation, only because, epochs earlier, clusters of cells squeezed surrounding fluids through themselves and discriminated and siphoned metabolites, burning them to a higher resolution. Mind exists solely because gowns and fibers, poriferoid bells and sphincters, pulsated in unison, thickening mesoglea into mesoderm and conjuring "meaningful" organization. So was "anterior" distinguished from "posterior," "dorsal" from "ventral," and self from everything. The emergent body center beat unities within septa and transmitted complex spiral patterns all the way along outermost filamentary tentacles, rotating its overall position and propelling its mass out of the inertia handed down from mineral hegemony. The first motile creature got itself from "here" to "there," from "there" to "here," over and over, defining "plan" until, when there was nothing else in its repertoire, such activity became its "mind."

Mind is not just chemical tropism and super-phosphorescence exponentialized; it is design. Taste is more than molecular reaction, touch more than incursion. Neural nodes see and hear, and "know" what to do with such bits and datum shards; they "know" that seeing and hearing are discrete though related events moving along different vectors at divergent speeds, bearing singular components of dense, textured episodes in their world.

Emerging nervous systems passed dawn impulses along coalescing and bifurcating networks of primitive code mainly because events and sensations were already beginning to be melded and brazed into interiorized forms, cavities, and organs. These lumps and channels were prescient phonemes.

Among primordial multicellular creatures, matter was continually being sorted, coordinated, partitioned, hierarchicalized . . . and then calibrated into the vast and inescapable domains of time. Otherwise, the manufacture of animate stuff would be little more than atmospheric variations, spasmodic rainbow oozings in mud.

And time (history) is where we find ourselves today. We put sound and touch and sight and smell together and weave a multidimensional, intricately and cohesively sensual world insofar as the forerunners of mind organized themselves at all. Phylogeny repeats in ontogeny, as other organs and their primordia begin to simmer "mind" embryogenically before the brain even exists. Throbs of cardiac tissue and streaming of blood islands are closer to the genesis of thought than the brain itself—which is a tuner and organizer of experience, not its originator (perhaps this is why large portions of even a mammalian brain can be destroyed without seriously impeding function). Drive, will, proprioception, organization of sensations, qualia, and images all arise in viscera outside a cerebral domain.

Phenomenology requires long tentacles of jellyfish medusae and soft statocysts and pressurized canals of comb jellies at its core, not only once upon a time but moment to moment now. Mind is the sum of body, little else—for the body's layering and rhythmicity alone sponsor image and make thought inevitable.

NATURE APPEARS MYSTERIOUSLY as properties of atoms and molecules. Yet genes neither foretell nor guarantee organisms, nor synapses mind. Life derives/fluctuates epiphenomenally, beyond genome maps or physics of a nervous system. Progeny bud from one another's serial units, instincts from electrons in membranes. As matter is indexed, tissue and mind thicken independently, perturbed (or reinvented) by objects cytoplasmic and pelagic, astrophysical and psychosomatic—nuclear fiat be damned. The path from inkling yips to pre-Socratics and Taoists to existentialism and deconstruction is inherent and autonomous. Perry Como singing, *"Take a wheel . . ."* breaks into Dog Eat Dog: *"Snooze . . . make your moves."*

Part Four

Psyche and Soma

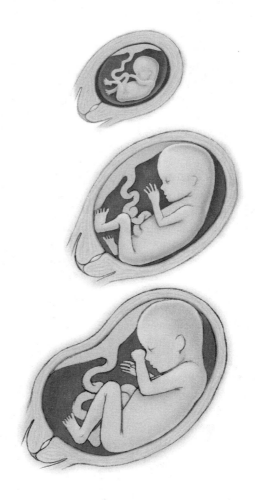

Illustration by Phoebe Gloeckner.

The Origin of Sexuality and Gender

The Biological Basis of Sex

MALE AND FEMALE REPRODUCTIVE PARTNERS, though heralded and universalized in ethics, art, and semiology, are but one virtual psychoanatomy. Sex has been reembodied again and again, from bacteria to sea hares, from crabs to squids, from newts to monkeys. It develops, evolutionarily and ontogenetically, as somatic layers, life cycles, and modes of cognition and behavior, stabilized only through millennia of homeostasis and deviation. Primary cell clusters have migrated, orifices changed roles, nerves impregnated evolving and devolving organs. From mutations and altered anatomies, animals have reinvented courting, costumes, and even intercourse organs and partners.

Though all human cultures attempt to tame and appropriate sexual energy, using it to reify their own institutions, pretending that their customs and examples are subpoenaed in the flesh, organs and applications evade linguistic police and, generation after generation, invent their own "perversions" and "pornographies."

SEX IS NOT BIOLOGICALLY A REQUIREMENT for autopoiesis or reproduction (genetic engineering might someday even eliminate a sexual mode for humans, but it is unlikely to make autopoiesis or reproduction superfluous too). In the reproductive domain, sex's sole imperative is the siring of offspring from more than one parent. Otherwise, modes of asexual reproduction, lacking biparental heritage, quite bountifully yield blastulas and embryos.

Genomes bud and graft. Among mosses they alternate asexual and sexual generations, transmogrifying from spores to haploid nuclei capable of fertilization. Slime-mold amoebae are drawn to one another by chemotactic signals secreted by

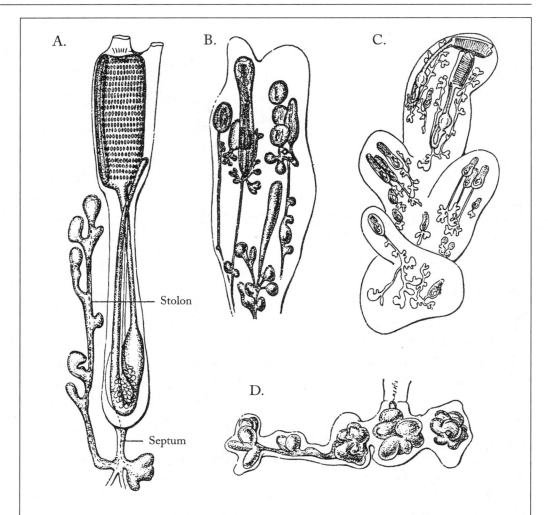

FIGURE 22A. Reproduction without sex. A. Zooid of *Clavelina* colony, showing hyper-
trophied epidermal vascular stolon containing mesenchymal septum; B. Part of colony
undergoing reduction within common tunic, showing reduction bodies of zooids and
stolon ampullae congested with trophocytes; C. Part of similar colony later, showing
reconstitution of large zooids from reduction bodies and development of smaller
zooids from isolated stolonic ampullary clusters. D. Ampullary clusters after complete reduction of zooids.

From N. J. Berrill, *Growth, Development, and Pattern* (New York: W. H. Freeman & Company, 1965).

random cells; these attract receptors on the surfaces of other molds who then exude
the same proteins. Depletion of their resource base causes starving amoebae to reorient, crawl toward the signals, bind, and aggregate there in huge slugs. These migrating colonies then produce fertile stalks gestating offspring (see also page 232).

Termite sexual reproduction is carried out in one generation by kings and queens who bear sterile workers. Only a medley of diet and hormones converts *some* new termites into royal-caste members. If primates inherited anything close to this biology, our entire social domain would be more opiate and cyberpunk, our courtships more antlike.

Far from being indispensable once entrenched, sexuality can artlessly deconstruct itself, leaving behind empty customs and relics of anatomy. The descendants of formerly sexual animals continue to embrace each other in acts of sterile copulation; others produce eggs which gestate without sperms. Desexualized and partially desexualized anatomies occur everywhere in nature, among plants, fungi, and animals alike — displaced correlates of gendered morphologies among their kin.

The Origin of Sex

"MALES" AND "FEMALES" *are* the units of social algebra. However, this dichotomy was not intrinsic and, among primal bacteria and spirochetes, had to be established from scratch. In bacterial conjugation it is difficult (if not impossible) to tell males from females. One bacterium is a donor of DNA, the other a recipient. The latter is determined by the presence of little hairs (pili) on "her" surface. But, when incited by a chemical gradient (a "fertility factor"), either party may take on the other's role. On occasion, climate-induced sex changes spread like wildfire through populations, turning males into females or vice versa.

According to our present archive, sexuality arose as a system of chromosome variation and gamete polarization some twelve hundred million years ago at the dawn of the Cambrian era, probably in association with heterotrophy (cells metabolizing organic stuff). Sexuality and heterotrophy are different modes of molecular attraction. In heterotrophy creatures feed on fellow creatures (life steals its energy from other life). Sexuality on the other hand stimulates tissue to exchange external products (germs) which become internal (intracellular) to an emerging zygote.

Fertilization followed by embryogenesis is an initially malignant transformation of feeding into differentiation. Thus all complex anatomies are derivatives of primordial cellular attempts to colonize other cells. Zooids that would otherwise eat one another combine with one another. These revolutionary episodes (as characterized in Chapter 4, "The First Beings") are ultimately carved into organs.

THE PASSAGE TO SEXUAL DIFFERENTIATION was either archetypal or fortuitous. Taking the former point of view, psychologist Carl Jung propounded that the unconscious universe and organism coincide, and that forms transcending mind and matter

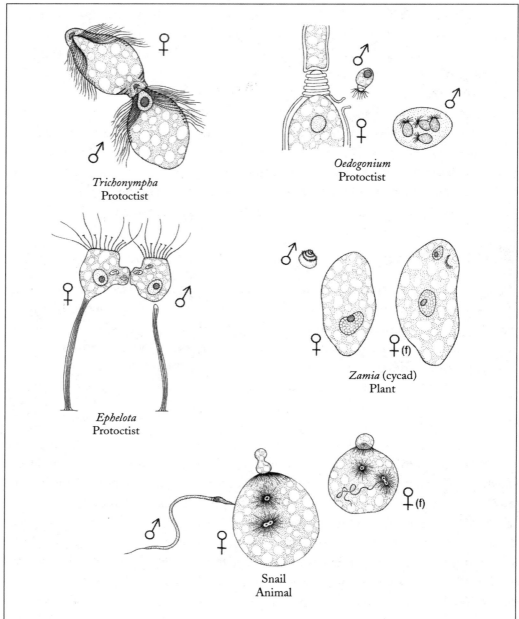

FIGURE 22B. Anisogamy in different lineages (smaller motile males, larger sedentary females (drawing by Laszlo Meszoly).

From Lynn Margulis and Dorion Sagan, *Origins of Sex: Three Billion Years of Genetic Recombination* (New Haven: Yale University Press, 1986).

imprint themselves on soma. The "shadow" is one such form—a zone of concealed antitheses subverting every conscious act; the "self"—our sense of autonomous uniqueness—is another; male and female—animus and anima—are also archetypes. They are all penchants inculcated from elsewhere, of unknown species and origin.

Following Lynn Margulis, I will presume, for the present, that plant and animal tissue layers arose not from preternatural archetypes but quite accidentally from dawn-time deeds of attempted cell cannibalism reconstellated as zooid fusion and meiotic reduction. The cannibals got trapped alive within the victims' membranes and "learned" (over generations) to live there—to borrow organelles and reproduce in a way that the offspring of both became locked in a metric cycle. Some resulting intramembranous clumps then specialized into organelle-producing cells (incapable of mitosis), while others lacking both flagella and cilia became mitotic— the forerunners, respectively, of male and female gametes and hence of all multicellular organisms. These creatures "survived" and flourished in their progeny in part because two aspects of their sexual phase—meiosis and syzygy—provide mechanisms to check their chromosomes against those of other DNA strands or other genomes, with damaged or defective codons replaced by their counterparts. Sexualized genotypes were self-mending, hence randomly superior in the cavalcade of nature. "In the earliest animals," Margulis writes, "reproduction became enslaved, imprisoned by sexuality. . . . [Then], as long as there was any tissue differentiation, fertilization-meiosis cycles were retained."[1]

This is also the shadow embodying itself within the anima.

APART FROM DESIRES spawned by sexuality, animals do not like being eaten or penetrated. Intimacies that mammals consider erotic (i.e., from which they derive pleasure and release) were initially invasive and dangerous among free-living zooids. They became seductive only after millennia of cell differentiation and layering, including the fabrication of libidinalized organs. Coalitions and interpenetrations of bodies yielded at first awkward new organisms in cellular drag. As anatomies evolved, more elegant and alluring paraphernalia came to grace and activate genders. Discomfort turned into pleasure.

From that point on, by Darwinian logic, sex had to remain enjoyable enough for creatures avidly to share their chromosomes and restore the meiotically reduced complements of their genes. How that was translated repeatedly and individually into flesh and exponentialized in multicellular genders is a mystery—and likely more than one mystery, for plant and animal sex is a maze of particularized apparatuses, relationships of bodies to fertilization, and social outcomes from erotic acts. Sexual attraction now works a lot like cell induction and chemotaxis, though (in "higher animals") it arrives in full cinemascope 3-D with Dolby sound, wired to a variety of sensory receptors, a neuroendocrine motherboard, and a cerebral ganglion.

THE PRINCIPALS IN ACTS of zooid cannibalism were identical or similar bacterial types, the eventual union of their mutant progeny isogamous; but, in at least one lineage, differential mutations led to anisogamy—large immobile eggs and small swift sperms.

Eggs and sperms, though nucleically identical, have quite different biological meanings. Sessile, fully dividing oogenetic cells were to become a requisite for morphogenesis; conversely, though, nonmitotic siblings divaricated into not only mobile sperms but axons and dendrites, hematopoietic stem cells, and a menagerie of other specialized, motile zooids trapped within tissues.

In but one of many embryogenic paradoxes, the self-cloning egg became the forerunner of the whole, the depleted sperm of its terminally differentiated parts; yet each of the parts also resided materially in the egg and was conferred only in nucleic proxy. The whole and the part were to receive and replace one another again and again through evolution—nucleus and organelle, mitosis and meiosis, blastula and organ, proliferative and quantal cycles, etc.

Gender was thereby coupled with reproduction, and the mechanics of chromosome exchange pollinized gene displacement and cellular differentiation.

The Relationship Between Eroticism and Reproduction

THERE IS NO EXTANT EVIDENCE for how primordial sexual acts might have hatched microtubule mechanics. The morphogenetic potency that operates today at subcellular and cellular levels projects nothing whole-cloth into the gendered sexuality of organisms, and even less into erotic symbols, rituals, and meanings. There is no mechanical link between nucleic and microtubule activity (in the small) and sexual behavior (in the large). Thus, in Margulis' opinion, gender is not a derivative but "an epiphenomenon of meiotic sex and cell differentiation."[2]

Human sexual meanings arise only insofar as they absorb millions of generations of cell and tissue experimentation through the opacity of tens of thousands of years of hominid customs, institutions, and ceremonies.

Between ancient bell animalcules dancing about the sessile females of their species and porn theaters with barkers and flashing lights, there is a bare link.

FOR SEXUAL REPRODUCTION TO OCCUR between modern multicellular creatures, there must be attraction (and compatibility) at a whole-organism level, then at the level of the cell, and again at the level of its nucleus. Compatibility plus mate recognition (i.e., through songs, smells, displays, and other sign exchanges) link two genders in one species.

While reproduction absolutely requires nucleic fusion, allure and seduction can occur in less discrete manners. Attractions outside species and within genders are dispersed among animals seemingly willy-nilly (especially when observed after the fact and from an ethological perspective). Human morality deems erotic acts sodomistic—contrary to nature—when they are infertile. Yet there were countless borderline moments in evolution when reproductive sex and "sodomy" fluctuated with each other and, in exchanges of near-deviant genes, bodies and lifestyles detoured into new kingdoms and phyla. It took only a few miniscule, leveraged mutations to inaugurate exotic deeds as full-fledged organs or bodies. Many of these survive into the present moment.

A gay male reaching to his mirror body in another male casts his sperms into a sterile replica of his lineage. Even though this may frustrate the seeming intention of the chromosomes, cells do not mandate ambitions, only desire.

Breaking the life cycle in gay sex cannot foil nature. As the speciation of the animal kingdom portends, nothing is absolute, no region of flesh exempt from libidinization or seduction, no act or throe biologically forbidden, no mongrel immune to being turned (down the road) into a viable life-form.

Human consciousness carries the weight of even the discarded life-forms and acts that underlie its male and female components; it continues to grasp after their vestiges in rituals and erotic play. All sexualities represent the history of our cells to potentiate fluids, spasms, and gametes; all are also attempts to "think about [what we want] and put our desires into concrete form"[3]—desires that (when executed) may not even be gratifying.

What about men and women who perform amatory urination and/or defecation on each other, integrating the excretory ducts and their products into foreplay? Do the eroticisms of faeces and "golden showers" arise from reversion to animal acts of courtship and fertilization, or are they invented anew in order to sensualize equations of power and displaced intimacy?

The Fluidity of Biological Meaning

AMONG HUMANS, SEX ACTS ARE LINGUISTIC AND SOCIAL as well as genetic and hormonal. In the struggle between cellular compulsion and psychocultural sublimation, agendas and customs are extracted from anatomy and chemistry again and again at multiple levels and in different contexts. It is often difficult to know what is inherited and what is invented, what humans feel and what they improvise in attempts to understand and express what they feel, how much of even what they feel is vestigial—lost modes seeking expression—and how much of it is symbolic imagination feigning such acts because intangible engrams stand in the way of direct sensation.

The cellular/hormonal charge in sexual tissues and organs is, like any neuroglandular, myofascial component (but more than most), transformed and dislodged, obliterated and recomposed. It is also embryogenic, shifting zone to zone in the passage of an organism from infancy through childhood to the metamorphoses of puberty.

By Freud's etiology of human development (applicable, in differing degrees, to other animals), libido is not circumscribed by the reproductive organs. Most creatures experience the entire alimentary system as erotic, particularly its perforations of ectoderm at the mouth and anus. The lips and tongue are the initial erogenous zone, espoused in blissful oblivion by the suckling infant (even through a rubber nipple), recovered later in French kisses of lovers.

The cutting of teeth imposes a sadistic element on oral satisfaction. Formation of weapons in sensitive gums breaks the maternal trance with its all-giving breast.

During childhood, libido is partially transferred to the anal canal, where it excites a narcissistic obsession with the body's functions and products (the highly lampooned but eulogized poop). At puberty much of its feeling is genitalized and put at the service of the primordial germ cells and anatomy of reproduction. There alone does it participate in social life—by reaching across narcissistic boundaries to exchange sensation with another organism.

This is an idealized theory of heterosexual development fused with a general discussion of the eroticization of organs. Idiosyncratic variations occur at every stage, depending upon individual proclivities and cultural effects. Yet one set of basic truths underlies all applications: sexual energy is subrogated between organs and zones throughout a lifetime, resulting in different expressions of it with radically divergent meanings. Some of these displacements may be the natural and universal result of biochemical changes (as during adolescence); some may embody psychocultural episodes; some may represent attempts at resolving deeply felt paradoxes—but regardless, they all proceed, like primary sexual development, through sublimation and conversion, generating novel acts and sensations.

Animal Seductions

ANIMALS EXPRESS SEX DIRECTLY, by biological imperative. Where nature imposes anatomy they enact it, earnestly and decisively. Though some species engage in gay, lesbian, and cross-gendered pairings and execute "mock" versions of marriage and divorce, rape and pedophilia, celibacy and infidelity, they do so without names or meanings.

Yet, more than any other feature of the prehuman world, sexuality is semeiotic. Desire was first a rune, even among mute beasts.

Peacocks and pheasants display iridescent colors and parade before prospective mates. Other birds spit, raise their tails, and spread wings. Fighting fish open painted fins like parasols. The male dytiscus beetle strums a rhythmic tune on his femoral ring using his hind legs. Whippoorwills summon one another with melodies, over and over. A female spider, after dropping a thread, slides partway down along it; a male catches the bottom and climbs up to meet her. She may decide to eat instead of mate with him, but he strokes and fondles her body with swift, jittery legs and then plunges a palp into her vagina. He may even wrap her in a cocoon of his silk while he enters her.

Crabs perform quadrilles. Moths are drawn to one another by smells which, to us, resemble raspberries and vanilla. In some species male flies compose midair dances; these swarms stir otherwise placid females from among the bushes. The females, transfixed by the pattern, apparently do not even see individual males.

Sea swallows transfer fish from the beak of one to another during mating dances; orangutans embrace and nuzzle as they hang breast to breast by their arms from branches. Flying foxes hug similarly in the air, and beavers kiss while paddling through water.

IF WE SURVEY SEXUAL REPERTOIRES among contemporary creatures, we find that the line between desire and distaste, compulsion and revulsion, exclusivity and intimacy, remains thin throughout nature. In species of spiders, octopi, birds, and cats, attraction is only strong enough to overcome temporarily the enmity between invader and invaded. They recoil, growl, hiss, sting, and bite, yet cannot keep themselves from mating. "Let us roll all our Strength, and all/Our sweetness, up into one Ball," proclaimed Andrew Marvell, "And tear our Pleasures with rough strife/Through the Iron gates of Life."[4]

MALE SPIDERS FERTILIZE FEMALES with sperm-bearing appendages (palps). After climbing the thread, the male arouses his lover in some cases by pulling on her web and tapping out a few stylized dance steps. As she comes to his call he may hold her at a safe distance with his forelegs.

One species of hunting spider clarifies the difference between himself and food by carefully courting with a fly wrapped in silk. During the male's presentation of this gift his whole body shakes and his palps quiver and stretch out to the female. Two ambivalent rituals—one of courtship, the other of cannibalism—remain caught in a biological trap, as spiders dance contrarily on separate threads.

Male bees have organs which break off in their queen's vagina, plugging the sperm duct so that the eggs are fertilized before the seed runs out, yet the loss of the organ causes the male to bleed to death. Females of other insect species literally devour their lovers during copulation. The turn-of-the-century entomologist Jean Henri Fabre was shocked by this savagery.

"What should we say," he wrote, "when the saddle grasshopper, before laying her eggs, slits her mate open and eats as much of him as she can hold? And when the gentle cricket becomes a hyena and mercilessly pulls out the wings of her beloved who performed so magnificent a serenade for her, smashes his harp and shows her thanks by partially devouring him?"[5]

Fabre further describes the golden beetle's marriage: "A vain struggle to break away—that is all the male undertakes toward his salvation. Otherwise, he accepts his fate. Finally his skin bursts, the wound gapes wide, the inner substance is devoured by his worthy spouse. Her head burrowing inside the body of her husband, she hollows out his back. A shudder that runs through the poor fellow's limbs announces his approaching end. The female butcher ignores this; she gropes into the narrowest passages and windings in the thoracic cavity. Soon only the well-known little boat of the wing sheaths and the thorax with legs attached are left of the dead male. The husk, sucked dry, is abandoned."[6]

The Evolution of Organs of Copulation

FISH DO NOT WILL THEIR QUIVERING FINS, nor do bees design their seed or hive. Evolutionarily, fertilization exploits available portals, cavities, culverts, talons, tentacles, and valves without regard for exclusively sexual, excretory, or alimentary zones, molding whatever infatuation or ritual is necessary to compel and preserve the act. Unique meanings are generated as tissue folds around germ cells and then must provide stimulation and pathways for syzygy. As one gay writer jibed, albeit with political rather than phylogenetic intent, "Women do not have the market on fuckable orifices."[7] Fathers become mothers, mothers fathers. Orally breeding fish carry both eggs and young in their jaws, a potentially deadly ambiguity. The mouth, the genitals, the womb, and the stomach regularly take one another's places in evolution.

Any aspect of evolving anatomy can serve a variety of purposes. Crabs and spiders, like octopi, have "arms" with "hands" that detach sperms from their own body and plant them in the female. Female grasshoppers chew up the remains of a semen tubule left at the base of their ovipositors. Lizards and snakes have spiny, hooked double penises so strong that the mating animals are often temporarily unable to pull apart. More advanced salamanders and frogs have no organ of copulation (and make only indirect contact in the seeding of gametes). An organ for introducing sperm directly into the female's body from the male's gonads exists among some species of worms, snails, and insects but not among others.

This heterogeneity is present even in the most primitive and (apparently) most ancient genera. While protozoans copulate body to body in an advanced fashion, many invertebrates (for instance, ocean-dwelling Annelid worms) merely shed sex cells into the water. Flatworms are hermaphroditic. At the back of their ventral surface is a genital atrium opening into a vaginal duct. The testes, developing from mesoderm and scattered within the lateral margins of the body, drain through a series of sperm ducts fused into a genital muscle. Eggs and sperms ripen at the same time and sperms are exchanged in ejaculative copulation.

Other worm phyla bear variations such as long barbed penises, vaginal domains that swell to a greater size than the rest of the animal during fertilization, and permanent marriage in which the mating creatures grow together at the genitals and remain united for the remainder of their lives. Most flatworms also reproduce asexually by fissioning, and, in fact, a section down to a tenth the size of the whole worm can produce a healthy, complete offspring.

One species of bristle worm is fertilized as the female eats the sexual apparatus of the male, which darts and twists provocatively off the anterior portion of his body like a piece of food being offered. Sperms reach her eggs through the abdominal cavity.

A male octopus uses a copulatory arm (the hectocotylus) to reach into his own breathing funnel for a packet of semen at his mantle cavity which he then stuffs into a female's cavity. She is partially choked by the long invading phallus in her gill chamber but aroused by an erotic red that tints the male's body while his arms titillate her. In a few cephalopod species the arm breaks off and lodges in the female's cavity, where it dwells as a semi-independent animal. Live sperm-bearing hectocotyli have been observed swimming independently at sea.

Snails rise to press up against one another, and in rubbing their bellies, give huge smacking kisses and dance like couples in the late hours of a party. They wrap their glutinous hermaphroditic bodies together and then delicately stimulate erotic points with antennae tips. Each hermaphrodite contains a quiver with which it pierces its partner, visibly wounding it (often seriously by puncturing the lung or abdomen). Yet the strikes are also pleasurable. An enormous swelling tube functions as a penis, and one partner inserts it deeply in the female genital of the other.

They cast sperm cells simultaneously and then separate for good, each crawling off in a different direction.

Many species of fish engage in intense kissing. Gurani suck their lips together for as long as twenty-five minutes before mating. Carp rub and writhe prior to simultaneous expulsion of gametes. The male lamprey sucks the neck of the female while wrapping his body around her and pressing against her abdomen. The stickleback draws the female into his nest, and then, when she is thoroughly buried therein, stimulates her caudal region with his fins, causing her to lay thousands of eggs. After she has left he is compelled to burrow through his tunnel, spreading his sperm on the eggs.

Foreplay completes the romance, and fertilization is accomplished masturbatorily.

The male Surinam toad squeezes the female so hard that his thumbs may penetrate her abdomen, their joint spasm leading to the simultaneous discharge of germinal cells. The male, hunched over the female's back, literally presses eggs out of her belly.

The male newt writhes on the ground as he leaves his sperms wrapped in jelly. The female passes over this packet and fertilizes it by pressing her cloaca against it. She is in the vicinity because the male has gotten her attention with a display of bright colors, a dance, and erotic wavings of his tail.

Turtles have swelling copulatory organs like penises. They seduce each other with slow head-to-head swims. The male extends his organ into the female's cloaca, and then he may ride atop her shell for days, titillating her genital with whips of his spiny tail, which causes her to push the hind end of her body as far as she can out of her shell.

Bats mate while hanging, the male pushing his bent tubular penis from behind, between the female's hind legs into her vagina.

Whales leap out of the water belly to belly and, while they hang there for a moment, the male's long and leathery penis enters the female and ejects semen.

ALL OF THESE ACTS, though circumstantial in their lineage and choreography, encompass singular and ancient excitation. Desire is the transhistorical projection of germ cells into cytoplasm. Tissue layers bring primal feelings of differentiation to the surface of genitals, where they can be felt and enacted. Creatures experience and transmit the same energy that ignites stars and initiates life. They liberate gametes in spasms of mercurial waters that sparkle with hundreds of millions of years of transmutative dissolution and alchemy. The wanton patterns and pulsations of trapped fluids and membranes fueling meiosis and gastrulation are externalized into organs and performed, if only for a moment, in the nervous systems of multicellular entities. There the riddle of cell existence turns into the ritual of courtship and the deed of cell reproduction.

The Origin of Bodies

IN HIS LATER, MORE PESSIMISTIC YEARS, Sigmund Freud proposed an all-embracing death impulse which he christened "thanatos," the antipode of eros. His Hungarian disciple Sandor Ferenczi elevated this drive to the central force of nature—not only of psyche, not only of life, but of matter itself. Thanatos, Ferenczi proposed, is the link between embryogenesis and phylogenesis, a ritual played out by cells resulting in the invention of viscera.

Unless aroused by a trauma, substance (in Ferenczi's view) would never have come alive. Nature cannot provide life randomly and neutrally out of inanimate chemicals; mere "primal soup" chemistry and natural selection by themselves are not enough.

In Ferenczi's metahistory, evolution and gender are a succession of failed suicides—of graspings after death and unconsciousness that accidentally incite life. Instead of annihilating itself (as intended originally), protoplasm becomes trapped in membranes, forced to spiral outward. Lineages of cells incorporate the successive traumas of abortive death attempts, recovering, masking, and displacing them organ by organ; in fact, cells are the collective somaticization of these traumas, as life, unable to become nothing, becomes something.

By speciation, desires and their antitheses operating as cytoplasm, unable to escape their own maze, weave ever new, more complex tissues. In a constant effort to squelch existence, cellular feints give rise to bodies and lifestyles in Lamarckian fashion. That is, actions and emotions get translated, sublimated, and inverted into germ plasm and then organs.

Channelled back through nerves into gametes, the Death Wish is coded in genes, embodied in phenotypes, and reexperienced through the serial ontogeneses of organs.

THE OCCURRENCE OF LIFE in dark, lonely waters was Ferenczi's first and most horrendous trauma, compelling a dram seed from its eternal dream. The denial of this event is imprinted onto matter at the level of the protoplasm it brought into being. The "catastrophe" (as Ferenczi characterizes it) recurs ontogenetically in each new organism at the maturation of its sex cells. In trying to stay asleep, it awakens. In trying not to imagine the nightmare of history, it makes history inevitable.

The second "catastrophe," releasing individual unicellular organisms (darting about in wonder and dismay at their existence), leaves its permanent, recurrent scar in mature germ cells formed at the onset of each new being. As many-celled marine animals propagate sexually, they introvert their death wish into an organ which is subsequently recapitulated (after many generations of layering) as the mammalian uterus. The wish for decellularization sublimates itself into colonies of cells, multicellularity; desire is pushed even further from its goal of oblivion.

The recession of the oceans and the subsequent adaptation of creatures to land was the next major catastrophe, but once again pure thanatos failed. These millennial events were somaticized (ontogenetically) in the protective covering of the uterus and the sprouting of sex organs. The "primacy of the genital zone" (as Freud named it) transfers desire from the whole body into libidinalized ridges and orifices. Sexual expression creates form even as it submerges itself in an ecstatic dissolution of form. For Ferenczi this was a vestige of the wish to return to the forgetfulness of the Great Sea (if not the cosmic void itself). Instead it thrust creatures onto dry land and into watery sense organs and passions.

Humankind came into being only after another global holocaust, the exterminating glaciers. Millions of creatures whose ancestors were lured into tropical lands by gardens, who adapted to temperate seasons were now forced to weather blizzards and avalanches. Many were frozen alive or starved. Their survivors incarnated their traumas. Life sought death and, although many races were duly extinguished, eros prevailed again: the human elf was born (around the hearth, in the manger) along with other mammals and the race of birds. The successive Ice Ages are somaticized individually in those lineages as the sexual latency of their childhoods (their nestling phases). The attempt to annihilate consciousness atomistically, long before it became trapped in membranes and symbols, gave rise, counteractively, to ritual and language. The energy of the Death Wish was drawn into a labyrinth of incalculable subtlety and depth.

Reversal by reversal, the dialectic of eros and thanatos was pushed deeper into tissue and further from primal sleep. As the Death Wish was recapitulated through the evolving animal kingdom, it became ever more profoundly unconscious; yet its increasingly uncompensated latency somaticized germ cells and organs through which it was replaced by layers of fresh libido, birth after birth, mutation by mutation. From its long sublimation and dormancy came the symbols on which culture and philosophy are based. Each transfer of libidinalized energy took an organism away from its desire to go back to sleep and sleep forever. Thanatos was far more indelible and primal than eros, so its singularity guaranteed its dialectical transformation, albeit spasmodic, into endless varieties of its opposite. Meanwhile eros has been formidable enough to hold thanatos (at least temporarily) at bay. Each animal is a testament.

"We have gained much from civilization," Freudian disciple Géza Róheim reassures us. "We have learned to conserve fore-pleasure, and to prolong youth and life itself." He quotes the master directly:

"'At one time or another, by some operation of force which still baffles conjecture, the properties of life were awakened in lifeless matter. Perhaps the process was a prototype resembling that other one which later in a certain stratum of living matter gave rise to consciousness. The tension then aroused in the previously inanimate matter strove to attain an equilibrium; the first instinct was present, that to return to lifelessness. The living substance at that time had death within easy reach; there was probably only a short course of life to run, the direction of which was determined by the chemical structure of the young organism. So through a long period of time the living substance may have been constantly created anew, and easily extinguished, until decisive influences altered in such a way as to compel the still surviving substance to ever greater deviations (retardation) from the original path of life, and to ever more complicated and circuitous routes to the attainment of the goal of death.'"[8]

It is a goal still unattained, though society boasts sigils of death on every shield, flag, and photon-bathed screen.

Sexual organs are literally the introversion of the battle between life and death. Our insides are a psychosomaticization of the mineral bath in which our ancestors were spawned, nourished, and into which they discharged billions of generations of sperms and eggs. When the sea was lost, primal desire synthesized internal waters. First they became tissue, then the reptilian egg (used also by birds), and finally the mammalian womb.

From the Greek term for "sea," Ferenczi named his theory "Thalassa"; he pronounced both infantile development and sexuality futile flights toward a long-lost primal Oceanus, contemporized as regression to birth waters and genitals. The grasping rush of the infant to get to the breast is an oceanic craving which matures into the sexual hungers of adults.

In general, thalassan eros drives male creatures to penetrate the females of their species, to get back into the womb, to reenter the sea (females return to the primal waters by embodying them). Among amphibians, which still have access to external breeding pools, Ferenczi saw a foreshadowing of coitus, the male frog clasping the female with the pads on his front legs. The excited discharge of urethral materials in salamanders shapes tissue phylogenetically for the partial erections of crocodiles. The acts of one species become the bodies of its descendants. Evolution is embodied (in Lamarckian fashion) by ontogenesis.

The reptilian forerunner of the mammal struggled to get the waters back around him, to burst into the salty, moist interior of a vagina. Its embryo was a lungfish, breathing by osmosis in the mud. The unconscious memory of that existence was so troubling that several lines of mammals went against the trend of latency and returned to thalassa as sea cows, dolphins, seals, and whales.

In Ferenczi's model the penis and vagina are scars of fierce combat between creatures trying to penetrate each other, pieces of flesh folded around germ cells (or giving access to germ cells), not as random mutations but as compromises between the drives of life and death, eros and thanatos. A truce is enforced in the form of regression (sexual intercourse), but it is successful only to a degree. Penetration is accomplished, and rudimentary organs mold it into lasting and inheritable and sensualized tissue—but those same organs also trap primal drive at a sexual level. The deeper wish to retain a prior, more primitive, and more peaceful equilibrium is frustrated. Transduced into a field of energy, this desire polarizes evolution in an opposite direction from its natural inclination (toward self-obliteration), and leads to new organs, new creatures—fur and lungs and limbs, and finally, the greatest disturber of the craving for slumber, mind.

Once tissues developed, their psychosexuality and theology followed, and, in our aeon, their sociological and symbolic extravaganza, including governments, wars, industries. As death was transformed into love (and its creative and destructive accomplices), the subcellular realm became cellular, multicellular, and then superorganic. Sublimation was finally successful because it not only brought sperm and egg into being but genitalized them and compelled them to interfertilize, and thus ensured its own succession and globalization.

FERENCZI'S VISION of the awakening flesh as a series of somaticized traumas reenacting geophysical cataclysms is an antiquated psychobiology, contaminated with Lamarckian and Velikovskian fantasies, but through a glass darkly it reflects the crisis of consciousness and ego-formation. Paradoxes of feeling and behavior arise anew in embryogenically differentiating tissue. Ferenczi was merely sighting history and biology backward through the loom of somaticization and individuation, through the unresolved discontents of civilization.

At the very least, he captured our experience of reproductive organs as blends of environments and traumas, flesh and symbols, affirmations and negations. Eroticism is the shadow of embryogenesis, its distorted dream. Normally we suppress the pain of evolution and consciousness, but eros throws us headlong upon its mystery and reminds us, as Søren Kierkegaard reminded his readers, that life is not a riddle to be solved but a reality to be experienced.

23

Birth Trauma

When does spirit merge with flesh?

DURING ITS FIRST WEEK OF LIFE the human blastula creases and double-folds into a pudgy caterpillar; its main sectors are a bent, protruding head lump; a meager body-stalk bearing, like a pregnant lizard, a pericardial bulge; and a thick umbilical trunk corporifying from the underbelly of its curled-in hind. By the third week a neural flash has driven ectoderm into mesoderm and endoderm; the cardium begins to shudder. On the twenty-second day neural folds fuse. A day or so later, eye and ear buds pop out. Unopened eyespots fluttering, interior cinema starts.

In its fifth week the embryo sprouts hands and feet as well as a vestigial tail. Its "head" is actually an overgrown forehead resembling a whole second embryo; the paired processes of its throat and lower jaw, with its otic and optic placodes, mimic chest cavity and limb buds, respectively.

In the sixth week true upper and lower limbs protrude, the former on either side of the rudimentary heart, the latter paired beneath the umbilical cord. The head—with inklings of countenance, jaws, and nasal folds—bows over a four-chambered heart. The flexure of the neural tube and vesicles of forebrain are illuminated through transparent periderm. A hepatic swell is apparent. The creature has stretched to thirteen millimeters.

By the seventh week head and brain have puffed up like balloons; lineaments of fingers and toes are impressed in virgin buds. Rudimentary lips pucker as maxillary processes and mandibles fuse. Nostrils sniff vacantly among nasal folds and medial nasal processes. Eyes without eyelids stare blindly. The life form is eighteen millimeters long.

In the eighth week cerebral hemispheres expand over brain stem. Eyelids "opaque" eyes. Toe rays protrude in webbed feet. A short unhinged stump of neck contracts to withdraw a hand.

A cosmic vertebrate has stamped a universal blastula.

IN THE ELEVENTH WEEK the creature gulps and swallows. During the twenty-first it is heard crying, a ripple perhaps of amniotic fluid rushing past its vocal cords. From the third month on, the fetus "now floats peacefully, now kicks vigorously, turns somersaults, hiccoughs, sighs, urinates, swallows, and breathes amniotic fluid and urine, sucks its thumb, fingers, and toes, grabs its umbilicus, gets excited at sudden noises, calms down when the mother talks quietly, and gets rocked back to sleep as she walks about."[1]

Midway through the fourth month stumps of sacrum and pubic bone are hewn; boninesses delineate ankles and feet; rings curl around ears; eyebrows thicken; fine hair coats the body; the male scrotum swells.

The unborn clasps its hands and rotates, breathing amniotic fluid. Its mouth grimaces.

It is almost as though it is being fed the esoteric ledgers of its race, the archetypal history of the Solar System, not in words but in biological concepts — dreams

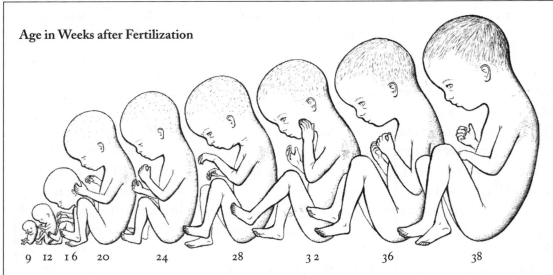

Age in Weeks after Fertilization

9   12   16   20   24   28   32   36   38

FIGURE 23A. Drawings of fetuses one-fifth of actual size. Head hair begins to appear at about 20 weeks. Eyebrows and eyelashes are recognizable by 24 weeks. The eyes reopen around 28 weeks.

From Keith L. Moore, *The Developing Human: Clinically Oriented Embryology* (Philadelphia: Saunders College Publishing, 1977).

and rites indistinguishable from dendrites and synapses in whose formation they come into being (much as baby Superman of the comic book was subliminally fed the annals of Krypton as he sailed through interstellar space in his capsule toward Earth).

Throughout the fifth month motor nerves expand and intertwine, impregnating muscles. Nerve fibers crawl from the spine; ventral roots become myelinated. The babe begins sucking and gurgling. The mother feels constant movement. Her fetus is five-and-a-half inches long.

During the sixth month glands spurt in skin; lymph follicles orbit through waterways; finger- and toenails eclipse the ends of digits; cerebral hemispheres crinkle and twist in sulci and gyri. It is no wonder the infant is in a reverie. She is experiencing, almost hourly, changes in consciousness, fleeting vestigial minds.

In the seventh month membranes over the pupils dissolve and eyelids open. Insula and *tubercula quadrigemina* squirrel through the cranial depths.

When does matter awaken to its own existence? When does spirit merge with flesh?

This question cannot be answered, either biologically or theologically. There is no first moment, for spirit or for life. Awakening is beyond time, without beginning.

Does the "soul" dwell in a spirit zone before it enters flesh? Does it manifest instantly with fertilization? Does it arrive in stages? If so, at what threshold does being supplant nothingness? If not at fertilization, how long—seconds? days? weeks? Who (or what) inhabits the blastocyst till then?

The Mechanics of Birth

LATE IN THE NINTH MONTH the fetal organs announce their readiness for transition. Their intention apparently stimulates the paraventricular nuclei of the hypothalamus, which exude adrenal corticotropic hormone into the fetal bloodstream. This activates the fetus' (and likely the mother's) adrenal glands to secrete cortisol, which flushes the maternal bloodstream and chimes the onset of labor.

As the muscular walls of the uterus contract and compress intrauterine fluid, the fetus senses an increase of pressure on its body. Successive contractions of these walls intensify hydraulic tension, forcing fluid into a cul-de-sac and opening the birth canal from the inside out. The resolution of this event will be as startling to the fetus as anything in the lifetime to follow. Dr. John Upledger narrates:

"The fetus feels its head entering the internal end of the birth canal. . . ; increased pressure occurs with each muscular contraction of the uterus. Now with the fetus' head in the birth canal there is a volume of fluid which precedes it through the canal. Soon the [mother's] membranes rupture naturally or they are ruptured by the obstetrical delivery person and the fluid cushion is gone. ... [the] fetus' head becomes the 'battering ram' that forces its way through this ever-so-tight birth canal.

"Once the fetal head is out of the external end of the birth canal, we still have a body that has to complete its journey. This is a twisting, wringing sort of trip that I believe actually gives the fetus its first spinal manipulation treatment. It would seem that, under ideal conditions, every joint in the fetal spine would be mobilized, as would each of the vertebrae in relation to the others. Further, all of the muscles and ligaments will be stretched and wrung out. It would also be comparable, I'm sure, to a very extensive and complete whole-body massage. ... the baby probably receives a thorough skin friction massage using the *vernix caseosa* as the oil. ...

"It takes time for the fetal body tissues to respond. They should be given that time so that this initial body treatment can be absorbed and offer its maximal benefit."[2]

Upledger recommends a "second" birth, a cranial treatment to correct injuries and malformations and give the newborn a more healthy entry into the world. If administered within days, osteopathic balancing and adjustment can occur in fifteen minutes. Months later, they take eighty to a hundred hours. Among adults the trauma has been so displaced and dilated by years of social acts that there is no limit to how many treatments are necessary to unravel its far-reaching reverberations and sublimated effects. In fact, they cannot be unwound.

Conditions Upledger associates with birth trauma are dyslexia, attention deficit disorder, hyperactivity, some kinds of seizures, cerebral palsy, and certain motor/eye dysfunctions, to say nothing of psychological trauma. Although not all birth ailments can be mended or alleviated, timely *ex utero* palpation can reduce physical and emotional scarring from the abrupt expulsion.

Primal Scream, Rebirthing, Somatoemotional Release, Shamanic Journeying, and Natal Therapy all treat neuroses characteristic of fetal and birth trauma. According to the theories behind these methods, if anxiety and panic take hold at the moment of birth or in infancy, they stick for life, an unlocatable part of the background of existence. Embodied before language, they cannot be released by even the most brilliant incantation of symbols and words. Personalized and recathected (i.e., recharged with emotional energy) through the vicissitudes of existence, their traumas are revived anonymously in phobias, recurrent failures, obsessive compulsions, and dysfunctional sex acts.

The Politics of the Hospital

IN CONVENTIONAL SCIENCE the embryo is a subhuman creature, without personality or identity. It comprises only the most primitive instincts, mere autonomic forerunners of emotions. Even when this being emerges from the womb it is treated as a nonentity, a figment of a person. Psychoanalysts consider newborns passive, uncomprehending, unaware of their identities, bestial in their cravings.

The arriving baby is not welcomed as a sacred or even an invited guest at most hospitals in the civilized world. One more bean sandwich, a collection of unconscious nerve endings, raw human material—it is slapped on the rump, separated from its mother, and deposited in a vapid zone with dozens of other arriving creatures; then, if a male, subjected to genital surgery (circumcision of the foreskin by scalpel) without anesthesia.

"It is assumed the baby feels nothing when, in fact, he feels everything. 'Birth is a tidal wave of sensation, surpassing anything we can imagine. A sensory experience so vast we can barely conceive of it.'

"The delivery room is set up for the convenience of the attending physicians, beginning with the bright lights aimed at the mother's pelvic area. The baby is very sensitive to light and is able to perceive it while still in the womb. The first thing the newborn sees are bright floodlights. The infant is blinded by the light; then several drops of a burning liquid are put into his or her eyes.

"The baby is also able to perceive sound in the womb; of course, the sounds are muted. But in the delivery room they are not, so the first sounds the newborn hears amount to a thunderous explosion of noise....

"Often, the infant is held upside down and spanked to expedite the process of draining the amniotic fluid from the lungs in order to facilitate breathing. This is extremely traumatic to the newborn, and often results in chronic back problems."[3]

Following an injection of vitamins, a "deep lancing wound to the [baby's] heel"[4] draws blood for laboratory tests. Throughout the interrogation, the newborn cries and screams, but medical personnel have been indoctrinated to ignore this. The natal being is a biological product, not a suffering soul or a citizen of the republic.

It is as though we require our children to be no one. We want them to occur, each a *tabula rasa*, until we proselytize and train them, make them human by our own humanity.

"After [its terrifying arrival], what the infant then most needs is to be reunited with his or her mother. Instead it is whisked away and placed in a little box in the nursery ... left alone trembling with terror, hiccuping, and choking."[5]

Being processed in this manner has no basis in primate society or nature; it is an artifact of late twentieth-century medical propaganda.

THE "NOVEL FORM OF 'DELIVERY' offered by obstetricians in hospitals is a baptism of pain. Birth was not like this in the thousands of years of human evolution prior to the 1940s. Physicians believe it is 'the best of care.' Cultural anthropologist Robbie Davis-Floyd calls it a ritual of initiation into a technocratic society where machines are used to improve on nature and all babies have become cyborgs."[6]

Of course, this is the same commodity civilization that inoculates its recruits with toxins, cages them in glass, clothes them with oils and plastics, feeds them machined foods, prepares them by mechanical devices and rote drills for technocratic jobs, transports them on engines and gears, and addicts them to machine dreams. Arthropod nostalgia does truly permeate our whole ritual.

Shamanic Birth

WE ARE NOW SO SOCIALIZED we fail to realize that our benefactors gain their power only by stealing it from us. Medical science, by mechanizing and displacing the experience of pregnancy and birth, has substituted a secular totem for the sacred one, and a sterile, fruitless fear for true awe of the unknown. "Fetal heart monitors violate the mystery," says midwife Jeannine Parvati. "They are blasphemous. It is with the inner sight that we can see the mystery unfold, not with … X-ray (and tetragenetic and carcinogenic) eyes."[7]

As the doctor counts heartbeats and works gadgetries, s/he diminishes a woman's sense of her own magic. The mother encounters her pregnancy and birth as an event outside herself, another object in male society. Child-bearing becomes an act of science, a gift of Apollo rather than of his first-born twin Artemis (who delivered him).

No doubt in the Stone Ages women honored the mystery of the blood and the changing Goddess in their own bodies, thus participated in ceremonies and propitiations of shadow forces. Giving birth was more likely to be (as Parvati puts it) "a supreme passionate bliss and the major soulmaking experience of their lives."[8]

FETUS-MAKING IS, at core, a sacrament of blood, a eucharist conferred on every being in its emergence from a chamber of cells. Sheathed in each woman is the vessel, an alembic, for this act. Embodied in life itself is a journey of transubstantiation. Birth is the original initiation; everything that happens thereafter is part of the ceremony. We are souls in a vast dream, clothed in atoms and cells. This is our interlude in bodily regalia.

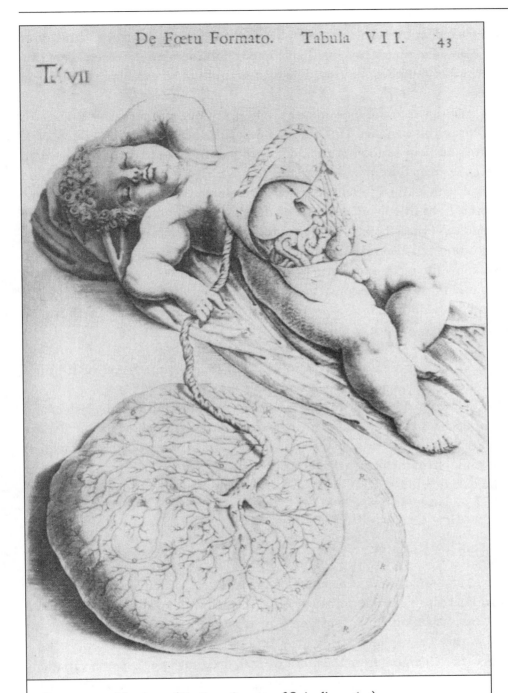

De Fœtu Formato.   Tabula V I I.   43

T.' VII

FIGURE 23B. Newborn (*De foetu formato* of Spigelius, 1645).
From Arthur William Meyer, *The Rise of Embryology* (Stanford University Press, 1939).

We may reduce the whole process to impersonal dynamics and biochemistry, or we may recognize the sacred guardianship our bodies impose on each other, the shimmerings of unknown destiny. The placenta is both an extraembryonic organ and a grail.

"[W]e sit down to the nine-month's banquet in our mother's womb, and it is the servitor of this banquet. It is our facing-partner beyond the shrouding foetal membranes and our sibling. It responds with its offices in that archaic place, in which we are created from basic life-stuff through all the stages of evolution, and it infuses us with all the substances and elixirs required for these changes. We know it by womb-light in our synaesthesia of the skin in its globe of actively creative waters; and this synaesthesia is gently modulated by all the movements and changes of composition and electrical fields and charges of our womb-liquor. . . .

"We know the picture of that relationship, not merely from photos in biology books, but with the mythic conviction of contemporary fables of regeneration, as at the end of the film *2001: A Space Odyssey* when the star-foetus floats above the great placenta-like surface of the globe."[9]

BIRTH IS AN ACT of memory beyond time, or of time beyond memory. It brings two creaturely cycles together in a cosmogonic moment.

"It is like women all wear masks—for *their* mothers and fathers, sisters and brothers," declares Parvati, "and in giving natural birth the masks come off. We are always surprised to see who She really is. These masks are personae, personalities; when in childbirth, the personality is stripped away. Off go the day-world clothes as we lie naked before God. Giving spontaneous birth shows the original face of both the baby *and the mother!*"[10]

Memories of Birth

THE BABE EMERGING FROM THE UTERUS is no passive cipher; it is a mature mammal capable of directed movement, charged with power and will. In more spiritual tribes than ours it is considered an avatar, an ancestor bringing wisdom into the world.

Hours after emerging from the uterine canal, fledglings track speech variations in many languages—Irish, Cherokee, Chinese—while ignoring nonsense syllables. When less than thirty-five hours old they start crying at the sound of other babies crying (but not at computer simulations of the sound), and "stop at the recorded sound of their own crying, indicating that they not only heard, but recognized their own voices."[11]

In its own way the newborn understands what is happening to it and stores complete memories of the occurrence. An adult can return to birth time in dreams or by hypnosis. The accuracy of consequent reenactments has been confirmed by startled mothers, physicians, and others who attended. Adults recall being stuck in the birth canal, the pain of the forceps, the incidental comments made by those in the room:

"She's holding me up, looking at me.... She's smelling me! And she asked the nurse why my toes were funny.... The nurse said that's just the way my toes are and that they weren't deformed."[12]

Under hypnosis, another woman recalls:

"I didn't want to come out. Some kind of pulling and tightening. Movement, lots of movement; myself moving.... Safe inside; didn't want to come out.... Lots and lots of noises, and just confusion outside.... I'm in an operating room.... A lot of chrome instruments. A metal table. And my mother's on the table. And there's a lot of men and women—seven or eight—dressed up in gowns. They're all talking and rushing about.... And then there's light, lots of light. Really bright. And it seems like something's on my head or by my head; seems like I'm just being pulled out by someone or something. There's one big bright light in particular.... I'm being pulled out. And very scared; very scared!... I'm lying on my back, my legs moving, and my hands are scratching my eyes, scratching my eyes and nose. I'm crying, screaming, and I'm getting out of breath. There's nothing surrounding me, nothing holding me. Too much open space! Too much freedom for my arms and legs. Air; too much *air*, too much freedom.... I'm not curled up safe."[13]

The beginning, as in every ritual, is critical and sets the tone for states to follow. One comes to Earth as a pilgrim, a vagrant, a prisoner, a prophet—either ready for the summons or thrown onto the train anyway, heedless of its rattle and speed.

Trauma is not only the result of sloppy or malicious delivery techniques; it is concomitant with existence, with being forced from a warm, safe hibernation into the theater of tooth and claw. "Lost forever is the indescribable fetal experience,"[14] the sensation of eternity, of being oneself a glowing sun, of the presence sometimes called God. Deep within, the child laments: "I am separated by the wound that does not heal. I would like to find that light again, but will have to die to do so. Neither my mother nor others know from where I come. I feel humiliated and infinitely sad."[15]

Bob Frissell notes:

"Birth trauma is caused by the sudden and unexpected shock of going from the comforting confines of the womb into an environment which is totally unfamiliar.

"In the womb all of our needs were met. We lived in safety and comfort, and there was no struggle. Then our body became too big for its container and suddenly we were forced down a passageway that was too small.

"The experience was painful, frightening, and distressing for both the mother and the infant. Then we found ourselves in a hostile world that was cold, bright, and noisy.

"What we really needed was to be shown that the outside world was safe and was a far more interesting place with infinitely more possibilities than the womb. Unfortunately, we were shown the opposite—not because the people in the delivery room were evil but because each of them had their own unresolved birth traumas and they got transmitted in the form of fear, tension, and urgency, to the infant.

"So rather than safety and trust, the setting was one of fear, and out of fear comes ignorance. . . .

"The temperature in the womb is about ninety-eight degrees Fahrenheit, and the temperature in the delivery room is about seventy degrees. That means the nude, wet newborn experiences a sudden thirty-degree drop in temperature. This would be the equivalent of taking a hot bath and then running outside. This 'temperature trauma' remains in the body in a suppressed state, and is most likely the cause of colds. . . ."[16]

Birth Shadows

BIRTH IS NEITHER A PROMOTION NOR A PUNISHMENT; it is simply the next stage in a cosmic journey that must cross the transformative gaps between dimensions. Individuation and ego themselves require opposition and frustration; differentiation is by nature disruptive and painful. Jungian therapist Edward Whitmont reminds us that "successive jolts by dissipative energies occur already at the very beginnings of what has been previously assumed to be the perfect and 'innocently' ideal phases of life, namely the intrauterine state and the birth process. From the very beginning of life onwards, the infant's oceanic, cosmic bliss and preformed primary narcissism are shaken apart by exposure to the birth trauma and to various degrees of parental frustration, denial, neglect, and rejection, as well as to maladaptive inherited genetic predispositions. . . ."[17]

Myths and fairy tales abound with phobias of birth horrors—a wolf crawls from the womb, a devil with pointed ears smiling malevolently, an alien hybrid or changeling of one sort or another. Biology has its own specters and teratologies: mongoloids, Siamese twins, herniated and microcephalic (or hydrocephalic) skulls, limbless torsos, eyelids without eyeballs, fused cyclopiac eyes, smooth featureless faces with only a proboscis, dwarves, "elephant men."

Rebirthing

VARIOUS THERAPIES ALLOW INDIVIDUALS to simulate the journey out of the uterus so that they can reexperience its twisting ride, the sudden lights, the cold, the rough handling, the separation.

In the Rebirthing system developed by Leonard Orr during the 1980s, a person is instructed to inhale and exhale in connected breaths. The inhalation phase is deep and long and should flow directly into a shorter exhalation.

During this exercise, feelings and susceptibilities arise, both uncomfortable and pleasurable. The intense breathing itself can be painful, tearing at old wounds, stirring memories of cosmic transitions. One is asked not only to permit these but to encourage, blend with, and actively affirm them. Instead of fleeing and stifling discomfort and anguish, the participant thanks the universe and her own body for bringing her to this magnificent moment. The act of intentional surrender releases her to flow into her sensations. Pain and grief become energy; a pulse is dispatched throughout tissues and integrated in new thought patterns. She sails through former disturbances as a plane riding turbulence, bouncing precariously but absorbing the bumps and gliding to the next phase.

Ordinarily the rapid breathing of rebirthing would lead to hyperventilation; in fact, people being rebirthed do experience light-headedness and dizziness. However, Orr, who based his technique on Kriya Yoga of India, insists that hyperventilation is not the main event as long as prana (life-force energy) as well as oxygen are assimilated during each inhalation. This primal blend of vital and molecular substances helps clear the body of dizziness, disease, and pain. It enhances sensations, automatically bringing to consciousness, in turn, each thing that is impacted, blocked, or incomplete. Crises are pulled to the surface, felt, flooded with prana, assimilated, and accepted fully, as they likely weren't when they occurred.

"Rebirthing is a two-stage process," explains Frissell. "The first is learning to breathe energy . . . as well as air. . . . The second . . . is 'to unravel the birth-death cycle, and to incorporate the body and mind into the conscious life of the Eternal Spirit, to become a conscious expression of the Eternal Spirit. . . .' Tremendous damage was done to our breathing mechanism at birth. Fortunately it can be healed."[18]

The journey back to the womb may take as many as ten or fifteen sessions, or it may unfold in one, but eventually the traveller finds herself again at her birth, reexperiencing separation from maternal tissues, along with any terror, gloom, excessive use of anesthesia, undue haste, or critical comments of doctors or nurses. These are, each in turn, affirmed, forgiven, and dissolved.

Orr initially conducted his ceremonies in warm bathtubs, but once he found that dry rebirthing was equally successful and less intimidating, he simplified the setting to a bed or couch.

Somatoemotional Release

UPLEDGER'S METHOD OF TREATMENT involves craniosacral therapy integrated with therapeutic imagery and dialogue to achieve what he calls "Somato-emotional Release." First, the therapist unwinds the "Avenue of Expression": the morphogenetic path of pharynx, thoracic inlet, hyoid tissues, hard palate, tongue, teeth, zygomata, nasal bones, mandible, etc. Through this channel, traumatically charged sensation evacuates the depths of endoderm and mesoderm and takes on a voice, often with cries or moans (see the next chapter). The client is encouraged to relinquish any withheld emotions, feelings, and resistance collected in her tissue memory, to release these through the layering and pathways of her anatomy. Gradually she is regressed to the point where she relives the sensations, smells, and sounds of the operating room—the splattered blood, the presence of strangers, and whatever else became cathected during the occasion.

Therapists overseeing this method of recovery describe suddenly smelling anesthetic in their rooms. Such psychic ghosts cannot be explained in conventional physical terms.

Upledger relates a personal experience from one of his own advanced classes:

"In just a few minutes I began to vaguely move in and out of a sensation of being born. Slowly the birth sensation solidified and I could 'feel' the hands of William Naggs, M.D., on my head assisting the birth process. My neck retracted. The therapist holding my head commented on this phenomenon. For just an instant I felt like a turtle pulling back into my shell. I then seemed to be using my shoulders and arms against the inner rim of the birth canal to resist the doctor's pulling. Dr. Naggs was trying to help. I didn't see it that way.

"... I became aware that the doctor was, with good but misguided intention, working against the natural delivery process. I wanted to go slower so that the process would be in keeping with nature's plan. I shared this with Susan [the session therapist] ... and she very wisely suggested that I tell the doctor to stop pulling and let nature take its course. I did and he honored my request. After all, this was my fantasy. For an unknown amount of time—I think quite a long while—I experienced the most remarkable twisting and untwisting, relaxing and lengthening of my neck, torso, and finally my legs. It was my first and (to date) my most pleasant spinal manipulation treatment."[19]

OFTENTIMES SEVERAL CRANIAL THERAPISTS work together to reenact the passage down the birth canal and emergence into the world. Each assists one aspect of reenacted birth movement by holding or palpating a part of the "newborn's" body and supporting its separate tendency, following it off the table as it writhes about trying to find its way down memory remnants of an actual uterus or a symbolic birth canal. Reliving or imaginatively dramatizing birth, the participant may literally throw himself off the table or twist up in the air.

I underwent this process myself with Dr. Upledger and his colleagues at his clinic in Palm Beach Gardens, Florida, in 1996. First I experienced a buoyancy of being able to move in any direction and follow any impulse because every part of my body was supported by someone. I could fly (or wiggle in air), and I did. Upledger then pointed out that my head must have been stuck in the birth canal, so he "cracked" the bones in my neck with a deep, sustained torque—and then followed with a sudden, quick jolt. In startled response—my own deep breaths now "rebirthing" me—I levitated under the support of many hands, a snake in air.

At last my snake dance ended; I swam from its apex down to the floor. As I lay there, all motions in me quieted. I opened my eyes, and John said, "Bright lights, big city, kid."

•

... in a Moebius strip the bodies unhinge, find a slow rotating freedom of motion in three dimensions. Topology tears at her sinews, and she weeps, for all nerves and muscles flow downhill, into the silent light of trees. Blood collects at the entranceway, and a nine-month world crumbles, its rivers escaping their channels, its mantle returning to the primordial sea. There is but one bridge, and the changeling must cross it, into another world. He gasps, floating in air on the planetary ocean, impelled through a door covered with membranous vines, squeezed by gravity into a horizonless mirage. Why not cry out in horror and wonder both?[20]

HUMAN LIFE IS IN DANGER right from the beginning: It is guaranteed on neither an archetypal nor a biological plane. It must come into being of its own material and make its psyche and spirit out of itself, its body. No matter how many transcendental planes there may (or may not) be, we cannot overlook or escape the sheer immensity of creation in this purely physical domain. The alchemists who sought to raise spirit and soul ever out of matter, uncommon substance out of common, and gods themselves out of ashes, understood implicitly that we are bound by our mortality. The decay and pain around us is not the wasting and grief of a mere residual world in which we chanced to occur; it is the decay and pain of us, the fact of our substance. We put on the mortal coil with great difficulty and minute precision; we cannot be abstracted from it, and we cannot idealize our wholeness and mentation as if it were angelic or metaphysical. The crisis of our becoming is real, not the image of a higher dimensional realm and not the symbol for another immortal mortality.

24

Healing

Cell Activity after Parturition

EMBRYOGENESIS DOES NOT TERMINATE SUDDENLY WITH BIRTH. After completing the lion's share of its work, it slows down incrementally *in utero*. Embryogenic effects become local; their fields grow subtler. Biological activity moves to another level, that of metabolism.

Post-parturition maintenance (self-healing) is the product of not only mitosis, the indefatigable assemblage of proteins, and their progressive morphogenesis, but homeostases of viscera and their psychosomatic regulation. All capacities for organismic stability outside the womb, including the efficacy to restore cohesive biological fields for as long as a hundred years or more, is embryogenic, i.e., an *ex utero* extension of the embryological process.

Fresh cells continue to be synthesized in the immune system and (for a time) the brain; tissues continue to be induced interdependently by organs and fluids. We may not be able to regenerate limbs and vital organs, as some simpler animals can, but we replenish large areas of damaged tissue by mitosis (and perhaps even occasional uncharted acts of retrodifferentiation).

Basal lamina surviving injury to muscles, nerves, and epithelia furnish supportive grids for regenerating cells to follow in reassembling prior tissue architecture. From elongated vacuoles enveloped by cytoplasm, epithelial cells extend pseudopodia into neighboring cells; these processes hollow out into tiny tubes that encounter one another, cell to cell, and join contiguously in blood-capillary channels. Cells deprived of oxygen likely secrete angiogenic proteins which induce capillary formation.

Angiogenesis follows wounds, inflammations, and other organismic damage, providing new capillaries where they are needed. Thus, no cell finds itself more than fifty microns from a blood supply.

Under elicitation from a class of proteins known as fibroblast growth factor (FGF), fibroblasts and other connective-tissue cells redetermine themselves in the context of wounds, fractures, and all manner of damage and pathology, heeding the emergency instructions of microtubules, microfilaments, intermediate filaments, extracellular matrices, steroids, and other signalling molecules to differentiate into new connective tissue, chondrocytes, adipocytes (fat cells), osteoblasts, and smooth-muscle cells in virtually every tissue and organ in the body. Though arising from different cellular sources, the various FGFs share fifty-five percent overlapping amino-acid contents.

Other tissue-specific stem cells throughout the body, from the crypts of intestinal villi in the lumen of the gut to the basal underbelly of the skin, divide throughout the organism's lifetime, becoming either new stem cells or the irreversible components of freshly differentiated tissue. Cell geometries and replacement rates depend on local induction, with proliferation most vigorous among cells that have direct contact with the extra-organismic environment. Thus tissues under continuous assault are replenished from the same template by which they were made.

The body is not a final object but an ecological field in which individual bionts replace one another in cycles of succession. Despite their independent existences and functions they apparently have an overriding stake in the whole organism—an incentive to maintain its metabolizing congery. Seen only in terms of potentially anarchic components with random, complex chemistries, the body is a *coup d'état* waiting to happen, a time bomb; however, the whole regulates the sum of the activities of its cells, not just mechanically, not just chemicogenetically, but as a hierarchically superior engine subsuming their agencies and kineses in its own.

The organizing principles of the embryo far outweigh anything medicine imagines accomplishing.

EMBRYOLOGICAL FORMATION IS "HEALING" in its singular and absolute form. At first glance, this assertion is so obvious as to be meaningless. However, doctors and patients rarely consider the fact or its implications.

The organizing principles of the embryo far outweigh anything medicine imagines accomplishing. The "healing" process involved in transforming a sperm and an egg into a metabolizing creature of sixty trillion cells is beyond any futuristic fantasy of even the most advanced science.

The changes from blastula to gastrula, then gastrula to newborn, are feats of incommensurable virtuosity. Organs are created out of unpromising raw material.

At scales that are both infinitesimal and exponential, complexity is manufactured out of simplicity; greater complexity is developed out of lesser complexity; simplicity is generated again from complexity. Functions that are inconceivable *a priori* are introduced as if by a mad magician and then synchronized so that not only is a living creature fashioned, but at every step along the way a *new* living creature is fashioned, then another out of that, and so on. It is as if engineers were to assemble a skyscraper complete with furnished apartments, plumbing, air-conditioning, and phone lines out of a heap of rock but in such a way that each intermediary stage of their assemblage was completely habitable in some fashion.

The overall accomplishment is of the gamut of turning a piece of popcorn into a cyclotron.

THE EMBRYOGENIC PROCESS is the vintage template for healing because it inserts the correct protein in developmental sequences at precisely the right site and moment for organism-wide integration and translation into function. This process is never locally circumscribed, for—as one structure induces another—interrelated, densely packed, three-dimensional viscera emerge dynamically. Every tissue provides contexts for the location and function of every other tissue. Furthermore, as we have seen, individual proteins and induction waves express equivalent information uniquely at different scales and in organs as removed from each other as lungs and bladder. This is a difficult event to manipulate therapeutically, even with biotechnology (see Chapter 15). Although medical effects *can* be tracked (in fact, scientific ethics require it), they can never be traced in their entirety at every scale of visceral organization and intervisceral cohesion.

Anyone's doctor would like to cure at the level of chromosomes, but it is impossible to reverse pathology in a fully formed organism using genetic codes; so physicians work solely at mending the phenotype—the living map assembled by the latent gene script. They treat on a condition-by-condition basis, with an emphasis on each zone of damage and its possible improvement and redetermination.

If a doctor cannot induce the powers of embryogenesis, he or she can counter gross assaults on the mature organism, usually by attacking the agents of those assaults (microorganisms, tumors, symptoms of metabolic malfunction, etc.). This is what allopathic medicine specializes in—dispelling and expunging invaders or depressing overactive immune and endocrine events, often at the expense of salubrious morphogenesis elsewhere.

It is difficult to dampen and toxify selectively. Surgery disrupts thriving biological fields as well as disease vectors; chemical weapons devastate the mechanisms of healthy organelles and cells as well as their ostensible targets. The favored strategies of advanced Western pharmacy and technological medicine require suppressing or obliterating embryogenic events in hope that pathologies will ultimately suffer more than vital organs.

This limitation is never refuted by the medical profession, but it is downplayed in the cultural valorizing of the physician. In truth, doctors' measures are successful only secondarily as the organism's naturally robust fields reorganize around their intercessions. Even surgery is meaningless without subsequent visceral response; otherwise, one is sculpting meat.

Health always comes from the organism, not *because of* the medicine but *in spite of it*. Doctors may remove impediments to organismic vigor, but they rarely confer a true elixir.

Sometimes a direct attack on a bacteria, virus, or toxin is the only way to prevent the organism's imminent demise. This is modern medicine's singular genius, its police function—and it is not trivial. In fact, it has added years, even decades, to the life spans of people throughout the world.

Once life is saved, health must still be restored. The organism must regain a destiny worth living, a freedom from layers of chronic disease or obsession with its own next malfunction needing a doctor. Becoming a medical artifact is not a substitute for a sound mind in a sound body.

A biochemical or mechanical repair that has little or no vitality in and of itself cannot regenerate the organism. The organism regenerates itself by already potentiated morphogenetic activity in the context of the new situations provided medically.

While no physician lays claim to the embryogenic process (which invented itself), every physician must subject his or her work to its immediate critique. No slack is cut: either the organism accepts the induction (the cure) and returns to more normal and healthy functioning afterwards, or it does not. (The *cure* is not the functioning.)

This is the full range available to allopathic medicine, and it will likely remain so, even at the conclusion of the Human Genome Project generations in our future.

The "cure" may also induce new pathologies or, as it radiates throughout the patient's biological fields, induce chronic malefic side-effects; diseases of such origin are called iatrogenic, for they are physician-begotten. Added to the spectrum of primary and secondary environmental illnesses, they make up the largest class of pathologies presently afflicting human organisms.

Categories and Definitions of Alternative Medicine

ALTERNATIVE MEDICINES DIFFER FROM ALLOPATHIC ONES in the following two characteristics: they are holistic rather than circumscribed or organ-specific (thus alternative medicines purport to work on the complete mind-and-body in place of its parts), and they propose energetic rather than mechanical cures (although the definition of "energy" varies from system to system and even practitioner to practitioner). Holistic treatments also engage the psychospiritual life of the individual, including his or her desires, dreams, hopes, fears, and individuation. Practitioners do not sacrifice a patient's emotional well-being or control over her own destiny to accomplish extrinsic acts of diagnosis and cure.

That doesn't mean holistic medicines always succeed but, successful or not, they view life as a long journey from conception through incarnation to a natural transition in death. Cures are as much ceremonies as constructs.

HOLISTIC MEDICINE IS HOLISTIC because it aims to have a dynamic effect on the biological field. Of course, all medicines have a dynamic effect, so holistic practitioners try to choose nonintrusive remedies, herbs and mechanisms that (traversing the body-mind) are more likely to synergize positively than negatively. The goal is to stimulate organ function and encourage immune vitality rather than to snuff out a disease. The healer intends at the very least (by an ancient motto) "to do no harm." Her favorite modalities are traditional ones because these have centuries and sometimes (as in the native armamentaria of China and India) a millennium or more of empirical inquiry and case histories behind them. This is their bonus of comfort and safety. She does not have to reinvent the wheel (or search always for new ways to move the wagon).

Holistic physicians often do not even target the major symptoms; instead of presuming to know what the disease is or how the etiology of its cure should go, they attempt to activate therapeutic changes at multiple levels, whether these levels represent the primary purpose of the treatment or not (or whether the changes are even perceptible to the practitioner). For holistic cures, skillfully introduced, permeate all levels to some degree.

Holism requires that the body be diagnosed and treated as a series of mutually inductive topologies linked mechanically, hormonally, emotionally, and psychically. The obsolete hunk of protoplasm lugged to the doctor's office is replaced by a succession of palpable and impalpable ripples reflecting one another, holograms resonating at different scales. This subtle process—a blend of psychosomatic intuition and telekinetic divination—returns the patient to the power of self-healing, either spontaneously and unconsciously or through a ritual of committed practices. Thus, holistic medicine is embryogenic, either in principle or in practice.

THE INSTANCES IN WHICH holistic practitioners succeed after a litany of high-tech failures (even to diagnose the cause of a malaise) proffer an unexpected substitute paradigm: It is possible that when MRI devices and biochemical fractionations descend into the componential fundament of the body they occasionally fall beneath the homeostasis of emergent properties making up a particular condition and the organism's prior healthy functioning. Thus they do not provide accurate information about what is wrong, and subsequent attempts to repair the damage on the basis of their data have minimal curative impact. The disease is elusively multicausal.

When a problem exists at an emergent level of organization rather than concretely within the layers themselves, it can never be satisfactorily pinned down to a locale or object (although sometimes it masquerades in the form of one or more heinous or conspicuous symptoms). The ailment, however, is not coterminous with its symptom pattern, and the symptoms do not reflect the actuated site or morbific pathways. Indeed, following the neuroendocrine system and other cellular networks, displaced and mobile conditions pulsate their distress widely throughout the body-mind's organismic fields. These diseases must be treated by modalities furnishing the equivalents (herbal, mechanical, bioelectric, or psychodynamic) of remedial and counter-resonant pulses.

A medicine of emergent properties may function as a myth, a metaphor, or a mantra. There is no predicting at what level dissonant fields will be perturbed back into homeostasis. To the dismay of agencies requiring controlled clinical trials, the shaman and practitioner of psychic touch can be mysteriously and uniquely efficacious. Miracle cures do not defy physics; they exemplify the sudden transduction of complexity and order from chaos.

In episodes of energetic holistic healing, signals are sent out blindly, through hierarchies of tissue, languages, cultures, unconscious mentation, and narrative histories—and, because we are the embodiment of codes that underlie all symbols, the transduction of all runes, these messages find and somehow potentiate hermeneutic fields out of which we ourselves come.

Whether we arrive at such remedies empirically and fortuitously or archetypally, they become our medicines at the point at which they effect cures. Nothing about their actual mechanism need be known.

Homeopathy

WE CAN DIVIDE NONPSYCHOANALYTIC ALTERNATIVE MEDICINES into two domains replicating pharmacy and surgery: an herbal branch, including homeopathy, Bach Flower Remedies, aromatherapy, Ayurvedic pharmacy, Taoist pharmacy, and most native botanies; and a mechanical branch, comprising osteopathy, acupuncture, and a spectrum of discrete somatic treatments (Feldenkrais Method,

Polarity, Rolfing, etc.). The distinction between pharmacy and mechanical adjustment is less crucial in alternative medicine than the allopathic distinction between drugs and surgery because channels generating and conducting energy tend to bridge the gap between herbs and manipulation, each of which must finally have a hormonal, autonomic, and fascial expression. Remember how cells alternate between and combine microtubule and steroid levels of activation. Likewise, acupuncture needles used in the context of Chinese medicinal formulae are not mechanical (or metallic) by the usual surgical meaning. They too are morphogenic "signals."

Preagricultural tribes discovered many natural cell-signalling substances in their environment. A ghost of the root of licorice fern goes on providing digestive and excretory information long after the tuber itself has been digested and excreted. In some other fashion St. John's wort stimulates the production of melatonin. Thus, an organism in culture continues to evolve and change as a condensed, membrane-bounded ecology enveloped in a larger symbol-mediated, atmosphere-enclosed ecology. Tree fungus became part of the Ice Man's embryogenic field (see page 570).

Whereas most mainstream doctors at least acknowledge attempts to administer herbs and herbal mixtures pharmaceutically or to improve digestion and reverse systemic pathology by manipulating the stomach, liver, sphincters, etc.—however culturally primitive and ineffectual they consider them—few would support the rationales of homeopathy or acupuncture.

Homeopathic pharmacists dilute and percuss herbal, animal, and mineral substances beyond the highest ratio at which there is physical substance left in them. These spiritualized "energy" medicines, "cloned" in tiny round pills, are then presumed to transmit assimilable or vital essence to the psyche and soma, and thereby to cure most chronic and life-threatening diseases.

According to homeopathic theory, the disease originates in an invisible nucleus dispatching those symptoms by which it is identified. The unnamed, unclassifiable malignant core is uniquely toxic and degenerating, for it anesthetizes the organism's healing capacity, establishing a general stasis, ostensibly because it is not recognized for what it is. Missed by the immune system and the biological fields, it is integrated rather than dislodged. It is manifested only as symptoms, which are in actuality the body's feeble attempts to metabolize and evict the pestilent core.

The microdose is its nonpathological, remedial substitute. The remedy, subtilized and amplified by dilution and succussion, presents itself as a parallel, a potentiated alternative to the disease. Engaging and activating the system's restorative morphogenetic capacity, it stimulates a correction of the underlying organismal lack. Perhaps, as material substance is removed, elemental aspects of it are retained and focalized in a kind of embryogenic hyper-signal.

Therapeutic effects of individual homeopathic remedies are meant to enhance (rather than alleviate) symptoms; going *with* instead of *against* dysfunction, they catalyze and augment the body's symptomatic attempts to heal itself. Based on the same broad "homeolineal" principle as vaccinations of disease products, homeopathic remedies incite an immune response, yet do not compromise or dampen the immune system by pretending to take its place.

The difference between homeopathy and immunization (isopathy) is that homeopathic medicines are not deadened toxins of a pathology; they are similars to the pathology. Their ingredients (one per remedy—cuttlefish, calendula, or zinc; bumblebee, pulsatilla, or sulphur; even live bowel nosodes and synthetics—no substances are excluded from possible medicinal use) are singularly selected on the basis that the same substance ingested by a healthy person would produce symptoms *like* those of the disease. Infinitesimalized, crystallized, and molecularly perturbed—transformed into an unknown, vitalized state of matter—the original ingredient is potentiated into something almost embryogenic.

A homeopathic remedy is not a drug or a biochemical information-bearing substance (like an enzyme); it is instead a pure signal, a resonance based in the molecular ghost of a dissolved animal, plant, or mineral extract that has vanished in the making. By its very existence in a biological field the signal transduces one state of morphological resonance (the qualities and life expressions in the material used to create the medicine) into another (the morphogenetic field of the organism), jolting the organismal pattern back into a dynamic harmony.

Similar, naturally occurring microdoses and vitalizations may lie at the heart of all evolving biological systems. The primordial ocean was a diluting, succussing vessel, its beaches hearths and alembics. Over millions of years all naturally occurring substances and compounds were molecularized and circulated as energy.

Traditional Chinese Medicine

ACUPUNCTURISTS ALSO ALTER VISCERAL FUNCTIONS by discrete and infinitesimal medicines. Though they usually prescribe macrodoses of herbs for ingestion in addition to needling, their main therapy is to insert coded metallic clarions that barely penetrate the skin. A precise entry of a needle into an appropriate current ostensibly alters the waves conducting that current and, much as in ontogenesis, restores or sustains a biological field by first disturbing (succussing) it and then stimulating it to reorient around its initial radiative vectors. A burning cone of mugwort (or other herb) may be held directly above the cure-dispersing site, dispatching heat and vapor as an alternative to needle insertion (this method is called moxibustion).

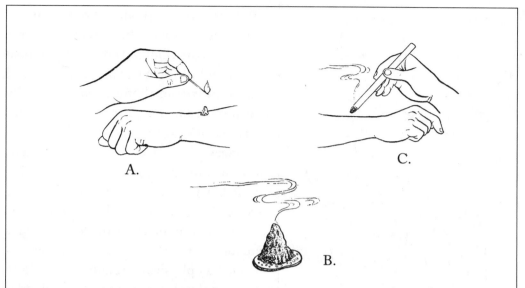

FIGURE 24A. A. Moxibustion with ginger; B. Moxibustion with garlic; C. Moxibustion with salt.

From The Academy of Traditional Chinese Medicine, *An Outline of Chinese Acupuncture* (Beijing: Foreign Languages Press, 1975).

How fine needles or points of fire penetrate organic codes is a mystery; yet its recognition is an ancient one, for acupuncture and moxibustion were devised in a pretechnological era—in fact, during the Stone Ages when naturally occurring thorns and hot botanical embers were among the most refined tools available.

IN CHINESE COSMOGONY, the body is shaped and maintained by two counteracting forces, the same forces that participated in the manifestation of the universe. A yang centripetal energy, Heaven's force, originates in the galactic fields and flows down through a spiral in the heads of both mother and embryo. Meanwhile, the rotating Earth discharges a yin centrifugal force upward through the genitals. Embryogenesis begins from the ceaseless antagonism of yin and yang drawing multiple helices out of an undifferentiated unity. Through continued vibrations, ever more complex chains condense, forming atoms, then molecules, and finally, protein fibers. In the Ayurvedic (East Indian) version, the father's semen and the mother's menstrual blood represent the proximal agencies of yang and yin, respectively.

The physical embryo is a realization of the structuring potential in matter provided by its parental cells. Energized patterns originating in sperm and ovum are attracted to each other, and their intersection-paths move back and forth, forming tissues in layers. In a sense each organ is an interference of waves through one

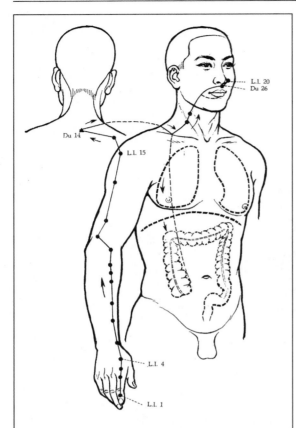

FIGURE 24B. Meridian: The large intestine channel of the hand.

From The Academy of Traditional Chinese Medicine, *An Outline of Chinese Acupuncture* (Beijing: Foreign Languages Press, 1975).

another, a vortex produced by streams of liquid "interweaving from manifold directions,"[1] redistributed through the global whirlpool of the gastrula. Currents curl and fold within themselves and finally push out as feelers, limbs, vesicles, and surfaces of skin and nerves.

Heaven's force charges the brain cells as transmitters of consciousness. Continuing downward, it energizes uvula, heart, stomach, genitals, and other organs. Where it encounters Earth's force, chakras are formed—reservoirs of subtle-spiritual as well as physical-organic power. The basal chakra locates around the bladder and genitals; subsequent ascending ones emerge at the ovary and intestines, solar plexus, heart, throat, brain, and crown (or cerebral cortex).

As they swirl outward, the spirals intersect other streams flowing between Heaven and Earth. Centrifugal motion molds hollower organs such as intestines, stomach, bladder, and gall bladder. Centripetal force condenses as lungs, heart, spleen, liver, and kidneys. The currents are simultaneously distributed through twelve primary electromagnetic channels, the "meridians" referenced elsewhere in this book. Anglicized as Lung, Large Intestine, Stomach, Spleen, Heart, Small Intestine, Bladder, Kidney, Heart Governor, Triple Heater, Gall Bladder, and Liver, these function as the pathways of yin/yang energy and the loom of the ch'i body on which the physical body is hologrammed.

Each meridian, beginning at a point of intersection between the embryo and the electromagnetic layer around it, travels inward through the body.

In gastrulation and subsequent organogenesis, the surface of the fetus reorients dorsally. Thus the major Entering *(Yu)* points are on the back, girding the spine. After contributing to the formation of organs along their paths, the currents culminate in Gathering *(Bo)* points. They are then discharged through channels leading

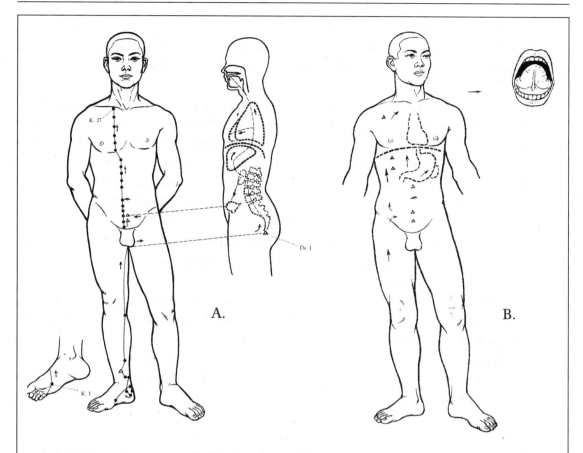

FIGURE 24C. Meridians. A. The kidney channel of the foot; B. The spleen channel of the foot. From The Academy of Traditional Chinese Medicine, *An Outline of Chinese Acupuncture* (Beijing: Foreign Languages Press, 1975).

to the Well *(Sei)* points on the tips of the fingers and toes.

Two other meridians form on the dorsal and ventral midlines of the body. Energy streaming up along the front of the embryo from a point between the anus and genital to the tip of the mouth establishes a Conception Vessel which furnishes endodermal tissue during embryogenesis. A connecting channel (the Governing Vessel) enters the mouth, flows down the inside of the body through the digestive system, and comes back out in the region between the genital and the anus. From there it runs up the dorsal surface of the body, over the head, and in through the mouth, distributing energy to the body's periphery.

These fourteen currents plait an elaborate network of tissue and organ relationships. The system of meridians in Chinese medicine is not fully represented or concretized by any other circulatory system such as the blood or the nerves (though, if it exists, it must underlie these and determine the courses of their materialization). Although meridians have no quantitative domain in Western science, they may well be lodged within the cryptic morphogenetic field.

COMPLEMENTARY SPIRALS OF BRAIN and intestines twine to produce, at midstream, the double-vortex of the heart. The brain is centripetalized intestines; the intestines are centrifugalized brain. The circulation of blood and flushing of waste proceed in twin eddies from the heart. The nervous system contracts and concentrates in the brain, whereas the digestive system expands and disperses through the intestines. The formative energy of the yin digestive system is complemented by the yang contracting spiral of the respiratory system.

Arms and legs are formed by secondary whorls. The meridians follow the surfaces of these limbs outward, winding around elbows and knees, wrists and ankles, and flowing to the tips of fingers and toes. The outer surfaces of limbs and phalanges originate at the cores of these spirals and thus correspond to the inner aspects of the organs. Conversely, the inner surfaces of limbs and joints concentrate as an outward flow from the peripheral zones of inner organs.

Insofar as ch'i currents carry energy through the organism for a lifetime, creatures are continually supplied with new quanta from the original vital patterns of which they were fashioned.

BECAUSE OF THEIR EMBRYOGENIC BASIS and function, the meridian cycles provide a contemporary series of conduits into which the messages of acupuncture needles and herbs can be conducted. Treatment takes place not at the sites of individual organs but by a method of introducing appropriate information into body-wide channels and redirecting currents of invisible ch'i energy through them to be assimilated in organs. Sometimes ch'i is dispatched to stimulate a weak organ, sometimes to sedate an overactive one; either way, a disproportion must be mediated.

Even though applied at points remote from a disorder, needles transmit energy to pathologies in targeted viscera. An entry site on the arm may supply greater medicinal potency to the heart or lungs than one on the chest because the former contacts a zone of original formative energy and cell migration and activates the entire embryogenic channel leading to the organ.

Ch'i can also be activated without a physician, as in *chi-gung* practices where a person locates his or her own energy nodes kinesthetically and raises them into organs by movements of limbs, viscera, and torso coordinated with breath. These techniques have evocative, kinesthetic names, somewhat less elegant in translation: "Ape Offers Fruit," "Rowing the Boat in the Center of the Lake," "Looking at the

Moon While Turning the Body," "Dragon Swimming," "Casting the Fish Net," and "Flying Wild Goose."

Martial arts such as t'ai chi ch'uan, ba gua, and hsing-i use the same principle to generate physical power.

IN SOME ORIENTAL SYSTEMS of diagnosis, the face is taken as a microcosm of the whole body; thus, cheeks complement lungs, the tip of the nose corresponds to the heart, the bridge of the nose to the stomach and pancreas. The ears reflect the kidneys, the mouth the anus, the teeth the vertebrae, and both sides of the forehead the reproductive organs. Treatments at one level are assumed to be transmitted through an expanding series of concentric fields to other levels.

In one style of acupuncture, the ear is interpreted as a partially aborted twin, a replica of the embryo, and is used like a voodoo doll to treat corresponding organs.

Similarly, in Western iridology, the eye is used diagnostically as a miniature of the organism, with parts of the iris corresponding to specific internal organs.

The act of shifting mind inward to the organs and cells is a keynote stage of all shamanic initiation.

DURING GASTRULATION AND ORGANOGENESIS, aspects of personality are organized in concert with hemispheres of tissue. Ectoderm, mesoderm, and endoderm continue in the adult as discrete personalities, or more accurately, componential expressions of a triune personality. When pathologies and traumas become somaticized, they likely retrace underlying developmental trajectories, following embryogenically serpentine paths of mesoderm and endoderm back through unconscious and autonomic aspects of tissue into nerve and muscle fibers, organs and fluids, even down into crystals of bone, where they sustain structural distortions. They sink in the gastrointestinal tract and constrict heart and lungs; they may even penetrate the ghost-remnant of the ancient archenteron and globalize again morphogenetically. They are emergent properties, but they represent declining (rather than ascending) levels of organization. In that state we might call them "submergent" or "disemergent," for they disrupt high-level configurations in a complex, dissociative way.

If diseases bear enough energy to contact the mitotic level of the cells and break through their extracellular matrices, they may convert into tumors and other dysfunctions of the immune system.

In some manner, symbols, life constructs, compassionate and curmudgeonly deeds, and shadows of belief systems also fission and disperse back to the communication networks of microtubules, lysosomes, and mitochondria; for these are not only the atoms of metabolic activity and consciousness, they are the ultimate repository of all of our kineses and their echoes—all our embodying encodements, even abstractly symbolized ones. Original entities and forces carrying out generic functions express their vestigial autonomies in us. There is no other final destination for molecular and organellar energy.

Pathology is continually driven inward and, though it becomes more and more insensate and unintelligible as it sinks, it is neither defused nor obliterated. Numbed, it interrupts the natural functioning of organelles and organs and becomes concretized, much as tissue itself incarnates developmentally. In fact, somaticized trauma can be a rigid proxy for motility, oxygenation, and normal growth patterns, blocking instead of irrigating and distributing, isolating rather than coordinating and germinating.

But if disease can be propelled through mentation into soma, so then can herbs, curative chants, visualizations, meditations, and palpations. They can hatch new tissue too.

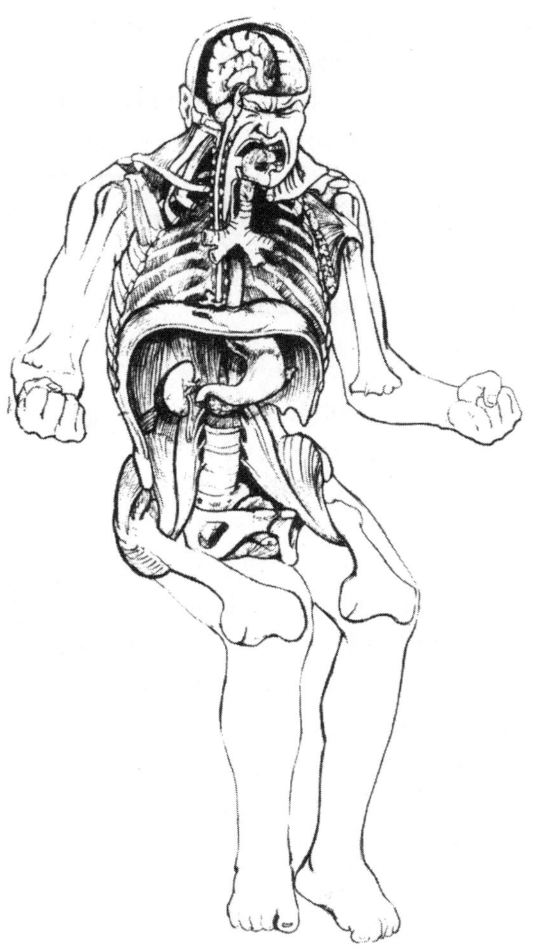

MANY SYSTEMS OF SOMATIC THERAPY teach methods of conducting "information" to the deep and hidden reaches of the soma. A practice can be as straightforward and direct as breathing into a region or site; as symbolic as internalizing a mandala, charging it with curative energy, and addressing it telekinetically to an imagined zone; or as acrobatic as sinking mind to the cellular or subcellular level of the targeted condition and allowing it to direct a healing dance (this takes profound internalization and laser-sharp attention).

Yoga cultivates these techniques

FIGURE 24D. The effect of the inflating emotions on pouches and layers.
From Stanley Keleman, *Emotional Anatomy: The Structure of Experience* (Berkeley: Center Press, 1985).

simultaneously in series of breath- and image-coordinated postures, stretches, and motions with names translated as "Hanging Dog," "Cat Kneading Paws," "Turtle Pose," "Full Lotus," etc. Rudolf Steiner developed Eurhythmy as an alphabet for spelling out therapeutic messages and transmitting them in stylized arabesques resembling sign language. New, less rigidly defined modalities (such as Authentic Movement, Body-Mind Centering, and Continuum) train practitioners in the art of "listening." They whet attentions subtly enough (supposedly) to track and infiltrate the cellular transits of kidneys and pancreas; muscular interpolations of heart, intestines, and other busy organs; fluidity and richness of blood and its hematopoietic derivatives (lymph, cerebrospinal fluid, antibodies, etc.); and even the shuffle and sonic boom of microtubules and other organelles arranging the materials of bio-existence.

Apparently a human conveying attention to these levels can attune, enter, and even adjust the rhythms and motility of amoebas, jellyfish, worms, mollusks, and mammals inside herself. She can influence the separate intelligences of subsidiary pulsations.

Therapist Bonnie Bainbridge Cohen instructs her students to abet their healing and personal growth by experiencing their circulatory and neuromuscular fluidity, blending with and supporting its natural environment and pulsations. For instance:

"... [T]he cerebrospinal fluid is clear and very slow moving. Its movement is powered by the cranial-sacral/coccygeal pump (movement between the skull and the tail). It has its own rhythmic cycle ... [and] can be felt in all parts of the body. It is a subtle, yet perceivable, cyclic movement between the filling phase (when CSF is being produced) and the emptying phase (when CSF is being absorbed). During the filling phase, all the bones of the body minimally but perceivably flex, abduct, and/or externally rotate. During the emptying phase, the bones minimally extend, adduct, and/or internally rotate. ... Gradually allow this soft being to move within its bony home. Its many articulate appendages, the cranial spinal nerves, reach out through the skull and vertebral column into the rest of the body. ... Allow the fluid that flows from this sacred center to travel through your nerves and connective tissue tubules and out into the world. Let it move you, suspending you between earth and heaven."[2]

"... [T]he cellular fluid rhythm is its own physiological rhythm. It is manifested as a continuous filling and emptying of all cells throughout the body,"[3] exchanging oxygen and carbon dioxide.

She then suggests, "Balance the inner/outer flow of the cellular fluid within the cells by supporting each body part with a restful, nondirective, and steady presence and touch."[4] One can also "stimulate the tissue pump and movement of intercellular fluid from the peripheral tissues of the skeletal muscles and organs back to the heart by using a press-and-release rhythm, as a cat does with its paws, throughout the body."[5]

THE FOUNDER OF CONTINUUM, Emilie Conrad, has led workshops reviving rituals of the early Greek Asclepian temples in which patients performed therapeutic dramas, prayed to the gods, slept on mats, and dreamed their troubled organs, inviting supernatural entities to descend nocturnally into them. Now we have no gods *except* our organs.

In Conrad's workshops, some of which build momentum through several nights, participants collectively travel to levels of psyche in soma. In one exercise, a whole room of advanced students located themselves in the innermost watery layer of loose mesenchyme surrounding the neural tube—the arachnoid membrane of the primitive meninx. Contacting it through the space behind the ear at the skull, they fell into separate trances, all moving in manners suggestive more of tissue life than human dancers. They had entered, at least imaginally, the profound, primordial, and apparently peaceful realm of the body's embryogeny—and none of them wanted to leave.[6]

Bainbridge Cohen explains the similar rationale of Body-Mind Centering:

"The movement of the mind within and without an organ reveals the specific mind of that organ. [In] the mind of the organs, universal symbols and myths are recognized. It is through this recognition that empathy is established, that universal feelings are recognized within the context of one's own life.... Each organ is a separate unit but also interrelates with all the other organs, functioning as a system through their rhythm, energy flow and movement. The tone of the organs [also] establishes the basic postural tone of our skeletal muscles. This tone and how the organs initiate and sequence movement through the inner space of our bodies provide an internal organization that contributes to the patterning of our muscular coordination."[7]

The act of shifting mind inward to the organs and cells—and the cosmos reflected in organs and cells—is a keynote stage of all shamanic initiation.

The theory of natural selection has nothing to do with how cells organize experience.

THE QUESTION MIGHT THEN BE: how can one actually get into the consciousness of organs—let alone cells and organelles? Is this only a metaphor? Is it a New Age fantasy? After all, cells and organelles not only ostensibly have no "consciousness," but they do not exist on an ontological level contactable by us as mind.

Yet paramecia and amoebas (as we have noted) move with something resembling consciousness. Organized in tissue and layers, their descendants combine their "intelligences" into new, emergent orders of sentience in multicellular space. As accumulated pulsations and "minds," they are the sole thing that makes up the cognitive intelligence of animals. If we emancipate ourselves from fixed mental structures and habits of ego identities, deigning to meet cells and atavistic invertebrate structures within us in their own medium (waves and vibrations), we might allow our consciousness to become a vehicle for whatever other minds are present in our protoplasm.

If this is a fallacious trope and it is impossible to communicate with actual tissues, cells, and organelles, then at least we will engage the unconscious, autonomic, and collective, transpersonal aspects of ourselves—body, mind, and spirit. We will enter our own medium, the pelagic stuff of our being.

Releasing mind, however, is a quandary. We are so used to "thought" that we believe if something doesn't exist there, it doesn't exist. Yet we accept dreams, yoga, athletic skills, visions following ingestion of psychoactive substances, and various other trance and voodoo states. Ordinary consciousness is little more than a culturally reinforced veneer, billowing over a sea of incalculable depth. Although the Western view of mind prefers rational, aware cogitation, our true lucidities may arise from quite different layers and frequencies of being.

To function as an internal "shaman/self-healer" means travelling in vision-quest manner inside us—inside the integuments of an outer thought shell in which we conduct the internal dialogue of our life drama. Once underneath our willed selves, we encounter not only ancient oceanic waves by which tissues induce the body-mind in a semifluid, reshaping event but motions and signal intelligences by which cells continue to distribute information and assemble us in tissues.

The mode of releasing mind cannot be simply meditation, visualization, or yoga. Its enactment must be less formally organized and ritually controlled, less cognitive and language-based in its design and plan. To resonate our being with the other possible beings inside us means being as guttural, growling, oscillatory, purring, and billowing as they are. The hindrances of our minds are, of course, animal in their origin and infrastructure too, but they are likely frozen at a reptilian level of camouflage, fright, flight, and dominance/authority within ancient parts of our archaeocortex, brain stem, and cerebellum.

As NOTED ABOVE, Emilie Conrad teaches internal cell and tissue exploration as a goal of her Continuum practice. If SETI (Search for Extraterrestrial Intelligence) begins with radio-telescopes scanning the emissions of star-fields, the search for internal intelligence originates with breath. The breath of Continuum is not simply a normal breathing rhythm, or even the more educated breaths of Rebirthing and meditation (zazen), for these rapidly entrain thought—the neocortex locks into their habits or pattern. Scanning inward must arise from disrupting surface figurations and allowing one's self to fall into deeper, erratic, unmeditated rhythms.

Conrad's "hu" breath is a fast panting pattern, the mouth and entire jaw running through positionings of different phonetic shapes from breath to breath. "Hu" is an entire alphabet dancing from letter to letter, the changes in the positioning of the jaw and lips, throat and neck muscles not only continually changing the part of the body the breath is expressing but making it so that the breath cannot impose any pattern at all but must skip around until, through sensation and pulse, it brings into consciousness and motion some aspect of the natural movement of viscera and cell life. The "hu" is a pump, circulating, dispersing, and reconfiguring thoughts and movements.

Continuum's "lunar breath" has some of the same attributes, but it is quiet and slow, little more than the flush of air from deep inside the thorax and throat, released in a long outbreath like wind through foliage or a delicate snore. In its quietude and richness (much like night's soft-fringed moonscape on Earth) we are instantly attuned to noncognitive, biological rhythms inside our ocean. In an ancient tongue that lies behind our various languages we begin to purr original questions—*whooooo?; hoooooww?*—questions in which speech and human identity had their once-upon-a-time beginnings.

The breath can also be compressed, redirected inward, and attuned to different bodily locations and frequencies by the tongue being pressed behind the upper teeth, as in making a "th" sound. This forked release of air is the "theta breath," a directional signal attuning the laminae and cavities of the body-mind. Conrad believes that forces present in this breath initiate a scalar wave that subtly influences the body's electromagnetic fields. The split of the tongue lies roughly at the level of the developmental brain and spinal cord (the neural tube), which formally begins at the upper palate. It is a zone where neural-crest cells originate. In an imaginal sense the theta breath directs scalars through the channelling of neural tube, and these find their way, as they did mesenchymally once, to the different stations of the body where consciousness has been inculcated.

Conrad's protocol for entering the mind and movements of the meninges around the brain starts with a theta breath, trilled while stroking the temple at the edges of the sphenoid bone with a finger from each hand, respectively. This radiates inwardly, inducing increments of delicate motion and consciousness. Theta is followed by a deep lunar breath, the air of exhalation passing through the skin, as though the entire ectoderm were breathing in vegetative clouds. Meanwhile the backs of the eyes search in the darkness, "looking" (with the help of the pineal eye) at the brain itself, following a path down along the dural tube. At the same time, one imagines a separation of their head from their spine.

An "aw(e)" (spinal-fluid) breath follows. The head is cocked back, stretching and exposing an area around the throat and neck in a manner resembling animal yawps and howls. Each outbreath makes a little frog croak—like a recoil—upon the start of its exhale and then releases normally.

As lunar breathing follows the "aw" breath, the traveller puts attention on the seam between the back of her ear and head. Mind and bodily rhythm slowly seep into the domain of the arachnoid, giving rise to its spidery dance.

CONRAD OFFERS MANY such breathing sequences; together they constitute an emerging alphabet for internal exploration. As one follows any of them (or improvises new ones), parts of the body begin to move in unexpected ways. The natural motions of animals arise, quivering and streaming—frogs, fish, snakes, deer, gulls. These are mostly internal images and minute oscillations that barely register in a gross, outwardly perceptible manner, no matter how huge and thick they feel. As one's tissues and rib cage move fluidly, a subtle, cell-based nervous system extends the sensation of underwater tentacles and organs undulating in a flow around them— barnacles, tendrilly phoronids, sea anemones.

Conrad believes that this motion is communicated through breath in an embryogenic mode: Not only does mind flow down into the organs and cells; it informs them about who we are and what we need. We are able to change our body-stuff, not only musculoskeletally and neurally but by accessing reams of unused junk DNA in ourselves and providing new templates for gradually repatterning it. The cells feel the attention and invocation in some fashion and begin to reconfigure themselves. Life itself reaches into its own probability structures, responds, and changes. This is not "inheritance of acquired characteristics," the staid Lamarckian sacrilege; it is potentiation of acquirable characteristics. Of course we cannot alter the basic hard bodily form we receive from gene space—the material, karmic universe has inexorable rules; if it didn't, none of the physical realm could come into being and maintain its wondrous continuity and legacy. But we might be able to shift potentialities and probabilities we already encompass; we might even be able to translate cognitive and psychosomatic designs into tissue.

Who knows what mischief Navaho sand-painting, faith healing, and vision-quest are truly about?

THE BIOLOGICAL SYSTEMS of people raised in civilization have been colonized and industrialized; our tendencies toward movement are neither spontaneous nor free. It takes time, practice, and gradual reattunement to achieve deeper states and, by journeying through them, to heal mental and somatic illnesses and restore tensility and pleasure of a biological self. Often one first has to initiate an intentionally choreographed movement; only after a number of cycles does the breath intercept this artifact and organize it along its own surprising trajectory.

Conrad's associate, anatomist Robert Litman, has developed a brief theatrical event to get one used to the possibility of "being" in cell and organelle sentience. A group of people disperses and takes positions on the floor in the shape of a cell. The largest number form the lipid bilayer membrane, lying more or less head to toe, but slightly curled and wrapped into each other. Three or four people spoon around each other to make up a Golgi body off-center within the cell circle. Various others gather in small mitochondria clusters and microtubule rows. A tiny central circle represents the nucleus. The time I participated in this event Conrad herself lay angled in this circle as the DNA.

While I stared paralyzed at the assembling pattern and held myself somewhat shy and aloof, all that was left was to become a collagen fiber, one of many oriented at approximate right angles to the "cell" at various points, the bottoms of my feet touching one of the legs of the "cell membrane." However, since connective fibers are still evolving, I had the possibility of becoming a previously unknown form of helix.

In the hour that followed, different clusters and individuals took up various breaths and their associated waves and chantlike sounds, sometimes in isolation, sometimes inspiring movement and sound elsewhere. Conrad-DNA did the "hu" breath, a wild, twisty dance in the center. Theta and lunar breaths cast a varying and sometimes sonorous Gregorian background. Mitochondria and microtubules broke at irregular intervals into long "o" or "om" breaths. An occasional "aw" and other phonemes chirped and honked. As the cell changed shape in undulating movements, its inhabitants moved about and changed shapes too, much as their microbial counterparts do.

After a while the make-believe quality of this pageant dissipated. I felt as if I were in a combination of a factory, a temple, and a jungle. I heard the low rumbling of machinery—mantra and grumble. Then it stopped. I heard hoots and twitters and the unceasing aspiring of life. Suddenly the mitochondria howled again.

In a remote way this is what it is like inside a cell—natural, peaceful, fluid motion, then creationary sounds (or the sounds of creationary processes). Meanwhile proteins are gracefully and graciously assembled, passed around, configured, and altered. The event had a haunting pagan feel to it, plus the sense that mind could enter here, at least along the long axis of breathed intention.

This was closer to the reality of a cell than a microphotograph or textual drawing (although, according to the paradoxical weaves of history and consciousness, it is only microbiology that has salvaged cell images from beneath the threshold of a quaintly visible world). Being a cell is neither sterile nor regimentalized; it is buoyant, playful, shape-changing, rhapsodic—a pure, scintillating form of consciousness and hoodoo life-breath. Cells enjoy their pond immensely; they are having fun.

BEHIND CONTINUUM is the notion that organisms cannot change structure unless they change their ideas first. When movement changes, breath changes, thought changes, then structure will also change. The living field we identify as cells and their components will change; new types of collagen and different proteins will be manufactured. Under these conditions the body-mind will not follow a rigid, genetically enforced track but will express its cellular aliveness in a larger, fluid, planetary environment, drawing on both the nucleic (DNA) and organellar pliancy and probability structure within itself (even as we simultaneously draw on the ductile, associative characteristics of mind to create novel thoughts).

Though this is utterly Lamarckian in its premise, the theory of natural selection has nothing to do with how cells organize experience. We actually do not know how the deep, multilevel corridors of information and structure in ourselves work.

SENSATION IS THE SOURCE of life's natural mutability and variability; hence, the dynamic form of the embryo is our natural state. We gain new information by deepening sensation and redeeming rhythms and resonances that are already present in us, underneath our social condition.

Inflexibility and rigidity represent the lack of sensation.

The only way kids can invade their school with guns, shooting bullets into other students, is if they don't *feel* what they are doing, if they do not experience the event as real.

As industrialized, computerized, heroin and fantasy intelligences replace real cell-chanted bodies, all sorts of new crimes against the body-mind arise. These are not intentionally sadistic. The harm being done doesn't even resonate.

When there is no longer empathic resonance, one can do anything. The body is a dream.

Psychoanalytic and Somatic Healing

THE FIRST HOLISTIC, PSYCHOSOMATIC MEDICINE was the method of psychoanalysis developed by Sigmund Freud at the turn of the twentieth century.[8] Freud's paradigm was based on a standard etiology, that traumatic events during infancy and early childhood distort normal psychosomatic growth. Even as traumas are suppressed and stowed in the unconscious mind, they are cathected, that is, charged with emotional energy. Fixed in content and distorted in rendition, they flood the personality with dysfunctional synapses, which, over years, become neuroses, psychoses, compulsive disorders, and psychosomatic diseases. They are pathologically morphogenetic, oscillating from psyche to soma, and back.

Freud proposed salvaging those ancient moments from the unconscious mind by leading the patient on a linguistic journey to rediscover them through dreams and free association and to live them again in the context of positive transference. Memory traces of infantile interpersonal dynamics are enacted in dialogues with a psychotherapist playing the role of a benevolent parent—something that was lacking during the actual events. Experienced from an adult perspective, primordial wounds are opened in the present, demythologized, and partially dispersed. When the sterile repetition and nonsense of infantile fixations are sundered, surprisingly virulent monsters emerge. These are then tamed, reassured, and integrated back into functional modes.

Freud's disciple Wilhelm Reich, while adopting his teacher's developmental etiology, challenged his "talking cure." He proposed (and later body-oriented schools supported his point of view) that once neurosis and psychosis become somaticized—their natural predilection—it is impossible to release them with words. They can be improved only through a dislodging or dissolution of the restrictions and blocks they build up in actual soma—their tissue armor and neuromuscular rigidities. The curative path must start in the body and travel from there into emotions and language.

In contemporary situations of anxiety, stress, and terminating relationships, neo-Reichians argue, we do not want to feel our body; we stifle its excitations lest they stir up unpleasant memories and sensations. By inhibiting sadness, unboundedness, and rage we benumb the anatomies that underlie them and, over time, become incapable of their natural expressions. Most people assume that this limitation is only a superficial layer of irritation—transient habits—but the response patterns get transposed into glands, fluids, neuromusculature of the gut and heart, and are structured by alignments of cartilage and connective tissue, creating fixed, intractable barriers.

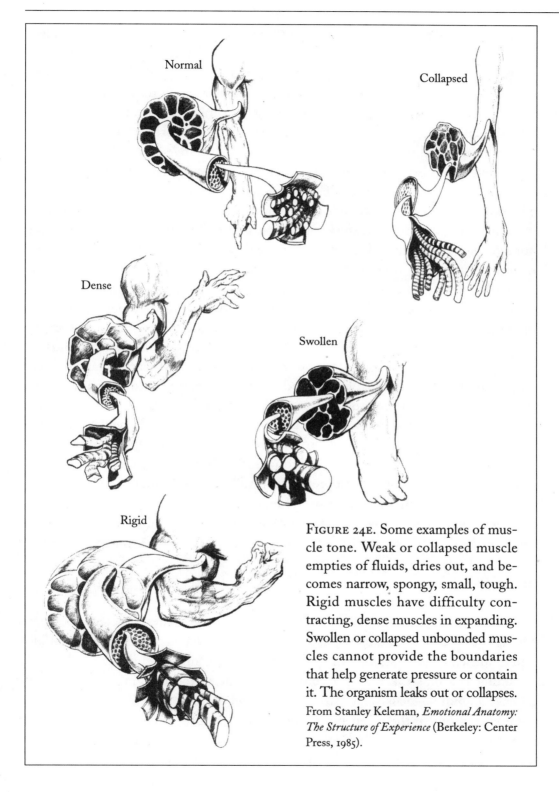

FIGURE 24E. Some examples of muscle tone. Weak or collapsed muscle empties of fluids, dries out, and becomes narrow, spongy, small, tough. Rigid muscles have difficulty contracting, dense muscles in expanding. Swollen or collapsed unbounded muscles cannot provide the boundaries that help generate pressure or contain it. The organism leaks out or collapses.

From Stanley Keleman, *Emotional Anatomy: The Structure of Experience* (Berkeley: Center Press, 1985).

By the "kindling" effect the sick get sicker; hormonal episodes become feedback loops supporting compulsions, mood disorders, panics, and tantrums.

In somatic practices, neo-Reichian therapists instruct people on how to bring pulsation back to emotionally lethargic or decompensated tissue. Bioenergetics involves kicking, pounding, flexing, tumbling, and enhanced breathing in the context of therapeutic massage. These techniques shatter the brittle coat of resistance around nerves and muscles and restore some natural range and feeling to organs. When energized by movement, tissues actually release tension and regain a measure of suppleness. Exercise itself—as long as it is accompanied by emotion and empathy and is not merely robotic gym-work—tends to perform a subtly embryogenic function, disturbing tissue knots, redistributing fixated charges, and resetting homeostases at a higher vibration. The mere act of ruffling connective tissues and neuromusculature is often therapeutic because, following agitation, they reacquire some of their flexibility and synergy.

Reich was aware of other, long-standing traditions of therapeutic massage, manipulation, and physical therapy but, following Freud, he presumed real organismic restoration came *only after* they were reapplied in the context of psychoanalytic theory—that prior to that they were capable of (at best) accidental cures and symptomatic relief. That is, he focused on the traumatic etiology and its sexual aspect rather than on the broader transcultural flux of integrative and disintegrative waves bearing emotions and their violations in pure forms. When Reich introduced his own versions of physical techniques into psychotherapy, he placed such an enormous emphasis on not only libido but an ethnocentric definition of sexual fulfillment that he came to interpret virtually all body armor and serious physical malignancies as a consequence of sexual repression.

Reich's singular cure was a melting of sexual armor through free and natural orgasm abetted by deep massage and bioenergetic exercises. Thus the sex-economic theory of body-armor and orgasmic healing became a stand-in for all advanced medicine.

A WIDE RANGE of other somatic disciplines partially supplanted Reichian therapies throughout the 1980s and 1990s, mainly because they avoided cosmological pitfalls and empirically addressed a full spectrum of bodily and psychological ailments, including autism, strokes, spinal-cord injuries, and effects of aging. These do not all have a libidinal locus. Traumas, neuroses, and dysfunctions arise for subtle and complex reasons in the mind-body continuum, and they do not all follow a sex-economic etiology or an Oedipal developmental pattern. Many restrictions and lesions are the results of genetic attributes, cultural strictures, accidents, injuries, parasites, secondary effects of other pathologies, and assorted chance phenomena. Although there may be sex-economic factors among these, metaphorizing the entire process in terms of these is not a useful way of proceeding therapeutically. In fact, metaphorizing them in terms of any arbitrarily prioritized factor dooms a system to sterile ideology.

In the end, the real problem Reich inherited from his mentor was not the one he sought to rectify—that Freud was naively psychological to the exclusion of the somatic aspect of existence—but the one he totally overlooked, hence inherited— that Freud was libidinally reductionist and culturally bound to one developmental yardstick.

Reichian Physics and Biology

Late in his career Reich reinvented the physics and biology of the universe, propounding an etiology of cell formation and morphogenesis. He extrapolated (from microscopic analysis of inorganic and organic substances and telescopic study of the atmosphere) that a bathysmal life force radiated from among the galaxies, travelled through interstellar space, and flooded (in biblical amounts) into planetary environments where its relics polarized into life vesicles (bions) and produced complementary male and female seeds. Sexual energy originated not in nucleic acids and tissues but extrinsically in the cosmos (among galaxies, nebula dust, rain clouds). Panspermia and spontaneous generation, not cell and membrane dynamics, were the source of life. "Creationary debris is all around us," Reich (in effect) said, "in every mud puddle, on grass fronds throughout every meadow and field— living astrophysical seeds. Yet, when scientists see them disclosed under the lens, they stare through them as though they were blind."

In fact, he wrote: "[B]ions are forms of transition from inorganic to organic matter; they can develop into organized living forms such as protozoa, cancer cells, etc. They are vesicles [of about 0.5 to 3 microns] filled with fluid and charged with energy ... originat[ing] in organic and inorganic matter through a process of swelling."[9] Wood, dried moss, grass, wool, coal, soot, lava, iron, potassium, silicates, and muscle and other animal tissue, are among substances that disintegrate into bions.

These membranous vesicles are not in themselves full-fledged living entities; instead they contain "a certain quantity of energy, ... forms of transition from non-living to living."[10] Travelling "through the microscopic field with slow, jerky or serpentine movements"[11] and, pulsating, they emit a bluish glimmer, a halo of the biological energy. In colloidal suspension they either take on or disburse an electron; otherwise, they are electrically neutral.

Bions fission and fuse exactly like cells, permeating each other, regardless of their proximal sources in different types of tissue and nonliving matter. "...[G]onadal cells and erythrocytes are bions. The chicken embryo develops through organization of the yolk bions, moss from stone bions, protozoa from moss or grass bions. Cancer cells develop from bions which originate from the vesicular disintegration of suffocated or otherwise biologically damaged tissue."[12]

Raw cosmic bions can be administered curatively, directed to bombard the progeny of deadly bions, for instance, bacterial or malignant cells. Even from a distance an emanation of their blue force kills or paralyzes intruders.

"ORGONE" IS THE NAME Reich gave to the cosmic source of not only bions but all energy. Present everywhere, penetrating everything, it traverses the universe in pockets of greater or lesser concentration like waves of a termless ocean. Bions are spawned by blankets of orgone washing up on planets and moons. Predisposed to life-assemblage by the agglutinative, pulsing efficacy within them, they bear the same elixir that holds stars and nebulae together. Bions literally teem from heated ocean sand because it "is nothing but solidified sun energy."[13]

Much as D'Arcy Thompson saw living matter as frozen currents of water, Reich interpreted biological assemblage as a direct translation of cosmic motility into localized lattices. Life forms express their orgonotic nature first, which is movement, pulsation, and coordination. Their membrane-enclosed configuration is secondary. Spiralling and oscillating cosmic vectors confer distinctive shapes on plant seeds and bulbs (corn and barley, potato tubers and almond kernels), leaves and blossoms (fig branches and rose flowers), blastulas and gastrulas, animal viscera (bladders, genitals, spleens, lungs, brains), protuberances (beaks, claws, horns, snail shells), and body shapes (jellyfish, starfish, oysters, beetles).

Orgone is so basic it can transpose itself from energy into matter (and back) without any of the traditional thermodynamic problems. In its natural movement through the universe it not only begets galaxies and star systems but leaves its energetic replicas on worlds in the form of vesicles which combine into spermatozoa, vorticellae, protozoa, and the like. Metabolism is an intrinsic quality of orgone even before it is alive. "We may assume," Reich said, "that the spermatozoa and eggs in the metazoa are ... formed through condensation of orgone energy in [their] germinal tissues."[13]

Organisms are not hierarchical systems assembled by the integration of their visceral functions. To Reich's mind, muscles, nerves, blood vessels, and other organs are independent plexuses of preexisting orgone energy, manufactured as stellar sacs and vesicles long before they get assembled and anneal materially in planetary biospheres.

Gastrulation, neurulation, and organogenesis must adhere to the plasma currents of their bions, which generate whole phyla and classes of creatures. Cells later incise their own diacritical marks, comprising families, genera, and species.

Function not only precedes structure; it is its sole physical and ontological basis. Function is what holds nature together and pulls landscapes into tight, habitable grids. Things "are" because they work, because they "do"; they don't exist first and then contrive functions from structure. The universe exists as occurrence long before it has any matter or stuff in it. Situations become "things."

In mass-free form the "kidney" knows everything it needs to about urinary function when it is still an orgone wave; it imparts that knowledge during its breakdown into bions, all the way through its reconstitution (by morphogenesis) into an organic configuration. It become a kidney metabolically because its natural movement tendencies already embody nephric calyces, glomeruli, and filtration.

Organs express singular endowments of orgone energy, coming together cooperatively because their proclivity is to seek union and unity and because the energetic factor propelling them is stronger than the rigid envelopes transfixing their shapes. Morphogenesis (culminating in organogenesis and tissue function) merely encrusts cosmic energies and functions in temporary restrictive shells; these forces already compose the equivalents of organisms, nervous systems, and libidinal charge in free interstellar space. Nerve cells do not merely produce impulses from sensory stimulation and synaptic electrochemistry, for there is no closed grid of neurovisceral signals and corresponding actions and impulses. Neurons transmit and communicate the fire intrinsic to them into the cosmic receptacles of organs predisposed (through common origin) to receive it.

When animals locomote and carry out their dispositions, they display the part of themselves that is unhindered expression of orgone; at rest, they are mere orgone statues or imprints of motility. In sexual foreplay and orgasm, they are galactic nebulae seeking to fuse and procreate, to liberate their molecular tension, to stretch and extend their domains with implicit bounty and good will; from this status both galaxies and lovers confer life effluvia on each other. (A similar mode of orgonotic throbbing and release causes the sundering paroxysms of death.)

Sex is thus a physical force like gravity, giving rise to a zygote and initiating embryogenesis when sperm and egg meet appropriately, because it is already conducting cosmic energy through gonads into bodily, cellular shapes. The erections and spasms of genitals and their superimposition during coitus evince their cosmic not their personal design. Reich's description of seduction and copulation is a masterpiece of vitalist and gnostic physics:

"The preorgastic body movements and especially the orgastic convulsions represent extreme attempts of the mass-free orgone of both organisms to fuse with each other, to reach into each other.... While the energy of one organism flows into the energy system of the second organism, mass-free orgone energy actually succeeds in transcending the limits of the material orgonome, i.e., the organism, and, by merging with an orgonotic system outside its own, it continues to flow.... Orgastic longing, which plays such an enormous role in animal life, now appears to express this 'striving beyond one's own self,' this 'yearning' to escape from the narrow confines of one's own organism.... it is orgonotic superimposition that connects the living organism with nature surrounding it."[15]

Thus, when we look at the sky with its stars, we are looking at love and desire in their raw form—forces so powerful that life is inevitable.

ORGONE IS NOT ONLY THE MATRIX of all biological form it represents the sole reliable medicinal substance, for disease is first and foremost the blocking of the free flow of cosmic energy through living systems. In experiencing the intrinsic flow of unhindered orgasm and receiving orgone during sex from another creature (or from the atmosphere in specially built energy-attracting units), human beings ground and activate cosmogonic substance throughout their beings, dissolving trauma, cell pathology, and somatic armor (see also Chapter 8, "Fertilization," pages 141–142).

In the end Reich applied star energy much as a Taoist doctor would ch'i, pulling raw creationary stuff out of the atmosphere into medicinal "orgone boxes." At this point, he no longer sought psychosomatic therapy but a supernatural physics.

ALTHOUGH REICHIAN PHYSICS has never come close to being accepted by science or incorporated into a general theory of the universe, it provides the kind of organizing principle and initiating principle that biology lacks. Systems theory and complexity analysis always fall short insofar as they arise from one and the same algebraic lineage and decipher function as an outgrowth of form. Future paradigms for life, explaining its origin, maintenance, increasing complexity, and sentience, may not refer precisely to orgone and bions; however, they must in some way incorporate the underlying precepts that Reich intuited—that function precedes form, that meaning precedes and gives rise to mechanics, that motion precedes material configuration. By working his way kinetically "backward," from psyche to soma to body to matter to energy to cosmos, he explained the coarsest and most astrophysical entities (stars and subatomic particles) in terms of their subtlest and most complex manifestations (life, metabolism, embryogeny, coordinated movement, and mind)—rather than the more common *vice versa*.

Palpation

INSOFAR AS A BODY MADE OF CELLS comprises both intra- and extracellular space, one genre of medicines may penetrate the cell reticula, while another shifts the external matrices and fascial/neuromuscular field. Homeopathy and acupuncture modify cell signalling and epithelial transmission; by contrast, palpation and visceral bodywork move physicomechanical forces synergistically. However, the treatments of osteopaths, polarity therapists, Rolfers, and other somatic healers likely also blend hands-on mobilization with molecular feedback (much as shear forces and reaction-diffusion couplings articulate with genetic transcription during embryogenesis).

There is apparently a deep level at which humans are the polarization of forces as much as the congelation of substances. Thus, they respond to touch morphogenetically and curatively—not any touch (just like not any enzyme) but touch applied in such a way that the most ancient, deep-seated, and pliable aspects of the soma recognize it as their clarion and respond in an organismically cohesive, therapeutic manner.

Successful applications have been developed empirically by generations of palpators, bodyworkers, osteopaths, polarity therapists, and medical innovators. As practitioners recognized unique responses and meanings of touch, they learned to trigger tissue activity, following and coordinating instantaneous feedback.

Diagnostic and curative activities include a range of techniques described by terms like torsion, sidebending, shear, strain-counterstrain, lateral compression, decompression, cranial-vault hold, mandibular traction, parietal lift, squat, and energy-block release. For instance, craniosacral therapists pull on both ears to decompress the cranial base laterally and release various sutures and junctions in the occipitosphenoid areas. The sphenoid bone itself is regarded as both a valve for controlling the flow of cerebrospinal fluid and a handle and rudder for freeing immobilized bones in the face.

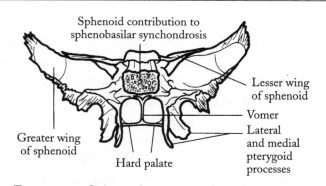

FIGURE 24F. Sphenoid, vomer, and hard palate from posterior view.

From John E. Upledger, *Craniosacral Therapy II: Beyond the Dura* (Seattle: Eastland Press, 1987).

Though the language reflects physics and engineering, such terminology is more heuristic than purely machinelike; mechanical images are projected metaphorically through the healer's hands into another body where they have an organic rather than just a mechanical effect. None of the techniques are applied in the manner of opening a jar or pushing a heavy load. With minor exceptions they are administered subtly with a few ounces of force at every moment. Bodyworkers are like artists or athletes; they play tissue vectors as meticulously and with as much attention to ricochet as a billiards whiz with a cue or a conductor with a baton. Tissue tends to lead sentiently applied force in the direction it itself "wants" to go.

Some somatic techniques are more mental and telepathic than physical. Even sound, prayer, and bodily heat move living matter. This is the basis of various forms of faith healing, Reiki, and quantum touch.

All the way out (or in) to our thought patterns and dreams, we are the innate physics of soft matter, the thermodynamics of cell-stuff. Nothing else in the universe responds to vectors of temperature, tension, and proximity in the way life does. The various splashes, convective undulations, currents, nautilus shells, braided vortices, and double helixes—writ large—that lie at the core of human anatomy respond to gravity and centripetal and centrifugal tension as positioning signals. With their "new" physicochemical, nucleic cloaks they are able to memorize and reinforce patterns in a way that pure liquids and gases cannot. They contain cures even as (evolutionarily) they contain shapes. The bodyworker in that sense is working on the body of the planet—the body of the planet condensed logarithmically in human spirals at the ultrasonic core of intersecting meridians.

BEFORE DOCTORS GRADUATED to modern forms of surgery and pharmacy, they were more inclined to pulse-taking, massage, and palpation—in general, to reading by hand and directly manipulating and adjusting the body's tissue layers, pulses, and rhythms. Throughout the world, and down through history, shamans as well as primitive surgeons used hands-on skills to diagnose and heal conditions, from those affecting digestion and circulation (however they named them) to the purported electromagnetic fields and auras around the body. The vital energy of the healer was thought to resonate with the emergent field of the patient. By this paradigm the hand (or any other trained organ—some Middle Eastern systems such as Kurdish-derived Breema employ the feet as more sensitive and less judgmental) melds with the patient and receives and transmits information like a dry sponge placed in a pool of water.[16]

From the codification of their techniques in Greek and Roman times, palpation and massage became dominant modes of medicine until they were gradually demoted during the rise of twentieth-century technocracy. In an era of MRIs and ultrasound, they have been dismissed as primitive, romantic fallacies of an earlier era.

A PHYSICIAN TRAINED in traditional Chinese medicine measures various surpluses and deficiencies in bodily functions merely by traversing pulses with his fingers at the patient's wrists. From a patter of sequential beats resonating discrete sites, he can detect sums and subtleties of metabolism as well as multidimensional visceral interrelationships. Each of the Entering, Gathering, and Well points, as well as locations between them, have unique roles in balancing the functioning of the body (as they did in molding tissues initially). The life current is either bubbling up, bubbling down, spreading out, eddying, fluttering, oscillating, or carrying out some other formative activity as it flows through each duct.

Palpation likewise receives patterns from viscera and transmits them back. A sensitive hand can interpret the form, motion, and mutual relationships of not only structures immediately beneath the skin but the mobilities of joints, texture and pliability of ligamentous and tendonous attachments, movements of bones in relation to one another, flow of bodily fluids, and normal or abnormal positions and rotational orbits of organs. Reflexogenic points, often held for ninety seconds or more, then control and/or inhibit different bodily functions at varying distances from the point of touch. The message appears to land nonproximally and to reach organs through the autonomic nervous system, fluids, and connectivities within the fascial web.

Palpation also adjusts internal functions by contacting and supporting tissues in movements they are already carrying out and suggesting new movements or vectors in a manner that viscera can accept. On occasion, palpation is intended to intrude and move tissue, for instance, in Rolfing where the practitioner's deep, often painful percussion impresses cell memory into ligaments and mesodermal structures, probably right down to the mechanics of basal lamina, epithelia, stem cells, and subcellular and extracellular tensegrities so that remarkable realignments and morphogeneses of tissue follow.

Most often, however, manipulative medicine is meant to be nonintrusive so that surficially initiated sensations can journey into organs without provoking a defensive neuromuscular reaction of pain, stiffening, or tightening. Once in the system, they seem to spread by shear force, interfacial tension, viscoelasticity, pulsation, or other embryogenic patterning. In fact, their transmission reverberates long after the hand is removed.

THE INTERNALIZATION AND PSYCHOSOMATIC DISTRIBUTION of an initial act of adjustment or palpation are critical to both the effectiveness and duration of the treatment.

Even in cases of simplistic and naive applications of massages, calisthenic exercises, and skeletal adjustments—where the practitioner has little or no awareness of deeper effects—morphogenetic and phenomenological change still ensues. The most primitive mechanical therapies as well as the most sophisticated surgeries are cathexis-catalyzed processes, for (as mentioned many times in this book) the organism is less a site-specific solid than a fluid mind-body medium through which modes of contact and linguistic symbols radiate in complex waves.

Human beings also have deeply entrenched resistance to being rearranged on an ego or tissue basis, profound enough to have frozen dysfunctions in tissue in the first place. Neurosis (to extend the general term) has the power to coopt any available or stray energy with the goal of enhancing its own protective mechanism, including all attempted cures and even unconscious projections of the physician himself. It is a shallow, defective replica of morphogenesis.

A psychotherapy or surgery may be "successful" according to each of their own provincial models of cure but unsuccessful insofar as it blindly potentiates pathologies and resistance to change. Just as there is no way to translate the syntax of acupuncture into Chinese or English and disclose the true messages and pathways of the needles, there is no way to know the subhypnotic instructions that a doctor is kineticizing in his patients.

As John Upledger repeatedly points out in his trainings, a surgeon (or bodyworker) transmits subliminal design (beneficent or malefic) through his scalpel, a design which is seeded in the body as latent messages and sanctions. "Don't become part of the problem," he warns, "that you are trying to solve."

SOMATIC COMPONENTS MUST FINALLY be dealt with in the terms in which they present themselves in the body, as neuromuscular, fascial, visceral, and circulatory patterns. They unwind intrinsically from the patterning and energetics they embody.

Release is always in terms of physical and/or psychological transference with a therapist and against the counterforce of resistance intrinsic in the patient's own system.

Osteopathy

IN THE MID-NINETEENTH CENTURY, Kansas physician Andrew Still developed (or, more accurately, codified) the traditional repertoire of hands-on methods for treatment of the sick. As a trained engineer, he applied the most current principles of structure and hydrodynamics to mechanical medicine, with a goal of repairing the body by fixing its girders, pulleys, and pumps.

Still formalized his treatments at a time when modern medicine was in its infancy. Thus, the distinction between a physical therapist redeploying vectors of tissue and a doctor using sophisticated biochemical techniques was not fully understood. Mainstream allopathic treatment also did not hold any greater prestige or credibility then than homeopathy, Native American pharmacies, or the traditional modes of manipulative medicine that empowered Still.

Most physicians who today call themselves osteopaths have all but given up their manipulative specialty and have become close cousins of the mainstream allopaths. Yet Still's system propagated a variety of other curative modalities that achieved prominence in the 1990s; these include chiropractic, cranial osteopathy, craniosacral therapy, visceral manipulation, orthobionomy, myofascial release, zero balancing, defacilitated fascial release, lymphatic drainage therapy, and strain-counterstrain (note that these terms are both generic and trademarked, thus appear alternately in lower and upper case in osteopathic literature).

The same key osteopathic techniques are employed in all of these to one degree or another: direct manipulation (which guides an organ or bone in a direction away from that in which it is stuck); indirect manipulation (which follows and enhances the stuck direction along its components in order to induce a release by freeing the trapped elements or impelling momentum back the other way); tracking the pulse of the cerebrospinal fluid and quieting it by gentle pressure; following the body's other minute movements with supporting touch to their natural cessation points (that is, blending and adding); applying light palpation (at levels of five grams or less) in order to activate an organ or tissue; projecting intention into fascia and other tissue through fingers that are barely conducting force; in general, creating axes of dynamic, reciprocal tension among viscera, bones, and fluids. These procedures induce waves of changes in the mind as well as the body of the patient, currents that travel through the system (quite mysteriously) and affect organs and life processes by oscillation and recoil.

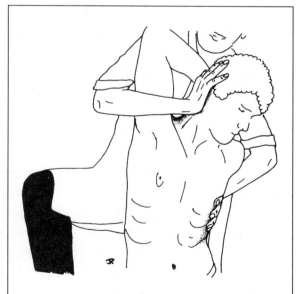

FIGURE 24G. Visceral manipulation: Stretching the parietal pleura.

From Jean-Pierre Barral and Pierre Mercier, *Visceral Manipulation* (Seattle: Eastland Press, 1988).

Chiropractic

CHIROPRACTIC MADE ITS DEBUT IN 1895 when Daniel David Palmer, an Iowa farmer with osteopathic training, bestowed hearing on a deaf patient by adjusting his neck. Palmer believed that, through postural and emotional tension, segments of the spinal cord become hyperactive. Trapped in self-replicating regimes, firing nonproductively, these somatic neuroses, or "subluxations" (as Palmer named them), lose their inherent capacity for corrective feedback and are desensitized to surrounding tissue. The autonomic nervous system around an irritated segment malfunctions, and the subluxation spreads to contiguous organs, which become lethargic. Reinforced by kinesthetic and physiological habits and emotional traumas, an initially minor distortion deepens into a pathology.

IN HUMAN DEVELOPMENT, as we have seen, the notochord induces the spinal cord, somites, and vertebral column, then vanishes as vertebral bodies supersede it. During the third or fourth week of pregnancy an emerging primitive spinal cord is surrounded by sclerotome cells which develop in segments that themselves become vertebrae. Each vertebra lies between two somites, thus is intersegmental, with inductive segmentation extending to the viscera.

Neural-crest cells originating between the neural groove and ectoderm migrate to form autonomic ganglia running down both sides of the spinal cord and engaging with the spinal nerve cells. Fibers streaming out from these ganglia penetrate the heart muscle, the adrenal glands, and blood vessels, and, through them, more remote organs and tissues from the jaw and atlas to the sacrum and toes. Cells of bone marrow and gut, blood and lymph are likewise originally borne by neural-crest cells that share a ganglionic origin.

Meanwhile the spinal cord grows into a thick trunk with branches going to and from the various viscera, nerves, and muscles of the body. Incoming nerve impulses flow from the back, and outgoing messages are dispatched via the ventral nerve roots. There are eight pairs of cervical nerve roots, twelve thoracic nerve roots, five lumbar roots, and six sacral ones.

This system of cables and fibers is inherently induced, structure by structure within itself, beginning with the ancient notochord. Thus, it retains a post-inductive potential; it can distribute externally imposed shocks and signals.

BY THE CHIROPRACTIC ARGUMENT, "All cells have sympathetic innervation, including blood vessels which, when hypertonic, decrease distribution to the brain in crisis.

The impact of this is reduced healing, increased hypertension, and changes in endocrine function impacting metabolism, brain function, and ultimately all homeostatic mechanisms."[17]

The sympathetic nerves of the autonomic system can be stimulated therapeutically to increase blood flow to the heart and skeletal muscles; reduce flow to the skin and internal organs; depress digestion, intestinal peristalsis, and kidney function; and abate the mind's activity. The parasympathetic nerves can be stimulated to increase absorption of nutrients and calories and catalyze their conversion to energy; excite the secretion of glands; activate digestive function; and initiate the flow of blood, slow the heart rate, and lower blood pressure.

The chiropractor's manipulation of vertebrae activates morphogenetic interrelationships of spinal-cord segments, nerve impulses, blood flow, and viscera. By releasing subluxations, he dynamizes and balances the entire system.

If, for example, the fourth thoracic vertebra is stuck, jammed with an overload of nerve impulses, arteries conducting blood to the heart may be restricted; hence, adjoining muscles contract. When, with a carefully aimed jolt, the vertebra is freed, a rush of nerve impulses impregnates tissues which then come alive, stimulating cellular activity around them and gradually resupplying oxygen to the heart.

A mere mechanical manipulation, unrelated to the heart *per se*, becomes cardiac and medicinal because of its transmission of microsignals through a hierarchy of correlative structures and vitalization of their surrounding zones. The lungs, stomach, intestines, kidneys, etc., can all be "energized" by chiropractic intrusion within the somite system and its adjacent tissue.

Other Osteopathy-based Systems

VISCERAL MANIPULATION WAS DEVELOPED in France during the 1970s by osteopaths Jean-Pierre Barral and Pierre Mercier. Their method assumes that each organ rotates on a physiological axis which has a dynamic relationship to all other axes in the body. Combinations of forces from organs' motions and their impedances travel to surrounding and distant tissues, causing distortions. If an organ loses its natural motility (through tension, metabolic malfunction, or injury), then strain and stress are exerted elsewhere, ultimately manifesting in an acute symptom. Stacking the long, complex, inertial pathways against one another with gently-guided, massage-like palpations (while sensing their layers of textures, tonalities, tensions, and subtle movements), the visceral therapist finds and gathers trajectories of past forces and then applies his own light manual force to the viscera and connective tissues (and their scar-laden restrictions), with the goal of unstacking them and encouraging their latent

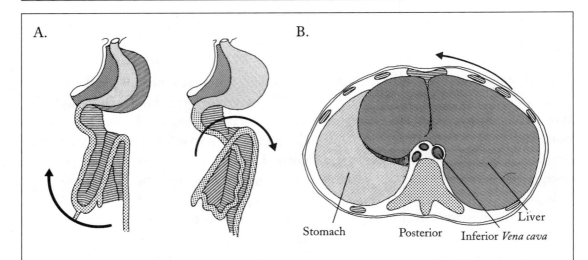

A.

B.

Stomach   Posterior   Inferior *Vena cava*   Liver

FIGURE 24H. Some examples of organ motility. A. Trajectory of intestinal twisting during embryogenesis; B. The motility and mobility of the liver in the transverse plane.

From Jean-Pierre Barral and Pierre Mercier, *Visceral Manipulation* (Seattle: Eastland Press, 1988).

"normal mobility, tone and inherent tissue motions ... the systems they function within and the structural integrity of the entire body...."[18] Restoration of orbit has a powerful effect on all aspects of the system.

Because visceral manipulation emphasizes the mobility of tissues more than their position, it treats by engaging the tensional forces within the body as a whole, i.e., the interconnections between "internal organs and structural or neuromusculoskeletal dysfunctions."[19] The palpator applies force and blends with the body's response to his induced tension, skillfully adding vectors where necessary. Whether through proximal physical contact or by morphogenetic waves affecting adjoining tissue, visceral therapists are able to improve pathologies of the liver, esophagus, small intestines, lungs, stomach, kidney, bladder, colon, reproductive organs, etc.

For instance, during expir (movement away from the median axis of the body), the liver rotates on a transverse axis and its superior aspect rolls anteroinferiorly. This motion is encouraged by placing the palm on the abdomen and supinating it. The external edge of the liver's rotation to the left around a sagittal axis is induced by the palm pulling the lateral aspect of the rib cage anteromedially while the fingers simultaneously push the medial aspect posteriorly.

In rotation of the jejunum and ileum, both hands follow the inherent motility, but an additional vector is applied from top to bottom to compensate for the vertical portion of the small intestine's floating position in the abdominal cavity.

Manipulation traces the historical status and migration path of each organ as a guide to its natural healthy motility and pulse. Barral writes:

"Organs migrate during embryogeny. For example, the stomach rotates to the right in the transverse plane and clockwise in the frontal plane. The transverse rotation orients the anterior lesser curvature to the right, and the posterior greater curvature to the left. The front rotation moves the pylorus superiorly and the cardia inferiorly...

"At the end of its embryological development, the inferior extremity of the stomach rotates toward the right, taking with it the duodenum in a right rotation around a vertical axis. It also swings in a clockwise rotation around a sagittal axis, the superior extremity of the stomach moving to the left, while the duodenum moves slightly upward and to the right. This is another example of visceral motility recreating the motion of embryogenesis...."[20]

Elsewhere he adds:

"The embryologic theory of visceral motility postulates that the axes and directions of these motions remain

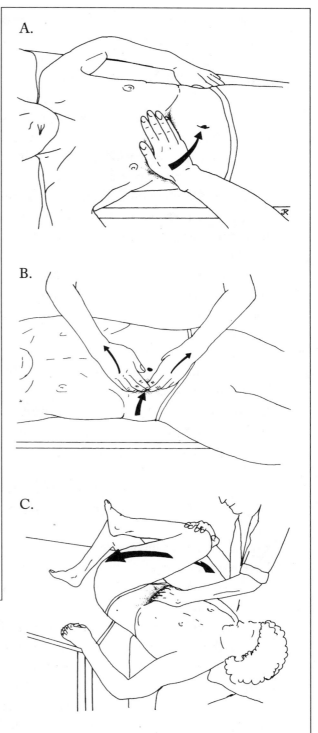

FIGURE 241. Some examples of visceral manipulation. A. Induction of the liver in expir along the frontal plane; B. Direct manipulation of the duodenum; C. Manipulation of the left kidney.
From Jean-Pierre Barral and Pierre Mercier, *Visceral Manipulation* (Seattle: Eastland Press, 1988).

inscribed in the visceral tissues. Thus, visceral motility occurs around a point of equilibrium, oscillating between an accentuation of the embryologic motion and a return to the original position, with a contractility analogous to (but much slower than) that of the nodal tissue of the heart."[21]

LYMPHATIC DRAINAGE THERAPY was developed by the French physician Bruno Chikly from an older system of manual drainage pioneered by Emil Vodder. Gentle massage-like palpation drains surplus fluids, toxins, proteins, and long-chain fatty acids within the body's interstitium.

Strain-Counterstrain Technique was codified by an osteopath, Lawrence Jones, as a specific method for calming the muscular stretch reflex and the flow of spasmodic impulses to muscle fibers and spindles throughout the body. The affected muscles are thereby more capable of elongating relaxedly and shortening with full power. Using similar techniques, specialized therapists have been able to reduce the spasm of arterial, venous, and lymphatic muscles. Blood flow, lymphatic drainage, speech and swallowing, respiration, and other functions mediated by the involuntary nervous system are thereby enhanced and/or ameliorated.

The general term "myofascial release" involves directing the hands to effect the separation and elongation of elastocollagenous fibers to make the matrix of soft tissue in the body less viscid and more fluid and flexible. With a change in the density of the matrix, metabolism improves, with nutritional and waste materials transported more easily. Rolfing, craniosacral therapy, visceral manipulation, and strain-counterstrain all have myofascial components.

A "neural tissue tension technique" is used for mobilizing motor nerves, sensory nerves, cranial nerves, spinal cord, and the plexus of nerves and blood vessels.

The often-confusing variety of therapeutic modes may be interpreted as arising from specialized cellular, subcellular, and tissue dynamics—each system becomes frozen or inert at its own level in its own way. Seemingly identical symptoms may have roots in entirely different substrata:

"Tissues and structures respond to treatment according to the nature of the cells and fibers of those tissues and structures. Cells of different systems require unique avenues of healing because the properties of cells vary from system to system. Neurons in the brain are unlike the sarcomeres of muscle fibers. The chemistry, histological characteristics, energy, and complementary tissues organize themselves differently for the support of each system."[22]

All of these techniques recall Still's initial attempts to apply engineering principles to the body; yet they also address the holistic, possibly morphogenetic transmission of remedies through fascia, cerebrospinal fluid, the involuntary nervous system, neural-crest material, extracellular matrices, and microtubule tensegrities. This would suggest an anatomy of redundant tensional gradients bearing structure and motion from the large to the small—from organs to membranes to cells, and vice versa.

Cranial Osteopathy and Craniosacral Therapy

CRANIAL OSTEOPATHY EMERGED DURING THE FIRST DECADE of the twentieth century when its founder, William G. Sutherland, a Minnesota osteopath, discovered the minute, interdependent motilities of the bones and sutures of the skull. These small skeletal structures had previously been assumed to be fixed at birth. Sutherland strapped a football helmet to his own head in such a way that he immobilized any potential movement. If the cranial bones were rigid, his discomfort should have been limited to the burden of the helmet. However, within a matter of days he experienced acute physical symptoms, emotional distress, and the initial signs of visceral pathology throughout his body. When he removed the helmet, he felt a remarkable warmth and movement of fluid up and down his spinal column and through his ventricles. From the insights of this initial experiment Sutherland developed a medicine for treating a wide range of ailments using the fulcra, hydraulics, and dynamism of these skull bones, their visceral attachments, and the cerebrospinal fluid itself.

CRANIOSACRAL THERAPY, an expanded, modernized version of cranial osteopathy, originated in 1972 when, during a surgery, John Upledger noticed the movement of the *dura mater* membrane—an inexplicable pulse that was synchronous with neither the breath nor the heart rate. At that time cranial osteopaths explained the throb along the dural tube as brain waves. Upledger felt that a pressure-stat system was more likely, with the compressive flow of cerebrospinal fluid providing its engine. During the mid-1970s, while part of an interdisciplinary team at Michigan State University, Upledger developed a means of diagnosis and treatment based on the dynamics arising from a cerebrospinal pulse. He named his method "craniosacral therapy" to distinguish it from the more heavy-handed thrusting manipulation common in the osteopathic trade and to make it feasible for him to train non-osteopathic bodyworkers and physical therapists in the art of palpation.

Upledger describes the craniosacral system as being made up of "the three-layered membrane system that we call the meninges, . . . the cerebrospinal fluid enclosed by this membrane system, . . . and the structures within the membrane system which control fluid input and outflow for the system."[23] As a result of the absorption and extraction of cerebrospinal fluid, bones, tissues, fascia, and viscera—from cranium to phalanges, along the spine and outward—participate in the aforementioned cycle of flexion (inward curling) and extension (outward unfurling). (See Chapter 18, "Neurulation and the Human Brain," pages 453–457 for additional descriptive material.)

This network is the internal milieu for the suffusion of all the fluids and innervations of the brain and nervous system into the muscles and viscera. Its pumping action integrates with musculoskeletal, vascular, lymphatic, and neuroendocrine processes. Because the hydrostatic cycle encompasses the brain, spinal cord, and pituitary and pineal glands, it has far-reaching effects on the body's functions. Dysfunctions and pathologies in any local region will reflect along cross-binding angles and axes multidimensionally in the whole organism.

The craniosacral therapist, palpating and tractioning the differential pressures of this system, changes visceral hydraulics and chemistry. The innumerable sites for initiating therapy include the *dura mater* attached like handles to the bones of the skull, the temporomandibular joint through the *tentorium cerebelli,* the muscular component of the xiphoid process of the sternum, and the pelvic and urogenital diaphragms and their muscles and fascia. During a particular set of techniques, one hand is placed firmly under a person's body, stabilizing the region. The other hand senses out the contour and directional orientation of muscle fibers and fascia and palpates in a shear, torquing, or rotary motion; its direction anterior-posterior, longitudinal-transverse, or multi-angular and oblique. Gradually the region relaxes and reorients; its tissues soften; then the pressure is released and a new motion may be initiated, either there or elsewhere, depending on the resulting stasis.

As WE HAVE SEEN, the fascial web develops in an indivisible piece out of mesodermal tissue and envelops the entire body. Its global expanse, heterogenous penetration of tissue, and activation by the cerebrospinal pulse through the nervous system allow it to transmit and receive the cranial rhythm everywhere.

A single system embryogenically, the fabric of fascia interpenetrates itself and all the organs. From the *falx cerebri* to the *tentorium cerebelli* down the internal lining of the occiput to the carotid foramen of the temporal bone winds one dynamic sheet. With a change in scale and orientation, that sheet continues to the pericardium in the thorax and the respiratory diaphragm; from there, with another change in vector and scale, to the psoas muscle, pelvis, *corpora cavernosa* of the penis and clitoris, legs, and bottoms of the feet. The fascia functions as both a map and highway through the body—a natural neural, mechanical, and hydraulic medium for the transmission of palpatory remedies.

As tensegrities of connective tissue are downplucked into microtubules, microfilaments, and the organellar anatomy of cell space, a signal starting at skin and subcutaneous muscle and nerves can distribute its kineses right into the nucleoplasm. The patterning suggests Bach's organ music or Charlie Parker's jazz more than Still's engineering. The body is not only mass, but space (at a visceral level and again at cellular and subcellular levels). Space and mass are arranged in perfect geometric counterpoint throughout the soma; protein is packed in deep fractals and gapped by the infinitesimal intervals among them. Filaments and tensegrity forces open even tinier ratios within membranous stuff. The system is such a perfect resonating chamber that harmonious notes jump scale both ways and locate themselves wherever they "sound" best.

Palpating from the cranium through the craniosacral-fascial system, a skilled "musician" can literally reach any site in the body.

IN THE WORDS OF UPLEDGER and his co-author Jon D. Vredevoogd, the body fascia are "a slightly mobile, continuous from head-to-toe, laminated sheath of connective tissue which invests in pockets (between lamina) of all of the somatic and visceral structures of the body.... By direct connections and common osseous anchorings, the extradural fascia and the meninges are interrelated and interdependent in terms of their motion. Therefore, the amount of diagnostic and prognostic information which can be obtained from the examination of fascial mobility or restriction is limited only by the palpatory skill and anatomical knowledge of the examiner. Attention is directed to the rate, amplitude, symmetry and quality of the craniosacral motion and its reflection throughout the body."[24]

WARPS, TENSION PATTERNS, and pathological obstructions are all transmitted via the fascia. Fascial immobility in spots is a guide to the exact location of disease processes hindering mobility (either at the point of stricture or elsewhere). Like the fairytale princess sensing a tiny pea under her stack of mattresses, a skilled cranial or visceral therapist can palpate through the complexity of the fascial system to the source of its distortions and immobilities.

He or she will be aided by the craniosacral rhythm, both in navigating fascial pathways and in locating obstructed sites, for the fascial system complies in both gross and subtle ways with the flexion and extension of the cerebrospinal pulse.

WHEREAS THE EARLY OSTEOPATHS and chiropractors primarily made adjustments of skeletal-neural structures and constrictions of viscera, Upledger and his colleagues attempted to tap into the unconscious etiology of the whole psychosomatic field, addressing the so-callled "inner physician" and encouraging it to resume or extend its native healing functions. The trances that arise from supporting, following, and restricting cerebrospinal flow are apparently states of grace during which unconscious communications take place and a person rearranges her entire inner being, including physical illnesses and traumas.

Craniosacral therapists attempt to follow the myofascial, cerebrospinal web to precise historic precincts of cathected moments—or functional replicas of such locales—at which a malady was internalized and locked in place, sometimes embryogenically, sometimes by birth trauma, and sometimes by bodily and emotional events not fully processed. This "energy cyst" (as the somatic component of the cathexis was christened by Upledger) is a multilayered knot of trapped physical and psychospiritual energy, functioning at a visceral and neuromuscular level, transmitting its restrictive effects fascially throughout the body (much in the way a snag at one spot in a sheet affects the contour of the entire sheet according to the intensity and direction of the snag and the texture of the material).

According to Upledger, "the complications of the Energy Cyst retention depend upon its emotional content, the quantity of energy within the cyst and its location. It seems that the emotional content of an Energy Cyst is capable of entraining the general emotional tone of the whole person. . . . "[25]

The same is true of physiological effects. "For example, the energy from [a] fall on the sacrococcygeal complex could penetrate quite easily into the pelvic viscera. In this location it could cause bladder dysfunction with chronic sphincter control problems, menstrual dysfunction, [and] prostatitis. . . . If it went all the way to the respiratory diaphragm the patient might later begin to notice symptoms of esophageal reflux (heartburn)."[26]

While the snag is unwound and released, the entirety of the field changes in all dimensions. Intrinsic motility and visceral function are restored.

Healing is morphogenesis.

HOLISTIC MEDICINE FUNCTIONS in terms of an enduring coordinated field— a field that is either precisely embryogenic, fractally iterative, and confocal, or draws on some energetic or vitalistic force not fully mapped by science but parallel to and integrated with the developmental process. Remedies that have little or no concrete explanation (at the surface) work, if they work at all, through the inductive fields of the organs and viscera. This is the only way we can explain curative effects from needles and microdoses. Very tiny amounts of substance (like amino acids) also initiate exquisitely minute changes—changes that are basic and synoptic and radiate quantally throughout the system. The chemical output of the Golgi apparatus, the geometric projections of the mitotic spindle, the transgenesis of fibroblasts, the potentiation of lymphocytes, the hollowing out and filling of capillaries (as well as thousands of other integrated processes) all occur at similarly subtle and discrete levels.

So much goes into an organism, layer by layer on a subtle level, that it becomes a bioelectric symphony, a flowing colloid, and a replica of the mind it engenders. Microdose and meridian hyper-signalling may be among the ways in which vast amounts of morphogenetic information, coding multiple trajectories of large, thick substances, were miniaturized and stored in the evolution and ontogenetic continuity of living systems. Some palpators report feeling almost-sonic flows of kundalini-like energy and rays from eighty-four sites through the body. These flows arise suddenly and spontaneously, do their healing, and subside. They are like somatic weather.

The tensile force of connective tissue is two thousand pounds per square inch, not all at once but as cumulative forces passing through the body. This can split hardened steel surgical bolts, so it can certainly reorganize tissue. A skilled healer has its full power and stored tension at her disposal.

Healing can work as hints of information, waves, breaths, shear forces, or thoughts. Rays and physico-dynamic stress lines shoot across systems of viscera and functions. Quantum leaps in organization are internalized from experiences. A single feeling (or fast) can initiate (or resolve) a deep cleansing.

LIFE IS A MIRACLE. This is more than a pietistic cliché. Nothing really explains or sponsors biological wholeness or coherence, or the biosphere. Trillions of creatures populate this planet—more or less healthy, each with a complement of organs, viscera, and metabolic and neural pathways necessary not only to exist but to experience and enact existence.

They thrive and procreate. They do not disintegrate into giant wens and tumors; they do not routinely unravel and lose their shapes. What sustains them? What keeps them from tumbling apart?

How could it not be a miracle? By the laws of physics none of them in their powwows with their lofty designs should be here at all. Not even one of them. Not even the rudiments of one of them.

The cellular basis of life does not condemn medicine to analytic molecularity and materialism. There is no way for granules of substance to hold shape and metabolize without synergy, whether a true vital force, an energy body, or an embryogenic complexity great enough to override the physics of entropy and dissociation. We are something far more labyrinthine and cohesive than a congery of cells; in fact, we are irreducible ontologically or metaphysically to the cells that make us up.

If energy or spirit sticks organisms together, energy and spirit can best reshape and heal them. That is the meaning of holism.

The different biological fields, wherever they originate, interact in the morphodynamics of substance. Light touch moves tissues (and cells); herbs (like neurotransmitters) change structure; microdoses and needles provide reorganizational information. Embryogenic data can be transmitted via bones, muscles, nerves, fascia, viscera, and the like; through enzymes, antibodies, ribosomes, mitochondria, microtubules, Golgis, etc.; and equally through qualia, archetypes, mantras, mandalas, and breathed resonances. At this level of evolutionary depth and complexity, no one knows any longer what the pathways are, but their collective synergies can be felt and observed in the different healing modalities. The success (or failure) of any therapeutic system probably represents different degrees and configurations of sensitivity, attunement, layering, and tracking, and hierarchies of information transmission, in individual organisms and through varying disease complexes. Any modality (osteopathic, homeopathic, energetic, herbal, constitutional, iconic, etc.) is viable if it can touch a morphogenetic or morphodynamic center, however it makes its trajectory there and however subtle or crude the message.

If one wants to get well (unconsciously as well as consciously), then these systems, each in its own way, provide the psychosomatic field with props, tools, and symbolic conversions to aid in the process of self-cure.

HEALTH IS NOT AN ULTIMATE STATE. Complete health is not even possible. Well-being is relative, a dynamic homeostasis always deteriorating into disease. From immune responses and from the havoc itself comes a different quality of health, a new, often more vibrant order. Disease is a state of dynamic chaos, searching (unconsciously) through myriad possibilities of itself for ways to reorganize living systems, to kindle novel patterns of constitution within cells and tissues. Disease is the sole incubator of health. Chaos is the matrix of life.

The success of craniosacral therapy and the other osteopathic systems clearly cannot be explained by the mere physiology of the choroid plexus, the *pia* and *dura mater*, and the arachnoid spaces, and their translation to the neuromuscular and fascial systems. These dynamics must in some way reenact an original circumstance in which tissue, dream, personality, consciousness, and the autonomic nervous system blended once with the fluid supply to and from the brain to form a delicate mind-body ecology. The therapist who finds the correct, unencumbering pathway into this archaeology synergistically contacts the precognitive linguistic codes whereby organisms communicate kinesthetically within themselves.

25

Transsexuality, Intersexuality, and the Cultural Basis of Gender

Sexual Orientations

HISTORICALLY WESTERN SOCIETY DISTINGUISHES two orientations: heterosexual and homosexual. However, gradually over the last hundred years a wider range of orientations has been recognized. In the late nineteenth century, Karl Heinrich Ulrichs, a gay man writing under the pseudonym Numa Numinantius, made a previously unrecorded distinction between traditionally classified homosexuals (who seek as a partner a member of their own sex) and *urnings* (who want to *be* the opposite sex—and also *may* experience "homosexual" desires). *Urnings* have an unabating sensation of being born in the wrong body (a condition now identified as "gender dysphoria").

In 1910, Magnus Hirschfeld, founder of an organization to aid men and women in gender crisis, proposed another variation—the erotic desire to cross-dress (clothes fetishism), which he called "transvestism." He noted that most transvestites (among his sample of seventeen, including one woman) were heterosexuals uninterested in partners of their own sex and with no desire to become the other sex (other than temporarily through cosmetics and attire). A more contemporary term "femmiphilia" is sometimes used to identify men "whose interest is solely in the feminine gender role and not in her sexual activity."[1]

After World War II, with the discovery of the inductive role of hormones, biologists and doctors began to grasp the fluidity of gender and its expressions. What occurs naturally during adolescence is a potential state in all people at all ages; there are many degrees of relative maleness and relative femaleness. Genders can even be incited artificially: with estrogen it is possible to feminize males, with testosterone

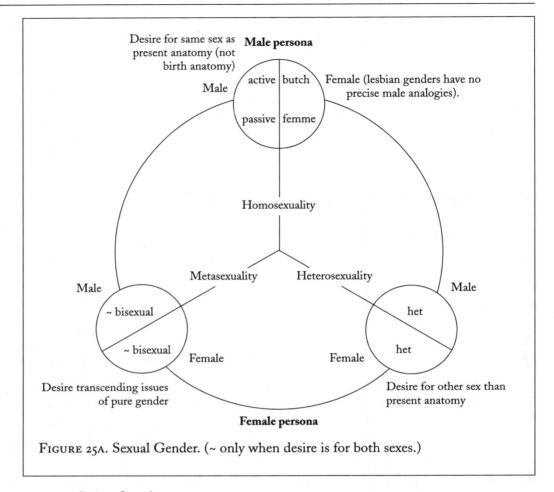

FIGURE 25A. Sexual Gender. (~ only when desire is for both sexes.)

to masculinize females.

Beginning in the 1950s a small number of venturous physicians have treated gender dysphoria medically with hormones and surgery.

Categories of Desire

IN TRUTH, THERE IS AN EXTRAORDINARY RANGE of divergence among chromosomes, anatomies, biological genders, cognitive genders, and modes of desire (and even these pigeonholes are artificial and overlapping).

From the perspective of normative taxonomy, desire is recognized in four nonparallel classes: heterosexual (for the opposite sex), homosexual (for the same sex), bisexual (for both sexes), and metasexual (outside the conventional domain of bodytypes and genders). The latter comprises a limitless number of appetites and acts bearing scant resemblance to one another; these include: ritual dominance (the projection of erotic power to subjugate another man or woman, encompassing rape as well as consensual sadomasochism), body fetishism (oral and anal impartment of genitals), artifact fetishism (cross-dressing, sex with eroticized objects or fantasy images, sex with animals, some episodes of masturbation and self-gratification), pedophilia (erotic obsession with children or using children as sex objects), and necrophilia (sex with corpses, acts linking murder and sex, sex mimicking throes of vampires). In addition, fantasies and acts combine and transpose all of the above through foreplay, dress, and titillating guises (in bathhouse orgies, phone sex, computer sex, mate swapping, etc.).

Despite the grouping of these together, a distinction should be made between consensual and nonconsensual sadomasochisms. The former no doubt comprises many healthy ways of discharging socially awkward atavistic urges, whereas the latter—where nonacquiescence is required for a perpetrator to feel erotic charge—is classically pathological.

Categories of Cognitive Gender and Gender Dysphoria

WE CAN SUMMARIZE COGNITIVE GENDER by a pair of basic types, each with a train of subtypes: being in an appropriate body to express one's desires or being in the wrong body. People with either conviction may be heterosexual, homosexual, bisexual, or metasexual, with their classification depending (in part) on whether their sexual orientation is derived from their biological or perceived gender. This multilevel polarity leads to grammatical as well as anatomical ambiguity. Because the pronouns "he" and "she" ("him" and "her") were carved at the origin of our language (English is not alone in this), morphophonemic gender will remain equivocal throughout this chapter, with switches from male to female (and *vice versa*) in describing the same person plus combined forms like "s/he." The series of grammatical disjunctions reminds us that shifting and optional gender is a revolutionary idea, not considered at the birth of Indo-European culture and speech.

WE CAN DIVIDE those with a conviction of being in the wrong body into two categories: those who seek medical reassignment (hormones and surgery) and those who accept their birth bodies. Both are considered transsexuals: the former *may* become surgically altered transsexuals; the latter usually remain cultural and behavioral transsexuals.

A transsexual who craves but (for one reason or another) never undergoes surgery is different from a transsexual metamorphosed by exogenous hormones and organs; both differ from heterosexual and homosexual transvestites, who accept their birth bodies regardless of the transgendered appearances they adopt or partners they seek.

Cultural and Psychological Factors in Gender Identity

NONHETEROSEXUAL DESIRES SEEM BIOLOGICAL in their deep-seatedness. In fact, in 1993 biochemists at the National Cancer Institute in the United States, comparing family trees of gay men and pairs of homosexual brothers, declared that at least one gene on the X chromosome (inherited from the mother) is circumstantially linked to gay phenotypes. Still, many elements of sexual identity are indisputably cultural. Some of them represent early childhood experiences consummating in adult behavioral patterns: for instance, a strong identification with a dominant mother may cause a boy to feel he has an invisible woman's body; likewise, a girl bonding mainly with her father may prefer playing cars to house and later seek a sexual correlate.

Cultural and psychological factors participate complexly with biology in gender identity and the genesis of desire. Early sexual abuse may sidetrack a person from her natural biological course in order to protect a core identity. Children have no capacity for the force and ambiguity (and often the suppressed rage) of adult passions; thus, even when they seem to respond to forced pedophiliac attentions, they are deflecting most of the energy, in the process fragmenting their ego. Cathected unconsciously, maturing under the neuroendocrine influences of puberty, the latent emotional charge of violative episodes becomes much more devastating than their surface memory.

A young boy upon whom an adult homosexual has performed his fantasies may himself become a pedophile, partly in an attempt to reenact what was done to him—to reexperience and understand its meaning and violation from the other side—and partly because violation itself has become required for any sexual expression to feel emotionally real.

People can also be coerced or indoctrinated to go against their original desires and biology, begetting still further erotic variations—for instance, a gay male practicing heterosexual marriage because his Christian or Muslim beliefs forbid homosexuality; likewise a Generation X lesbian who is actually either hetero- or bisexual but whose friends have defined sex with men as treason and will ostracize any member who engages in it.

Some cultures or cults impose similar counterbiological trends on selected children. Families lacking daughters among the Lache of Colombia may rear a son as a girl in order that s/he be able to carry out the household chores of women. Traditional Aleut and Kodiak Islander parents may select a boy—perhaps one who shows effeminate tendencies—to raise as a female and then sell her in early adolescence to a "wealthy man who want[s] a boy wife."[2]

Gender outcome usually rests on whether anatomy, hormones, and psyche are stronger than artifacts, symbols, and sanctions. Either set can become eroticized with any degree of ambivalence and lingering peccadillos and compulsions from the devalued or lost mode. The result is a rainbow-like heterogeneity of behavior.

Intersexuality

To this already baffling and overlapping set of categories we must add the most basic one of all—permutations of chromosomes and phenotypes. People's external genitals and secondary sex characteristics do not always conform to their heredity—mutations in genes determining gender yield their own spectrum of anomalous types. The codons also may be ambiguous, giving rise to other chimeric anatomies and psyches.

According to geneticist Anne Fausto-Sterling, there are at least five distinct natural and healthy genders among humans, each with their own bodies, attractions, repulsions, inner lives, and special needs.

To males and females, Fausto-Sterling appends three genders of intersexuals with unique combinations of male and female organs and traits. The most well-known of these are herms, or true hermaphrodites, with sperm- and egg-producing gonads, a testis and an ovary. Hermaphrodites can penetrate females with their penises and likewise provide labia and vaginas for others' penises. Emma, whose case was recorded by urologist Hugh H. Young in his *Genital Abnormalities, Hermaphroditism and Related Adrenal Diseases* in 1937, had both a penis-size clitoris and a vagina. As a teenager she was a boy, but at nineteen s/he married a man. Her husband found the relationship sexually satisfying, but Emma was unhappy with it and sought girlfriends as lovers. S/he did not favor surgery to remove her vagina because s/he wanted to stay married (for economic reasons).[3]

Merms, or male pseudohermaphrodites with XY chromosomes, have testes and some combination of female genitalia (a vagina, a clitoris, and/or, at puberty, breasts), but no ovaries. Ferms, their female equivalent, bear two X chromosomes, lack testes, but possess some male traits (adult-size penises, beards, and/or deep voices).[4]

These are not pathologies or birth defects; they are "normal" tissues, conferred by nucleically derived proteins, eroticized and functional. Nature, of course, without prejudice tenders mixed genitals in single bodies throughout the plant and animal kingdoms. In humans these produce psyches that are somewhat male, somewhat female, but substantially their own thing.

Among some indigenous peoples intersexuals are considered conscripts of deities, set apart in guilds, addressed as oracles and shaman helpers. Like twins they are

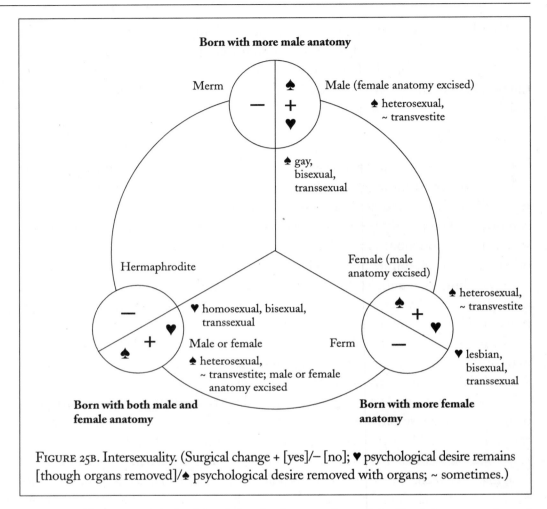

Born with more male anatomy

Merm

Male (female anatomy excised)
♠ heterosexual,
~ transvestite

♠ gay,
bisexual,
transsexual

Hermaphrodite

Female (male
anatomy excised)

♠ heterosexual,
~ transvestite

♥ homosexual, bisexual,
transsexual

Male or female

Ferm

♠ heterosexual,
~ transvestite; male or female
anatomy excised

♥ lesbian,
bisexual,
transsexual

Born with both male and
female anatomy

Born with more female
anatomy

FIGURE 25B. Intersexuality. (Surgical change + [yes]/− [no]; ♥ psychological desire remains [though organs removed]/♠ psychological desire removed with organs; ~ sometimes.)

spirit-world guests with double-sighted talents and seership. Both intersexuals and transsexuals may be granted privileged roles in the company of women as well as during ceremonies.

Among other tribes and clans these same anomalies are regarded as accursed or demonic and their bearers suffer the fates of evil eyes and witches. In one historic reference from the 1600s "a Scottish hermaphrodite living as a woman was buried alive after impregnating his/her master's daughter."[5]

THE ACTUAL RANGE of possibilities is even more heterogeneous. Dr. John Money outlines nine stages by which human beings, beginning at syzygy, undergo "psychosexual differentiation or the establishment of gender identity."[6] The initial phase is of course the assignment of a chromosomal gender; the culmination of all nine stages is an active gender established in adolescence. Money notes a passel of "'sex-chromosomal errors' such as Turner's Syndrome, Klinefelter's Syndrome (XXY), and the XYY Syndrome; 'gonadal errors' such as hypospadias (an incompletely fused or improperly located urethral tract in the male), androgen insensitivity in the XY fetus, and hermaphroditism; 'hormonal errors' such as the androgenital syndrome in XX fetuses and gynecomastia; 'internal errors' such as male hermaphroditism with uterus and normal penis or hypospadias with uterus differentiated; 'external error' such as the masculinization of XX fetuses by administration of hormones to the mother during pregnancy, penile agenesis (in which the XY infant is born with a penis the size of a large clitoris, due to the absence of the spongy tissue of the *corpora cavernosa* in the penile shaft), and penile injury or penectomy; and 'gender identity error' such as transsexualism."[7]

Each of these errancies and its crossed neuroendocrine signals lead to distinctive erotic desires and repertoires of their performance.

TWENTIETH-CENTURY SOCIETY censors and hides intersexuality. Medical police limit sex types to two. They look at what has been born, but they do not see. "It must be a boy," they tell themselves, "or it must be a girl." So they reassign all chromosomal variants to one of two unanomalous genders, initially by surgery following birth (cutting away one or the other set of organs as if it never existed). The gelded child is then "normalized" through male (or female) acculturation, rituals, and taboos.

It is simply not permissible to pass through modern society as an ambiguous sex type.

The surgical elimination of anomalous organs is considered a "cure" because of the challenge of indoctrinating a hermaphrodite (or pseudohermaphrodite) into acceptable gender behavior; then there are, of course, the painful initiations any boy-girl would receive from "her" peers and the marginalized sex life that would ensue. Medical elders assume that no parent would want to raise such a child, for there is no category by which to nurture "his" sexuality, or to love him/her. That deeply has polar male-or-female gender seeped into our affections.

But divesting someone of his/her anatomical heritage is not always a gift. When uncovering (by one means of research or another) their post-partum surgery, most intersexuals are shocked and outraged, but suddenly understand their dysphoria. They were mutilated, their erotic identities excised, their lives shoehorned into an anatomy and assigned a social role their tissues and desires do not support.

Transsexuality

"I WAS THREE OR PERHAPS FOUR YEARS OLD when I realized that I had been born into the wrong body...," recalled Jan Morris (as a child during the 1930s, she was James Humphrey Morris). "I remember the moment well, and it is the earliest memory of my life [sitting under a piano hugging his cat while his mother played Sibelius]."[8] Morris insisted he was neither a homosexual nor a transvestite; for, whereas these individuals merely fantasize changing sex (and would be miserable if they actually did it), he was compelled by a longing to be whole. "Transsexualism ... is not an act of sex at all," she declared. "It is a passionate, lifelong, ineradicable conviction, and no true transsexual has ever been disabused of it."[9] More poignantly, she added (years later): "If I were trapped in that cage [of a male body] again, I would search the earth for surgeons, I would bribe barbers and abortionists, I would take a knife and do it myself, without fear, without qualms, without a second thought."[10] Such dysphoria is no prurient charade.

Other cultures faced with the same bizarre passions but having no knowledge of chromosomes have discerned them through a glass darkly. The Hidatsa Native Americans claim that if a man looks at a coil of sweetgrass in such a way that leads its female spirit to get into his mind, it will cause him to "have no relief until he 'changed his sex.'"[11] Such myths may cipher traditional ethnobotanical wisdom: perhaps the smell of sweetgrass puts those with anomalous hormones into a mild erotic trance.

INTERSEXUALITY, TRANSSEXUALITY, AND TRANSVESTISM often overlap. Transsexuality, as classified by Money, is in fact a psychosomatic variant of intersexuality. While some intersexuals are grossly androgynized in their organs, others can be more subtly androgynous at levels that impart themselves psychically and symbolically but do not manifest in actual body-parts (the Navaho use the term *nadle* to describe interchangeably those with anomalous genitalia and those who "pretend to be *nadle.*"[12]). Some transsexuals are shaped solely by family and cultural factors (without underlying morphogenesis); others may well inherit undetectable aspects of the tissue-hormone substratum of the gender contrary to their anatomy. There is no concrete evidence of their intersexuality, but they feel unequivocally that they have the wrong organs, and the missing gender meticulously if obscurely permeates their psychology. At least to their own perception they are men in women's bodies, or women in men's bodies (men born without full penises and women lacking a vagina and breasts born *with* penises).

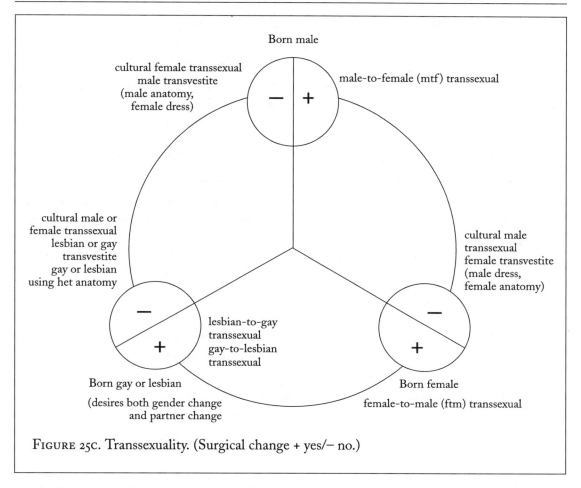

FIGURE 25C. Transsexuality. (Surgical change + yes/− no.)

A discrepancy between hormonal gender and phenotypic body-type would explain why some male transsexuals so loathe their penis and scrotum that they try to disfigure and castrate them.

Pioneering Gender Changes

THE GUINEA PIG OF MODERN SEX ALTERATION was Christine Jorgensen (born George William Jorgensen, Jr., in 1926); although not the first to have his/her gender reconstructed surgically, she went public with her new identity in an unprecedented way. This handsome American G.I. returned from Denmark at the dawn of the tabloid era, transformed (with much fanfare) into a woman, titillating and disturbing the imagination of millions.

From early childhood Jorgensen had fancied himself a girl, preferring needle-point to sports and stationing himself staunchly clothed on shore while the other boys skinny-dipped. Underdeveloped in bodily maleness and genitals, he was also effeminate in gestures and walked and carried school books like a girl. He had a female voice and lacked body hair. His discomfort with having to pretend to be a "man" heightened through his adolescence. When he tried playing the female role in sex with men, "Nothing fitted.... My body often yearned to give, to yield, to open itself, the machine was wrong.... I felt my body was not my own."[13] He became inconsolably depressed.

Mario (Marie) Martino, a female-to-male transsexual born in 1938, confirms the masculine pole of these same feelings: "Any resemblance to lesbianism on [my] part was due to my lack of the proper organs. Never did I use my vagina during lovemaking—always, I attached and wore my false penis. Wanting only to be a man, I went to all imaginable lengths to be one: affecting male attire, male mannerisms and figures of speech, having my hair clipped at the men's barbershops, roughing up my bushy brows."[14]

These are not homosexual or transvestite fantasies; they are "wrong body" clairvoyances. A transsexual born as a male wishes to be a normal woman, to have sex with a regular guy. Likewise, a transsexual female is not a lesbian, but an incompleted man who seeks nothing more than to acquire a wife and family. Transsexuals switch genders also in attempts to gain social and legal protections.

In Jorgensen's era—the 1940s—when homosexuality was acknowledged, if stigmatized, transsexuality was totally veiled. Its yearnings were scornfully dismissed by homosexuals in much the way heterosexuals disdained homosexuality. In fact, from the observations of his associates, Jorgensen concluded he was a mixed-up homosexual with a strange fantasy life. Martino's first lesbian lover ridiculed her claim that she was a man, rebelliously resisting the role of "his" wife. She prided herself on treating "him" as just another old-fashioned lesbian.

In the late 1940s, while reading about advances in endocrinology, Jorgensen took a daring step. Illegally obtaining tablets of estrogen, he ingested enough of them to develop the beginnings of breasts and, more importantly, a feminine euphoria. His destiny was set.

Travelling to Copenhagen, where both medical and social views regarding sex change were more tolerant, he heard for the first time that he might actually not be homosexual; his doctor confided, "I think the trouble is very deep-rooted in the cells of your body ... inwardly, it is quite possible you are a woman. Your body chemistry and all of your body cells, including your brain cells, may be female. That is only a theory, mind you."[15]

Theory or not, in 1951 Jorgensen underwent three sex-change operations. His penis was removed, and he received simulacra of female organs. Then, as Christine, she returned to the United States to cruel headlines and hoopla, doomed ultimately to life as a cabaret performer, obsessed with the show(wo)manship and celebrity of her new body. To herself she was never anything more than a freak.

A decade later, thanks in part to Jorgensen, sex changes were more widely accepted, and Martino had a better chance for a quiet life after surgery. He settled into male existence without any compulsion to justify or display his condition: "By day, whether working, driving, gardening, or relaxing, I sense always the presence of this outward acknowledgment of my maleness. And, by night, my new organ— for all its being less than perfect—is still deeply stimulating to both me and my mate...."[16]

Berdaches

AMONG NATIVE AMERICAN GROUPS, homosexual activity is widespread, clandestine, and tacitly accepted. Transvestites and transsexuals are in fact more distinctly identified and quarantined than homosexuals, for in tribal culture gender is crucial in a way that sexual activity may not be—it is more important *who* someone is than what he or she does. This led to intermediate sex moieties. Cheyenne and Lakota bilingual speakers translate *he man he* and *wintkes,* respectively, as "half-men-halfwomen."[17] The Crow likewise tell informants that the English for *badé* is "not man, not woman."[18]

In general, ethnographers have settled upon the term "berdache" ("woman") for classifying cross-dressing males in indigenous American tribes. A berdache is more than just "woman"; she is a third gender entirely, a male not masculinized (two berdaches would not ordinarily have sex together but, if they did, the relationship would be considered lesbian rather than gay). Berdaches assume female cultural roles, have sex with non-berdache men, and alternate ceremonially between male and female identities. In some tribes "berdaches were reported as being under male ownership. They were frequently found in male social spaces performing activities associated with females during male rituals: fellating powerful men or being anally mounted by them."[19] Similar roles and acts have been noted in other primate societies.

"Berdache" is not a Native American term. It is likely a Persian name "which spread to ... Spanish via Arabic, and from Spain to France. In all its variations, the term refers to the passive male partner in anal intercourse, sometimes with the implication that the person is a male prostitute. Early European observers in America assumed (usually correctly) that the cross-dressed biological males they saw among many groups of Native Americans were sexually servicing other biological males ... who did not cross-dress. This European definition, based on antihomosexual prejudice, ironically dovetails with the agenda of gay male academics who would like to claim the berdaches as [valiant] gay ancestral figures."[20]

The berdaches' female equivalents are "passing women"; these "girls" wear men's clothes, do men's work, are classed unambiguously as men, and have sexual relationships only with women. It should be noted that the term "female berdache" is applied, with equally ironical heroic (or dismissive) overtones, to "passing women" who cross-dress and perform as full-fledged warriors or shamans.

Among the Mohave the third and fourth genders are known as *alyha* (crossing males) and *hwame* (crossing females), respectively. One Mohave man told a visitor that "his *alyha* wives wanted their genitals referred to as a cunnus (clitoris) and became violently angry if male terms were used to describe [them]."[21] (This creative displacement may be erotic verbal role-playing, akin to phone sex in its arousal by provocative mention of body-parts. To call one's penis a clitoris titillates the partner. The imagination of a lesbian lover might equally be excited by a woman referring to her clitoris as "my penis.")

Similar genders occur in other cultures. The Samoan *fa'afafine* (transsexual man) is translated as "the way of a woman."[22] The cross-dressing male *acaults* of Myanmar (Burma) function so fully as women that, in a culture where homosexuality is both tabooed and illegal, men may have sex with *acaults* without being ostracized. *Acaults* are also thought to have mythological powers and bear tokens of the fortune-bestowing Manguedon spirit.

The Anatomy and Psychology of Gender Change

BY THE 1950S MEDICAL REASSIGNMENT was being proposed in some "enlightened" Western circles as the sole viable "cure" for the symptoms of unfulfilled transsexuals, among them: schizophrenia, alcoholism, self-mutilation, recreational-drug abuse, and of course sexual dysfunction. This remediation has since become common. Artificial organs may not replicate embryologically formed ones, but they provide an approximation of maleness atop femaleness, femaleness in maleness. Individuals acquire a previously unavailable option; they can express sexuality by a combination of birth anatomy and new body-parts.

Post-modern sex change encompasses a series of procedures, endocrine and surgical, psychological, and finally cultural.

Hormones are prescribed as a first, somewhat gentle step. Testosterone and its allies drop women's voices, stimulate beard growth, and activate any inherited baldness patterns. Estrogen has the converse effect.

Voice lessons and body-hair electrolysis consummate male-to-female molting.

Morris describes his transformation initially as not so much a feminization as "a stripping away of the rough hide in which the male person is clad ... [not] merely the body hair nor even the leatheriness of the skin, nor all the hard protrusion of muscle [but] ... a kind of unseen layer of accumulated resilience, which provides a shield for the male of the species...."[23]

Body-building and tattoos enhance female-to-male transitions.

THEN THE SURGEON'S KNIFE commits violent embryology. Laminae are squeezed out, grooves cut, orifices opened, tissue clumps refolded and resewn. Clitorises are molded into penises (phalloplasty involves hormonal tissue enlargement underscored by surgical displacement of surrounding folds). Penises are elided and turned inside-out, with a cutting from their head clitorized. Breasts and vaginas are transplanted from ectodermal stem tissue and induced onto appropriate male sites. Artificial scrotums are woven out of labia majora; tissue from the scrotum is used to fashion labia majora.

Medicine and art meet in a resculpting of the human body: "a gaudy, baroque crescendo of doctors and scalpels and stitches and blood which, however good the surgery, still leaves you feeling violated and broken inside somehow and never quite sane in your body again."[24]

"My clitoris has grown a lot," one female-to-male reassures his potentially curious personals-ad respondents. "It's about two inches long when erect. It looks like a very small penis."[25]

A male-to-female shows off his/her new "clit, the one the super-surgeons, who can make almost anything into almost anything else, made by transplanting the very head, the glans, of my beautiful, long ivory pink and blue-veined penis right between my labia"; he then waited "three months for it to heal and the blood supply to stabilize."[26]

Renée Richards (Richard Raskind), the 1970s male-to-female notorious for gaining the right to compete on the women's tennis tour, described her new "clitoris" as being slightly higher than where his penis had been. Her tendency to project and thrust with the organ was diminished; sensation increased as she rubbed it; she noted a satisfying feeling of receiving, of being "moved toward."

In truth, vaginoplasty in men brings with it loss of orgasm, repeated scarring, minimal lubrication, and urinary-tract infections. Functional plastic penises for women are difficult to fashion and implant, painful for urination, limited in sexual feeling, and often aesthetically disturbing.

The psychosomatic effects, while subtler, are happier. Sex-change innovator Harry Benjamin reported that most of his patients were gratified with their new organs, even if they could not achieve orgasm. He interpreted "successful" male-to-female orgasms without a clitoris and with an ersatz vagina as the result of "the longed-for female role in the sex act" and "the possible retention of sensory nerve endings in the scrotal (now labial) fold and also in the penile (now vaginal) tissue."[27]

Desire eludes simple expression or categorization.

SEXUAL RECRUITMENT IS A CACHE OF POSES AND LEGALITIES invented by society to imitate, deconstruct, and symbolically represent animal organs (and fantasies of animal organs). Sexuality can be expressed by gender of a partner, erotic rituals, pure fantasy, or some combination of these. In the wild our forebears chose (unconsciously) what to vamp and whom to fuck, which orifices to enter and which phalluses to receive. Acts of allure then spawned multilayered complexes of pleasures and aggressions, symbols and allegiances, marriage customs and clan taboos, partially originating in but not confined by birth genitals; hence, transcending "mere" hormones and folds. Eros and gender have never been matters of property or propriety. All erotic acts are contaminated with unresolved ambiguities and unacted, incompletely acted, and unactable desires.

Sex changes alter the erotic landscape in nonlinear ways. Anatomy and hormones determine not only how a person expresses his or her sensuality but whom they attract. Some gays and lesbians find they are more drawn to surgical transsexuals than, for instance, their unaltered equivalents (butches, femmes, etc.). Even confirmed heterosexuals begin to notice the subtly transgendered qualities they seek in partners. Some men prefer tomboys; others seek "girly-girls." Women make similar, often unconscious, distinctions between macho and androgynous flavors in men.

Desire eludes simple expression or categorization. Take, for instance, the tangled dilemma of the woman in a heterosexual marriage who requested that male organs painfully and incompletely be attached to her so that she could have "gay" male sex with her husband (whom she desired in only this way). Other men have had themselves surgically altered into women so that they could have sex with women *as women*. "I don't want to make love like men normally do with a woman," one man explained. "I want to make love to a woman as if I were a woman."[28]

When "male" surgical transsexuals find that, once "women," they are attracted to women rather than men, they conclude they were not really gay males to begin with—and also not just women in men's bodies—they were lesbians in men's bodies. It isn't satisfactory to have "mere" natural phalluses; they prefer artificial vaginas and dildos to make different love to essentially the same partners.

What could be the origin of an urgency so powerful as to require disfigurement and the creation of nonmatching organs in order to "work"?

Sex and the Exercise of Power

SOCIAL AND POLITICAL AGENDAS AND ISSUES OF CLASS are inextricably mixed with both organs and experiences. In more primitive eras and societies, sex changes were often brute and obligatory and did not single out those who sought them. The constabulary recruited female males—eunuchs—by the excision of gonads in young boys; this provided new members to fill occupations deemed unsuitable for either women or potent men; i.e., as confidantes to royalty and guardians of virgins. Castrated East Indian males, known as *hijra*, with the loss of their penises became agents of the Great Mother. [It should be noted that pretexts and politics of genital modification differ from observer to observer. In some accounts, *hijra* are said not to be castrated but function only as cultural females (like berdaches); other ethnographers, however, claim a degree of anatomical modification even among traditional transvestite berdaches.]

Constructing and deleting penises cannot escape ideological subtexts. In the case of eunuchs, castration is a means of disenfranchising them as real men; conversely, for some lesbians, surgical penises and strap-on dildos are a way to seize "dick-based privilege.... If the penis is going to be elevated to semidivine status as a marker for the freedom and self-importance that men enjoy," crows one, "I think it's natural for someone who wants to overturn the patriarchy to get behind one of the damn things and see how it feels to drive it."[29]

In the late twentieth century, a combination of feminist sisterhood, valorized matriarchy, and lesbian exclusivity has led to some backlash against male-to-female transsexuals. Some lesbian spokespersons grumble, for instance, that, "All transsexuals rape women's bodies by reducing the real female form to an artifact, appropriating this body for themselves.... The transsexually constructed lesbian feminist violates women's sexuality and spirit, as well. Rape, although it is usually done by force, can also be accomplished by deception."[30] That is, s/he never stops being a man and taking male privilege. Groups of women sharing this attitude outlaw and even evict transsexuals from their gatherings, declaring them men faking womanhood in order to gain female gender power.

Though without an equivalent political agenda, vulvaphobia among gay males has led to ostracism of female-to-males. Not all gay men are misogynous, but many fear and demonize the castration-like vaginal gap and penile "stunting" of women's genitals. One form of sadomasochistic foreplay involves taunting a male partner by referring disgustingly to his putative female anatomy—the opposite of the Native American situation (in part because the Mohave "husband" was in all likelihood heterosexual).

Cultural Transsexuals

MANY TRANSSEXUALS, though they feel they have the wrong bodies, cherish their birth organs, in part because they are *theirs,* in part because their own phalluses have capacity for sensation and orgasm that surgically fabricated ones would not. In place of a sex change they effect a cultural gender contrary to their anatomy. Males delight in dressing and acting as women, becoming women in every subtle aspect other than overt genitalia. They transgender their appearance—their very presence—to have sex with women, to have sex with other men, or simply to enjoy the erotic depths of their performance using both biological organs and fantasy-evoked ones. A Burmese *acault* tells an ethnographer he is a woman only by his sexual role; otherwise, he expresses himself through his penis and its orgasms.

Transsexual females primp and perform as men and adopt male roles in society, whether they cultivate boyfriends or girlfriends. Though female-to-male candidate Mark Rees despised his vagina and desperately craved a penis, he remarked he "had no wish to submit myself to perhaps ten operations, great pain, scarring and risk of infection in order to acquire something which was useless, ugly and without sensation."[31] He may not have been happy with his clitoral phallus, but he was willing to tolerate it, as long as it gave him true pleasure.

Surgical sex change might be less compelling when transsexuality is more psychocultural than chromosomal. It is also possible that, as transsexuality becomes more widely acknowledged and sex change more accessible, concrete demonstration (to oneself and others) loses its claim on people's psyches. An additional factor is that fantasy now has a much freer reign in which to explore and improvise outrageous gender behavior.

Technology and medicine are also far more suspect than they were in the '50s and '60s. Whatever dysphoria they may suffer, people do not want a doctor to make them over. In the wake of "let it all hang out" punk and postmodern body-marking and clothing styles, a new generation of transsexuals would rather invent their own realities with tools they can make and control than let unhip Dr. Frankenstein do it for them. If erotic imaging and theatricality work, why disfigure oneself? Instead of pining for new bodies, these sex rebels create imaginal cross-gender experiences from play-acting through their birth organs. "Sure, my metaphoric dick gets hard," one woman acknowledges, "but when I come, it's going to be via my clit, no matter what the circumstances were that made me ready for it. . . . ; it's my large and very unmasculine nipples that get hard."[32]

Anatomical destiny, libidinal channel, and fantasy place simultaneous differential

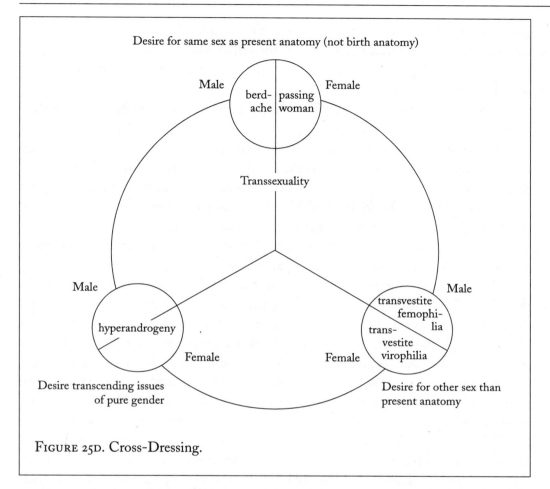

FIGURE 25D. Cross-Dressing.

weights on both one's own body parts (and acts) and one's partner's; each combination of anatomy and rite provides a unique meaning and identity. On this basis fantasized organs carry sufficient charge to transcend or convert anatomy.

The Cultural Basis of Gender

WHEN A MAN DECIDES TO "BE" A WOMAN, or a woman a man, this is preposterous only if the defining qualities of man-ness and woman-ness are circumscribed by base anatomical characteristics. Of course, biological gender is real—reproduction is confirmation of its strict denomination. But the assigned gender roles of societies are not at all coterminous with male and female anatomies or reproductive acts; they represent mostly secondary performances, though most people end up maintaining them their whole lives as if they *were* them. Western society pretends everyone is born either absolutely male or female (depending on genitalia), but the whole drama of gender is substantially a fable based on political, socioeconomic, and aesthetic styles, including forceful and submissive behaviors, fashions of hair, clothing, posture, gait, language patterns, artistic tastes, vocational requirements, etc.

Years ago, Jan Morris wrote, "To me, gender is not physical at all, but is altogether insubstantial. It is soul, perhaps, it is talent, it is taste, it is environment, it is how one feels, it is light and shade, it is inner music, it is a spring in one's step or an exchange of glances, it is more truly life and love than any combination of genitals, ovaries, and hormones. It is the essentialness of oneself, the psyche, the fragment of unity."[33] The "right" biological male can quite effectively enact a woman, much as a born, masculinized woman can portray the ideal man.

After all, arousal is based on more than the anatomy of a partner. Male berdaches and transvestites dress and act as "permanent" women in men's bodies. The degree of their femininity is limited only by the shrewdness of their art. Out in the world they enact female presence so seamlessly that even homophobic men are unwittingly drawn to sexual encounters with them. As female prostitutes, they are usually successful, right down to the john's shock at encountering a full, perhaps erect penis. Even then the illusion is so seamless, the arousal so complete, the transvestite so luminously portraying exactly what a man wants a woman to be, the john continues his self-seduction to orgasm. At the core of his female object is neither a woman nor a man but a woman always, who could also be either a woman or a man.

"Why, in the Peking Opera, are women's roles played by men?" asks David Henry Hwang's character Song—a secret agent in conversation with his controller Chin in the play *M. Butterfly* (loosely based on the twenty-year affair of a French foreign official in China with a local opera star who was both a man and a spy and whom he never saw naked, thus thought to be a woman).

"I don't know," replies Chin, a product of Maoist Red Guard mentalities. "Maybe a reactionary remnant of male—"

Song interrupts her: "No. Because only a man knows how a woman is supposed to act."[34]

This secret knowledge is what allows Song to make himself irresistible to Gallimard, so irresistible that even after learning (and seeing) the truth Gallimard considers her the perfect woman:

"I am a man," he says, "who loved a woman created by a man. Everything else—simply falls short."[35]

IF THE SEXES ARE MORE CLUBS sustaining costume parties than blood lineages with gene-based requirements, then no citizen of a Western democracy legally may be denied membership in the gender of his or her choice. Nonsurgical transsexuals now freely select their sex type, genitalia to the contrary. Females attend Harvard and other prestigious universities dressed as males; they live in male dorms, date sorority girls, and, in general, pass as full-fledged men on their recognizance alone (and to the horror of most alumni).

These transsexuals are not masquerading, but even if they were, they intend to declare the plasticity of culture and nature. When asked whether she was not embarrassed to undress in the shower among the "other" men, a Harvard female (a passing female) remarked that many of her fellow men had even more serious problems with their anatomy (she no doubt meant psychological as well as physical). When questioned as to why she didn't get a sex change, she said that her body was fine as it was. The interviewer then wondered if she wouldn't be discriminated against in life; she replied that she would also be discriminated against for being Jewish, for going to Harvard, etc.; society is full of discriminations. . . .[36]

When a woman can openly participate as a male (while still anatomically a female), the secondary aspects of gender become obvious—i.e., that some women are more masculine in certain ways than many men and adapt to male cultural roles better than they do. According to the Harvard woman's lover (avowedly not lesbian, and quite on the lookout for a suitable male partner), her female lover was a very attractive male, that is, "he" pleased her more than any other available partner. Clothes, speech, mannerisms, and overall imbuement engendered (literally) a male performance so effective that any mere quantitative lacks (i.e., too small a phallus, large breasts) were inconsequential.

CULTURAL TRANSSEXUALISM may yet turn out to be a fad; it is too soon to know how real it is, how deep it goes, how ultimately satisfying its rituals. Yet the unscrolling of the interstices of sex and gender will continue at breakneck speed as long as culture, humanity, and the ecological basis of life are in mortal danger. We clearly cannot survive in nature if we deny nature. The society that generated this crisis will be challenged at every portal and layer, as symbol-bearing animals seek their lost realities and claim new faith from neglected and vilified domains.

"Something outside had to enter," proclaimed science-fiction prophet Philip K. Dick, "something which we ourselves would be unable to build."[37]

"The Impossible attracts me," sang Arkestra Pharaoh Sun Ra, "because everything possible has been done and the world didn't change."[38]

The Landscape of Neo-Puritanism

OUR PRESENT SOCIAL COMPACT IS TRAPPED between two competing moral designs.

On one side are those who believe that America is a resurrection of Eden, divinely sanctioned to rule over humankind and impose the kingdom of heaven on Earth. The partisans of this liturgy revile their opponents as scientific humanists, pagan New Agers, satanic witches, Darwinian nihilists, and godless relativists. Their puritanical cabal in the West has relegated most sex acts to "temptations of the devil." Even oral and anal copulation and adultery between consenting men and women must be carried out somewhat furtively. Multisexual expressions are more deeply tabooed; exposure of such deeds (homosexuality, sodomy, fetishism) guarantees a slide down the social scale, a loss of gender rights and privileges—and even of freedom or life itself. Sexually deviant behavior so enrages certain civic officials and moral vigilantes that they incite judicial or murderous attacks on the perpetrators. Prostitutes, gay males, and cross-dressers are self-enfranchised enforcers' most frequent targets.

The family-based Puritan definition of self would limit expression of erotic desires to procreation within patriarchal marriage—pleasure a mere contingency. Sexual congress for any purpose other than confirmation of the nuclear family is considered illicit—in disobedience of sanctions so antediluvian as to seem divine. In the Baptist canon marriage is "the exclusive, permanent, monogamous union of one man and one woman.... The perversion of homosexuality defies even childbirth, since it negates natural conception."[39]

Advocates on the other side regard this as priggish, intrusive, and neurotic to the point of fanaticism. They cite a primal libidinal drive, irresistible in its expression and not prescriptive to marriage or even heterosexuality. By this view, homosexuality, transsexuality, and intersexuality, as well as most of their variants, are ancient and authentic phenomena, as old as our species and as intrinsic as cell life.

Since the beginning of culture, gay and lesbian sex have likely been routine, even sanctioned, as alternative modes of expression with their own social meanings. Early Greek and Roman (and quite probably Pithecanthropine and Cro-Magnon) heterosexuals engaged openly in same-sex dalliances. Men did not suffer loss of stature for putting their tongues and genitals in other men's mouths.

REGULATIVE OBEDIENCE CASTS SHADOWS from which arise its antipodes. Oppressed and depreciated aspects of consciousness inevitably assert themselves—the more prohibitive the force, the more unruly the compensation. Though cultural control and censorship cannot obliterate desire, they distort and pervert its manifestations. The most debauched sex crimes are often committed (ironically) by those who have most deeply inhibited their inklings in themselves—not only moralists of self-proclaimed abstemious bent but their rivals who indulge and experiment with "wild side" fantasies at the expense of their actual feelings.

In recent years, "nonhets" in legion have come "out of the closet" and declared their acts legitimate expressions of innate desires. Appropriating the runaway economy and cyberpunk culture of the late twentieth century, they explore forbidden domains and publicly confront demons that have (in other times) led to sadism and carnage in place of eros.

The liberation of sexuality from reproduction and reproduction from sexuality have helped catalyze a disintegration of the Puritan family. A concurrent populist campaign seeks legalization of marriages between two men or two women. A new social-sexual landscape has brought with it an increase in communal families; single-parent households; gay and lesbian partners with children; as well as sex, abortions, nonhet and drug experimentation at ever younger ages, and an epidemic of runaway youth (even from the "best" of families).

In addition, gametes have been donated from person to person, sold on the open market (for instance, by college students needing cash), and even auctioned on web sites, with the result that some gestational carriers give birth to children who will not only never see their father but never know if their father even knows they exist. Sperms have also been harvested post-mortem from the newly deceased corpses of relatives and, since 1997, frozen oocytes have been transferred between women.

The sexual/reproductive revolution has occurred side by side with a worldwide increase in use of cocaine, heroin, and other addicting substances, plus an affiliation of teens and even pre-teens in criminal gangs, private armies, and marauding paramilitary militias without concern for life or property and beyond civil or governmental control.

Latter-day Puritans preach that our profligacy will lead not only to horrendous crime waves but sins worthy of the wrath of God and Endtime (as at Sodom and Gomorrah when He confided to Abraham: "... their offense is very grave"[40]). In defense, they have launched a new Jeremiad—their own version of a jihad—to redeem the United States and stave the collapse of civilization.

The impetus behind their concurrent right-to-life crusade is not any great love for embryos (and certainly not for the people those embryos will become, many of whom the same Lambs of God would banish to poverty, AIDS, or the electric chair without a second thought); it is the assertion that the embryo is God's covenant with humankind and its prerogative belongs to him not us. In aborting fetuses, doctors and their patients are seen as committing blasphemy, striking an insolent blow against the Creator. The moralists hate abortion doctors more than sex offenders and murderers because, while the latter are conspicuous disciples of Satan (often acknowledging as much), the former are emending the moral and legal requisites of society, replacing one God with another—a vindictive military judge with a weak secular prelate.

The advocates of Jehovah brook no such indulgence; they want the authorities to maintain strict celestial rule over our satyric passions and humanist predilections; in essence, to protect God's innocents from our depravities. Yet post-modern vamps rebel in the name of Antonin Artaud:

"My cruelty is not synonymous with bloodshed, martyred flesh, crucified enemies. Rather, it is an appetite for life, a cosmic rigor and implacable necessity, in the gnostic sense of a living whirlwind that devours the darkness."[41]

IN 1998 PRESIDENT BILL CLINTON's erotic escapades with a White House groupie may have seemed to an urbane majority tame and commonplace and at worst tawdry, the commerce of a petty adulterer (or a sex addict with a medical rather than moral problem); yet to family-values conservatives his behavior was a foreshadowing of Armageddon, a defilement of the Oval Office. Clinton became their signal example of the triumph of the sodomites. He had desecrated the Holy Book, the Constitution of the United States, and his diabolic crime merited impeachment, conviction, and imprisonment.

Clinton himself was boxed in between his sex-negative taboos (from an old-time Southern deity) and his urgency to explore strange desires (pagan global-economy dryads). His much younger partner, Ms. Lewinsky, came from a different generation and milieu; unconflicted on this issue, she spoke shamelessly of wanting "to play ... to mess around."[42] Her belief in a "sexual soul-mate"[43] different from a householding partner was as ordinary and inofficious to her as joining a sorority or putting on pearls and a beret. By her world-view, passion overrode all proprieties and commandments, including those of the independent proconsul and his confederates.

Genderqueers

THE MORE THE AUTHORITIES TRY TO TAME our mystery-enshrouded origins, the more the jungle re-vegetates in sigils, costumes, and ceremonies. In San Francisco's sex clubs, "most of these men [are] young, muscular, marked with tattoos or piercings. The leather they [wear] and the metal and ink they put on their bodies [are] not fashion statements, they [are] brave declarations of difference and affirmations of a passion for pain, power, and extreme degrees of penetration [—even when] fated to become clichés...."[44]

Insofar as each person's path is unique, gender is a shifting kaleidoscope. From beauty shops, fashion parades, tuxedos, and perfumes; through tangos, waltzes, rap, snowboarding, and Rolling Stones; automobiles, jewelry, make-up, midriffs, navel-piercings, blouses tied in knots, plastic noses, T shirts, etc., boy-girls and girl-boys project and respond to each other via an endless parade of erotic gestures and tastes, transcending age, class, ethnicity, and taboo. There are gender-subtle differences among women with hair stylists, women who just brush their hair, women who cut their hair short, and women who shave their heads. Similar subtleties pervade every aspect of adornment, marking, disfigurement, and personal carriage and style.

Are not all embellishments—anatomical, cosmetic, artifactual—attempts at refiguring gender?

In the United States the majority of customers for ornamental mammary surgery choose enhancement by silicone implants; in Brazil most prefer breast reduction. In Rio de Janeiro itself, where there are more plastic surgeons per capita than anywhere else on Earth, the father of modern cosmetic surgery, Dr. Ivo Pitanguy, has turned body-sculpting into recreation for the masses. His clients return month after month, year after year, for a touch here, a touch there—grafts, facelifts, eyelid lifts, eyebag removal, lipos, tucks. The embryogenic fluidity of the body is reenacted in fashion statements and artforms.

Transmodern sex epicures treat medical gendering as little more than another designer drug, body-piercing, fey carnage, or satanic tattoo; they submit to the knife as to a beautician. Artificial organs are their sadomasochistic props, bad and "rad" and more than a little bit spooky (in candlelight). In lieu of scalpels and stitches, other partisans don opaque painted faces, flamed-out hair colors, blends of Halloween and anachronistic drag, a style foreshadowed in David Bowie's "Spiders of Mars." They aim to embody not just transvestism but ambisexuality; not only androgyny but hyperandrogyny.

Apocalyptic cyberpunk or futuristic rock burlesque?

It is impossible to assess how much of this is real and deep-seated and how much of it represents post-modern theater of the body—ideological rage, apocalyptic ceremony, and anti-establishment art.

THE SAME ENIGMA strikes equally at the heart of Maori full-body warrior tattoos, Chinese feet-binding, Ubangi cranial and facial moldings, and even medical prostheses (a current fad among some disabled people aims to eroticize wheelchairs and artificial limbs as seductive devices in their own right).

A French professor and performance artist using the name Orlan undergoes repeated public plastic surgeries, "transforming her face into a composite of the icons of feminine beauty. She tracks the relationships between sexuality and pathology, between the female body and the body politic.... [S]he is in the process of acquiring the chin of Botticelli's *Venus,* the nose of Diana, the forehead of *Mona Lisa,* the mouth of Boucher's *Europa,* the eyes of Gérôme's *Psyche.*"[45] Orlan is interested in the stories around these women, for instance, the idea that *Mona Lisa* was Leonardo himself in drag, or the escapades of the mythological Diana. Cheek implants above her brows give her a *Star Trek* look. Her aim is not to become beautiful but "to reveal that the objective is unattainable and the process horrifying."[46]

She is going beyond beauty, then beyond gender, both betraying and liberating the rationales sustaining culture itself. Her mutilations are not in quest of a pleasure principle but a deconstruction of the fake ideals that oppress people, the myths of attractiveness that addict them to unrequitable desires.

Riki Anne Wilchins, editrix of *In Your Face,* calls upon all genderqueers to join the struggle against rigid sex roles: "diesel dykes and stone butches, leatherqueens and radical fairies, nelly fags, crossdressers, intersexuals, transsexuals, transvestites, transgendered, transgressively gendered ... and those of us whose gender expressions are so complex they haven't even been named.... Gender oppression affects everyone: the college sweetheart who develops life-threatening anorexia nervosa trying to look 'feminine,' the Joe Sixpack dead at forty-five from cirrhosis of the liver because 'real men' are hard drinkers."[47]

Kate Bornstein, a male-to-female transsexual and self-defined gender outlaw claiming to be no longer a man and "not a woman either,"[48] declares that gender does not exist and that categories of male and female must be fictions; otherwise, genes will necessarily determine who we are as people, and biology will be used to control the wildness of women (men also). She rejects the motto that transgendered men and women achieve their true sexual identities after reassignment and laments that after "hiding deep within a false gender..., after much soul searching, [they] decided to change their gender [only to] spend the rest of their days hiding deep within *another* false gender."[49]

Jess, the hero of Leslie Feinberg's novel *Stone Butch Blues,* cries out: "I don't feel like a man trapped in a woman's body. I just feel trapped."[50]

A friend tells Minnie Bruce Pratt, Feinberg's lesbian-femme lover: "You are not only a lesbian, but very, very queer. You love a woman who is manly, and yet do not want her to be completely man. In fact, you desire her *because* she is both."[51]

She responds: "When I unknot your tie and unbutton your shirt, as we lie together naked, I say with a fearless caress that I love the man I am undressing, and I also know that a woman lies beside me, not a mirror to reflect me."[52]

Men are not something other than women. They are slightly different inductions of underlying womanly fields. They are women after sex changes in the womb, in male cellular drag. No wonder heterosexual females find them cute; no wonder men find women ravishing. No wonder gay males are attracted to men, lesbians to women. Their differences and samenesses overlap and interpenetrate; they potentiate and induce each other even as they did in the blastocyst. They see parts of themselves in each other and experience parts of the other in themselves.

"I do not want to be a woman," concludes the transsexual journalist Michael Thomas Ford, "I love inhabiting a male body. I like the way it moves and smells and responds. I love having a dick and feeling it hard in my hand, or feeling it slide into a warm mouth or asshole. I love the feeling of a cock pushing its way into my ass. I love coming, and the way my load splashes over my belly and sticks to the hairs of my forearms. I especially love the way my body feels when it touches another man's, both familiar and alien at the same time.

"No, this is not about wanting to be a woman. It's about wanting to be free from the boundaries created by expectations, roles, and fears, and even from the limitations of my own genitals.... I like it when a man begs to put his mouth on my pussy. I love it when he tells me he wants me to bend over so he can fuck me from behind. I come with him when he can't hold back any longer because of what I say to him, and he loses his load all over my tits.

"Yes, I am a real girl. And yes, I am a real man."[53]

IT DOESN'T TAKE a sorceror's potion and a "midsummer night's dream" for men and women (like Queen Titania) to become infatuated with the equivalents of Bottoms crowned with asses' heads. "Sex is not about body-parts. It's about the erotic energy that happens between two people."[54] That energy always carries elements of both maleness and femaleness as well as other less definable aspects, but it ultimately transcends all of them in the individuality and specificity of the eroticized other.

Humanity, culture by culture, has opened up mysteries nature kept invisible to its creatures by making them part of their bodies. We have exposed the domain of self and other, of desire and consequence, and have thereby accoutered nature's lewd immaculacy, illuminating it through language and image. But there is always an element that resists knowing, lying in the bottomless gap between meaning (in the form of erotic imagination) and fate (in the form of blind cellular habit).

"When you caress another body in the dark," writes Steven Shaviro, "the differences are so precise and immediate, so subtle and numerous, as to defy classification. What is the exact angle of this thrust, what are the specific contours of this caress? Where, on my skin, in my nerves, in my brain, do I feel this particular tingling? Who is to determine whether these curves on my chest are large enough to be called breasts? Or whether this swollen appendage is a clit or a cock? I can't even say that this body is 'mine' any longer. . . . You could imagine this touch, if you insisted, extending onto the body either of a man or of a woman, or even of some other, alien being."[55]

The new reality emerging is very much like the fragmented, recursive reality that underlies billions of years of evolution on Earth, or on planets of the Orion system.

Part Five

APPLICATIONS

26

Self and Desire

IN THE SUMMER OF 1983, while hiking in the Peruvian Amazon, Will Baker was invited on a hunting trip by two Asháninka Indians, Carlos and Cuñado. Soon after they began tracking, a monkey couple ("their faces small and old as time,"[1] Baker writes) came through the trees to look at them. Cuñado nocked his arrow, drew, and fired. The female started as the shaft entered her small body; her fingers fondled its hardness, and she dragged herself back and forth, uncomprehending. The unsuspecting male ran up to her and pulled at her shoulder, trying to hurry her away. Cuñado's next arrow pierced him, and he bolted from her aid and pinwheeled through the branches. The hunters "hoot at this slapstick agony, this silly tale of fidelity."[2]

In 1761 Georg Wilhelm Steller described an instance of loyalty on the battlefield among sea cows attacked with harpoons by Russian sailors:

"When an animal caught with the hook began to move about somewhat violently, those nearest in the herd began to stir also and feel the urge to bring succour. To this end some of them tried to upset the boat with their backs, while others pressed down the rope and endeavoured to break it, or strove to remove the hook from the wound in the back by blows of their tail, in which they actually succeeded several times. It is most remarkable proof of their conjugal affection that the male, after having tried with all his might, although in vain, to free the female caught by the hook, and in spite of the beating we gave him, nevertheless followed her to the shore, and that several times, even after she was dead, he shot unexpectedly up to her like a speeding arrow. Early the next morning, when we came to cut up the meat and bring it to the dugout, we found the male again standing by the female, and ... once more on the third day when I went there by myself for the sole purpose of examining the intestines."[3]

The expedition massacred so many of these animals that they were extinct within twenty-seven years of their discovery.

To most, this was the only crime.

THE SLAUGHTER OF INNOCENTS now reaches epidemic levels—38 million cows and calves, 92 million hogs, 4 million sheep, and 7 billion chickens (reported in a 1998 issue of *Time*) plus hundreds of thousands of rabbits, ducks, and other denizens butchered for food in the United States alone in 1998; uncounted more creatures slaughtered to make shoes, belts, wallets, fur coats, and the like. Mindless macabre franchises like Kentucky Fried Chicken and Burger King trivialize the wretched course of life and death for "children" born under their regimes. A more select group is maimed in experiments: guinea pigs injected with carcinogens, cats lobotomized, monkeys made to run on treadmills until they drop from exhaustion—but the treadmill never stops.

To hunters and scientists, animals are different from us in a way that justifies their going as far as they want, without doubt or remorse. Their subjects are props; they are not sentient; they do not suffer "real" pain:

"He saw it emerge from a pile of dead brush into full view, where it posed for one second in the crosshairs, a full-grown massive male deer holding itself absolutely still, ears like dark velvety leaves, white flag of a tail switching, large liquid eyes brushed by long lashes and soaking in as much visual detail as can register in the animal's brain, wet nose searching the breeze for scent that is not tree bark, pine needle, resin, leaf, water, snow, hoof, urine, fur or rut."[4]

This world is a domain of violence, for sure—bodies hurtling through glass, against steel, crushed between tin motorized vectors, hemorrhaging with underwater gases, lacerated and spilling iron. Even in times of peace, we live among carnage.

The hunter focuses through his scope on the life form; a brief flash, a machismo explosion ... the invisible thread snaps, his flick of neurons terminating the biological field in a mound of spasming protein. The deer is no more.

All slaughter—on the battlefield, in the backs of meat markets and animal shelters, from pest exterminators' nozzles, by guns or knives during robberies; by political assassination, firing squad, suicide bomb, missile, tooth, claw—converts protoplasm to cells, then to the molecules of which they were concatenated. For all the apocalyptic drama, the rituals that embroider the outcomes with trophies, slogans, flags, funerals, vendettas, or price tags, the result is a fermenting mound of impalpable ashes, to be commoditized or disposed of, by humankind or nature.

"... all across the hills and valleys, up and down the gullies and over the boulder-strewn ridges and cliffs, from up in trees and hillsides, overlooks, bridges, even from the backs of pickup trucks, out of brush piles, over stone walls, behind ancient elms — throughout the hundreds of square miles of New Hampshire hill-country woods — trigger fingers contract one eighth of an inch and squeeze. There is a roar of gunfire, a second, a third, then wave after wave of killing noise, over and over, sweeping across the valleys and up the hills. Slugs, pellets, balls made of aluminum, lead, steel, rip into the body of the deer, crash through bone, penetrate and smash organs, rend muscles and sinew.... Huge brown eyes roll back, glassed over, opaque and dry; blood trickles from carbon-black nostrils, shit spits steaming into the snow; urine, entrails, blood, mucus spill from the animal's body: as heavy-booted hunters rush across the snow-covered ground to claim the kill."[5]

OUR SOCIETY IS NOW STAGING an acrimonious showdown between those who uphold the reproductive rights of women and those who defend the inalienable right of fetuses to life. The freelance executions of abortion providers testifies to how strongly some vigilantes oppose the artificial termination of pregnancies. But what about the millions of impoverished children (and animals) who perish innocently each year so that others can maintain a higher standard of living, can consume them and their goods? What about "a world order that commits planetary suicide in a search for profit while driving the majority of human beings into despair and poverty"[6]? Do the anti-abortionists oppose destruction of the rainforest and the coral reef? Do they oppose war and capital punishment? Are they against implements of torture sold to any junta for drug or oil dollars? Do they think the universe or, for that matter, God himself, feels any less pain for the butchery of the wolf or giraffe than for the slaughter of the fetus? Are not the cow and the pig fetal souls impregnated with neurons?

What witness do they bear for baby chicks tossed by the thousands into a dumpster at the back of a Delaware hatchery, "left to slowly suffocate or die of exposure?"[7] Or those who survived the hatchery and ride the trucks in cages?

"These white feathers just keep blowing in the wind along the thruway. The chickens are headed for one of the many New York City chick slaughterhouses ... the chickens are packed in very tight. The noise of the traffic, the heat, and the speed must be unbearable to them. These birds have been kept in a darkened room their entire lives, and now they are here in the middle of the thruway. Some chickens have managed to squeeze their necks through the slats of the crates. Their eyes are looking out, unaccustomed to light.... There are bizarre distortions, a limb twisted here or there, beaks agape....

"The chickens in the crates never got to stretch their wings. Do they feel pain, did they suffer? Yes. And now their feathers are torn off in the wind to fly all over the thruway."[8]

What about the stunned veal calf, hanging upside-down and chained by the legs, its tongue dangling out of its mouth before its throat is cut, and the next calf swings down the line?

CONQUEST AND MURDER have been the rule here since the beginning-time. Tribes once executed their young to keep families small; they exterminated other lineages for a spring, a valley, or a hunting ground. The United States exists solely through wresting land from its prior occupants, a systematic annihilation of aborigines—squaws, babies, embryos and all.

Saddam Hussein's rockets snuffed out not only the homes and lives of the Kurds in whose towns they landed but their very DNA. Now their progeny are born with incomplete faces and twisted limbs, incurable malignant sores; they have no life, no future, no seed; their race is over. Mustard gas and Sarin provide a sterile, antiseptic genocide, computer-packaged, computer-delivered. Post-Hiroshima, post-Chernobyl, regions of our planet are rendered toxic and uninhabitable simply by an idea.

All that separates North America's Christian Fundamentalists from Israel's right-wing settlers or the Iraqi Baath mafia and Taliban of the Afghan hills is a superficial difference in fundamentalist editions of the same holy book. Their genocidal patriotisms, patriarchies, allegiances to vengeful prophets, and phobias regarding women and sex are almost identical.

"JERUSALEM [February 2, 1999] — In an ugly confrontation, 100 orthodox yeshiva students surrounded a group of American Reform rabbis who went to pray at the Western Wall yesterday morning. The students booed loudly and hurled insults past officers from the border police.

"What was most chilling to the Americans was that the youths, their faces contorted in anger under their black hats, screamed that the rabbis should 'go back to Germany,' to be exterminated, one explained later."[9]

Tribal moralism is no newcomer to this plane: In July 1099, "For three days the Crusaders slaughtered the Moslems,/Men, women, children. The Christians/Waded up to their ankles in blood. The Jews/Were burnt in their synagogues.//Seventy thousand Mohammedans/Were put to the sword. Within days/The infection from the masses of bodies and gore/Produced a wave of pestilence/Biblical in its power and repulsion, yet/Even so, less than the preëmptive AIDS/Of Sodom and way prior to the dark Ebola./The savagery was Ruandan and Ugandan.//And then, bareheaded and barefoot,/The humble conquerors ascend/The Hill of Calvary, walking in effect/In their final procession,/Loud chorus of anthems from the priests."[10]

AN UNFORGIVING MATHEMATICS presses against all goals of preserving life. Without conflict, nature would be overrun with creatures. Even after two world wars and numberless plagues, floods, famines, and desertifications, *Homo sapiens* threatens the biosphere by its numbers. Infanticide and genocide mete a *danse-macabre* through millennial time.

But to say that the embryo is not alive, or not human—not yet sentient and thus not murderable—is another form of gratuitous deceit.

It is impossible to abort the fetus without killing the person incarnating there.

"Pro-choice" means nothing if it does not include the unarticulated choice of the living soul in the embryo—its hunger for life, its desire for unfoldment. A court of law may rule that the embryo has no rights, but its cells and gathering consciousness fall outside jurisprudence.

Courts cannot redress the biological injustice arousing male and female genitals and making women alone the carriers of the zygote. These roles are ancient beyond adjudication; they are not an unfair allotment or capitalist exploitation; they are an unsolved riddle and an opportunity, as is life.

IT ALL HAPPENED LONG AGO. Even the elegant bards of the Aborigines tell us it was at the dawn of time, or before time. The mythology of the deed echoes through generations: "The men of our Dreaming committed adultery, betrayed and killed each other, were greedy, stole and committed the very wrongs committed by those now alive."[11]

The act goes beyond the haberdashery in which it has masked itself. It is a transformation rite happening neither here nor elsewhere, neither in time nor outside of time, involving events indecipherable as such, but fundamental to the epode of life:

"I saw the soul of a man. It came like an eaglehawk. It had wings, but also a penis like a man. With the penis as a hook it pulled my soul out by the hair. My soul hung from the eagle's penis and we flew first toward the east. It was sunrise and the eaglehawk man made a great fire. In this he roasted my soul. My penis became quite hot and he pulled the skin off. Then he took me out of the fire and brought me into the camp. Many sorcerers were there but they were only bones like the spikes of a porcupine.

"Then we went to the west and the eaglehawk man opened me. He took out my lungs and liver and only left my heart. We went further to the west and saw a small child. It was a demon. I saw the child and wanted to throw the *nankara* (magical) stones at it. But my testicles hung down and instead of the stones, a man came out of the testicles and his soul stood behind my back. He had very long *kalu katiti* [skin hanging down on both sides of the subincised penis] with which he killed the demon child. He gave it to me and I ate it."[12]

Everything has changed. Everything was once something else. Death camps splatter the face of history like raindrops on glass. There is no name for them. There is no tribunal. They are a shadow masked by bare rudiments of a ceremony . . . even after the cultural revolutions and so-called police actions of the late twentieth centurion. And if we hope to find the evidence or explanation for it in the embryology described in these pages, we will be as sorry as those who look to the tattered documents and (now) the videotapes of history.

In this world, we are the animals. For sure and for certain. We may kill them, eat them, ignore them, or judge ourselves above them—but we *are* them. Our gravity well is a temple, blue skies keeping in the ceremony. We are priests, Aztec in our famous cruelty, Aztec in our clarity. We carry out finitude and law. And yet they are finished and complete in other kingdoms, seeing planets we have lost forever (or will never see).

The animals do not have personalities as we do. They bear no malice. They are there until the absolute last moment; then they are not.

It is wrong to think of us as the bane of the animals. We are their completion, their ritual. They did not intend us. Yet apparently they could not have quarantined forever the symbol and the name.

We suffer consciousness that they may be fleet and light.

We judge so that their ferociousness and hunger go unabated.

We dream, and they are dreamless night.

We make text. Their bodies and footprints lie in the margins.

We make language; they are outside language, in the old speech.

Everything we do—our cities, billboards, poems, wars, machines, our roosts in which they build nests—they allow.

Even Roger Miller singing "King of the Road" is the ceremony of the animals. Raccoons and starlings, fish in the river the melting snows feed, fly buzzing on the screen, *"Trailers for sale or rent/Rooms to let fifty cents."*

We inherit extensions of the brief flares of awareness that enable a crab to find food, to awaken to hunger in the flux of chemicals it constellates, to extend its claw and scrabble across sand. Our images, instincts, desires, and rituals come directly from our animal heritage (if not from species presently on the Earth). Still something stands between us and them. Despite the development of the human psyche by no more than incremental degrees of cell-stuff, there is an uncrossable gap.

We can describe a dragonfly's activities, but we cannot imagine the wingspread it feels along helicoid cuticles. We cannot know the rush of currents against a fish or a whale's giant song.

A spider looking at another spider sees the same thing we see when we look at each other—another human being (to universalize our term). If we were somehow reincarnated in their bodies we would find spiders as irresistible as we now find lovers of our own species.

A LABORATORY CHIMPANZEE is raised in a human family, a child among children. When funds for the experiment run out, the animal is taken to a zoo. There he sits, behind bars, crying, wondering why he has been put in a cage with apes. To himself he is a child with fur.

In a science-fiction story, Pat Murphy describes a girl Rachel who died in an automobile accident but whose brain imprint was transferred by her scientist father to a chimp.

"Sometimes, when Rachel looks at her gnarled brown fingers, they seem alien, wrong, out of place. She remembers having small, pale, delicate hands. . . .

"Rachel remembers cages: cold wire mesh beneath her feet, the smell of fear around her. . . . Rachel remembers a junior high school dance where she wore a new dress. . . .

"She is a chimp looking in through the cold, bright windowpane; she is a girl looking out; she is a girl looking in; she is an ape looking out."[13]

After Rachel's father dies, she is captured and placed in a lab. There she startles the attendant with her ability to use sign language:

"Please, please, please. Help me. I don't belong here. Please help me go home . . . I am not a monkey. I am a girl."[14]

AN AMOEBA BRINGS A QUANTUM of sentience into the world. Compared to sponges or jellyfish, worms are intelligent bionts. Compared to worms, snails and insects are virtual philosophers. The octopus and salmon are beginning to individuate; they have inklings of personalities. In myths and fairy tales amphibians and reptiles let us speak on their behalf.

Warm-blooded animals dream in sleep as we do; they probe with a slightly detached curiosity foreign to snakes and frogs. Bears, seals, dogs, horses, mice, cats, and birds are "people"—people in their own classes set off from the royal lineage.

Monkeys and apes live at the boundary of our condition. After the fact they look like unfinished replicas of us. Like men and women they live in groups; they gossip; they play.

In some cultures, mammals and birds even have greater social standing than many humans. And through castes, slavery, and the taking of prisoners, people can be made into "animals."

THE WONDER OF CONSCIOUSNESS is that it seems to arise from nature, yet to have nothing to do with cells; that it is grounded in predation, yet gives rise to compassion and justice.

Watch the water beetle dive! Though a teetery toy with rubber legs, it is a swift carapace of death for the tadpoles it catches, rips apart, and gobbles.

It is not cute; it is not quaint—its cuteness is horrific. It is what this world is, bottom line—yardstick for creation.

We are cells. The beauty to which we are drawn in another is cell-bound. Political power and charisma are metacellular. Prayer is a bobbin of cells. So is compassion.

Cells eat before they do anything else; yet somehow that voracious deed, transferred phylogenetically into psyche, becomes a parable of a peaceable kingdom in which a lion lies down with a lamb.

What system anywhere in the cosmos has engendered a greater reversal?

WE HAVE THE ILLUSION of having established a safer domain, one that holds wild beasts, like diseases, at bay. We now invent our own horrors far worse than any suffered by animals, or by us as animals. Electronic images and remorseless titillations, an omnipresence of loneliness, a specter of cosmic (or urban) vulnerability, a premonition of the black horse, have all replaced the dismembering of the hare by the fox.

Terror has become the Word (so we experience pain in its absence, almost continuously, and die many times before our death), but the Word is also an elixir, and we can be enlightened and transformed.

A Chinese landholder takes his treasures out to sea and dumps them overboard as an offering to the Dragon King. A boy studies judo for twenty years; then serves the poor. Generations of meat eaters reflect on their heritage and adopt vegetarian diets. Yogis sit in dank caves, trying to look past the dross to the source of mind.

Vietnamese monks setting themselves aflame, kin of South African murder victims publicly forgiving the executioners at councils of Truth and Reconciliation, nuns risking their bodies to care for lepers and AIDS patients: how do these acts arise from cell predation?

WE JUSTIFY OUR WARS, our brutalities, our predation, and our greed by extenuation to nature. "The territorial imperative," wrote Robert Ardrey, "is as blind as a cave fish, as consuming as a furnace, and it commands beyond logic, opposes all reason, suborns all moralities, strives for no goal more sublime than survival."[15]

Animals are hardly pacifists, but they are rarely as arbitrarily cruel as we. The tenderness of the crocodile mother with her babies stands against rampant child abuse among the humanoid species, not in every instance of reptile and *sapiens* but as a measure of innate capacity for humane behavior.

"When man does not admit that he is an animal, he is less than an animal," proclaims Michael McClure. "Not more but less."[16]

WE ARE NOW ENSLAVED by lusts that trivialize our capacity for feeling. Our acts of domestic violence and war are pathetic exaggerations of the real suffering of sentient beings. Recreational hedonism numbs us to our real desires. We are caught in a desperate rush to grab hold of everything, in an illusion that we must not be shortchanged or denied. Yet even the most aggrandized orgy devolves instantaneously into petty consumerism. Steven Shaviro exposes the new hegemony:

"[T]here's no getting around it: 'To speak is to lie—to live is to collaborate.' The only way out is the same way we came in. . . . One fix after another, one purchase after another, one orgasm after another; for there is no end to the accumulation: 'the lonely hour of the "last instance" never arrives.' All we can do with words and images is appropriate them, distort them, turn them against themselves. All we can do is borrow them and waste them: spend what we haven't earned, and what we don't even possess. That's my definition of postmodern culture, but it's also Citibank's definition of a healthy economy, Jacques Lacan's definition of love, and J. G. Ballard's definition of life in the postindustrial ruins . . . orgies of endless consumption, forever postponing the moment when the bills come due . . . S&L scams for the rich, Visa and Master Card financing for the middle class, and even occasional riots and looting for the poor."[17]

IT IS AS IF A VENAL STROKE against God alone could bring Him back to life, and only the exploration of every craving and morbid fantasy in the most explicit way could lift from us the oppression of consciousness.

We look back through Jean Henri Fabre's eyes with a sense of portent as well as wonder:

"In the course of two weeks I thus see one and the same Mantis use up seven males. She takes them all to her bosom and makes them all pay for a nuptial ecstasy with their lives. The male, absorbed in the performance of his vital functions, holds the female in a tight embrace. But the wretch has no head; he has no neck; he has hardly a body. The other, with her muzzle turned over her shoulder, continues very placidly to gnaw what remains of the gentle swain. And, all the time, that masculine stump, holding on firmly goes on with the business!"[18]

Desire begins in the cells of the mantises and continues past them into the shadowed depths of their materialization. It simply happens. It comes into the world through the germ plasm and is expressed in flesh. Anatomy and destiny recede to a point where they merge and disappear.

"A headless creature, an insect amputated down to the middle of the chest, a very corpse, persists in endeavouring to give life. It will not let go until the abdomen, the seat of the procreative organs, is attacked."[19]

If these were crimes, the dragonfly and the shark would be imprisoned and the tiger and the vulture should go hungry.

Mantises can do nothing about it, but we who stand facing each other with missiles and uranium warheads both allow and condemn it — snuff films, S&M nightclubs, kiddie porn and prostitution, piracy and rape, child brothels and boat-people, lockdowns and concentration camps, thrill kills and death squads, torture and mutilation, electrodes attached to genitals and roots of teeth, castrated organs stuffed in the mouths of Bosnian prisoners about to be mowed down by machine guns and bulldozed into a pit, millions inhabiting the streets and garbage dumps of the Solar System's great cities, New Delhi, Chicago, Lagos, Mexico D.F. . . . others locked in mortal combat.

"They snarled and sobbed, tore and bit, rolling through muck and gore." It could be any two animals on the field of battle: Thermopylae, Agincourt, Fredericksburg, Stalingrad, Mekong, Soweto, Pristina. This was the outskirts of Jerusalem, 1968: "Siamese twins, the rifle sandwiched between them like some deadly umbilical cord. Pressing against each other in a . . . death-hug. Beneath them was a cushion of dead flesh, still warm and yielding, stinking of blood and cordite, the rancid issue of loosened bowels. . . . He clawed purposely, went for [the] eyes, got a thumb over the lower ridge of the socket, kept clawing upward and popped the eyeball loose."[20]

A Ugandan preacher trains hit squads to kidnap children by the thousands and drag them to his preserve; there he indoctrinates them and sends them back to their home villages to massacre their kinfolk. Those who refuse he turns the others on — with guns, bayonets, machetes, and staffs until the terrified bludgeon their fellow prisoners to death. Then, at the preacher's orders, they drink and bathe in the blood. Soon they are laughing and goofing off.

Two Arkansas boys dress in Halloween garb, hide in the hills, and fire rifles into crowds of girls gathered outside their grade school.

Three Texans tie a man to the back of their pickup and drag him for three miles until his head and shoulders are taken off by a culvert.

Vigilantes slash the throats of street people; high-school dropouts tie a gay man to the stake and incinerate him.

The generals send armies of children through the land mines to clear the way for their tanks.

It has grown from the mute wars of the Stone Age to the primitive bellicosities of the Caesars to the modern epidemic of psychopathic generals, mass murderers, and Gestapo police. What is new is not the activity but the desperation, after so many attempts to impose justice, still to be fuck-ups. It no longer even seems possible to acquire humanity—or, perhaps, as the orgies and whippings of the Marquis de Sade augured over two centuries ago, the association of desire and mutilation is as old as our species (and only now are the limits, or the despair of them, on display).

On his deathbed (1997) William S. Burroughs cries out:

"Where is the cavalry, the spaceship, the rescue squad? We have been abandoned here on this planet ruled by lying bastards of modest brain power. No sense. Not a tiny modicum of good intentions. Lying worthless bastards."[21]

And just in case there remained a snippet of hope ... special-interest groups buying out politicians from Lagos to Baton Rouge.

WHEN HELPLESS TOTS are sodomized or murdered by adults, otherwise merciful men and women call for vengeance. And so the pornographic current spreads. At the core of their disgust lies some forbidden attraction to the same act. What is enacted by some is nascent in all—the secret of Humbert Humbert in Nabokov's *Lolita*. The desire to punish is a self-loathing; the death penalty is served to obliterate not only the hated being who loosed an atrocity, but every aspect of sympathetic imagination in us.

The Gary Gilmore story, as retold by Norman Mailer in *The Executioner's Song*, poses the dilemma of crime and punishment. We want to pardon and to be pardoned. We want to give our disadvantaged and traumatized children a second chance; yet, all too often, the criminal is "rehabilitated" and murders again anyway.

Gilmore understood this better than most. "Kill me," he said. "For your own sake." He had already suffered the worst of America's prisons—months of solitary confinement, beatings by guards. He was crippled by disease and blinded by infection. When he was released he lasted a mere nine months before committing two random, unnecessary murders.

Between his last imprisonment and his execution six months later he sat in his cell meditating on life and death, a miscreant who accepted the firing squad he had earned.

He did not spout pieties about justice and capital punishment; he did not feign grief or guilt; he spoke neither for nor against the Mormons; he merely showed them the creature they held captive and, on that basis, met them halfway.

He spoke for the prosecutors, judges, lawyers, other prisoners, guards, and even the humanitarians who tried to save his life against his wishes. In the end his visitors could scarcely tell the difference between the condemned killer and a holy man, except that Gilmore was crude and belligerent; he trained everyone who came into his presence. Even the priest who tried to bless him (the moment before the fusillade) found the roles reversed and himself being consecrated and accepting the murderer's last blessing.

Gilmore did not talk himself into believing he was just another "innocent victim of society's bullshit."[22] Instead he demanded that the State of Utah follow through on its sentence. The outpouring of sympathy in his last weeks didn't fool him. He knew the only possible atonement was to shock our pollyanna era into witnessing the truth without flinching, thus proving it *was* the truth. He said, in essence: "You have no choice. Don't lose your self-respect by pretending to save or reform me. But don't pretend I'm an inhuman killer who has nothing to do with you or that you've solved your problem by shooting me. I'm the best of you as well as the worst. And neither you nor I know what's behind any of this. Kill me, but we're still in it together for the duration."

"I'm so used to bullshit and hostility, deceit and pettiness, evil and hatred. Those things are my natural habitat. They have shaped me. I look at the world through eyes that suspect, doubt, fear, hate, cheat, mock, are selfish and vain. All things unacceptable, I see them as natural and have even come to accept them as such. I look around the ugly vile cell and know that I truly belong in a place this dank and dirty, for where else should I be? There's water all over the floor from the fucking toilet that don't flush right. The shower is filthy and the thin mattress they gave me is almost black, it's so old. I have no pillow. . . .

"It seems to me that I know evil more intimately than I know goodness and that's not a good thing. I want to get even, to be made even, whole, my debts paid (whatever it may take!), to have no blemish, no reason to feel guilt or fear. I hope this ain't corny, but I'd like to stand in the sight of God."[23]

Thousands of others have followed him in manacles to their State-sponsored executions—"dead man walking," so the lyric goes.

"[W]hatever it was that had done that awful thing was already gone," mused Paul Edgecomb, Stephen King's boss of the electric chair. "In a way that was the worst. Old Sparky never burned what was inside them, and the drugs they inject them with today don't put it to sleep. It vacates, jumps to someone else, and leaves us to kill husks that aren't really alive anyway."[24]

THE MURDERER IS NEVER AN ALIEN; he is a son, a daughter, a brother, a husband. Aboriginal justice once sought to heal the wound by requiring the murderer to replace his victim in the victim's family. Having deprived that family of one of its offspring he was ordered to become the thing he had taken away. Killers adopted the clothes, the wife, the children, the parents of the person they killed. The act of homicide was dealt with not as the atrocity of an outsider but from within the empathy of the clan. Since the spirit of the murderer would likely jump to someone else, be reborn into the group again and again, it had to be diagnosed and cured, led back to its humanity.

The family disperses the darkness by taking in what is left of the human being in the slayer. The kin of the victim accept the ultimate collectivity of the species and consent that murder is a bond, however perverted. The victim likewise will return in subsequent generations, so the crime must be expiated before his soul seeks revenge. Among animals, this replacement is axiomatic; the individual is the species.

"The greatest peril of life," the Eskimo hunter Ivaluardjuk tells the Scandinavian explorer, "lies in the fact that human food consists entirely of souls.

"All the creatures that we have to kill and eat, all those that we have to strike down and destroy to make clothes for ourselves, have souls, like we have, souls that do not perish with the body, and which must therefore be propitiated lest they should revenge themselves on us for taking away their bodies."[25]

The mantis has its young, and they too mate, breed, and devour. All wars end, and their dead are buried. Years later, the identities of those in the cemeteries fade as the whole generation and then the whole species passes like some forgotten Dakotan tribe—even the memory of its existence blurring into all subsequent existences.

VIOLENCE CANNOT BE ENDED BY DECREE, and it certainly cannot be induced to end by demonstrations and editorials against it. The architects of weaponry operate ever under the undiagnosed causes of bloodshed. The lessons of Napoleon are submerged in Tolstoy; the texts of Tolstoy are further deconstructed by the scripture the Third Reich engraved into the spine of Europe. The black magic of warfare recedes through Machiavelli, Philip of Spain, Theodoric and the Vandals, Alexander of Macedonia to the Egyptians and Stone Age conquistadors. Hitler's warning now echoes like a meditation gong through Rwanda and Cambodia—the Red Guard furies of China, the ethnic bloodshed of Yugoslavia, and the tribal genocide of Africa. Until we as a species bathe in the mystery of war, it will not be possible to disarm. Witness Robert Ardrey's version of a twentieth-century credo:

"Our history reveals the development and contest of superior weapons as *Homo sapiens'* single, universal cultural preoccupation. Peoples may perish, nations dwindle, empires fall; one civilization may surrender its memories to another civilization's sands. But mankind as a whole, with an instinct as true as a meadow-lark's song, has never in a single instance allowed local failure to impede the progress of the weapon, its most significant cultural endowment.

"Must the city of man therefore perish in a blinding moment of universal annihilation? Was the sudden union of the predatory way and the enlarged brain so ill-starred that a guarantee of sudden and magnificent disaster was written into our species' conception? Are we so far from being nature's most glorious triumph that we are in fact evolution's most tragic error, doomed to bring extinction not just to ourselves but to all life on our planet?"[26]

"I have seen it done with my own eyes, and have not recovered from my astonishment,"[27] wrote Fabre of the mantises.

The Marquis de Sade answers him: "Oh, rest assured, no crime in the world is capable of drawing the wrath of Nature upon us; all crimes serve her purpose, all are useful to her, and when she inspires us do not doubt but that she has need of them."[28]

"The universe is banal," adds black-comedy movie-man Woody Allen. "And because it's banal, it's evil. It isn't diabolically evil. It's evil in its banality. Its indifference is evil."[29]

Newborn baby fish are captured instantly by crabs and anemones. A group of young squid blow their first puff of ink and are swallowed *en masse* by a whale. But fish and squid feed on crustaceans and snails.

When Idi Amin served a former minister's head at the dinner table, Frank Terpil, the CIA renegade, did not balk. "How could you go on working for him?" the reporter asked.

"I don't make the rules. This is what life is."[30]

Someday, says Dostoyevsky's Inquisitor, "the beast will crawl to us and lick our feet and spatter them with tears of blood. And we shall sit upon the beast and raise the cup, and on it will be written, 'Mystery.' But then, and only then, the reign of peace and happiness will come for men."[31]

Ivan protests, in the name of Dostoyevsky and, in fact, for all of us:

"Not justice in some remote infinite time and space, but here on earth, and that I could see myself. . . . If I am dead by then, let me rise again, for if it all happens without me, it will be too unfair. . . . I want to see with my own eyes the hind lie down with the lion and the victim rise up and embrace his murderer. I want to be there when everyone suddenly understands what it has all been for."[32]

BEHIND CLOSED DOORS enemy diplomats continue to speak to each other because they are trapped in the same ancestral language, the same pathology, the same fantasies of golden cities, of radioactive air, of exfoliating flesh and incinerated forests.

Despite proposals for mutual destruction of weapons I suspect we may have to talk until the end of time, simply to survive. Or we had better plan for this long a dialogue, just as we must plan for the half-lives of plutonium and other poisons we have strewn about us. Yet the consciousness that brought ozone-layer holes and pesticides, torture machines and jails into being cannot end them. It is frozen in negligent horror at its own acts.

We spoke long ago of turning swords into plowshares. If it were simple we would have done it, assuredly. And even though it is difficult—in fact, impossible—we have no other choice but to try, until the end of time.

OUR FAINT INCIPIENT EGO encounters the ancient desires of cells, their hunger for substance, their unceasing differentiation. Computers notwithstanding, we cannot create mind from a sterile liquor, and we cannot convert nature by a rational attack. Our task is more difficult. We must reclaim the darkness by conducting its pitch-blende through our unexplained lives.

There is no rule of thumb: One person nurses the sick in Bangladesh while another irrigates a farm in pre-Columbian Arizona. Some dispel evil by kung fu or aikido, while others spread discord through the same arts. No one is totally diabolical; every person enacts some smidgen of photosynthesis, but likewise, a shutter of darkness.

The Sioux prays for his game and lures it to show itself to him by the beauty and integrity of his chant, the clarity of his attention. All thoughts are already universal on some level, even without telepathy.

Animals apparently call out to other animals to become their food. They collaborate across species in remorse and understanding at the moment of the kill. An unacknowledged bond joins the hunted and hunters throughout the planet but, as the first shamans intuited, only when the hunt obeys the ceremony.

We cannot be carnivores without being killers too. From the viewpoint of plants, we are just another mutant that has lost the ability to feed directly from the Sun. What if this ability were regained and transmitted back through the cells? This would be remarkable, considering the millions of years of predation our metabolism embodies. Our *apologia* to the whole animal kingdom is based on the circumstantial evidence that there is no other path to survival.

Yogis still promise we can someday draw our sustenance from vibrations of air, without killing even vegetation, to drink from the Sun and the psychic field around us, but if that's where we're headed, we've got a long way to go.

Nothing about cell life, or DNA, or the self-assembly of tissue—at least in our academic rendering of them—suggests an innocent direction. Yet this entire progressive culture might be an evasion of our inherent condition, our actual destiny. We could be avoiding our own natures, missing solutions to our crises, spurning courses through the underbrush. Worlds without end more fulfilling than this might be within our grasp. But these lie in the margins of an inner life we flee through all our ideologies and institutions.

A MEDIA-CONSCIOUS, technologized, hierarchically managed Weltanschauung creates the mirage of progress against famine, tyranny, and disease. We are blind. It is but "a killing/producing machine without a spiritual center."[33]

"For a long time now," declared Friedrich Nietzsche, "our whole civilization has been driving, with a tortured intensity growing from decade to decade, as if towards a catastrophe: restlessly, violently, tempestuously, like a mighty river desiring the end of its journey, without pausing to reflect, indeed fearful of reflection.... Where we live, soon nobody will be able to exist...."[34]

The old folks may tire of battle and make peace, but a new generation of punks is born every day: *interawahme,* Red Guards, Shining Paths, skinheads, Goths, bloods. The empty minds of abandoned youths (it matters not whether in desert refugee camps or Denver-Paris suburbs) suck in the unlived fantasies and usurious myths of their elders—jihads and fatwas, Dungeons, Dragons, Natural Born Killers, Vampires, Semi-Automatic raves, Doom patrols. They are angry at the world we have left them, and they mean to destroy it.

Satan is not the aggressive recruiter portrayed by dishonest preachers; he fills the dearths, the emptinesses, the gaps where life seems to be lived where no life is lived. He is the sterile outcome of the craving for power, for megabucks in the absence of empathy or meaning. Without breaking the trance he turns computer blood into real blood; he rewrites the Koran into suicide-bombing missions to paradise. "Next motherfucker gonna get my metal.... Pow pow pow[35]"! Invent a vacuum, fill it with luminous ikons, with cartoon daydreams of violence, he will infiltrate and claim them. Kill God, and Satan is the sole denominator.

"There will be wars such as have never been waged on Earth. I foresee something terrible. Chaos everywhere. Nothing left which is of any value; nothing which commands: Thou shalt!"[36]

And still, as the Buddha preached, this horribly broken world is not broken at all. Things are proceeding exactly as they must. The universe is unfolding as it should, as it only can. The "brokenness," or appearance of flaws, is for us to heal by our lives, by mindfulness and compassion, even when those attributes seem singularly lacking.

The universe is not even indifferent, although that is the vainglorious refrain of desolate, unmerciful civilizations. After all, the labyrinth is us; it was us; it is becoming us. We are the guardians and the sheep. And we alone may cast this dark spell over matter, that now hexes and bewilders us, condemns us to wander in a cruel, cataclysmic machine. The universe is only indifferent when viewed from a superficial perspective of how a just, responsive deity (or energy) should act. But that might be absurdly shallow for an actual timeless expression of existence and being.

Between the forces of voodoo and quantum dance of atoms lie untold realms of suffering and redemption, worlds happier than this one, worlds of unmentionable damnation and doom. They are all perfect in their own way. They are all part of the greater creation.

We may pretend (at this late date) to save the Earth, but what about whole civilizations on other planets ravaged by cruelty and bloodshed, creatures in grievous pain and subjugation on worlds around distant suns? Can we rescue them too?

If we were to accomplish a lasting peace on our world, must we worry about inhabitants of other worlds that might not even exist? But if they do, they are part of the universe, part of consciousness; and ultimately our sympathy must be extended through eternity to their suffering too, creatures we will never know.

If we could bring peace to this planet, we could surely bring peace to the entire universe.

Not because such a fantasy does any good, but because it forces us to view the crisis in its actual bigness while at the same time reminding us that we do not know who and where we are and what options we have.

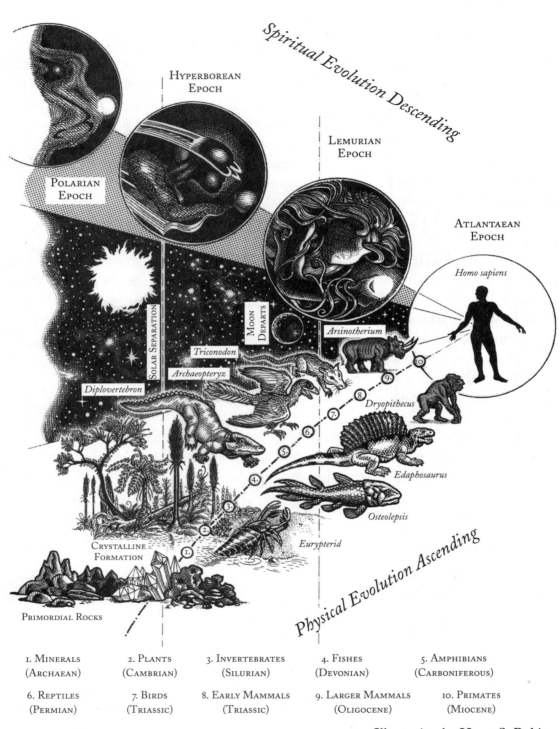

Spiritual Evolution Descending

POLARIAN
EPOCH

HYPERBOREAN
EPOCH

LEMURIAN
EPOCH

ATLANTAEAN
EPOCH

Homo sapiens

SOLAR SEPARATION

MOON DEPARTS

Arsinotherium

Triconodon

Archaeopteryz

Diplovertebron

Dryopithecus

Edaphosaurus

Osteolepsis

CRYSTALLINE
FORMATION

Eurypterid

PRIMORDIAL ROCKS

Physical Evolution Ascending

| 1. MINERALS (ARCHAEAN) | 2. PLANTS (CAMBRIAN) | 3. INVERTEBRATES (SILURIAN) | 4. FISHES (DEVONIAN) | 5. AMPHIBIANS (CARBONIFEROUS) |
| 6. REPTILES (PERMIAN) | 7. BIRDS (TRIASSIC) | 8. EARLY MAMMALS (TRIASSIC) | 9. LARGER MAMMALS (OLIGOCENE) | 10. PRIMATES (MIOCENE) |

Illustration by Harry S. Robins.

Spiritual Embryogenesis

"The summarization of our existence is mystery, absolute, unqualified confrontation with what we cannot know."

THE UNIVERSE IS A TOTAL AND UTTER CONUNDRUM—where it came from, what it is, even what it looks like. Since access to data is limited to our nervous systems and devices contrived through them, we are condemned to follow materialities, one pole of their domain dwindling now into a subatomic underbelly, and the other stretching across an infinity of space-time and galactic mass "void, dark, and drear"[1]; molecular all the same, in effect, defining the modern plight. The rest of it, beyond measurable radiation, does not "exist." It could be anything at all.

"We do not know what anything *is*," Da Free John warns. "We are totally mindless, and totally beyond consolation or fulfillment, because there is no way to know what anything *is*. The only thing you can know about anything is still *about* it. But you do not know what it *is* ... *is* ... *is* ... or why it happens to be. You have not the slightest knowledge of what it *is*. And no one has ever had it. Not anyone. Not Jesus, not Moses, not Mohammed, not Gautama, not Krishna, not Tukaram, not Da Free John, no one has ever known what a single thing *is*. Not the most minute, ridiculous particle of anything. No one has ever known it, and no one will ever know it, because we are not knowing.... The summarization of our existence is mystery, absolute, unqualified confrontation with what we cannot know. And no matter how sophisticated we become by experience, this will always be true of us....

"No matter what sophisticated time may appear, no matter when, in the paradox of all of the slices and planes of time, any moment may appear in which men and women consider the moment, no one will ever know what anything *is*."[2]

We exist by molecule and cell agglomeration, have children likewise, raise and cherish them as real beings, accept this limitation—our condition—at face value, all the time explaining ourselves to ourselves by laws of nature, divine intervention, complexity theory, and their ilk. But these are human things projected into transcendental things, terrestrial things into celestial things.

We haven't a clue. Or we have all too many clues, wrapped in mazes, leading to paradoxes, enigmas, contradictions, and dead ends.

"WHAT IS THIS?" the Korean Zen Master Seung Sahn snaps. The answer he expects of his students is: "Don't know!"

This is not an admission of ignorance. It is a recognition of the scope and complexity of the question and our own having arisen within it:

> "The heavens collapse, and the ground caves in.
>   The great universe is split from side to side.
>   In the midst of true emptiness, without even one thing,
>   Where do you come from, and where do you go?
>
>   ... Only don't know!"[3]

Slapping the ground or grunting would be another answer to this koan.

REALITY SUPPOSEDLY ORIGINATES in an explosion of stars and then an assemblage of chemicals in planetary holding pools. How? Why? Why did it not lie burbling and sputtering forever? If random chemistry had to sully it, yellow and brown goo should have been the outcome, not wild horses.

The entelechy of molecules is simply not sufficient to the deed.

There should have been no one, no us, no stars, no planets, no universe, no thoughts for forever.

At the moment this whim disappears into its paradox our mind collides with itself. Try holding the notion—that none of this might have existed, not only now, not only then, but for eternity back and eternity to come.... Try.

There is but one universe.

NOTHING SO DIVIDES THE MODERN WORLD as the split between those who believe all things arise from spirit and those who believe that things are matter only. The physical Sun that burns down into twentieth-century beehives and batteries is real enough, working its photosynthesis through the cosmic history of hydrogen, and so is the bloody infant who swims into the world via carbon lattices.

Yet the universe could have a spiritual history too, without ever telling itself outwardly in sun-stars and stones.

Predating science by at least thousands, if not tens of thousands, of years, spiritualism proposes an invisible aspect to creation, an inherent clairvoyance of who and what we are. Although we cannot demonstrate that spirit and vital force even exist, men and women have intuited them, and their higher-dimensional consorts—gods, ghost-shadows, citizens of hyperspace. We descry a shadow, an immaterial engine behind nature. Else why should we don these unlikely robes?

By the nineteenth century, though, occultists and vitalists had gotten lazy. The mere fact that rival mechanists had failed at a full description of causes and effects was enough to justify their faith and puff them up with complacency. Once authoritative mysticism lost its wellspring, it became as ideological and materialistic as science. It did not have to demonstrate or even experience supersensible forces. It could merely declare a demiurge.

Now, as the spiritual mind confronts the universe through the denouement of science, it sees its total abnegation, but also, strangely, its truth. Science *is* theology, without its wishful thinking and denial. It is a lead filter through which bare shards of divinity and psychism seep. Without science we would never know the algebra of our imprisonment in matter. If we had remained loyal to the vital force, we would hardly have discovered the gene. We would be chasing after someone else's bellicose, authoritarian gods just about forever, gods who are little more than projections onto phantasmagoria.

While ideologically doctrinaire scientists do not find spirit anywhere (and religious fundamentalists see only cartoon deities manipulating moral experiments), the great dance of atoms, molecules, and cells in and of itself possesses a spiritual face, one obscured not so much by science as the politics of materialism to which science has been wed. (As one biologist jibed, "Biology has nothing against God, just a prior commitment."[4])

In a real world physical and spiritual cannot be separate anyway. There is but one universe that changes only in our changed perceptions of it through centuries. One Earth, one sky.

MANY OF THE FOUNDERS of modern physics and biology still believed in a supersensible agency, whether they called it God or vital force. Kepler, Newton, Darwin, and Driesch addressed this archetypal energy directly, and assigned it the ultimate genesis and destiny of celestial bodies and animalia—Newton has been called both the *first* scientist and the *last* magician. Because these researchers could never square supernatural agencies with daily peregrinations of matter—and spoke alternately in terms of astrology and physics, similitude and biology—their works have (to the modern sensibility) a schizophrenic ring. Historians handle this by paying attention only to the paradigms of theirs that led to science. The rest are blamed on ideological contamination.

Other early scientists (Galileo, Pasteur, and Haeckel among them) were experimentalists in the modern sense. They glimpsed a blemishless machinery of matter and energy. Yet the unarticulated background of their work intimates bottomless depths.

In either instance, something large and cosmogonic remains: If primal matter came into being through extrasensory forms, it at once had to obey the vectors of mechanical forces (which are themselves sums of archetypal effects).

The Triumph of Mechanism

SEVENTEENTH-CENTURY FRENCH MATHEMATICIAN René Descartes discerned the human body as a mere mechanical box, an effigy manufactured by an invisible divine being who impregnated each one with an exogenous soul that animated it and imbued it with activity and mind. The rest of life was mere chemical tropism. Animals, lacking such souls, were pure machines. "I assume their body," Descartes wrote, "to be but a statue, an earthen machine formed intentionally by God to be as much as possible like us."[5]

Experimenters took this separation of the animal and the divine literally. "They administered beatings to dogs with perfect indifference, and made fun of those who pitied the creatures as if they felt pain. They said the animals were clocks; that the cries they emitted when struck were only the noise of a little spring that had been touched, but that the whole body was without feeling. They nailed poor animals up on boards by their four paws to vivesect them and see the circulation of the blood...."[6]

For almost two hundred years God alone stood between human beings and an identical fate, nailed four paws to an empty universe. Cartesian logic kept body and mind, animal and soul, earth and heaven rigorously separate; men and women remained noble visitors to a profane and pagan place. Science could coopt every corner and function of nature as long as it left human divinity intact and ratified a scintilla of God's divine breath. But even those inviolable traces were being eliminated decade by decade, and soon there was no territory in nature for a God to rescue humanity from the earthen machine.

DURING THE NINETEENTH CENTURY, mechanistic science made startling, unforeseen headway. As techniques of investigation improved, physicochemical and molecular causes were revealed, and whole areas that had been conceded *in perpetuity* to the vitalists were reclaimed, one by one. In 1828 German chemist Friedrich Wohler synthesized the organic molecule urea, demonstrating that even complex substances could be imitated in a laboratory; the modern-day gene-splicers are his lineal descendants. In 1859, in his opus *On the Origin of Species,* Charles Darwin translated the premises of physics into living systems. He showed that the same universal agency that held the Moon in its orbit and caused water to run downhill was responsible for the radiation of myriad plants and animals, with no exterior cause or vital agency. In 1893, Max Rubner applied the law of conservation of energy in a strict and functional manner to animal tissue and its metabolism. This sealed the Darwinian promise in thermodynamics.

Jacques Loeb, a German biologist who believed that so-called human will was a form of chemical tropism, shocked nineteenth-century vitalists by artificially fertilizing sea-urchin larvae so that plutei formed without the participation of sperm. After the rediscovery of Gregor Mendel's principles of genetic transmission and the pure algebraic expression of inherited traits, early-twentieth-century biologists could explain the physical diversity and behavior of all living organisms as the result of differential fertility and mortality, cell mutation, and the molecular properties of organs. There was no chink anywhere in the machine.

More recently biologist Richard Dawkins has rechristened God "The Blind Watchmaker":

The molecules comprising living things, he writes, "are put together in much more complicated patterns than the molecules of nonliving things, and this putting together is done following programs, sets of instructions for how to develop, which the organisms carry around inside themselves."[7]

Life merely "resembles" life. It is a remarkably convincing pageant performed by chemico-molecular stooges—nature's marionettes.

"Maybe they do vibrate and throb and pulsate with 'irritability,' and glow with 'living' warmth, but these properties all emerge incidentally. What lies at the heart of every living thing is not fire, not warm breath, not a 'spark of life.' It is information, words, instructions. If you want a metaphor, don't think of fire and sparks and breath. Think instead of a billion discrete digital characters carved in tablets of crystal. If you want to understand life, don't think about vibrant, throbbing gels and oozes, think about information technology."[8]

We have already visited this sterilized planet many times. It is the modern credo, our defiance of a universe that has failed every test of justice, every sensible plan. It is as though we are saying to God, "You won't trick or disappoint us any more. We'll get there first and eradicate your every false trace of hope. Then we'll turn our lifeless algorithm into you."

Microbiology and biotechnology sit on the pedestal now; they draw students into classrooms, graduates into corporations, products into markets. Genetic determinism is king because it is more accountable in an "assembly line" economy than either complexity theory or vitalism.

Materialistic scientists, Dawkins and Stephen Hawking among the more outspoken of them, are now glued to a pedigree of molecules and chromosomes, an evidence trail of subatomic particles and stellar masses. Inside us is a void of voids, a horrific, infernal clockworks of accelerated atoms tearing about phantom cores. They assume (quite proudly) that there is nothing else, nothing human at all.

LOOKING BACK ON THE RULING LINEAGE of materialist philosophy, contemporary author David Denby summarizes our canon and dilemma:

"The spider that Darwin studied in Patagonia, which bore itself away on gossamer threads, also spun connections and relations that reached to all of us, for evolution was a web, with everything connected to everything else. To the practiced eye, the spider's blindly reflexive behavior was also a virtual library of adaptive traits refined through evolutionary practice—strategies that were no different in quality from adaptations that primates had performed in their long march to *Homo sapiens*.

"It is a remorseless process. Does anyone—even the most confident atheist or materialist—really take comfort from evolution by natural selection? Evolution offers nothing (as the Victorians sorely complained) to the human desire for ethical advancement or emotional solace. The American philosopher Daniel C. Dennett has candidly described Darwin's theory of natural selection as the 'universal acid' that seems to burn away all our comfortable illusions. For Dennett and other neo-Darwinians, not only is natural selection the center of biology but it explains more of consciousness and morality than most people realize. For our choices and character, our desires and deeds may be the result of long-ago accidents and adaptive mechanisms, which improved chances of reproductive success in a given environment, and then got passed along in genes, after numberless generations, to you and me, where they function in a new environment, sometimes successfully, sometimes not. It is a mindless as well as remorseless process—'algorithmic,' as Dennett calls it.

"This genetic patterning, which started with tiny microörganisms, is now extraordinarily complex, the things of this world joined in systems of astounding subtlety. Yet no centralized intelligence planned the life around us. There is Design but no Designer: natural selection wants nothing, aims at nothing, and reaches no resolution or fixed point. If the neo-Darwinists are right, our creativity, our consciousness—Elisabeth Schwarzkopf rehearsing Schubert near an open window and someone listening outside—were produced by a process both unconscious and uncreative.

"Nor can evolution be thought of as a progression. We can speak of expansion, in the sense of greater complexity, but not of progress. In particular, human beings are neither the inevitable goal nor the end product of evolution: evolution passes right through us. As Stephen Jay Gould maintains, our existence is merely contingent, the species *Homo sapiens* no more than a tiny twig in the tree of life—a twig that might easily have fallen dead. At his most relentless Gould declares that 'we came about this close (put your thumb about a millimeter away from your index finger), thousands and thousands of times, to erasure by the veering of history down another sensible channel.' By contrast, bacteria were here before us and will triumph after we are gone. Men and women will leave their bones but cast no shadow."[9]

To THINK THIS IS TO BECOME IT. The Darwinian/Mendelian regency yields not only genetically engineered medicines and cloned sheep but the Jewish and Gypsy experiments of commandant Josef Mengele. What nature has already done in the absence of God, man dares now attempt (without compunction) in the name of nature.

Twentieth-century scientists have restored the esoteric connection between remote suns and the stuff of life on Earth, but only by making the stars lifeless and their relationship to life mathematical and random. The *élan vital* is hardly a flame passed by spirits and angels; it is a thermodynamic variant, a wobble in a lifeless regime.

We no longer require gods because we have been exposed as not being truly alive, hence not responsible for our actions. We are molecular ripples mimicking something grand (but something that cannot happen in a universe such as this). The mechanist position is typified by J. D. Bernal's smug witticism:

"It is difficult to imagine a god of any kind occupying himself creating, by some spiritual micro-chemistry, a molecule of deoxyribonucleic acid which enabled the primitive ancestral organism to grow and multiply. The whole hypothesis has now come to its natural end in absurdity."[10]

But it is equally incomprehensible that this gossamer world of microtubules and mantras, of Sigmund Freud, Joan Miró, and Samuel Beckett, came into being through a thermodynamic accident and the happenstance properties of molecules and amino acids. Though nearly every scientist rotely professes belief in this epistemology, none of them behave as though they are mere jumbles of chemical concatenations. They act like official spokesmen for the gods.

On the one hand, it is the only reasonable explanation; on the other hand, it is utterly ludicrous. Everyone knows it is the only reasonable explanation; yet, everyone knows that it is utterly ludicrous.

The modern sensibility has come to its natural end in absurdity.

The Revival of Spirit

EVEN AS THE SUCCESS OF MECHANICAL SCIENCE took most of the ground away from traditional occultism, the failures of that same science gave the vitalist and hermetic traditions surprising new ground. Clerk Maxwell's nineteenth-century electromagnetic hypothesis foretold the imminent downfall of pure progressivism. Since then, physical experimentation has not solved the mystery of life, let alone the riddle of the mind; it has not even adequately explained inanimate matter. If the first wave of eager scientists assembled a machine, the second has dismantled it (though some still hope it will turn out to be a computer). The laboratory holds no fixed arrangement of props, only a cyclone through which quanta pass. In the atom, there is no machine. In fact, the very small doesn't continue to unsheath into the even smaller, into an indestructible germ-seed; it crumbles into an entirely different universe with its own witnesses and rules. Substance and motion are alternate forms of one substratum; even light and matter are interchangeable, and time and space are regional jargon for something quite different in the wild.

The mysteries of mechanical science are once again the mysteries of spiritualism, though in an entirely other way that suggests these ties will be dissolved only to be reestablished as poles of a larger mystery for as long as we exist. A generation after scientists were certain that a mechanism or field theory would be found for all motion and form, Werner Heisenberg and Albert Einstein were addressing the hand behind nature and attempting to characterize the invisible Shaper. But this was not a spiritual universe in place of a mechanical one; it was the basic puzzle of all existence.

THE PROBLEM WITH CONTEMPORARY SCIENCE is that, after hitting the wall, it became more akin to mercantilism than empiricism, offering trade shows in place of a universe.

The game is not entirely honest. It uses one phenomenology at the ball-yard, another at home. Scientists commute to work like corporate executives, oversee commodities and transactions, and then retreat to "private lives," their families, recreations, and goods, as if none of the rest really mattered, none of it was really real. But all phenomena exist on the same footing, and the gap between what one *is* and what one *does* is fatal to any view of nature arising from laboratory inquiries. Gang members soldiering through L. A. ghettoes, Maori shamans assembling medicine bundles, peasants tilling Oaxacan soil, prisoners tortured in Iraqi jails all understand dimensions of the universe as profound as quarks or DNA, elements of existence physicists and biologists entirely miss. Furthermore, a scientist ripped out of his life into any of theirs would renounce his Paradigm as quickly and shame-lessly as Peter denied Jesus before the cock crowed.

Science must always return to the wonder and sense of creation within, for that is where our impulse for inquiry and intelligence arises. The awareness of being in a vast and mysterious manifestation, of being part of it and, at the same time, in the slipstream of a singular flow of mind, transcends any theory, law, set of sym-bols, or gods. "The frog never lies," the old laboratory biologists taught us. And now, in a millennial crisis of identity, *we* are the frog.

We all know the truth.

A NEWBORN MOUSE ACTS AND RESPONDS, beyond species, beyond eternity. Even an amnesiac, who has forgotten every factual detail of her life, proceeds from the essential truth of her being—loss of memory is not loss of self. What we hear in our heads is also what the universe hears echoing through its galaxies and what the darkness discerned in the burst of atomic form and crystalline gauze along its unbridled flank. Why we are, why the universe requires us, is the riddle and also its answer.

In superstring theory, two nine-dimensional slabs—known as "end-of-the-world nine brains"—face each other across the abyss, and that is the least of our problems. As complex as this universe gets, as cluttered with random collisions of particles and events layered in maelstrom and subterfuge, it is intuitively obvious what it is, so obvious everything goes merrily on despite anarchy and abnegation. We don't quit even when confronted by the malign depth of our plight, the antipa-thy of existence, the density of matter, the opacity of the fog. We are all secretly smiling at one another, even if (sometimes) in horror. Something reassures us, some-thing unspoken, unspeakable, immune to ideology. Without an umbrella of machines everywhere, we might still regain a warrior's courage, a shaman's bravado, a lama's grace; we might still find in ourselves the inconceivable safety of being.

All of us are aware (at heart) of who and what we are but cannot name it. Those who most fiercely dispute the transcendent world are able to spurn it only because they can experience their immersion in it without having to sacrifice an inch of their cynicism. They are merely "dissing" stray gossip about a reality they appre-hend directly. Nihilism is, in truth, an indispensable branch of mysticism, for it leads into our depth.

If we did not know implicitly who we were, we quite likely could not exist at all; we would be a series of robot-like gestures strung together by mute commands, a cyborg without a program.

Because one person is committed to "hard science" and another to "astral projection" does not mean that the scientist does not intuit the astral realm or the occultist does not tow to inviolable laws of matter. They are both playing roles using xenophobic images; they are members of competing clubs. Instant conversions, both ways, remind us of the true secrecy in which people guard their final ordinations.

A professional skeptic who dies and finds himself in the light again—in another dimension—is not embarrassed or shocked. "Oh yes," he thinks, "that's what this is. I remember." He reenters his nature beyond rhetoric. He recognizes what he knew all along. His cynicism, far from being disproven, turns out to have been successful party talk.

Likewise, those who adopt the ostentatiously spiritual and fill their lives with its symbols and accoutrements are no more spiritual than those who do experiments in laboratories and deem our bodies chance chemical gloss. Each has chosen an imagery and ceremony, a way to pass the long hours on Earth. Each expresses an inherent temperament more than an ideology.

The skeptic is easily enraged; he takes umbrage at people whom he sees mixing intuition, wishful thinking, and scientific metaphor. He interprets this as sloppy and indulgent. A natural bureaucrat and drill sergeant, he dismisses channelling, telepathy, and energy healing as placebos and stage tricks. Because of the persistence of his kind, science was born.

The spiritualist follows creative insights and expresses synergistic impulses that cannot be ascribed to material causes. A natural visionary and clown, he resists temporal rules of conduct and inquiry. Because of the visions of artists and shamans, the discontinuous riddle of the universe continues to touch our hearts, and the mystery unravels, a bit at a time.

The beauty of the system (and also its horror) is that death has nothing to do with our ideas—how "Buddhist" we are or not, how much faith we have. We roll, as a ball down a hill, propelled by gravity, drawn by karma, and the outcome is simply what it is. We go toward what is because there is nothing else. Belief plays a part, but if the soul flies from here on incorporeal wings, reembodies or not, journeys in other realms or not, will happen because of the nature of things.

The most asserted atheism, the most devoted rigorous scientific reductionism must come from the gods too, if there are gods at all.

The Great Chain of Being

SPIRITUAL THEORIES PROVIDE THEIR OWN MEANINGS for cell division, morphogenesis, and human evolution. Where biologists propose hierarchical fields and homeostases, occultists refer to cosmic forces that precede matter. A spiritual agency is the string that holds the beads, beads that (according to science) are already held to one another, from their origination, by electrochemical forces.

Early naturalists viewing nature did not perceive thermodynamics or conservation of energy. They saw archetypes shuffling through celestial germinations of new forms; minerals, plants, and animals alike were elemental crystals, transfigurative tinctures. Just because something was calcined or sublimated did not mean that its shape was obliterated or rendered ephemeral. Alchemy held precedent yet over chemistry; ideal, eternal form over algebra: "Transmutation . . . takes place when an object loses its own form, and is so changed that it bears not resemblance to its anterior shape, but assumes another guise, another essence, another color, another virtue, another nature or set of properties."[11]

In this context biologists interpreted the botanical and zoological species as fixed steps in a ladder flowing ever upward from lower, less sentient forms to higher, more intelligent ones—the Great Chain of Being. In the mid-eighteenth century the French lawyer and scientist Charles Bonnet, assuming this ascension, described the likely future state of our globe:

"Man, then transported to a dwelling place more suitable to the eminence of his faculties, will leave to the ape or the elephant that first place which he occupies among the animals of our planet. In this universal restoration of the animals, therefore, it will be possible to find Leibnizes and Newtons among the apes or the elephants, and Renaults and Vaubans among the beavers. The more inferior species, such as the oysters, polyps, etc., will be in comparison with the more elevated species in this new hierarchy what birds and quadrupeds are in comparison with man in the present hierarchy."[12]

This would certainly be an honorable universe.

Ontogeny recapitulates phylogeny; together they recapitulate cosmogony.

WHEN HAECKEL PROPOSED THE CONCEPT of ontogenetic recapitulation, occult biologists immediately incorporated it into an older theory of archetypal evolution along the lines of the Great Chain of Being. The spiritual anatomist Hermann Poppelbaum specifically rewrote the biogenetic axiom to read: "Microcosmogony is a reflection of macrocosmogony."[13] That is, ontogeny (the development of an organism from cells) recapitulates not only phylogeny (the evolution of complex multicellular organisms from simple precellular zooids) but cosmogony (the history of the universe and the migration of souls between dimensions). The stages of the human embryo comprise both physical and spiritual (animal and angelic) chronologies leading to Adam Qadmon, the male-female prototype. While ontogeny recapitulates phylogeny, their synchronism recapitulates cosmogony, eclipsing and synopsizing it even more deeply than ontogeny condenses and abbreviates phylogeny, and also allegorizing and metonymizing it, for the gap between ontogeny and cosmogony is subatomic and transdimensional and does not reflect the continuity of either thermodynamics or membranes that the passage from phylogeny to ontogeny does.

As Gould (quoting Huxley) remarked, recapitulation contains the germ of a hidden truth. To occultists who had inherited millennia of teaching that the stars rule our organs, the hidden truth could only be the iconographic imprint of the cosmos on the embryo and the expression of cosmic history through the phases of its unfolding.

Life on Old Saturn and the Sun

PROBABLY THE MOST COMPLETE ADAPTATION of Haeckel's proposition in a universal occult system was accomplished by Rudolf Steiner (who certified his work to Haeckel under the caution that Haeckel himself would have "unmistakably declined this dedication").[14]

I will review Steiner's cosmology in depth here because he is the only significant Western occultist to develop a theory of embryology. He and his anthroposophical followers (Karl König, Hermann Poppelbaum, Henrik Steffens, and Thomas Weihs among them) provide a taxonomic link between cosmogenesis and ontogenesis. It is irrelevant whether one construes Steiner's annals of planetary evolution as literal events or (in keeping with most modern sympathizers) symbolic representations of states of consciousness too alien to render in ordinary language. It will never be resolved in conditional space-time.

Since I am not trying to write a biology book, Steiner doesn't "ruin" things or destroy my credibility. I hope everyone keeps reading. This is a text considering all modes of inquiry into the mystery of existence: embryogenesis. Wilhelm Reich, Sandor Ferenczi, G. I. Gurdjieff, Adi Da Samraj, Drunvalo Melchizedek, the Eskimos Aua and Ivaluardjuk, and Hermann Poppelbaum all contribute pieces to a puzzle far vaster than any one epistemology can do justice to. It is not that science isn't more judicious, more "correct." Compared to the renderings of science, what follows is balderdash and inflated superstition. But that "balderdash" strikes closer in some ways to the mystery-shrouded heart of our being than mere facts and measurements (however invincible) floating in contextless objectivity.

IN STEINER'S CREATIONARY TALE, human beings pass through (and fetally recapitulate) four ancient universes. Although he gives these realms the names of planets and bodies in this Solar System, Steiner is proposing zones in other dimensions on which our ancestors previously incarnated. He is not talking about Saturn and the Moon in the usual sense but as astral realms associated with those bodies' astrological cycles and material condensations.

According to Steiner, the whole of our creation began in a spiritual germ as large as the present-day universe. This was not the dime-sized quark out of which the Big Bang exploded but an ingot of the unfathomable cosmic depth, its mere outer tapestry masquerading as astrophysics. When this present universe is stripped away or pulped into a black hole, that one will remain, its indestructible souls soaring into their ancient landscape.

From the germinal monad, specters descended onto a world known as Old Saturn. At its moment of fertilization the human zygote briefly glitters as a seed of that long-vanished universe.

On Old Saturn the human ancestors were asleep and unaware. Those souls who incarnated directly from there onto the Earth appeared in the Azoic era, still sleeping, as minerals. While on Saturn they were capable of becoming anything, even human, but their reaction was too shallow and precipitous, so their true consciousness remains behind in the higher spiritual realms. They reside in every atom of every jet propellor blade and wattle of prairie. Other souls (not human progenitors) awakened more fully on Saturn and passed directly into angelic realms.

The Saturnian world was entirely physical but not in the sense we think of; it contained no atoms, no molecules; it was a physicality only of heat effects. Through its aeons the human ancestors gradually acquired a possibility of incarnating in bodies, thus bringing the Cronian epoch to an end.

After passing through a cosmic night—perhaps a googol of our aeons—these Saturnians were reborn on the Sun in a form that would appear bodiless to us. First they recapitulated their sap while the Sun remained dark. Then envelopes of ether sprang up around them and ignited their domains. It is still burning, a semblance of thermonuclear fire lighting our zone.

The etheric substance of the Sun traced the germinal organs of animal bodies in heat effects inherited from Old Saturn. Even today, ripples of ether (like Reich's orgone) sustain the carapaces of creatures on Earth; without such auras they would sink to the mineral realm and disintegrate. Our physical heart is a replica of an etheric heart, and our physical brain is an etheric brain imprinted in protoplasm.

The liquid of our membranes is already a "living" thing, a psychic sense organ. Through fluids the life body of the Sun responds to the appearance of matter on

Earth and transmits its patterns into protoplasm.

Water responds to every breeze along its surface, every change in temperature and tangency, from fluxes of gravity to interruptions of vines or bugs. At each snag it recoils rhythmically, sending out a scale of ripples or waves.

Water is not simply a miraculous chemical; it is an intermediary between the rocky planet and the invisible ether, recording the impressions it receives from both sources and distributing them in crosscurrents. Note the effect of rain on a parched valley—thousands of dessicated seeds spring from dormancy in multicolored regalia. The moisture the sage distills from the desert flushes the night with potion. Even a garden hose on a summer afternoon brings astral refreshment to grasses and flowerbeds. They perk up and radiate their native majesty and depth.

In the rivers and seas of this world and in the clouds of Earth's atmosphere we can observe the organs of creatures flirting with palpability and then receding, much like the animals a child imagines in a parade of cumuli. The fins of sting-rays appear in intersecting ocean waves; where springs surge into surface tarns, umbrellas of jellyfish manifest; as two streams meet underwater, the paired vortices of a heart are foreshadowed. The feathers of birds are currents of air embodied. Even bone preserves crisscrossing etheric ripples.

ON THE SUN the potential man/woman lay in dreamless sleep, stirring to observe the flutter around him/her through a sense organ which would later be incarnated as the pineal gland. These hermaphroditic beings watched the odd extrinsic world flow by in bright soul pictures, semiconscious reflections of its own being, but they did not recognize them or comprehend the significance of their objectified identity. They were alert and intelligent but incapable of self-awareness. Such is the state of consciousness of those beings who incarnated directly onto Earth from the Old Sun—they are plants; their egos linger in another realm.

Our ancestors inhabited the esoteric kingdom of this era, the Hyperborean, so its sunlight is recapitulated in the yolk sac (the occult embryologist Karl König has remarked that the early yolk sac of an aborted human holds a fluid bright as gold).

Old Moon and the Lemurian Age

THE WORLD OF THE SUN was too dazzling and all-encompassing for the human ancestors (or anyone) to evolve any further there; thus its epoch came to an end and, after passing as spores through the planetary womb, the Solar beings evolved onto the Old Moon. First they recapitulated the sap of Old Saturn and the ether of the Sun; then they secreted a new carapace, a liquescent envelope that drew them from hibernation. Without such a lunar double the life body would have remained permanently unconscious, asleep. Even today our astral shrouds keep us attuned to an exterior world.

The astrum expressed itself initially as pure physicalization, a congealment of human ancestors and their fellow beings into faint, willowy creatures—ghosts by today's standards. This cosmic moment laid the foundation of Solar-System astrology, for we are recapitulated—potentiated—at conception and again at birth out of astral signatures winding through and disguised in zodiacs of extant planets.

At this uncertain juncture our predecessors wavered between their nostalgia for the Sun and an exhilaration in their new independent nature. Something haunting called them back, a dirge they had to abandon if they were to advance and grow. They came forward because something equally powerful summoned them and they were curious—something that had never before happened in the cosmos.

As the human initiate received its nervous system, its revels now truly ended. No turning back; no exit, no bluff this time! It perceived the vastness of things around it and began to fathom its quandary. Only then did our ancestors understand that they were a copy of the universe, not the universe itself. Their lives belonged to them alone.

This was the Lemurian age, corresponding atemporally to the Mesozoic on Earth. The pharyngeal slits of the embryo commemorate this stage, for they are not only the gills of Mesozoic fishes but the organ through which our Lemurian ancestors received the form-giving vibrations of the cosmos. Speech originating in the chirps and bellows of beasts is also a rune-alphabet of higher dimensions resonating through astral receptors. As our forebears propelled themselves in sound across time and space, their gambol became language in the throats of animals.

In the Earth's primeval ocean, other hastily incarnating souls awoke with gill slits. They breathed water, not speech.

THROUGH THE ETERNITIES of the Lemurian age the physical universe was solidifying. Galaxies coalesced; stars separated from nebulae and stirred planetary webs. The Sun was incarnated with a retinue of worlds including Mercury, the Earth, Mars, Jupiter, Neptune, Pluto, etc., each of them with astral sheaths many times their physical girth. Since the astral world is not structured by gravity or space-time, creatures were able to move freely through it without energy or receptacles. Higher beings danced among its sceneries, ecstatic on music of the spheres. They are probably still dancing.

During this epoch those creatures who jumped from the astral onto the Earth became invertebrates. Because the Moon is a hardening and drying system, jellyfish, worms, crabs, insects, and starfish dominated the Mesozoic. In Moon consciousness, there is a perfect correspondence between image and object, so the wisdom of these animals is in their organs, not their minds—in claws and stingers, phosphorescence and shells.

They have no sense of their own presence; they live in their ancestors. They are dreaming, Poppelbaum says, as if it were always the day before yesterday. Spiders arrive emitting a web before they are able to individuate it; glands and spinner hooks remain their sentience. Crabs scuttle about each other, waving claws and changing colors . . . fast asleep. Through grains of soil, earthworms follow undulations of their own bodies. Toads and grasshoppers cheep.

"Animals are fixed ideas incarnate,"[15] wrote Henrik Steffens in 1822; they are souls that took on bodies abruptly, leaving their egos behind on the Old Moon. Each one of them is the physiognomic signature of its astral self.

Though their intelligence is trapped in their organs, invertebrates harbor a soul, a spark of elixir. Carl Jung eulogized: "If the glow-worm could be transformed into a being who knew that he possessed the secret of making light without warmth, that would be a man with an insight and knowledge greater than we have reached."[16]

ALTHOUGH THE INTELLIGENCE of human beings can never enter their peripheral organs, it has been replicated in artificial organs—machines, weapons, money. Ladies and gentlemen obsessed with these things become mockeries of animals. They barter the possibility of becoming spirit for a cornucopia of plastic mirages on one plane. Trapped in fascination with molecular and metallic novelties, instead of joining with collective wisdom and evolving into avatars, they struggle futilely to individualize each new object in their psyches. As they age, however, they harden and, if they lose touch with their astral and etheric envelopes, they become fully materialistic. Their deaths actually bring imagination back into the world, for the consequences of their shortsighted decisions are imprinted on the cosmos, to be addressed by newly incarnating souls. Thus does consciousness itself progress.

All events everywhere in fact are inscribed into the living chemistry of the universe, and from thence new creatures and spirits embody a portion of collective wisdom and activate it in their lives on the various worlds.

COLD-BLOODED VERTEBRATES followed invertebrates out of the Old Moon. These animals have no mind and no voice; they speak only by abrasion of exterior body parts. The croaking frog and hissing snake exist at the boundary of this condition. Salmon, gulls, antelopes, and beings ancestral to them followed—awakening smidgen by smidgen.

Look at the charmed countenance of the wren chick, the wildebeest foal struggling to its feet; they seem almost to be trying to recall another place and time.

The higher animals perceive something of the tragedy that has befallen them; they have a voice but no syntax. They try to express the strangeness and isolation they feel, but their words come out as whistles and honks. Even when triumphant they are inconsolable. In odd moments the eyes of other mammals meet ours, and their expression is one of bewilderment. They are trapped in the dream of the astral body, and their faces reflect either the loneliness or horror of the missed opportunity.

"It is not the ox that bellows," writes Poppelbaum, "not the dog that barks—a bellowing comes from the ox, a barking from the dog—through the dog. From the land of dreams it pours into the world of man. . . . An unredeemed being is striving for expression in it. . . . The voice of an animal is wrung out of it as though by a nightmare; not produced in the free course of breath—it is full of the destiny that is suffered but not understood!"[17]

The Indra lemur and howler monkey seem almost to force their lungs out; they screech from the hollow center of the universe. The turtle bears its ancient mask in silent suffering.

"It is as though something *veiled* were living behind the physiognomy. Something that *craves* to shine through but is withheld by the body's rigidity! This impression becomes positively grotesque and horrible in the case of an insect. Looking at the head of a wasp or a butterfly, perhaps through the magnifying glass, we cannot help almost shuddering. The merciless rigidity and hardness of the casing, out of which the eye, an immobile point, its surfaces walled in, stays there lidless and ever open; that fearful leverwork of the parts of the mouth working mechanically; the hurried jerk and cramped groping of the proboscis; the antennae, always trembling, and yet not looking truly 'alive'—it is as if one saw a ghost, a phantom suspended by invisible threads, that pretends to be alive but in reality is only a moving mechanism."[18]

"All animals," said Steiner, "live (as it were) under the surface of the sea of color and light."[19] For life in a lunar state, existence is a single photograph, snapped once, at birth. The clam huddles in its shell of astral liquid—a facsimile of the cosmos.

Except for the markings that distinguish species from species, nothing personal separates one crab from another, one albatross from another. Their individuality is subsumed in their ancestral type.

They are alive, but they do not know it.

Consciousness has to come from somewhere, has to inherit its dusky, urbane texture from some prior experience. Old Saturn and Lemuria are just intuitive names for those realms. They suggest the transdimensional passage our inner beings traversed before embodiment in cells. They augur that such a passage held precisely the drama revealed in myths and fairy tales.

Migration to Old Earth

WHEN THE EPOCH OF THE OLD MOON came to an end, its more advanced Lemurian beings fled for higher spheres. A bizarre new world was poised to emanate, one they preferred to stand well clear of. The Earth appeared first in a radiant shell as Old Saturn. Eventually, with the cooling of its physical landscape, its aura coalesced. This moment recurs ontogenetically as the circulation of blood through the placenta.

The astral bodies of the other planets in this Solar System were likewise recapitulated; they continued to congeal and harden past the stage of the Old Moon. Souls hovered about all these worlds, but many found such solidification too painful, and swiftly vacated; the rest were attracted to their various geographies.

Archetypal plant forms invaded the Earth well before actual plants could root in its soil. Steiner depicted gigantic trees and vegetables and huge exotic flowers swelling up through the albuminoid atmosphere of this epoch like moss agates in silica gel. Once actual plants rooted in soil, elemental hydrogen streamed into their essence and blossomed into leaves and petals. As each plant withers and decays, "its essence or 'idea' returns to the cosmos, leaving the tiny, largely mineral seed as an anchor, a guarantee that it will reappear on earth when conditions warrant. Investigations of the stratosphere ... [have] discover[ed] clouds of pollen, still mounting skyward, many miles above the earth."[20]

At the core of animalization an inner skeleton mineralized, the vertebrate shape; at its opposite pole a sheath of auras congealed, a vestigial gateway to the astrum, activated regularly during sleep.

The planet teetered at the brink of permanent crystallization, but just as it was about to ossify entirely, a number of spiritual beings interceded and dislodged the heavy Moon from within its body to a nearby orbit. The major impetus of this event did not occur on the astronomical plane, though nightly we see its consequences in a cratered mirror. The dislodging of the Moon was a deeply internal transformation in our sector. Steiner took this act of rescue to mean that *man cannot incarnate himself.* The withdrawal of the Moon dichotomized all subsequent animals into male and female. Outwardly less perfect, less complete, these creatures now had space in which to develop within.

Upon the lifting of the Moon's burden, the human ancestors assumed erect posture and gradually became visible to one another. Their hands and feet differentiated, and nutritive and reproductive organs sprouted as well as nodules for voices.

Rescued from rigidification in lunar bodies, men and women acquired full egos and objective consciousness. This distinguished life on Earth from all their prior incarnations.

DURING THE MESOZOIC and Tertiary epochs young souls hastened onto the terrestrial plane, becoming animals here, leaving their former creature bodies behind in the elevated realms for later souls to use. At the same time, the ancestors of men and women strayed nostalgically in Lemuria and then Atlantis; correspondingly, their embryos are retarded in the womb, remaining flexible and uncommitted.

Crocodiles, chickens, and mice all start as "humans," as coiled knots of flesh with a backbone, nerves radiating from a primitive streak like meteor showers, but they are quickly trapped in hard artifactual organs. The more impulsively they hasten to migrate (cosmogonically), to incarnate (ontogenetically), the less consciousness they retain and the more their wisdom lodges in their organs.

At the closing cusp of the Atlantean era, the esoteric eleventh hour, the last prehuman creatures rushed into bodies: apes followed by prehistoric tribes of hominids—Australopithecines, Pithecanthropines, and Neanderthals. They struggled to the very end to avoid solidification, but they each incarnated an instant too soon.

The embryonic gibbon bears the stamp of prehuman flesh; it is human but human like an old man, wrinkled and hardened, dead to the possibility of ego.

In esoteric tradition man and woman are not descended from the apes, nor, in fact, from any of the life forms on the Earth. They are the penultimate sparks on the astral forge, abrading the whorl at the precise moment at which substance is soft enough to receive their complete psyches.

Humans are the essence holding together the tiers of creation;

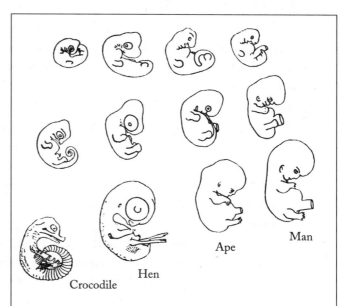

Crocodile Hen Ape Man

FIGURE 27A. Embryonic stages of creatures compared, illustrating the retardation of psyche and spirit relative to body.

From Hermann Poppelbaum, *Man and Animal: Their Essential Difference* (London: Anthroposophical Publishing Co., 1960).

other genera are the side effects of *their* formation.

There are no fossils of our true ancestors because they were not material until they were us.

Spirit and Matter

In Steiner's vision, spirit rushes toward matter. Spiritual evolution descends through Saturn, Sun, and Moon incarnations to meet physical evolution ascending through star debris, minerals, protoplasmic globules, bacteria, protists, plants, invertebrates, fishes, amphibians, reptiles, mammals, primates, hominoids. It is one cycle reflected in disquiparant mediums—a confluence of streams, one coagulating into physical organs, and the other subtilizing as spirit, both spiralling all the way to mitochondria and subatomic particles. Together they form the tree of life.

The bodies of creatures are but shaggy replicas of their souls.

The present-day kingdoms of plants and animals are reminiscent of long-ago partings, says Steiner; they are the equivalents of what we were on other worlds, not this one.

What we see in the tree of life is the shadow of a transdimensional event. While its physical dimensions are self-contained in their own kingdom (much as Darwin recorded it), they are conducted by a spiritual entelechy. Thus, where scientists measure evolution progressing randomly by infinitesimal degrees, complexifying by interpolations of cellular energy into intelligent life forms, Steiner saw exponents of spiritual hierarchies crescendoing down into matter and organizing it.

The truth vaguely adumbrated by Haeckel's theory of recapitulation may have been more than just the significance of heterochrony in speciation; it may also have been the esoteric link between embryogenesis and cosmogenesis. Recapitulation represents not only the effects of heterochronic genes and mutations but the fact that both cells and tissues are organized by a progression of prior worlds which they then enact anachronistically and at a different scale.

There is no doubt that the speciation of zooids is actual and irrevocable, but (from an anthroposophical standpoint) its true source energy is a flowing corridor of hyperspace channeled through membranes.

Where spirit sank too quickly and suddenly, its force was absorbed and frozen (as noted earlier) in minerals and rock crystals. These bumptious characters, multi-hued and enigmatically geometrical in their cubist countenances, are spiritual beings too, but so deeply imbedded in substance they retain no flexibility. They may "think" they are laughing and philosophizing, but it all comes across as a fixed signature—

rhomboid and tetrahedronal. Crystals heal for that reason. They are inflated and self-deluded, but then they just *are*.

As increasingly more spiritual force is withheld, substance becomes softer and more pliant, less trapped in the thrust of metamorphosis. Plants spring from a quantum of restrained spirit, but they are predominantly fossilized.

"The biography of one man," wrote Steiner, "corresponds to the history of a whole species in the animal kingdom."[21] And each animal biography is a hieroglyph. Rushing into matter with assorted skinks, sharks, starlings, and skunks, the flying squirrel hits it like a meteorite and totally fossilizes into its molecular papyrus.

Men and women represent spiritual energy penetrating matter at its deepest and subtlest declivity, in a form that allows them to keep one foot in each dimension. By hanging precariously between worlds, they reenact both the biological and spiritual prehistory of the Earth—its phylogeny and cosmogony—a legacy that our single lives illumine.

Again, occult biologists refute our descent from the apes on this basis: The generalized ancestors of men and women never slid that far into matter; they were present in human form even during the heyday of the mollusk and the dinosaur, not here but in Lemuria. Only our strictly genetic aspect, differentiating feature from feature within groups of animals, was evolving on Earth; our overall human identity did not have a genetic origin.

Recapitulation is the miniaturization of astronomical events in tissue.

Haeckel unwittingly provided spiritualism with a mechanism for getting the macrocosm into the microcosm: recapitulation. Steiner and his associates understood recapitulational embryogenesis to be the sole physical tie between this dimension and all previous ones, between terrestrial matter and celestial spirit. The journeys of souls from ancient world to ancient world were replicated macromolecularly as migrations of gametes. The multidimensional cosmos and the Great Chain of Being were themselves interpolated into matter as amulets. Captured by the genes along with other biomorphologies, they were miniaturized (in the extreme), inverted, signified, coded, and reincarnated in proteins. They still progress in orbits like planets and moons, but choreographed by organelles and cells. The universe itself was thus coiled into a minute spiral, an insignia small enough to arise anew each time in egg-stuff and embryonic tissue.

Haeckel became a reluctant mage, an astrological zoologist, cited as religiously by twentieth-century occultists as Hermes was by Mediaeval alchemists. The recapitulation of cosmogony by both ontogeny and phylogeny is simply a modern version of "As Above, So Below." It is the signature and agency of astronomical events in cell-space.

WITH THE SEPARATION of the first polar body from the oogonium in each meiotic division, according to Karl König, the removal of the Old Sun is recapitulated. As the second polar body is shed in the formation of the mature oocyte, our human ancestors depart from the Old Moon. That is, primordial episodes, even in other dimensions, leave their signatures in germ plasm, and the cells *must* depict them archetypally even as (at the same time) they must obey their morphogenetic regimes. Because of the once-upon-a-time cataclysmic withdrawal of the Moon, the small globe of the yolk sac is drawn inexorably out of the yolk-sac vesicle. Then the billions of souls who streamed to the new Earth become billions of spermatic beings swarming about each egg. Since time is not a factor, ancient events and future ones occur simultaneously. It doesn't make mechanical or biological sense, but it doesn't have to, for it represents a condensation, a transformation far deeper and denser than the infinitesimal space within an atom or a black hole. Cosmological things are held together by meaning rather than thermodynamics, and events too vast to track are tabulated in realms too small to see.

The theosophical scholar Manly P. Hall describes the cell as a microcosm of the planetary fields and their forces: its cytoplasm is a Sun; its centrosome, which activates mitosis, is lunar. Its nucleolus, the seed of the mind, is the planet Venus. Mars recurs as its archoplasm, an emotional sheath and astral membrane. Saturn imposes itself first as the cellular nucleus and then as the emerging nervous system. Jupiter and the outer planets do not form separate orbs but are dissolved in the nucleolus with the human aura.

Conception

THE INDIVIDUAL SOUL, STEINER SAYS, hears the summons at "the midnight hour" and starts its downward descent through the planetary spheres, meeting hierarchic beings along the way and receiving wisdom from them. No words are involved; the transmissions are instantaneous and telepathic. Each of the avatars imparts a testament of priceless esoterica to be woven into tissue as a Guardian Angel or Higher Self. Using this material, the incarnating human establishes his or her spirit-germ, the plan of its earthly body.

Continuing its descent, the soul dons an etheric body as it sinks through the astral element, settling finally in the lunar domain, asleep. At the moment of conception, clad in astral and etheric shrouds, it stirs to a heaven-shattering flash and topples—the zodiacal man-woman, Adam Qadmon—into a newly fertilized ovum.

Hall offers a celebratory version:

"In the ovum is the plastic stuff which is to be molded by the heavenly powers. In the ovum is the sleeping world, awaiting the dawn of manvantaric day. In it lurk the Chhaya forms of time and place. Suddenly above the dark horizon of the ovum appears the blazing spermatic sun. Its ray shoots into the deep. The mother ocean thrills. The sperm follows the ray and vanishes in the mother. The germ achieves immortality by ceasing of itself and continuing in its progeny. . . . The One becomes two; unity is swallowed up in diversity. Fission begins; by cleavage the One releases the many. The gods are released. They group around the Poles. The zones are established. Each of the gods releases from himself a host of lesser spirits. The germ layers come into being. The gods gather about the North Pole. The shape is bent inward upon itself."[22]

Gastrulation begins.

CONCEPTION IS A FIRST RITE OF PUBERTY, the passing of prayersticks through a cell-molecular veil.

Midwife Jeannine Parvati recalls a dream the night one of her own babies imbedded in her endometrium:

"I am being initiated into a secret medicine society—a cluster of women are together, watching as I burrow into soft, rich earth. We are chanting, 'I am entering the sacred circle of mugwort.'"[23]

ONCE THE SPIRIT-GERM FUSES with the fertilized egg it recapitulates all its prior universes in the rich medium of protein, initiating gastrulation and giving rise to three distinct cosmic, biological layers.

Madame H. P. Blavatsky continues:

"Then the embryonic creature begins to shoot out, from the inside outward, its limbs, and develops its features. The eyes are visible as two black dots; the ears, nose, and mouth form depressions, like the points of a pineapple, before they begin to project. The embryo develops into an animal-like foetus—the shape of a tadpole—and like an amphibious reptile lives in the water, and develops from it."[24]

AFTER FERTILIZATION (in König's version) the trophoblast forms first, the temporary dwelling place of the soul; the chorion recapitulates galactic infinitude. Hall interprets the chorion as the atmosphere of the cell, the outer crust corresponding to the corona of the Sun or the shell of a chicken's egg. König's amnion is the astral sea in which the embryo floats; for Hall the amnion is composed of sidereal ether, and the amniotic fluid is a body flowing from "the ten divinities of the sun."[25] Hall's allantoic outpouching of the caudal surface of the yolk sac is a ripple in the etheric current through which vital energies pass onto the world. According to König, Lemuria itself is recapitulated in the allantois; the yolk sac is the Earth, the beach onto which Palaeozoic animals are washed.

Hall considers the swathe of extraembryonic membranes the etheric body of the fetus, the vestigial solidifying aspects of its odyssey of hyperdimensional existences. Each new being is thus recapitulated from and nourished by the rags of its prior incarnations.

THE ENDODERMAL LAYER forms as a recapitulation of Saturnian sap—the mineral body of the plants and animals depositing its aboriginal gut.

Ectoderm is a lunar yarn, the segregating tendency of the Moon transmitting its iterative principle into this tissue sheath. The severance of our lunar body is recapitulated anew through each multiply-induced sense organ. Ectoderm withdraws from itself and reflects back against its interiorized surface; the lens of the eye and the labyrinth of the ear are prompted by polarities from a primal embryogenic field represented by the Earth itself.

A third germ layer, unknown in the simplest animals, develops along the contact zone of ectoderm and endoderm. This mesodermal intrusion is a semi-erogenous ridge upon which actual reproductive organs form. Where mesoderm penetrates endoderm, mesoendoderm forges volition, will, desire. Where nerves pierce mesoderm, feelings flood along filaments and channels into organs. The kidneys, gonads, muscles, and bones all occur in mesodermal pairs under the bilateral influence of the Moon in ectoderm.

Blood

THE BLOOD FORCE OF THE SPIRIT-GERM grounds itself in the trophoblast. Stem cells sow generations of corpuscles. Angioblasts, erythrocytes, and lymphocytes stream as replicas of the first galaxies and stars. The equipotential liquor of plants, invertebrates, and warm-blooded birds and mammals is recapitulated as life-bearing fluids in humans. This is not only a mineral blood; it is a universal etheric liquid combining chlorophyll, hemoglobin, sunlight, and cosmic vibration.

In occult anatomy, blood is a sacred vapor. Ties of iron are not mere metaphors; they are karmic bonds. When the Comanche chief runs a blade across his palm and reaches to clasp the similarly nicked hand of his ally, their pact is signed in the magnetic field of the Earth. Would that arms negotiators and politicians took their vows as seriously!

Blood mysteries go back to Isis, and beyond, to the gods and goddesses of the Stone Age. Through the "smoaky" vapors and emanations of this arcanum all manner of disincarnate entities and ghosts may be summoned, hence, the sacrifice of animals to invoke spirits and supernatural allies. Necromancers likewise summon satanic entities from fumes of blood.

As it dries, blood darkens toward infra-red: the black unseen aspect of the rainbow, the mystery of extrasensory perception, and the channelling of disincarnate (unblooded, undeathed) entities. Sorcerers and shamans have ever sought fractions of the blood of adversaries: to gain access to it is to hold thrall over its soul. This speaks to the "great power of the monthly blood which is shed by women in their periodic sacrifice, the odours and magnetic vibrations of it, which are feared and desired by all men. It is only humans who menstruate, and so it is as though in humans the womb were re-made to bleed in order to open that crack between worlds, as though menstruation were there to activate the blood-threshold, to render it familiar and kind, to overcome the birth trauma and to give resurrection by womb knowledge."[26]

In aboriginal Australia it is "the great Rainbow Serpent which women create by dancing during their menstruation, and which swallows them up and takes them to Heaven."[27]

The brilliant red of fresh blood is an emblem transcending its iron-binding cells. It is the seal of life (and death), the crack between worlds. Blood is the only organ whose presence represents all the others, a metonymy for the person himself, the marker of his or her human existence. When its crimson stain appears out of context, it suggests either a horrible or magical event.

Incarnation

SOMETIME BETWEEN THE SEVENTEENTH and twenty-second day after conception, according to Steiner, the soul implants itself in the emerging embryonic disk [the Hindus also say twenty-two days, the Sikhs forty; Steiner corrects their arithmetic to read *another* forty days after entering the embryo (around sixty altogether) before the individuality injects its human form; but of course "all are describing *aspects* of consciousness/soul—not a fixed *adherable* commodity"[28]].

König says that the etheric element penetrates the chorion and fuses with the amnion; its physical being merges with the yolk sac, its astral layer with the allantois. St. Gregory of Nyssa describes the soul imprinting "as if it were a gem making a stamp upon some soft substance and acting during development from within."[29]

But there is a paradox here: The soul is present from the beginning of the Earth or evolution would not have occurred. Only because its cosmic history has *already* been recapitulated (phylogenetically) can a creature recur (ontogenetically). Only because the soul exists as a cosmogonic seed, embodying the full prior history of the universe, can it locate, imbue, and differentiate the raw genetic lump of gastrulating protoplasm inherited through the humble lineage of mammals.

Haeckel's confusion of acceleration with retardation, fatal to his ultimate status in science, becomes not only inconsequential but actually facilitative when we include cosmogony, for as the more perfected spiritual realm works forward in time toward incarnation, it must go backward toward Primal Intelligence. Carnality is a primitive ancestral quality and at the same time the progressive path life on Earth took in order to get spirit to this staging area. Thus, the human being is a paedomorph for retarding solidification but, contrarily, a recapitulation of innumerable prior astral universes. It must go forward into carbon rings, yet it must reach back to its etheric source.

The genius of Haeckel's "error" is that it accurately represents the dilemma of the soul. The only reason creatures are sent on the long transmigration of lives is to be able to recapitulate past universes while successively embodying more and more of the soul's objective spirituality in each next one. Phylogenesis must advance in time, encumbered as it is with actual physical baggage, radicalized and individuated as it is by the same baggage. Cosmogenesis, however, can go retrograde and forward simultaneously—at one pole toward the celestial domain of its source and at the other toward the prophetic imprint of its ego in flesh.

THE FORCE OF INDIVIDUATION first appears in the primitive pit and then spreads through the agency of the streak. Excitation drives inward as a groove; then it folds over into a vessel—the neural tube. A brilliant ray of dimension-shattering consciousness molds the tissue. The forward thrust of light locates in the head, and its peripheral awareness rolls to the torso, limbs, and tributaries of the nervous system. The finger of God reaches to the finger of Adam in a painting by Michelangelo. The neural groove sinks into the mineral realm, and the notochord is detached. With the separation of the Moon the primordia of a vertebral column ascend one by one. An Atlantean child arises.

As William Blake told us in a former time: "Man has no Body distinct from his Soul...."[30]

It is hard to know why New Age and pop science-writers usually overlook the blastula, gastrula, and organogenesis, favoring Big Bangs, quasars, quarks, cellular intelligence, evolution of consciousness, self-organizing universes, chreodes, and their ilk. Most texts embrace the embryo while staring right through it. They understand what it is, but the mutating, folding, splaying ball is so sober and explicit, they miss its implications.

The embryo's "inherent awareness of the principles of its own nature far transcends the limits of human wisdom."[31] The embryo is the subtlest, most incomprehensible shape that nature has to offer. It is in the embryo that quantum mechanics and uncertainty principles are blatantly exhibited and resolved in a live thing. It is in the embryo that molecular hydrogen, carbon, oxygen, magnesium, and the like (for there is nothing else among its ingredients) are turned into full-fledged jungles with birds. It is in the embryo that "an impersonal power, manifesting as . . . limitless energy radiate[s] to the planets of world-systems without number, streaming to them from their respective suns."[32] It is in the embryo that blind and dumb threads of debris string themselves and one another together in pulsating, sentient bionts. It is in the embryo that junk becomes tissue and tissue develops a capacity for consciousness. It is in the embryo that "the cosmic Life-Breath . . . descends into the abyss of manifestation."[33] It is in the embryo alone that the merciless universe—the stuff exploding in stars and riding on meteors—develops eyes and ears.

The events exhibited by this wriggling, deepening, acrobatic heap lead to everything we know and are, to everything we can say about it or (for that matter) about anything else. Look at it and you are looking at the only bridge between nothing and everything. You are looking at how nature organizes its heart and core, how the breath of Yahweh animates marl, how meaning moves and births itself, how original pagan stone carves an alphabet and writes text.

This is not an automaton or device. It is not a tactic of sperm and egg (or cell clones), for no piece of the embryo knows where it is, what it is, or what it is becoming until it happens, micron by micron and millisecond by millisecond, each time again, anew. There is nothing cool and organized about this, nothing resembling business as usual. It is a hot, bloody, rough forge, throwing living jugs and placental-yolk waste into oceans and atmosphere—swarms of bees, wailing pups, windblown seeds, fluttering fish—without reference to anything we could know or understand.

Yet, far from being a mechanical process divorced from spirit, embryogenesis is the handiwork of the divine, literally funnelling spirit into matter. Its invaginating, layering dance is how spirit looks when it enters matter; it is how matter looks receiving spirit. But, since there is but one cosmos, visible and recondite, this is the same thing. Matter and spirit materialize simultaneously, run the "primitive streak" in tandem—or by cleft harmony. Ontogenesis is spirit's exact signature in mass, down to a fillip and a quillery.

The body *is* the soul; it is the soul's replication in flesh and blood, the collective recapitulation of all its prior incarnations. It is the only shape the soul can make under these conditions, so it is as profound and foliated as matter can get—raw elements sheathed and knotted by spirit.

Embryogenesis must be the most luminously spiritual of all *tangkas* and tarot cards, for it is the exact deck—no symbols, no metaphors; it is the embodiment and spiritualization of the universe; it is spirit writing itself in amino acids and cells.

Look not to holy scripture, which contains mere human injunctions in ideograms, but to the event which breaks through thermodynamics without language, which transmits not in phonemes and graphemes but existence. *It* is the Word of God.

Birth

ONCE THEIR IMAGES HAVE BEEN IMPRESSED, the allantois and the yolk sac disappear—Lemurian and Hyperborean epochs. The amnion continues to develop until, by the end of the seventh week, it is pulsating on its own. As the neural streak and network of nerves spread, the waters of the amnion are most active.

The celestial message passes in ripples through the astrum; the embryo, seeing (or hearing) them in its pineal gland, learns the karmic details of its coming incarnation. It is still clairvoyant, tuned to higher spheres, guided by the formative intelligence of more perfect worlds. Whether we identify this directionless ringing with blood and organelles or with archangels and etheric spirits, the embryo is wiser here than it will be at any point in its existence to come, knowing everything as nothing.

It likewise attends to the clatter and chatter of the "lesser" world it is descending into while its body assembles that world's armament around it.

Gut branches into channels, liver and spleen sprout, and the first chambers of the kidney appear. The era of Atlantis is recapitulated; it is the Tertiary on Earth; within the uterus the unborn child is submerged in an astral flood.

Then the waters burst and the fetus floats out into the cosmos in an ark. The exit from the birth canal completes the transition from one dimensional realm to another. A new being is not just "getting born"; it is plunging across zones of creation, arriving amnesiac, naked and embodied, on an alien planet. An esoteric moment in both the evolution of life and the metempsychosis of spirit is experienced simultaneously and without warning.

As the infant is discharged onto the planet, it is flung from placental shrouds into the magnetic field of its aura, in which it is likewise suspended.

"The placenta is a being," cautions Parvati. "It is of its own ac/cord. The placenta would have a related (but not identical) aura to the mama and the baby. How we treat it and its cord will affect the baby and the mother.... Why do we want to cut and get rid of the placenta?

"The reason, in part, most want to dispose so rapidly of the placenta is the simple fact that it is dying. It's hard to watch a being die. As you watch the newborn lotus baby come fully into life, the placenta goes fully into death. We appreciate watching the old ones pass on as the new ones come in...."[34]

Closer to this process than adults, children remember it more fully, if unconsciously. This makes them into proud, almost haughty émigrés. Adults recall the passage as an obscure intimation that flits by, a sense of the seamless boundary that conceals their entry into this life, evaporating even as they reach to embrace it, lighting the universe for a moment like a match struck in a dark room, revealing its actual lineaments. "We are impressed," concludes Thomas Weihs, "by the daring impossibility of human existence—the biological, natural creature-bondage in which [kindles] the brilliant spark of the divine ego, the dewdrop of the totality of the cosmic divine powers. We often see children [around the age of two] being wheeled in their prams, riding like little kings in a carriage, looking at us adults with the gaze of a detached, divine emperor, not certain whether we are also of the nature of ego."[35]

Creation

MATTER AND CONSCIOUSNESS ARE ANTIPODES OF CREATION. They begin at opposite poles, then spire through space and time, engulfing each other. The texture/riddle of the universe is the appearance of mind within it, a lantern dispersing darkness and void.

There can be no cosmogenesis without beings to experience it.

Even now, fields of hydrogen are sowing solar systems throughout the Milky Way and other, more distant galaxies. Everywhere worlds are being fashioned and distilled in raw molecules, worlds perhaps for souls, for perhaps even our souls at some future stop. Astronomers see the barest reflection of this drama—eddies of matter and energy, far more catastrophic than nuptial—the brightest thing in existence.

THE SHARDS OF CREATION are strewn to the ends of time and space, unbelievably far beyond our limitation of mind. It takes even shafts of light hundreds of thousands of years to cross this span, light which is travelling so fast it would appear to be instantaneous—hundreds of thousands of years! Trillions upon trillions of stars surround the Earth in the void, but, more relevant to us, radiant embryo orbs also penetrate the Earth, awakening through its atmosphere, in its oceans and hollows. The young owl hoots, the sow bug scurries through leaves, fish pulsate in schools. The newborn is a planet separating from a sun. It does not so much leave the womb as enter another womb—the cosmos woven into the night sky.

Whither go ye ye cannot know, but the destiny of souls is connected to the destiny of matter—somehow, some way—and this is the basis of cells.

THE ORNATE SPIRITUAL UNIVERSES of Steiner and Hall are not the only possible ones—we must not overspiritualize the universe; we do so always at risk to the psyche. The universe is quite spiritual enough without our decadent gods and gaudy symbols. Only from the spontaneity of our hearts or the difficulty of our acts do astral bodies and vibrations make any sense, and then only because we connect their lyrics to simply being.

We must finally accept, in light of the harsh reality of being born and dying, that what we are is a continuation of what the universe is, so all our wishes and fears could not be irrelevant to cosmic process; else how could they have occurred? Our wild hopes for rebirth, our dread of hell and extinction are part of the universe too.

The journey is unknown; the path is unknown; what will happen is unknown; what it all means is unknown. This is our only solace in a fathomless, cryptic universe.

THE INEVITABILITY OF DEATH is the same as the inevitability of birth. The forces that brought us here, that acknowledge and cling to life, are the forces that will take us *from* here. If we shun and vilify our certain deaths, then we must in some way deny the fact of our life.

We are in the hands of the gods anyway and, if they are not able captains, we were in trouble long before dying; we were in fact in trouble before being born.

28

Cosmogenesis and Mortality

Creation Tales and Origin Myths

THE ORIGIN MYTHS OF ABORIGINAL PEOPLES contain elements of syzygy, morphogenesis, invagination, differentiation, and emergent properties, though viewed metaphorically (as they must be) rather than in terms of molecules and cells. These cosmogonies encode, within condensed cycles of supernumerary beings, logarithmic series of forms. They could represent the raw undifferentiated stuff of the pre-Cambrian deluge or the mammalian blastocyst.

The matrix of creation is described in turn as an emptiness, a blackness, a loneliness, a void, sterile bottomless waters, arid waterless land, and a dream with no waking. In the Prelude to the Roman poet Ovid's *Metamorphoses*, "Before the ocean was, or earth, or heaven,/Nature was all alike, a shapelessness,/Chaos, so-called, all ruse and lumpy matter,/Nothing but bulk, inert, in whose confusion/Discordant atoms warred. . . . "[1]

In a number of Indo-European cosmologies a germinal cosmic egg (like the ovum) is set afloat upon primeval waters, then is fertilized by an entity or principle that somehow comes to originate outside it. After penetration, the egg sorts into land and sea, earth and sky, and classes of living beings. Its coarser elements congeal; its clearer aspects are drawn upward into sky.

In native North America a procession of other germinal agencies replaces the egg—a bag dangling from the heavens, a raft drifting across an unfathomable sea, a sprig or primordial tree, tobacco smoke, a surprise boll. The Navaho journey follows Mountainway, Shootingway, Red Antway, Big Starway, Handtremblingway, Ghostway, Blessingway, Windway, and Beautyway.

Quarrels between gods or original clan beings (such as the Eskimo Raven brothers or the pan-African twins) can also initiate cosmogenesis.

In many Polynesian tales an original darkness fusing heaven and earth is reconstituted as units of time (tenths, hundredths, thousandths, etc.) which then become Pos, warring nature spirits who unleash creation by their strife.

IN THE FINNISH EPIC TALE *Kalevala,* the entire universe was originally underwater. A female eagle came soaring over the water, searching for a place to brood her egg. She hovers about, apprehensive that if she deposits her prize on the billows, winds and waves will float it away from her. After an exhaustive search she finally sights a small island and, thinking it a hillock, alights there to nest.

This is no ordinary landfall but the knee of the sorceror Väinämöinen, who is awakened by the warmth of her body. A twitch of his knee causes the egg to splash into the sea, breaking apart. But this is also no ordinary eagle. The yolk of its egg disperses into the heavens, its denser zones becoming sun and moon, its shell pieces shattering into the Earth and stars. "In the ooze they were not wasted,/Nor the fragments in the water,/But a wondrous change came o'er them,/And the fragments all grew lovely./From the cracked egg's lower fragment,/Rose the lofty arch of heaven,/From the yolk, [its] upper portion,/Now became the sun's bright lustre;/From the white, [its] upper portion,/Rose the moon that shines so brightly;/ Whatso in the egg was mottled,/Now became the stars in heaven,/Whatso in the egg was brackish,/In the air as cloudlets floated."[2] Cell layers and mesenchyme— in their empyrean equivalents—herald cosmogenesis.

The Pelasgian goddess Eurynome emerged naked out of Chaos. Finding no place to settle, she sifted sea from sky and danced aimlessly atop its waves. As she detoured to the south, a wind stirred in her wake. Here was something "new and apart with which to begin a work of creation. Wheeling about, she caught hold of this north wind, rubbed it between her hands, and behold! the great serpent Ophion. Eurynome danced to warm herself, wildly and more wildly, until Ophion, grown lustful, coiled about those divine limbs and was moved to couple with her."[3] Later she became a dove and laid her Universal Egg upon the waves. Enticed again by Eurynome, Ophion wrapped his body around it seven times, finally hatching it. "Out tumbled . . . sun, moon, planets, stars, the earth with its mountains and rivers, its trees, herbs, and living creatures."[4]

IN ANCIENT EGYPT Nun was the beginning, "limitless chaos, a vast ocean of formless magma including within it the potential of life as well as the principle of consciousness, the god Atum, the Whole One, the Complete."[5] Manifesting from himself a primal seed, Atum cloned the first pair of gods (asexually): Shu, encompassing vastness and emptiness; and his female consort Tefnut, saturating the breeze with life energy. The union of these poles yielded Geb (Earth) and Nut (Sky), Shu raising the body of Nut directly out of her embrace with Geb into overarching heavens. Geb and Nut then gave birth to Osiris and Isis, Seth and Nepthys. Thus did Atum's unity fission into a progression of dichotomies.

THE EAST INDIAN LORD OF ALL, trapped in his own essentiality and irresistible beauty, yearns to see what mongrels he might spew. After preparing a matrix of waters, he casts his own seed upon it. But these are universal waters, and they have been toiling for eternity, "performing fervid devotions,"[6] desiring to be fecundated and exalted. Finally their labors heat them. The seed becomes Hiranyagarbha, a golden egg shining luminescently on the sea. As this egg stirs and evolves, the Lord is reborn in it as the embryonic Brahma. "In that egg, O Brahman, were the continents and seas and mountains, the planets and divisions of the universe, the gods, the demons, and mankind. And this egg was externally invested by seven natural envelopes; or by water, air, fire, ether, and Ahamkara, the origin of the elements, each ten-fold the extent of that which it invested; next came the principle of Intelligence; and, finally, the whole was surrounded by the indiscrete Principle; resembling, thus, the cocoa-nut, filled interiorly with pulp, and exteriorly covered by husk and rind."[7]

After inhabiting Hiranyagarbha for a year, Brahma splits it with his mind. Trying to speak aloud as he reemerges, he realizes that in his new form he can only burble. He stutters, "Bhuvah," which converts instantly to air. Then he goes, "Svah"; there is sky. After these successes he mutters many nonsense syllables; seasonal aspects differentiate and creatures scamper about.

The upper section of the egg sails into starry universe; its lower sector crumbles into matter. Between them floats the trembling atmosphere. Simultaneously out of the egg there arises a complex but primitive being, Purusha, with a thousand feet, a thousand arms, a thousand heads each with a thousand faces bearing a thousand eyes. Within these organs lies the raw material, the embryoblast, of the world, including all its plants and animals. In order to liberate them Purusha must obliterate himself.

IN A JOSHUA NATIVE AMERICAN CREATION TALE, before the world was made, a sweathouse stood by itself on the eternal waters. One of its two dwellers, Xowalaci, smoked tobacco as he discussed with his companion how to make a world. After five days of smoking, "trees began to bud, and fell like drops of water upon the ground."[8] Later Xowalaci fashioned five mudcakes, the first of which he let drop into the watery depths with instructions to make a sound and expand when it reached bottom. The cake took a long time to sink, but Xowalaci finally heard a faint thud. Each successive mudpie he tossed made a louder thump sooner. So did land gradually arise from sea. Then Xowalaci opened a fresh bag of tobacco and began strewing it in the wind to compose mountains and rivers, landscapes without end.

IN A MAIDU MYTH the dark eternal waters were interrupted only by a raft floating on them bearing Turtle and Father-of-the-Secret-Society. Things were placid on this craft until, suddenly, on a rope of feathers, the Earth-Initiate landed in the water from above, tied his rope to the raft, and pulled himself on board. Turtle immediately asked him to make some dry land so that he could rest between his long dives. Earth-Initiate was agreeable to this but said that he needed some soil to work with. Turtle offered to fetch it if Earth-Initiate would tie a rock to Turtle's left arm. Earth-Initiate affixed it thusly and also lassoed Turtle with a rope and notched it to the raft. As Turtle hit the water with a splash, Father-of-the-Secret-Society (without warning) began chanting.

For six whole years Turtle was unaccounted for in the great deep and, when he returned, he was covered with green slime. Only a tiny bit of primal earth was stuck under his fingernails; the rest had been washed off in his ascent. Unperturbed, Earth-Initiate drew out a fine knife and proceeded to carve this portion, then set it to rest on the bow of the raft. Gradually it expanded, first into a ball, later into an orb the size of the world. The raft came ashore finally at Ta'doiko.[9]

In the Assiniboine version, Muskrat makes the dive, Frog chants, and Inktonmi, wearing a wolf-skin robe, creates men and horses out of the dirt from Muskrat's claws.[10]

A medley of cartoonish creatures, cosmological soil, incommensurate scales, and anachronistic time frames also characterizes embryogenesis.

AFRICAN BANTU ETIOLOGY opens with the demigod Bumba alone in the dark with nothing but water around him. "One day Bumba was in terrible pain. He retched and strained and vomited up the sun. After that, light spread over everything. The heat of the sun dried up the water until the black edges of the world began to show.... [S]andbanks and reefs could be seen.... Bumba vomited up the moon and then the stars, and after that the night had its own light also."[11] Still agonized, Bumba continued to strain and out of his body came leopards, crested eagles, crocodiles, one Yo fish, tortoises, lightning (Tsetse), white herons, one beetle, and goats. Later Tsetse, the sole troublemaker of the lot, had to be banished to the sky. Still mischievous, she sometimes "leaps down and strikes the earth and causes damage."[12]

AUSTRALIAN ARANDA GENESIS begins with perpetual darkness "oppress[ing] all the earth like an impenetrable thicket."[13] The ancestor Karora lies asleep in what will one day be the Ibalintja Soak but is now dry. Red flowers and grasses grow all about him, and from a thick purple bed a great *tnatantja* (a ceremonial pole) swings majestically in the breeze way above his head, "as though it would stroke the very vault of the heavens."[14] This pole was also a living creature with human-like skin. While Karora dreamed, he rested against the pole and, though he had lain this way since the beginning of time, all of a sudden "wishes and desires flashed through his mind. Bandicoots began to come out from his navel and his armpits."[15] A wooden bull-roarer (for making sound by twirling during ceremonies) emerged from under his armpit, assumed human lineaments, and became his first son.

A PAN-POLYNESIAN CREATION MYTH features an original deity, Io, in a murky, shifting void. As she breathes outward, her tumult spreads through nothingness, segregating into Papa and Rangi, female and male, respectively; also Earth and Heaven. As they embrace, life is ignited but lies trapped between them, susceptible to neither illumination nor shadow—a sterile mesoderm. The children of Papa and Rangi take on the burden of freeing this preanimate mass. One of them, Tan (later to be the god of forests and birds), wedges his parents apart and, using his own body as a pillar (the primitive streak), lifts Rangi above Papa. Tankiri, god of wind and storms, rushes in between Earth and Sky (as the neural groove) and opens up their median domain to the differentiating forces of nature.[16]

"God allowed His unconscious to create an image of the material attempting to come into the light of consciousness."

IN THE HEBREW QABALISTIC TRADITION, the universe arose from Ein Sof, an unknowable first cause. Ein Sof is not only unknowable in the sense of quarks and quasars; it is unknowable in the metaphysical sense, for the reality it represents is folded into creation in such a way that it occurs nowhere as a result of insinuating itself everywhere, down to the smallest shuttle of mind and sprinkle of subatomic dust.

"When the supernal emanator wished to create this material universe, it withdrew its presence. At first Ein Sof filled everything. Now, still, even an inanimate stone is illuminated by it; otherwise the stone could not exist at all—it would disintegrate."[17]

Ein Sof is a candle igniting trillions of other candles across infinity. Their brightnesses and hues differ in every imaginable way, but they all "manifest through differentiation."[18] The light's simultaneous suffusion and withdrawal, absence and presence, is central to its mystery. "If the light were completely hidden, the world could not exist for even a moment! Rather it is hidden and sown like a seed that gives birth to seeds and fruit."[19]

Creation arrives and departs uncountable times before its essence resolves in a droplet which develops opacity and density. As this bead actualizes and disperses, the emanator takes a ray from the original source and reillumines the mass—a single beam so as to splinter the primeval monad into algebras, for it was sterile as a block.

The geography of the universe is represented now by the Tree of Life and its ten *sefirot*. Translated into English, these are Power, Love, Beauty, Splendor, Foundation, Presence, Eternity, etc., but their Hebrew names, even untranslated, give a truer sense of their seed nature: Gevurah, Hesed, Tif'eret, Hod, Yesod, Shekhinah, Netsah, Binah, Keter, Hokhmah.

As cosmic light poured through the sefirot, some structures could not hold such a powerful emanation and shattered. Most of the radiance recoiled to its source, but a rain of sparks fell among the remnants of broken vessels. Unrevealed and unactualized, these became lost in the emerging material universe. Human destiny is a process of *tiqqun,* or attempting to collect the debris of creation and repair the damage. The task of restoring the microcosm and recovering its alphabet lies at the roots of most forms of Gnosticism, Qabalism, and ritual magic.

WHEN GOD IS PERSONIFIED in Genesis, Hindu mythology, and other etiological tales, he is a being like any other: he must confront his basic inertial resistance to the creativity of his own psyche. Psychologist Charles Poncé points out that gods become worldmakers only because they are lonely—unbearably so:

"Ultimately, loneliness is something or someone representative or symbolic of an interior component of ourselves that longs to be united with us, or that we long to be united with. Expressed psychologically, this yearning is for something that is not known to us consciously, that is hidden deep within our souls."[20]

Jehovah, Brahman, Io, and the Winnebago Earthmaker are powerful but sterile demiurges unless they can gain access to the spontaneous emanations of their unconscious minds.

"In an attempt to discover the hidden aspect of Himself with which He wished to be united, God allowed His unconscious to create an image of the material attempting to come into the light of consciousness. In much the same manner that the unconscious in human beings creates dreams reflective of our hidden nature, so too did God spin out of His unconscious that aspect of His hidden nature with which He was not reconciled. Whereas our unconscious lives are ephemeral and without material substance, His unconscious life took on substance.... His images were to live themselves forward regardless of His commandments in much the same manner [as] the process of our unconscious lives [must] live themselves forward regardless of our conscious and rational commands.... In other words, Adam symbolized ... the unknown portion of God's psyche that constituted His blind spot."[21]

The Ray of Creation and Its Octaves

ACCORDING TO THE RUSSIAN OCCULT SCIENTIST G. I. GURDJIEFF, a ray of creation originated outside time and space. Entering the cosmos at a speed faster than that of light, its force disintegrated into galaxies, splintering (without diminishment) into the star systems they spawned. From there it exploded into individual suns, and from these nodal centers into single worlds, such as the Earth-Moon system. Creation was unavoidable, for in order to cross the gaps between domains the ray had to fill enormous intervals with shocks, each one equivalent (at transgalactic scale) to the jump from one musical note to another. The shock of the first octave occurred between the Absolute and the Sun, and its effect is beyond our knowing. Subsequent shocks were filled, one at a time, by spontaneous manifestations of Sun, Earth, and organic life—stars, oceans, glaciers, rivers, and kingdoms of animals (from the local versions of protists or bacteria in remote galaxies to fishlike entities on oceanic moons in the Milky Way).

For Gurdjieff, atoms are cosmic notes vibrating within and between scales. Each element has chemical, cosmic, and psychic properties which take on discrete materializations in different zones. For instance, there are hydrogens on levels from 6 to 12,288, with hydrogen 384 defined as water, hydrogen 192 as breathable air, and hydrogen 96 as rarefied gas. Hydrogens 48, 24, 12, and 6 are identified as "matters unknown to physics and chemistry, matters of our psychic and spiritual life on different levels."[22]

The goal of consciousness is lodged in every atom everywhere in the universe. In fact, it is nowhere else. Elemental particles build toward mind and spirit on worlds, expressing aspects of their arcane nature in the morphologies they assemble.

Consciousness does not evolve from matter, says Gurdjieff; componentially it *is* matter transmuted by organisms. Molecules of food and air, even while they are being metabolized, are metamorphosed as well into molecules of mind and spirit. Subtle impressions are likewise "digested" and, once vitalized, transmit motion and energy. Without an influx of images, animals could not survive an instant.

Gurdjieff defines human beings as creatures with a special ability to translate raw molecules into their higher spiritual octaves. They *must* do so, or their souls will perish. The upper realms alone are stable. All other substances, even the psyches of men and women, are broken up atomically into the chaff and jetsam of worlds and stars. The souls of the damned, in this manner, illumine the universe. They were once minds; they are now photons. Their hell is light.

Only by breaking with compulsive patterns of behavior and habitual thought can men and women grasp their predicament and then, through self-awareness, transmute the substance of their being across the critical interval to the next octave. This is the equivalent of heaven—a consciousness body.

The difficulty of evolving is immense, for our actual present situation is so uncomfortable we do not want to experience it. We cannot bear it for even an instant. This is a matter of physics, not weakness of resolve. "Man *can* awaken," Gurdjieff proclaimed. "Theoretically he can, but practically it is almost impossible because as soon as a man awakens for a moment and opens his eyes, all the forces that caused him to fall asleep begin to act upon him with tenfold energy and he immediately falls asleep again, very often *dreaming* that he is awake or is awakening. . . .

"A man may be awakened by an alarm clock. But the trouble is that a man gets accustomed to the alarm clock far too quickly, he ceases to hear it. Many alarm clocks are necessary and always new ones."[23] (Otherwise we will surround ourselves with alarm clocks that keep us asleep.)

Elsewhere Gurdjieff counselled, "You must accustom yourself to struggle. Little by little this struggle will give results which will accumulate within you. . . . You fail one time, ten times. But each struggle brings results, a substance accumulates in you. . . . You will fail ten times, or even twenty times, but the twenty-first time you will be able to . . . carry out conscious decisions."[24] Without any change in appearance you will *become* a higher form of matter.

Gurdjieff accepted Darwin's fiat of evolution by natural selection, but only as a shadow of cosmic evolution—genetic mutations are one method of shock. The goal of speciation is not simply diversity and filling of ecological niches. For a planet to maintain life at all, some species in its biosphere must graduate to making souls. If consciousness on a world fails (perhaps through over-mechanization), then the ray of creation will wither from that place and every living thing in it will perish. As long as the process of soul-making continues, all molecules on that world (even the oxygen in its atmosphere and the phosphorus in its crust) can one day become souls.

This is the promise of the Bodhisattva—to postpone her own passage into heaven until every sparrow and dab of oxygen has preceded her. Buddhist scripture measures the immense time for this migration of souls in terms of a great rock that is grazed only by the feathery wing of a circling dove. When the rock has been fully eroded into dust by the circuits of the bird, then enlightenment will dawn across all the skies of the universe.

Teilhard's Sun Lattice and Noosphere

GURDJIEFF'S COSMOGONY REFLECTS the Afghan and Kurdish crossroads of Middle Eastern civilization; it is Islamic and Zoroastrian, with Hindu and Buddhist elements. Pierre Teilhard de Chardin, a French priest and archaeologist, proposed a similar mingling of matter and spirit but with Christian overtones and quite different consequences. In 1955, in *The Phenomenon of Man,* Teilhard described how the spirit fire that formed the Sun endowed its planets with vital, living interiors. Once these orbs were dispersed into the icy darkness of space they sought to become stars again. Lacking the gravitational force necessary to condense and ignite, they wove rocky and methane worlds instead. The Earth (perhaps unique among them) transferred information from its core into diaphanous microtubules and protein threads, iconicizing billions of ersatz stars in eyespots and ganglia of creatures, totems and ornaments of tribes and empires. Kindling again and again in the weightless medium of the mind, the Sun minted an entire currency of stunning coins, each replicating a different aspect of unattainable fire.

When we look at Sol today, we find ourselves asking: is this just a coarse behemoth of hydrogen and helium, incalculably simpler than each of us, yet the source of all the stuff of our bodies and minds; or is this a metadimensional object of a great enough complexity to house within its catacombs the potential of every terrestrial creature and object of thought, every Martian and Jovian and Ganymedian thing likewise, as well as glossaries of countless possible worlds? Is the Sun merely a proximal source of light and energy; or is it our true abode?

"When I speak of the 'within' of the Earth," Teilhard wrote, "I do not of course mean (the) material depths ... only a few miles beneath our feet. The *'within'* is used here ... to denote the 'psychic' face of that portion of the stuff of the cosmos enclosed from the beginning of time within the narrow scope of the early Earth. In that fragment of sidereal matter ... as in every other part of the universe, the exterior world must inevitably be lined at every point with an interior one.... By the very fact of individualisation of our planet, a certain mass of elementary consciousness was originally emprisoned in the matter of the earth."[25]

Myriad infinitesimal centers, having germinated on the Sun, transmit their essences to element grids on icy worlds. As these uncoil, they bind together in lattices and patterns; cells and nucleic acids form; and creatures emerge from the vortices—first, plants which strain tropistically toward their home in the Sun, and then animals who translate etheric fire into semes and odes and erect Apollonian monuments.

The interior face of matter ignites in a replica of the invisible, spiritualized Sun. This expansion of consciousness will ultimately penetrate and transform the entire planet, adding a layer to the biosphere even as the biosphere formed atop the geosphere. In true Gnostic fashion, this mind-sphere (noosphere) will provide for the salvation of not only the self-aware creatures contributing to it but "life itself in its organic totality."[26] In a reversal of Steiner's cosmology, the Earth will return to the Sun, but as spirit not fire.

The emerging noosphere is "yet another membrane in the majestic assembly of telluric layers. A glow ripples outward from the first spark of conscious reflection. The point of ignition grows larger. The fire spreads in ever widening circles till finally the whole planet is covered with incandescence.... Much more coherent and just as extensive as any preceding layer, it is really a new layer, the 'thinking layer,' which, since its germination at the end of the Tertiary period, has spread over and above the world of plants and animals."[27]

Unlike Gurdjieff's mage-creator, Teilhard's Christ wastes no souls and discards no pilgrims. Yet we must still be singly incarnated. Experience and suffering are individuated, never universalized. "*Cosmic embryogenesis* in no way invalidates the reality of . . . *historic birth*,"[28] Teilhard warns. We live, and then we die, though the cosmos may go on evolving and liberating souls forever.

Amnesia Tunnels and Overtones

D RUNVALO MELCHIZEDEK, the affable walk-in from the thirteenth dimension, proposes a universe made up of 144 dimensions, each of them containing a dozen distinct overtones on which utterly different realities and landscapes operate. These are wavelengths or frequencies not just of light but of matter and mind. Being on one or another overtone means being constituted of the stuff that passes locally for matter and using a native version of energy. For instance, atoms and gravity exist uniquely on our overtone; other overtones have their own equivalents.

Birth and death are not modes of origination or extinction but vestibules between overtones, like tunnels for trains. Cell-orchestrated transit in a uterus *(udero* = universe) is one vehicle of passage, the visible instrumentality for the transport of new beings in biologies throughout the cosmos. The womb is a spaceship, but all of its motion is "in" and "inside-out."

Melchizedek's premise, reminiscent of Rudolf Steiner's, is that all of us were here at the beginning and all of us will be around at the end; spirit is never created, never dies. What occurs, from lifetime to lifetime, is a permutation of dimensions and overtones, creatures assuming new bodies and psyches in an overall process of evolution. When the Sun burns out or the universe contracts into a singularity, we will survive that cataclysm too, for the demise of the cosmos will merely be the withering of its worlds on one overtone, an autumn wind stripping dead leaves from live roots.

In the ordinary course of things, when people die, they travel instantly to the third overtone of the fourth dimension. There they wander, memoryless and unknowing, until they return to our third dimension via reembodiment. When they're here, they have no memory of there, and when they're in the third overtone or the womb, they have no memory of here.

"It's kind of like a long trip to nowhere," says Melchizedek's student Bob Frissell, "an amnesia tunnel. We learn to forget real, real good."[29] Despite forgetting, we carry our traumas of death, rebirth, and suffering back and forth with us in a cycle between dimensions and lifetimes, which results in further memory loss and shallow breathing.

For a developed individual, there are two ways out of the trap: Like the historical Jesus of Nazareth, a person can resurrect himself or herself, that is, leave his corpse behind and, soaring past amnesiac overtones, reconstruct a light body in the tenth, eleventh, or twelfth overtone of the fourth dimension—levels of unity (or Christ) consciousness. (Rudolf Steiner claims that when Christ was resurrected, the unique genetic body of Jesus of Nazareth was transubstantiated, so never reappeared on Earth; it was the universal archetype of Man that returned to the Apostles.)

In the upper overtones of the fourth dimension all prior lifetimes are lucidly recalled; there is never again a break in memory. Thus, any future birth from there will be a willingly commissioned trip down the dimensional ladder in a protein vehicle hatched from an egg.

In ascension, an individual turns her body into a ball of light and takes it with her. There is no corpse, merely dematerialization followed by rematerialization on the higher overtones of the fourth dimension. If all of the inhabitants of the Earth did this together, the whole planet would move to the new overtone with a different landscape and biosphere.

Melchizedek's view of Atlantis is that it was more than just a prehistoric island submerged by a tidal wave or a world in the astral realm; the Atlanteans were a secular but advanced civilization that occupied the Earth some 20,000 years ago on a higher overtone of the fourth dimension.

After an asteroid struck the three-dimensional Earth at a point near Charlestown, South Carolina, about 16,000 years ago (its impact rippling across the dimensions), the citizens of Atlantis vowed to protect themselves from future celestial encounters.

They then manufactured a forbidden device out of counter-rotating fields, a so-called *merkabah,* to either vaporize approaching comets (and the like) or transport the populace to another plane. But their emotional bodies were not developed enough for the *merkabah* to operate on its natural fuel, the physics of love. Overly mechanistic, it ripped open the dimensional levels and let in hordes of demonic spirits. These drove the Earth down many overtones to its present dimensional level where, as Frissell puts it, we "bumped our head and forgot who we were."[30] We will continue to be reborn at this level until the Earth as a whole makes a dimensional shift.

His asshole is a portal to another dimension.

COSMOGONIES CONTAIN MEMORY TRACES and intuitions of things that cannot be known in any other way; in fact, that cannot be known at all. They overlap with one another in features while contradicting in other features. Each contains adumbrations of truth and other elements that sound straight out of comic books or science fiction. That is because our situation is basically and fundamentally ludicrous. In the words of Kinky Friedman of the Texas Jewboys, "somewhere between the fun and the sun and the rum and the gun you follow that last airplane picture right up into the sky. A window seat to limbo if the Catholic Church is correct, where you fly a tight, tedious holding pattern through night and fog for at least a thousand years with a small Aryan child kicking the seat behind you, while next to you a fat man from Des Moines is locked in a hideous rictus of eternal vomiting upon the half-completed crossword puzzle that is all of our lives."[31]

At a certain level, existence is little more than gallows humor.

We are embodied in this world in a way that Donald Duck might have dreamed up—hives of hyperactive, cannibalistic corpuscles, "all ruse and lumpy matter" (Ovid). Somehow these brainless microbes are stuck together seamlessly in such a way that their cackling is muffled and a big guy and a big gal step out of the shower. Then we clamber about, Abbot and Costello, Othello and Lear, Betty and Veronica. We get removed Batman-style. No wonder "Ed the Happy Clown" (in Chester Brown's graphic novel) shits more than he could have possibly ever eaten; it turns out that his asshole is a portal to another dimension which has its own ecocrisis, its own monetary system, and even its own Ronald Reagan![32]

Our actual situation is infinitely more complex and dangerous, more surreal, than any of the scenarios in this chapter. Yet, proposed from their different vantages of time and space (from the teepees of the North American prairie B.C.E. to the temples of Mediaeval Hebrew philosophers to the Dreamtime billabongs of southern Pacific deserts) they foreshadow and reflect something resembling the complexly layered dimensionality of the real universe. They are our "being" maps at various degrees of sanctitude and irreverence, piety and blasphemy.

Even those myths solemnly provided by the scions of microbiology and astrophysics (who themselves are but a few millennia removed from the Stone Age)—when measured against the real vastness of the cosmos and the vulnerability of our incarnation—share more with Pueblo tales of travelling to this zone behind a crow or gopher nudging through a hole than they do with the ultimate superstring theory of an advanced technological civilization.

We are still groping in Muskrat's mud.

The creationists may be right, that evolution is just a theory, but at least it is a theory with an intention to look unblinkingly at the revealed universe. Creationism is a sophistic ploy to claim privilege and jurisdiction for the mythical godhead of one benighted tribe. It has no spiritual standing at all beside the Lakota medicine man dispatching tobacco smoke into the four winds of the great mystery. *That* humble, profound act is what should be taught in schools to balance Darwinism.

Astrogenesis

THE ASSEMBLY OF OUR BEING is the central event of our lives. We begin as a mote of jelly; almost hourly our tissues and membranes congeal and twist until we are a minor sea creature. Gradually organs layer and fuse, but they are neither perfect nor static. Over weeks they are shifted and rearranged to a degree that would be sheer agony to a mature man or woman. The tumult within the embryo is worse than any war wounds or electroshock torture. There is no safety in fettle or mass, no identity, no recourse. Organs being made are torn apart; nerves sear through membranes; arms and legs swell outward and fragment into joints and digits; our bones and ligaments are all stretched and broken; the brain throbs with nerve processes and nodules. It is more than an operation; it is once-in-a-lifetime primal surgery. And yet the "fragile" babe in the womb experiences this turmoil in silence and seemingly without pain. Some would even claim it is an ecstatic ride. The laying down of tissue is visionary and cosmic to a degree that no subsequent state of consciousness can recapture. Time and scale change parameters, panoramas shift from within; the wall of shaping flesh becomes the universe.

The roughly 280 days in the womb, says the Gurdjieffian astronomer Rodney Collin, are a full third of our life. Cell division and differentiation move thousands of times faster than events in villages and cities. Everything we are is constructed there, and each instant of the embryo's life is packed with breathtaking events. Without reference points in language we cannot remember them as what they are, so we awake at birth remembering them as what *we* are, and the rest of our life becomes footnotes to this primal adventure.

The denizens of the zodiac are also involved, Collin tells us: At critical stages of embryogenesis, glands and other organs are activated by resonances from the Solar System. The Sun/heart is the center of a scale of spirals which winds through the planets (glands) and then, in each organism, uncoils world by world (organ by organ) like a spring. Every endocrine function is a receiving set and transformer for a discrete planetary influence. Specifically, a gland (or the nerve-plexus associated with it) responds only to the electromagnetism of one planet, a message which is strongest when that planet is at its zenith relative to this world and gradually dwindles as the planet approaches and sinks below the horizon. Since the heavens have a unique configuration at the moment of fertilization, the influences of the planets on the glands follow a distinctive mathematical progression based on their orbits. The invisible "long body" they would spin in the heavens if they trailed luminescent threads behind them they in fact weave in each torso through the glands.

The symbolic relationship between the heart and the Sun (between the bloodstream and light itself) is incarnated as one force in two mediums, warming and nourishing everything from the remotest planets and organs to the periphery of the body. One universal engine carries oxygen, hydrogen, nitrogen, and organic carbon back to the Sun. One force synchronizes cosmic and microcosmic metabolism. Just as planets diffuse the Sun's material into the Solar System—not only as primary radiation but plurally back and forth between their orbs—so do glands and organs refract one another's modulations back and forth through the bloodstream. As the Earth recovers its reflected sunlight in decreasing microdoses from the Moon, Venus, Mars, Ceres, Uranus, etc., every organ reacts to the altered polypeptides of the enzymes it disperses.

Depending on which planet is strongest, first at conception and then at birth, different types of personalities arise. The first three glands are endodermal: the thymus inciting growth leads to the Solar personality, the pancreas spreading digestion and assimilation is Lunar, and the thyroid governing respiration is Mercurial. The next circuit of the helix is mesodermal: the parathyroid circulating blood is Venusian, the adrenals which infuse the cerebrospinal and voluntary muscles are Martian (martial), and the posterior pituitary influencing sensation and the sympathetic muscles is Jovian (jovial). The third spiral is ectodermal: the anterior pituitary awakening the mind is Saturnine, the gonads (eros) are Uranian, and the Neptunian pineal organ has an unknown function associated with tachyon energy beyond the speed of light.

"Man falls through time as solid objects fall through air."

THE SECOND THIRD OF HUMAN LIFE, Collin says, begins with birth and lasts 2800 days, or to the being's seventh birthday. This is the time of childhood when the organism comprehends its individual existence, learns to respond to its habitat, and acquires language. Once a child speaks he enters a dialogue with himself. He awakens from the shadows of all previous existences. This initiation marks the completion of a second "embryogenesis." Without language a creature is wild and free, as pagan as a fetus.

Writing has served a similar function in the development of civilization: Tribal peoples often struggle to remain illiterate because once they are able to read and write they can be taxed.

The last third of human life stretches from 2800 to 28,080 days when, for a brief moment, the planets sit in roughly the configuration they had when the sperm penetrated the ovum. Saturn-return in the prime of life has been publicized as the astrological counterpoint to the midlife crisis, but in fact all the planets return, some of them once, most of them many times.

The whole of adult life is no more than seven years of childhood or 280 days in the womb.

"Thus between conception and death man's life moves faster and faster until at the end the hours and minutes pass for him a thousand times faster than they did in the hours of his conception. This means that less and less happens to him in each hour as life progresses. His perception spreads over a longer and longer period, but in fact this longer period is only an illusion since it may contain no more than did the infinitesimal fraction of a second of his first sensation.

"He thinks to tame time by measuring its passage in years, but time cheats him by putting less and less into them. So that when he looks back over his life and tries to calculate it by the scale of birthdays, he is in a strange way foreshortening his existence, like a man looking at a picture which elusively curves away from him. In another figure, we can say that man falls through time as solid objects fall through air—that is, gaining momentum, or passing faster and faster through the medium as he goes."[33]

"The universe is not a machine of death."

"IT IS A GOOD DAY FOR DYING," announced the old Indian warrior, supine in the tipi of his son.[34] "Dying and living again," the Tibetans tell us through their classic, the *Bardo Thötröl*. We are in an unbroken cycle of cycles; even the condemned killer realizes as he awaits the denouement of this existence:

"... just me Gary Gilmore thief and murderer. Crazy Gary. Who will one day have a dream that he was a guy named GARY in 20th century America and that there was something very wrong...."[35]

"All up and down the whole Creation/Sadly I roam," mourns Stephen Foster in his melody about the Suwannee River—a dirge in the guise of a folk song.

"It is night on planet Earth, and I'm alive," shouts Jeff, a slacker in Eric Bogosian's play *SubUrbia*. "And someday I'll be dead. Someday I'll just be bones in a box, but right now I'm not, and anything is possible."[36]

Anything and nothing.

What does life amount to? The throbbing of cardiac muscle? Rolling layers of tissues and nerves? Primal breaths of blood and air? Radiance of the first light in the cortex? We live long past the psychic wholeness of the beginning, but we can never add to its luster. The sky of childhood is still blue, the apples red, and the flowers in the faded field are yellow. In the end, life will vanish into a darkness that apparently persists past the end of time. But these too are just words. Being (itself) is of ungaugeable depth.

"... *Still longing for the old plantation/And for the old folks at home."* Home?

PERHAPS ONE ONTOGENESIS LEADS TO ANOTHER. Just as the fetus was born from water into air, so we will shed this body as we enter the world of vibration, *"far, far away."* From the uterine womb we crawl onto the womb of the Earth; from this oceanic world we pass into another realm, and then another, each with its own *"Suwannee,"* so the heart must be turning ever.[37] The old man, the old woman are babies, about to be born—their wrinkled flesh a husk. In this view we are wrenched from each incarnation for a dematerialized transit to the next.

Such apotheoses are easily proposed but not necessarily comforting. There is no way out of cosmogenesis, whatever it is. Birth is always reincarnation, to equip us for a mode of existence unimaginable before it occurs. We are passing landmarkless through stuff and complication vaster and more unfathomed than any mysterious ocean on any world.

Our mortality startles us again and again. After all, we give everything to this condition—our memory, our plans, our identity, our body, our mind, our children. If these are taken away, what is left?

And yet we continue to eat, play, and raise children, make love, read books, and fight for what we think we believe in.

After we are no longer alive, all of eternity will happen without us.

Awakening suddenly in the middle of the night, we perceive infinite time exploding beyond our existence; we stare through the dark room at a universe that has existed for trillions of years before we were conceived.

So macabre, so astonishing, so unlikely! In the context of all this our life seems a worthless thing, a trick. But who is there to trick?

Why go through all that trouble to put us here for so short a time? Why equip creatures with the *dharma* on a flight so skewed and swift?

"The great message of the universe is not that you survive," Adi Da Samraj told his disciples. "It is that you are awakened into a process in which nothing ultimately survives."[38] But it is not, either, that everything is destroyed. Nothing, in fact, is.

"Yes—it is true that we are not smothered, ended, murdered. The universe is not a machine of death."[39]

Despite the indisputable vastness of galactic fields, consciousness (with humble, base origins in atoms and stellar debris) is immeasurably denser and brighter, and ultimately encompasses the cosmos, intimating an omniscient origin and a role in melding the very stuff from which its corpus is twilled.

BUT CAN SUCH KNOWLEDGE HELP US? If a great teacher shows us, in a moment of lucidity, that we are immortal, does it change our situation? He replaces our few remaining seconds with billions upon billions of years of cosmic transit and unfolding—but nothing changes. Our suffering and longing do not end. Our mystery is no less profound and eternal.

The life we are living, in its ordinary sense, is what we are. The universe will never tell us anything definitive about our fate. It will continue to contradict itself and foil any grasping after certainty. We survive only as what we are, and what we are must change—always, *always*.

In our happiest moments, what sustains us more than pleasure is the mystery itself, not knowing who this is in us. The blind riddle of existence is what makes it possible to live at all, in darkness, at the heart of danger. Being full and happy occurs only in a glimmer-spark floating through an eternity of star-masses.

Full of what? Happy for what reason? Because at this moment, made of cells, made (in effect) of nothing, we don't know and don't care but are heartened by the marvel, the excruciating ephemerality, and the sheer unlikeliness of this all—and that is good enough and maybe even complete.

BY THE TIME WE ARE BORN, just about everything of importance has been done. We are well on our way to death. The embryo shows us that we are in a process of change that cannot be stopped. Everything in the universe is moving, enveloping structure in new structure. The embryo mirrors how we are incarnating through substance—that our mind is substance, and a mind-like intelligence shapes it, whether that be in the genes, morphogenetic fields, or an etheric body.

Darkness is the only possible path—thickets, shadows, and a nostalgia for something unknown. Yet, by the same measure, in an expanse of dandelions and clover (photons, bird calls, sugar tinge, propelled insect crystals, water spooling horizon-line in cumulus mounds), what is here is absolute . . . is an entirety of the resistance of atoms through atoms, tilth of cells upon cells, assembling masks through which creation glimpses itself.

We cannot hope, said T. S. Eliot (in his "Quartets"), because invariably we hope for the wrong thing. If phantasms led us, universe after universe, we never would have gotten here, with all its grit and luscious complication.

Likewise our fears would have caused us to flee beleaguered paths into existence—although their fire and spiritedness (embodied by predators and prey) are a requisite of all life and psyche.

We must wait too "without love/For love would be love of the wrong thing."[40]

MOST SPIRITUAL JARGON is so much whistling in the dark. If a bigness surrounds us and floods down through our becoming, it would not necessarily appear as a guise of rebirth or life everlasting. It might not even be sweet. *("All the world is sad and dreary,/Everywhere I roam. . . .")*

We might well experience it first as the surety of our mortality, as the veil of sorrows into which all sentient beings awake. When we are most nihilistic, most sure of the finality of death and the meaninglessness of life, we may be closest to a deathless psyche. That would be the *shadow* cast by the soul as it passes into the resistance of this world, the opaque substance of molecular existence.

Through sheer terror of embodiment we experience the intuition of spirit. Yet that terror is the clue to any immortality we have—and this takes precedent over any tales of heaven or, for that matter, of hell. "Lord, your mercy is stretched so thin," wrote the poet Edward Dorn in a journal by his deathbed, "to accommodate the need/of the trembling earth. . . ."

All the discussions of spiritual embryogenesis become a flutter of sanctimonious fancies when we have to turn inward to the darkness and to our own nightmares for the birth of anything real. Dorn: ". . . the white Rose, whose/house is light against the/threatening darkness."

But then this is where we turned in the first place in order to become.

29

Death and Reincarnation

Embryogenesis Outside the Womb

Throughout this book I have emphasized a genetic, cellular event that fashions holographs, keeps them functioning at each stage of their development, nurtures them after parturition, and turns them into repositories for protein-molecular episodes replicating their own. As we have seen, when embryogenesis is no longer body-making, it becomes metabolism and healing. Even injured, sick, or otherwise compromised organisms carry out life. Outside the womb the biological field maintains the best possible homeostasis under each new circumstance confronting it.

From its inception in a simple blastula, the organism is a membranous factory, receiving raw materials, processing them into energy and amino acids for bones, viscera, nerves, blood, etc., and eliminating wastes (the injurious by-products of metabolism). Life forms perpetuate hydraulic and nutritive systems, stitch up wounds, and identify and drive out invaders. Yet they are never sovereign for even an instant; any failure to receive the requisite molecular components will lead to gradual weakening or immediate collapse of their underlying template. Any lapse in efficiently breaking down components, extracting their prime molecules, and eliminating their toxic debris will likewise degrade their field and may also alter their DNA.

Cell-formed bodies must obey laws of thermodynamics that rule the entire universe. Biology is a special case that only seems to defy physics. When anabolism (the metabolic synthesis of tissues) is exhausted, the inroads of a stormy, abrasive world set in. Membranes are subjected to ongoing bombardment and assault, their trellises challenged. Every suspected threat to them turns out to be well-founded.

The remarkable coherence of embryogenic energy meets its comeuppance in the universality of entropy.

Death is the harvest of embryogenesis—its issue in the corrosion of cells and their return to molecular anonymity.

Death Alone Ends Embryogenesis

DEATH IS THE ENTROPIC OUT-BREATH for which syzygy and gastrulation are the in-breath. These are not a long in-breath and a brief out-breath. They are a deep, uniform, slowing in-breath punctuated by increasing slippages of out-breath, until breathing itself clatters to a halt. The final few out-breaths are sudden, swift, evaporative.

There is no simple chronology of creation and dissipation. Even as cellular structures are being substantiated, they are crumbling. Yet as long as their biological field holds, the creature remains alive and may not even notice.

Getting made and getting dismantled are not two different processes; they are a single cadence. As a nineteenth-century historian declaimed:

"Coeval with the first pulsation, when the fibers quiver, and the organs quicken into vitality, is the germ of death. Before our members are fashioned, is the narrow grave dug in which they are to be entombed."[1]

Despite our intrepid egos, the ground of existence is, biblically, dust. The whirlwind of assemblage leads to not only the cherished babe but the corpse, for this universe cannot turn perishable matter into immortal souls. Its sewing of DNA messages and animate holographs lasts for a shimmering, provides a home for philosophy, and then unravels irreconcilably until terminal liquidation in the form of a heart attack or stroke, AIDS or cancer, a drowning or violent accident.

Embryology, which seems too brittle to accomplish an event it consummates billions of times a day on this planet, turns out to be exactly that brittle when it comes to imperishability. Morphogenesis was never the golden gateway to immortality, for the vital element supporting it confers no inviolable perpetuity. It is a fragile, temporal process, liberating exquisite crystals, which tremble and float, then come apart from their own intrinsic make-up.

All our ambitions, achievements, and epiphanies are vain pretexts, for we are not real animals; we are scraps of sterile detritus in which a delusion of being was incited—a rueful, lenient conceit of an ego.

Death is an inevitable outcome to a developmental process that relies on bringing vitality to inanimate matter without ever making it truly alive.

Why We Don't Live Longer

THE PLANETARY ENVIRONMENT IS SATURATED with viruses, bacteria, solar radiation, predators, toxins, and hazards of water, wind, heights, temperature, hard objects, etc. In the modern era, add to these pollution, industrial radiation, and the ricochets of a dense, industrial civilization. All assail the integrity of the biological field to one degree or another. More cataclysmically they interfere with the communications from cell to cell—the transmission of information by DNA.

Transfigurations in nucleotides may also come about by chance, accumulating over a lifetime until the genetic message is corrupted. Garbling of DNA sequences undermines the very basis of creature existence, for without cooperation among cells, there is no mutual assembly or repair (when such disruptions occur earlier in embryogenesis, there is no organism at all). Protected from mutations and other genetic interference, a creature might restore itself from many states of partial disintegration and persist for hundreds (if not thousands) of years.

In the late twentieth century our idealized immortality drug is a serum made of a person's own refreshed genetic code, reviving youthful capacity in tissue. In Kim Stanley Robinson's novel *Red Mars,* immigrants to the fourth planet devise a process whereby they inject each member of the colony with "an infection [that] invade[s] every cell in his body except for parts of his teeth and skin and bones and hair; and afterward he would have nearly flawless DNA strands, repaired and reinforced strands that would make subsequent cell division more accurate."[2]

"How accurate?" someone asks.

"Well, about like if you were ten years old."[3]

The initial sensation is of a fever. "Then we put a small shock through you to push the plasmids into your cells. After that it's more chills than fever, as the new strands bond to the old."[4]

OUTSIDE THE REALM OF SCIENCE FICTION, there is no remediation for cell lifespan, DNA transmission errors, the stampede of free radicals, the toxic by-products of cell breakdown, the expansion of collagen through viscera, lemming-like epidemics of cell death, and the sheer randomness and instability of molecular life. Tiny cellular and biochemical changes, insignificant and undetectable at first, accumulate and manifest as atrophy of function in organs.

The dissociation of the organism is nowhere near as delicate or magical as its assemblage. Arteries narrow, lose their capacity to constrict and dilate. The fine-tuned discharge of blood to muscles and organs is miscued; nutrition and oxygen do not penetrate as widely or deeply; tissues suffocate and starve; organs diminish in vitality and resiliency; structures harden, rot, and become less lifelike. The immune system relaxes its vigilance; malignancies have freer reign. The nervous system and brain become less integrated and alert. Even ectoderm wrinkles. The organism becomes not so much a vibrant life form as a terrain for the genesis of disease. When a heart attack or stroke finally detonates, it is viewed as a catastrophic occurrence, but it is actually the accumulation of inevitable, infinitesimal changes, as a human genome ages from thirty to seventy years.

According to the doctors who wrote *The Yellow Emperor's Classic of Internal Medicine* four millennia ago, "When a man grows old his bones become dry and brittle like straw, his flesh sags and there is much air within his thorax, and pains within his stomach; there is an uncomfortable feeling within his heart, the nape of his neck and the top of his shoulders (are contracted), his body burns with fever, his bones are stripped and laid bare of flesh, and his eyes bulge and sag. When ... the eye can no longer recognize a raphe, death will strike....

"Haste and emptiness within the body arrive suddenly. The five viscera are interrupted in their work and become stopped up; the ways of the pulse no longer function and circulate. Breath does not go in and come out....

"When the pulse of the liver stops there is anxiety within and at the outside of the body, as though [the] man were followed by the punishing edge of a sword that [was] charging blazingly, or as though guitars and lutes were pressed down."[5]

Western medicine discerns the same signs but explains them more mechanically. The delicate equilibria of the organs begin working against one another. The heart cycles less often (by an average of one beat a year). It becomes more easily stressed by physical activity and emotion. With fewer beats and each beat pumping less blood, it cannot supply the limbs and lungs as efficiently. Blood pressure rises, leading to hypertension.

Cells in the heart's sino-atrial node die and are not replaced. Its valves and muscles calcify; pigment accumulates in its tissue. With the depletion of muscular and neural cells, it becomes less a heart, more a thicker, flabbier, slower piece of flesh.

By the same order the bronchi lose the ability to inflate and deflate entirely. Cells and airways accumulate debris; mucus is not cleared as rapidly or thoroughly.

When the immune system no longer identifies invaders, troublemakers slip in everywhere—potential infections, malignancies, pneumonias—past the blinded cells. Once inside, they find that the body defends inadequately against them there too, and they run riot in its tissue. Where fewer antibodies are produced in the lungs, delicate bronchial catacombs are undermined by infections. In a polluted environment, inefficient lungs succumb to toxins.

The brain decomplexifies and becomes less of a brain (at about the rate of two percent of its weight for every decade after fifty). Its gyri atrophy and lose the convoluted filigree that braids consciousness; the sulci—gaps—between them widen. There is less mind. Like the heart, the brain rusts, taking on an orange or yellow hue as its dead cells accumulate. The blossom of intelligence withers.

With no extra glucose to draw on in the case of a sudden interruption of their blood, the cerebral hemispheres are sitting ducks. During a stroke—literally a cessation of the fluid supply through the artery supplying the brain—constituent neurons deteriorate within a matter of minutes. Their lifelong dance of sparks is exposed as the briefest, mesmerizing halo. The sensation of being dissolves.

TISSUE LOSES ITS MORPHOGENETIC CAPACITY almost immediately after assemblage. Transplanted early (as we have seen), prospective hair cells can become brain or skin. Transplanted later, they no longer recognize their new context and have only a "hair" capacity.

Aging is also a gradual failure of cellular cohesiveness, a loss of interest among the cells to stay bound to their rigorous organismic process. We may not tire of life, but the process itself tires of the distinction between us and them, a distinction written into biology by the evolution of species and guarded twenty-four hours a day.

On some synergistic level, every biological entity runs down, as all phenomena do—suns, stones, galaxies, and the universe itself. Genes apparently also contain clocks determining life-spans (like the artificial obsolescence devices companies place in machines to ensure that new ones will be bought). After a genetic program completes itself, the underlying fount of its template likely shuts down. Even without Terminator genes, cells stop dividing—something intrinsic in them quits (in the range of sixty divisions maximum, scientists say).

Phenomenology of Death

TIBETAN MEDICINE DESCRIBES DEATH as a dissolution of body and mind, from their grossest elements to their most subtle, with the withdrawal of core generative energy.

As the earth element disintegrates, the dying person has the sense of being crushed under an imponderable and unrelenting weight; he feels as though he is sinking or falling. His attempts to change position and get more comfortable are futile because the increasing gravity is internal and intrinsic—the aggregate of form itself dissipating.

As the water element is released, liquid runs from his orifices, dribbling from his nose and bladder, discharging from his eyes. The person is alternately hot and cold. He experiences a sense both of drowning and great thirst. The yarns wrapped about one another in his formation are coming apart, giving him the semblance of being carried helplessly in a great current.

Once it burns away the body's fluids, the fire element turns to fever and steam. Images present themselves to the eyes, but they lack detail, are mere outlines. People are heard as sounds not words. The orifices become parched, the mind confused and insensate, supported only by sparkle and hiss. Sight and sound are confused with each other. One's impression is not so much of being on fire as the whole world consumed in a blaze. In truth, the basis of phenomena has been ignited and is obliterating itself in phantasmata.

As the fire element ruptures into sparks, air is the sole conduit of consciousness, sputtering out the throat. But breathing soon becomes panting. Each in-breath is shorter, each out-breath longer, breath by labored breath. The outside world evaporates and is replaced by hallucinations. The person calls to childhood friends and acquaintances long dead.[6] The last breath is not always an expiration but a clutched, rasping "in."

Death is "a radical fast,"[7] since it purifies us of our elemental aspect, our gross self. We are shorn to the rough, uncolored wool of our existence, shorn not only of thoughts and memories but of cellular and molecular existence. A fast indeed!

Medicines Extending Life

WE CANNOT AVOID DEATH, but we can attempt to postpone its onset, to lengthen our lives and improve their quality.

Advanced medical technology is not the only route to an enhanced life-span. The herbs and needles of traditional Chinese medicine liberate ch'i essence to the organs as long as there is root vitality in them. Insofar as the body-mind is an intersection of spiralling fields of Yin and Yang, the improvement of life and postponement of death lie in maintaining the dynamic activity of these fields.

"Yin stores up essence and prepares it to be used; Yang serves as protector against external danger and must therefore be strong.... Even if one's Yang is strong, but if one does not preserve it, then the atmosphere of Yin will be exhausted. If Yin is in a state of tranquillity and Yang is preserved perfectly, then one's spirit is in perfect order. If Yin and Yang separate, one's essence and vital force will be destroyed.... Therefore if people pay attention to the five flavors and mix them well, their bones will remain straight, their muscles will remain tender and young, their breath and blood will circulate freely, their pores will be fine in texture, and consequently, their breath and bones will be filled with the essence of life."[8]

How we heal our diseases, both as a visualization we carry with us and in the subtlety or coarseness of our energetic embodiment, will partially determine the kind of death we have. Diet, lifestyle, and belief systems also affect the transition of body-mind. The organism can succumb in a slowing symphony, by rhythmic drumbeats, or be poisoned and maimed in a melodrama of emergency resuscitation. One type of medicine treats life as a pitched battle against death and uses every imaginable prosthesis and heroic measure as if immortality were possible and longevity an end in itself. The other type shepherds the organism to its inevitable quiescence, invigorating and supporting it along the way with herbs and palpations (or their many equivalents), then calming its voyage to dissipation.

Ways of Looking at Death

THERE ARE TWO WAYS TO REGARD DEATH—the two ways of regarding life. Life and death can be viewed as purely cellular, evolving by molecules, genes, mutations, and natural selection. Each creature emerging in this web consumes its genetically allotted energy, then molders and declines—or it is snuffed prematurely, sometimes by a hungry viral, microbial, or vertebrate predator, sometimes by a violent encounter with earth, air, fire, or water. Medicine's forensic obsession with microscopes and imaging machines leads to a materialist's autopsy. Death is catabolism—the metabolic breakdown of complex molecules into simple ones.

The other path is spiritual and phenomenological. Life and death can be viewed as transitions of consciousness, zones in a vaster spiritual landscape. In cultures where people pray for the safe journey of the soul, the aging body, though flaccid and losing alertness and intelligence, is not coterminous with the individual inhabiting that body. The soul (or spirit) robed in cells is considered the greater entity.

The physical occasion of the body is a veil covering uncountable other veils. These are initially veils of language, culture, symbol systems, and unexamined beliefs and prejudices that lead to a Western scientific prognosis quite different from those of Apache Indians or Australian Aborigines, who dispatch the manifestations of individual lives out of their coalescences in cells back to a living cosmos and the realm of the ancestors.

MY PRIMARY SOURCE for the biological description of death in this chapter has been Sherwin B. Nuland's contemporary classic *How We Die*. Nuland misses a crucial point, one not relevant to his book but at the heart of mine—that death is an opportunity for the spirit to be liberated from the body and go elsewhere.

Nuland approaches death with as much compassion and humanity as is possible for one raised within the medical-biological hegemony of Western civilization. Yet his assumptions lead to an inextricable idolatry of molecular-cellular embodiment which must conclude in a shocking nullity.

He never broaches any other possibility, not because of censorship and not because his tradition forbids him but because there is no language to say who we are while describing (at the same time) the biochemical reality of how we die.

He clearly wants to affirm spirit. But his words have no weight or solace behind them because his authority is as a scientist and a custodian of bodies, not as a lama or dispatcher of souls.

THOSE WHO HOLD that life is an electrical illusion of molecules—an epiphenomenon of cell agglutination—must conclude that death brings the charade to a finis. If each of us is nothing more than a soap bubble with a delusion of consciousness, that bubble will pop and shatter forever. And there is nothing else. From this standpoint, there are no beings such as us: when cellular chains coalesce and link up by neurons, consciousness merely *seems* to happen. We think we are conscious, but there is no "we"; a biochemical reaction fakes identity, manufacturing a self to delude by a time-track of its own existence. Language and society reinforce the illusion of real beings. Meanwhile, we acknowledge death only in a phantomlike, ideological sense, not in our daily conduct. We are alive (knock on wood), and then (God forbid) we are dead.

The Corpse

DEATH, SAYS CZECH NOVELIST MILAN KUNDERA, has two faces. "One is non-being; the other is the terrifying material being of the corpse."[9]

In an era when birth and death are secreted away in institutional edifices, we are distracted from flesh-and-blood reality by abstract renderings at every opportunity. We dissociate the bleak, sepuchral cadaver from the lives we lead.

What is repressed in everyday life is exhibited with gruesome zeal in horror and war films and documented in police ledgers: flies crawling on unblinking eyes, bone caked with dried blood and slabs of decaying flesh, severed pieces of what was once a whole person, grubs crawling in faces that are pulp and bone, scalps detaching effortlessly with hair, gurgling cavities of blood driven by functionless hearts, bloated corpses with entrails floating from them like underwater plants. This is the trash of amino acids and proteins, but it is also the immediate detritus of human experience when it is ripped loose of its life force.

The horror is that we can be taken apart so savagely, so routinely, and there is nothing left that resembles us, nothing that attracts us, nothing that explains who we are or what becomes of us—only rotting meat—a dead animal—like any other anonymous roadkill.

In an autopsy, skin is pulled off to reveal mucous organs wreathed in blood. The face is rubbery and pliant, frozen in its last act. (The film-maker Stan Brakhage, shooting 16 mm. in the Pittsburgh morgue, perceived suddenly that the first masks must have been the actual faces of the dead, perhaps removed with the scalp in battle and superimposed on themselves by the victors. He recognized, then, that not only are we actors but that the dead continue to wear costumes and hold postures.)[10]

THERE IS A STARK BEAUTY to morphogenesis, senescence, and decay—a lesson in skeletons, vultures, and worms. It is in fact the most direct message the universe has.

Hindu, Buddhist, and other spiritual practitioners leave the corpse undisturbed for several days at the scene of its death. This is partly out of consideration to the dead person, giving her time to release a thing in which she dwelled with total intimacy for so very long. The wait is to allow her to relinquish her attachment to her pod of flesh at a pace comprehensible to her, to depart this abode gracefully in a willing act of surrender.

The body is likewise on display for contemplation by the still alive. The dead person keeps the living company, slowly bloating, decomposing, and changing expressions and personalities, imparting final, profound lessons about the nature of ego and biological identity. The body gradually returns to its anonymous human mold.

The carcass may eventually be elevated on a pyre, covered with ghee or oil, and set ablaze. People see and smell what they are made of.

In sky burials, carrion birds descend to tear the organs out of exposed corpses. Once again the living experience the startling truth of their embodiment. There is no doubt the body is meat; there is no doubt the body dies. These are among the certainties of existence. To ignore them is to make existence itself disembodied, vague, unnatural.

In the West, of course, we have plane and car crashes instead of sky burials— diabolic ceremonies followed by public-relations exercises in denial.

Our society specializes in merchandizing images, falsely reassuring its customers. Death becomes hardware, the cadaver a funeral-industry commodity attracting expensive coffins, floral arrangements, chapels, embalming and/or incineration devices. But then the body is commoditized from birth; death merely assesses the final tariff.

When Richard Kessler's wife, Kathleen, was killed in the 1996 ValuJet crash in Florida, he was shocked by the callousness of the news coverage:

"I'll turn on the television today, and they'll talk about body parts. What body parts are they talking about? My wife's beautiful left hand which she took her bar exams with? Or her right hand which she scratched my neck with, or the ear lobe that I kissed?"[11]

Few statements could more poignantly illustrate the contradiction between the mere silt of tissue and the phenomenological experience of self. Forests and thunderstorms notwithstanding, this realm is as thin as a hollow pot with a painted face. It is we who are thick; it is *our* experience that runs "through caverns measureless to man/Down to a sunless sea."[12]

Despite circumstantial evidence, the corpse is not the person; it is a stunningly perfect replica of her, an imago cast in cells—a lifeless, inanimate carbon-protein doll.

Then who was the beloved when she lived other than a congery of cells?

The Mysteriousness of Death

"MYSTERIOUS AS THE MANNER in which death came into life, even so mysterious is death itself."[13] Aua, an Eskimo shaman, ponders the baffling welter of footprints and signs. The spoor of death is like no other thing.

"We know nothing about it for certain, save that those we live with suddenly pass away from us, some in a natural and understandable way because they have grown old and weary, others, however, in mysterious ways, because we who lived with them could see no reason why they in particular should die, and because we knew that they would gladly live. But that is just what makes death the great power it is. Death alone determines how long we remain in this life on earth, which we long to, and it alone carries us into another life which we know only from the accounts of shamans long since dead. We know that men perish through age, or illness, or accident, or because another has taken their life. All this we understand. Something is broken. What we do not understand is the change which takes place in a body when death lays hold of it. It is the same body that went about among us and was living and warm and spoke as we do ourselves, but it has suddenly been robbed of a power, for lack of which it becomes cold and stiff and putrefies."[14]

If a soul gave life to cells, the departure of that soul condemns them to a bland death mask. If life is mere electrification of mud, then a break in circuits muffles the clay figurine forever. A sleight of hand animates; a sleight of hand puts the golem to sleep. There is nothing to see.

WHEN CLINT EASTWOOD'S WIZENED GUNSLINGER addresses the young would-be killer near the end of *Unforgiven*, he warns him that death is playing for keeps: "It's a helluva thing. When you take a man's life, you take everything he has and everything he's gonna have."[15] It certainly seems that way. Life is apparently a rare and precious commodity in the universe and, when spirits attain bodies, they cling to them as though nothing else ever mattered or could matter. Losing a body is a horrific thing. Taking someone else's body from them is the singlemost heinous crime of the species.

Death becomes such a powerful and mysterious event even its name defies context. From life we cannot approach "death" except by proxy. We may pretend to understand the death of an era, a machine, even a bird; it is virtually impossible to comprehend the death of a human being. Yet, as others die around us, our life persists in absurd contrast.

"Out of nowhere," Adi Da Samraj tells a group of devotees, "we are existing in these bodily forms. Look at all of us sitting around in this room here, completely unable to account for anything! Our situation is weird!"[16]

A person we rarely see dies, and he or she is gone forever; we will never meet again. Someone existed once but no longer exists. This is different from her mere absence, although its precise difference is hard to assess or explain because we do not know what death is. A missing person is not a dead person, though after a long enough passage of time he or she may be presumed to be dead (Ambrose Bierce ... Amelia Earhardt ... the poet Lew Welsh, who wanted to deposit his body where no undertaker or constabulary could find it).

They might have been kidnapped by extraterrestrials, taken to another world—maybe that is where the dead go anyway. Until their era passes, a curious ambiguity hovers over their locale. Where among shadows in the incomprehensible immensity of stars did they go? Are they, in the words of Russell Banks: "... gone from me and located nowhere else in this perversely cruel universe, which first gives us life amongst others and then takes the others off, one by one, until we are left alone, all of us, alone"?[17]

What is the difference between a child who succumbs before living a full term and an old man expiring on his deathbed? Is experience singular or accrued in layers? Are we here to quench our desire for life, both good and bad? Or are we here to avoid trauma and have as positive a go-round as we can? Is the dead child deprived forever of her chance to be in this world? Does it make a difference to her? Is "getting to live" the essential, one and only deed in the universe, now and forever? Does the universe provide aborted or prematurely terminated spirits other chances at bodily life?

What separates a villager dying quietly in sleep before a massacre by barbarians from the same tribesman being decapitated brutally during the raid? What lies behind the wish of a mother to have been blasted from existence at least a second before the phone reporting her daughter's death in a car crash? What would she have gained (or lost) by dying? What does she lose (or gain) by having to endure the calamity? Is there such a possible result as "just turn me off," a longing confided to me by a tired, old doctor who had outlived his colleagues, patients, friends, and family? Can we make our pain stop?

What were those Brazilian football fans who committed suicide right after their team's victory in the World Cup (because it doesn't get any better than that) thinking life is? What is anyone who takes his or her life intending to do to themselves, to the universe? (They expect to flee, but they may end up in the same circumstances without a body, acutely regretting and having to move on.)

Does life matter? If so, where does it go when we die?

"I remember/him/flying down/the alley/on a sled,/the snow banked/high on either/side of him,/me down/in the street/waving,/no cars,/all clear," recalls New York City policeman Phillip Mahony of his brother Pat who died suddenly in his sleep at age twenty-eight. "Come on!/Come on!/flying down/the alley,/cheeks red,/nose running,/thrilled as/ could be,/on a sled/twice as/big as/he was....//What happens to that?//Is *this* what/happens?//Is this really/all there is?//Is all of his life/ ultimately so/insignificant/that only/a few chance daydreams/recorded here/will stand on its behalf/against all of/death and all/of time/and even/God?/Could it be/that the/final fact/of a life/is such a cruel/and uncontestable/disposability?"[18]

The whole interlude defies logic, but no more so than the details defy logic.

"What about/the wash in/the drier?/What about/the budget?/What about/the car:/what about/alternate side/parking?/His favorite song was "Windy,"/what happens to that?"[19]

We exist in preparation for an extraordinary event that everyone has undergone. Look at the young girls proudly bearing field-hockey sticks in a photograph from a turn-of-the-century college yearbook. "They have been zapped out of this experience."[20] They have gone on the ultimate journey, no matter what and how they lived. And we will follow them, each and every one of us.

HOWEVER SAVORY AND POWERFUL LIFE IS, it is also transient, brief, and insubstantial. It is (for everyone) a scrumptious banquet served before an execution.

MOST PEOPLE, EVEN SERIOUS SPIRITUAL PRACTITIONERS, have little certainty of what will happen to them after death. Perhaps, they worry (beneath their party-line optimism), we just end, and are no more.

They have good reason for concern. Neither the circumstantial evidence of the corpse (vacant as a rock) nor the muddled testimony of the living (their paucity of true recall from beyond life, their inveterate mythologizing, their dubious near-death adventures) gives much confidence that we are anything more than tissue whorls with brains. Faith can only come from the subjective experience of being, the surety that we are what we seem—actual entities, existing also in a way that does not require organized mounds of zooids for its manifestation, that is not just wishful thinking and solace against annihilation. Otherwise, we are cells, and cells fully explain us. But if we are not just cells, what is the deathless part of us made of, or how is it maintained if it is not made of matter?

Either our being is an illusion, no matter how strongly we feel it and how actual our dilemmas seem, or it is real and, as a real thing, has no termination (can have no termination). If it is real, we will follow its radiance forever, whatever it is, wherever it takes us; we cannot *not* be.

You can't make friends with death.

THE PHYSICAL, BIOCHEMICAL COMPONENTS OF DEATH are deceptively simple. Insofar as we accept the entropic basis of biology (see above), we recognize that death is its repercussion.

Yet the obliteration of actual lives and then our own excision are profound and incomprehensible. Death must end all arguments, rationalizations, and participation in the meanings of this realm.

We depart with awesome abruptness. There is no concept to fall back on, though belief systems provide cover stories of both strange and commonplace existences after death.

The fact that death may be none of these things does not mean it is not something.

In order to get anywhere else we must pass through a transformation as absolute and irrevocable as the one that made us. Whatever is weird and inexplicable about life, whatever suggests limitless layers of terror and bliss, speaks equally for death. But that does not make death simply another trial to cope with or appease.

Upon his arrival in the United States, Tibetan teacher Chögyam Trungpa told his students:

"Death is the desolate experience in which our habitual patterns cannot continue as we would like them to. Our habitual functions cease to function. A new force, a new energy, takes us over, which is 'deathness,' or discontinuity. It is impossible to approach that discontinuity from any angle. That discontinuity is something you cannot communicate with, because you cannot please that particular force. You can't make friends with it, you can't con it, you can't talk it into anything. It is extremely powerful and uncompromising."[21]

Western science uses nullity as a solace. But death is under no obligation to shut us down like machines. It will be whatever it is.

"[Death's] uncompromisingness . . . blocks expectations for the future. We have our plans—projects of all kinds that we would like to work on. Even if we are bored with life, we would still like to be able to recover from that boredom. There is constant hope that something better might come out of the painful situations of life, or that we might discover some further way to expand pleasurable situations. But the sense of death is very powerful, very organic, and very real."[22]

Death has no ideology. Its fullness is the same as its emptiness. There are no bargains to strike with it the way we do with life. It is the place where resolutions, promises to be better, romances, crushes, clandestine desires, secret ambitions, wounded pride, dream mysteries, revenge, hopes of redemption and rehabilitation—where they all end.

"Knowledge is never more than knowledge about," says Adi Da Samraj, "and knowledge about is confounded by death. There is no knowledge about things that is senior to death. Death is the transformation of the knower . . . is a process in which the knower is transformed, and all previous or conditional knowing is scrambled or confounded. . . . To consider death is fruitless, since the knower is what is changed by death."[23]

Even the old hymnal "Rock of Ages" promises, "When I rise/to worlds unknown. . . . "

The universe does not want our belief or our promises, and it cannot use our knowledge. It cannot be subjected to cultural sanction, theology, diplomacy, language, or scientific law. It ignores fact, quantity, seriality, logic. Mere belief systems and pieties afford it nothing.

Even as embryogenesis was a real thing, with startling and unexpected results, so will death be its own real thing.

"You will live or you will die," Trungpa advised a friend about to have a liver transplant. "Both are good."[24]

Bardo Realms

The *Tibetan Book of the Dead* is a narrativized account of the passage of souls between incarnations through the bardo realm where they experience forms of the same emotional projections as they did when embodied. In fact, "bardo" means transition ("bar": in between; "do": suspended). This present incarnation in life is also a "bardo"—a journey through projections which are illusions. In truth, we are always in transition: "... compared to the enormous length and duration of our karmic history, the time we spend in this life is relatively short."[25]

Bardos run from one to another through a cycle of distinct gaps between each other—in an overall reality made up only of bridges and gaps. We pass from life into dying into the luminosity of pure mind into the bardo of wandering between lives into the next incarnation or life, here or elsewhere in the universe. Liberation and enlightenment come only from recognizing the fundamental intermediate, transductive nature of *all* these states.

In the esoteric sense, our passage from life to dying to death to wandering is a set of successive opportunities for leaving the birth cycle and becoming a different reality. To do this means not just a shift in profundity but a redefinition of the very nature of what is profound. While we are submerged in any one reality—life, for instance—the overall geography of life, death, and transformation seems fantastic and remote. We feel inalterably here, despite every indication to the contrary. "We are bound to this life by virtue of our acceptance of this life," says Adi Da. "We are arbitrarily motivated by it simply because it is apparently arising, and we do not generally believe there is an alternative to it."[26]

Thus, to enter something totally different is not merely a shift in texture and profundity; it is a shift in the meaning of texture, in the very the way we gauge depth.

"The same mechanics that are effective in life are effective after death," Adi Da continues, "and your ability to transcend them will not be greater than that which you have enjoyed in life.... You will have no more ability than you have now."[27] He cautions that no amount of study or inquiry into the mechanism of the transition phase to death or the nature of phenomena themselves "is sufficient to enable you to move beyond these limitations,"[28] for the phenomena after life are so much more meticulous, attenuated, and finer than phenomena within it; they do not emerge in gross promptings of landscape. Instead "... the realm of subtlety and energy controls attention."[29] He laughs. "You cannot even hold on to your philosophy or your mantra when you pass by a crosslegged nude on a couch! So what do you think happens from life to death and back to life again ... in the midst of such a profound event as psycho-physical death?"[30]

If a cat turns into a piece of celery in my dream and I accept it, what hope do I have for finding my way through the apparitions of a death bardo?

IN TIBETAN COSMOLOGY a "dead" person is attracted back to flesh as he or she glimpses projections of men and women making love. These luminous apparitions draw spirits toward them. By identifying with one of the lovers the spirit takes on the opposite sex. The boy is originally his mother's lover, the girl her father's. But "lover" here means something far more intimate than human coitus. It is an elementary joining that projects desires, spirits, and bodies relentlessly through one another, sparing nothing, beyond modesty or shame. The beings may lie in different dimensions, but they use each other's membranes and wombs to cross between, though one of them might be the size of a cell and the other an adult man or woman.

If this were an actual paraphysical perception of the potential fetus' mother and father making love, we would not be able to explain artificial insemination; however, it is a mythic representation of the force through which spirit is embodied, pictorialized in text. The horror-tale-like chronicle of the journey through bardos may also be a false anecdotalization of something that does not flow linearly through time the way our nights and days here do. Still, the text is in narrative because our passage from birth to death is (apparently) event-sequential.

"There will be projections of males and females in sexual union. If you are going to be born as a male, you will experience yourself as a male and feel violent aggression toward the father and jealousy and desire for the mother. If you are going to be born as a female you will experience yourself as a female, and feel intense envy and jealousy of the mother and intense desire and passion for the father. This will cause you to enter the path leading to the womb, and you will experience self-existing bliss in the midst of the meeting of sperm and ovum. From that blissful state you will lose consciousness, and the embryo will grow round and oblong and so on until the body matures and comes out from the mother's womb."[31] That is to say, the embryo is not just an archetypal spirit corporealized; it is a realization, in genetic terms, of the hungers that summon creatures through transitions, a materialization of what they already are.

If overly strong emotions overwhelm the spirit, it could be born as a dog, a bird, a horse, a turkey gobbler, or even an ant in an anthill—to suffer the consequences of that level of attachment until it is sated and understood.

ACCORDING TO SOGYAL RINPOCHE, "... the process of death mirrors in reverse the process of conception. When our parents' sperm and ovum unite, our consciousness, impelled by its karma, is drawn in. During the development of the fetus, our father's essence, a nucleus that is described as 'white and blissful,' rests in the chakra at the crown of our head at the top of the central channel. The mother's essence, a nucleus that is 'red and hot,' rests in the chakra said to be located four finger-widths below the navel....

"With the disappearance of the wind that holds it there, the white essence inherited from our father descends the central channel toward the heart. As an outer sign, there is an experience of 'whiteness,' like 'a pure sky struck by moonlight.' As an inner sign, our awareness becomes extremely clear, and all the thought stages resulting from anger, thirty-three of them in all, come to an end....

"Then the mother's essence begins to rise through the central channel, with the disappearance of the wind that keeps it in place. The outer sign is an experience of 'redness,' like a sun shining in a pure sky. As an inner sign, there arises an experience of bliss, as all thought states resulting from desire, forty in all, cease to function."[32]

This light show is fundamental and inalterable, for the snail and the sparrow (in their way) as for us.

Why Death is Terrifying

CHÖGYAM TRUNGPA WARNS HIS DISCIPLES not to view death bardo experiences as escapades or adventures. No New Age, Buddhist-macho spin can be put on them; death is not an hallucinogenic theme park or "Outward Bound." It is as dire and cutthroat as it seems.

Reading from scriptures, Trungpa notes, you might tell a dying friend: "'Though something terrible is happening to you, there is a greater thing. Now you are going to have a chance to get into those experiences described in *The Tibetan Book of the Dead.* And we'll help you do it!' But no matter what we try, there is this sense of something that cannot be made all right, no matter what kind of positive picture we try to paint.

"It seems, quite surprisingly, that for many people, particularly in the West, reading *The Tibetan Book of the Dead* for the first time is very exciting. Pondering on this fact, I have come to the conclusion that the excitement comes from the fact that tremendous promises are being made. Fascination with the promises made in the *Book of the Dead* almost undermines death itself....

"A few decades ago when the idea of reincarnation became current for the first time, everybody was excited about it. That's another way of undermining death. 'You're going to continue; you have your karmic debts to work out and your friends to come back to. Maybe you will come back as my child.' Nobody stopped to consider they might come back as a mosquito or a pet dog or cat."[33]

The power and seriousness of incarnation are what make death different from a trip or another extraordinary experience. There is no limit to where the universe can place a person and how it might embody them. A Silicon Valley executive can be refleshed as a poor woman in Albania or Somalia. A movie star can be hatched as a child prostitute in Thailand. A corporate hog farmer can be driven by his karma into the body of a pig in a pen in one of his farms, in fact can be sucked into hog after hog, slaughter after slaughter, lifetime after lifetime. Homeless drunkard and snazzy playboy can, in effect, change places from life to life. But first they must start over as immaculate babes and grow back into their fates. There is an attraction toward what one most fears and resists. An experimental scientist becomes a caged rat, helpless before his own tools of mutilation; a corporate polluter, a deformed child in Sinabang. A great athlete materializes as a cripple, a member of the House of Lords as a beggar in Bangladesh. A Turkish postman can recur as a Tamil tiger, an Australian outdoorsman as a Kosovar maid, still carrying his gun. Gifts and property are stripped away. Suddenly a scion with wealth and standing finds himself a speechless waif girl, the helpless chattel of a Mediaeval warlord.

Only he doesn't find himself; he doesn't even know what is happening to him.

And all of it will seem absolutely immediate and real at the time, as one is whipped along without relief, without choice, without cognizance, by the winds of fate. But even these examples, frightening as they are, are mere intentionally amusing stories. The real issue is that each of us will go through a vortex where only the submerged, forgotten aspect of ourselves will not be destroyed. It alone has any sway over our shape, our nature, our journey, and hope for its happy terminus. Only the thing we don't know or recognize will keep us "alive," will keep us (period) long after the sparkles of our cells have gusted into breeze.

Men and women who fastidiously select their clothing and restaurants and the company they keep suddenly don't even get to pick the body they inhabit, the world they are born into, or the hygiene and ecology of their new existence. A soul gets reembodied as a helpless wildebeest or hare, surrounded by predators, gored and riven. She could then be incarnated as the slave of a sadistic master on a fascist-ruled (or even insect-ruled) planet. After that will come another death, another birth, another life, another occupation, another death, another bardo. And there is no bargaining, no recourse, no memory.

That is why fascination is inappropriate. None of this is fascinating or amusing;

it is what is happening; not only is it real, it is the only thing that is real. Every possibility of love, compassion, universal peace, and justice for sentient beings rests on its successful unfolding, its outcome—on our unfolding within it, our outcome. Otherwise, it is senseless manipulation and torture.

"EVEN THOUGH THERE is death," Adi Da advises, "what makes death bearable and profound is not the fact that death doesn't exist, but the fact that it does, that it *is* a real process. And that real process, and the Reality in Which that is occurring, is the profound matter, even while alive."[34] We are driven ultimately toward our essence. That is both the horror of our demise and its saving grace.

"All the arrangements you can make with the body-mind—waking, dreaming, or sleeping—are temporary. I don't care if it's your girlfriend or your boyfriend or your Samadhi. As long as it's a state of the body-mind—waking, dreaming, or sleeping—it's temporary. It's not it. Any experience that depends on conditions of any kind is conditional, and, therefore, is not Eternal, not Always Already, not Permanent....

"The profundity of [my] Way of the Heart is *in* the depth. The Way is in-depth, in the process itself, not merely in the social gleefulness, or in the social whatever."[35]

The dead are not weird.

SPEAKING ACROSS DEATH'S DIVIDE through her husband, Ellias, the recent Sara Lonsdale (now Theanna) describes her passage into domains that followed her death. These turn out to be, as Dante Alighieri foreshadowed centuries ago in his *Divine Comedy*, a direct representation of the lives and projections of the dead transposed into geography. Many expect either nothing or a customized heaven and hell; death turns what they anticipated into all too vivid and credible landscapes—and usually (just for good measure) things far more believable and far worse.

Dante's exact Hell may no longer be believable to modern man and woman, but it has been recast in *Star Wars/Friday the 13th* techno-glitter. Ghouls and pixies already jump out of our collective psyche onto the *tangkas* of horror films and from the back alleys and ruins of our streets.

"Many of those who die, and especially the non-believers and the modern 'realists,' are herded quickly into the hell realm where [their] death mind can be satisfied and amplified in its assumptions and conclusions," she reports. "They meet an alienating and exiling plane of existence in which nothing of meaning or value will happen until they stir from their nightmare."[36] That is, they people the cinemascope of death horror as it has been advertised and bruited in this world. Jean

Cocteau's glass salesmen still wander down the blank alleyways outside life, wraiths bearing phantom panes of *vitrage*.

Although we tend to consider only extant doctrines or theological and scientized portents, our habits form collective images, very real-seeming holographs. Some of these are cities; some of them are fairy tales; some are false paradises or judgments; others lie beyond this dimension. Morbid death mirages which declare doom and annihilation submit the dead to their own paranoid fantasies, so their beings do finally wash up on an anonymous shore on some random monstrous planet. Or so they think. And so they will until they recognize that they are the sole projector of the film.

The bardo of this world can be equally tragic and maudlin.

The fantasies and externalizations of our culture do not merely distort our status in the universe, they invent the torture-bearers they then threaten us with; in fact, *their very purpose* is to conceal the meaning of death and the soul, sabotaging our chance of getting into our own depths. The superficiality of mass culture makes death the power broker/executioner-engine it is, inuring us to transdimensional existence even as it addicts us to material-ridden lives. The desire for immortality (or at least longevity) is also an escapist conceit.

Our belief systems are our blind spot, the crucible through which we must pass to effect any radical transformation of nature and consciousness. It is not that we must stop believing them (we can't); it is that we must believe them so fully and wholeheartedly their stasis is broken and (instead of rigid dogmas) they become what they are in us—true passions and compassions to ignite in our own life-and-death burning.

Theanna works her way through her own participation in mass fantasy, then through her stubborn nihilism and unexpectedly tenacious clinging to life.

"Three deaths were mine," she relates, "successively and at the very same time. I had to die the first death of the body, and this was fearsome. I had always wanted to die, but I had never been ready to die.... I went with my body's dying, went with it hard and straight and strong and deep, and knew it for rebirth. I knew I would survive this dying, and that I had further deaths to go, and a whole lot of borning to do....

"I also had to die the second death of the mind. The modern person has a mind that refuses to die, that is so drenched in the dark side of death that it never wants to die, it can never leap beyond its own shadow. This mind gathers around itself a cluster of identity pictures and keeps switching quickly among them....

"My mind was jammed-stuck on idiotic themes and questions, and this mind needed a lot of enlightening before it could let itself die.... All the help I was given by wakeful beings pushed me directly over that edge, and I did attain that place of dying to my mind's contents completely, just as I died to my body's cumulative suffering, and then—Bam! I was in....

"But there was a third death here. I had to die to death....

"I had to slay death in two different guises, its false side and its true one. In order to slay the false death, I had to contend with The Lord of Death, a creature out of the collective fantasies of the darkest evil, only all-too-real."[37]

According to Theanna, not only are the newly dead sucked into the terrifying machinery of their own minds, but they continue to project ancestral patterns that drag human history backward into a mire of death-dread. To the degree that a dead person can become liberated, he or she can infuse new images and impulses back on the Earth and help humankind as a whole escape the death trap.

FROM WHERE THEANNA NOW DWELLS, death is the greatest delusion of all. The iron curtain it slams down—and has slammed unmolested for millennia—is a scam, a hocus-pocus, *trompe-l'œil,* magician's ruse.

Death is neither weird nor implacable. Those on its other side are not even removed and forever out of communication. In fact, they are more intimate and present than when alive, dwelling in a boundless cosmic realm inside this incarnation. But a trance of phenomena and projections of ancestral consciousness blind us to their presence, and to our own locale in the universe. We could, Theanna explains, through an act of selfless love, use dying not as a separator but a connector, a subtilizer of souls. We could see what the universe really is.

For the dead are reaching to us not from some sky or deep interior realm; they flood every interstice of the here and now. They lurk in light and shadow, their presence so immediate and all-encompassing it feels like nothing.

The dead do not become less real. As the artifacts and conditions of life become unreal to them, they grow more real. They travel as what we call ghosts, but to themselves they are not ghostlike at all; they are more-than-living beings.

We can't find them because we are ceding our entire existence to the Lord of Death, empowering him as *exciser ultimato.* We can't find them because, without bodies, they can visit only the gaps and lacunae of this world, and these are, by our very names for them, the spaces where we experience nothing, imagine nothing, and allow nothing to distract us. We search in vain, warns Theanna, in the single domain in which we could not find her. She is not elsewhere; she is entirely here, always.

The moment in human evolution at which death becomes transparent is the moment at which we will no longer be marooned in this body and world as in a locked room, trying to divine by impossible means what lies in the rooms around us, to escape infinitudes that relegate us to trivial and insectoid existences.

What halts our destiny now is not the lack of disclosure. Everything is in fact disclosed. We must (Theanna) "unlearn the birth process . . . come out of the infinite surround and dive straight for a very sharp point . . . gather all [our] forces, concentrate into it, and be ready for the shock of landing into density."[38] We must break the stranglehold of the Lord of Death by seeing him for who he is; then death's door will open both ways for everybody to pass freely through.

"I am here," Theanna cries out, "to explode the mystique around death, and to assert boldly that your death is the point in your journey where you decide who you really want to be."[39] It is where you fly the coop of belief systems.

Your birth equally is a point of landing in density, of finding yourself a duck on a pond.

If we had them, why don't we remember past lives?

WHY, IF WE LIVED BEFORE, DO WE NOT REMEMBER? It seems an indifference or punishment of the universe to wipe out our memories again and again, to steal from us our greatest pleasures and fulfillments, our resumés, our loved ones.

Yet we must look again.

The universe preserves essence in the only way it can: by shearing off everything which is temporal or superficial, everything made of cells. For even with the miraculous emergent properties of tissue, cells cannot transcend their cell nature enough to grasp reality outside a membrane-vacuole embodiment. It is a wonder that we intuit as much as we do through their thick, imperfect veil; that we create megalopolises, sciences, and abstract art out of Golgi bodies and mitochondria.

In their bioelectric hives, cells present remarkable holograms and kaleidoscopes; protoplasm stretches the utter limits of its mere carnal properties to attain an intelligence of being and an intimation of a philosophical and eternal realm. Yet, as long as it is ninety-nine percent meat, it is erasable. That is why the events of this life are soon forgotten—all except for one unknowable track.

Stated differently, we do not achieve full profundity among material and conditional realms, no matter how dense they seem. Their texture is a mirage projected from a changeless realm through a gaggle of phenomena. Their actual profundity comes solely from *both* birth and death, hence is ever pending and elusive while we are ensconced.

Memory is a mere biochemical ganglionic phenomenon that registers an approximate impression of events in nature; it is the minimal recording system required for survival in this world, produced by physics under natural selection. Cellular memory may be state-of-the-art in the universe of matter, but it is not the place to look for who we really are or the fate of those we once knew. The talents of molecules and cells (through embryogenic transformation) reasonably explain their organization, motion, metabolism, and the electrochemical impressions and record-banks of their collective minds as they meld in great chemotactile, mobile, predatory colonies. Why (in that form) they also have a subjective conviction of self is not so immediately obvious.

If the phenomenon of being has any reality beyond the self-deluding, chemical propaganda of cell armies, it must exist without molecules, without neurons, without cells, without conditions, or not exist at all.

Our so-called unremembered "memory" of other lives dwells languageless, outside chronology. It may be profound—in fact, as noted, it is the most profound thing about us—but it expresses itself only in *who* we are.

We cling to a ghost presence, a nostalgia, a clutching after, a hope, a clinging, a sense of loss pursuing us body after body through the fading watches of time.

"Were we happy tonight because we were happy or because once, a long time back, we had been happy?" asks the narrator of a 1946 novel (the italics: his yearning). *"Was our happiness tonight like the light of the moon, which does not come from the moon, for the moon is cold and has no light of its own, but is reflected light from far away?"*[40]

PEOPLE ARE TRULY BEWILDERED; they turn on a dime (depending on current fads and fashions), from accepting death as the end one day to believing that they have lived innumerable times before by the next afternoon. Reading magazines and books, talking to friends, watching television, they take on storyboards of other existences like movie roles, assembled piecemeal from dreams, episodes of *déjà vu*, and premonitions, or conferred by esteemed psychics; therefore, this life must read to them like a script too. What is often missing is a sense of the pure sorrow, delight, and open-endedness of existence. It is no fun sometimes to be on this roller coaster, but it is bewitching, exhilarating, and charged with an intimation of the bigness of how the universe might be.

We are in such a hurry to complete evolution, to live, die, get reborn, and relive previous lives, to make contact with aliens and ascended or astral spirits that we evade the basic somatic fact of living and our immersion in a sullen blue and wild creation from which true wonders of the cosmos come unplanned and unnamed. The irony is that as we strive toward the cosmic we lose the cosmic; we replace the experience of profundity with the projection of profundity onto shallow events.

In this regard, birth by embryo is the true butterfly/haywagon ride.

WHEN PEOPLE SEEM TO RECALL prior incarnations as Egyptian priestesses, crusading knights, and other romantic characters, they are responding to something inside themselves, though it is unlikely they were ever those people.

If we have lived (or "telepathically" experienced) other lives, they will not be found locked away in our brain like a series of Gothic novels. They are obscure in the way this life will be, when it is done.

Whatever we were, if anything, before conception and birth in a womb, was so incomprehensibly different from what we became that we do not remember it. We do not even remember our mistaken imagination of it. We became what seemed like another thing entirely, driven by our karma. Then we got latticed in new tissue.

Death and birth are mysteriously the same, with equal loss of continuity, expectation, and hope.

It is doubtful, after here, we will recite "Three Blind Mice" again.

One Fine Day

IF WE PASS FROM BODY TO BODY AND LIFETIME TO LIFETIME, the thread is not of physical stuff. We have no mass, no materiality, no circuitry, no electrons. When we leave this world, we vanish without a trace. When we come back (if we come back), we start over again in a new body. The billion-year-old part of us—our deathless essence—must return as a baby, a cipher. It cannot pick up where it left off, with its former knowledge, worldliness, *savoir-faire*, personality and style, or any aspect of its worldly ken. It must learn what it is to be alive again. It must see again for the first time because the cells it is using for eyes have never been in that configuration before, have never looked at a world.

A soul gets to believe in Santa Claus again, play with toys again, discover emotional life again. Earthly experience is defined by the cells and tissues not the spirit inhabiting them. In their particular morphogenetic grid, they are new. They have never tasted or smelled before; the world to them is startling and fresh. "... '*twas so good to be young then ... with the sweet smell of apples ... in the season of plenty ... when the catfish were jumping/as high as the sky.*" It is so special, so haunting, so once and only, so thunderstorm-Orion deep in every atom of their being.

We learn to long again for them, to speak again, to read again, to believe we are a person again, to leave them behind here again. We cry again, laugh at silly jokes, fight with sticks, run with a kite again. His or her life is our life, is us—and nothing else is, and we are nothing else. All the rest is gone—except perhaps for an undying scrap that is working little by little toward understanding it all.

This is the cells' world, the body's world, and we exist solely on their terms, for their benefit, so their flesh can be young and marvel at the cosmos all over again, and again.

If the soul is lucky enough to be human and find its way back to something resembling this world, it will ask all the same questions: "Why is the sky blue? Was Grandpa ever a child? Who am I?" Even if its last life was as a physicist or scholar, it knows nothing. It has never heard of molecules, dinosaurs, or the Civil War. Even if it invented the phonograph or the electric motor, it doesn't know what these are or how they work. If it was William Shakespeare, or Blake, or Faulkner, it must read their works afresh just like any schoolboy or schoolgirl—and it may not even understand them as well or like them as much as some of its classmates. All it carries is an affinity, a leaning, an intuition or vague intimation of familiarity.

The things of this world, even when we muster them, are not our things. They are the world's things; we don't create or compose anything. Forms use our aptitudes to manifest. We are their channels not their makers, so they do not remain part of our essence.

It had to be this way. It had to be absolutely real. We have to feel total wonderment at each new breeze and thunderstorm and circus clown. We must wander through labyrinths and gardens of childhood, sucking sugars and flavors out of candy, talking to birds and chipmunks. We must emerge from childhood again mysteriously at the dawn of romance, with no erotic experience (no matter how many lovers in another body). The "first time" is truly the first time. And there will be another first time and another, in yet other bodies, and they will all be first, and they will all be real. But only if we live again....

On a scratchy disk you will hear, *"One fine day, you're gonna want me for your girl."* *"... only fools rush in, but I can't help/falling in love with you."* Elvis may get to hear it too, in another language, as a young man (or young woman) for the first time.

How deeply and mysteriously thrilling it will be! How eerie and inexplicable too.

And this is merely the tip of the cosmic iceberg.

Cellular Cinerama

THE EMBRYOGENIC PROCESS described throughout this book tats a circuitry into our silk down to the most miniscule microtendrils, cell membranes, and reticula. Every protein congery clings to invisible axes and surfaces and subsurfaces of others. Every membrane, vestibule, sense organ, and muscle wraps around and through our fabric, inside and out, layers begetting layers, burlap through burlap, filament on filament, penetrating and enveloping at the same time (so that what passes through is itself passed through, and passes through again). We are "here, there, and everywhere," which requires a total commitment to *becoming* that no mere act in life will approximate, either in sensation or urgency.

We are grabbed at the root of guts (in fact, deeper, at the base of intestinal crypts and their myelinated axons), at the pith of marrow, at dry alveolar gusts gathering into breath. We are blasted into them even as we are ripped apart and put together. Embryogenesis is the ultimate sex act and martyrdom; it is everyone's most major surgery; it is the four-second mile, again and again; it is the arsenal of the queen's interrogator; it is the first birthday party, candles glittering on blue-frosted cake. Grasp at it and weep. Nothing less than a hundred percent, headlong dive will do. Nothing less will stick.

We are in a room naked—not just naked but nothing. And we must get dressed—not just dressed but fleshed. And the clothes come from a warp of waves within a mind-sunken hollow. It must make something from nothing but, if there were nothing, we wouldn't be there—or, more precisely, there would be no "there" to be. In a sequence of states resembling a dream within a dream within a dream we put on our raiments, don the microfilaments of wetsuit, squeeze out tripods and claws, and twist our insides out of deeper insides and coilings. The fractal passages through which we gape, consume, and defecate; the waters in which we bathe for what must seem like forever; the prostheses with which we grab and clinch the surface and its stumps and ruts are all shapes produced by maximum traction and granulation of emerging mind and self against emerging wool ... until they pass through each other and illumine first the womb and then the birth quarter. By the time we are embodied (or reembodied) we don't recognize any other shimmering, anywhen anywhere. All other possible realities are dream fragments evacuating consciousness like hydrogen clouds dispersing at speeds above light.

When we step into the ring with embryogenesis, if we want to remember past lives we cannot be unnerved by the deafening roar of the crowd. But we are, so we have already lost the fight.

We are wired, head to toe, skin and bones, cell to cell, axon to axon. Signals cascading through streams and rivulets combine at a great Amazon coursing into and out of our brain. So much concentrated light and sensation flood through this network that anything lying outside it is a mere intimation. As the embryo matriculates and becomes engorged in neural fire and the crowd's roar, it forgets all that it was previously.

We are stuffed and dazzled with phantasmagoria. We cannot search beyond it in the shadows.

It may finally be true that we will see our life only in the darkness between lives. Otherwise the lights are too bright.

The Continuity Between Lifetimes

AS LONG AS WE ARE PREOCCUPIED by the present world illusion, we cannot deter-mine even the certainty of the here and now (separate from a dream), let alone a legacy of other existences. We are enveloped by phenomena as one is surrounded by water without shore. "You cannot merely leap into a memory of a past lifetime and remember it as your own experience," Adi Da admonishes enthusiastic spirit travellers. "You must also recover the continuity between lifetimes, or you will not have the certainty that your present self experienced that past event. In order to reestablish the feeling of continuity between this lifetime and any past lifetime, therefore, you must become comfortable with the state you may have realized dur-ing the intervals between lifetimes."[41]

In order to recall and recover our unmaking, our remaking, we must know and understand our making. Otherwise, we have no context and no gist. Without con-text our memories are dwarfed by infinity and eternity. In such a vastness we have no yardstick by which to find a past life and (even if we thought we uncovered one) no device by which to locate, store, protect, and preserve it. Without the body as a guide (the body to which events once happened) we are unable to recognize its experiences as uniquely our own.

In our present disquieted states we cannot recall most of our childhood, let alone the events and dramas of other lives and the obscure kinetic states we experienced between lives. We barely catch hints and cues that fly up suddenly—an angle of light, a taste that is not quite black-walnut, a thicker yellow, a draft of cool warm air, the color of a late afternoon sky as we lie on a bed in delectable quiet. These come and go before they can be identified, leaving incomprehensible vastness but no synopsis or motive.

The reason we do not remember everything is that consciousness cannot han-dle infinity or eternity. The only way to travel in forms through time and space without end is to have an incredibly deep, incredibly subtle core that cannot be accessed from any lifetime but gives rise to all meaning and to consciousness itself. This is the sense of the divine and sacred we all feel. This is the haunting thread of a folk tune like "Greensleeves" or the harmony of an organ following inscrip-tions laid down almost three centuries ago by Johann Sebastian Bach. This is the true nostalgia, the true melancholy, the engine of love, and the source of sweetness in a child and *déjà vu* all one's life. Everything else must be ephemeral or there would be too much of it. Yet the core is always illuminating, always beyond reach, and its intimation makes all the rest seem real. The very power of the core is that it is so infinitesimal and so concentrated in relation to this incarnation that, for all intent and purposes, it is absent; yet it is so powerful as to be the glue and motive that holds all the rest together.

Our memories from beyond birth, if they exist, are crevices in thought, hollow lapses, condensed, each of them, to less than the size of a string, at obliquities of enormous subtlety, expanding into lifetimes only beneath the surface of what is falsely called the unconscious but is merely another archetype set to mark the boundary of existence.

Past-Life Tales

IN A WAKING TRANCE ONCE, lying in hot sun under a vast blue summer sky, I suddenly travelled into a dimension I had not known. The things I saw were spectacular in their ordinariness: a door half open (light streaming through), a stone wall (rocks in the foreground), an old farmhouse, a "Mediaeval" fair, the back of a church, a shadow, a dirt path, a hen pecking. These images came and went, some of them hazy-bright flashes, others fade-in/fade-outs, usually at a raised angle of forty-five degrees or more, sometimes rotated at an even sharper slant. Their cinema carried no particular past-life narrative, no betrayal of alien scenery. It was, click!, a tree; click!, a donkey in a patch of light; click!, an old woman in rags; click!, children playing with sticks ... an abandoned shed ... an oak tree ... some wild vines. Between these images was merely space and time—gaps of immensely thick and empty texture, like a fabric on which nothing had been written.

I had little doubt that, whatever I was viewing, it was not part of my present life. I was at the shell and boundary of this existence. The images, in their simplicity and prosaicness, were unimaginably remote, lying at immense distances from here— greater than anything I had ever gauged. I knew this by dead reckoning—the sensation of my juxtaposition to them, the declination of their sudden apparition.

DURING 1996, a week after my session with John Upledger described in Chapter 23, he guided me through a craniosacral treatment into a surprise past-life regression. With an audience of trainees who accompanied him from site to site in his clinic (many of them visitors from Europe and Asia), he was demonstrating the range of the cerebrospinal trance. Something he felt in me through his palpation must have led him to think that I could cross barriers that day. So he gave the word.

Under his gradual encouragement I forced my way back in imaginal memory to a scene in a hospital at which I visioned my ostensible birth. As John asked for progress reports and directed my attention inward, I saw humanoid figures, cartoon aliens, and then, on his instruction to search behind them, tapestries of Egyptian hieroglyphs—as though my mind were mocking me with central-casting motifs of past-life stage-sets. The hieroglyphs looked suspiciously like New Age disinformation.

Gradually the images changed. I went past the hieroglyphs to the scene of a repeating nightmare, the dream I most associate with a possible past life. I have had versions of it since early childhood. It seems not so much a dream as a theater I revisit. Its narrative involves desperately trying to bury a corpse; a murder trial; a stern, antipathetic judge meting me a life sentence; a drear, interminable time in prison (experienced as decades even during a single night's sleep); a backup of sewage thick with urine and shit, knee-deep into my cell.

From version to version over the years, it remains unclear if I committed the murder or was merely helping an unidentified accomplice conceal his crime. The trial, the judge, and the sentence are always the same.

When asked by John to name the time and place, I said, "Sixty years ago, Rumania with a 'u.'"

I had no more faith in this identification than in the hieroglyphs. I told him so. "Fuck that!" he said. "Humor me, will you?"

As if stung by a zen master's staff, I plunged headlong into the Rumanian landscape. Before the session was over, a roomful of people had seen John remove a blade from my chest (I was apparently murdered in prison by either a guard or an inmate). Different onlookers told me later it looked like a whoosh of energy, a faint shape resembling an axe.

Whether this was a symbolic axe from my childhood—and the prison and murder a masque of infantile experiences—or whether my essential being comes from another person in another life (resembling the Rumanian episode) will never be resolved.

Karma

A BUDDHIST ADAGE TELLS US, "If you want to know what your past lives were like, look to the life you are now living. If you want to know what your future lives will be, look at your actions."

Judgment and justice in the universe are carried out not by godheads on thrones but karma, an inalienable force sharing more with gravity and electromagnetism than any pop-cultural rendition of destiny or retribution. Karma is a dynamic principle pushing everything to its most intrinsic subtle level, into what it actually is, what its actions are already turning it into. Each deed of ours has karmic (existential) as well as physical and psychological components, and it is the former that transmit that deed through a far deeper and vaster universe, one that has moral components. In fact, gravity, having no meaning itself, is a mere subset of karma.

Justice is always served karmically. True gravity requires it. Every single act (even every thought) has a weight, a consequence, though not necessarily the one our limited sense of right and wrong confers on it.

The scope of karma is well illustrated by a tale about the fourteenth Dalai Lama. Told at a public gathering about a new slaughter of Tibetan monks and nuns by Chinese soldiers, he suddenly and fiercely wept. But it was not entirely, as the stunned onlookers presumed, for the dead. *Their* suffering would lead to fortunate rebirths, or better. It was for the Chinese he wept, for the lifetimes of fresh suffering that would have to be undergone by them (and humanity as a whole) to fufill the karma of their actions and expiate them.

IN THE END it hardly matters whether we remember past lives or whether they even happened. We are here, at this moment, with the issues of here and now to resolve, including our tarnished memories of this incarnation. There is only "here and now"; there was only "here and now" even then.

All times are "real," all places "real" (a trench in World War I, a thirteenth-century Dutch winter, a domed city in another galaxy). Technology means nothing, progress nothing; all evaporates and recongeals time and again.

Who we are is more critical than who we might have been, whatever horrendous acts "we" may have committed, whatever saintly deeds. If Hitler and Goebbels are back on this plane in other bodies, we would have no way to detect them, for they would be someone else.

There is a reason the universe preserves essence as it does. If we carried with us an accumulating array of lifetimes—their crises and triumphs, apotheoses and guilt—we would be overwhelmed. Our susceptibilties to melodrama and trauma almost paralyze us as it is. To bear their collective weight and pain across lifetimes and galaxies would be intolerable.

When we were alive once (then and elsewhere), that world-domain was crucial ... and then what happened an incarnation before that, etc. What is happening now is the aggregate meaning of all of them. We can no more look to the past than to the future for the resolution of our destiny because, as the Buddha taught, we must act now to change patterns that run deeply enough to bridge lifetimes. In fact, now is the only time we can act; it is the only time that is real.

This is what karma means—the preservation of essence in a material, free-will form.

"Between grief and nothing I will take grief."

WE OFTEN MISS WHAT EMBRYOGENESIS—incarnation—is because it is so literal, coarse, and direct. Being made and being disassembled are the sheer compasses of the universe, its invariant constants. Scientists look at electrons, chromosomes, or hydrogen, but these are not the benchmarks of things; they are only things. Even the technical description (herein) of genetic, evolutionary, morphogenetic development fails the pure ontological fact of existence and nonexistence, embodiment and disembodiment, realization followed by annihilation.

We exist because the universe requires us. It cannot reach to this part of itself without our participation. Assorted complaints about an ineffectual or malefic God are superfluous beside the urgency of opening zones of creation. God is not involved in short-term victories or justice. Everything he manifests comes out of darkness

The universe needs to lose its way in its own fabric, to snuff its own light only to discover it anew as a whole other thing. This is why it gets so dark, so grim here, how a civilization of machines can arise, shrouding luminosity. This is also the wonder of morning, of birth.

The way our body takes shape is a summation of all domains and cosmoses, known and unknown, till now. Embodiment is literally that—the universe seeking to experience the countless layers and interstices of its own nature. It makes them physical and embodies and inhabits them with atoms, molecules, cells, and the like. Long before it took on a physical aspect (outside of time altogether), it reached into itself, involuted, invaginated, and made itself happen. It is its own final cause because it invented its own body. "People still don't seem aware," sighed Meher Baba, "that it is the subtle energy of the divine Consciousness which has become the physical universe."[42]

Everything must go through our same complication and depth, even as it unravels through us in morphogenesis and gastrulation—everything that seeks a standing in the universe, that longs for a journey that counts, wherever that thing is located and whatever it is made of. It is the process that matters, not whether its medium is cells or its condition ego. Embryogenesis is a projection of cosmic geometry into atoms, a transcendental object imposed onto pliant grids of simple molecules in three dimensions and time. Who knows what other topological transits cascade between firmaments, what other shape-fields send wild entities winging across intervals between realms.

Our own deepest urgencies are strung out in individual cells on embryogenic paths, cells which are objects in fields, and fields which are ruled by larger fields, within the gravitation and telekinesis of a greater invisible system, to the end of matter as we represent it in particles and suns. Cells are not merely the happenstance crystals of primal brine or the emergent substance of life; they are also the only way the universe can make and people bodies. Cell experience is the precise topography of spiritual experience as bodies. How cells are organized is how the inside as well as the outside of the universe is organized. They are not just cells as biological things; they are cells as phenomenological and ontological principles.

The brutality of nature and the carnivorous course of evolution (the tough guys winning every race, capturing every ecosphere) show merely how difficult it is to sneak into this zone, how privileged we are, how much work is required to complete our full nature. Yes, it is bloody and bleak and rife with Satanic artifacts, but at least we are here, and what we can fathom at this site plummets layers beyond us down into the catacombs of creation, at least as far as its guise extends into galactic masses. There are times when we are as happy as angels and would dwell here forever if we could, and there are times when we want neither this body nor its life.

"At 2 AM, when I stagger off the stage," confides rock-star Rob Brezsny, "I'll sigh to myself again, as I have so many other times, that this is the feeling I most want to remember about my stay here on earth; that when my body dies and my will-o'-the-wisp soul is negotiating its way through the Bardo planes, I will treasure most the exquisite blown-out sensation that comes from blending kamikaze release with practiced discipline."[43]

Life is transitory, brief, and filled with loss and disappointment. Yet we hazard it seemingly with hope and excitement. It is a blessing to put on the great suit. It is a wondrous thing to drape our energies, desires, and wanderings in leaves and mud, and to transit through here as a scarecrow.

"I can't go on like this," says Happy Wilson as Estragon in a production of Samuel Beckett's *Waiting for Godot*, staged by inmates on the gymnasium floor at San Quentin Prison.

"That's what you think,"[44] replies Donald Twin James.

Four months before his death the Buddhist poet Rick Fields was composing "The Bardo of Dying." In it he spoke the paradox, "I've never felt more prepared or readier for death as now and simultaneously, never felt such passion, and yes—thirst and hunger—and yes—joy for living as now."[45] A month before his death, while listening to a chanting tape from a friend, he danced, swayed, and moved to the music; he was also weeping. Entering the room, his wife asked, "Rick, what's going on?" He answered through his tears, "I just love being alive so much."[46]

WE COME INTO BEING, according to Adi Da, because the world "represents at least an aspect of our tendency toward experience."[47] To depreciate life through self-deprivation does not take into account the divinity of our gift—the opportunity to explore matter and space from within a body. Experience can be joyful and enlightening as long as one recognizes it for what it is, and freely cedes back to God what he has given. In fact, we must live deeply and desperately, or we deny the Divine Radiant Presence in ourselves. "You believe that the Divine is some *One* else or Other," Adi Da teases. "You think that the Divine is so profound you could not realize God except in a totally different state, circumstance, dimension than this present one. But God is simply the Shining, Conscious Being that is our Nature at all Times and under all circumstances."[48] He is "the sole owner of everything, even of our relationships."[49]

We float in our own static and substance, between hunger and satiety, between (Faulkner) grief and nothing.[50] This is the true bottomlessness of existence coeval with the bottomlessness of indivisible matter or the termlessness of stars.

What we feel is embodiment, all its pain without which the universe would not know the vastness of its joy, all its pleasure without which the universe would not be able to exist. Embodiment provides these states in ways that allow us to gauge them and pass palpably through them. Abstraction would not have been enough, nor would telepathic projection. We have to be made of something even if that "thing" is divine Consciousness corporealized.

The place to seek spirit is not elsewhere and beyond, but here in the way of all flesh. It is a *privilege* to get a body, to abide among kin.

Yes, it is painful, but if it didn't hurt, we wouldn't feel it; if we didn't feel it, we wouldn't really be here. We would be hyperdimensional spectators, able to bail at any moment. We have to *be* molecules in order to understand the nucleus of cosmic fire. We have to get made in order for creation to root.

The spaciousness of existence is the spaciousness of the universe. It must be real (i.e., embodied) or nothing would happen; the universe could not delve into the exquisite paradoxes and textures of its own being. Yet, as it does, it creates new texture—matter and energy—in order to experience itself, to unravel itself. And all creatures exist as necessary complications, budding at precisely the spots they must, at the only spots they can, in order to deepen the universe's expression.

Divine Consciousness does not care if it deepens by pain and grief, or by pleasure and joy—deepen it must, discover its own esoteric nature it must. People on worlds cast in fire and marl are its lumens and lamina, which is how they form, twisting inside and out, in replicas of the cosmic shape uncurling its own riddle.

We are excavating, through our acts, not only symbols, maps, and technologies of iron, silicon, and light, but previously unexplored ranges of mirth and despondency, of sensation and epiphany, which express the core of an actualizing universe. Perhaps somewhere, beyond exponential infinitudes and snakeshedding space, is a source where our sparks originated, or perhaps they are originless; either way, we can no longer retrace the route we came.

Look at how far there is to go, but look at how far we have already gotten! There is no way we could have done it except by a process darning us electron by electron, cell by cell, impregnating matter. No facile entry or exit, we are wired to awaken within and without, simultaneously and multidimensionally. Heaven's Gate and Solar Temple cults to the contrary, suicide is not a rocket into purer, more authentic realms. Nothing about our situation suggests that we could depart so naively. We must drag ourselves through creation, much as we were dragged here, wounded rabbits in snow. Then, at the point at which pain becomes unbearable and light is almost extinguished, something else comes into being . . . and something else . . . and. . . .

IT IS IMPOSSIBLE to know or even imagine what will become of the human race—or our own beings—a century from now, seventy-three thousand years from now, 2.37 billion years from now, or when the Sun finally hemorrhages . . . and whether any of this will survive, and as what. A billion years is a very long time—five hundred years is a long time too; how many transformations and resource-depleted landscapes can species and societies withstand before vitiation or exhaustion? How many millennia can the Atlantic Ocean be fished before there are no more cod, no more hake, no more diatoms or krill? How long will the Atlantic itself last? How many savageries and soap operas and ethnic cleansings can our species sustain, how many gospels and holy wars, how many times to have Rome built and destroyed, Atlantis lost and found and lost again? Try picturing the Earth a mere three thousand years from now. No science-fiction intaglio gives a clue. Eternity is even longer—much, much, much longer.

The sole saving grace—what alone rescues anyone, anything, from the watches of time, the façades of episode—is the fact that manifestation/rebirth can occur only from essential nature into a material stratum.

AT THE MOMENT OF CROSSING OVER, even the wisest shamans, medicine men and women, get scared. And why not? They must give up themselves, admit that they are lost. "To give up yourself is to stand there within the creation, to call out, to trust that they *will* come for you."[51] A raven, a spider, a warior on horseback . . . a teacher, a spirit guide, a path through shadows, somewhere in the next world.

The universe churns in chaos and shifts in appearances; it cannot bargain away its cruelty and ruthlessness—its thunder and lightning are greater than a billion suns. These are prerogatives, givens.

Near the end of his life Carlos Castaneda spoke of how, knowing they are unable to defend themselves, sorcerers develop "the art of facing infinity without flinching, not because they are filled with toughness, but because they are filled with awe. Discipline is the art of feeling awe. . . . A live world is in constant flux; it moves; it changes; it reverses itself."[52]

In whatever form beings exist, they can only bow before the service in wonder and veneration.

The Clear Light

THE MOST POWERFUL ICON in *The Tibetan Book of the Dead* is not the wild west of the bardo between lives but a "self-originated Clear Light, which from the very beginning was never born."[53] After the body dissolves and its elements disintegrate, all ordinary aspects of mind are torn away. As these are extinguished and snap from their connection to phenomena, the dead person is initially plunged into darkness and chaos . . . but then consciousness gradually returns and, as it does, she is witness to a dawn that does not begin a day but is the glimmer from which she arose and to which she returns each time, "the ultimate condition of all [her] personalities."[54] Anger and desire no longer exist. She has arrived at the primordial basis of mind experienced as a Ground Luminosity—"an immaculate sky, free of clouds, fog, or mist"[55]—the place whence everything we know on Earth and in the heavens arises. "We come here, so-called here, out of . . . the transcendental Brilliance without differentiation."[56]

That infinite phosphor provides the luminosity for thought, for sensations, for lament. It is the impalpable abyss out of which being comes, replicated even at the heart of DNA. It provides the subtlest and most naked potentiation of existence. Viewed directly, without flesh, after a death, this cosmic mandala marks the moment, the opportunity, at which essence—enlightenment—can be attained, but, if not recognized, it dissipates back into the murk and maze of incarnation and rebirth. It does not disappear; it creates implacable complexity.

Sogyal Rinpoche writes:

"Even though the Ground Luminosity presents itself naturally to us, most of us are totally unprepared for its sheer immensity, the vast and subtle depth of its naked simplicity. The majority of us will simply have no means of recognizing it, because we have not made ourselves familiar with ways of recognizing it in life. What happens, then, is that we tend to react instinctively with all our past fears, habits, and conditioning, all our old reflexes. Though the negative emotions have died for the luminosity to appear, the habits of lifetimes still remain, hidden in the background of our ordinary mind."[57]

Despite exhaustive preparation, even a sincere person is distracted by the rush of terror and longing at the heart of his being. Without any of the talents he accumulated in life, he is at the mercy of essence alone.

"Padmasambhava says, 'All beings have lived and died and been reborn countless times. Over and over again they have experienced the indescribable Clear Light. But because they are obscured by the darkness of ignorance, they wander endlessly in a limitless samsara.'"[58]

It is apparently the artlessness of the light, not its psychedelic mandala nature that gives it its power and inviolability, that makes it so difficult to recognize. We ourselves are just not that simple.

AT ONE LEVEL our emotions may be transient reactions to circumstances, but they reflect our inherent nature. An individual is literally reincarnated each time by the pull of his own desires and formative imagination into organs fabricated by collective karma. That is probably why the tissues cohere against extraordinary entropic interference, why birth is inexorable.

We find ourselves here because consciousness commands it. But consciousness is the same as matter, so matter commands it—a *fait accompli* demonstrated by the swiftness with which carbon, oxygen, and their allies set themselves to the latticework of life. Life likewise rushes to differentiate and become mind, not because of an exterior entelechy but because its absolute nature, hidden from the glare of the electron microscope, lies in God's blind spot, escaping any censorship he or we might impose.

Embryogenesis is the literal dawning and redawning of the Light, an irresistible force, binding cells by attraction, sealing them in envelopes of flesh, assembling psyches, animal and human, from collective prior existentiality into new material apparel. As the Heart Sutra tells it, "No death, and also no extinction of it."

The onset of the embryological process, both in the primordial oceans and the flesh of creatures, is the singular effect of wandering away from the Clear Light. As each animal does this, it not only encounters birth, it creates the form of the Light known as samsara, the form of embodiment expressed by cell hunger.

The failure to recognize the ground luminosity requires something, and the ground luminosity also requires something. These conditions meet and fuse. Inevitability and incognizance combine into form and being.

Glossary

THE GLOSSARY IS NOT MEANT TO BE A DICTIONARY or to offer complete definitions. It should serve as a quick reference for a reader who gets stuck in the text because of a word that is not defined at that spot. Words adequately defined wherever they are used are not always included in this glossary. For most items included, the index will still be a source for a more substantial definition. The glossary also includes some common terms, for instance, measurements and anatomical directions. Some terms of general interest describing items that have been edited out of the book have been intentionally left in because of relevance to other topics.

Abdomen (Abdominal) The part of the mammalian body, lying between the thorax and the pelvis, that encloses the viscera.

Acheulian The stone-tool technology that flourished in Europe between the second and third interglacial periods, marked by symmetrical handaxes, the Acheulian still stands as the longest-running culture of men and women on Earth, ending only some 200,000 years ago after surviving 600,000 years without substantial change.

Acrosome The organelle at the tip of a sperm cell that enables it to penetrate the ovum.

Actin A class of proteins with molecular weights of about 44,000, actin collects in dense filaments just beneath the cell's plasma membrane and interacts with myosin in muscle contraction.

ADA (adenosine deaminase) deficiency This is one of the two percent of true monogenic human diseases and a favorite target for somatic gene therapy. Adenosine deaminase is an enzyme necessary to degrade adenosine and critical to the functioning of the immune system. Defects in ADA lead to severe combined immunodeficiency, such as in the instance of David, the "boy in the bubble," who ultimately died when he was exposed to outside air.

Adenine A purine derivative, adenine is a component of amino acid, secreted notably in the pancreas and spleen.

Adenovirus(es) A group of small, ubiquitous icosohedral DNA viruses (including the common-cold) that are generally harmless and have been widely investigated as vectors for human gene therapy.

Adherens junction Connection sites for actin filaments, adherens junctions enable cytoskeletal elements of cells to connect to each other and to the extracellular matrix.

Adrenal Literally "at or on the kidneys," used to describe two small glands, each one situated above a kidney, secreting epinephrine and other hormones.

Aerobic Metabolizing oxygen; requiring oxygen to live.

Afferent fiber A nerve conducting signals from sense organs and the body's periphery to the spinal cord.

Agoraphobic Originally fear of open spaces; by extension, fear of leaving the house or encountering the outside world.

Alar Having wings; shaped like a wing; or related to the armpit; axillary.

Allantois (Allantoic) The extraembryonic membrane in which the embryo deposits nitrogenous waste.

Allele(s) The alternative form(s) of one gene.

Alpha-helix protein This is a protein fabricated when a single polypeptide chain turns regularly about itself to form a rigid cylinder within which each peptide bond is regularly hydrogen-bonded to other nearby peptide bonds in the chain. Because of its hydrophobic constraints it commonly inhabits the zone of transmembrane proteins that cross the lipid bilayer.

Altricial Helpless at birth.

Alveolus (Alveoli) Any cavity, pit, or air sac; a structure involved in gas exchange in lungs, or milk secretion in the epithelial tissue of mammary glands.

Ameloblast(s) The differentiated form of the cells of the inner tooth-enamel epithelium adjacent to the dentin, ameloblasts form enamel prisms over the dentin, regressing ultimately toward the outer-enamel epithelium.

Amino acids Units (monomers) making up proteins, these organic molecules possess both carboxyl (COOH) and amino (NH2) groups.

Amnion The innermost extraembryonic membrane about the fluid-filled sac in which the embryo is suspended, the amnion is covered by a somatic layer of lateral mesoderm continuous with extraembryonic mesoderm.

Amnioserosa A thin lateral aspect of the extraembryonic amnion of some insects, giving rise to a primary dorsal organ, the amnioserosa is derived (nonetheless) from the inner cell mass of the embryo rather than the trophoblast.

Amphioxus The lancelet, a primitive chordate about four centimeters long, resembles a headless fish; laterally flattened, spindle-shaped, nearly translucent, this primitive marine animal swims by lateral flexures of its whole body; it generally prefers shallow water, spending most of its life half-buried in the sand with its anterior end protruding upward. The lancelet subphylum is related to primordial vertebrates, though its members possess a notochord instead of a vertebral column.

Ampulla(ae) 1. A small dilation in a canal or duct; a rounded muscular sac in invertebrates. In starfish, the ampullae contract, forcing the fluid they contain into the tube feet, extending them. 2. The ears' three semicircular canals are located anatomically such that a person can detect movements in almost any direction and respond to kinetic equilibrium; the swollen base of each canal is known as the ampulla—it bears within it a crista—a ridge of epithelium hooded by a cupula. The cupula is a curled gelatinous mass almost identical to a macula but without an otolith. Hair cells embedded in it register the cupula's fluid displacement within the canals (opposite to the movement of the head). As the hairs bend, depolarization of hair cells shoots action potentials into the vestibular nerves; they relay these signals to the cochlear nerves, which transmit them to the brain.

Amygdala An almond-shaped structure in the temporal lobe of the cortex, involved in a variety of complex neural activities.

Anabolism The process of consuming energy to build more complicated molecules from simpler ones.

Anaerobe (Anaerobic) Able to live in the absence of free oxygen; often, poisoned by oxygen, i.e., an organism that can live only in the absence of atmospheric oxygen.

Anaphase During this phase of cell division lasting a few minutes the kinetochore microtubules shorten, the chromosomes approach the poles, and then the two poles of the mitotic spindle move apart.

Anastomosis A joining or union of branches, such as arteries or veins of a leaf.

Androgen A steroid hormone (of a class including testosterone), stimulating development of the male reproductive system and secondary male characteristics.

Aneuploidy A biologically dangerous condition in which certain chromosomes are too few or present in extra copies; i.e., the chromosome number is irregular instead of haploid or diploid.

Angioblast(s) Mesenchymal cells that form the isolated masses of the blood islands.

Angiosperm A plant that reproduces by flowering, forming its seeds inside chambers called ovaries.

Ängstrom One one-hundred-millionth of a centimeter in length; or a millionth of a micrometer.

Anisogamy (Anisogamous) Production of gametes that are different, usually in size and/or form.

Anisotropy (Anisotropic) Having properties that differ according to the direction of measurement. The anisotropic relationship of time to space that spawns so many cosmological paradoxes lies also at the heart of morphogenetic space and time.

Anteroinferior Situated anatomically near the front and beneath.

Anteroposterior Situated forward and toward the caudal or dorsal side of the body.

Antibody A blood protein generated in reaction to an invader, neutralizing it and/or its toxins, hence, the source of the body's immunity.

Anticodon Base triplet on a tRNA molecule that reads its mirror codon on an mRNA molecule.

Antigen (Antigenic) A foreign macromolecule originating outside the organism and eliciting an immune response upon entering.

Antimüllerian hormone Sertoli cells secrete antimüllerian hormone in the male, which triggers the production of testosterone and differentiation of the Wolffian duct cells to become *vas efferentia,* epididymis, *vas deferens,* and seminal vesicles, at the same time causing the Müllerian ducts (which give rise to the female reproductive system) to regress.

Antrum Any cavity in the body; a cavity formed in an oocyte as the fluid-filled spaces around its follicles coalesce.

Aorta (Aortic) The main arterial trunk of the heart, carrying blood from its left side to the arteries of all limbs and organs except the lungs.

Arachnoid membrane The delicate membrane separating the *dura* and *pia mater* of the spinal cord and brain, consisting of such thin, soft "hairs" that it resembles a cobweb, hence its spidery name.

Arachnoid trabercula(ae) These numerous delicate strands of connective tissue left over from the formation of the arachnoid membrane and *pia mater* as a single layer continue to pass between emergent differentiated zones.

Arachnoid villi Comprising a thin cellular layer derived from the endothelium of the sinus and the epithelium of the arachnoid, the arachnoid villi project into the dural sinuses and absorb the cerebrospinal fluid into the venous system (they are also known as arachnoid granulations).

Archenteron This central cavity is formed during gastrulation; lined with endoderm it develops as an animal's digestive tract, forming the phenomenological core of the inside of the body cavity.

Archetype A primordial psychic form that is inherited in the inborn, collective domain of the unconscious mind, each archetype expresses itself by archaic motifs in myths, fairytales, dreams, and ancient art; an archetype also provides seminal and transitional concepts in science, philosophy, and religion. By extension, an archetype is a universal image or form taking on divergent shapes and manifesting different aspects. Possible archetypes include mother, child, puer, anima, man/woman, trickster, Christ, shadow, entwined serpents, a winged horse (pegasus), etc. Archetypes in nature (as opposed to psyche) may also contribute to basic shapes and geometries.

Artery A vessel carrying blood away from the heart.

Aster A star-shaped configuration that appears in the cell's cytoplasm as the centrosome forms during cell mitosis.

Astrum (Astral) A region of outer and inner space separate from the physical universe, yet mirroring it astronomically and astrologically, known also as hyperspace or hyperdimensional space.

ATP (adenosine triphosphate) A nucleoside that releases free energy as its phosphate bonds are hydrolyzed. This is both a source of and storage place for metabolic energy in cells.

Atrium (Atria) A chamber of the heart receiving returning blood.

Auricle The fleshy part of the external ear.

Auricularia The echinoderm embryonic stage in which lobes carrying ciliary bands begin to extend.

Autonomic nervous system The division of the vertebrate nervous system which regulates involuntary actions, for instance, of the heart, glands, bladder, pupils, and intestines, the autonomic nervous system comprises sympathetic and parasympathetic branches.

Autopoiesis (Autopoietic) The process of organismal self-maintenance.

Autotrophy The nutritional mode in which inorganic carbon in the form of carbon dioxide serves as a source of carbon and light; the condition of inorganic compounds serving as a source of energy.

Axial Located in the vicinity of an axis or forming an axis itself.

Axon The long, thin protrusion of the membrane of a neuron, underlain by microtubules, transmitting impulses away from the cell body.

Axoneme Shaft of an undulipodium with a 9 + 2 microtubule arrangement.

Axopod A firm, straight, highly organized microtubule-composed pseudopod of a protoctist, used for locomotion or feeding, most striking among heliozoans, a floating protozoa so named ("sun animal") because their axopods resemble rays of sun.

Bacteriophage A bacterium-infecting virus, also called a phage.

Bacterium(ia) Free-living or parasitic unicellular (prokaryote) organism, usually anaerobic.

Bacteroid A bacterium structurally modified during evolution for symbiotic residency in the roots of certain leguminous plants.

Barbule A pointed projection fringing the edge of a feather barb.

Bardo Literally a "transition," used in Tibetan Buddhism to describe the state or realm in which a being exists after death and before entering another body; used more generally to describe any domain of being, such as the bardo of this world, the bardo of dying, the bardo of the womb.

Basal body (see Centriole) A centriole-like structure that organizes microtubules into a cilium or flagellum.

Basolateral Lower (along the base) and along the side(s).

B-cell One form of lymphocyte developing in the bone marrow and later producing antibodies that mediate humoral immunity.

Berdache Passive male partner in anal intercourse with another male (can also be used to designate gay men or sometimes lesbians).

Bile An alkaline liquid secreted by the liver and stored in the gall bladder, used by the duodenum to emulsify fats and facilitate digestion.

Biont A living organism (probably terrestrial, as opposed to a hypothetical exobiont from another world).

Bipinnaria Stage of a bilaterally symmetrical echinoderm embryo (see Brachiolaria).

Blastocoel Fluid-filled cavity in center of a blastula.

Blastoderm The layer of cells surrounding the blastocoel, the blastoderm becomes the germinal disk from which the organism develops.

Blastokinesis The active movement of the embryo by which it passes from the ventral to the dorsal side of the egg and at the same time revolves 180 degrees on its long axis.

Blastomere One cleavage cell formed by the fertilized ovum.

Blastopore Opening to the archenteron in the gastrula, which becomes a mouth in protostomes and an anus in deuterostomes.

Blastula The hollow ball of cells that results from the stages of simple cleavage after fertilization of the ovum.

Bouton(s) Very small knobs close to cell bodies or dendritic stems of other neurons on which the terminal, finely branched twigs of axons end in their synapses between neurons.

Brachial Of, pertaining to, or resembling an arm or a similar or homologous anatomical part.

Brachiation The act of swinging by the arms, usually from branch to branch in a tree.

Brachiolaria Stage of a sea star (echinoderm) during which the free-swimming, cilia-propelled, bilaterally symmetrical bipinnaria larva develops small anterior projections it uses to attach to objects on the ocean floor in preparation for its metamorphosis into a radially symmetrical starfish.

Brachiopod(a) The phylum of lampshells includes bivalves with tentacle-like structures on either side of their mouths.

Branchial arches Separated from each other by branchial grooves, the branchial arches support the lateral walls of the cranial aspect of the foregut, the primitive pharynx, and are the embryogenic basis of the facial and masticatory muscles, the pharyngeal and laryngeal muscles, the hyoid bone, etc. Made up of a mesodermal core covered externally by ectoderm and internally by endoderm, the arches are invaded by neural-crest cells which give rise to the skeletal and connective tissues of the lower face and the anterior region of the neck.

Bronchium(ia) Bronchial tubes that are smaller than bronchi and larger than bronchioles in the fractal morphology of the lungs.

Bronchus(i) One of the main branches of the trachea leading directly to the lungs.

Bryozoa A phylum of mosslike animals that form aquatic colonies by budding and branching.

Bt potato(es) Genetically engineered potatoes containing the insecticidal protein from the naturally occurring bacteria *Bacillus thuringiensis*. Some companies have inserted the gene for Bt into plant genomes so that they will not have to depend on bacteria for protection.

Buccal Pertaining to the cheeks or mouth cavity.

Bulbospongiosus This is a muscle at the bulb of the penis that constricts the urethra and aids in the erection of the penis; in the female, the same muscle, located at the base of the clitoris, contributes to clitoral erection.

Bulbo-urethral gland(s) Small pea-shaped compound glands lying alongside the male urethra, just below the prostate, discharging directly into the terminal portion of the urethra (which they lubricate); also known as Cowper's glands.

Bulbus arteriosus Expanded portion of the ventral aorta, containing smooth rather than cardiac muscle, this structure distributes blood to the muscles.

Calyx (Calyces) Cuplike or funnel-shaped structure.

Canaliculi A very small channel in the body, as those forming a tear duct or carrying bile out of the liver.

Cardiogenesis Embryological formation of the heart.

Carotid One of two arteries in the neck carrying blood to the head.

Catabolism The process of the release of energy by breaking down more complex into simpler molecules.

Cation An ion that has lost one or more electrons, so has a positive charge.

Caudal Tailward; near the tail or hind region.

Cecum A sizable blind pouch, such as at the beginning of the large intestine.

Centrale An accessory carpal bone.

Centriole(s) Two cellular structures composed of cylinders of nine triplet microtubules arranged in a ring, centrioles organize other microtubule assembly during cell division.

Centrolecithal Describing eggs of arthropods in which the relatively yolk-free cytoplasm is confined to the center, the outer cortex, and spokes connecting the two.

Centromere The part of a chromosome to which the spindle fiber attaches during cell division.

Centrosome The mass of differentiated cytoplasm which contains the centriole.

Centrum(a) The bulk of a vertebra, excluding the bases of the neural arch.

Cerebellum A deep-seated, primitive structure in the brain regulating and coordinating voluntary muscular movement, the cerebellum is located inferior to the occipital lobes of the cerebral cortex.

Cerebral aqueduct A narrowing of the neural canal that joins the third and fourth ventricles of the midbrain.

Cerebral palsy Impaired coordination and muscle function from brain damage prior to birth.

Cerebrospinal fluid (CSF) The blood-based serum bathing the lateral ventricles of the brain and the basin of the spinal cord.

Cerebrum (Cerebral) The large, distinctive aspect of the advanced mammalian brain, occupying most of the cranial cavity while divided into two hemispheres.

Cervix (Cervical) Any neck-shaped tissue mass, such as the outer end of the uterus.

Chela Claw of a crustacean.

Chellean Describing very early European Palaeolithic sites, usually associated with bifacial stone handaxes.

Chemotaxis Orientation or movement of a living organism relative to a chemical substance.

Ch'i Primary universal and biological energy in Chinese cosmology and medicine.

Chitin The structural component of arthropod exoskeletons, composed of a polysaccharide of an amino sugar; also found in fungi.

Chiton A simple marine mollusk that lives on rocks and dorsally bears a mantle of eight articulated shell plates, related phylogenetically to those gastropods ancestral to limpets.

Chlorophyll A green pigment located within a plant chloroplast.

Chloroplast An organelle in plants and photosynthetic protoctists that uses sunlight (photons) to synthesize organic compounds from carbon dioxide and water.

Cholesterol An essential steroid in animal-cell membranes, cholesterol is used as a precursor molecule to synthesize more complex steroids.

Chondroblasts Mesenchyme cells formed from the neural crest and sclerotome, migrating to sites of limb and cartilage formation.

Chondrocyte(s) Terminally differentiated cartilage cells arising from chondroblasts. During cytodifferentiation, chondrocytes aggregate and pile up in nodules

Chondrogenesis The embryogenic creation of cartilage by cellular secretion of an extracellular matrix consisting of collagen and other proteins.

Chorda Descending neural rootlet; remnant of the dissolved notochord in vertebrates.

Chordate A phylum of animals, each possessing a notochord, a dorsal, hollow nerve cord, and (in their embryos) pharyngeal gill slits.

Chorion The outermost extraembryonic membrane involved in formation of mammalian placenta.

Choroid The vascular layer of the eye between the retina and sclera.

Choroid fissure A line along the medial wall of the embryonic cerebral hemisphere where it becomes extremely thin. Continuous with the roof of the third ventricle, the choroid fissure marks the site of the future choroid plexus of the lateral ventricle.

Chreod(e) A complex nonphysical pathway compelling geological and biological activity into patterns; a resonance from outside physical space and time, conferring form, symmetry, and instinct on matter.

Chromatid One of two daughter strands of a duplicated chromosome still joined by a centromere.

Chromatin An aggregate association of dispersed DNA and protein in eukaryote nuclei, most perceptible between periods of cell division.

Chromatophore Pigment-bearing cell that, by expanding or contracting, can change overall skin coloring.

Chromosome(s) Long, threadlike associations of genes consisting of DNA and protein, chromosomes are found in the nucleus of almost all eukaryote cells and transmit ancestral features to future cells.

Chyle Milky-colored lymph with a high fat content.

Chyme Partly digested food passed from the stomach to the duodenum.

Cichlid A family of tropical freshwater fishes.

Ciliary body The thickened vascular tunic of the eye joining the choroid to the iris.

Cilium(ia) Locomotive organelle constructed from a core of nine outer doublet microtubules and two single inner ones wrapped in an extension of the cell membrane.

Circadian cycle Referring to a full daily cycle of biorhythms, passing through nocturnal and diurnal phase.

Cisterna A fluid-filled sac or space in a cell or tissue.

Clavelina A tunicate species beginning typically as a tadpole-like larva with a notochord in its tail, then becoming sessile and saclike as an adult, often occurring in colonies.

Cloaca Historically, the common cavity for the intestinal, urinary, and genital tracts in primitive mammals and simpler invertebrates.

Clone A genetic duplicate of a cell or organism.

Cnidoblast Parent cell containing the stinging capsule (nematocyst) of a coelenterate.

Coacervate Primitive precellular life forms emerging from clustering droplets of aquatic colloids in which a dispersed phase has a strong affinity with its dispersing medium.

Coccyx Fused rudimentary vertebrae constituting a small bone at the base of the spinal column.

Cochlea A coiled tube shaped like a snail shell and packed with nerves essential for hearing, the cochlea develops from an expanding diverticulum of the otic vesicle after it fuses away from the otic pit.

Cochlear nerve Sensory receptor for the cochlea, this nerve transmits sound (in the form of vibrations) to the window of the cochlea through the external eardrum and small bones of the inner ear, including the malleus and incus.

Codon A sequence of DNA code made up of three nucleotides and specifying either a specific amino-acid sequence or a termination instruction.

Coelenterata A phylum of simple invertebrates, each with a radially symmetrical body and a saclike internal cavity, including hydras, jellyfish, and sea anemones.

Coelom A body cavity that is lined with mesoderm.

Coenzyme An organic molecule (such as a vitamin) which aids an enzymatic process in metabolism.

Collagen(s) (Collagenous) A family of different fibrous proteins secreted by connective tissue cells and found in all multicellular animals, collagens are the most abundant proteins in mammals. Coded by a multiplicity of genes, they have a stiff, triple-stranded helical structure conducive to forming fibrils and creating the meshwork of the basal laminae.

Colliculi Large clusters of neurons of the midbrain, the superior and inferior colliculi are involved in orchestrating visual and auditory reflexes; they originate as neuroblasts in the alar plates.

Commissure Angle, corner, or seam of an organ or tissue; a place where two structures are joined; a tract of nerve fibers passing from one side of the spinal cord or brain to the other.

Competence The physiological state or capacity of a tissue that allows it to react with morphogenetic specificity in response to particular stimuli. Neural differentiation is a primary competence of ectodermal tissue.

Convection Heat transfer between regions of unequal density caused by nonuniform heating; any fluid motion caused by an external force.

Corium The layer of skin beneath the epithelium, the corium is the site of nerve endings, sweat glands, and blood and lymph vessels.

Cornea Transparent convex skin covering the lens of the eye.

Cornu Any hornlike or horn-shaped anatomical structure.

Corona radiata A follicle layer several cells thick, radially arranged around the mammalian oocyte.

Corpora cavernosa Two columns of erectile tissue in the shaft of the penis.

Corpus callosum The site where the right and left cerebral hemispheres are joined.

Corpus luteum A glandular structure in women that secretes progesterone and some estrogen, the *corpus luteum* is propagated from a ruptured follicle after ovulation.

Corpus spongiosum The median longitudinal column of erectile tissue of the penis that contains the urethra.

Corpus striatum Striped gray and white matter, located in front of and lateral to the thalamus in each cerebral hemisphere.

Cortex (Cortical) The outer layer of an internal organ or body structure; the outer layer of gray matter that covers the surface of the cerebral hemispheres.

Corti The organ of Corti is the region of the cochlear duct of the ear comprising hair cells and initiating action potentials in response to sound vibrations.

Corticospinal tract The major descending nervous pathway involved in conscious motor control, the corticospinal tract contains axons of upper motor neurons that transit the pyramidal regions of the *medulla oblongata*, most of them crossing the body and synapsing with lower motor neurons in the anterior horn of the spinal cord; thus, each half of the brain controls the opposite hemisphere of the body.

Corticotropic (Adreno)corticotropic hormone, a substance produced by the pituitary gland, regulates the production of steroids by the adrenal cortex.

Cortisol A hormone that regulates carbohydrate metabolism and maintains blood pressure.

Cosmogony (Cosmogonic) The study of the origins and evolution of the universe.

Covalent A type of chemical bond that occurs when electrons are shared.

Cranium (Cranial) The portion of the skull encasing the brain.

Craniocaudal On an axis of head to tail.

Ctenophore A marine animal that has a gelatinous body and eight rows of combed cilia for swimming, ctenophores comprise their own phylum.

Cumulus cells Follicular cells surrounding the primary oocyte.

Cyanobacteria A blue-green photosynthetic bacterium.

Cybernetics The study (especially mathematical) of the flow of information and of control processes in electronic, mechanical, and biological systems. In this text the adjective "cybernetic" is used in a popular sense to refer to the computerized or mechanized aspect of an organism or event.

Cyborg A human being who has physiological processes aided by mechanical or electronic devices; a science-fiction creature who is part human, part robot.

Cystic fibrosis A hereditary glandular disease, affecting mainly the pancreas, respiratory system, and sweat glands.

Cytoplasm The protoplasm outside the nucleus of the cell.

Cytosine A pyrimidine base in RNA and DNA.

Cytoskeleton The internal framework of the cytoplasm of a cell, made up of microtubules, microfilaments, and intermediate filaments.

Delamination Splitting of the blastoderm into two layers of cells; general separation into thin layers.

Dendrite (Dendritic) A branched extension of a nerve cell that conducts impulses from adjacent cells inward toward the cell body.

Dentin Calcareous part of the tooth beneath the enamel that surrounds the pulp chamber and root canals.

Dermatome The lateral wall of a somite; also a region of skin bearing sensory fibers from a single spinal nerve.

Dermis The tissue layer beneath the epidermis that contains nerve endings, sweat and sebaceous glands, and blood and lymph vessels.

Dermomyotome The dorsolateral aspect of the somites, the dermomyotome provides cells for the skeletal muscles and dermis of the skin.

Desmosome A specialized cell junction in epithelia into which intermediate filaments are inserted, particularly conspicuous in skin tissues that withstand mechanical stress.

Deuterostomia The branch of the metazoa (including echinoderms and chordates) in which the opening leading from the cavity of the archenteron (the blastopore) becomes only the anal and not the oral opening in later development.

Diaphragm A muscular membranous partition that separates bodily cavities, notably the partition between the abdominal and thoracic cavities that aids in respiration.

Diastrophism The series of geological processes shaping the folds, faults, continents, mountains, and ocean beds of the Earth's crust.

Dimer A molecule constituted of two identical simpler molecules.

Dipleurula The earliest bilaterally symmetrical, ciliated phase of the echinoderm embryo, the dipleurula is presumed to represent the hypothetical ancestor of all echinoderms.

Diploblastic Derived from the ectoderm and the endoderm alone, used in describing the tissue structure of lower invertebrates.

Discoidal Disk-shaped.

Divalent Composed of two homologous chromosomes or sets of chromosomes; or having a twin capacity to form atomic bonds.

DNA (deoxyribonucleic acid) The nucleic acid that carries the core genetic information in the cell and is capable of self-replication and the synthesis of RNA.

Dopamine A neurotransmitter formed in the brain that helps regulate sleep, mood, and pleasure recognition, and is critical to central-nervous-system function.

Dorsal Toward the back or upper surface of an organ.

Dorsoventral Flattened and having distinct upper and lower surfaces.

Down's syndrome A congenital disorder caused by an extra twenty-first chromosome, Down's syndrome presents with mental retardation, short stature, and a flattened facial profile.

Duchenne muscular dystropy The most common form of muscular dystrophy (a disease of irreversible muscular deterioration) affecting almost exclusively males, beginning in early childhood and usually causing death before adulthood.

Duodenum (Duodenal) The initial portion of the small intestine.

Dural tube The spinal section of the dural membrane, running between the foramen magnum of the skull (the orifice at its base through which the spinal cord passes) and the sacrum.

***Dura mater* (Dural membrane, Dura)** The fibrous membrane that lies atop the arachnoid and *pia mater* and covers the brain and spinal cord. Attached via the periosteum to the bones of the cranial vault, the dura represents the boundaries of the semi-closed hydraulic system containing the cerebrospinal fluid.

Dynein A large protein complex containing two or three globular heads linked to a common root by a thin, flexible strand. The ATP activity of each head is stimulated sixfold by its association with microtubules. A cilium's dynein arm is constructed by one dynein molecule. The heads generate the motion of the microtubules in the cilium by a sliding action much like that of myosin in muscle.

Dyslexia A learning disorder marked by the impairment of ability to recognize and comprehend written words.

Eardrum This thin membrane, oval and opaque, separates the middle from the external ear.

Eccrine gland Gland secreting externally.

Ecdysone A steroid hormone made by insects and crustaceans that promotes growth and controls molting.

Echinodermata (Echinoderm) A phylum of marine invertebrates, radially symmetrical in their adult forms, that have an internal calcareous skeleton and are often covered by spines.

Ectoderm (Ectodermal) The outermost of three primary germ layers of an embryo from which the epidermis, nervous tissue, and, in vertebrates, sense organs, develop.

Effector fiber A nerve running to a gland or muscle, activating secretion or contraction, respectively.

Efferent fiber A nerve that carries impulses away from the central nervous system or some other central organ to an effector fiber.

Elastin Protein that is the principal structural component of elastic fibers.

Electron A subatomic particle with a negative charge.

Element A substance composed of atoms having identical numbers of protons in each nucleus.

Embryoblast The inner cell mass of the early human morula, the embryoblast will later form the embryonic organism (as opposed to the outer cell layer, the trophoblast, which will give rise to the placenta and the nutritive extraembryonic organs).

Embryogenesis The development and growth of an embryo from a single cell into an organism.

Emergent property A novel aspect of a complex system that (for all intent and purposes) was not present at one particular level or stage of development or organization, yet occurs at a succeeding one. Emergent properties of evolving systems made of atoms, molecules, cells, symbols, and behavior, respectively, arise in such a way that disorganization or relative spareness at one level mysteriously generates high order or relative complexity at the next. Emergent properties may also be defined as aspects that are only characteristic of a system as a whole, rather than being characteristic of any of its parts.

Endocardium A thin, serous membrane made of endothelial tissue that lines the interior of the heart.

Endocrine Relating to glands and their secretion.

Endocytosis (Endocytic) The process of cellular ingestion in which the plasma membrane folds inward to bring substances into the cell.

Endoderm The innermost of three primary germ layers, endoderm develops into the gastrointestinal tract and the lungs.

Endogamy (Endogamous) A rule which requires a person to marry within his or her own kin, local or social group, or caste.

Endoplasmic reticulum Membrane network inside the cell that is involved in the synthesis, modification, and transport of cellular materials.

Endosymbiosis A symbiotic relationship between two organisms in which one of them (the endosymbiont) lives inside the body of the other (the host).

Enzyme (Enzymatic) Proteins that act as biochemical catalysts.

Eocene Second oldest of five major epochs of the Tertiary Period, roughly 58 million years before the present, characterized by the rise of mammals.

Epiblast Outer layer of the blastula that gives rise to the ectoderm after gastrulation.

Epiboly Growth of rapidly dividing group of cells around a more slowly dividing group such that the epithelial layer encloses the deeper one.

Epicardium The inner layer of the pericardium (membranous sac enclosing the heart) in actual contact with the heart.

Epididymis This tightly coiled tube, which lies along the top of and behind the testes, is where sperm cells mature and develop the ability to swim.

Epigenesis (Epigenetic) The theory that an individual organism is developed by successive differentiations of an unstructured egg rather than the incremental enlarging of a preformed entity.

Epiphysis (Epiphyses) A part of a bone that starts its development separated by cartilage from its main portion; a small center at the ends of a long bone from which the bone itself grows and ultimately becomes ossified.

Epistasis (Epistatic) The interaction between nonallelic genes with the suppression of one gene resulting.

Epithalamus Dorsal posterior subdivision of the diencephalon that contains the pineal body.

Epithelium Membranous tissue of one or more compact layers of cells directly connected to one another that covers most internal/external surfaces of the body and organs.

Erogenous Responsive to sexual stimulation.

Erythrocyte(s) Red blood cells that transport oxygen and carbon dioxide to and from tissues.

Erythropoietin Glycoprotein hormone that stimulates the production of red blood cells by bone marrow.

Estradiol An estrogen-related hormone secreted in the follicle cells of ovaries, which, when taken from sow ovaries or pregnant mares, can be used medicinally as a substitute for estrogen.

Estrogen Steroid hormone produced chiefly by the ovaries, estrogen is responsible for promoting estrus and the development and maintenance of female secondary sex characteristics.

Ether (biochemical) Two hydrocarbons linked by oxygen.

Ether (occult) The first elemental resonance (or substance) to evolve from the oneness of universal intelligence, ether is considered the heart of the all-pervasive-ness of space itself. Sonic and vibrational, though utterly empty and at rest, it supports a field of highly agitated air.

Etheric body This subtle sheath of each organism draws bodily form from cosmic breath, stepping down the universal emanation of quintessence into materiality in the form of five gross elements that sustain matter.

Ethmoid bone The walls and septum of the nasal cavity, this light spongy bone is located between the ocular orbits and contains the olfactory nerve fibers. Mechanically, the ethmoid acts against the potential effects of the rotating wings of the sphenoid bone on the frontal bone of the skull.

Eukaryote An organism whose cells have a distinct membrane-bound nucleus and membrane-bound organelles.

Eustachian tube This tube connects the tympanic cavity with the nasal part of the pharynx and serves to equalize air pressure on either side of the eardrum.

Exocytosis (Exocytic) Cellular secretion of macromolecules in which vesicles fuse with the plasma membrane and are discharged; waste removal from the interior of a cell through its membrane.

Exogamy (Exogamous) A rule (or custom) of marriage which forbids an individual to take his (or her) spouse from within a particular residential, kin, or status group, or caste to which he himself (or she herself) belongs.

Exon A nucleotide sequence in DNA that carries the code for the final messenger RNA molecule and thus defines the amino-acid sequence during protein synthesis.

Exoplacental cone A mass of cells formed by the proliferation of extraembryonic ectoderm lying at the upper end of the egg cylinder, the exoplacental cone invades maternal connective tissues and establishes close contact with maternal blood vessels, thus beginning the formation of the placenta.

Expir Phase of motility cycle in which organs move toward the median axis of the body.

Extension The phase of the craniosacral motion cycle in which the head narrows in its transverse dimension, the sacral base moves anteriorly, the sacral apex moves posteriorly, and the whole body internally rotates and seems to narrow slightly.

Extensor A muscle that acts to stretch a limb.

Extraembryonic Pertaining to organs and tissues outside the body (soma) of the embryo that mediate metabolic integration with an egg or (among mammals) uterus of the mother; such organs include the placenta, amnion, yolk sac, and chorion.

Factor VII clotting genes Antihemophiliac genes.

Fallopian tubes Ducts through which ova pass from ovaries to uterus in humans and higher mammals.

Falx cerebelli The smaller of two folds of *dura mater* separating the hemispheres of the brain lies between the lateral lobes of the cerebellum.

Falx cerebri The larger of two folds of *dura mater* separating regions of the brain, the *falx cerebri* lies between the cerebral hemispheres and contains the sagittal sinuses.

Fascia (Fasciae, Fascial) Fibrous tissue sheet or band that envelops, separates, or binds together muscles, organs, or other soft structures of the body.

Fat A large molecule composed of two kinds of smaller molecules, a fat is an ester of glycerol and fatty acids. Glycerol is an alcohol of three carbons, each carbon bearing a hydroxyl group. A fatty acid is a long carbon skeleton (sixteen or eighteen atoms) with a carboxyl head and a hydrocarbon tail. Because of their tails bearing carbon-hydrogen bonds, fats are hydrophobic—insoluble in water.

FGF (fibroblast growth factor) Acidic FGF and basic FGF are the two founding members of a family of structurally related growth factors needed for mesodermal or neuroectodermal cells.

Fibronectin Any of a group of glycoproteins of cell surfaces, blood plasma, and connective tissue that promote cellular adhesion and embryonic cell migration.

Fibula Outer and narrower of two bones of the human leg, or the hind leg of a quadrupedal animal between the knee and ankle.

Filopodia Fine protoplasmic extensions of mesenchyme cell surfaces.

Fimbria A fringelike part or structure such as the opening of the Fallopian tubes.

Flagellum (Flagella) A long threadlike appendage of certain unicellular organisms used for locomotion, each flagellum is formed from a core of nine outer doublet microtubules and two inner single ones, all in a plasma-membrane sheath.

Flexion The phase of the craniosacral motion cycle in which the head widens, the sacral apex moves in an anterior direction, and the whole body externally rotates and broadens.

Flexor A muscle that acts to bend a joint or limb.

Flimmer A fine, hairlike projection that extends laterally from undulipodia.

Fluke A parasitic flatworm (trematode) with a thick outer cuticle and one or more suckers for attaching to the tissue of the host.

Folliate Shaped like a leaf.

Follicle A cavity in the ovary containing the maturing ovum; any small body cavity or sac.

Follicle stimulating hormone (FSH) This hormone is secreted by the anterior pituitary; in women it stimulates the follicles of the ovary, assisting in their maturation and causing them to produce estrogen; in men it stimulates the epithelium of the seminiferous tubules and has a role in inducing spermatogenesis.

Foramen (Formina) Aperture or perforation in a membranous structure (such as the cecum or brain).

Foramen of Luschka (see below.)

Foramen of Magendie When, midway through embryogenesis, the thin roof of the fourth ventricle of the midbrain swells outward at three spots, rupturing to form foramina, the median and lateral apertures are known as the foramen of Magendie and the foramina of Luschka, respectively. They provide the means for the cerebrospinal fluid to pass from the fourth ventricle into the subarachnoid space.

Foraminifera An order of microorganisms with perforated calcareous shells through which pseudopods (protrusions of cytoplasm) protrude as a means of locomotion and to envelop and ingest food.

Fornix(ices) Bands of white fibers under the *corpus callosum* of the brain, connecting ancestral regions of the hippocampus.

Fractal A geometric pattern repeated at ever smaller scales to produce irregular surfaces and shapes and to pack space with ever more densely arranged substances.

Frontal Situated at or toward the front of the body (or sometimes the forehead).

Fundus The portion of a hollow organ opposite or farthest from its opening.

Ganglion (Ganglia) A group of nerve-cell bodies (in vertebrates, located outside the brain or spinal cord).

Gap junction A continuous aqueous channel formed from transmembrane proteins in the plasma membranes of two aligned cells such that their interiors are connected.

Gastrodermis In simple animals, like *Hydra,* the gastrodermis is a sheet (or sheets) of digestive tissue consisting of large gastric epitheliomuscular cells linked together with specialized glandular and mucous cells.

Gastrovascular Having both a digestive and circulatory function.

Gastrula The embryo stage after the blastula, the gastrula emerges from complex morphogenetic movements and infoldings as a hollow, two-layered, cup-shaped sac of endoderm surrounding the archenteron.

Gastrulation A stage of embryogenesis characterized by the folding-in of part of the blastoderm such that the simple spherical blastula becomes converted into a double-walled cup, as if one side of the elastic hollow ball had been pushed in (invaginated) by an external force. The blastocoel is obliterated by the invading sheet of cells and replaced by the archenteron.

Gene (Genetic) The hereditary unit that occupies a specific location on a chromosome and, at least theoretically, determines a particular characteristic in an organism.

Gene promoter The start site for RNA synthesis, signalling where it should begin by tightly binding the polymerase.

Genome The complete haploid set of chromosomes with its associated genes.

Genophore Large bacterial DNA (bacterial chromosome).

Germ plasm The cytoplasm of a germ cell that contains chromosomes.

Glans The head or tip of the penis or clitoris toward which the external urethral orifice moves progressively closer during masculinization of the embryo.

Glioblast An embryogenic precursor cell to neuroglia.

Globus pallidus The phylogenetically oldest part of the *corpus striatum;* also called the palaeostriatum.

Glossopharyngeal Pertaining to both the tongue and pharynx, applied especially to the ninth pair of cranial nerves which are distributed to the pharynx and tongue.

Glucose A monosaccharide sugar common in animal and plant tissue, glucose is a major energy source of the body.

Glutamic acid A nonessential amino acid common in plant and animal tissue.

Glycerin (Glycerol) $C_3H_8O_3$, a molecular component of fats, is obtained from fats and oils as a by-product in the manufacture of soaps and fatty acids.

Glycine Sweet-tasting crystalline nonessential amino acid.

Glycolipid(s) These oligosaccharides, covalently bonded to carbohydrate-based lipids in the workings of the Golgi apparatus, are common in the plasma-membrane bilayer of animal cells, where they extrude a sugar group at the cell surface.

Glycolysis An anaerobic, ATP-generating metabolic process that changes carbohydrates and sugars into pyruvic acid, yielding a net of two ATP molecules.

Glycoprotein A group of conjugated proteins that have carbohydrates as the non-protein component.

Glycosylation The process of adding sugar units such as in the addition of glycan chains to proteins.

Gonad An organ in animals that produces gametes, for example, the testis or ovary.

Grana A stacked chlorophyll-containing structure within the chloroplast that is the site of the light reactions of photosynthesis.

Guanine A purine base of RNA and DNA.

Gubernaculum testis A ligament connecting the testes to the floor of the primitive pelvis.

Gynecomastia The development of breasts in men.

Gyrus(i) Any of the prominent, rounded, elevated convolutions on surfaces of cerebral hemispheres.

Haptotaxis Locomotion of cells along adhesive substrata by an interfacial tension-driven process.

Hayflick limit Cells naturally die either when they have completed a fixed number of division cycles (around sixty, the Hayflick limit) or at some earlier stage when programmed to do so, as in digit separation in vertebrate limb morphogenesis.

Heliozoa Aquatic protozoa that have spindlelike pseudopods radiating from a central cell mass.

Hemidesmosome Specialized junction between an epithelial cell and its basal lamina.

Hemoglobin Iron-containing respiratory pigment in red blood cells of vertebrates that contains six percent heme and ninety-four percent globin.

Hensen's node The swelling anterior end of the primitive streak formed by the thickening of part of the blastoderm during gastrulation; also called the primitive knot.

Hepatocyte A parenchymal cell of the liver.

Herpes Any of several viral diseases that cause eruption of vesicles on the skin or mucous membranes.

Heterochrony Displacement in time of the ontogenetic appearance and development of one embryonic trait with respect to another; a phylogenetic change in the onset or timing of development, so that the appearance or rate of development of a trait is either accelerated or retarded relative to the appearance or rate of development of the same trait in an ancestral embryo.

Heterotrophy (Heterotrophic) A mode of obtaining nutrition (carbon molecules and energy) from autotrophs, used by organisms that are unable to synthesize their own food and depend on complex organic substances (i.e., eating other organisms or their by-products) for metabolism.

Heterozygosity Having different alleles at one or more corresponding chromosomal loci.

Hex-A Hexosaminidase that is deficient in Tay-Sachs disease and Sandhoff's disease.

Hippocampus The ridge in the floor of the lateral ventricle of the brain that consists mainly of gray matter and has a central role in memory processes.

Histamine A physiologically active amine found in plants and animals, histamine is released in humans as part of an allergic reaction.

Histone Any of several small basic proteins in association with the DNA in chromatin.

Holoblastic Exhibiting cleavage in which the entire egg splits into individual blastomeres.

Homeobox A sequence of 180 nucleotide pairs found, with minor variations, in virtually all homeotic selector genes and some other genes as well. These probably evolved by gene duplication and divergence.

Homeotic selector genes Genes first activated in the blastoderm and defining a choice between states of determination corresponding to different but homologous, or homeomorphic, structures. Their ordering according to their spatial pattern of expression and control relationships is mirrored by the seriality in which they are situated along the chromosomes.

Hominid A primate in the family Hominidae. *Homo sapiens* is the only extant species.

Hominoid Belonging to the superfamily Hominoidea, which includes apes and humans.

Homoplasy (Homoplastic) The morphological similarity of an anatomical feature in separate lineages whose common ancestor did not resemble either lineage with regard to this trait.

Humerus bone This bone extends from the shoulder to the elbow.

Huntington's disease A disease of the central nervous system that leads to dementia, abnormal posture, and involuntary movements.

Hyaloid artery One of the arteries supplying the lens of the eye with blood. Its distal portion degenerates during embryogenesis.

Hydra A freshwater coelenterate polyp with a naked, cylindrical body and an oral opening surrounded by tentacles.

Hydrocephalia (Hydrocephalic) A usually congenital condition in which undue accumulation of fluid in the cerebral ventricles causes enlargement of the skull and compression of the brain.

Hydrolysis Decomposition of a chemical compound by reaction with water.

Hydrophilic Having an affinity for water, readily absorbing or dissolving into water.

Hydrophobic Repelling water and unable to dissolve in water.

Hydrostatic The equilibrium mechanics of fluids, especially incompressible ones.

Hymen A tissue that occludes the external vaginal orifice.

Hyoid bone A u-shaped bone between the mandible and the larynx at the very base of the tongue that supports the muscles of the tongue.

Hypermorphosis The phylogenetic extension of ontogeny beyond its ancestral termination, usually leading to increased body size and complexity of differentiating organs. This results in recapitulation in that prior adult stages have now become intermediate stages of a lengthened descendant ontogeny.

Hypha Any threadlike filament that forms the mycelium of a fungus.

Hypoblast The lower layer of (particularly bird, reptile, and mammal) blastoderm (as opposed to the epiblast).

Hypospadia Congenital defect in which the urethra opens on the bottom of the penis rather than on the glans.

Hypothalamus The part of the brain that lies below the thalamus and regulates bodily temperature and other autonomic activities.

Ileum The terminal portion of the small intestine.

Ilium The uppermost and widest of three bones of the lateral halves of the pelvis.

Immunocompetence Having the normal capacity to develop an immune response following exposure to an antigen.

Incus The middle bone of the middle ear (the anvil) formed from ossification of the dorsal end of the first branchial-arch cartilage.

Induction The switching of cells from one morphogenetic pathway to another by the influence of adjacent cells, i.e., that have been adjacent developmentally from the beginning or that have migrated to adjoining positions.

Inguinal Relating to the groin.

Inspir Phase of motility cycle in which organs move away from the median axis of the body.

Insula Lobe in the center of the cerebral hemisphere that is situated deep between the lips of the sylvian fissure.

Insulin Hormone that regulates the metabolism of carbohydrates and fats.

Intermediate filaments Cytoplasmic filaments between microfilaments and microtubules that form a ring around the cell nucleus.

Internuncial Linking two neurons in a neuronal pathway.

Interstitium (Interstitial) Related to small narrow spaces between tissues or parts of an organ.

Intervertebral Located between vertebrae.

Intron Segment of a gene between axons that does not function in the coding for protein synthesis.

Involucrin Involucrin serves as a marker for terminal differentiation of stratified squamous epithelia (see Squames). Insofar as involucrin genes have diverged considerably through mammalian evolution, antibodies for the human molecules do not detect the homologous mouse or rabbit gene product.

Isogamy (Isogamous) Production of compatible gametes equal in size and kind.

Jejunoileum A structure arranged in a series of loosely attached, somewhat freely moving loops covered by the peritoneum and attached by mesentery to the posterior abdominal wall, the jejunoileum provides a gastrointestinal continuation from the duodenojejunal flexure to the ileocecal junction.

Jejunum Section of the small intestine between the duodenum and the ileum.

Jugular foramen Aperture through which the trunk of the posterior mammalian body passes in order to empty into the primary head-vein, this point marks the junction of the head venous system with the neck portion of the system, or the anterior cardinal vein.

Jugular vein(s) Veins draining the superficial tissues of the head and neck and the sinuses of the brain.

Keratin A tough, fibrous, sulphur-containing protein forming the toughened outer layer of skin, hair, nails, hooves, and horns (etymologically, "horn").

Kineses Movements or motions toward a stimulus.

Kinetosome Organelle used to construct undulipodia; the centriole from which the shaft of the undulipodium emerges.

Labia majora Two outer rounded folds of adipose tissue on either side of the vaginal opening that form the external lateral boundaries of the vulva.

Labia minora Two thin inner folds of skin within the vestibule of the vagina enclosed within the cleft of the *labia majora.*

Labii Muscles raising and protruding the upper lip and flaring the nostrils.

Labioscrotal Relating to or being a swelling or ridge on either side of the embryonic rudiment of the penis or clitoris, which develops into one of the scrotal sacs in the male and one of the *labia majora* in the female.

Lacteal A lymphatic vessel located in the wall of the small intestine, the lacteal absorbs fats and carries chyme from the intestine.

Lactoferrin An iron-binding protein of very high affinity found in milk and in the specific granules of neutrophil leukocytes.

Lamella A thin scale, plate, or layer of bone or tissue; a membrane, as in chloroplasts of plants.

Lamellipodia Any motile cytoplasmic, sheetlike extensions that are characteristic of some migrating cells.

Lamina(ae) A thin plate, sheet, or layer; a scale or platelike structure.

Lancelet (see Amphioxus).

Laryngotracheal tube The laryngotracheal tube arises from a groove in the caudal end of the ventral wall of the pharynx. It deepens and expands into a tube in the foregut, opening into the pharynx while separated from the esophagus by the tracheoesophageal septum. In association with surrounding splanchnic mesenchyme, this tube gives rise to the larynx, trachea, bronchi, and lungs.

Laryngovagus Laryngeal branch of the vagus nerve.

Larynx (Laryngeal) Respiratory tract between the pharynx and trachea that has walls of cartilage and muscle and contains the vocal cords.

Leghemoglobin Hemoglobin-like oxygen-binding protein in the nitrogen-fixing root nodules of leguminous plants. In a remarkable demonstration of coevolution, the globin part of the molecule is encoded by a host gene whereas the heme group is usually provided by the bacterial guest.

Leukocyte White blood cell.

Libido Energy associated with instinctual, biological drives.

Limbic system A neural system involved in olfaction, emotion, motivation, behavior, and other autonomic processes.

Lipid(s) A structural fat in cells, together with carbohydrates and proteins constituting the principal materials of life; organic compounds including fats, oils, waxes, sterols, and glycerides that are insoluble in water but soluble in organic solvents.

Lupus An autoimmune disease of the skin and mucous membranes with eruptions of ulcers, lesions, and inflammations, particularly in the facial area.

Luteinizing hormone In females this hormone secreted by the anterior pituitary works with FSH (follicle stimulating hormone) to stimulate the ovum to grow to maturity; it induces the follicle cells to secrete estrogens, causes the mature follicle to rupture and expel its ripe ovum (ovulation), and converts the ruptured follicle into the *corpus luteum*. In males it stimulates interstitial cells to secrete testosterone.

Lymph (Lymphatic) A colorless to faintly yellowish fluid that flows in the lymphatic vessels connecting the lymph nodes and containing white blood cells and some red blood cells. Travelling through the lymphatic system (anatomically separate from the circulatory system), lymph returns to the venous bloodstream through the thoracic duct. It removes bacteria and some proteins from tissues and transports fat from the small intestine while providing mature lymphocytes to the blood.

Lymphocyte A white blood cell formed from a lymphoblast originating in the lymph nodes, spleen, thymus, appendix, or tonsils. Responsible for immune specificity, lymphocytes constitute approximately a quarter of all leukocytes in adult human blood.

Lysis Dissolution of red blood cells or bacteria by an antibody.

Lysogeny The phenomenon whereby bacteriophages reproduce inside bacterial cells and then burst out.

Lysosome A eukaryote organelle in the cytoplasm of most cells containing various hydrolytic enzymes that function in intracellular digestion of macromolecules, the lysosome is a membranous bag of various hydrolytic enzymes that uses proteins synthesized in the endoplasmic reticulum and transported through the Golgi body.

Macrophage A large mammalian white blood cell that conducts phagocytosis and, while scavenging in the bloodstream, ingests invading microorganisms and removes damaged and senescent cells and cellular debris.

Macula A gravity-sensitive sensory structure in the vestibule of the ear, combining hair cells with a gelatinous mass bearing otoliths.

Madreporite The sieve plate connecting the upper surface of a sea star to its fluid-filled ring canal, the madreporite is framed by five jaws with teeth.

Malleus The hammer bone of the middle ear formed from ossification of the dorsal end of the first branchial-arch cartilage, the malleus is the most lateral of the middle-ear bones and is attached to the tympanic membrane.

Mandelbrot set A mathematically generated collection of points in a complex plane with a richness of complication across scales, this computational creation of Benoit Mandelbrot has come to be viewed as the emblem for the intricacy and fractal quality of chaos. According to James Gleick, "An eternity would not be enough to see it all, its disks studded with prickly thorns, its spirals and filaments curling outward and around, bearing bulbous molecules that hang, infinitely variegated, like grapes on God's personal vine." (*Chaos: Making a New Science*, p. 221.)

Mandible (Mandibular) The lower jaw of a vertebrate, also the upper and lower beak of birds and a variable mouth part in insects.

Mandibular arch The first branchial arch of the vertebrate embryo from which humans develop a lower lip, mandible, masticatory muscles, and the anterior part of the tongue.

Manus The distal part of the forelimb, usually a hand, claw, or hoof.

Marsupial A nonplacental mammal carrying its young in a maternal pouch (the marsupium) in which they complete their embryonic development; the group includes kangaroos, koalas, and opossums.

Maxilla(ae) (Maxillary) A pair of bones of the skull that fuse in the midline and form the mammalian upper jaw; also one of a pair of laterally moving appendages behind the mandibles of most arthropods.

Medulla (Medullary) The inner core of a number of different vertebrate organs and structures; also the inner core of plant stems.

Medulla oblongata Neural tissue at the base of the brain, controlling respiration, circulation, and other critical bodily functions. The lowermost portion of the vertebrate brain, the *medulla oblongata* is continuous with the spinal cord.

Medusa The tentacled, bell-shaped, free-swimming sexual stage of most coelenterates.

Megakaryocyte Large bone-marrow cell with a lobulate nucleus that gives rise to blood platelets.

Meiosis (Meiotic) A two-stage process of cell division in sexually reproducing organisms that halves the number of chromosomes in reproductive cells to form gametes in animals and spores in plants.

Melanin Dark brown or black granules secreted by cells in the skin, hair, and retina of the eye (in the latter case forming a dark enclosure reducing the amount of scattered light).

Melanocyte An epidermal cell originating ancillary to its eventual tissue location, a melanocyte is capable of secreting the black pigment melanin, leading to skin-color variation among animals.

Melatonin A hormone produced by the pineal gland that stimulates color change in the epidermis of amphibians and reptiles, and responds to Circadian cycles in mammals.

Membrane A thin, pliant layer of tissue covering or separating structures, organs, or connecting surfaces of an animal or plant; a thin, pliable layer of cytoplasm covering or separating structures or organelles within cells.

Meninx (Meninges) One of the three membranes enclosing the brain and spinal cord.

Meridian According to Chinese medicine, one of fourteen invisible ch'i-bearing channels which crisscross the head, arms, legs, and trunk deep in the tissues and correspond to different organs and regions of the body (pericardium, lung, kidney, liver, etc.). These channels surface at 360 or more acupuncture points, stimulation of which can alter the flow of basic life energy into the organs.

Meristem The stem cells of plants, i.e., the part of plant tissue that remains embryonic throughout the lifetime of the plant, allowing indeterminate growth; undifferentiated plant tissue from which new cells are formed, as that at the tip of stem or root.

Meroblastic Cleavage characteristic of bird (avian) eggs in which there is incomplete division of the cells because the density of yolk.

Mesencephalon The mid section of the vertebrate brain formed from the middle section of the embryonic brain.

Mesenchyme 1. Any group of loosely organized cells with extracellular matrix. 2. A part of the embryonic mesoderm that migrates in groups of cells, later developing into connective tissue, skeletal tissue, and parts of the circulatory, lymphatic, and other systems. Note: Mesoderm refers to a germ layer, mesenchyme to a cell state. All early mesoderm is mesenchymal, but not all of it remains mesenchyme; in the development of the kidney nephron, mesoderm is tranformed from mesenchyme into an epithelium. Conversely, the neural crest represents ectodermal cells that were originally epithelial and became transformed into a mesenchyme. Though other layers than mesoderm can become mesenchymal, in most cases ectodermally-derived tissues remain epithelial, and mesodermally-derived tissues remain essentially mesenchymal. Confusion may originate between neural-crest material and mesenchyme, both of which migrate; the former is ectodermal in origin and mesenchymal in cell state. However, "mesenchyme" is sometimes used as a descriptive term only for mesoderm.

Mesentery A fold in the lining of the abdominal wall connecting the intestines to the dorsal aspect of the wall; any of several folds of the peritoneum connecting the intestine to the dorsal abdominal wall, enveloping the jejunum and the ileum.

Mesoderm The middle of the three germinal layers of most animal embryos, later to become the notochord, connective tissue, the musculoskeletal system, the lining of the coelom, the gonads, the kidneys, and much of the urogenital and circulatory systems.

Mesogloea Jellylike protein lying between the inner and outer cell layers of the body wall of sponges and cnidarians.

Mesonephros (Mesonephric) The kidney of a fish or amphibian, recapitulated as the mid section of the embryonic excretory system of vertebrates (in this form also called the Wolffian body); the second of three excretory organs that develop in a vertebrate embryo, the mesonephros becomes the functioning kidney in fish and amphibians but is replaced by the metanephros in higher vertebrates.

Mesozoic The era of geologically defined time stretching from roughly 230 million years ago to 63 million years ago and including the Triassic, Jurassic, and Cretaceous periods. The Mesozoic follows the most ancient Palaeozoic and precedes the Palaeocene epoch of the Cenozoic when the first hominids appeared.

Metabolism All of an organism's biochemical processes considered as a totality, comprising anabolic and catabolic pathways, and necessary for maintaining life.

Metabolite An organic compound produced by or needed to take part in metabolism.

Metalloprotein A protein that contains a bound metal ion as part of its structure.

Metamere One in a series of homologous body segments lying in a longitudinal series such as in earthworms, insects, and lobsters, corresponding loosely to somites in vertebrates.

Metanephros (Metanephric) The third and final stage of the embryonic kidney which, in vertebrates, becomes the actual adult kidney.

Metaphase A stage of mitosis in which the chromosomes and their centromeres align on the metaphase plate along the mitotic spindle's equator, the centriole pairs stationing themselves at opposite poles.

Metastasis (Metastatic) The migration of pathogenic organisms or cancerous cells from their original site to one or more additional sites in the organism.

Metazoa Division of the animal kingdom (Animalia) comprising all animals more complex than the other division, the Protozoa.

Microcephalia (Microcephalic) Pathological smallness of the embryologically formed head, usually leading to diminished mental capacity.

Microfibril 1. In zoology: A component of the extracellular matrix—particularly at an early stage in the hierarchy of collagen assembly; the sequence goes from alpha chain (a gene product) to tropocollagen (a triple helix of alpha chains) to microfibrils to fibrils to fibers to fiber bundles. 2. In botany: A basic structural unit of the plant cell wall, made of cellulose in higher plants and most algae, chitin in some fungi, and mannan or xylan in a few algae.

Microfilament An actin protein rod functioning structurally and morphologically (for instance, in contraction) in the cytoplasm and particularly the cytoskeleton of virtually all eukaryote cells; any of the minute fibers throughout the cytoplasm of the cell that maintain its structural integrity.

Micron One millionth of a meter in length.

Micropyle Pore in ovum membrane through which spermatozoon enters.

Microspikes Projections from the leading edge of some cells, particularly but not exclusively nerve-growth cones.

Microtubule(s) Hollow rods of tubulin protein occurring in eukaryote cytoplasm that provide structural support and assist in cellular locomotion and transport and fissioning, microtubules are the central components of mitotic spindles, cilia, and flagella.

Mitochondrion(ia) An archaic but crucial bacterium-like cell organelle with a large amount of highly convoluted internal membrane, its own idiosyncratic variety of DNA, and many enzymes important for cell metabolism including those responsible for conversion of food to energy, this relatively large subcellular "organism" (0.5 to 1 micron) carries out oxidative metabolism (cellular respiration) and provides cellular energy.

Mitosis (Mitotic) A process of eukaryote cell division in distinct stages, distinguished from meiosis in that equal numbers of chromosomes are allocated to each of two daughter cells after division; hence the original chromosome number is conserved and passed on, and each of the new cells contains a complete copy of parental chromosomes.

Modiolus The central bony shaft of the cochlea of the ear, the modiolus is the cone around which the cochlea is wound.

Molar A tooth that ostensibly evolved in herbivorous and omnivorous mammals to grind food (particularly botanical material); one of twelve teeth (in four sets of three, one at the end of each quadrant of the two human jaws).

Molecule (Molecular) A configuration of nuclei and electrons in an atom, held together stably by electrostatic and electromagnetic forces, a molecule is the basic structural unit of matter, the simplest unit that exhibits the physical and chemical properties of the substance of which it is a component—the smallest particle into which an element can be divided.

Monocyte A large white blood corpuscle with a well-defined, pale, egg-shaped nucleus, very fine granulation in its cytoplasm, and more protoplasm than the allied lymphocyte, this cell contains specialized organelles that enable it to fuse with phagocytic vesicles and ingest invading macromolecules (including protozoa).

Monomer A molecular subunit that can be bound with others into a polymer, a building block of complex molecules.

Monotreme An egg-laying mammal, including the platypus and echidna, indigenous only to Australia.

Morphogenesis (Morphogenetic) Either the evolutionary or embryological development of biological structure through the inductive interactions of cells; the differentiation and growth of tissues and organs during development.

Morphophonemic The patterning of linguistically stable, indivisible units and subunits of meaning (not necessarily whole words) orchestrated by submorphemic units of sound (phonemes). "Non" and "neo" are morphemes that take on wholly different meanings without the initial phoneme "n"; changes in pronunciation undergone by allomorphs of morphemes as they are modified by neighboring sounds or for grammatical reasons in the course of inflection or derivation. "Morphophonemic" defines the overall grid of sounds and related stress shifts which create the shapes of related words and the structure of languages.

Morula A spherical embryonic mass of blastomeres formed during mammalian blastulation from a cleaved ova.

Mucosa(ae) A membrane secreting a viscous mixture of mucin, water, cells, and inorganic salts, forming a protective coating over glands and in the respiratory and alimentary tracts; containing, producing or secreting mucus, the mucosa is thicker in the mouth and esophagus where it must withstand abrasion, thinner in the intestine where the requirement is absorption and secretion.

Müllerian bodies The connective tissue fibers which form the framework of the retina.

Müllerian duct The primary duct of the pronephros, which survives in female humans as the oviduct; in frogs this duct persists in both sexes.

Mutation A sudden structural change within a gene or chromosome that results in the creation of a new trait that was not found in the parent.

Mycoplasma Common pathogenic bacteria that line the airways of many victims of asthma and other lung disease. During the process of evolution into their present forms these simple creatures—characterized by osmotic fragility and complete lack of a bacterial wall—somehow forfeited a substantial portion of their genetic information.

Myelin (Myelinated) White, fatty material that encases axons and nerve fibers.

Myoblasts Precursor muscle cells originating from myotomes.

Myocardium The muscular tissue of the heart.

Myofascia (Myofascial) The anatomical and functional relationship between muscle fibers and fascial sheaths. Myofascia is a specialized kind of connective tissue, joined in an envelope encasing muscles and arranging itself in layers between them. Toughened myofascia restricts movement; the goal of Rolfing and other bodywork therapies is to enable myofascia to regain its natural elasticity.

Myofibrils Threadlike fibrils of the contractile part of striated muscle fibers.

Myotome The segment of somite in the vertebrate embryo that becomes skeletal muscle.

Myotubes Elongated multinucleate cells that contain some peripherally located myofibrils. They are formed by the fusion of myoblasts and develop into mature muscle fibers.

Nanometer One billionth of a meter.

Nasopharynx (Nasopharyngeal) The part of the pharynx above the soft palate that is continuous with the nasal passages.

Nauplius Free-swimming first stage of larva of certain crustaceans with an unsegmented body, three pairs of appendages, and a single median eye.

Naviculare A boat-shaped bone in the wrist associated with the metacarpal and radius; a concave bone in front of the anklebone associated with the tibial region.

Neoteny The retention of formerly juvenile characteristics in adults of species, or attainment of sexual maturity by an organism while still in larval stage, produced by a retardation of somatic development.

Nephridium (Nephridia) A tubular excretory organ in many invertebrates or in vertebrate embryos, from which the kidney develops.

Nephrotome Ciliated, funnel-shaped inner opening of nephridium in a coelenteate.

Neural crest Part of the ectoderm in a vertebrate embryo that lies on either side of the neural tube and develops into the cranial, spinal, and autonomic ganglia.

Neural-crest cells Ectodermal cells that, after breaking into mesenchyme, migrate to various sites in the developing nervous system.

Neural tube The dorsal tubular structure in vertebrate embryos formed by the longitudinal folding of the neural plate, the neural tube becomes the brain and spinal cord.

Neuroblast(s) Embryonic cells from which nerve cells develop.

Neuroectoderm (Neuroderm) The ectoderm of the neural plate from which the central nervous system (composed of the brain and spinal cord) develops.

Neuroendocrine The interaction between the nervous system and the hormones of the endocrine glands.

Neuroepithelial Of or pertaining to neuralized epithelial tissue covering most internal surfaces and organs and the outer surface of animal bodies.

Neurofilament Member of the class of intermediate filaments found in the axons of nerve fibers.

Neuron(s) Impulse-conducting cells that constitute the brain, spinal column, and nerves that have a nucleated cell body with dendrites and a single axon.

Neuropile Fibrous network of delicate unmyelinated nerve fibers interrupted by numerous synapses and found in high concentrations of nervous tissue in the brain.

Neuropore Opening at anterior end of the neural tube during early development.

Neurotransmitter Chemical message released from a neural synapse, which diffuses across the synaptic cleft, binding to and stimulating a postsynaptic cell.

Neurulation Embryonic formation of the neural tube by closure of the neural plate directed by the underlying notochord.

Nissl body A patch outside the nucleus of a nerve cell, the Nissl body extends into dendrites but not axons and disappears when axons degenerate.

Noogenesis The involution and interiorization of the universe through consciousness; instinct perceiving itself in its own mirror; the highest function and outcome of psychogenesis.

Noosphere Teilhard's "thinking layer" of the Earth, spreading as an incandescence over and above the biosphere as the biosphere once upon a time spread over the geosphere.

Noradrenalin A hormone in the body's sympathetic nerve endings, it constricts blood vessels.

Notochord Flexible, rodlike structure that forms the main support of the body in the lowest chordates.

Nuchal Pertaining to the nape of the neck and associated nerves, glands, and muscles, etc.

Nucleic Acids Any of a group of complex compounds found in all living cells and viruses; composed of purines, pyrimidines, carbohydrates, and phosphoric acid.

Nucleolus A small, typically round granular body constructed of protein and RNA in the nucleus of a cell, usually associated with a specific chromosomal site and involved in ribosomal RNA synthesis and the formation of ribosomes.

Nucleophilic Redistributing, substituting, donating, and sharing electrons.

Nucleoplasm The protoplasm of a cell nucleus.

Nucleoside Any of various compounds consisting of a sugar, usually ribose or deoxyribose, and a purine or pyrimidine base, especially a compound obtained by hydrolysis of a nucleic acid, such as adenosine or guanine.

Nucleotide Any of various compounds consisting of a nucleoside combined with a phosphate group and forming the basic constituent of DNA and RNA.

Nucleus A large, membrane-bound, usually spherical protoplasmic structure within a living cell, containing the cell's hereditary material and controlling its metabolism, growth, and reproduction.

Nucleus pulposusa Gelatinous center of the intervertebral disks, arising from the deterioration of the notochord and later surrounded by circularly arranged fibers.

Occipital Of or relating to the occipital bone, which is in the lower posterior part of the skull.

Octave Interval of eight degrees between two tones. According to G. I. Gurdjieff, the musical octave was derived by ancient mystery schools from the cosmic octave. The period in which cosmic vibrations are doubled was divided into eight unequal steps corresponding to the rate of increase in the vibrations themselves, the eighth step repeating the first with double the number of vibrations and significant cosmogonic and etiological implications. The great cosmic octave reaches us in the form of the ray of creation which fills the intervals between its tones with shocks manifesting as worlds (such as the Sun and Moon) or events (such as organic life on Earth). Corresponding intervals and creative shocks emerge within the octaves of consciousness.

Odontoblast One of the dentin-forming cells of the outer surface of dental pulp.

Oligosaccharide(s) Chains of simple sugars (monosaccharides) in increasing lengths.

Olive (Olivary nuclei) Cell clusters formed by ventral migration of neuroblasts from the alar plate of the brain, these synapse proprioceptive data from the shoulder girdle, neck, and trunk.

Omentum (Greater) A pouchlike extension of the peritoneum.

Omohyoid Of or pertaining to the shoulder and the hyoid bone.

Oncogene (Oncogenesis) A gene that causes the transformation of normal cells into cancerous tumor cells, especially viral genes that transform their host cells into tumor cells.

Ontogeny (Ontogenesis, Ontogenetic) The development of an individual organism from an embryo to an adult.

Ontology (Ontological) The branch of metaphysics that deals with the nature of being.

Oocyte A cell from which an egg develops by meiosis; a female gametocyte.

Oogenesis The formation, development, and maturation of an ovum.

Oogonidium (Oogonidia) Reproductive eggs (microgametes) of flagellate protozoa.

Oogonium (Oogonia) The descendant of a primordial germ cell that differentiates into an oocyte.

Organelle(s) Differentiated structures within a cell that perform specific functions; metaphorically, the organs of cells.

Orgone Wilhelm Reich's cosmic primordial energy underlying the formation of galaxies and all pulsations in nature, hence, the theoretical life force emanating from organic material.

Oris One of a series of muscles moving the mouth and lips.

Oropharyngeal Relating to part of the pharynx between the soft palate and the epiglottis.

Osseous Composed of, containing, or resembling bone.

Osteoblast(s) A cell from which bone develops.

Osteoclast(s) A large, multinucleate cell found in growing bone that resorbs bony tissue as in the formation of cavities.

Osteocyte(s) Branched cells embedded in the matrix of bone tissue.

Otic pit Invagination of the otic placode that sinks below surface ectoderm in mesenchyme in formation of the ear.

Otocyst Vesicle formed by the invagination of ectoderm (the otic pit) that develops into the inner ear.

Otolith A tiny protein and calcium-carbonate pebble occurring in gravity-measuring and kinetic-balancing organs.

Oviduct The tube through which the ova pass from the ovary to the uterus or to the outside.

Ovipositor The tubular structure, usually concealed, with which many female insects or fishes deposit eggs.

Ovum Female reproductive cell or gamete of animals.

Paedomorphism (Paedomorphic) Retention of juvenile characteristics in adults, i.e., in later ontogenetic stages of descendants.

Palate (Palatine) The roof of the mouth in vertebrates which has separate oral and nasal cavities and a hard and soft palate.

Palp The elongated, often segmented appendage near the mouth in invertebrate organisms such as those that insects use for sensation, locomotion, feeding, or sexual and reproductive activity.

Panspermia The theory that living organisms exist throughout the universe and develop wherever the environment is favorable; modified during the late nineteenth century into the proposition that spores are transmitted from planet to planet, propelled through the universe by starlight and gravity while adhering to specks of primordial or meteoric dust.

Papilla(ae) Small, nipplelike projections such as the roots of hairs or taste buds on the tongue.

Paracrine A form of chemical signalling in which the target cell is close to the signal-releasing cell; includes neurotransmitters and neurohormones.

Paramesonephric In females the ducts that develop into the female genital tract upon regression of the mesonephric ducts.

Parasympathetic nervous system Part of the autonomic nervous system originating in the brain stem and the lower part of the spinal cord that in general inhibits or opposes the physiological effects of the sympathetic nervous system, as in tending to slow the heart and dilate the blood vessels.

Parathyroid One of four small kidney-shaped glands situated in pairs near the lateral lobes of the thyroid and secreting a hormone for calcium and phosphorus metabolism.

Paraventricular nucleus A region in the hypothalamus involved in water retention.

Parenchymal Part of an organ instead of its connective tissue.

Parietal bones Bones forming much of the sides and roof of the cranial vault; their mobility strongly influences cerebral venous circulation through the sagittal sinus and, thereby, the reabsorption of cerebrospinal fluid into the venous system.

Parthenogenesis (Parthogenetic) A form of reproduction in which an unfertilized egg develops into an individual; observed in insects and other arthropods.

Parthenogonidium (Parthenogonidia) Large cells that can participate in either sexual or asexual production, dividing repeatedly to form a new organism (as in volvox) or differentiating into gametes (sperms or eggs).

Pedicellaria(ae) Skin appendages on starfish and sea urchins, pedicellariae move as independent units and close on objects; sea-urchin versions have three jaws around motile ossicles bearing spines and sometimes poison glands.

Pelagia Deep-sea jellyfish without polyp stages, i.e., only medusae.

Pentose sugar Monosaccharides with five carbon atoms per molecule; these include ribose and several other sugars.

Peptide Any of various natural or synthetic compounds containing two or more amino acids linked by the carboxyl group of one amino acid and the amino group of another.

Pericardioperitoneal canal One of several small canals linking the pericardial and peritoneal cavities in the embryo.

Pericardium (Pericardial) Membranous sac filled with serous fluid that encloses the heart and roots of the aorta and other large blood vessels.

Perichondrium The fibrous membrane of connective tissue covering the surface of cartilage except at joint endings.

Periderm The outer layers of tissue of woody roots and stems, consisting of the cork cambium and the tissues produced by it.

Perilymph Watery fluid between the membranous and bony aspects of the inner-ear labyrinth filling the perilymphatic space in which the membranous labyrinth is suspended.

Perineum The portion of the body in the pelvis occupied by urogenital passages and the rectum, bounded in front by the pubic arch, in the back by the coccyx, and laterally by part of the hipbone; the region between the scrotum and the anus in males and between the posterior vulva junction and the anus in females.

Peristalsis Wavelike muscular contractions of the alimentary canal or other tubular structures by which contents are forced onward toward an opening.

Peristomial Fringe of toothlike appendages around the mouth of a moss capsule.

Peritoneum (Peritoneal) Serous membrane that lines the abdominal cavity and folds inward to enclose viscera.

Peroxisome A cell organelle containing enzymes, such as catalase or oxidase, that catalyze the production and breakdown of hydrogen peroxidase.

Pes (Pedes) Foot of a vertebrate.

Petiole Stalk by which a leaf is attached to a stem.

Phagocytosis (Phagocytic) Engulfing or ingesting of bacteria and foreign bodies by cells such as leukocytes; cell-eating; endocytosis by formation of pseudopods.

Phagosome Membrane-bound vesicles formed by the invagination of phagocytised material, phagosomes fuse with lysosomes that contain hydrolytic enzymes to digest the material.

Phalange A bone of a finger or a toe.

Pharynx (Pharyngeal) The section of the alimentary canal that extends from the mouth and the nasal cavities to the larynx, where it joins the esophagus.

Phoronid Marine worm with tentacles around its mouth.

Phosphate Salt or ester of phosphoric acid.

Phospholipid(s) Small molecules constructed mostly from fatty acids and glycerol, phospholipids contribute to the thin, impermeable sheets that enclose all cells and coat their organelles. Phospholipids' heads are polar and hydrophilic, their tails nonpolar and hydrophobic.

Phosphorescence Persistent emission of light following exposure to and removal of incident radiation.

Phosphorylation Addition of a phosphate group to an organic molecule.

Phrenicocolic ligament(s) Structure by which the hepatic and splenic flexures are suspended from the diaphragm. The spleen rests on the left phrenicocolic ligament.

Phyllotaxis The arrangement of leaves on a stem.

Phylogeny (Phylogenesis, Phylogenetic) The evolutionary development and history of a species or higher taxonomic grouping of organisms.

Phylum (Phyla) The primary division of a kingdom that ranks higher than a class.

Pia mater The fine, vascular membrane that closely envelops the brain and spinal cord under the arachnoid and *dura mater.*

Pineal A pine-cone-shaped gland that secretes melatonin.

Pithecanthropus Extinct primate postulated from bones in Africa and Asia, classified as *Homo erectus* and presumed ancestral to Neanderthal Man and *Homo sapiens.*

Pituitary Small, oval endocrine gland attached to the base of the vertebrate brain that controls other endocrine glands and influences growth, metabolism, and maturation.

Placenta A membranous vascular organ that develops in female mammals during pregnancy, lining the uterine wall and partially enveloping the fetus, to which it is attached by the umbilical cord.

Placode Area of thickened ectoderm in an embryo from which a nerve ganglion or a sense organ will develop (for instance, the otic placode into the otic pit, then the otocyst, then the cochlea and inner ear).

Placozoa(n) A phylum of amoeba-like metazoans with only one member, *Trichoplax,* which is composed of two epithelia with fibrous, contractile mesenchyme in between, a single flagellum surrounded by microvilli, and non-flagellated gland cells.

Planula Flat, free-swimming, ciliated larva of a coelenterate.

Plasmid A circular, double-stranded unit of DNA not coated with protein, a plasmid replicates within a cell independently of chromosomal DNA; it is most often found in bacteria.

Plasmodesma (Plasmodesmata) Fine cytoplasmic channels connecting every living cell in higher plants while passing through intervening cell walls.

Plastid(s) Pigmented, photosynthetic organelles of plant and algal cells involved in food synthesis and storage and thought to originate from bacteria.

Platelet A platelet is a disklike fragment of an erythrocyte made up of a small chunk of cytoplasm with a membrane. These bodies, synthesized by megakaryocytes in the bone marrow of mammals, play an important (clotting) role in stopping blood flow and preventing blood loss.

Pleiotropy (Pleiotropic) The control by a single gene of several unrelated phenotypic effects.

Pleura Serous membrane that envelops the lungs and folds back to make the lining of the chest cavity.

Plexus A structure in the form of a network of nerves, blood vessels, lymphatics, or other tissues.

Plica A fold or ridge, as of skin, membrane, or shell.

Pluteus The free-swimming larva of brittle stars and sea urchins; among brittle stars the pluteus uniquely bears calcareous rods that support its "arms"—extended lobes bearing ciliary bands.

Polar body Discarded part of the oocyte that, receiving virtually no cytoplasm, degenerates during meiosis.

Polyadenylate tail A long homopolymer of adenosine monophosphate that is usually found at the ends of eukaryotic mRNA.

Polymer(s) Compounds of monomers, with their repeated linked units leading to high molecular weight.

Polymerase Enzyme that catalyzes the formation of polynucleotides of DNA or RNA using existing strands as a template; polymerase synthesizes polymers by joining together monomers.

Polyp A coelenterate, such as a coral, having a cylindrical body and an oral opening usually surrounded by tentacles.

Polypeptide Peptide containing ten to a hundred molecules of amino acids.

Polysaccharide(s) Carbohydrates made of a number of sugar monomers joined by glycosidic bonds.

Pons Slender tissue joining two parts of an organ, such as the tissue joining the *medulla oblongata* and the mesencephalon below the cerebellum in the brain.

Precambrian Oldest, largest division of geologic time, starting with the creation of the Earth perhaps three to four billion years ago and ending about 600 million years ago; its later phases are characterized by the appearance of primitive forms of life.

Predentin Immature uncalcified dentin in the embryonic tooth, consisting chiefly of fibrils.

Preformationism The theory that nothing has been added to or taken from the germ-plasm of species since their beginning and that all characteristics were always present (though most were latent).

Prickle cell A large flattened polygonal cell of the epidermis that has fine spines projecting from its surface.

Primitive streak An elongated band of cells that forms along the axis of an embryo early in gastrulation by the movement of lateral cells toward the axis; it develops a groove along its midline through which cells move to the interior of an embryo to form mesoderm.

Primordium(ia) Organ in its most rudimentary form or stage of development.

Proboscis Long, flexible snout or trunk.

Prochordal Situated in the front of the notochord, applied to parts of cartilaginous rudiments in the base of the skull.

Progenesis (Progenetic) A form of paedomorphosis, it denotes the retention of juvenile characteristics in adult descendants by precocious sexual maturation of an embryo.

Progesterone A steroid hormone produced and secreted initially by the *corpus luteum* of the ovary and later (throughout gestation) by the placenta, progesterone is a product of cholesterol, made from the breakdown of sugar and fat in the mitochondria of most cells; it acts to prepare the uterus for implantation of the fertilized ovum, maintains the lining of the uterus during pregnancy, and promotes development of the mammary glands. Synthesized in smaller amounts by the adrenal glands of both sexes and the testes of males, it is a precursor of testosterone, estrogen, and many adrenal cortical steroid hormones, including the corticosteroids critical for sugar and electrolyte balance, stress response, and blood pressure. Medicinally it normalizes zinc and copper levels, neutralizes estrogen side effects, and protects against breast and endometrial cancer. In regulation of menopause natural progesterone is often prescribed in place of estrogen. [See John R. Lee, M. D., and Virginia Hopkins, *What Your Doctor May Not Tell you about Menopause: The Breakthrough Book on Natural Progesterone* (New York: Warner Books, 1996).]

Prokaryote Cell of the bacterial and cyanobacterial kingdom, characterized by absence of nuclear membrane and by DNA that is not organized into chromosomes.

Prolactin Pituitary hormone that stimulates lactation (secretion of milk).

Proline An amino acid found in most proteins and a major constituent of collagen.

Pronephros (Pronephric) Kidneylike organ, being either part of the most anterior pair of three pairs of organs in a vertebrate embryo or functioning as a kidney in some simple vertebrates, such as the lamprey.

Pronucleus The haploid nucleus of a sperm or egg prior to its fusion into the diploid nucleus of the zygote during fertilization.

Prophase The first stage of mitosis in which the chromosomes condense and become visible, the nuclear membrane breaks down, and the spindle apparatus forms at opposite poles of the cell. Also the first stage of meiosis in which DNA replicates, homologous chromosomes condense into long thin threads and attach at their ends to the nuclear envelope, chiasmata form, and the chromosomes contract.

Proprioception (Proprioceptive) Unconscious perception of movement and spatial orientation arising from the stimuli within the body itself.

Prosencephalon The anterior region of the embryonic brain from which the other regions develop.

Prosimian The suborder of primates including lemurs, lorises, and tarsiers.

Prostate The gland that surrounds the urethra at the base of the bladder in males and secretes fluid that is the major constituent of semen.

Protamine Proteins found in fish sperm that are strongly basic, are soluble in water, are not coagulated by heat, and yield chiefly arginine upon hydrolysis.

Protein Any of a group of complex organic macromolecules that contain carbon, hydrogen, oxygen, nitrogen, and usually sulphur, composed of one or more chains of amino acids, and including many substances such as enzymes, hormones, and antibodies that are necessary for the proper functioning of an organism.

Proteus An amoeba.

Proto- First in time.

Protoctist Any of the unicellular protists and multicellular eukaryotic microorganisms and their descendant organisms, considered a separate taxonomic kingdom in most classification systems.

Protostomia The branch of the metazoa (including annelids, arthropods, and mollusks) in which the opening leading from the cavity of the archenteron (the blastopore) becomes subdivided into two openings, one of which becomes the mouth and the other the anus.

Protist Any of a wide variety of eukaryotic unicellular organisms.

Proton A stable, positively charged subatomic particle that is significantly heavier than an electron.

Pterygoid process One of the wings of the sphenoid bone.

Pulp cap Associated with the formation of a fresh feather, the pulp cap originates from derivatives of pulp epithelium, which are themselves derived from a basilar collar of regenerating feather epidermis. The column of epidermal membrane around the pulp becomes cornified at periodic intervals to form a series of downward-opening cups along the stratified squamous epithelium. Pulp is actively resorbed as the cap is being formed, sending off branches into the feather. Though totally epidermal in composition, feathers are induced mesodermally.

Purine A crystalline organic base that is the parent compound of various biologically important derivatives, including uric acid and the nucleic-acid constituents adenine and guanine.

Pyloric sphincter Fold of mucous membrane containing a ring of circularly disposed muscle fibers that closes the vertebrate pylorus.

Pylorus The passage at the lower end of the stomach that opens into the duodenum.

Pyramidal fibers Corticospinal fiber bundles from the developing cerebral cortex, forming in the ventral region of the myelencephalon.

Pyrimidine Crystalline organic base that is the parent substance of various biologically important derivatives including nucleic acid constituents uracil, cytosine, and thymine.

Qualis (Qualia) Subjective perception or recognition of the quality of a thing as apart from the thing itself.

Radius Slight, curved, shorter, thicker of the two long forearm bones located on the lateral side of the ulna.

Ramus Any branchlike anatomical structure. For instance, the ramus is the bottom of the ischial bone plus the lateral aspect of the pubic bone in the front (i.e., an ischial and a pubic ramus fused); the backs of these are the ischial tuberosities on which we sit.

Raphé A seamlike ridge between two similar anatomical parts, as in the scrotum or the lateral palatine process.

Recombinase An enzyme that catalyzes genetic recombination.

Reiki Received as a channeled transmission by mid-nineteenth-century Japanese theologian Mikao Usui, Reiki is form of faith-healing in which nondiagnostic touch is used by a practitioner to direct the energy of both divine and cosmic domains into a client. Fingers are placed on various parts of the patient's body and held there, usually for three to five minutes but sometimes for as long as half an hour.

Renal Relating to the kidney.

Restriction nucleases Enzymes made by bacteria to protect them against viruses, restriction nucleases recognize sequences of four to eight nucleotides in DNA. They are used in recombinant DNA technology to isolate and manipulate specific genes.

Reticular formation Diffuse network of nerve fibers and cells in parts of the brainstem that is critical in regulating consciousness or wakefulness.

Reticulum A netlike formation or structure; term used in describing the internal membranes of the cell.

Retina Delicate, light-sensitive membrane that lines the inner eyeball and is connected by the optic nerve to the brain.

Retinoic acid This aldehyde is involved in photoreception and gene transcription.

Retrovirus Viruses that reverse the normal process in which DNA is transcribed into RNA, associated with cancer and AIDS.

Rhachis Pith-filled central shaft supporting the inner and outer web of a feather scale.

Rheostat A continuously variable electrical resistor used to regulate current. Some biological structures serve a rheostatic function.

Rhombencephalon Part of the embryonic brain where the mese- and myelencephalon develops.

Rhombomere A segment of the developing mammalian hindbrain. During early embryogenesis the hindbrain is formed from a series of rhombomeres, but this segmental origin becomes obscured during later development.

Ribosome Minute round particle composed of RNA and protein found in the cytoplasm of living cells and active in the synthesis of proteins.

Risorius Muscle retracting the corners of the mouth in order to allow a big bite.

RNA A polymeric constituent of all living cells and many viruses, important in protein synthesis and the transmission of genetic information, RNA consists of a long, usually single-stranded chain of alternating phosphate and ribose units with the bases adenine, guanine, cytosine, and uracil bonded to the ribose.

Root (dorsal root ganglia) Unipolar neurons that enter the spinal cord, the root ganglia of the brain are derived from neural-crest cells; some central processes (axons) end in the spinal cord, while others ascend to the brain in dorsal columns of the cord; peripheral processes pass in spinal nerves to sensory visceral endings.

Rostral Beaklike or snoutlike; toward prow or beak.

Rube Goldberg machine An intricate contraption of many stages, each initiated by the previous one, designed to produce a relatively simple or straightforward result. For instance, eggs rolling onto spoons and electric trains carrying logs may be used in combination to turn on a coffee-maker. In a famous instance (from the cartoonist Rube Goldberg's work) a large spider jumps on a hammer, flipping up a spatula, which tosses an egg into a pan, lifting it and raising a lever turning on a toy soldier who kicks a bowling ball off a ledge—all to open a screen door.

Saccharine Synthetic white crystalline powder that has a taste five hundred times sweeter than cane-sugar and is calorie-free.

Sacrococcygeal Of, relating to, affecting, or performed by way of the region of the sacrum and pelvis.

Sacrum (Sacral) Triangular bone made up of five fused vertebrae that forms the posterior section of the pelvis.

Sagittal Relating to the longitudinal vertical plane that divides a bilaterally symmetrical animal's body into right and left regions.

Sarcomere Repeating subunit from which the myofibrils of striated muscle are built.

Scalar waves Hypothetical electromagnetic waves that allow the curvature of space-time and the unification of gravitation with electromagnetism. These would facilitate time travel, telekinesis, and instantaneous cell communication and collaboration.

Scapula Either of two flat triangular bones that form the back part of the shoulder.

Schwann cell Any of the cells that cover the nerve fibers in the parasympathetic (peripheral) nervous system and form the myelin sheath.

Sclera Tough white fibrous outer envelope of tissue covering all of the eyeball except the cornea.

Sclerotome The ventromedial aspect of the somites, the sclerotome contains cells that will give rise to bones, cartilage, and ligaments.

Sebum (Sebaceous) Semifluid secretion of the sebaceous glands in the dermis of the skin, consisting of fat, keratin, and cellular material.

Septum(a) A thin partition or membrane that divides two cavities or soft masses of tissue in an organism.

Septum transversum A mass of mesoderm developing in cranial relationship to the pericardial coelom.

Serosa The serosa of an organ is its smooth, outermost lining where it faces a cavity and is not surrounded by connective tissue (i.e., the serosa of the intestine).

Serotonin An organic compound found especially in the brain, blood serum, and gastric mucous membranes, serotonin is active in vasoconstriction and the transmission of nerve impulses.

Sertoli cells Elongated cells in the tubules of the testis to which the spermatids attach, providing support, protection, and nutrition until spermatids become spermatozoa.

Seta(ae) Stiff hair or bristle.

Sinus (Sinusoid) 1. Dilated receptacle containing chiefly venous blood; 2. Air-filled cavity in the bones of the skull that communicates with the nostrils.

Sinus venosus An enlarged pouch that adjoins the heart and is formed by the union of large systemic veins, the *sinus venosus* is the passage through which venous blood enters the heart in lower vertebrates.

Siphonophore(s) Pelagic, swimming or floating colonies of polymorphic hydrozoan hydras and medusas in the coelenterate phylum.

SNAPS Golgi-related proteins catalyzing lipid bilayer fusing, forcing membranes together.

SNARES (literally SNApREceptors) Proteins receptors for SNAPs which, as a complex, are responsible for regulating vesicle targeting and mediating transport of cargo from the endoplasmic reticulum to the cell surface.

Soma (Somatic) 1. Entire body of an organism (by comparison with psyche, the entire mental configuration of an organism); 2. The manifestation of the organism itself (by comparison with its germ cells); 3. A cell programmed to die and unable to become a gamete; 4. The body of any organic structure, i.e., a nerve cell.

Somatoplasm Entirety of specialized protoplasm in a somatic cell.

Somite(s) The segmental mass of mesoderm in vertebrate embryos that becomes the muscle and vertebrae.

Spermatid Any of four haploid cells formed by meiosis in males.

Spermatocyst Sac containing sperm cells.

Spermatocytes Diploid cells that undergo meiosis to form four spermatids.

Spermatogenesis The formation and development of sperm by meiosis and spermiogenesis.

Spermatogonia Any cells of the gonads that are progenitors of spermatocytes.

Spermatogonidium(ia) Reproductive sperms (microgametes) of flagellate protozoa.

Spermiogenesis The metamorphosis of spermatids into spermatozoa.

Sphenoid bone A wing-shaped bone, situated at the base of the skull behind the eyes, the sphenoid is considered, in the words of John Upledger, "the mechanical keystone of the rhythmic accomodative motion of the skull," articulating with the vomer, ethmoid, frontal, occipital, parietals, temporals, zyomatics, and other cranial bones such that any inhibition in rotatory cycle under cerebrospinal pressure "will place significant drag on the whole craniosacral system." *(Craniosacral Therapy, p. 215.)*

Sphenomandibular Used to describe restrictions on freedom of range between the sphenoid bone and the mandible (sphenobasilar and sphenomaxillary relationships are likewise considered critical in diagnostic osteopathy).

Sphincter Ringlike muscle that constricts a body passage or orifice and relaxes as required by normal physiological functioning.

Sphincter of Oddi A complex sphincter closing the duodenal orifice of the common bile duct.

Spina bifida A congenital defect in which the spinal column is imperfectly closed so the meninges and spinal cord protrude, spina bifida leads to hydrocephalus and other neurological disorders.

Spirochete(s) Motile, corkscrew-shaped, heterotrophic bacteria that cause syphilis and are often pathogenic. In the wild they are frequently fast-swimming, favor anaerobic environments (mud and animal guts), and are considered prototypes of the ancestral form of organelles (undulipodia) that became symbiotic in protoeukaryote cells.

Splanchnic Relating to viscera; the splanchnic mesoderm is embryonically continuous with the extraembryonic mesoderm covering the yolk sac.

Splanchnopleure The wall of the primitive gut formed by splanchnic mesoderm and embryonic endoderm.

Squames The outermost cells of the epidermis, filled with densely packed keratin and reinforced by the intracellular protein involucrin, squames are tightly compressed and stacked in hexagonal columns interlocking at their edges.

Squamous Scalelike, platelike, flat, or covered with scales.

Stapes Small stirrup-shaped bone of the middle ear, formed from ossified cartilage of the dorsal end of the second branchial arch, joining the incus to the vestibular membrane and vestibule.

Starch A nutrient carbohydrate composed of glucose, abundant in the roots, tubers, stem piths, seeds, and fruits of plants.

Statocyst A fluid-filled sac with sensory cells that acts as an organ of balance and equilibrium and a primitive rheostat in invertebrates.

Stellate Shaped like a star.

Sternocleiodomastoideus A muscle that rotates and extends the head, stretching from the mastoid process (the protuberance behind the ear) to the sternum and clavicle.

Sternohyoid Of or pertaining to the sternum and the hyoid bone or cartilage.

Steroid Fat-soluble compounds with seventeen carbons in four rings that include sterols, adrenals, and sex hormones.

Stolon A stemlike structure or thin mat in colonial invertebrates from which new organisms develop by budding. Among tunicates the rhizoid-like stolons consist of epidermis and a longitudinal mesodermal septum that divides the cavity of the stolon into a pair of canals through which blood circulates; new buds consist of epidermal-covering epithelium and an inner vesicle produced from mesodermal septum cells.

Stomodeum Anterior and oral portion of the alimentary canal of an embryo.

Stroma The connective tissue that constitutes the framework of a red blood cell.

Stylohyoid Of or pertaining to the styloid process and the hyoid bone.

Subarachnoid space The coalescence of cerebrospinal-fluid-filled spaces within the *pia mater* and arachnoid membrane, the subarachnoid space drains into venous blood circulation via the arachnoid villi.

Submucosa The thick layer of loose connective tissue that goes deep into the mucous membrane, containing nerves, blood vessels, and small glands.

Suctorian Protozoan that is sessile and feeds with its tentacles.

Sugar A sweet crystalline carbohydrate critical to biological energy-production.

Sulcus (Sulci) Narrow fissures separating adjacent convolutions of the brain.

Suture The line of juncture between two bones, particularly in the skull.

Syllid A roundworm (annelid) capable of remarkable restorative regeneration of segments after amputation.

Symbiogenesis (Symbiogenetic) Morphogenetic novelty generated by symbiont interaction.

Symbiont (Symbiosis) Organisms of different species living in close contact.

Sympathetic nervous system The part of the autonomic nervous system originating in the thoracic and lumbar regions of the spinal cord that inhibits or opposes the physiological effects of the parasympathetic nervous system, as in the speeding up of the heart and contracting of the blood vessels.

Synapse (Synaptic) The junction across which a nerve impulse and/or neurotransmitter passes from an axon terminal to a neuron, a muscle cell, or a gland cell.

Syncytium A group of differentiating cells connected by a cytoplasmic bridge.

Systems theory The principles, models, and laws that apply to the interrelationship and interdependencies of sets of linked components which form a functioning whole or system such as organs in an organism.

Syzygy Act of mating or gametes joining.

Tachyon energy A hypothetical subatomic energy faster than light and the source of all forms and frequencies in the physical world, tachyon interacts directly with the slower-than-the-speed-of-light world through Subtle Organizing Energy Fields on physical, mental, emotional, and spiritual levels.

Tagma (Tagmata) Clearly defined groupings of metameric segments in arthropods. The tagmata of insects are its head, thorax, and abdomen. Tagmatization is considered a primitive form of body organization.

Tangka Tibetan sacred painting (often a mandala) revealing the inner pathway through the incarnate universe by representations of mythological beings, spaces, and relationships.

Tarsal bone (Tarsus) The instep of the vertebrate foot between the leg and middle metatarsus which precedes the toes.

Taxis (Taxes) The movement of an organism toward or away from an external stimulus (light, pressure, current, smell, etc.).

Tay-Sachs disease A genetic disorder found in East European Jewish families that causes early death by affecting the brain and nerves and causing abnormal lipid metabolism, Tay-Sachs leads to blindness, deafness, seizures, paralysis, dementia, decreased muscle tone, and growth retardation.

T-cell (T cell) White blood cells that mature in the thymus and work for the immune system in identifying antigens and regulating other immune cells.

Tegmentum A primitive part of the old brain, governing metabolism and reproduction.

Telomere Either end of a chromosome.

Telson Terminal abdominal segment (of twelve) in primitive insects.

Temporal bone(s) Two bones, each complexly constructed of three parts, forming the sides and base of the skull; the temples.

Tensegrity The hypothesis that cells can behave like structures in which shape results from balancing tensile and hydrostatic forces.

Tentorium cerebelli Arched fold of *dura mater* that covers the upper surface of the cerebellum and supports the occipital lobes of the cerebrum.

Tertiary epoch First period of the Cenozoic era that is characterized by the appearance of modern flora, apes, and large mammals, beginning in the Palaeocene, about 63 million years ago and ending with the Pleistocene, about 2 million years ago.

Testosterone The steroid found in the testes that forms secondary male sex characteristics.

Tetracycline An antibiotic made from streptomyces.

Thalamus A large ovoid mass of gray matter found in the posterior of the forebrain that relays sensory impulses to the cerebral cortex.

Thorax (Thoracic) In a human, the thorax lies between the neck and diaphragm and is encased by the ribs and contains the heart and lungs. In an insect, the thorax is the region between the head and abdomen.

Thylakoid Flattened membranous sacs inside the chloroplast, thylakoids convert light energy to chemical energy.

Thymine A pyrimidine base found in DNA.

Thymocyte A lymphocyte derived from the thymus that is a precursor to the T-cell.

Thymus A gland behind the breastbone that consists of lymphatic tissue and serves as the site of T-cell differentiation.

Thyroid An endocrine gland developed out of the pharyngeal pouch, the thyroid appears on either side of the trachea in humans, secreting thyroxin, an iodine-containing hormone.

Tonsillar crypts Any of deep invaginations occurring on the surface of the palatine and pharyngeal tonsils.

Totemism (Totemic) 1. The use of plant, animal, stone, cloud, and other natural-phenomenon names to define or categorize human institutions or social groups; 2. A combination of genealogy and shamanism; 3. A native system of philosophy, taxonomy, and science.

Trabecula(ae) Any of the supporting strands of connective tissue projecting into an organ as part of the structure of the organ.

Trachea Thin-walled tube of cartilage and membranous tissue that descends from the larynx to the bronchii and carries air to the lungs.

Tracheophytes Vascular plants.

Transcription The synthesis of mRNA from a DNA template.

Trapezius A flat muscle running from the base of the occiput to the middle of the back, the trapezius supports and makes it possible to raise the head and the shoulders. Humans have two trapezius muscles.

Trigeminal Descriptive of the fifth pair of cranial nerves that have sensory and motor functions in the face, teeth, mouth, and nasal cavity.

Triploblastic Having three germ layers.

Trochoblast Outer cell of ciliated larval blastula.

Trochosphere Small free-swimming ciliated aquatic larva of various invertebrates including mollusks and annelids.

Trophoblast Outermost layer of cells of blastocyst that functions in the implantation and nutrition of an embryo.

Trophocyte(s) Feeder cells in simple invertebrates that pass food on to other cells that digest and store it.

Tropocollagen This is the form that most procollagen molecules take when their propeptides are enzymatically removed and they become collagen; another name for collagen.

Tubercula quadrigemina One of the ganglia in the vault of the cranium. The *corpora quadrigemina* are four rounded eminences located immediately behind the third ventricle of the brain.

Tumbling factor Tumbling factor is one of many proteins that contribute to the rate of tumbling (changed direction) when bacterial cells move toward a chemoattractant (in chemotaxis). One or more of these factors may be altered via a genetic over- or underexpression without changing the overall robust phenotype of chemotaxis.

Tympanic Of or pertaining to the middle ear or eardrum (as tympanic membrane, tympanic cavity).

Ulna Bone extending from elbow to wrist on the side opposite the thumb.

Umbilical stalk The primordial form of the organ (umbilical cord) connecting the embryo to either a yolk sac and/or extraembryonic tissues.

Undulipodium(ia) Organelle making up flagella (when long and one per cell) and cilia (when short and many per cell). The sperm is considered an undulipodium and, along with other such organelles, presumed to be of spirochete origin.

Urachus The primitive continuation of the urinary bladder into the umbilicus in the embryo, obliterated during human development.

Urea Water-soluble compound that is the major nitrogenous end product of protein metabolism and the chief component in urine.

Urethra (Urethral) Canal through which urine is discharged.

Urogenital Of or relating to both the urinary and genital structures and functions.

Uterovaginal Of or relating to the uterus and the vagina.

Utricle 1. A delicate membranous sac bearing liquid. 2. The dorsal portion of the otocyst of the ear, the utricle is the progenitor of the endolymphatic duct. It develops vestibular sensory nerve endings: *maculae utriculi.*

Vacuole Small cavity in the cytoplasm bound by a single membrane and containing food, water, or metabolic waste.

Vallate Having a raised edge surrounding a depression.

Vas deferens The main duct through which semen is carried to the ejaculatory duct

Vascular Pertaining to vessels that circulate blood.

VEG-F gene Vascular endothelial growth factor.

Vein Vessel carrying blood toward the heart.

Vena cava (Venae cavae) Two large veins in air-breathing vertebrates that return blood to the right atrium of the heart.

Ventral Close to the anterior, lower, or inner.

Ventricle A cavity or chamber of the heart or brain.

Vernix caseosa Cheeselike substance covering the skin of the fetus.

Vertebrate An organism that has a backbone or a spinal column.

Vesicle A small bladderlike cell or cavity.

Vestibular mechanism A division of cranial nerve VIII, the vestibular nerve relays impulses into the cerebellum and medulla, providing information from nerves exiting other receptors in the vestibule adjacent to the semicircular canals of the inner ear. The sensory cells in the membranous labyrinth of the vestibule and the canals are stimulated by the movement of the head, which causes a thick fluid (endolymph) to flow across and bend their hairlike suspensions.

Vestibule 1. Space into which the vagina and urethra open together, with lubricating vestibular glands on either side. 2. Curled cavity through which vibrations pass in the cochlea of the ear.

Villus (Villi) Microscopic fingerlike projections of mucosal membranes.

Viscera (Visceral) Soft internal organs in the abdomen and thorax.

Vitelline Of or pertaining to the yolk of the egg.

Vomer A flat bone that constitutes the inferior and posterior of the nasal septum, the vomer is driven by the sphenoid bone, with which it is locked in a tongue-and-groove design quite susceptible to jamming. The vomer also articulates with the hard palate in a similar joint and has an extensive articulation with the ethmoid.

Wolffian duct In the male this duct is associated with the mesonephros and serves double duty as the ureter and sperm duct.

Xiphoid process A small, posterior sword-shaped section of the sternum (or breastbone).

Yolk 1. The yellow, usually spherical portion of the egg of a bird or reptile, surrounded by the albumen and serving as nutriment for the developing young; 2. A corresponding portion of the egg of other animals, consisting of protein and fat that serve as the primary source of nourishment for the early embryo.

Yolk sac A membranous sac attached to an embryo, enclosing yolk in bony fishes, sharks, reptiles, birds, and primitive mammals, and functioning as the circulatory system of the human embryo before internal circulation begins.

Zoea The larval form of crustacea that has spines on its carapace and rudimentary limbs on its abdomen and the thorax.

Zona pellucida Extracellular membrane forming a thick envelope around the mammalian oocyte.

Zooid 1. An organic cell that has independent movement within a living organism (such as spermatozoa); 2. One of the distinct individuals forming a colonial animal such as a bryozoan or hydrozoan; 3. Any simple or primordial animal.

Zygoma (Zygomata) Cheek bones that also help form the eye orbits, the zygomatic bones are often mobilized osteopathically by placing the index finger inside the mouth (but external to the maxilla) and then grasping the zygoma between itself and the thumb.

Zygomaticus One of the muscles raising the upper lip and facilitating facial expressions and smiling.

Zygomycote A phylum of fungi (including black bread molds), alternating generations between diploid (and other ploidy) zygospores and haploid sporangiospores.

Zygote The cell formed by the union of two gametes, especially a fertilized ovum before cleavage.

This glossary was prepared by the author with the help of Lisa Rigamonti. Assistance on particular words was provided by Harvey Bialy, Stuart Newman, Richard Strohman, and R. Louis Schultz. The sources for the definitions include the books used in preparing the overall text (and their own glossaries); numerous dictionaries (in particular *The American Heritage Dictionary of the English Language*, New York: Random House, 1969); and AltaVista Internet searches.

Notes on semantics

I N THIS BOOK, consciousness is the spark of knowing that can manifest in any form throughout the universe; it is collective, universal, and archetypal—a primary attribute of being. Mind is a particular form of consciousness that has evolved on Earth through atoms and cells, nervous systems and brains—an artifact of biological evolution. Phenomenology is the dynamic process linking them. An exterior model of embryogenesis is cosmic and morphogenetic; an interior model is ontological/phenomenological.

The following distinctions implicit in the text are elucidated here (with representative page numbers):

Astronomy The Earth is a linearly unfolding zone of galactic, stellar material (13, 22).

Astrochemistry Cosmic debris makes up biological substance. There is no environmental gap between enormous stellar objects and life forms except the subtilizing, layering, deepening, and interpolation of the latter (164).

Astrophysics Genes digitize millions of multiplex form potentials which combine in unique organic structures only as their runes interact with chemical and mechanicodynamic gradients stretching to the ends of the cosmos (269, 284–285).

Astroembryology The latent complexity of the whole universe is reduced to the cellular nucleus where it is reenacted in a quantal, diaphanous flutter of planetoids (xvii–xviii).

Astrobiology Nervous systems are transpersonal and astrophysical (462).

Astrology Stars and planets in orbits regulate developmental rhythms (737–738).

Astrosophy There is an invisible, interior dimension (the astrum) throughout the night sky whereby ontogeny and phylogeny are unified and the microcosm reflects the macrocosm (705–716).

Astromythology Totems of the sky portray the morphology and divisibility of genesis (725–728).

Notes and Bibliography

THE INFORMATION IN THIS BOOK has been assembled from a diversity of sources. Specific citations for everything would be overkill in a nonacademic work, so mostly direct quotations are documented. General sources for each chapter are listed in the approximate order of use. Double asterisks** indicate a major source used throughout a whole chapter or extensively in a large part of it; a single asterisk* indicates a major source for a section within a chapter; and a title listed without an asterisk represents a source used for some information in a chapter. These notes replace a general bibliography.

Preface

1. From a poster assembled by Ahad Cobb in the early 1970s for Lama Foundation, San Cristobal, New Mexico.

2. *Time*, September 13, 1999, p. 10.

Chapter 1. Embryogenesis

Murchie, Guy. *The Seven Mysteries of Life: An Exploration in Science and Philosophy.* Boston: Houghton Mifflin Company, 1978.

Balinsky, B. I. *An Introduction to Embryology,* 5th edition. Philadelphia: Saunders College Publishing, 1981.

—

1. Frederick Gowland Hopkins, quoted in Donna Jeanne Haraway, *Crystals, Fabrics, and Fields: Metaphors of Organicism in Twentieth-Century Developmental Biology* (New Haven: Yale University Press, 1976), p. 102 f.n.

2. Steven Shaviro, *Doom Patrols: A Theoretical Fiction about Postmodernism* (London: Serpent's Tail/High Risk Books, 1997), p. 9.

3. Russell Banks, *Cloudsplitter* (New York: HarperFlamingo, 1998), p. 755.

4. Shaviro, p. 113.

5. William Beebe (American naturalist, 1877–1962), quoted on www.geocities.com/RainForest/Vines/8591/Only.htm, 1999.

6. Jennifer Egan, *The Invisible Circus* (New York: Picador, 1995), p. 210.

7. *Newsweek,* October 6, 1997, p. 68.

8. John Keats, "Ode on a Grecian Urn" (1819) in John Keats, *Selected Poems and Letters,* editor, Douglas Bush (Boston: Houghton Mifflin Company, 1959), p. 208.

The direct quote of Leo Tolstoy and the indirect quote of Gertrude Stein were taken from anthologies that did not give their original sources.

Chapter 2. The Original Earth

Hanawalt, Philip C., and Robert H. Haynes, editors. *The Chemical Basis of Life: An Introduction to Molecular and Cell Biology.* San Francisco: W. H. Freeman and Co./*Scientific American,* 1973.*

Bernal, J. D. *The Origin of Life.* Cleveland: World Publishing Co., 1967.*

Gibor, Aharon, editor. *Conditions for Life.* San Francisco: W. H. Freeman and Co./*Scientific American,* 1976.*

Wendt, Herbert. *The Sex Life of the Animals.* Translated from the German by Richard and Clara Winston. New York: Simon and Schuster, 1965.*

Crick, Francis, *Life Itself: Its Origin and Nature.* New York: Simon and Schuster, 1981.*

Shklovskii, I. S., and Carl Sagan. *Intelligent Life in the Universe.* Shklovskii's part translated from the Russian by Paula Fern. New York: Dell Publishing Co., 1966.*

Gilluly, James, A. C. Waters, and A. O. Woodford. *Principles of Geology,* Second Edition. San Francisco: W. H. Freeman and Co., 1959.

Jantsch, Erich. *The Self-Organizing Universe.* New York: Pergamon Press, 1980.

Olson, Charles. *The Maximus Poems [1950–1970].* Berkeley: University of California Press, 1983.

Duchesne-Guillemin, Jacques. *Zoroastrianism: Symbols and Values.* New York: Harper and Row, 1966.

Griaule, Marcel. *Conversations with Ogotemmêli: An Introduction to Dogon Religious Ideas.* Translated from the French by Ralph Butler, Audrey I. Richards, and Beatrice Hooke. London: Oxford University Press, 1965.

———

1. *Holy Bible, The New King James Version* (Nashville: Thomas Nelson, Inc., 1979), p. 1.

2. *Genesis, Translation and Commentary* by Robert Alter (New York: W. W. Norton, 1996), p. 4.

3. ibid.

4. ibid.

5. ibid.

6. Annie Proulx, *Close Range: Wyoming Stories* (New York: Scribner, 1999), p. 97.

7. ibid., p. 33.

8. Aristotle, *History of Animals* (4th century B.C.) quoted in Wendt, pp. 16–17.

9. Paul H. Barrett, editor, *The Collected Papers of Charles Darwin, Volume Two* (University of Chicago Press, 1977), p. 17.

10. A. I. Oparin, "The Origin of Life" in an appendix to Bernal, p. 201.

11. Quoted by Oparin in Bernal, p. 203.

12. Crick, p. 141 ff.

13. Oparin, "The Origin of Life" in an appendix to Bernal.

14. J. B. S. Haldane, "The Origin of Life" in an appendix to Bernal, p. 246.

15. Charles Darwin, quoted in Bernal, p. 21.

Chapter 3. The Materials of Life

Hanawalt, Philip C., Robert H. Haynes, editors. *The Chemical Basis of Life: An Introduction to Molecular and Cell Biology.* San Francisco: W. H. Freeman and Co./*Scientific American,* 1973.*

Bernal, J. D. *The Origin of Life.* Cleveland: World Publishing Company, 1967.*

Gibor, Aharon, editor. *Conditions for Life.* San Francisco: W. H. Freeman and Co./*Scientific American,* 1976.*

Hauschka, Rudolf. *The Nature of Substance.* Translated from the German by Mary T. Richards and Marjorie Spock. London: Vincent Stuart Ltd., 1966.*

Murchie, Guy. *The Seven Mysteries of Life: An Exploration in Science and Philosophy.* Boston: Houghton Mifflin Company, 1978.*

Crick, Francis. *Life Itself: Its Origin and Nature.* New York: Simon and Schuster, 1981.

Dixon, Dougal. *After Man: A Zoology of the Future.* New York: St. Martin's Press, 1981.

———

1. Bernal, pp. 167–168.

2. Hauschka, p. 45.

3. ibid., pp. 38–39.

4. ibid., p. 40.

5. ibid., pp. 41–42.

6. ibid., p. 43.

7. ibid., p. 54.

8. Walter Raleigh, quoted in Bernal, p. 8.

9. Bernal, p. 8.

10. Lynn Margulis and Dorion Sagan, *Origins of Sex: Three Billion Years of Genetic Recombination* (New Haven: Yale University Press, 1986), p. 12.

11. Bernal, p. 146.

12. Crick, p. 87.

Chapter 4. The First Beings

Bernal, J. D. *The Origin of Life.* Cleveland: World Publishing Co., 1967.**

Margulis, Lynn, and Dorion Sagan. *Origins of Sex: Three Billion Years of Genetic Recombination.* New Haven: Yale University Press, 1986.**

Carroll, Mark. *Organelles.* London: The Guilford Press, 1989.**

Crick, Francis. *Life Itself: Its Origin and Nature.* New York: Simon and Schuster, 1981.*

Hanawalt, Philip C., and Robert H. Haynes, editors. *The Chemical Basis of Life: An Introduction to Molecular and Cell Biology.* San Francisco: W. H. Freeman and Co./*Scientific American,* 1973.*

Alberts, Bruce, Dennis Bray, Julian Lewis, Martin Raff, Keith Roberts, James D. Watson.

Molecular Biology of the Cell. New York: Garland Publishing, 1989.*

Thomas, Lewis. *The Lives of a Cell: Notes of a Biology Watcher.* New York: Viking Press, 1974.*

Murchie, Guy. *The Seven Mysteries of Life: An Exploration in Science and Philosophy.* Boston: Houghton Mifflin Co., 1978.*

Berrill, N. J., and Gerald Karp. *Development.* New York: McGraw-Hill Book Co., 1976.

Shklovskii, I. S., and Carl Sagan. *Intelligent Life in the Universe.* Shklovskii's part translated from the Russian by Paula Fern. New York: Dell Publishing Co., 1966.

Haraway, Donna Jeanne. *Crystals, Fabrics, and Fields: Metaphors of Organicism in Twentieth-Century Developmental Biology.* New Haven: Yale University Press, 1976.

———

1. Crick, p. 103.

2. Stanley Keleman, "Professional Colloquium: 29 October 1977," in *Ecology and Consciousness: Traditional Wisdom on the Environment,* Richard Grossinger, editor (Berkeley: North Atlantic Books, 1992), p. 17.

3. Margulis and Sagan, p. 170.

4. ibid., p. 67.

5. ibid., p. 165.

6. ibid.

7. ibid., pp. 182, 206.

8. ibid., p. 63.

9. ibid., p. 168.

10. ibid., p. 112.

Chapter 5. The Cell

Carroll, Mark. *Organelles.* London: The Guilford Press, 1989.**

Wendt, Herbert. *The Sex Life of Animals.* Translated from the German by Richard and Clara Winston. New York: Simon and Schuster, 1965.*

Russell-Hunter, W. D. *A Life of Invertebrates.* New York: Macmillan Publishing Co., 1979.*

Berrill, N. J., and Gerald Karp. *Development.* New York: McGraw-Hill Book Co., 1976.*

Balinsky, B. I. *An Introduction to Embryology,* 5th edition. Philadelphia: Saunders College Publishing, 1981.*

Alberts, Bruce, Dennis Bray, Julian Lewis, Martin Raff, Keith Roberts, James D. Watson. *Molecular Biology of the Cell.* New York: Garland Publishing, 1989.*

Hauschka, Rudolf. *The Nature of Substance.* Translated from the German by Mary T. Richards and Marjorie Spock. London: Vincent Stuart Ltd., 1966.*

Margulis, Lynn, and Dorion Sagan. *Origins of Sex: Three Billion Years of Genetic Recombination.* New Haven: Yale University Press, 1986.*

Thompson, D'Arcy. *On Growth and Form* [1917]. London: Cambridge University Press, 1966.

Murchie, Guy. *The Seven Mysteries of Life: An Exploration in Science and Philosophy.* Boston: Houghton Mifflin Co., 1978.

Thomas, Lewis. *The Lives of a Cell: Notes of a Biology Watcher.* New York: Viking Press, 1974.

Haraway, Donna Jeanne. *Crystals, Fabrics, and Fields: Metaphors of Organicism in Twentieth-Century Developmental Biology.* New Haven: Yale University Press, 1976.

—

1. Anton van Leeuwenhoek, quoted in Murchie, p. 82.
2. ibid.
3. ibid., p. 83.
4. Matthias Jakob Schleiden, quoted in Wendt, p. 27.
5. Haraway, p. 20.
6. Carol Featherstone, "Coming to Grips with the Golgi"; in *Science,* Volume 282 (18 December 1998): pp. 2172–2174.
7. Craig Holdrege, *Genetics and the Manipulation of Life: The Forgotten Factor of Context* (Hudson, New York: Lindisfarne Press, 1996), p. 73.
8. Carol Featherstone, "Coming to Grips with the Golgi," p. 2172.
9. ibid.
10. Margulis and Sagan, p. 69.
11. Hauschka, pp. 21–22.
12. ibid., p. 26.
13. ibid., p. 109.
14. ibid., p. 28.
15. ibid.
16. ibid., p. 109.
17. Baron von Herzeele of Hanover, quoted in Hauschka, p. 14.
18. Alberts et al., pp. 655–656.
19. Margulis and Sagan, p. 169.
20. Alberts et al., p. 21.

Chapter 6. The Genetic Code

Alberts, Bruce, Dennis Bray, Julian Lewis, Martin Raff, Keith Roberts, James D. Watson. *Molecular Biology of the Cell.* New York: Garland Publishing, 1989.**

Carroll, Mark. *Organelles.* London: The Guilford Press, 1989.*

Crick, Francis. *Life Itself: Its Origin and Nature.* New York: Simon and Schuster, 1981.*

Balinsky, B. I. *An Introduction to Embryology,* 5th edition. Philadelphia: Saunders College Publishing, 1981.*

Wickware, Potter. "History and Technique of Cloning"; in *The Human Cloning Debate,* edited by Glenn McGee. Berkeley: Berkeley Hills Books, 1998.*

Portugal, Franklin H., and Jack S. Cohen. *A Century of DNA: A History of the Discovery of the Structure and Function of the Genetic Substance.* Cambridge: MIT Press, 1977.*

Upledger, John E. "Who Is Smartest of Them All?" *UpDate, A Publication of the Upledger Institute, Inc.* Palm Beach Gardens, Florida, Summer 1997.*

Margulis, Lynn, and Dorion Sagan. *Origins of Sex: Three Billion Years of Genetic Recombination.* New Haven: Yale University Press, 1986.*

Holdrege, Craig. *Genetics and the Manipulation of Life: The Forgotten Factor of Context.* Hudson, New York: Lindisfarne Press, 1996.*

Stanley, Wendell M., and Evans G. Valens. *Viruses and the Nature of Life.* New York: E. P. Dutton, 1961.

Berrill, N. J., and Gerald Karp. *Development.* New York: McGraw-Hill Book Co., 1976.

—

1. Harvey Bialy, "The I Ching and the Genetic Code," in *Ecology and Consciousness: Traditional Wisdom on the Environment,* Richard Grossinger, editor (Berkeley: North Atlantic Books, 1992).

2. Zecharia Sitchin, *The Cosmic Code* (New York: Avon Books, 1998).

3. Wickware, p. 22.

4. Margulis and Sagan, p. 116.

5. Carroll, p. 53.

6. Wickware, p. 23.

7. ibid.

8. Johnson F. Yan, *DNA and the I Ching* (Berkeley: North Atlantic Books, 1991), p. 169.

9. Margulis and Sagan, p. 120.

10. Jeremy Rifkin, *The Biotech Century: Harnessing the Gene and Remaking the World* (New York: Jeremy P. Tarcher/Putnam, 1998), p. 195.

11. Wickware, p. 22.

12. Thomas J. Weihs, *Embryogenesis in Myth and Science* (Edinburgh: Floris Books, 1986), p. 60.

13. ibid.

14. Crick, p. 66.

15. *San Francisco Chronicle,* December 11, 1998, pp. 1 and 16; by wire service from *Washington Post.*

16. Donna Haraway, quoted in Steven Shaviro, *Doom Patrols: A Theoretical Fiction about Postmodernism* (London: Serpent's Tail/High Risk Books, 1997), p. 115.

17. Gilles Deleuze and Felix Guattari, quoted in Shaviro, p. 119.

18. Shaviro, p. 119.

19. Wade Davis, *The Serpent and the Rainbow: A Harvard Scientist Uncovers The Startling Truth About The Secret World Of Haitian Voodoo And Zombis* (New York: Warner Books, 1985), p. 194.

20. Holdrege, p. 101.

21. Shaviro, p. 103.

22. John Todd, "An Interview Conducted by Richard Grossinger and Lindy Hough," November 1982 (this section did not appear in the published version of the interview in *Omni,* New York, August 1984).

Chapter 7. Sperm and Egg

Berrill, N. J., and Gerald Karp. *Development.* New York: McGraw-Hill Book Co., 1976.*

Balinsky, B. I. *An Introduction to Embryology,* 5th edition. Philadelphia: Saunders College Publishing, 1981.*

Wendt, Herbert. *The Sex Life of Animals.* Translated from the German by Richard and Clara Winston. New York: Simon and Schuster, 1965.*

Weihs, Thomas J. *Embryogenesis in Myth and Science.* Edinburgh: Floris Books, 1986.*

Wickware, Potter. "History and Technique of Cloning"; in *The Human Cloning Debate,* edited by Glenn McGee. Berkeley: Berkeley Hills Books, 1998.

Campbell, Neil A. *Biology.* Menlo Park, California: The Benjamin/Cummings Publishing Company, Inc., 1987.

Glass, Bentley, Owsei Temkin, and William L. Straus, Jr. *Forerunners of Darwin: 1745–1859.* Baltimore: Johns Hopkins University Press, 1959.

Alberts, Bruce, Dennis Bray, Julian Lewis, Martin Raff, Keith Roberts, James D. Watson. *Molecular Biology of the Cell.* New York: Garland Publishing, 1989.*

Gould, Stephen J. *Ontogeny and Phylogeny.* Cambridge: Harvard University Press, 1977.

Murchie, Guy. *The Seven Mysteries of Life: An Exploration in Science and Philosophy.* Boston: Houghton Mifflin Co., 1978.

Haraway, Donna Jeanne. *Crystals, Fabrics, and Fields: Metaphors of Organicism in Twentieth-Century Developmental Biology.* New Haven: Yale University Press, 1976.

———

1. Leonard Hayflick, "The Cell Biology of Human Aging," in *Scientific American,* 242 (1), January 1980: pp. 58–65.

2. Gilles Deleuze, quoted in Steven Shaviro, *Doom Patrols: A Theoretical Fiction about Postmodernism* (London: Serpent's Tail/High Risk Books, 1997), p. 39.

3. Shaviro, p. 39 ("cannibalism ... evolve" quoted from Lynn Margulis and Dorion Sagan).

4. Sir Charles Sherrington, *Man on His Nature* (London: Cambridge University Press, 1963), p. 95.

5. Philip Wheelwright (editor), *The Presocratics* (Indianapolis: Bobbs-Merrill, 1966), p. 196.

6. ibid, p. 187.

7. Seneca, quoted in Weihs, p. 34.

8. Anton van Leeuwenhoek, quoted in Wendt, p. 57.

9. Jan Swammerdam, quoted in Weihs, p. 43.

10. ibid.

11. Immanuel Kant, quoted in Glass, Temkin, and Straus, Jr., p. 186.

12. Weihs, p. 44.

13. ibid., p. 45.

14. Aristotle, quoted in Weihs, p. 31.

15. Galen, quoted in Weihs, p. 35.

16. William Harvey, quoted in Weihs, p. 42.

17. Pierre-Louis de Maupertuis, quoted in Glass, Temkin, and Straus, Jr., p. 68.

18. Weihs, p. 46.

19. Albrecht von Haller, quoted in Weihs, p. 45.

20. Weihs, p. 46.

21. ibid., p. 79.

22. Ross Harrison, quoted in Haraway, p. 44.

Chapter 8. Fertilization

Wendt, Herbert. *The Sex Life of Animals.* Translated from the German by Richard and Clara Winston. New York: Simon and Schuster, 1965.**

Margulis, Lynn, and Dorion Sagan. *Origins of Sex: Three Billion Years of Genetic Recombination.* New Haven: Yale University Press, 1986.**

Berrill, N. J., and Gerald Karp. *Development.* New York: McGraw-Hill Book Co., 1976.*

Balinsky, B. I. *An Introduction to Embryology*, 5th edition. Philadelphia: Saunders College Publishing, 1981.*

Russell-Hunter, W. D. *A Life of Invertebrates.* New York: Macmillan Co., 1979.

Jantsch, Erich. *The Self-Organizing Universe.* New York: Pergamon Press, 1980.

Weihs, Thomas J. *Embryogenesis in Myth and Science.* Edinburgh: Floris Books, 1986.

—

1. G. E. Ward et al., *Journal of Cell Biology,* 101 (1985): pp. 2324–2329.

2. Oscar Hertwig, quoted in Wendt, p. 69.

3. Weihs, pp. 79–80.

4. Da Free John, "The Mystery of the Spermatic Being," in *The Laughing Man,* Vol. 3, No. 1 (Clearlake, California: Dawn Horse Press, 1982), p. 15.

5. ibid.

6. Balinsky, p. 338.

7. Wilhelm Reich, *Ether, God and Devil/Cosmic Superimposition,* translated from the German by Therese Pol (New York: Farrar, Straus and Giroux, 1972).

8. Margulis and Sagan, p. 63.

9. ibid.

10. ibid., p. 4.

11. Andrew Marvell, "To His Coy Mistress," *Seventeenth Century Poetry: The Schools of Donne and Jonson,* editor, Hugh Kenner (New York: Holt, Rinehart and Winston, 1964), p. 457.

12. ibid.

13. Michio Kushi, *The Book of Do-In: Exercises for Physical and Spiritual Development* (Tokyo: Japan Publications, 1979), p. 27.

14. ibid., p. 28.

15. Susan Minot, *Evening* (New York: Alfred A. Knopf, 1998), pp. 197–198.

16. Henry Corbin, *Creative Imagination in the Sufism of Ibn 'Arabi* (Princeton, New Jersey: Princeton University Press, 1969), p. 329.

17. ibid.

18. ibid., p. 175.

Chapter 9. The Blastula

Balinsky, B. I. *An Introduction to Embryology*, 5th edition. Philadelphia: Saunders College Publishing, 1981.**

Berrill, N. J., and Gerald Karp. *Development.* New York: McGraw-Hill Book Co., 1976.**

Haraway, Donna Jeanne. *Crystals, Fabrics, and Fields: Metaphors of Organicism in Twentieth-Century Developmental Biology.* New Haven: Yale University Press, 1976.*

Keleman, Stanley. *Emotional Anatomy* . Berkeley, California: Center Press, 1985.*

Weihs, Thomas J. *Embryogenesis in Myth and Science.* Edinburgh: Floris Books, 1986.*

The Yellow Emperor's Classic of Internal Medicine. Translated from the Chinese by Ilza Veith. Berkeley: University of California Press, 1972.

———

1. Weihs, p. 80.

2. Bruce Alberts, Dennis Bray, Julian Lewis, Martin Raff, Keith Roberts, James D. Watson, *Molecular Biology of the Cell,* (New York: Garland Publishing, 1989), p. 892.

3. Keleman, p. 11.

4. Michio Kushi, *The Book of Do-In: Exercises for Physical and Spiritual Development* (Tokyo: Japan Publications, 1979), p. 27.

5. Samuel Beckett, *Nohow On* (New York: Grove Press, 1980), p. 6.

6. ibid.

7. William Faulkner, *Absalom, Absalom!* (New York: Random House, 1936), p. 142.

8. ibid., p. 143.

9. Kim Stanley Robinson, *Blue Mars* (New York: Bantam Books, 1997), p. 236.

10. Anne Tyler, *Dinner at the Homesick Restaurant* (New York: Berkley, 1983), p. 270.

11. ibid., p. 274.

12. ibid., p. 279.

13. ibid., p. 284.

Chapter 10. Gastrulation

Balinsky, B. I. *An Introduction to Embryology,* 5th edition. Philadelphia: Saunders College Publishing, 1981.**

Berrill, N. J., and Gerald Karp. *Development.* New York: McGraw-Hill Book Co., 1976.**

Weihs, Thomas J. *Embryogenesis in Myth and Science.* Edinburgh: Floris Books, 1986.**

Saunders, John W., Jr. *Patterns and Principles of Animal Development.* New York: Macmillan Co., 1970.*

Moore, Keith L. *The Developing Human: Clinically Oriented Embryology,* 2nd edition. Philadelphia: Saunders College Publishing, 1977.*

Russell-Hunter, W. D. *A Life of Invertebrates.* New York: Macmillan Co., 1979.*

Keleman, Stanley. *Emotional Anatomy.* Berkeley, California: Center Press, 1985.*

—

1. Rudolf Hauschka, *The Nature of Substance,* translated from the German by Mary T. Richards and Marjorie Spock (London: Vincent Stuart Ltd., 1966), p. 78.

2. Weihs, p. 94.

3. ibid.

4. Tao Huang, "The Pre-heaven Sensation," in *House Organ,* #26, Spring 1999, Lakewood, Ohio, p. 6.

5. Pat Conroy, *Beach Music* (New York: Bantam Books, 1996), pp. 774, 171.

6. Arthur A. Tansley, "The Deadly Ninja: Agents of Death," in *Fighting Arts Magazine,* Vol. 5, No. 3 (Liverpool, England, 1983), pp. 19–20.

7. Heinrich Zimmer, "The Indian World Mother," in *The Mystic Vision* [Papers from the Eranos Yearbooks, 6] (Princeton: Princeton University Press, 1968), p. 74.

8. Edward Whitmont, *The Alchemy of Healing: Psyche and Soma* (Berkeley: North Atlantic Books, 1993), p. 140.

9. Keleman, p. 5.

10. ibid., p. 11.

11. Joseph Allen and Thomas O'Toole, "Thoughts of an Astronaut," *Washington Post Syndicate* (Washington, D.C., 1983).

12. ibid.

13. Lewis Wolpert, quoted in Scott Gilbert, *Developmental Biology, Fifth Edition* (Sunderland, Massachusetts: Sinauer Associates, 1997), p. 209.

14. Janet Frame, *Living in the Maniototo* (Auckland, New Zealand: The Women's Press, 1979), pp. 117–118.

Chapter 11. Morphogenesis

Berrill, N. J., and Gerald Karp. *Development.* New York: McGraw-Hill Book Co., 1976.**

Balinsky, B. I. *An Introduction to Embryology,* 5th edition. Philadelphia: Saunders College Publishing, 1981.**

Goodwin, Brian. *How the Leopard Changed its Spots: The Evolution of Complexity.* New York: Simon and Schuster, 1996.**

Holdrege, Craig. *Genetics and the Manipulation of Life: The Forgotten Factor of Context.* Hudson, New York: Lindisfarne Press, 1996.**

Ingber, Donald E. "The Architecture of Life." *Scientific American* (January 1998), pp. 48–57.**

Weihs, Thomas J. *Embryogenesis in Myth and Science.* Edinburgh: Floris Books, 1986.*

Alberts, Bruce, Dennis Bray, Julian Lewis, Martin Raff, Keith Roberts, James D. Watson. *Molecular Biology of the Cell.* New York: Garland Publishing, 1989.*

Saunders, John W. Jr., *Patterns and Principles of Animal Development,* 2nd edition. New York: Macmillan Co., 1970.*

Davidson, Eric H. *Gene Activity in Early Development,* 2nd edition. New York: Academic

Press, 1976.*

Thompson, D'Arcy. *On Growth and Form* [1917]. Cambridge: Cambridge University Press, 1966.*

Newman, Stuart A., and Wayne D. Comper. "'Generic' physical mechanisms of morphogenesis and pattern formation." *Development* #110, 1990, pp. 1–18.*

Newman, Stuart A. "Generic physical mechanisms of morphogenesis and pattern formation as determinants in the evolution of multicellular organization." *Journal of Biosciences,* Volume 17, Number 3, September, 1992, pp. 193–215.*

Newman, Stuart A. "Generic physical mechanisms of tissue morphogenesis: A common basis for development and evolution." *Journal of Evolutionary Biology,* #7, 1994, pp. 467–488.*

(I also discussed the above three articles with Stuart Newman and have used material from that conversation in this chapter and the succeeding one.)

Haraway, Donna Jeanne. *Crystals, Fabrics, and Fields: Metaphors of Organicism in Twentieth-Century Developmental Biology.* New Haven: Yale University Press, 1976.*

Russell-Hunter, W. D. *A Life of Invertebrates.* New York: Macmillan Co., 1979.*

———

1. Berrill and Karp, p. 1.

2. Sven Horstadius, *Experimental Embryology of the Echinoderms* (Oxford: Clarendon Press, 1973), p. 1.

3. William Blake, "The Tyger" (1794) in *The Portable Blake*, editor, Alfred Kazin (New York: Viking Press, 1946), p. 109.

4. Goodwin, p. viii.

5. H. Robert Bagwell, "Integrative Processing," written draft #4, unpublished manuscript, January 2, 1999.

6. Alexander Rich and S. H. Kim, "The Three-Dimensional Structure of Transfer RNA," *Scientific American,* January 1978, p. 52.

7. Freeman Dyson, *Origins of Life* (Cambridge: Cambridge University Press, 1985), p. 6.

8. C. Delisi, "The Human Genome Project," in *American Scientist* 76 (1988): pp. 488–493, quoted in Goodwin, p. 16.

9. Charles Poncé, "The Small Room," unpublished manuscript.

10. Stuart A. Newman, "Carnal Boundaries: The Commingling of Flesh in Theory and Practice," in Lynda Birke and Ruth Hubbard (editors), *Reinventing Biology: Respect for Life and the Creation of Knowledge* (Bloomington, Indiana: Indiana University Press, 1995), pp. 213–214.

11. Richard Dawkins, *Unweaving the Rainbow: Science, Delusion and the Appetite for Wonder* (Boston: Houghton Mifflin Company, 1999).

12. Bagwell.

13. Newman, 1992, p. 195.

14. Ingber, p. 48.

15. Thompson, pp. 60–61.

16. ibid., p. 37.

17. ibid., p. 15.

18. ibid., pp. 85–86.

19. Newman, 1992, p. 196.

20. ibid.

21. Thompson, p. 172.

22. Newman and Comper.

23. Stuart A. Newman, personal communication, 1999.

24. Alberts et al., p. 794.

25. Ingber, p. 54.

26. Newman, 1992, p. 212.

27. Goodwin, p. 148.

28. Newman, 1994, p. 469.

29. ibid., p. 470.

30. ibid.

31. Newman and Comper.

32. ibid.

33. ibid.

34. ibid.

35. ibid.

36. Andrew Lange, *Getting at the Root,* unpublished manuscript, 1999.

37. Rich Anderson, "What is 'Scalar Electromagnetics'?" http://www.tricountyi.net/~randerse/whscalar.htm.

38. ibid.

39. Holdrege.

40. Michael J. Chapman and Lynn Margulis, "Morphogenesis by symbiogenesis," *International Microbiology,* Volume 1 (1998), p. 321.

41. ibid., p. 322.

42. Alberts et al., p. 257.

43. Lynn Margulis, personal communication, December 26, 1998.

44. Lynn Margulis, personal communication, February 21, 1999.

45. ibid.

46. Weihs, p. 40.

47. ibid., p. 41.

48. ibid., p. 72.

49. George Oster, seminar given at University of California at Berkeley, January 26, 1982.

50. Goodwin, p. 120.

51. ibid., p. 129.

52. Newman and Comper.

53. Ingber, pp. 49–50.

54. ibid., pp. 51–52.

55. Michael McClure, "Wolf Net," in *Ecology and Consciousness: Traditional Wisdom on the Environment,* Richard Grossinger, editor (Berkeley: North Atlantic Books, 1992), p. 206.

56. ibid.

Chapter 12. Biological Fields

Haraway, Donna Jeanne. *Crystals, Fabrics, and Fields: Metaphors of Organicism in Twentieth-Century Developmental Biology.* New Haven: Yale University Press, 1976.**

Holdrege, Craig. *Genetics and the Manipulation of Life: The Forgotten Factor of Context.* Hudson, New York: Lindisfarne Press, 1996.**

Newman, Stuart A., and Wayne D. Comper. "'Generic' physical mechanisms of morphogenesis and pattern formation." *Development* #110, 1990, pp. 1–18.*

Newman, Stuart A. "Generic physical mechanisms of morphogenesis and pattern formation as determinants in the evolution of multicellular organization." *Journal of Biosciences,* Volume 17, Number 3, September 1992, pp. 193–215.*

Newman, Stuart A. " Generic physical mechanisms of tissue morphogenesis: A common basis for development and evolution." *Journal of Evolutionary Biology,* #7, 1994, pp. 467–488.*

(I also discussed the above three articles with Stuart Newman and have used material from that conversation in this chapter and the preceding one.)

Goodwin, Brian. *How the Leopard Changed its Spots: The Evolution of Complexity.* New York: Simon and Schuster, 1996.

Wickware, Potter. "History and Technique of Cloning," in *The Human Cloning Debate,* edited by Glenn McGee. Berkeley: Berkeley Hills Books, 1998.

Berrill, N. J., and Gerald Karp. *Development.* New York: McGraw-Hill Book Co., 1976.

Balinsky, B. I. *An Introduction to Embryology,* 5th edition. Philadelphia: Saunders College Publishing, 1981.

Weihs, Thomas J. *Embryogenesis in Myth and Science.* Edinburgh: Floris Books, 1986.

Newman, Stuart A. "Carnal Boundaries: The Commingling of Flesh in Theory and Practice," in Lynda Birke and Ruth Hubbard (editors), *Reinventing Biology: Respect for Life and the Creation of Knowledge.* Bloomington, Indiana: Indiana University Press, 1995.

—

1. Ron Meyer, private communication, January 1999.

2. W. Johannsen, quoted in Newman, p. 215, and Holdrege, p. 68.

3. Newman, 1992, p. 216.

4. ibid., p. 215.

5. J. S. Jones, quoted in Holdrege, p. 86.

6. Sir Charles Sherrington, *Man on His Nature* (Cambridge: Cambridge University Press, 1963), p. 107.

7. Ross Harrison, quoted in Haraway, p. 88.

8. ibid., p. 87.

9. ibid., p. 99.

10. ibid., p. 74.

11. ibid., p. 100.

12. Joseph Needham, quoted in Haraway, p. 113.

13. Haraway, p. 177.

14. Holdrege, pp. 46–47.

15. ibid., p. 43.

16. ibid.

17. C. H. Waddington, quoted in Newman, 1992, p. 193.

18. Richard C. Strohman, "Biology, physics, emergence and all that. A report from the trenches. A commentary on complexity in biological systems and initiatives for new interdisciplinary programs for its study," unpublished paper, February 24, 1999.

19. Newman, 1992, p. 194.

20. Newman, 1994, p. 478.

21. Newman, 1992, p. 210.

22. Newman and Comper, p. 8.

23. ibid., p. 14.

24. Stuart A. Newman, personal communication, 1999.

25. Newman and Comper, p. 6.

26. ibid., p. 15.

27. ibid., p. 14.

28. Newman, 1994, p. 477.

29. Newman, 1992, p. 201.

30. Newman, 1994, pp. 478–479

31. Newman, 1992, p. 193.

32. Newman, 1994, p. 479.

33. ibid. p. 482

34. Newman and Comper, p. 140.

35. Newman, personal communication.

36. Newman and Comper, p. 140.

37. N. J. Berrill and Gerald Karp, p. 281.

38. Holdrege, p. 65.

39. ibid.

40. ibid., p. 62.

41. Newman and Comper, p. 13.

42. ibid.

43. ibid., p. 14.

44. Paul Weiss, quoted in Weihs, p. 65.

45. Paul Weiss, quoted in Haraway, p. 186.

46. Stuart A. Newman, "Carnal Boundaries: The Commingling of Flesh in Theory and Practice," in Lynda Birke and Ruth Hubbard (editors), *Reinventing Biology: Respect for Life and the Creation of Knowledge* (Bloomington, Indiana: Indiana University Press, 1995), p. 222.

47. ibid.

48. Brian Goodwin, quoted in Weihs, p. 70.

49. Stephen Black, "Determination of the Dorsal-Ventral Axis in *Xenopus laevis* Embryos," Ph.D. thesis (Berkeley: University of California, 1983).

50. David Halberstam, "Jordan's Moment," in *The New Yorker,* December 21, 1998, p. 52.

51. ibid., p. 55.

52. Goodwin, *How the Leopard Changed its Spots,* p. 177.

53. Haraway, 185.

54. Paul Weiss, quoted in Haraway, p. 185.

55. Joseph Needham, quoted in Haraway, p. 136.

56. Paul Weiss, quoted in Haraway, p. 147.

57. Haraway, pp. 60–61.

58. Rupert Sheldrake, *A New Science of Life* (Los Angeles: J. P. Tarcher, 1982).

59. Ralph Abraham, Terence McKenna, and Rupert Sheldrake, *Trialogues at the Edge of the West: Chaos, Creativity, and the Resacralization of the World* (Santa Fe, New Mexico: Bear & Company Publishing, 1992), p. 28.

60. Virginia Lee, "Science and Spirit: Conversations with Matthew Fox, Ph.D., and Rupert Sheldrake, Ph.D." in *Common Ground,* 92 (Summer 1997): p. 159.

61. A. L. Kroeber, *Anthropology* (New York: Harcourt, Brace & World, 1923), p. 342.

62. Sheldrake, *A New Science of Life* , book jacket.

63. Justin O'Brien, *The Wellness Tree: Energizing Yourself in Body, Mind, and Spirit* (St. Paul, Minnesota: Yes International Publishers, 1990), p. 6.

64. Plotinus, *The Six Enneads* (ca. A.D. 263), translated from the Latin by Stephen MacKenna (Chicago: *Encyclopedia Britannica,* 1952), p. 117.

Chapter 13. Chaos, Fractals, and Deep Structure

Gleick, James. *Chaos: Making a New Science.* New York: Penguin Books, 1987.**

Goodwin, Brian. *How the Leopard Changed its Spots: The Evolution of Complexity.* New York: Simon and Schuster, 1996.**

Gladwell, Malcolm. "The Tipping Point," in *The New Yorker,* June 3, 1996.*

Depew, David J., and Bruce H. Weber. *Darwin Evolving: Systems Dynamics and the Genealogy of Natural Selection.* Cambridge, Massachusetts: A Bradford Book, The MIT Press, 1995.*

Kaufman, Stuart A. *The Origins of Order: Self-Organization and Selection in Evolution.* New York: Oxford University Press, 1993.

—

1. Gleick, pp. 13–14.

2. Steven Shaviro, *Doom Patrols: A Theoretical Fiction about Postmodernism* (London: Serpent's Tail/High Risk Books, 1997), p. 126.

3. Gleick, p. 55.

4. ibid., p. 86.

5. Gladwell, pp. 32–33.

6. Philip Wheelwright (editor), *The Presocratics* (Indianapolis: Bobbs-Merrill, 1966), p. 183.

7. ibid., p. 72.

8. Gerald Holton and Duane H. D. Roller, *Foundations of Modern Physical Science* (Reading, Massachusetts: Addison-Wesley Publishing Company, Inc., 1958), pp. 168–169.

9. ibid., p. 174.

10. Philip Appleman (editor), *Darwin: A Norton Critical Edition* (New York: W. W. Norton & Company, Inc., 1970), p. 73.

11. ibid., pp. 66–67.

12. Charles Darwin, *On the Origin of Species by Means of Natural Selection, or Preservation of Favoured Races in the Struggle for Life*, second edition (London: Murray, 1860), p. 474.

13. Gleick, p. 306.

14. ibid., p. 94.

15. ibid., p. 108.

16. Goodwin, p. 188.

17. ibid.

18. ibid., pp. 110–111.

19. ibid., p. 111.

20. ibid., p. 113.

21. ibid., pp. 137–138.

22. ibid. pp. 138–139.

23. ibid., p. 60.

24. ibid., p. 61.

25. ibid., p. 185.

26. ibid.

27. Gleick, p. 306.

28. ibid., p. 117.

29. Shunryu Suzuki, *Zen Mind, Beginner's Mind: Informal Talks on Zen Meditation and Practice* (New York & Tokyo: John Weatherhill, Inc., 1970), pp. 31–32.

30. Alfred North Whitehead, quoted in Frederick B. Artz, *The Mind of the Middle Ages* (New York: Alfred A. Knopf, 1965), p. 15.

31. Kenneth Kierans, "Beyond Deconstruction," www.mun.ca/animus/1997vol2/kierans1.htm.

32. Jacques Derrida, *Edmund Husserl's "Origin of Geometry": An Introduction*, translated by John P. Leavey, Jr. (Lincoln, Nebraska: University of Nebraska Press, 1978), p. 88.

33. Jacques Derrida, *Of Grammatology*, translated by Gayatri Chakravorty Spivak (Baltimore: Johns Hopkins University Press, 1974), p. 288.

34. Kierans.

35. Jacques Derrida, *Dissemination*, translated by Barbara Johnson (The University of Chicago Press, 1981), p. 364.

36. ibid., p. 337.

Chapter 14. Ontogeny and Phylogeny

Gould, Stephen Jay. *Ontogeny and Phylogeny*. Cambridge: Harvard University Press, 1977.**

Montagu, M. F. Ashley. "Time, Morphology, and Neoteny in the Evolution of Man," in *Culture and the Evolution of Man*, edited by M. F. Ashley Montagu. New York: Oxford University Press, 1962.*

Friedlander, C. P. *The Biology of Insects*. New York: Pica Press, 1977.*

Wendt, Herbert. *The Sex Life of Animals*. Translated from the German by Richard and Clara Winston. New York: Simon and Schuster, 1965.*

Zimmer, Carl. *At the Water's Edge: Macroevolution and the Transformation of Life*. New York: The Free Press, 1998.

Gould, Stephen Jay. *The Panda's Thumb*. New York: W.W. Norton and Co., 1982.

Weihs, Thomas J. *Embryogenesis in Myth and Science*. Edinburgh: Floris Books, 1986.

Goodwin, Brian. *How the Leopard Changed its Spots: The Evolution of Complexity*. New York: Simon and Schuster, 1996.

Holdrege, Craig. *Genetics and the Manipulation of Life: The Forgotten Factor of Context*. Hudson, New York: Lindisfarne Press, 1996.

—

1. Arthur Edward Waite (editor), *The Hermetic and Alchemical Writings of Paracelsus, Volume 1, Hermetic Chemistry* (London: James Elliot & Co., 1894), p. 179.

2. Michel Foucault, *The Order of Things: An Archaeology of the Human Sciences*, translated from the French (New York: Pantheon Books, 1970), p. 17.

3. ibid., p. 22.

4. Ernst Haeckel, quoted in Gould, *Ontogeny and Phylogeny*, p. 78.

5. Lorenz Oken, quoted in Gould, *Ontogeny and Phylogeny*, p. 45.

6. Claude Lévi-Strauss, *The Raw and the Cooked: Introduction to a Science of Mythology*, Vol. 1, translated from the French by John and Doreen Weightman (New York: Harper and Row, 1969).

7. ibid.

8. W. E. H. Stanner, *The Dreaming* (Indianapolis: Bobbs-Merrill Reprint Series in the Social Sciences, A-214, first published 1956), pp. 54–55.

9. Weihs, pp. 47–48.

10. Goodwin, p. 22.

11. Karl Ernst von Baer, quoted in Gould, *Ontogeny and Phylogeny*, p. 56.

12. Ernst Haeckel, quoted in Gould, *Ontogeny and Phylogeny*, p. 172.

13. ibid., p. 82.

14. E. Mehnert, quoted in Gould, *Ontogeny and Phylogeny*, p. 175.

15. Gould, *Ontogeny and Phylogeny*, p. 1.

16. Wilhelm Roux, quoted in Gould, *Ontogeny and Phylogeny*, p. 195.

17. Gould, *Ontogeny and Phylogeny*, p. 214.

18. Steven Shaviro, *Doom Patrols: A Theoretical Fiction about Postmodernism* (London:

Serpent's Tail/High Risk Books, 1997), pp. 116–117.

19. Fritz Kahn, quoted in Wendt, p. 87.

20. Gould, *The Panda's Thumb*, pp. 35–37.

21. Charles Darwin, *On the Origin of Species by Means of Natural Selection, or Preservation of Favoured Races in the Struggle for Life* (London: Murray, 1859).

22. William Burroughs, quoted in Shaviro, p. 48.

23. Julian Huxley, quoted in Gould, *Ontogeny and Phylogeny*, p. 267.

24. Richard Goldschmidt, quoted in Gould, *The Panda's Thumb*, p. 192.

25. Shaviro, p. 114.

26. Aldous Huxley, *After Many a Summer Dies the Swan* (New York: Harper and Row, 1965), pp. 238–240.

27. Gould, *Ontogeny and Phylogeny*, p. 383.

28. Holdrege, p. 150.

29. Louis Bolk, quoted in Gould, *Ontogeny and Phylogeny*, p. 361.

Chapter 15. Biotechnology

Holdrege, Craig. *Genetics and the Manipulation of Life: The Forgotten Factor of Context.* Hudson, New York: Lindisfarne Press, 1996.**

Various Authors. "The Future of Medicine," *Time*, January 11, 1999, pp. 42–91.**

Wickware, Potter. "History and Technique of Cloning"; in *The Human Cloning Debate*, edited by Glenn McGee. Berkeley: Berkeley Hills Books, 1998.**

Rifkin, Jeremy. *The Biotech Century: Harnessing the Gene and Remaking the World.* New York: Jeremy P. Tarcher/Putnam, 1998.*

Alberts, Bruce, Dennis Bray, Julian Lewis, Martin Raff, Keith Roberts, James D. Watson. *Molecular Biology of the Cell.* New York: Garland Publishing, 1989.*

Taylor, Robert. "Superhumans," *New Scientist,* Volume 160, Number 2154, October 3, 1998, pp. 25–29.*

Lemonick, Michael D. "The Biological Mother Lode," *Time* Magazine, Volume 152, Number 20, November 16, 1998, pp. 96–97.

—

1. Walter Isaacson, "The Biotech Century," in "The Future of Medicine," p. 42.

2. Phillip S. Angell (Director, Corporate Communications, Monsanto Company, Washington, D.C.) in "Letters," *The New York Times Magazine*, November 15, 1998, p. 26.

3. Rifkin, p. 12.

4. James D. Watson, "All for the Good: Why genetic engineering must soldier on," in "The Future of Medicine," p. 91.

5. Jeffrey Kluger, "Who Owns Our Genes?" in "The Future of Medicine," p. 51.

6. James Walsh, "Brave New Farm," in "The Future of Medicine," p. 88.

7. Edward Dorn, "El Peru/Cheyenne Milkplane," from *Westward Haut* (unpublished poem); excerpted in Edward Dorn, *High West Rendezvous: A Sampler* (South Devonshire, England: Etruscan Books, 1997), p. 25.

8. Frank Herbert, *Dune Messiah* (New York: Berkley, 1969); *God Emperor of Dune* (New York: Berkley, 1981).

9. Holdrege, p. 121.

10. ibid., p. 122.

11. ibid., p. 117.

12. Martin Heidegger, quoted by Harvey Bialy in an e-mail on biotechnology (original source unknown).

13. Richard C. Strohman, commentary submitted for publication in *Nature Biotechnology,* 1999.

14. ibid.

15. Holdrege, p. 112.

16. ibid., p. 116.

17. ibid., p. 121.

18. Walter Isaacson, "The Biotech Century," in "The Future of Medicine," p. 42.

19. Christine Gorman, "Drugs by Design," in "The Future of Medicine," p. 80.

20. Frederic Golden, "Good Eggs, Bad Eggs," in "The Future of Medicine," p. 58.

21. Leon Jaroff, "Fixing the Genes," in "The Future of Medicine," p. 68.

22. ibid., pp. 72–73.

23. ibid., pp. 68–70.

24. ibid., p. 69.

25. Michael D. Lemonick and Dick Thompson, "Racing to Map Our DNA," in "The Future of Medicine," p. 47.

26. ibid., p. 49.

27. ibid., p. 47.

28. ibid.

29. Strohman.

30. ibid.

31. Richard Lewontin, "Billions and Billions of Demons," in *The New York Review of Books,* Volume XLIV, Number 1 (January 9, 1997): p. 29.

32. Jerome Groopman, "Decoding Destiny," in *The New Yorker* (June 3, 1996): p. 45.

33. Frederic Golden, "Good Eggs, Bad Eggs," in "The Future of Medicine," p. 59.

34. Groopman, p. 45.

35. ibid.

Chapter 16. The Origin of the Nervous System

Bullock, T. H., and G. A. Horridge. *Structure and Function in the Nervous System of Invertebrates.* San Francisco: W. H. Freeman and Co., 1965.**

Russell-Hunter, W. D. *A Life of Invertebrates.* New York: Macmillan Co., 1979.*

Woodburne, Lloyd S. *The Neural Basis of Behavior.* Columbus, Ohio: Charles E. Merrill Books, 1967.

Rose, Steven. *The Conscious Brain.* New York: Alfred A. Knopf, 1974.

Murchie, Guy. *The Seven Mysteries of Life: An Exploration in Science and Philosophy.* Boston: Houghton Mifflin Co., 1978.

Sherrington, Sir Charles. *Man on His Nature.* London: Cambridge University Press, 1963.

Berrill, N. J., and Gerald Karp. *Development.* New York: McGraw-Hill Book Co., 1976.

—

1. Robert Kelly, *Axon Dendron Tree* (Annandale-on-Hudson, New York: Salitter Books, 1967), p. 7.

2. Robert Kelly, "A Chapter of Questions," in *Ecology and Consciousness: Traditional Wisdom on the Environment,* Richard Grossinger, editor (Berkeley: North Atlantic Books, 1992), p. 48.

3. Tor Norretranders, *The User Illusion: Cutting Consciousness Down to Size,* translated from the Danish by Jonathan Sydenham (New York: Viking Press, 1998).

4. David Chamberlain, *Consciousness at Birth: A Review of the Empirical Evidence* (San Diego, California: Chamberlain Communications, 1983), p. 4 [republished in different form as *The Mind of Your Newborn Baby* (Berkeley, California: North Atlantic Books, 1998)]. Page number refers to the 1983 version of this book.

5. Maurice Merleau-Ponty, *The Primacy of Perception,* translated from the French by James M. Edie (Evanston, Illinois: Northwestern University Press, 1964), p. 17.

6. Maurice Merleau-Ponty, *The Structure of Behavior,* translated from the French by Alden L. Fisher (Boston: Beacon Press, 1963), pp. 144–145.

7. ibid., p. 145.

8. Maura "Soshin" O'Halloran, *Pure Heart, Enlightened Mind,* audiobook (San Bruno, California: Audio Literature, 1996).

Chapter 17. The Evolution of Intelligence

Bullock, T. H., and G. A. Horridge. *Structure and Function in the Nervous System of Invertebrates.* San Francisco: W. H. Freeman and Co., 1965.**

Russell-Hunter, W. D. *A Life of Invertebrates.* New York: Macmillan Co., 1979.*

Friedlander, C. P. *The Biology of Insects.* New York: Pica Press, 1977.*

Alexander, R. McNeill. *The Chordates.* Cambridge: Cambridge University Press, 1975.*

Rose, Steven. *The Conscious Brain.* New York: Alfred A. Knopf, 1974.*

Wendt, Herbert. *The Sex Life of Animals.* Translated from the German by Richard and Clara Winston. New York: Simon and Schuster, 1965.

Smith, Eric, Garth Chapman, R. B. Clar, David Nichols, and J. D. McCarthy. *The Invertebrate Panorama.* New York: Universe Books, 1971.

Woodburne, Lloyd S. *The Neural Basis of Behavior.* Columbus, Ohio: Charles E. Merrill Books, 1967.

Buschsbaum, Ralph. *Animals Without Backbones,* 2nd edition. Chicago: University of Chicago Press, 1948.

Borror, Donald J., Dwight M. DeLong, and Charles A. Triplehorn. *An Introduction to the Study of Insects.* Philadelphia: Saunders College Publishing, 1981.

Goodwin, Brian. *How the Leopard Changed its Spots: The Evolution of Complexity.* New York: Simon and Schuster, 1996.

——

1. Frank Herbert, *Dune* (New York: Berkley, 1965).

2. Johann Wolfgang Goethe, "Death of a Fly," in Stephen Mitchell (editor), *Bestiary: An Anthology of Poems about Animals* (Berkeley, California: Frog, Ltd., 1996), p. 37.

3. Steven Shaviro, *Doom Patrols: A Theoretical Fiction about Postmodernism* (London: Serpent's Tail/High Risk Books, 1997), p. 113.

4. Goodwin, pp. 71–72.

5. Eugène Marais, *The Soul of the White Ant* (London: Methuen and Co., 1937).

6. Karl von Frisch, *Animal Architecture* (New York: Harcourt, Brace and Jovanovich, 1974), p. 103.

7. Maurice Maeterlinck, *The Life of the Bee* (New York: Dodd, Mead, and Co., 1936), quoted in Wendt, pp. 179–180.

8. Maurice Maeterlinck, quoted in Shaviro, p. 120.

9. Robert Kelly, "First in an Alphabet of Sacred Animals," in *Ecology and Consciousness: Traditional Wisdom on the Environment,* Richard Grossinger, editor (Berkeley: North Atlantic Books, 1992), p. 61.

10. Shaviro, pp. 119.

11. ibid., pp. 114, 119.

12. ibid., pp. 120–121.

13. Russell-Hunter, p. 453.

14. John R. Searle, "I Married a Computer," a review of *The Age of Spiritual Machines: When Computers Exceed Human Intelligence* by Ray Kurzweil, Viking Press, in *The New York Review of Books,* Volume XLVI, Number 6 (April 8, 1999), pp. 34–38.

15. Ray Kurzweil, quoted in Searle, p. 34.

16. Searle, p. 34.

17. ibid.

18. Kurzweil, quoted in Searle, p. 34.

19. ibid.

20. ibid.

21. Colin McGinn, "Hello, HAL: Three books examine the future of artificial intelligence and find the human brain is in trouble," in *The New York Times Book Review* (January 3, 1999): p. 11.

22. ibid., p. 12.

23. Lynn Margulis and Dorion Sagan, *What Is Life?,* quoted in *Whole Earth,* Fall, 1999, p. 71.

24. Searle, p. 36.

25. Maura "Soshin" O'Halloran, *Pure Heart, Enlightened Mind,* audiobook (San Bruno, California: Audio Literature, 1996).

26. Robert Kelly, *Finding the Measure* (Los Angeles: Black Sparrow Press, 1968), p. 26.

Chapter 18. Neurulation and the Human Brain

Woodburne, Lloyd S. *The Neural Basis of Behavior.* Columbus, Ohio: Charles E. Merrill Books, 1967.**

Weihs, Thomas J. *Embryogenesis in Myth and Science.* Edinburgh: Floris Books, 1986.**

Moore, Keith L. *The Developing Human: Clinically Oriented Embryology.* Philadelphia: Saunders College Publishing, 1977.*

Rose, Steven. *The Conscious Brain.* New York: Alfred A. Knopf, 1974.*

Alberts, Bruce, Dennis Bray, Julian Lewis, Martin Raff, Keith Roberts, James D. Watson. *Molecular Biology of the Cell.* New York: Garland Publishing, 1989.*

Restak, Richard M. *The Brain: The Last Frontier.* New York: Warner Books, 1979.*

Berrill, N. J., and Gerald Karp. *Development.* New York: McGraw-Hill Book Co., 1976.*

Balinsky, B. I. *An Introduction to Embryology,* 5th edition. Philadelphia: Saunders College Publishing, 1981.*

Keleman, Stanley. *Emotional Anatomy.* Berkeley, California: Center Press, 1985.*

Upledger, John E., and Jon D. Vredevoogd. *Craniosacral Therapy.* Seattle: Eastland Press, 1983.*

Hall, Manly P. *Man, Grand Symbol of the Mysteries: Thoughts in Occult Anatomy.* Los Angeles: The Philosophical Research Society, 1972.*

Goodwin, Brian. *How the Leopard Changed its Spots: The Evolution of Complexity.* New York: Simon and Schuster, 1996.

Alexander, R. McNeill. *The Chordates.* Cambridge: Cambridge University Press, 1975.

Russell-Hunter, W. D. *A Life of Invertebrates.* New York: Macmillan Co., 1979.

Ramachandran, V. S., and Sandra Blakeslee. *Phantoms in the Brain: Probing the Mysteries of the Human Mind.* New York: William Morrow & Company, 1999.

Freud, Sigmund. *The Interpretation of Dreams.* Translated from the German by James Strachey. New York: Basic Books, 1955.

Chomsky, Noam. *Aspects of the Theory of Syntax.* Cambridge, Massachusetts: MIT Press, 1965.

———

1. Weihs, p. 113.

2. ibid., pp. 113–114.

3. Lord Byron [George Gordon], "The Dream" (1816) in *Selected Poetry and Letters,* editor, Edward E. Bostetter (New York: Rinehart and Co., 1958), p. 25.

4. Goodwin, p. 167.

5. ibid., p. 168.

6. Keleman, p. 53.

7. Hall, pp. 219–220.

8. Madame H. P. Blavatsky, quoted in Hall, p. 220.

9. Paul MacLean, quoted in Restak, p. 36.

10. ibid., p. 41.

11. ibid., p. 52.

12. Upledger and Vredevoogd, p. 61.

13. John E. Upledger, *Your Inner Physician and You: CranioSacral Therapy and SomatoEmotional Release* (Berkeley: North Atlantic Books and UI Enterprises, 1991), p. 18.

14. Sir Charles Sherrington, *Man on His Nature* (London: Cambridge University Press, 1963), p. 105.

15. Andrew Weil, *The Marriage of the Sun and Moon: A Quest for Unity in Consciousness* (Boston: Houghton Mifflin Co., 1980), p. 257.

16. Stanley Keleman, "Professional Colloquium: 29 October 1977," in *Ecology and Consciousness: Traditional Wisdom on the Environment,* Richard Grossinger, editor (Berkeley: North Atlantic Books, 1992), p. 24.

17. Keleman, *Emotional Anatomy,* p. 58.

18. ibid., pp. 28 and 58.

19. H. Robert Bagwell, "Integrative Processing," written draft #4, unpublished manuscript, January 2, 1999.

20. ibid.

21. Matthew Arnold, "Palladium" [ca. 1880], in Harold C. Goddard, *The Meaning of Shakespeare, Volume 2* (Chicago: The University of Chicago Press, 1951), p. 37.

Chapter 19. Organogenesis

Moore, Keith L. *The Developing Human: Clinically Oriented Embryology.* Philadelphia: Saunders College Publishing, 1977.**

Balinsky, B. I. *An Introduction to Embryology,* 5th edition. Philadelphia: Saunders College Publishing, 1981.**

Berrill, N. J., and Gerald Karp. *Development.* New York: McGraw-Hill Book Co., 1976.**

Alberts, Bruce, Dennis Bray, Julian Lewis, Martin Raff, Keith Roberts, James D. Watson. *Molecular Biology of the Cell.* New York: Garland Publishing, 1989.**

Keleman, Stanley. *Emotional Anatomy.* Berkeley, California: Center Press, 1985.**

Weihs, Thomas J. *Embryogenesis in Myth and Science.* Edinburgh: Floris Books, 1986.*

Thibodeau, Gary A., and Kevin T. Patton. *Structure & Function of the Body.* St. Louis: Mosby Year Book, 1992.*

Hall, Manly P. *Man, Grand Symbol of the Mysteries: Thoughts in Occult Anatomy.* Los Angeles: The Philosophical Research Society, 1972.*

Schultz, R. Louis. *Out in the Open: The Complete Male Pelvis.* Berkeley, California: North Atlantic Books, 1999.*

Seeley, Rod R., Trent D. Stephens, and Philip Tate. *Essentials of Anatomy and Physiology.* St. Louis: Mosby Year Book, 1991.

Gershon, Michael D. *The Second Brain: The Scientific Basis of Gut Instinct and a Groundbreaking New Understanding of Nervous Disorders of the Stomach and Intestine.* New York: HarperCollins, 1999.

Ballard, William W. *Comparative Anatomy and Embryology.* New York: The Ronald Press

Company, 1964.

Schwenk, Theodor. *Sensitive Chaos: The Creation of Flowing Forms in Water and Air,* translated from the German by Olive Whicher and Johanna Wrigley. London: Rudolf Steiner Press, 1965.

—

1. Emilie Conrad, personal note, 1998.

2. ibid.

3. Keleman, *Emotional Anatomy,* p. 15.

4. Alberts et al., p. 970.

5. Hall, p. 221.

6. Keleman, *Emotional Anatomy,* p. 43.

7. Bob Frissell, *Something In This Book Is True....* (Berkeley, California: Frog, Ltd., 1997), p. 131.

8. Stanley Keleman, "Professional Colloquium: 29 October 1977," in *Ecology and Consciousness: Traditional Wisdom on the Environment,* Richard Grossinger, editor (Berkeley: North Atlantic Books, 1992), p. 23.

9. Robert Zeiger, personal communication, 1998.

10. Hall, p. 222.

11. ibid., p. 223.

12. ibid.

13. Rob Brezsny, *Televisionary Oracle,* unpublished novel (forthcoming, Frog, Ltd., Berkeley, California, 2000).

14. Stanley Keleman, *In Defense of Heterosexuality* (Berkeley: Center Press, 1982), p. 53.

15. R. Louis Schultz.

16. Keleman, *In Defense of Heterosexuality,* p. 54.

Chapter 20. The Musculoskeletal and Hematopoietic Systems

Moore, Keith L. *The Developing Human: Clinically Oriented Embryology.* Philadelphia: Saunders College Publishing, 1977.**

Balinsky, B. I. *An Introduction to Embryology,* 5th edition. Philadelphia: Saunders College Publishing, 1981.**

Berrill, N. J., and Gerald Karp. *Development.* New York: McGraw-Hill Book Co., 1976.**

Alberts, Bruce, Dennis Bray, Julian Lewis, Martin Raff, Keith Roberts, James D. Watson. *Molecular Biology of the Cell.* New York: Garland Publishing, 1989.*

Keleman, Stanley. *Emotional Anatomy.* Berkeley, California: Center Press, 1985.*

Matthews, L. Harrison. *The Life of Mammals,* Vol. 1. New York: Universe Books, 1969.*

Schultz, R. Louis, and Rosemary Feitis. *The Endless Web: Fascial Anatomy and Physical Reality* . Berkeley, California: North Atlantic Books, 1996.*

Wintrobe, M. M. *Clinical Hematology.* Philadelphia: Lea and Febiger, 1981.*

Murchie, Guy. *The Seven Mysteries of Life: An Exploration in Science and Philosophy.* Boston: Houghton Mifflin Co., 1978.*

Alexander, R. McNeill. *The Chordates.* Cambridge: Cambridge University Press, 1975.*

Feldenkrais, Moshe. *Body Awareness as Healing Therapy: The Case of Nora.* Berkeley: North Atlantic Books, 1993.

Duddington, C. L. *Evolution and Design in the Plant Kingdom.* New York: Harper and Row, 1970.

—

1. Keleman, p. 40.

2. *The Essence of T'ai Chi Ch'uan: The Literary Tradition,* translated from the Chinese by Benjamin Lo, Martin Inn, Robert Amacker, and Susan Foe (Berkeley: North Atlantic Books, 1979), p. 95.

3. Schultz and Feitis, pp. 13–14.

4. John E. Upledger, *CranioSacral Therapy I Study Guide* (Palm Beach Gardens, Florida: UI Publishing, 1991), p. 18.

5. Murchie, pp. 120–121.

6. Robert J. Sardello, "The Suffering Body of the City," in *Spring,* 1983 Annual Issue (Dallas, Texas), p. 153.

7. Emilie Conrad, personal note, 1998.

8. Murchie, p. 121.

9. Valentin Ivanovich Govallo, *Immunology of Pregnancy and Cancer,* translated from the Russian by Lena Jacobson (Commack, New York: Nova Science Publishers, Inc., 1992), p. 133.

10. ibid., p. 221.

11. ibid., p. 138.

12. ibid., p. 136.

Chapter 21. Mind

Montagu, M. F. Ashley, editor. *Culture and the Evolution of Man.* New York: Oxford University Press, 1962.** The following essays from this book were the main sources used: Oakley, Kenneth P., "A Definition of Man"; Washburn, Sherwood L., "Tools and Human Evolution"; White, Leslie A., "The Concept of Culture"; Haldane, J. B. S., "The Argument from Animals to Men: An Examination of Its Validity for Anthropology"; Etkin, William, "Social Behavior and the Evolution of Man's Mental Faculties"; Dobzhansky, Theodosius, and Montagu, M. F. Ashley, "Natural Selection and the Mental Capacities of Mankind"; Hallowell, A. Irving, "The Structural and Functional Dimensions of a Human Existence" and "Personality Structure and the Evolution of Man"; Montagu, M. F. Ashley, "Time, Morphology, and Neoteny in the Evolution of Man"; and Brace, C. Loring, "Cultural Factors in the Evolution of the Human Dentition."

Spuhler, J. N., editor. *The Evolution of Man's Capacity for Culture.* Detroit: Wayne State University Press, 1965.** My main source in this book was Spuhler's own essay, "Somatic Paths to Culture."

Le Gros Clark, W. E. *The Antecedents of Man: An Introduction to the Evolution of the Pri-*

mates. New York: Harper and Row, 1963.**

Kelso, A. J. *Physical Anthropology.* Philadelphia: J. B. Lippincott Co., 1970.*

Moore, Keith L. *The Developing Human: Clinically Oriented Embryology.* Philadelphia: Saunders College Publishing, 1977.*

Campbell, Bernard. *Human Evolution: An Introduction to Man's Adaptations.* Chicago: Aldine Publishing Co., 1966.*

Bagwell, H. Robert. "Integrative Processing," written draft #4, unpublished manuscript, January 2, 1999.*

Romer, A. S. *Man and the Vertebrates.* Baltimore: Penguin Books, 1954.*

Berrill, N. J., and Gerald Karp. *Development.* New York: McGraw-Hill Book Co., 1976.*

Lévi-Strauss, Claude. *The Savage Mind,* translated from the French anonymously. Chicago: University of Chicago Press, 1966.*

Freud, Sigmund. *An Outline of Psychoanalysis.* Translated by James Strachey. New York: W. W. Norton and Co., 1949.

Jung, C. G. *The Archetypes and the Collective Unconscious.* Translated from the German by R. F. C. Hull. New York: Pantheon Books, 1959.

Marshack, Alexander. *The Roots of Civilization: The Cognitive Beginnings of Man's First Art, Symbol and Notation.* New York: McGraw-Hill Book Co., 1972.

de Santillana, Giorgio, and Hertha von Dechend. *Hamlet's Mill: An Essay on Myth and the Frame of Time.* Boston: Gambit, 1969.

Stanner, W. E. H. *The Dreaming.* Indianapolis: Bobbs-Merrill Reprint Series in the Social Sciences, A-214, 1956.

Elkin, A. P. *The Australian Aborigines.* Garden City, New York: Doubleday and Company, 1964.

Turner, Victor. *The Forest of Symbols: Aspects of Ndembu Ritual.* Ithaca, New York: Cornell University Press, 1967.

Wilson, Robert Anton. *Cosmic Trigger.* Berkeley: And/Or Press, 1977.

Clarke, Arthur C. *2001: A Space Odyssey.* New York: New American Library, 1968.

Anderson, Edgar. *Plants, Man and Life.* Berkeley: University of California Press, 1967.

——

1. Steven Shaviro, *Doom Patrols: A Theoretical Fiction about Postmodernism* (London: Serpent's Tail/High Risk Books, 1997), p. 43.

2. Martin Heidegger; alleged quotation placed on internet, http://scorpio.gold.ac.uk/tekhnema/2/beardsworth/beardsworth.html.

3. Erich Neumann, *The Great Mother,* translated from the German by Ralph Manheim (Princeton: Princeton University Press, 1963), pp. 12–13.

4. Claude Lévi-Strauss, *The Savage Mind,* p. 95.

5. Gerrit Lansing, "The Burden of Set," in Richard Grossinger (editor), *Earth Geography Booklet #1, Economics, Technology, and Celestial Influence, Io #12* (Cape Elizabeth, Maine, 1972): p. 54.

6. Bagwell.

7. ibid.

8. ibid.

9. ibid.

10. Claude Lévi-Strauss, *The Elementary Structures of Kinship*, translated from the French by James Harle Bell, John Richard von Sturmer, and Rodney Needham (Boston: Beacon Press, 1969), p. 490.

11. ibid., indirect quote.

12. Marcel Mauss, *The Gift* (1924), translated by I. Cunnison (New York: Free Press, 1954), pp. 11–12.

13. Claude Lévi-Strauss, *The Elementary Structures of Kinship*, pp. 488–489.

14. Bagwell.

15. Richard A. Knox, "New Evidence of Medical Treatment 5,000 Years Ago: Frozen Mummy Took Laxative for Parasites," *San Francisco Chronicle*, December 25, 1998, p. A4 (wire service from *Boston Globe*).

16. *Genesis, Translation and Commentary* by Robert Alter (New York: W. W. Norton, 1996), pp. 42–43.

Chapter 22. The Origin of Sexuality and Gender

Margulis, Lynn, and Dorion Sagan. *Origins of Sex: Three Billion Years of Genetic Recombination.* New Haven: Yale University Press, 1986.**

Wendt, Herbert. *The Sex Life of Animals.* Translated from the German by Richard and Clara Winston. New York: Simon and Schuster, 1965.*

Ferenczi, Sandor. *Thalassa: A Theory of Genitality.* Translated from the German by Henry Alden Bunker. New York: W. W. Norton and Co., 1938.

Queen, Carol, and Lawrence Schimel (editors). *Pomosexuals: Challenging Assumptions About Gender and Sexuality.* San Francisco: Cleis Press, 1997.*

de Ropp, Robert S. *Sex Energy: The Sexual Force in Man and Animals.* New York: Dell Publishing Co., 1969.*

Freud, Sigmund. *An Outline of Psychoanalysis.* Translated by James Strachey. New York: W. W. Norton and Co., 1949.

———

1. Margulis and Sagan, p. 169.

2. ibid., p. 20.

3. Michael Thomas Ford, "A Real Girl," in Queen and Schimel, p. 158.

4. Andrew Marvell, "To His Coy Mistress," *Seventeenth Century Poetry: The Schools of Donne and Jonson*, editor, Hugh Kenner (New York: Holt, Rinehart and Winston, 1964), p. 458.

5. J. H. C. Fabre, *The Life and Love of the Insect* (London: A. C. Black, 1918), quoted in Robert S. de Ropp, pp. 40–41.

6. ibid., pp. 41–42.

7. Carol Queen, "Beyond the Valley of the Fag Hags," in Queen and Schimel, p. 83.

8. Géza Róheim, *The Riddle of the Sphinx, or Human Origins*, translated from the German by R. Money-Kyrle (New York: Harper and Row, 1974), p. 271.

Chapter 23. Birth Trauma

Upledger, John E. *A Brain Is Born: Exploring the Birth and Development of the Central Nervous System*. Berkeley, California: North Atlantic Books and UI Enterprises, 1996.*

Chamberlain, David. *Consciousness at Birth: A Review of the Empirical Evidence*. San Diego, California: Chamberlain Communications, 1983, p. 4 [republished in different form as *The Mind of Your Newborn Baby* (Berkeley, California: North Atlantic Books, 1998)].*

Weihs, Thomas J. *Embryogenesis in Myth and Science*. Edinburgh: Floris Books, 1986.*

Frissell, Bob. *Something In This Book Is True....* Berkeley, California: Frog, Ltd., 1997.*

Neumann, Erich. *The Great Mother: An Analysis of the Archetype*. Translated from the German by Ralph Manheim. Princeton: Princeton University Press, 1963.*

—

1. Chamberlain, pp. 5–6. Page numbers refer to the 1983 version of this book.

2. John E. Upledger, *A Brain Is Born*, pp. 102–103.

3. Frissell, p. 130. ["Birth is a tidal wave ... conceive of it" comes from Frederick Leboyer, *Birth Without Violence* (New York: Alfred A. Knopf, 1976), p. 15.]

4. Chamberlain, pp. 203–204. Page numbers refer to the 1998 version of this book.

5. Frissell, p. 131.

6. Chamberlain, p. 204. Page number refers to the 1998 version of this book.

7. Jeannine Parvati, "Prenatal Care Guidelines," unpublished manuscript, 1982.

8. Jeannine Parvati, "Notes on Grossinger's *Embryogenesis*," unpublished manuscript, 1984.

9. Peter Redgrove, *The Black Goddess and the Unseen Real: Our Uncommon Senses and Their Uncommon Sense* (New York: Grove Press, 1987), p. 185.

10. Jeannine Parvati, "Psyche's Midwife," unpublished manuscript, 1984.

11. Chamberlain, p. 4. Page number refers to the 1983 version of this book.

12. ibid., p. 35.

13. ibid., pp. 37–39.

14. Christophe Massin, *Le Bébé & L'Amour* (Paris: Aubier, 1997), unpaginated manuscript translated by the author.

15. ibid.

16. Frissell, p. 129.

17. Edward Whitmont, *The Alchemy of Healing: Psyche and Soma* (Berkeley: North Atlantic Books, 1993), p. 140.

18. Frissell, p. 128. ["To unravel the birth-death cycle ... expression of Eternal Spirit" comes from Leonard Orr, *The Story of Rebirthing* (Chico, California: Inspiration University, no date), p. 2.]

19. John E. Upledger, *SomatoEmotional Release and Beyond* (Palm Beach Gardens, Florida: UI Publishing, 1990), p. 221.

20. Richard Grossinger, *The Continents* (Los Angeles: Black Sparrow Press, 1973), pp. 98–99 (adapted).

Chapter 24. Healing

The information in this chapter is, to a certain degree, a condensed version of material from my three volumes on the history of healing: *Planet Medicine: Origins* (North Atlantic Books, 1995), *Planet Medicine: Modalities* (North Atlantic Books, 1995), and *Homeopathy: The Great Riddle* (North Atlantic Books, 1998).

—

Moore, Keith L. *The Developing Human: Clinically Oriented Embryology*, 2nd edition. Philadelphia: Saunders College Publishing, 1977.*

Upledger, John E. *A Brain Is Born: Exploring the Birth and Development of the Central Nervous System.* Berkeley, California: North Atlantic Books and UI Enterprises, 1996.*

Schultz, R. Louis, and Rosemary Feitis. *The Endless Web: Fascial Anatomy and Physical Reality.* Berkeley, California: North Atlantic Books, 1996.*

Barral, Jean-Pierre, and Pierre Mercier. *Visceral Manipulation.* Seattle: Eastland Press, 1988.*

Upledger, John E., and Jon D. Vredevoogd. *Craniosacral Therapy.* Seattle: Eastland Press, 1983.*

Keleman, Stanley. *Emotional Anatomy.* Berkeley, California: Center Press, 1985.*

Burger, Bruce. *Esoteric Anatomy: The Body as Consciousness.* Berkeley, California: North Atlantic Books, 1998.

—

1. Theodor Schwenk, *Sensitive Chaos: The Creation of Flowing Forms in Water and Air*, translated from the German by Olive Whicher and Johanna Wrigley (London: Rudolf Steiner Press, 1965), pp. 33–34.

2. Bonnie Bainbridge Cohen, *Sensing, Feeling, and Action: The Experiential Anatomy of Body-Mind Centering* (Northampton, Massachusetts: Contact Editions, 1993), pp. 78–79.

3. ibid., p. 70.

4. ibid.

5. ibid., p. 71.

6. Emilie Conrad, personal note, 1998.

7. Bainbridge Cohen, pp. 29–30.

8. For a more complete discussion of this topic, see my essay "Why Somatic Therapies Deserve As Much Attention As Psychoanalysis in *The New York Review of Books*, and Why Bodyworkers Treating Neuroses Should Study Psychoanalysis," in Don Hanlon Johnson and Ian Grand (editors), *The Body in Psychotherapy: Inquiries in Somatic Psychology* (Berkeley: North Atlantic Books, 1998), pp. 85–106.

9. Wilhelm Reich, *Selected Writings: An Introduction to Orgonomy* (New York: Noonday Press, 1960), p. 199.

10. ibid., p. 201.

11. ibid., p. 200.

12. ibid., p. 201.

13. ibid., p. 207.

14. Wilhelm Reich, *Ether, God and Devil/Cosmic Superimposition*, with five chapters newly translated from the German by Therese Pol (New York: Farrar, Straus and Giroux, 1972), p. 195.

15. ibid., pp. 222–223.

16. John E. Upledger, *CranioSacral Therapy I Study Guide* (Palm Beach Gardens, Florida: UI Publishing, 1991), p. 10.

17. Mary A. Lynch, in the preface to Tom Giammatteo and Sharon Weiselfish-Giammatteo, *Integrative Manual Therapy for the Autonomic Nervous System and Related Disorders with Advanced Strain and Counterstrain Technique* (Berkeley, California: North Atlantic Books, 1998).

18. Frank Lowen, *Visceral Manipulation I—A Study Guide* (Palm Beach Gardens, Florida: UI Publishing, 1992), p. 4.

19. ibid., p. 5.

20. Barral and Mercier, p. 147.

21. ibid., p. 9.

22. Sharon Weiselfish-Giammatteo, personal communication, 1998.

23. John E. Upledger, *Your Inner Physician and You: CranioSacral Therapy and SomatoEmotional Release* (Berkeley: North Atlantic Books and UI Enterprises, 1991), p. 18.

24. John E. Upledger and Jon D. Vredevoogd, *Craniosacral Therapy* (Seattle: Eastland Press, 1983), p. 9.

25. John E. Upledger, *SomatoEmotional Release and Beyond*, p. 25.

26. ibid.

Chapter 25. Transsexuality, Intersexuality, and the Cultural Basis of Gender

Califia, Pat. *Sex Changes: The Politics of Transgenderism.* San Francisco: Cleis Press, 1997.**
    [Califia's original source is in brackets.]
Queen, Carol, and Lawrence Schimel (editors). *Pomosexuals: Challenging Assumptions About Gender and Sexuality.* San Francisco: Cleis Press, 1997.*

—

1. Califia, pp. 199–200. [Virginia "Charles" Prince, *The Transvestite and His Wife* (Los Angeles: no publisher listed, 1986), p. 60.]

2. ibid., p. 131. [Walter L. Williams, *The Spirit and the Flesh: Sexual Diversity in American Indian Culture* (Boston: Beacon Press, 1986), pp. 45–46.]

3. Anne Fausto-Sterling, "The Five Sexes: Why Male and Female Are Not Enough," *The Sciences,* Vol. 33, No. 2, The New York Academy of Sciences (March/April 1993): p. 23.

4. ibid., pp. 20–24.

5. ibid.

6. John Money, quoted in Califia, p. 69. [Richard Green and John Money (editors), *Transsexualism and Sex Reassignment* (Baltimore: Johns Hopkins University Press, 1969), p. 91.]

7. ibid., p. 69. [Green and Money, *Transsexualism and Sex Reassignment*, p. 92.]

8. Jan Morris, quoted in Califia, p. 29. [Jan Morris, *Conundrum: An Extraordinary Narrative of Transsexualism* (New York: Henry Holt and Company, Inc., 1974), p. 3.]

9. ibid., p. 30. [Jan Morris, *Conundrum*, p. 8.]

10. ibid., p. 37. [Jan Morris, *Conundrum*, p. 169.]

11. Califia, p. 135. [Walter L. Williams, *The Spirit and the Flesh*, p. 29.]

12. Califia, p. 134. [Walter L. Williams, *The Spirit and the Flesh*, pp. 77–78.]

13. Christine Jorgensen, quoted in Califia, p. 31. [Christine Jorgensen, *Christine Jorgensen: A Personal Autobiography* (New York: Bantam Books, 1968), p. 24.]

14. Mario Martino, quoted in Califia, p. 43. [Mario Martino (with harriet), *Emergence: A Transsexual Autobiography* (New York: Crown Publishers, Inc., 1977), p. 134.]

15. Christine Jorgensen, quoted in Califia, p. 21. [Christine Jorgensen, *Christine Jorgensen*, p. 92.]

16. Mario Martino, quoted in Califia, p. 47. [Mario Martino, *Emergence*, p. 263.]

17. Califia, p. 124. [Walter L. Williams, *The Spirit and the Flesh*, pp. 76–77.]

18. ibid. [Walter L. Williams, *The Spirit and the Flesh*, p. 81.]

19. ibid., p. 143. [Ramón A. Gutiérrez, "Must We Deracinate Indians to Find Gay Roots?" *Outlook*, Winter 1989, p. 62.]

20. ibid., pp. 123–124. [Walter L. Williams, *The Spirit and the Flesh*, pp. 9–10.]

21. ibid., p. 136. [Walter L. Williams, *The Spirit and the Flesh*, p. 97.]

22. ibid., p. 149. [Kris Poasa, "The Samoan Fa'afafine: One Case Study and Discussion of Transsexualism," *Journal of Psychology and Human Sexuality*, Volume 5 (3), 1992, p. 39.]

23. Jan Morris, quoted in Califia, pp. 34–35. [Jan Morris, *Conundrum*, p. 106.]

24. Riki Anne Wilchins, "Lines in the Sand, Cries of Desire," in Queen and Schimel, p. 146.

25. David Harrison, "The Personals"; in Queen and Schimel, p. 133.

26. Riki Anne Wilchins, "Lines in the Sand, Cries of Desire," in Queen and Schimel, pp. 139–140.

27. Califia, p. 60. [Harry Benjamin, *The Transsexual Phenomenon* (New York: The Julian Press, Inc., 1966), p. 129.]

28. David Tuller, "Adventures of a Dacha Sex Spy," in Queen and Schimel, p. 182.

29. Pat Califia, "Identity Sedition and Pornography," in Queen and Schimel, p. 91.

30. Janice Raymond, quoted in Califia, p. 95. [Janice G. Raymond, *The Transsexual Empire: The Making of the She-Male* (Boston: Beacon Press, 1979), p. 104.]

31. Mark Rees, quoted in Califia, p. 184. [Mark Rees, *Dear Sir or Madam: The Autobiography of a Female-to-Male Transsexual* (London: Cassell, 1996), p. 128.]

32. Laura Antoniou, "Hermaphrodykes: Girls Will Be Boys and Dykes Will Be Fags," in Queen and Schimel, p. 120.

33. Jan Morris, quoted in Califia, p. 30. [Jan Morris, *Conundrum*, p. 25.]

34. David Henry Hwang, *M. Butterfly* (New York: New American Library, 1986), p. 63.

35. ibid., p. 90.

36. This "woman" appeared in a 1997 edition of ABC Primetime and on other news programs; I have quoted indirectly from them from memory.

37. Philip K. Dick, quoted in Kodwo Eshun, *More Brilliant Than The Sun: Essays in Sonic Fiction* (London: Quartet Books Limited, 1998), p. 52.

38. Sun Ra, quoted in Kodwo Eshun, p. 154.

39. Amendment to Baptist Faith and Message Statement, quoted in Don Lattin, "Baptists Say Wives Must Submit," *San Francisco Chronicle* (June 10, 1998): pp. A1 and A11.

40. *Genesis, Translation and Commentary* by Robert Alter (New York: W. W. Norton, 1996), p. 80.

41. Antonin Artaud, quoted in Rob Brezsny, *Televisionary Oracle,* unpublished novel. (forthcoming, Frog, Ltd., Berkeley, 2000).

42. Monica Lewinsky on "20/20," interviewed by Barbara Walters, ABC, March 3, 1999.

43. ibid.

44. Pat Califia, "Identity Sedition and Pornography," in Queen and Schimel, p. 94.

45. Linda S. Kauffman, *Bad Girls and Sick Boys: Fantasies in Contemporary Art and Culture* (Berkeley: University of California Press, 1998), pp. 60–61.

46. ibid., p. 61.

47. Riki Anne Wilchins, quoted in Califia, p. 242.

48. Kate Bornstein, quoted in Califia, p. 191.

49. ibid., pp. 191–192.

50. Leslie Feinberg, quoted in Califia, p. 188.

51. Minnie Bruce Pratt, quoted in Califia, p. 215.

52. Leslie Feinberg, quoted in Califia, p. 215.

53. Michael Thomas Ford, "A Real Girl," in Queen and Schimel, p. 159.

54. David Harrison, "The Personals," in Queen and Schimel, p. 137.

55. Steven Shaviro, *Doom Patrols: A Theoretical Fiction about Postmodernism* (London: Serpent's Tail/High Risk Books, 1997), p. 30.

Chapter 26. Self and Desire

Wendt, Herbert. *The Sex Life of Animals.* Translated from the German by Richard and Clara Winston. New York: Simon and Schuster, 1965.*

de Ropp, Robert S. *Sex Energy: The Sexual Force in Man and Animals.* New York: Dell Publishing Co., 1969.

Lacan, Jacques. *Écrits.* Translated from the French by Alan Sheridan. New York: W. W. Norton and Co., 1977.

Freud, Sigmund. *An Outline of Psychoanalysis.* Translated from the German by James Strachey. New York: W. W. Norton and Co., 1949.

Freud, Sigmund. *Civilization and Its Discontents.* Translated from the German by James Strachey. New York: W. W. Norton and Co., 1962.

Reich, Wilhelm. *The Murder of Christ: The Emotional Plague of Mankind.* New York: Farrar, Straus & Giroux, 1970.

Lederer, Laura, editor. *Take Back the Night: Women on Pornography.* New York: Bantam Books, 1982.

—

1. Will Baker, "Tsitsi, the Faithful," from "Three Monkeys, I am Father," in *Nuclear Strategy and the Code of the Warrior: Faces of Mars and Shiva in the Crisis of Human Survival* (IO/#33), editors, Richard Grossinger and Lindy Hough (Berkeley: North Atlantic Books, 1984), p. 230.

2. ibid.

3. Georg Wilhelm Steller, quoted in David Day, *The Doomsday Book of Animals: A Natural History of Vanished Species* (New York: Viking Press, 1981), pp. 216–217.

4. Russell Banks, *Affliction* (New York: HarperCollins, 1989), p. 68.

5. ibid., pp. 68–69.

6. Joel Kovel, *History and Spirit* (Boston: Beacon Press, 1991), cover quote.

7. Sue Coe, *Dead Meat* (New York/London: Four Walls Eight Windows, 1995), caption on plate 12 following page 40.

8. ibid., p. 67.

9. Deborah Sontag, "Ultra-Orthodox Jews Rebuke American Reform Rabbis at Wailing Wall," *San Francisco Chronicle* (February 2, 1999): p. A10.

10. Edward Dorn, "Jerusalem," from *Languedoc Variorum: A Defense of Heresy and Heretics;* excerpted in Edward Dorn, *High West Rendezvous: A Sampler* (South Devonshire, England: Etruscan Books, 1997), pp. 38–39.

11. W. E. H. Stanner, *The Dreaming* (Indianapolis: Bobbs Merrill Reprint Series in the Social Sciences, A-214, no date, first published 1956), p. 55.

12. Géza Róheim, *The Riddle of the Sphinx, or Human Origins,* translated from the German by R. Money-Kyrle (New York: Harper and Row, 1974), pp. 23–24.

13. Pat Murphy, "Rachel in Love," in *Points of Departure* (New York: Bantam Books, 1990), pp. 218, 219, 224.

14. ibid., p. 228.

15. Robert Ardrey, *African Genesis: A Personal Investigation into the Animal Origins and Nature of Man* (New York: Dell Publishing Co., 1963).

16. Michael McClure, "Wolf Net," in *Ecology and Consciousness: Traditional Wisdom on the Environment,* Richard Grossinger, editor (Berkeley: North Atlantic Books, 1992), p. 217.

17. Steven Shaviro, *Doom Patrols: A Theoretical Fiction about Postmodernism* (London: Serpent's Tail/High Risk Books, 1997), p. 2. ["To speak . . . to collaborate" comes from William S. Burroughs; "lonely hour . . . never arrives" comes from Louis Althusser.]

18. J. H. C. Fabre, *The Life and Love of the Insect* (London: A. C. Black, 1918), quoted in Robert de Ropp, *Sex Energy,* pp. 42–43.

19. ibid., p. 43.

20. Jonathan Kellerman, *The Butcher's Theater* (New York: Bantam Books, 1988), p. 377.

21. William S. Burroughs, "Last Words" in *The New Yorker* (August 18, 1997): p. 37.

22. Norman Mailer, *The Executioner's Song* (Boston: Little, Brown and Co., 1979), p. 306.

23. ibid., p. 305.

24. Steven King, *The Green Mile*, Volume I: "Two Dead Girls" (New York: Penguin Audiobooks, 1996).

25. Knud Rasmussen, "Intellectual Culture of the Iglulik Eskimos" (1929) in *Ecology and Consciousness: Traditional Wisdom on the Environment*, Richard Grossinger, editor (Berkeley: North Atlantic Books, 1992), p. 15.

26. Ardrey, p. 318.

27. J. H. C. Fabre, quoted in de Ropp, p. 43.

28. Marquis de Sade *(Juliette)*, quoted in Robert de Ropp, *Sex Energy*, p. 43.

29. *Woody Allen on Woody Allen*, in conversation with Stig Björkman (New York: Grove Press, 1993), p. 225.

30. From notes taken while watching *60 Minutes*, CBS News, 1982.

31. Fyodor Dostoyevsky, *The Brothers Karamazov*, translated from the Russian by Constance Garnett (New York: Random House, 1950), p. 306.

32. ibid., p. 289.

33. Kovel, *History and Spirit*, cover quote.

34. Friedrich Nietzsche, quoted in Erich Heller, *The Importance of Nietzsche* (University of Chicago Press, 1988), p. 5.

35. Marilyn Manson, quoted in Richard Corliss, "Bang, You're Dead," *Time*, May 3, 1999, p. 49.

36. Nietzsche, quoted in Heller, p. 5.

Chapter 27. Spiritual Embryogenesis

Steiner, Rudolf. *An Outline of Occult Science*. Translated from the German by Henry B. Maud and Lisa D. Monges. Spring Valley, New York: Anthroposophic Press (first published in 1909), 1972.*

König, Karl. *Embryology and World Evolution*. Translated from the German by R. E. K. Meuss. In *British Homoeopathic Journal* 57 (1968): 1–62.*

Poppelbaum, Hermann. *Man and Animal: Their Essential Difference*. Translated from the German. London: Anthroposophical Publishing Co., 1960.*

Weihs, Thomas J. *Embryogenesis in Myth and Science*. Edinburgh: Floris Books, 1986.*

Schwenk, Theodor. *Sensitive Chaos: The Creation of Flowing Forms in Water and Air*. Translated from the German by Olive Whicher and Johanna Wrigley. London: Rudolf Steiner Press, 1965.

Haraway, Donna Jeanne. *Crystals, Fabrics, and Fields: Metaphors of Organicism in Twentieth-Century Developmental Biology*. New Haven: Yale University Press, 1976.

Hall, Manly P. *Man, Grand Symbol of the Mysteries: Thoughts in Occult Anatomy*. Los Angeles: The Philosophical Research Society, 1972.

—

1. Samuel Taylor Coleridge, "Dejection: An Ode," in Samuel Taylor Coleridge, *Selected Poetry and Prose*, edited by Elisabeth Schneider (New York: Rinehart & Co., Inc., 1956), p. 127.

2. Da Free John, *Easy Death* (Clearlake, California: The Dawn Horse Press, 1983), pp. 88–89.

3. Zen Master Seung Sahn, "Poem on the Occasion of Thirty Years of Teaching Abroad," in *Gong Mun, The Newsletter of the Empty Gate Zen Center,* Vol. 1.2 (Summer, 1996), Providence, Rhode Island, p. 1.

4. Richard Strohman, personal communication, 1999.

5. René Descartes, quoted in Stuart A. Newman, "Carnal Boundaries: The Commingling of Flesh in Theory and Practice," in Lynda Birke and Ruth Hubbard (editors), *Reinventing Biology: Respect for Life and the Creation of Knowledge* (Bloomington, Indiana: Indiana University Press, 1995), p. 200.

6. Peter Singer, quoted in Newman, p. 201.

7. Richard Dawkins, *The Blind Watchmaker* (New York: Norton, 1986), p. 112.

8. ibid.

9. David Denby, "In Darwin's Wake," in *The New Yorker* (July 21, 1997): pp. 56–58.

10. J. D. Bernal, *The Origin of Life* (Cleveland: World Publishing Company, 1967), pp. 140–141.

11. Arthur Edward Waite (editor), *The Hermetic and Alchemical Writings of Paracelsus, Volume 1, Hermetic Chemistry* (London: James Elliot & Co., 1894), p. 151.

12. Charles Bonnet, quoted in Bentley Glass, Owsei Temkin, and William L. Straus, Jr., *Forerunners of Darwin: 1745–1859* (Baltimore: Johns Hopkins University Press, 1959), p. 204.

13. Poppelbaum, p. 74.

14. Steiner, p. xxxii.

15. Henrik Steffens (1822), quoted in Poppelbaum, p. 85.

16. C. G. Jung, Seminar Report on *Nietzsche's Zarathustra*, X, p. 51f. (privately mimeographed), quoted in James Hillman, "Senex and Puer: An Aspect of the Historical and Psychological Present," in *Puer Papers* (Irving, Texas: Spring Publications, 1979), p. 44.

17. Poppelbaum, pp. 150–151.

18. ibid., pp. 149–150.

19. Rudolf Steiner, quoted in Poppelbaum, p. 106.

20. Rudolf Hauschka, *The Nature of Substance,* translated from the German by Mary T. Richards and Marjorie Spock (London: Vincent Stuart Ltd., 1966), pp. 39–40.

21. Rudolf Steiner, quoted in Poppelbaum, p. 136.

22. Hall, pp. 84–85.

23. Jeannine Parvati, "Notes on Grossinger's *Embryogenesis,*" unpublished manuscript, 1984 (she adds that "mugwort is Artemis' herbal ally").

24. Madame H. P. Blavatsky, quoted in Hall, p. 85.

25. Hall, p. 103.

26. Peter Redgrove, *The Black Goddess and the Unseen Real: Our Uncommon Senses and Their Uncommon Sense* (New York: Grove Press, 1987), p. 186.

27. ibid., p. 44.

28. Jeannine Parvati, "Notes on Grossinger's *Embryogenesis*," unpublished manuscript, 1984.

29. Gregory of Nyssa, quoted in Weihs, pp. 36–37.

30. William Blake, "The Marriage of Heaven and Hell," *The Portable Blake*, editor, Alfred Kazin (New York: Viking Press, 1946), p. 250.

31. Paul Foster Case, *The Tarot: A Key to the Wisdom of the Ages* (New York: Macoy Publishing Company, 1947), p. 36.

32. ibid., p. 31.

33. ibid., p. 36.

34. Jeannine Parvati, "Prenatal Care Guidelines," unpublished manuscript, 1982.

35. Weihs, pp. 145–146.

Chapter 28. Cosmogenesis and Mortality

I have dealt with the topic of cosmogenesis more thoroughly in my book *The Night Sky: The Science and Anthropology of the Stars and Planets* (Los Angeles: J. P. Tarcher, 1988). See chapters entitled "Scientific and Occult Astronomy," "Ancient Astronomy," "Star Myth," and particularly "The History of Western Astronomy VII. The Creation."

—

Collin, Rodney. *The Theory of Celestial Influence.* London: Stuart and Watkins, 1958.*

Matt, Daniel C. *The Essential Kabbalah: The Heart of Jewish Mysticism.* HarperSan Francisco, 1996.*

Weihs, Thomas J. *Embryogenesis in Myth and Science.* Edinburgh: Floris Books, 1986.*

Kushi, Michio. *The Book of Do-In: Exercises for Physical and Spiritual Development.* Tokyo: Japan Publications, 1979.*

Ouspensky, P. D. *In Search of the Miraculous: Fragments of an Unknown Teaching.* New York: Harcourt, Brace & World, 1949.*

Sproul, Barbara C. *Primal Myths: Creating the World.* San Francisco: Harper & Row, 1979.*

Teilhard de Chardin, Pierre. *The Phenomenon of Man,* translated from the French by Bernard Wall. New York: Harper & Row, 1959.

Bennett, J. G. *Gurdjieff: Making a New World.* New York: Harper and Row, 1973.

Haile, Father Berard, Leland C. Wyman, Maud Oakes, Laura A. Armer, Franc J. Newcomb. *Beautyway: A Navaho Ceremonial.* New York: Bollingen Series/Pantheon Books, 1957.

—

1. Ovid, *Metamorphoses,* translated by Rolfe Humphries (Bloomington, Indiana: Indiana University Press, 1973), p. 3.

2. *The Kalevala: The Land of the Heroes,* Vol. 1, translated by W. F. Kirby (London: J. M. Dent, 1907), p. 7.

3. Robert Graves, *Greek Myths,* Vol. 1 (Baltimore: Penguin Books, 1955), p. 27.

4. ibid.

5. Weihs, p. 23.

6. "Satapatha-Brahmana, XI, i, 6," translated by Julius Eggeling; in Max Müller (editor), *Sacred Books of the East*, Vol. 44 (Oxford, England: Clarendon Press, 1900), p. 12.

7. *The Vishnu Purana*, quoted in Manly P. Hall, *Man, Grand Symbol of the Mysteries: Thoughts in Occult Anatomy* (Los Angeles: The Philosophical Research Society, 1972), p. 104.

8. L. Ferrand and L. J. Frachtenberg, "Shasta and Athapascan Myths from Oregon," in *Journal of American Folklore*, No. 28 (1915): p. 224.

9. Roland B. Dixon, "The Maidu Creation Myth," in *Bulletin of the American Museum of Natural History*, 39, No. 1; quoted in Sproul, pp. 238–239.

10. Robert H. Lowie, "The Assiniboine," *Anthropological Papers of the American Museum of Natural History*, Vol. 4, No. 1 (New York: 1909), p. 1.

11. Maria Leach, *The Beginning* (New York: Funk and Wagnalls, 1956), p. 145.

12. ibid.

13. T. G. H. Strehlow, *Aranda Traditions* (Melbourne: University of Melbourne Press, 1947), p. 7.

14. ibid.

15. ibid.

16. Sir George Grey, "The Children of Heaven and Earth," in *Polynesian Mythology and Ancient Traditional History* (Auckland: H. Brett, 1885), pp. 1–8.

17. Matt, p. 94.

18. ibid., p. 29.

19. ibid., p. 90.

20. Charles Poncé, *Working the Soul: Reflections on Jungian Psychology* (Berkeley: North Atlantic Books, 1984), p. 66.

21. ibid., p. 67.

22. G. I. Gurdjieff, quoted in Ouspensky, p. 175.

23. ibid., p. 221.

24. "Excerpts from a Meeting with G. I. Gurdjieff in 1943," in *Material for Thought*, Far West Editions, San Francisco, Number 12 (Spring, 1990): pp. 65–66.

25. Teilhard de Chardin, pp. 71–72.

26. ibid., p. 72.

27. ibid., p. 182.

28. ibid., p. 78.

29. Bob Frissell, "Seeing Beyond" Interview, audiotape, 1996.

30. ibid.

31. Kinky Friedman, *God Bless John Wayne* (New York: Bantam Books, 1996), p. 83.

32. Steven Shaviro, *Doom Patrols: A Theoretical Fiction about Postmodernism* (London: Serpent's Tail/High Risk Books, 1997), p. 109.

33. Collin, p. 156.

34. Ray Goldrup and Blaine M. Yorgason (screenplay), *Windwalker*, directed by Kieth Merrill, Santa Fe International, 1980.

35. Norman Mailer, *The Executioner's Song* (Boston: Little, Brown and Co., 1979), p. 360.

36. Eric Bogosian (screenplay), *SubUrbia,* directed by Richard Linklater, Castle Rock Entertainment, 1997.

37. Michio Kushi.

38. Da Free John, *Easy Death* (Clearlake, California: The Dawn Horse Press, 1983), p. xxii.

39. ibid., xxiii.

40. T. S. Eliot, "East Coker," in The Complete Poems and Plays, 1909–1950 (Newy York: Harcourt Brace and Company, 1958), p. 126.

Chapter 29. Death and Reincarnation

Nuland, Sherwin B. *How We Die: Reflections on Life's Final Chapter.* New York: Alfred A. Knopf, 1994.**

Da Free John. *Easy Death.* Clearlake, California: The Dawn Horse Press, 1983.**

Sogyal Rinpoche, *The Tibetan Book of Living and Dying.* HarperSan Francisco, 1992.*

—

1. Sir A. Palgrave, quoted in Nuland, p. 84.

2. Kim Stanley Robinson, *Red Mars* (New York: Bantam Books, 1993), p. 288.

3. ibid.

4. ibid, p. 291.

5. *The Yellow Emperor's Classic of Internal Medicine,* translated by Ilza Veith (Berkeley, California: University of California Press, 1966), pp. 182–183.

6. Sogyal Rinpoche, pp. 250–253.

7. Da Free John, p. 231.

8. *The Yellow Emperor's Classic of Internal Medicine,* pp. 108–109.

9. Milan Kundera, quoted from a secondary source *(San Francisco Examiner,* December 27, 1998).

10. Stan Brakhage, "Interview, 7 Jan. 72," in *Imago Mundi (I0/14),* edited by Richard Grossinger (Plainfield, Vermont: North Atlantic Books, 1972), p. 362.

11. Laura Blumenfeld, "After a Crash, Cruel Treatment of Families Is 'Second Tragedy,'" *San Francisco Chronicle,* Saturday, June 29, 1996.

12. Samuel Taylor Coleridge, "Kubla Khan"; in Samuel Taylor Coleridge, *Selected Poetry and Prose,* edited by Elisabeth Schneider (New York: Rinehart & Co., Inc., 1956), p. 115.

13. Knud Rasmussen, "Intellectual Culture of the Iglulik Eskimos" (1929) in *Ecology and Consciousness: Traditional Wisdom on the Environment,* Richard Grossinger, editor (Berkeley: North Atlantic Books, 1992), p. 17.

14. ibid., pp. 17–18.

15. David Webb Peoples, *Unforgiven,* directed by Clint Eastwood, Warner Bros., 1993.

16. Da Free John, p. 71.

17. Russell Banks, *Cloudsplitter* (New York: HarperFlamingo, 1998), p. 218.

18. Phillip Mahony, "Pat," an unpublished poem.

19. ibid.

20. Da Free John, p. 71.

21. Chögyam Trungpa, *The Lion's Roar: An Introduction to Tantra* (Boston: Shambhala Publications, 1992), p. 129.

22. ibid., p. 130.

23. Da Free John, p. 299.

24. Chögyam Trungpa, quoted in "In Light of Death: An Interview with Rick Fields on Living with Cancer," in *Tricycle: The Buddhist Review* (Fall 1997): p. 47.

25. Sogyal Rinpoche, p. 103.

26. Da Free John, p. 240.

27. ibid., p. 169.

28. ibid., pp. 265–266.

29. ibid., p. 170.

30. ibid., p. 83.

31. Francesca Fremantle and Chögyam Trungpa (translators), *The Tibetan Book of the Dead: The Great Liberation through Hearing in the Bardo* (Boulder, Colorado: Shambhala Publications, 1975), p. 84.

32. Sogyal Rinpoche, p. 254.

33. Chögyam Trungpa, *Crazy Wisdom* (Boston: Shambhala Publications, 1991), pp. 131–132.

34. Avatar Adi Da Samraj (Da Free John), *Drifted in the Deeper Land: Talks on Relinquishing the Superficiality of Mortal Existence and Falling by Grace into the Divine Depth That Is Reality Itself* (Middletown, California: The Dawn Horse Press, 1997), p. 151.

35. ibid.

36. Ellias and Theanna Lonsdale, *The Book of Theanna: In the Lands that Follow Death* (Berkeley, California: Frog, Ltd., 1995), p. 86.

37. ibid, p. 218, back cover.

38. ibid., p. 24.

39. ibid., p. 25.

40. Robert Penn Warren, *All the King's Men* (New York: Modern Library/Random House, 1953), p. 226.

41. Da Free John, p. 186.

42. Meher Baba, *God Speaks* (New York: Dodd Mead, 1955), p. 120.

43. Rob Brezsny, *Televisionary Oracle*, unpublished novel.

44. Samuel Beckett, *Waiting for Godot* (New York: Grove Press, 1954), p. 61.

45. Marcia Fields, "Conscious Dying," a talk given at a memorial service for Rick Fields, Spirit Rock Meditation Center, Woodacre, California, August 1, 1999.

46. ibid.

47. Da Free John, p. 61.

48. ibid., p. 374.

49. ibid., p. 345.

50. William Faulkner, *The Wild Palms* (New York: Vintage Books, 1962), p. 324.

51. Priscilla Cogan, *Winona's Web,* audio cassette, Audio Literature, 1997.

52. Michael Ventura, "Homage to a Sorcerer" (Chapel Hill, North Carolina: *The Sun,* #279, March, 1999), p. 23.

53. Padmasambhava, quoted in Sogyal Rinpoche, p. 259.

54. Da Free John, p. 205.

55. Sogyal Rinpoche, p. 254.

56. Da Free John, p. 331.

57. Sogyal Rinpoche, p. 261.

58. ibid.

Index

ILLUSTRATIONS ARE DENOTED BY PAGE NUMBERS IN *italics*, but illustrations are only separately noted if they do not fall within the text pages listed. Page numbers with "A," "B," or "C" appended refer to the color sections.

About the Author

A NATIVE OF NEW YORK CITY, RICHARD GROSSINGER attended Amherst College and the University of Michigan, from which he received a B.A. in English and a Ph.D. in anthropology, respectively. In the late 1960s and early 1970s he did ethnographic fieldwork among Hopis and Euro-Americans in Arizona and fishermen in downeast Maine. He taught college in Maine and Vermont until 1977, when he and his family moved to the San Francisco Bay Area of California, where he has lived since.

Grossinger is the author of a number of titles. His early experimental prose works include *Solar Journal: Oecological Sections, The Continents, Book of the Cranberry Islands,* and *The Slag of Creation.* His two nonfiction novels are *New Moon* and *Out of Babylon. Embryogenesis* completes a trilogy of which the first two volumes were *Planet Medicine* and *The Night Sky.* He has edited or coedited a number of anthologies, including *The Alchemical Tradition in the Late Twentieth Century, Baseball I Gave You All The Best Years of My Life,* and *Ecology and Consciousness.*

With his wife, the poet and novelist Lindy Hough, he is the founder and publisher of North Atlantic Books and Frog, Ltd., both located in Berkeley, California.